In the Shadow of the Shield

In the Shadow of the Shield

The Development of Wireless
Telegraphy and Radio Broadcasting
in Kingston and at Queen's University:

An Oral and Documentary History,
1902-1957

By Arthur Eric Zimmerman, Ph.D.
© 1991

ISBN 0-9695570-0-0

Published with the assistance of the Ontario Heritage Foundation,
Ontario Ministry of Culture and Communications

Cover and book design:
The Brookview Group Inc.

Cover photograph:
Mark Shapiro

Dedication

To Cyril Kalfin and Judy Pike
and all those who made this project possible.

In memory of Professor Harold Stewart.

TABLE OF CONTENTS

INTRODUCTION

The idea for this project, to discover the origins of radio at Queen's University, seeped into my mind in May 1982, at the annual conference of the Association for the Study of Canadian Radio and Television in Toronto. Several delegates presented oral histories of the pre-commercial days of radio on the east and west coasts of Canada - personal reminiscences and stories that no one had ever bothered to write down. In fact, there is still no well-documented history of the development of wireless and radio in Canada. I began to wonder whether anybody knew the true details of the history of CFRC, Queen's Radio in Kingston, where I was a volunteer broadcaster.

It is part of Queen's legend that CFRC is one of the oldest radio stations in North America. Generations of CFRC Radio Club volunteers have been nurtured on the story that Queen's has the second oldest campus station on the continent, that it started in 1922 with experimental call 9BT and that the first broadcast was of a cornet solo followed, at some indefinite interval, by a football game. For the last 30 years, its official birthdate was celebrated as October 27, 1922. In fact, CFRC first came on the air on October 27, 1923, with the play-by-play of a Queen's-McGill rugby game. Station breaks during the 1970s referred to CFRC as "Canada's second oldest radio voice," after CFRB in Toronto. But the CFRB call was granted much later, in 1927.

I became more and more curious about the origins of Queen's Radio. How and why did radio come to Queen's University? Did it all start in 1922 or did it antedate that? Is there any extant documentation, with names? Might any original members of the station still be alive? Would an oral history still be possible? Where were the licences, the station logs, the purchase orders for equipment? As the ideas gelled, I made enquiries around Queen's campus, was steered to the little publications by Douglas Geiger and Professor Harold Stewart on the origins of CFRC, and was shown the log book begun in 1923.

So, with a place to start digging, a few names and some records available, an oral history radio program seemed an ideal project to celebrate the 60th anniversary of Queen's Radio in 1982. I got a small grant from the Vice-Principal (Services) for recording tape and set off in the middle of August to do my first interview, with Prof. Harold Stewart. By October, I had 40 hours of oral history on tape. The Queen's Radio Oral History program had to be postponed because I had too much information to sort and assemble.

My intention in the radio series was to stay in the background and just let people tell their own stories. I quickly faced the complication that each raconteur tells a slightly different version of any particular story. People's memories are usually pretty good, except for dates, but an individual experience of an event may represent only a part of one side of a story. More often, I was fortunate if I had even a single observer for an event or situation, with no independent corroboration possible. So collecting and organizing oral depositions was not entirely productive and satisfying. The radio series turned out to be anecdotal, unconnected and full of huge gaps that needed plastering over. The story was traceable, falteringly, only as far back as the winter of 1921-22 and from there I peered hopefully backward through the fog.

The breakthrough came in June 1984, when the Rev Dr W. Harold Reid wrote to tell me that Queen's radios were used at Barriefield War Camp in 1915-17. This marvellous gentleman supplied the key for unlocking 20 years of unsuspected pre-history of radio at Queen's. Of necessity, I turned to newspapers and archives for the particulars of these stories, which are now almost beyond living memory, and found gold. That got me searching through these resources for verification of dates and details of stories already collected, and the project evolved into a documented oral history. Although much more detailed and colourful oral evidence is available for relatively recent events, I persisted in documenting whatever I could.

It also seemed reasonable to try to fit the Queen's history into the larger context of radio in the Kingston area and in Canada. To maintain the local perspective, however, I looked for most of my documentation in local newspapers. Unfortunately, newspapers do not seem to have any memory, which makes tracking an extended sequence of events very difficult.

Many radio history books I've consulted have disappointed because they are very economical with the facts and dates but generous with the flavour enhancer. My scientific side insists that stories be examined and verified from several different approaches, that peripheral threads be checked for consistency and correctness of chronology. I was very concerned that telescoping or speculation would constitute a kind of fiction, so I struggled far beyond where it was sensible to stop in order to sort things out. Sometimes I was rewarded. Usually there was nothing more to be found and I had to suppress my instincts and try to be content with a surfacey kind of fiction.

I hoped I could produce a scientific history, a clothes-line of hard facts with the stories hung along it. But I now see that history is just not suited to a linear string of words, or even to being spread out in two dimensions like a mural. It's multi-layered, a polydimensional mosaic, and has to be put up in globs with links going out to every other glob and layer. This notion makes it difficult to keep an exposition entirely chronological. The reader will find many pages of apparent digression, as I tried to link up various layers to give the stories the depth they deserve, or to give some feeling for the times. Rather than ignore apparent contradictions, I preferred to include and discuss them since they will certainly surface again in future. The work also makes many circles, some small, some very large, as themes developed in one period return for further elaboration or for a wrapping together.

In the course of my research I destroyed some myths about CFRC, substantiated others and unearthed a deep foundation that nobody realized was there. CFRC itself is certainly not one of the very oldest radio stations in North America. It is one of about 11 of the oldest surviving radio stations in Canada and is the only one still held by its original licencee. I discovered the date and content of the first broadcast as 9BT and was shown the requisitions for the purchase of parts and licences for 9BT and CFRC. With help, I pushed the history of radio on the Queen's campus back 20 years before 9BT. I resurrected James Lester Willis Gill, the first Professor of General Engineering at Queen's, and fleshed out the life and contribution of Robert Leland Davis, co-founder with Prof. Jemmett of 9BT. I was led to the discovery of Gill's early wireless experimentation on campus, to details of the use of Queen's radios in wireless training at Barriefield in the First War, to Kingston's wireless amateurs of the teens and twenties, to an early Queen's thesis on wireless, and I followed threads to a unique aeroplane built at Queen's from 1908 to 1915 and to an early demonstration of Bell's telephone on campus. I have dug up original material on Marconi's Canadian chain of wireless coast stations, on the origins of Montreal's CFCF and on the first wireless time signal system in the world, on Canada's east coast. I have pieced together pictures of three early radios used on the Queen's campus between 1911 and 1923 and discovered that a famous aerial photograph of the campus, circa 1919, shows the antenna of the Queen's Wireless Club. And I've heard wonderful stories, about Queen's and about radio and people in Kingston and around the continent. I am the custodian of some of those stories now, because the tellers are no more.

The story of the development of radio in Kingston is especially interesting because the district is in a radio hole - they used to call it an ether or aerial pocket. The city sits in the radio shadow of the Canadian Shield and is pretty well isolated from the rest of the country by that great outcropping. Cross-Canada radio relays in the 1920s had to bypass Kingston to the north or south. Kingston hams struggled against that radio shadow from the beginning to establish links with Toronto, Ottawa and Montreal to be part of a trans-Canada network, but their main contacts were necessarily largely in and their influences from the United States. When CFRC became the CBC affiliate in 1936, it was the city's first reliable listening post on the rest of the country. For the rest, the Kingston story contains most of the elements of every story of radio development across Canada. It is local Canadian radio in microcosm, striving to establish itself under adverse conditions. When the history of radio in this country is eventually assembled, it will be made up of dozens of district histories like this one.

Many hundreds of individuals and institutions have assisted in this project and are gratefully acknowledged: Corinne Allan, Elizabeth (Black) Allen, James D. Allen, Dr Margaret Angus, Dr William Angus, Mabel (Hickey) Archer, Dr Lawrence Badash, Keith Baker, Mrs John R. Bain, Ian Baines, Harold W. Beardsell, John Bermingham, Lenore Black,

Dr E.J. Bond, Alan F. Bowman, Brian Brick, Dr R. Carl Brigham, Eric Bronskill, Clara (Farell) Brooke, Basil Brosso, Philip Brown, Joseph G. Burley, Duncan Campbell, Jack M. Campbell, Mary I. (Clark) Campbell, Dr John R. Card, H. Lincoln Chadbourne, Elspeth Chisholm, William Choat, Prof. Lewis D. Clark, Dr Ralfe Clench, Cecil Climo, Don Close, Dr Brian M. Cochrane, Harold A. Cohen, Dr Montague Cohen, Josephine (Donnelly) Cole, Robert A. Conway, Beatrice Cook, Donald A. Cooper, Florence (Daly) Courneya, William J. Coyle, Stuart Crawford, Harry Creagen, James Creighton, Dr Duff Crerar, Marion Cruse, Steve and Nancy Cutway, Arthur L. Davies, Michael Davies, Henry B. Davis, Dr Peter T. Demos, A.E. Diamond, Douglas Donnelly, Orton Donnelly, Robert Donnelly, William (Buster) Doubleday, Dr A. Vibert Douglas, Alan S. Douglas, Allan Dove, Len Dover, Mel Easton, Ronald Ede, Dr Ron Elliott, Frederick G. England, Rev Dr Victor H. Fiddes, Bill Fitsell, Eileen Fleming, Eleanor Foster, Sid Fox, J. Douglas Frame, John Fraser, Gerald R.M. Garratt, Dr Douglas W. Geiger, George F. Geiger, Goldie Bartels Geiger, Prof. Fred Gibson, Donald R. Gordon, James W. Grant, Dr K.W. Greaves, Gary Greer, Herb Hamilton, Fred Hammond, Dr G.B. Harrison, Herbert W. Hartshorn, Dr John L. Heilbron, Joan (Annand) Henderson, Stuart and Jean (Campbell) Henderson, Dr Clarence Y. Hopkins, John H. Houser, Dr Fred S. Howes, Dr Pauline Hughes, W.J. Hurn, the Jackson Sisters (Wilma Nichol and Louise Wittish), William Jaffrey, Harry Jemmett, Maud Jemmett, Dr V.I. John, Prof. Alfred W. Jolliffe, Ernest J. Jury, Jane (Sherman) Kaduck, Verna (Saunders) Kerrison, Ken Keyes, James E. Kirk, Cuth Knowlton, Dr Hilda Laird, Lareta Lamoureux, Al Lennard, David Lennick, Hugh A. Lightbody, Esse W. Ljungh, David Lockett, Lawrence W. Lockett, Georgina (Ettinger) Logan, Dr G.A. Lyons, Dr W. Donald MacClement, Art and Audrey Macdonald, Keith A. MacKinnon, John MacMillan, M.S. Macphail, Harry D. Marlatt, Grace E. Marrison, Andrew Marshall, Dr J.L. Mason, Ruth (Marrison) Mathes, Dr Donald I. Matheson, Lynne McAlpine, Lancely R.C. McAteer, Alphonsus P. McCue, Dorothy F. (Folger) McDiarmid, Alex J. McDonald, Kenneth McIntyre, Margaret McKay-Clements, Peter C. McLeod, Bill McUen, Florence Fraser McHugh, Charles Millar, Prof. and Mrs Norman Miller, Edith R. Millman, Jean (Bangay) Mills, Winnifred L. Moir, William Morley, Bud Moyse, John Murray, H. Gordon Mylks, Kevin Nagle, John B. Nayler, Agnes (Morrissey) Neale, William A. Neville, Bonnie Nichols, Ian Nicholson, Dr Michael Nolan, Audrey Norrish, Mr A.E. O'Kane, Mrs H.G. Ott, Joseph Paithouski, Larry Palef, Fred Paquin, George M. Parsons, Dr Sid Penstone, Catherine M. Perkins, Sidney V. Perry, Prof. Robert M. Pike, Dr Lewis Pyenson, Gary Racine, David Raymont, Rev Dr W. Harold Reid, Lois Baker Rich, Bruce Riggs, Bruce Robinson, Dr Donald C. Rose, Dr Wilfred A. Roy, John R. Rutledge, Kathleen (Whitton) Ryan, Dr Robert and June (Pryce) Sanderson, Dr W.B. Sargent, Marion Rose (Gill) Service, S. Daniel Shire, Jeffrey Simpson, John F. Sirett, Lillian M. Smith, Harold Sprott, Prof. Harold H. Stewart, Dr John D. Stewart, Dr John B. Stirling, George Stone, Dr Irwin Sugarman,

Noreen Sugarman, William A. Taylor, Barbara Teatero, Jack Telgmann, Lou Tepper, Douglas Thwaites, Joseph Thwaites, Jr, William R. Topham, Richard D. Travers, Bernard Trotter, Eric Truman, Bogart Trumpour, George W. Vosper, Verna (Tibbles) Vowles, Dr Ronald L. Watts, Freeman C. Waugh, Sandy Webster, Arthur Wehman, Erwin Wendholt, Mrs George T. White, Dr A. David Wilson, Dorothy (Redeker) Wilson, William J. Wilson, Prof. F. Morris Wood and J. Michael Woogh.

Special thanks to Susan Robertson, who obtained a SEED grant to work on the project in the summer of 1986 and contributed some important material; to Dr E.J. Bond, Judy Pike and many others for encouragement; to Cathy Perkins, George Laverty and Cyril Kalfin for reading the manuscript-in-progress; and to John Parry for editorial assistance.

I am also grateful to the Ontario Heritage Foundation and to Queen's Principal, Dr David C. Smith, and the Sesquicentennial category of the Principal's Development Fund, Queen's University, for making publication possible. A generous grant from the Davies Charitable Foundation made possible the production of this book in hard cover.

Thanks also to the staff of Queen's University Archives; Hilda Baerg and the staff of Queen's Alumni Affairs; the staff of Douglas Library, Queen's University, especially Interlibrary Loans; Queen's Documents Library; George Innes, Department of Geography, Queen's University; the staff of the National Archives of Canada, Ottawa; Phebe Chartrand and the staff of McGill University Archives; the Graduates' Society, McGill; Louise L. Cowing, Public Library, Meriden, Ct.; Boston Public Library; Carol Duthie of Canadian Marconi, Montreal; Nancy Hurn, Canadian National Exhibition, Toronto; Cornell University Library; Imperial War Museum, London; Royal Air Force Museum, Hendon, London; Vimy Wireless Museum, Kingston; Dr W.A.B. Douglas, Department of National Defence, Ottawa; Dr Charles A. Ruch, Westinghouse, Pittsburgh; C.A.H. James, Royal Commission for the Exhibition of 1951; Charles Schille, Harvard University Archives; C.J. Stafford, International Electrotechnical Commission; New York State Library, Albany; Archives, Massachusetts Institute of Technolgy Library; Morley K. Thomas, Atmospheric Environment Service, Toronto; Library of Congress, Washington, D.C.; Omer Lavallee, Corporate Historian, CPR; Ernest A. DeCoste, National Museum of Science and Technology, Ottawa; Elliot N. Sivowitch, Smithsonian Institution; National Film, Television and Sound Archives, Ottawa; Leslee Niethammer, City of Cambridge Library; Nancy Bartlett and Brian Etter, Bentley Historical Library, University of Michigan; Joseph E. Baudino, Westinghouse, Washington, DC; Bruce Kelley, Antique Wireless Association; John James, Biddle Instruments, Blue Bell, Pa; Canadian Communications Foundation; Ministry of the Attorney General of Ontario; Hamilton Board of Education; McMaster University Library; David A. Middleton,

CFCF, Montreal; Mrs E.H. Swanborough, Mohawk College, Hamilton; H.M. Tory Archives, University of Alberta, Edmonton; Alumni Department, Olivet College; US Department of Justice; Louise Gay, Associate Registrar, University of PEI; Frank Fluellon, Canadian Society of Military Medals and Insignia.

Arthur E. Zimmerman
February 9, 1991

Convention in Endnotes

Scientific papers are cited: author, title, journal, volume, issue, date, page.
Newspaper articles are cited: title, newspaper, date, page.
  (in old newspapers, some references are untitled squibs).
Books are cited: author, title, publisher, city, date, page.

Abbreviations

BAAS -   British Association for the Advancement of Science
DBW -   The Daily British Whig, Kingston
DMF/MF - Department of Marine and Fisheries, Ottawa
KDS -   Kingston Daily Standard
KDS/DSJ - KDS Saturday feature: Daily Standard Junior
KWS -   The Kingston Whig-Standard
MPS -   McGill Physical Society
NAC -   National Archives of Canada, Ottawa
OLRSC -   Official List of Radio Stations of Canada
PESQU -   Proceedings of the Engineering Society of Queen's
          University
PRSL -   Proceedings of the Royal Society, London
PTRSC -   Proceedings and Transactions of the Royal Society of
          Canada
QUA -   Queen's University Archives
QJ -   Queen's University Journal or Queen's Journal
QQ -   Queen's Quarterly
QR -   Queen's Review or Queen's Alumni Review
QST -   Official organ of the American Radio Relay League
SM -   School of Mining, Kingston
WBW -   The Weekly British Whig, Kingston

CHAPTER 1

*James Lester Willis Gill, B.A.Sc., Professor of General
Engineering (QUJ 28 #4, Dec. 7, 1900, opp. p. 93)
-courtesy Queen's Archives*

J. Lester Willis Gill and Wireless at Queen's

Professor James Lester Willis Gill mounted the first public exhibition
of wireless telegraphy in Kingston at a Queen's University convocation
lecture on April 28, 1902, only four months after Marconi's first successful trans-Atlantic transmission of the letter "S." This mysterious
new technology, by which words and ideas were flung into space on
tiny lightning bolts and then recaptured at a distance in a peculiar jumble of coils, had been shown to Canadians in just a few places by this
time, most notably at the Toronto Industrial Exhibition in 1899.

The work was delicate and tricky. To make the apparatus perform on demand required great technical skill, knowledge and steady nerves. It should have caused a sensation, but Gill's remarkable exhibition resulted in a one-column account in one of Kingston's dailies and a squib in the other - as much space as they would give to a local farmer falling off his barn roof or to the beltline streetcar jumping the track. There was no follow-up to the article and no editorial comment in either paper on the implications for the future of signalling through space without wires.

In those days Kingston had excellent communications with Toronto and Montreal - the two largest cities in Canada and the nearest major nuclei of commerce - by rail, road and water as well as by telephone, telegraph and post. From the meagre coverage given Gill's notable demonstration in the local newspapers, it appears that the possibilities of communicating instantaneously over long distances by wireless telegraphy did not excite Kingstonians very much.

Who was this progressive young academic gentleman with expertise enough that he would dare to attempt a public demonstration of the ethereal wireless art? Where did he acquire his equipment and his experience? What was the state of Queen's University and Kingston in those days, and why did Gill choose Queen's and Kingston for his demonstration? A comprehensive laying out of all the facts that could be discovered is in order, since Gill and his contributions to Queen's have been completely forgotten over the last seventy years.

The Situation in Kingston and Queen's

At the end of the nineteenth century, Kingston was a moderately important focus of wealth in Canada. It was a shipping centre and a transshipping port for grain proceeding up the St Lawrence to Montreal. It was at the entrance to the Rideau waterway, and on the Canadian National and Grand Trunk railway lines. With a population of approximately twenty thousand, Kingston could boast a complement of industry and institutions far out of proportion to its size. The city had a locomotive factory, shipbuilding facilities, two drydocks, ship salvage companies, grain elevators, a cereal factory, several brewing plants, a knitting mill, a piano factory, a tannery, Queen's University and the affiliated School of Mining, the Royal Military College, and a large Dominion centre for incarceration and penitence.

In 1900, Queen's University and the affiliated School of Mining had 72 teachers and a combined enrolment of 727 students. There were 72 students in Practical Science in the School of Mining.[1] In comparison with the physical plant of the 1990s, the campus was then quite small and bare, with only four permanent buildings: Summerhill (1839; purchased by Queen's College in 1853), Old Medical (1858), Old Arts (1879) and John Carruthers Science Hall (1890). The Mill and Professor Dupuis' Mechanical Laboratory (the "Tool Shed") were the only other buildings on campus, and these were frame structures. The four stone buildings were overcrowded and the classroom and research facilities were poor. This was a critical period, for little Queen's was faced with heavy competition from the other Ontario universities, at Toronto and London. If Queen's were to survive, constitutional change and expansion were essential.

As a denominational institution, Queen's itself was not eligible to receive grants from the Province of Ontario, and the Presbyterian Church steadfastly refused to take any financial responsibility for the maintenance of its university.[2] The School of Mining, however, had been created by Principal George Monro Grant in 1892 with an independent and non-denominational constitution in order that it be qualified to receive government support. Thus, the School of Mining supported all of the costly science teaching at Queen's, yet officially the School's affiliation with Queen's was only for the purpose of granting degrees. This relationship would have to change as well. Principal Grant had decided upon separation of Queen's from the Presbyterian Church in 1899,[3] but this did not go forward until 1912 and did not come to fruition until several years later. Grant did, however, succeed in implementing plans for expansion of the university. By the turn of the century he had commitments for doubling of the space available for teaching on campus. In 1900 the ratepayers of Kingston voted $50,000 for the construction of a new arts building, Kingston Hall, and the Government of Ontario was in the process of approving a grant of $122,500, spread over five years, for two new science buildings for the School of Mining. These would be named Ontario and Fleming Halls.

Grant had also attempted to induce Frontenac County to contribute $20,000 for the erection of a new convocation hall, to be named Frontenac Hall.[4] The by-law was defeated, according to J.M. Macdonnell, later Chairman of the Queen's Board of Trustees, because the good people of the county objected to financing a building in which dancing would take place.[5] The truth of the matter was conveyed privately to Dr Grant in an unsigned letter from one of his churchmen.[6] In the late 1890s the Principal had appeared in a public debate with Methodist ministers prior to a referendum on temperance. Grant correctly predicted the evils of prohibition and denounced the temperance liquor laws from the platform of Kingston City Hall.[7] He was in favour of a liquor licencing act! A scandalous position for a Minister of the Presbyterian Church to propound! By wide concensus, Principal Grant was in the pockets of the hotel keepers, and as far as the electorate was concerned he could go to the hotelmen for his money. The students of Queen's came to the rescue and raised $35,000 in subscriptions for the building. There were plans to dedicate the new hall to honour Grant on the occasion of his 25th anniversary as Principal, in November 1902. The great and beloved Principal was in declining health, however, having worn himself out promoting the welfare and growth of Queen's. He did not live to see the construction of the building bearing his name. Queen's University and the School of Mining were well launched by Grant's efforts into a building program that would produce most of the familiar buildings of Queen's central campus by 1910.

The Situation in Engineering

The teaching of engineering in Canada was changing rapidly at the turn of the new century. It had become clear from the examples of Toronto and McGill that the School of Mining at Kingston had to rapidly broaden its scope to remain competitive. The several areas of engineering were going through a radical shifting of their relative importance and attractiveness to students. Civil engineers were having difficulty finding work, and students were now flocking to mining, mechanical and electrical engineering. It was becoming obvious too that, in the future, mining engineers would need to know the theory and practice of mechanical and electrical devices for their work. The School of Mining did not have a full-time instructor in either area and, if the curriculum were not widened, students would go elsewhere to study mining engineering.

In 1899 the Director of the School of Mining, Dr W.L. Goodwin, urged his Board of Governors to engage a full-time professor of general engineering, in addition to the agreed need of an instructor in advanced drawing and mechanical engineering.[8] A year later, Goodwin changed his mind and presented the Board with a report proposing that they hire instead a professor of electrical and mechanical engineering.[9] Even so, Goodwin was having second thoughts about disregarding civil engineering.[10] Mining, mechanical and electrical were in demand for the moment. When conditions returned to normal, however, and engineers were able to go back to their proper jobs building railroads, bridges and other public works, all the mining, mechanical and electrical men might be thankful for a thorough training in civil engineering.

The New Professor of Engineering

A compromise was reached with Principal Grant. The School of Mining would hire the professor of electrical and mechanical engineering, an assistant in mathematics with a civil engineer's training, plus a demonstrator to help him. Mr McKelvey, Dr Goodwin and Principal Grant were appointed a committee to hire the new professor at a salary of $1,500 a year. Negotiations were carried on through the spring and summer of 1900 with Louis A. Herdt, M.A. (B.Sc., McGill, 1893; then Lecturer in Electrical Engineering at McGill), who had been recommended by Dr H.T. Bovey, Dean of Applied Science at McGill. When it became apparent, in early August, that Herdt was going to refuse the offer, Goodwin wrote to Bovey to ask for the name of another suitable man.[11] Whether through Bovey's agency or not, Goodwin received a letter from Mr L.W. Gill of Westinghouse in East Pittsburgh, Pa., dated August 11, 1900, enquiring about the proposed appointment.[12] Gill accepted the position in a letter dated August 21, 1900, and at a meeting of the Board on September 9,[13] Goodwin announced that L.W. Gill, B.A.Sc., had been engaged for one session, October 1, 1900, to May 1, 1901, at $1,200. It was too close to the beginning of the fall session for the hiring of a permanent occupant of the chair. Gill must have made it plain that he needed to have the summer free to continue his M.Sc. work at Westinghouse in East Pittsburgh, for a letter dated August 24, 1900, signed by Registrar and Treasurer George Y. Chown, indicates that the Committee felt it was increasing rather than decreasing Gill's salary in allowing $1,200 for seven months instead of $1,500 for the full twelve. For the full salary, Gill would have to be available during the summer of 1901 to supervise the construction of a new building on campus.[14]

Goodwin wrote to Gill in East Pittsburgh,[15] expressing his pleasure that he had accepted the position offered, and informing him that the new science building would not be available for that season and that the School had no equipment for electrical engineering (E.E.). Therefore, the course of instruction in E.E. would have to remain largely as it had been in the past. In short, Gill was to teach general engineering as an elementary introduction in second year, and in third year civil, mechanical and elementary electrical engineering for the miners, plus the steam engine. The fourth year would include materials of construction and design, plus practical mechanical and electrical engineering for mining students, using the apparatus available. He could devote as much extra attention as he wished to those few students interested in mechanical or electrical as a specialty.

Within a month of his hiring, Gill submitted to the Principal a list of electrical equipment to be purchased by the School: standard generators, motors, electrical measuring instruments and lecture room apparatus.[16] He proposed a total expenditure of $3,300 on equipment for E.E., and pointed out that, to teach efficiently, the Board of Governors would have to spend $8,000 to $10,000. The Board had never been subjected to such demands and had to go on faith about the more expensive pieces. Gill recognized that space in Carruthers Hall would permit installation of a DC generator only, but it could be used after the day's teaching to supply electricity at night. The School had a real live wire on its hands, and the Board must have been highly stressed. In the School of Mining Annual Report for 1901-02, Director Goodwin stated, "Mining engineers must know to use electrical and other machines. Professor Gill has organized this department of study with great energy and ingenuity."[17] Later, Gill's students said of him:

> He's a tiny little crank but makes things hum,
> And he says all smoking shall be stopped. By gum!
> But he's dreaming. But he's dreaming. Wake him up.
> We wish we knew him better. We do. We do.
> And as his classes are no jest
> We know Prof. Gill has done his best,
> And we smile when we think of the future.[18]

Forty years later, Gill was still remembered as "a man of tremendous energy and an indefatigable worker...He gave the Director no rest and pushed him into new demands on the Board of Governors...Regulations were tightened up or introduced where none existed, for example the attendance regulation...no doubt...much of this was due to Gill."[19]

Along with his list of apparatus for E.E., young Professor Gill recommended the installation of a combined central heating and electric power plant on campus. Heating campus buildings was a dirty, smokey, labour-intensive business, depending on individul furnaces and stoves tended by the janitors.[20] Gill argued that central heating would cut down on soot, fire hazard and labour, would conserve fuel and would provide cheap electric power for lights and ventilation to the whole campus as a fringe benefit. The Board deferred action until the matter of a proposed power plant for the new engineering building itself, Fleming Hall, had been decided.[21]

Gill was certainly involved in the planning of Fleming Hall in his first year in Kingston. After all, he was one of those who would have to stake out space in the new building for his two departments and fill his space with the appropriate equipment for his teaching and research - he had been hired for his special expertise, and had impressed upon the Board some very definite ideas about what he required to do his job efficiently. The final plans for the new building were approved in the spring of 1901 and then Gill left for his summer research in East Pittsburgh, confident that he would be re-appointed Acting Professor of Electrical and Mechanical Engineering for a further two year term at $1,500 per year. When Gill returned to Kingston in the fall, Fleming Hall and its power plant were well above the ground and the earth was soon to be broken for Ontario Hall. During the summer, however, the plans for Fleming Hall had been modified to save money. The building had been made narrower and now the corridors took up a disproportionately large amount of space. Gill was furious, but it was too late to do anything about it, and occupants of Fleming Hall have always had to live with rooms that are too narrow.[22]

Somehow, Gill got permission to build his power plant. He probably designed and surely helped supervise the construction of that steam heating and power plant attached to the east end of Fleming Hall. In his history of the Science Faculty, Dr A.L. Clark recorded an incident told him by Mr G.Y. Chown, Registrar and Treasurer of Queen's: "Neither (Chown) nor Gill had noticed that no provision had been made for taking the boilers into the building and a part of one side wall had to be taken down to allow them to go in. He said that he and Gill were called up on the carpet and were given a good 'wigging'."[23]

How Wireless Telegraphy Came to Queen's

J.L.W. Gill's Background

James Lester Willis Gill was born into a farming community near Little York Station, a few miles north of Charlottetown, Prince Edward Island, on December 2, 1871. He was the eighth and youngest child of Abraham, Jr, and Sarah (Henderson) Gill. Grandfather Abraham Gill, Sr, an Island pioneer, came from Devonshire, England, in 1819 when he was 21,[24] and built up a prosperous farm on lot 34, PEI. Abraham, Jr, farmer and manufacturer of farm implements, was a pillar of the community and was known throughout the Island for his intelligence, integrity and public spirit. He was a successful fruit grower, active in the management of the annual Provincial Exhibition, and a staunch Liberal-Conservative.[25] He taught all of his children "the habit of hard work, for (he) believed that no child should grow up in idleness. The girls as well as the boys in the family were taught every phase of farming, and because their father continually impressed on them that it required genius to farm successfully, they also learned to respect their work."[26] The children were certainly encouraged to explore and be creative. In later life, Lester's older sister Elizabeth Jane Duncan observed that most people lacked pep. She ran an Alberta farm almost single-handed after her young husband was blinded in an accident. She looked after her husband, raised four children, cultivated and marketed produce, and still had energy to keep a half-acre flower garden, make the family's clothes, paint in oils, work leather, carve in wood, knit and crochet, embroider, hook and dye rugs, read poetry and study nature. She collected "rocks and minerals, shells and sea weed, wild flowers and ferns as well as birds and animals that she...stuffed herself."[26]

Lester was also a frenetic dymnamo, but he never sat still long enough for anyone to record all the things that he was up to. As a boy, he was passionately interested in aviation and built kites to study their flight.[27] He must also have helped his father in the manufacture of farm implements, because he later proved to be remarkably adept at machining and all sorts of shopwork. He attended the country primary school, but if he did go to Prince of Wales College in Charlottetown, even briefly, there is no record of it.[28]

In 1887, Abraham Gill, Jr, died intestate and deep in debt.[25] Young Lester left home shortly afterward and landed in Boston with two dollars in his pocket.[29] He used to say that he ran away from home at 15, but in fact he had relatives in the Boston area. The family of his aunt Jane (Gill) Morey lived in nearby Milford, Mass., and there is reason to believe that he had some contact with them, because he later married Emma Elizabeth Morey, Jane's daughter.

According to Lester,[29] he saw a newspaper advertisement by a baker seeking a boy to look after his delivery horses. He had been brought up on a farm, so Lester applied and got the job. After about six months, satisfied with the boy's work in the stables, the boss let him go out to sell. Soon he was turning over $50 and $60 a day! Sunday mornings were special, because the wagons used to go out with Boston rye bread and baked beans, hot out of the oven, and people would make these the mainstay of their Sunday meals. In the next few years, he also worked as a streetcar conductor and then as a contractor, doing carpentry, painting and odd handyman jobs. His manual skills and zeal for work, ingrained in him by his father, are certainly evident even this early in his life.

Lester soon realized that he wasn't getting anywhere without an education, and he made up his mind to return to Canada and go to

McGill University. By the fall of 1892, when he was 21, he had saved up enough money to put himself through, and he went up to Montreal and registered for admission into engineering. But Lester had no high schooling, and the McGill authorities turned him down. Lester came back the next day and kept coming back, day after day, until they got tired of kicking him out. The Admissions Board finally settled on a deal with him. If he passed all of his examinations at Christmas time, he could stay at McGill. If not, Lester agreed that he would leave. This kind of arrangement was not so unusual, as can be seen in the McGill calendars of the 1890s. Certain students were allowed to enter the university without admission examination, pending the results of the Christmas examinations. Full standing was conferred only after successful completion of the spring examinations. If in no other way, Gill must have impressed the Board with his tenacity, because on paper he was by far the least qualified of any of the applicants.

That first term at McGill must have been a horrendous struggle for the young man. He knew nothing at all about many of the subjects. He had to assimilate all of the basic matriculation material and master the university courses at the same time. But he passed every examination at Christmas. His grades continued to improve through the second term and he stood fourth in his year in the sessional examinations. He also took the prize in descriptive geometry and shared the Fleet Prize in woodworking in shop.[30]

The McGill calendars show that through the next three years, Lester consistently stood at the top of the Mechanical Engineering group, and won awards for shopwork, mechanism, physical culture, physics, the theory of structure and for his work in the testing laboratory. In the spring of 1896, he graduated with a Bachelor of Applied Science with honours in the dynamics of machinery, hydraulics, thermodynamics and mechanical engineering,[31] and he took the British Association for the Advancement of Science Gold Medal in Mechanical Engineering.

Lester Gill entered graduate studies in physics at McGill in the fall of 1896, working under the supervision of Prof. Hugh Longbourne Callendar[32] in the new Macdonald Physics Building. Callendar, a well-rounded physicist, came from J.J. Thomson's Cavendish Laboratories in Cambridge and specialized in thermodynamics and heat measurement, investigating the thermal properties of water and steam. He was one of those old-fashioned scientists who designed and built apparatus for performing exquisitely accurate measurements of known physical constants. One of his important contributions was a new standard in thermometry, the platinum resistance thermometer.[33] Gill was a mechanical engineer, a machine builder, and he probably fit right in. He began "examining into the various methods of testing the magnetic qualities of iron in October of 1896."[34] Also working with Callendar at that time were two graduate students, R.O. King and R.W. Stovell,[35] and three senior demonstrators in physics, Howard Turner Barnes,[36] Henry Marshall Tory[37] and Frank H. Pitcher,[38] all engaged in some aspect of continuous flow calorimetry.[39] There is no evidence that Gill was involved in the thermal experiments, although he certainly worked with Barnes, Tory and Pitcher.

Gill soon demonstrated his analytical skills and ingenuity in his research project. He found that the textbook method of end-on ballistics for measuring hysteresis of an iron core could lead to large errors. He showed how the method of measuring the attraction between an iron core and a solenoid, round which a current is passed, might better be employed to measure the hysteresis of the iron core if a series of force measurements were made: first as the core was inserted more and more into the solenoid and then as the core was drawn out of the coil, and then repeated with the current around the solenoid flowing in the opposite direction. This was the beginning of the development of the Gill hysteresis tester.[40]

In April 1897, with Callendar's support, Gill applied for a Royal Commission for the Exhibition of 1851 Scholarship, on the basis of his work with the hysteresis tester.[34] A scholarship of £150 per annum was granted, and he spent the first year of his post-graduate studies in the Physics Department at McGill on the hysteresis project. That year, he published a paper on his hysteresis tester.[40] He was also reported to be engaged in research on the Lorenz method for determining the value of the Ohm in absolute measure, using a specially imported English-made apparatus which was set up on a free-standing pier in one of the magnetic laboratories.[39,41] Barnes and Stovell used it too. Lester was also doing research on the distribution of magnetism in cylinders.[39]

Gill exhibited his hysteresis tester at the meeting of the British Association (BAAS) in Toronto, in August 1897.[42] Lord Kelvin, Nicola Tesla and Prof. Reginald Aubrey Fessenden were among the distinguished scientists at that meeting. Fessenden would soon apply for the Macdonald Chair in E.E. at McGill, to be rejected[43] in favour of an American, R.B. Owens of the University of Nebraska.[44] It is entirely possible that Gill and Fessenden met that August in Toronto, for Fessenden had worked on hysteresis in 1891 and was constantly plagued by it in his wireless research. Gill's apparatus might have interested him. Incidentally, it was just after the BAAS meeting, as Fessenden was on his way to visit his uncle Cortez Fessenden at Chemung Lake near Peterborough, that he hit upon the notion that Hertzian waves are actually continuous electromagnetic waves, like radiating ripples in water, rather than whiplash disturbances of the ether, as in Marconi's theory of transmission. This was Fessenden's first real breakthrough in the process of inventing amplitude modulation wireless transmission, or AM radio.

Lester Gill and Wireless

It is evident from the preceding that Lester Gill was an uncommonly determined, energetic and capable young man who did not always go by the prescribed route. Having earned a degree with highest honours in mechanical engineering, he was expected to go out into the REAL world to earn a living and build up his infant country. That's what graduate engineers did. They built railways and canals, roads, power plants and power transmission systems. Instead, he entered graduate studies in physics in order to become proficient in electrical matters. His undergraduate courses had not included any electrical work, and electricity was the coming thing. In those days there was little in the way of graduate studies in engineering at Canadian universities. Faculty members had private consulting practices and no time or use for graduate students or research.[45] Gill was strongly motivated toward E.E., so he found a route through graduate work in physics, rather than through formal undergraduate studies in E.E.

It may appear natural, then, for such an enterprising, inquisitive young man to gravitate toward the new and exciting field of wireless telegraphy. There is circumstantial evidence which places him geographically very close to several primary centres of research in wireless between 1898 and 1902. At least three people who were very active pioneers in wireless could have come into his life during this period. Then there is the matter of challenge. Four decades later, Arthur L. Clark wrote of Gill that he was "a man of tremendous energy and an indefatiguable worker...one of the most dynamic and lovable characters one could imagine. Nothing ever seemed impossible to him. If something seemed so to others, he immediately did it to prove that it could be done."[46] Given his character, there was probably no way to keep him out of wireless.

Callendar's laboratory was an exciting place for a young scientist to be working. The new Macdonald Physics Building was the most up-to-date and best equipped of its kind anywhere in the world. Moreover, Callendar had contacts in England, then the centre of the English-speaking scientific universe, and access to the latest and best equipment and information coming from the Cavendish and elsewhere. And Callendar's scientific interests were much wider than his list of ongoing projects at McGill indicates. After all, Callendar was the Macdonald Professor of Physics and, unlike his senior, Prof. John Cox, was an active and productive researcher in the laboratory.

The McGill Physical Society

An excellent way for a busy laboratory chief to review work in progress, to keep abreast of the latest in the scientific literature, and at the same time to make sure that his graduate students are exposed to all of the currents in related fields and to teach them to read the literature critically, is to form a "journal club" within the group. The McGill Physical Society (MPS) was founded on Saturday, September 25, 1897, and was to meet every Saturday evening for a series of presentations.[47] The founders were all members of Callendar's group: H.M. Tory, F.H. Pitcher, H.T. Barnes, L.W. Gill, R.W. Stovell and R.O. King. Callendar was President, Prof. Cox was Vice-President and R.O. King was the Secretary who left the too terse minutes. Each member was assigned journals to read and dates for presenting relevant articles drawn from these journals. Gill was to review <u>Electrician</u> and other electrical journals. Others assigned were Prof. Nicholson, Mr Louis Herdt and N.N. Evans, and a committee was struck to draw up a constitution and by-laws.

At the first regular meeting of the MPS, Saturday, October 9, 1887, "Professor Callendar gave a demonstration of the principle of the new Tesla oscillator, which had attracted considerable attention at the 1897 meeting of the British Association" in Toronto.[48] The Tesla oscillator was a machine for the rapid generation of high tension sparks, using a sort of interrupter regulating the primary; that is, it was for wireless transmission of electromagnetic energy - a radio transmitter. Several Saturdays later, Cox performed an experiment in electrical oscillations and Callendar demonstrated a new X-ray tube in which one could regulate the vacuum by means of a spark gap. So, there was an interest in that group in things related to wireless.

At the third regular meeting, Saturday, October 23, Lester Gill "reported on a paper by Fleming in the Phil. Mag. and on the September Nos. of the Electrician." The Fleming paper was titled "A Method of Detemining Magnetic Hysteresis Loss in Straight Iron Strips,"[49] right in Lester's own area of research.

The articles in the September numbers of Electrician were not further described in the minutes by Mr King. The September 10 number contained a review of papers by Fleming and Dewar on electrical phenomena in liquids at low temperatures. September 24 contained Gill's first published paper[40] and September 17 had a review of Marconi's first patent on wireless telegraphy, No. 12039 of 1896, "Improvements in Transmitting Electrical Impulses and Signals, and in Apparatus Therefor." It was applied for on June 2, 1896, and granted on July 2, 1897, by the British Patent Office. Until that time, the details of Marconi's wonderful invention were kept a deep secret, since publication would have made the system ineligible for patent. The article reviewed the Marconi system in great detail and showed eight illustrations from the patent. An illustrated editorial note followed, "Dr. Oliver Lodge's Apparatus for Wireless Telegraphy," which made the point that, from a demonstration given by that gentleman to the BAAS at Oxford in 1894 and from his subsequent publications, almost anyone could have done what Marconi claimed as his own: "It is reputed to be easy enough for a clever lawyer to drive a coach and four through an act of Parliament. If this patent be upheld in the courts of law it will be seen that it is equally easy for an eminent patent-counsel to compile a valid patent from the publicly described and exhibited products of another man's brain. No longer is it necessary to devise even so much as a 'novel combination of old instrumentalities', and the saying ex nihilo, nihil fit evidently was not intended to apply to English patents at the end of the nineteenth century."[50] The editorial missed the subtle point, however, that although Lodge had shown that Morse code could be sent by wireless, neither he nor anyone else had proposed a practical exploitation of this method for sending messages. With Marconi's patent, the idea took fire in the public imagination and the wireless

Equipment bought by Prof. John Cox, McGill University, for the Macdonald Physics Laboratory. Pieces of this collection would have been available to Gill in his wireless demonstration to the McGill Physical Society in 1897. Clockwise from bottom centre: glass-enclosed piece, which may be an electrocope; electrolytic detector (?); large induction coil for transmitting or receiving; resistance box (?); two large rings on an appartus for demonstrating mutual induction and transformer effects; unidentified apparatus with horse-shoe magnet, and below it a telegraph key; Clarke's Machine for generating an electric current; apparatus resembling a single-slide tuning coil; torsion or strain gauge using a travelling light beam.
-courtesy McGill University Archives

age was, if not born, at least weaned.

Two weeks later, on Saturday, November 6, 1897, the minutes of the MPS stated tersely, "Mr. Gill gave a demonstration of Wireless Telegraphy."[51] There is no further indication of what took place or of the equipment employed. The Montreal daily papers, both English and French language, were silent, and so are the surviving papers of those in attendance. Perhaps the news was not intended to go beyond the walls of the Macdonald Physics Building. If Gill did present the Marconi patent article to the MPS, perhaps Gill himself may have suggested that there was enough information in the literature to enable him to take a shot at building the apparatus. A register survives in the McGill Archives listing equipment available in the Macdonald Physics Building at that time, and the components necessary for the construction of a Marconi- or Lodge-type wireless telegraph were certainly there for the asking: Ruhmkorff coils, Leyden jars, induction coils, telegraph keys and batteries.[52] It would have been a relatively simple matter for a mechanical engineer and skilled machinist like Gill to manufacture any of the other components needed, from coils to spark gaps to coherer. Fourteen days later, Gill was able to give a practical demonstration of wireless telegraphy. Where demonstrations at meetings of the MPS were not successfully concluded, the fact was noted by the Secretary, Mr King. There is no such entry in the minutes with respect to Gill's demonstration of wireless telegraphy.

Was Lester Gill's the First Demonstration of Wireless Telegraphy in North America?

William Joseph Clarke, of Trenton, Ontario, gave the first public demonstration of wireless telegraphy (W/T) in the United States on Thanksgiving eve, November 24, 1897, at the New York home of banker Jacob Schiff.[53] Clarke's apparatus was built for him by J.H. Bunnell and Company, based upon the descriptions in the literature of the Marconi patent, and he used it to send a simple "on" signal to close a relay. Clarke did not send a message, so it is doubtful whether the thing could properly be termed telegraphy. "Some weeks before" he had made a preliminary test in the offices of the United States Electrical Supply Company at 120 Liberty Street in New York, starting a small motor located across the hall in the offices of The Electrical Engineer. At some time between these two demonstrations he had done some experimenting at Lancaster, Pa. The dates are not recorded, and it seems that Clarke was not fussy about staking a claim of priority.

Gill's W/T demonstration at McGill did not seem to be very important to him either, as he never made reference to it in the bits of his recollections that were published. His demonstration at McGill may or may not have involved the sending of a real message in Morse code, but the date of November 6, 1897, could qualify it as the very first verifiable demonstration of W/T in all of North America. In any case, a Canadian was the first to do it.

Lester Gill Continues His Graduate Work

Gill worked at perfecting his method of measuring hysteresis loss in iron until June 1898.[34] The difficulties to be overcome necessitated a thorough investigation of the distribution of magnetic induction in straight iron rods, and this was the subject of his second scientific paper.[54]

In the middle of April 1898, Callendar was notified by cable that he had been appointed Quain Professor of Physics at University College, London.[55] He resigned his professorship at McGill in May and left almost immediately to take up his new position.[56,57] Exactly what Gill's options were at this point cannot now be discovered, nor is it possible to know whether he had already plotted his future course in anticipation of the break up of Callendar's group. Certainly, his work was not in Callendar's area of immediate interest and, in any case, Callendar did not take any of his McGill graduate students with him. Principal Peterson and Prof. Cox went back to J.J. Thomson at the Cavendish Laboratories in Cambridge that summer and sought another of his students to replace Callendar. On Thomson's recommendation they engaged a sharp young physicist named Ernest Rutherford, then 28 years old.[58] J.J. Thomson thought very highly of the young man.

Rutherford was disdainful of engineer types who constructed machines to help determine the next decimal place of a known physical constant.[59] To his way of thinking this was not useful or interesting. It was the discovery of new scientific truths that excited young Rutherford. Gill fell squarely into the category of machine-building engineer, but there is no evidence that the anticipation of problems with Rutherford as Junior Professor of Physics determined Gill's future. Certainly, other of Callendar's pupils later collaborated with Rutherford. Their expertise in heat transfer was essential in the early work on radioactive decay.[60] It is intriguing to note that Rutherford arrived at McGill in late September 1898,[61] and that Gill registered at Harvard on October 10, 1898, taking his Exhibition Scholarship with him.[62] One of the conditions of the Exhibitions is that the student study at a university other than his own, so Gill may have been fulfilling a requirement of the award by leaving McGill. His colleague, R.O. King, had spent the second year of his Exhibition at Harvard in 1896.

In his year at Harvard as a graduate student in physics, Gill studied mathematics and physics and worked in the Jefferson Physical Laboratory with Prof. Edwin Herbert Hall[63] of "Hall Effect" fame, attempting to work out a method of measuring permeability using his hysteresis tester. His long and detailed investigation showed that it was possible to determine the value of the permeability easily and quickly with his tester. These results were never published, and he did not write the final examinations in his courses at Harvard. The report of the 1851 Scholarship Commission concludes, "Mr Gill appears to possess not merely the power of working hard, but initiative in devising new methods and apparatus. His work is throughout distinctly good..."[34] During the period of his studies with Hall, Gill patented a mobile hydraulic extensible-bridge fire-fighting and -escape apparatus, probably under commission from a consortium.[64]

Possible Contacts with Other Wireless Pioneers

At Harvard, Gill may have become acquainted with a brilliant Texas farmer's son his own age, George Washington Pierce, who arrived in the same year to take up a scholarship to work with Prof. John Trowbridge in the Jefferson Physical Laboratories.[65] Pierce was interested in the behaviour of radio waves, and he devised a radio micrometer for measuring short radio waves. Later he was to study refraction of radio waves, resonance in wireless cicuits, crystal rectifiers or semiconductors, radio oscillations and the radiation characteristics of antennae, as well as piezoelectric crystal resonators and oscillators. Pierce went on to write two classic books on wireless,[66] led Harvard to a pioneering position in wireless communication, became President of the Institute of Radio Engineers in 1918-19, and succeeded E.H. Hall as Rumford Professor of Physics at Harvard. Pierce had spent four years doing odd jobs after leaving the University of Texas in 1894, and that work experience might have brought him together with Gill. They had one more thing in common - Pierce's mother's maiden name was Gill. No documentary evidence exists to confirm or refute it, but perhaps Gill practiced the new wireless art with Pierce at Harvard in 1898-99.

In 1899, Gill went to the Testing Department of Westinghouse in East Pittsburgh, Pa., with the purpose of "making a study of manufacturing methods and shop practice."[67] It is more accurate to say that Westinghouse gave its Machine Test Course to selected engineering graduates as a means of obtaining eager and intelligent workers at little cost.[45] The students would help with final testing of products - machines and switching gear - both prototypes and shop production for shipping.[68] They would also be encouraged to experiment to improve the accuracy and efficiency of the tests. This sort of practical work experience was essential for a novice engineer on the way to a steady job or to the M.Sc., and the company had the option of taking the best of these young men onto its employment rolls.

East Pittsburgh and Westinghouse were also the stamping grounds of the Professor of E.E. at the University of Western Pennsylvania, Reginald Aubrey Fessenden.[43] Fessenden had saved the Westinghouse contract to light the Columbian Exposition in Chicago by working around the Edison patents, and George Westinghouse was grateful. He wanted the young scientist as a local consultant, convinced the university to offer the new chair of E.E. to Fessenden, and gave him access to the Westinghouse factories. At this point, Fessenden would have been experimenting with alternating current (AC) generators for his wireless work, attempting to produce continuous waves with frequencies above audibility to carry voice and music through the atmosphere. It is reasonable to assume that Fessenden would have been invited to present his work to the students and staff at Westinghouse. This was five years before Westinghouse began publishing The Electric Club Journal, proceedings of the journal club of the students in the testing department, so there is no published record of his presenting a paper. By early 1900, though, Fessenden had gone to Cobb Island in the Potomac River to work for the US Weather Bureau, and it was there that amplitude modulation radio was born on December 23, 1900.

Although a Gill-Fessenden connection in wireless telegraphy is an enchanting possibility, the little evidence existing in Gill's publications on wireless does not support such a relationship. He never made written reference to Fessenden. The other notable circumstance of Gill's wireless activity, that he showed no interest in voice transmission, also tells strongly against a Fessenden connection.

A Rutherford connection, however, is not at all out of the question. Ernest Rutherford built a magnetic detector for sensing electromagnetic radiation while a student in New Zealand in 1895, and took the device with him to the Cavendish Laboratories. As Junior Macdonald Professor of Physics at McGill, Rutherford lectured to the Royal Society of Canada on wireless in 1899 and three years later experimented with the reception of wireless signals aboard a speeding train, in company with

H.T. Barnes, C.H. McLeod and H.T. Bovey. His first graduate student in Montreal was Harriet Brooks, whose M.A. thesis in 1901 was on the damping of electrical oscillations in the discharge of a Leyden jar across a spark gap. That year, she and Rutherford discovered the recoil of radioactive atoms, which led to their seminal paper on the atomic weight of radium emanation.[38,69]

Gill continued in some manner with his M.Sc. studies until 1904, well after his appointment to the School of Mining at Kingston. He spent the summers of 1901 and 1902 at Westinghouse in East Pittsburgh and possibly the next two summers as well. He may also have done further academic work toward his degree in Montreal, although when and what that work might have been is not known. Whatever the circumstances, Ernest Rutherford's signature is one of nine on Gill's M.Sc. sheepskin, suggesting that Rutherford may have been one of Gill's classroom instructors sometime between September 1898 and April 1904, or that Rutherford was a member of a rather large committee attesting to Gill's worthiness for the M.Sc. There is no evidence that Gill was involved in any of Rutherford's wireless experiments at McGill, or that he was in attendance at Rutherford's jam-packed lectures on wireless telegraphy.

Wireless at the School of Mining

Survival of materials documenting the day-to-day life at universities is a very haphazard kind of thing, even in the best of cases. Around the turn of the twentieth century, Queen's College and the affiliated School of Mining at Kingston were very small institutions. There were few administrators and most of the daily business was probably handled more or less informally, so there was little need to make detailed records. Much of what was put on paper just did not survive the periodic scrambles to find extra space or to dispose of potential flammables. What has survived at Queen's often comes out of personal files of deans and directors, from individual initiatives to preserve caches of old material bequeathed or found, or from hidden corners that nobody ever got around to cleaning out. Much of what survives does so as an entry, sometimes vague, in an unique register or letter book, or as a single slip of paper tucked long ago into an inappropriate file. The newspapers, campus and city, are some help but they have never had any memory, even for the recent past. The student papers are often sarcastic in tone and the writers take pains to make their references obscure, so these are not very useful in resurrecting the past. Thus, substantial allusions to wireless telegraphy in the Kingston press before the First World War are extremely rare. Unfortunately, most of the information about the early days of wireless activity at the University and in Kingston has perished with the witnesses. They kept no diaries or logs and evinced little sense that they were helping to break new ground.

The wireless activities at Queen's in the first decade of this century were entirely extra-curricular, and would have been confined to the select few who knew the Morse code and had a working knowledge of the latest advances in the electrical arts. As a subject of instruction at Queen's College or in the School of Mining, wireless can have been accessible only to those few students studying E.E. as a specialty. When he arrived in Kingston in 1900, Gill may have found one or two of these queer ducks struggling to find an electrical path through the School of Mining. "To these", instructed Director Goodwin, "you could devote particular attention and, in fact, give them all they can take of these subjects."[15] But the first batch of real electricals emerged in 1905 with Gill's developing program. Unfortunately, detailed curriculum records from this period do not exist, and no ephemera have surfaced to shed any light on Gill's classroom or extra-curricular wireless activities during his first 18 months in Kingston. There is, however, an intriguing entry in the School of Mining Cash Book for February 1, 1901, tersely noting:

J.G. Biddle   Engin 186 $118.46   (requisition #266)[70]

The James G. Biddle Company of Philadelphia was an electrical supply house which handled a wide variety of apparatus for industrial, educational and home use. The records at McGill Archives show that Callendar dealt with Biddle in 1897. In his first year or two at Kingston, Gill often ordered small items for his department from Biddle, the requisitions rarely exceeding $50, because there was not much money available to spend.

A decade later, Gill ordered a complete outfit of wireless telegraphy from James G. Biddle, so it is not at all unlikely that this major purchase of February 1, 1901, may indeed have been a wireless telegraph set. It cannot be proven though, because requisition #266 no longer exists. It is equally likely that Gill built a W/T set from scratch or from commercially available components. The incontrovertible facts are that he had access to a wireless transmitter and receiver by late April 1902, and that he knew how to send and receive Morse code messages with the rig, for he gave a public demonstration of wireless on the campus on April 28, 1902.

Professor Gill's Wireless Exhibition of 1902

The closing exercises of the 61st session of Queen's and the School of Mining took place on Sunday and Monday, April 27 and 28, 1902. On Monday the 28th, the Annual Meeting of the School of Mining was held in Carruthers Hall at 3 pm, and the associated Popular Science Lecture was delivered by Professor Gill at 8 pm in Convocation Hall in the Arts Building. No letters of proposal or invitation concerning this lecture have been found but, of course, things were quite informal then. It must suffice that it was plain to someone of influence that young Gill had a handle on a hot topic and that someone considered that this wireless business would make for an evening of wholesome and uplifting instruction, suitable for such a serious occasion. The topic, "Wireless Telegraphy," was announced without fanfare by the Daily British Whig on April 25.[71]

As inevitably happens with truly momentous events, Gill's wireless lecture was overshadowed in the press. First, there were the many newsworthy doings during Convocation Days at Queen's: the Rev Dr Clark of Toronto would preach the Baccalaureate Sermon on Sunday; Students' Day, Tuesday, would feature the unveiling of a portrait of the late Dr Mowat; and on Wednesday, there was Queen's Convocation proper and the laying of the cornerstone of the new engineering building by Sir Sandford Fleming, KCM, Chancellor of Queen's; not to mention the laying of the cornerstone of the new mineralogy-geology-physics building by the Hon Richard Harcourt, Ontario Minister of Education! A breathtaking whirl of excitement. So, there's little wonder that Gill's lecture on wireless was not able to attract attention in all of this fun and festivity. It was only after the event that Kingston learned of the miracles of modern science displayed at the College on Monday evening.

Then, on the very afternoon of the wireless lecture-demonstration, a shocking and senseless tragedy took place in Miss Boyd's Junior Fourth class at Frontenac School, and blew Queen's right off the front page. Beatrice Holland, aged 14, was shot in the temple and killed by a classmate with a .22 calibre revolver during a playful altercation in the cloakroom.[72] That ensured that science news was kept in the back pages of the Kingston papers.

Kingston News devoted just a few lines on page 4 to Gill's lecture-demonstration:

> Last evening Convocation Hall, Queen's University, was filled by students and friends to hear Professor Gill lecture on the Marconi System of Wireless Telegraphy. Chancellor Fleming presided, and a vote of thanks was passed to the lecturer.[73]

The Daily British Whig, uncharacteristically, splurged half a column-page on the wireless story,[74] and it is this lone article that provides the only evidence that Gill performed a practical radio demonstration that night during his lecture. Chancellor Fleming's diary contains no report of the proceedings. The text of the wireless lecture was published in the Queen's Quarterly in 1903,[75] including illustrations of the standard laboratory apparatus that Gill employed to show physical analogies of the behaviour of electromagnetic waves. No mention was made in the Quarterly of Gill's demonstration of the transmission and reception of wireless messages that evening, although he did it at least twice: first with wires, and then without wires.

There is enough information in The British Whig and in the Quarterly articles to make possible a reasonable reconstruction of Gill's lecture-demonstration at Convocation. On the basis of The Whig story, and some knowledge of the Marconi apparatus of the day, one can make some educated guesses about the apparatus that he was using. It is useful to reconstruct the event here in some detail because it gives a picture of the state of the wireless art of the day and of the under-

Reconstruction of Prof. Gill's wireless demonstration, April 28, 1902, in Convocation Hall, Old Arts. The metal frame with suspended copper spring, for demonstrating wave motion, is on table at left along with the receiving apparatus. Transmitting equipment is on the table at rear and suspended above is the large sheet metal radiator, which was part of the antenna.

pinning of theoretical comprehension.

Picture the intense little man with liver-coloured hair and handlebar moustaches holding forth from a low-rise platform under the rose window to the assembly in the solemn and small Convocation Hall. Toward the front of the platform, at the north end of the Old Arts Building, would have been a table bearing a long spiral-coiled copper spring suspended by elastic bands from a metal frame. Another table would have held one or perhaps two curious little contraptions, with wires and coils and brass rods, mounted on individual boards and connected by more wires to an array of batteries underneath the table. The whole scene would have been dominated by a large sheet of metal, perhaps ten feet square, suspended over the stage by stout ropes.

According to the <u>Queen's Quarterly</u>, Gill first defined wireless telegraphy as "the science or art of communicating over long distances without the use of wires. 'Ten years ago,' he continued, 'we heard practically nothing of this subject; to-day it is engaging the attention of all the civilized nations.'" The most ancient system of human communication, by means of sound, is naturally limited as to distance. Sound energy, he explained, is dissipated rapidly in all directions, so that in order to attempt to communicate over long distances by sound waves, the magnitude of the disturbance at the point of generation would result in "serious damage to person and property." One can send and receive messages by means of smoke signals or by means of light, for example by using a heliograph, but these signals will not pass through fog or opaque objects. The third system, the Marconi system, uses electric waves to carry messages. What all three of these systems have in common is that they all effect communication through the use of wave motion.

Gill illustrated sound wave motion by means of his long copper spiral spring, suspended from a metal frame by elastic bands every few turns. Oscillations set up by a pendulum connected to one end of the spring would result in a series of periodic waves, contractions and rarefactions, travelling along the spring. Attaching tiny paper discs to the spring made the motion visible to the audience, and he explained that the motion observed corresponds to the motion of the particles of the particular atmosphere propagating the sound waves. Less energy is required to send sound waves through a group of less massive particles, such as hydrogen gas, for less energy is required to impart the motion to the particles in that atmosphere. With light and electric waves, the transmission medium has practically no mass, so only a small amount of energy is needed to communicate by means of these waves over very long distances.

Next he discussed the proposed medium of transmission of light and electric waves, the ether, about which very little was then known: "It is not difficult to reason that such a medium exists, and that it penetrates all matter. For instance, if we take all the air out of a glass jar, sound will not travel through it, yet light will; if there were nothing left in the jar, how could the motion which constitutes light be carried through?" He went on to point out that magnets will attract or repel in a vacuum, so that force must be transmitted through some medium other than the atmosphere. And light travels millions of miles through space from the sun and the stars. Light is a wave motion, and so must have a transmitting medium. Light will penetrate all matter to a greater or lesser degree, so the ether must be omnipresent. It must possess the properties of inertia and mobility so that the waves can go through it. As inertia depends upon mass, the mass of the ether should be measurable. To that point, no instrument of sufficient delicacy had yet been devised to measure the mass of the particles of the elusive ether. Light waves, Gill continued, differ from sound waves in that the motion of the ether in the transmission of light waves is at right angles to the path of propagation, rather than parallel to it as with sound waves. He illustrated the propagation of light waves by modifying his copper spring apparatus, attaching to one end a weight suspended between two vertical springs. Vertical oscillation of the weight set up undulating transverse waves in the spring, and generated similar vertical oscillations in an identical spring-mounted weight at the other end of the copper spring.

He went on to explain that electric waves or "Hertz waves" are of the same character as light waves, and are transmitted by the same elusive etheric medium at the same velocity as light, 186,000 miles per second, but at 100×10^6 cycles per second rather than the 400 to 800 x 10^6 cycles per second of light. Just as in the spring model on the stage, an electric wave striking a conductor gives rise to oscillations in that conductor. If then, two wires are placed parallel to each other and electric oscillations are set up in one wire, the electric waves generated by these oscillations radiating in all directions strike the second wire and set up oscillations in it, by induction.

There were ways to detect the oscillations in the second wire, most commonly by means of a device called a "coherer," which was attached between the second wire and a ground. The coherer, attributed to Edouard Branly, Professor of Physics at the Catholic Institute of Paris, took advantage of the observation that the resistance of a coat or varnish of fine copper dust decreased markedly whenever an electric spark was produced in the neighbourhood. What had been almost an insulator suddenly became a fine conductor, and remained so until the "cohesion" (a term coined by Sir Oliver Lodge) of the particles was broken down by some mechanical tremor. Essentially, the coherer was a small (4 cm) glass tube with tightly fitting rods of brass or silver inserted into each end until the rods were a few millimetres apart. One terminal was attached to ground and the other to an antenna plate. Inside the glass tube between the ends of the two rods was a small quantity of fine metal filings, perhaps mixed silver and nickel, or iron, making a "bad" contact across the terminal rods. An electric wave anywhere in the immediate vicinity would reduce the high resistance of the metal filings, in proportion to the intensity of the disturbance, and the coherer would become a good conductor. If one then tapped the coherer tube, known as "tapping back" the coherer, the filings were disturbed and the apparatus once more assumed its high resistance.

Now, if a battery were connected across the metal plugs of the coherer, current would not flow in the initial state since the resistance is high. When electrical oscillations are set up in the antenna plate, resistance in the coherer drops rapidly and a local current will flow through the filings and the battery. A bell can be put into the battery circuit, and when the battery current flows through the coherer under the influence of an incoming electromagnetic wave (i.e. when the coherer is in the "on" state), the bell will ring. The filings will remain cohered even after the incoming oscillations have stopped, and therefore the bell will continue to ring. In order to shut off the bell, the coherer must be "tapped back" to de-cohere the filings, increase the resistance of the filings and stop the flow of battery current through the coherer. The tapping back can be performed most efficiently by putting an electric bell hammer into the battery circuit to tap the coherer tube when the battery current flows. Rapid tapping back by this apparatus would allow one to distinguish between a short pulse of waves (a Morse dot) and a long pulse (a dash). This system of wave detection is inherently slow, permitting a receiving rate of only 12 to 15 words per minute, and the operation of a delicate instrument like the coherer is a complex art in itself, requiring great patience and ingenuity.

In the Marconi system of wireless telegraphy, the ground and antenna are arranged vertically, each ending in a large, roughly square sheet of metal, called a "radiator," with the coherer wired in between the aerial wire and the earth.

An unexpected tone of skepticism then crept into Gill's account. The crude spark transmitters of those times put out an extremely broadband signal, so that each one took up the whole spectrum and interfered with everyone else. Only one transmitter could be on the air at a time. Signor Marconi had claimed to have perfected a syntonic or tuned system, so that the transmitter and receiver could be tuned to each other. In this scheme, each receiver would respond only to waves of a certain length, and would not respond to waves of any other length. Each transmitter could be made to radiate waves of any required length, and each transmitter and tuner could be continuously tunable to each other, thus permitting many transmitters to operate simultaneously without any mutual interference. A tuned system was then demonstrated with the copper spring apparatus, showing how a body could be set into vibration by a wave motion, if the natural oscillation frequency of that body were in tune with the incoming wave motion - a sort of sympathetic vibration. The wave motion would not, however, affect bodies whose natural frequency of vibration was not in tune with the impinging wave.

In 1902, the spark transmitters in general use were sending out signals as broad as a barn door, highly damped signals that contained an enormous range of frequencies, and no means was apparent for sorting out specific frequencies to send. Hence Gill's skepticism. When the tuned system was perfected in the future, Gill speculated, each person could be supplied with a tuned receiver and have his own "wave length" registered in a published call book - sort of like a telephone book. If you wished to communicate with a friend, you would go to the nearest transmitting station, look up your friend's wavelength in the registry, tune the transmitter and call up your friend. If you received no reply, you might conclude that he was dead! The new system would not replace the telephone or telegraph, but would popularize communication and would reduce the rates. The telephone did not replace the telegraph, nor did electric light reduce the use of gas. "In time," Gill predicted, "we might do away with ordinary speech, and communicate with electric mouths and ears." Citizens' Band radio and synthesized speech messages have brought his predictions into our lives. It is typical of the day, though, that the idea of wide public broadcasting was not even considered. Wireless was envisioned by everyone as being purely for interpersonal communication.

Gill did not appear skeptical of Marconi's most recently announced triumph, trans-Atlantic telegraphy, only four months before. This receiving of signals from across the Atlantic Ocean (even if it was only the code for the letter "S" - three dots) was not so remarkable, he thought. The distance is only two thousand miles. This is hardly as mysterious as sunlight or starlight waves travelling millions of miles to earth. Why is it so remarkable that electric waves will pass through stone walls and fog, when light will pass easily through glass? After all, X-rays pass through things opaque to light, and X-rays and light differ only with respect to wave length. Indeed, sound waves are transmitted through the walls of buildings. So why not electric waves? Gravity is also a profoundly mysterious force which operates through barriers, yet we take it for granted. "It adds another germ of knowledge which helps to impress upon us how little we know and how little we can ever know concerning the true nature of the things around us."

At some point during the proceedings, Gill sent messages across the platform of Convocation Hall by means of two wires attached to his apparatus. Whether the equipment used was telegraphic or telephonic was not disclosed by <u>The British Whig</u>, nor whether the means of transmission was conductance or inductance. Then came the <u>pièce de resistance</u>, which should have astonished the people there gathered to be edified (or rather, marconified): "The professor had an apparatus arranged by means of which signals were transmitted without wires, through two walls and down the length of Convocation Hall, and the message 'Success to Marconi' was spelled out by taps of a bell on the platform." The transmitter was probably located inside what is now the Theology Department Office, opposite the entrance of Convocation Hall on the south side of the building and two stone walls away. Gill must have had an assistant at one end or other of the system, but neither his existence nor identity is hinted at in the articles.

The transmission equipment would have been an adaptation of the Marconi apparatus of the day: a Ruhmkorff induction coil with interrupter, a spark gap, probably tipped by small brass balls, called "oscillators," across the secondary of the coil, a Morse telegraph key, a tray of batteries, and a large metal plate or radiator hung above the apparatus and connected to one side of the spark gap. Another wire led from the other side of the spark gap to a second large metal plate buried in the ground.

The <u>British Whig</u> and the text in the <u>Queen's Quarterly</u> make only oblique references to the receiving apparatus that Gill had with him on the platform in Convocation Hall. The former mentions a bell as the sounder, while the latter describes the coherer at some length and mentions that "in the Marconi and other systems the wires are placed vertical, and the coherer is placed between the wire and the earth." The wire referred to is the antenna, which in the Marconi system of the time was a single vertical wire likely ending in a large metal plate, or radiator. These references, plus the message sent during the demonstration, imply that a Marconi-type of receiver was being used as well as a Marconi-type of transmitter. By this time, the Marconi system was no secret in its general characteristics. A keen amateur could find instructions for building a complete Marconi-type wireless telegraph set in a growing number of magazines.[76] The standard receiver of those days had a coherer, sounder, tapper, battery, plus aerial and ground plates. The aerials were to be mounted upon 30 or 40 foot masts.

The students at the <u>Queen's College Journal</u> were as perverse and blasé as students anywhere, and reported: "During his lecture on Wireless Telegraphy the other evening, Professor Gill received a wireless message from Marconi congratulating him on the success of his experiments, and expressing his regret at not being able to be with us for Convocation."[77]

If we may believe that Gill had been conducting experiments in wireless in Kingston before his lecture-demonstration, then the arrival of radio at Queen's University obviously antedates April 28, 1902, and the reason for extending the invitation to present the Convocation lecture is no mystery. Finally, if he was indeed using the coherer as his wave detector, then all doubt subsides that he was conducting experiments in wireless telegraphy in Kingston. The coherer was such a temperamental and obstinate beast, requiring endless adjusting and tweaking to get it to perform, that only a practiced hand could have tamed it. In fact, in those days it was an act of bravery, or foolhardiness, to attempt a public demonstration of wireless using a coherer as detector.

Apart from the newspaper articles reporting Gill's lecture-demonstration of wireless telegraphy, there seems to have been no other manifestation of excitement engendered by the showing off of this startling new technique in communications, either within the College or the city. The last word on the matter appeared in 1903, when the Queen's Journal included a comment on the text of Gill's lecture as published in the Queen's Quarterly: "...Professor Gill's contribution on wireless telegraphy presents in popular form the salient features of this new means of communication and will help to explain the physical phenomena on which it is based..."[78]

Professor Gill's Wireless Activities, 1902-09

Gill's interest and active work in wireless must surely have continued through the period 1902-09, but little documentation survives. It was a time of feverish activity for him in many other areas of responsibility. In his first few years at the School of Mining, Gill was a prime mover in the rapid expansion of teaching. The School of Mining Report of 1903 stated that Gill's courses had developed so well that he would need a teaching assistant to free him for advanced work. There were now 25 registered in E.E. and 4 in Mechanical.[79]

As he built his two departments from scratch, developing courses and laboratory exercises, he always kept very much in mind the welfare and needs of his students. And they appreciated what he was doing for them. In 1903, the Science valedictorian tendered "to Professor Gill the assurance that his efforts have been more than appreciated and that his energies and labour have won the esteem and confidence of the students."[80] He established a study room for the students,[81] gave special lectures to the Engineering Society on electric cranes and derricks[82] and on pending legislation which he believed would be detrimental to the young engineer.[83] The students were "loud in their praise" of his efforts to give them practical experience in boiler tests in his new power plant on campus,[84] and of his organization of tours to important mechanical and electrical manufacturing industries in Peterborough, Hamilton,

Niagara Falls and Montreal.[85] He took his students to investigate the Kingston Locomotive Works and had them participate in tests on the Kingston light and power plant.[86] And finally, Gill arranged with a number of these firms to take his students on as apprentices during their second summer vacation.[87]

In the middle of all of this activity, Gill completed the requirements for his M.Sc. at McGill University, and received the degree in the spring of 1904.[88] No record survives to indicate the subject of the thesis, although it probably had to do with the substance of his two papers on hysteresis. There is, likewise, no record extant to indicate why Ernest Rutherford was a signatory to Gill's sheepskin.

Gill's restless activity continued. In 1905, he, Macphail and A.K. Kirkpatrick proposed to the Board of Governors that the School of Mining establish a national correspondence school.[89] This did not get past the planning stage. Gill also proposed that the University sell electricity from its steam power plant to Kingston at low cost.[90] Gill and Kirkpatrick were authorized to negotiate with the city,[91] but Queen's power requirements quickly expanded to take up the plant's entire reserve production and the scheme came to nothing.

Another of Gill's projects which did not see proper completion was a series of articles, titled "Elementary Electrical Engineering," published between March 25 and September 29, 1910 in The Canadian Engineer.[92] The last article was prefaced by a notice that the series would be continued for some months, but no explanation was proffered for its sudden disappearance. Late in the summer of 1910, the articles were published as a textbook of the same title by the Jackson Press of Kingston.[93] The book ends exactly where the uncompleted series of articles ends, with storage batteries.

At the same time, beginning about the fall of 1909, Gill began to design and construct a "safety first" aeroplane, engine, wings and all, out of welded tubular aluminum,[94] a project that literally failed to get off the ground and was abandoned in the spring of 1915 in favour of an active part in World War I. Earlier, Gill had patented a spinning machine,[95] possibly for the Kingston Knitting Mills, and had developed an improvement on the Prony brake for laboratory use.[96] That Prony brake was in use in Queen's E.E. through the 1950s.

Gill also had the time to serve as a member of the Canadian Bureau of Standards, and was Canadian representative at the Turin meeting of the International Electrotechnical Commission in September, 1911.[97] In addition, he was in charge of the University's steam and power plant, and of the installation of the heating, lighting, ventilation, telephones and low-power electrical systems in the new Gordon Hall in 1910. He turned up at the 24th Annual Meeting of the American Institute of Electrical Engineers,[98] and made a presentation to the Dominion Royal Commission on Technical Education in September, 1910.[99] References to Gill's activities on campus and in Kingston turn up in all sorts of places. He was a founding member and Secretary of the Kingston Branch of the Canadian Society of Civil Engineers, an active member of the

Kingston Board of Trade from 1912 and Superintendent of Ice (1906-08) and President (1909-10) of the Kingston Curling Club. He also belonged to the Kingston Canadian Club, and was elected President in January 1915.

Professor Alexander (Sandy) Macphail recalled the character of his old friend Lester Gill in an obituary tribute:

> Lester Gill died last August at the age of sixty-eight, after a life full of fun, of good fellowship, of loving regard from his students, his comrades, and his fellow citizens. He was of an affectionate nature, of a rare simplicity of character, and of brilliant achievement...Gill was a most engaging fellow. His irrepressible mirth, his delight in a good story, his infectious laughter were a joy to us. We used to meet often in Fleming Hall where nearly all "Science" was housed. We "worked" there every night, preparing our lectures for the next day's classes. We all had about twenty-five hours of instruction a week. A notable figure among us was Lester Gill with his short stature, his sturdy figure, his mop of red hair, which he kept to his last day, and not red either, but a lovely liver colour, or sometimes oxblood, or tinged with gold; his incisive speech, his accuracy of information...I had many a visit with him at his battery positions (in Flanders in World War I); in times good and bad, he was always the same old irrepressible Lester Gill, exploding in laughter, or passionate and obstinate in argument, or serious in discussion, as the boy I had played with forty years before...In all the years I knew Gill, I never lost my wonder at his boundless, overflowing energy, his capacity for increasing work, his clearness of thought, his directness of approach to any problem, and his complete honesty of purpose. He was one of the best teachers I ever knew; he demanded hard work from his students - and got it. The vivid impact of his personality hit the student between the eyes, and the student gave back of his very best...He had built up quite a department of electrical engineering, with a minimum of material equipment, but with a vast fund of enthusiasm and personality...Gill will long be remembered for his many engaging qualities of mind and heart.[27]

Gill was a work-a-holic, and cannot have had a great deal of time during his first decade at Queen's to experiment with wireless. All was quiet on the campus with respect to wireless until an unsigned article, commenting on current events in wireless telegraphy, appeared in the Queen's University Journal late in 1907.[100] The author commented upon the predictions of disaster for the Marconi system made some seven years before by the cable proprietors, warning of loss or unauthorized diversion of wireless signals, and of problems with plagiarism and theft of and atmospheric electrical obstructions to the "thought-laden current." But Marconi had persisted in perfecting his equipment, and on

October 17, 1907, had inaugurated his new commercial system for public use. The article concluded with comments on the power of this new technology and on its value to all phases of human activity and intercourse. A couple of months later, a "Scientific Note" from Gill appeared in the Queen's Quarterly, expressing skepticism about the Marconi Company's announcement that trans-Atlantic wireless service was to begin on November 16, 1907:

> ...but very few commercial messages have as yet been transmitted, and indeed there are many people who are in doubt as to whether a single complete message has been transmitted....and, when it is remembered that this company has failed more than once to meet its announcements since 1901, it would appear that the transmission of messages across the Atlantic on a commercial basis is a problem yet unsolved. In making this statement, the writer has no desire to depreciate the work of Marconi, which is monumental, even if his trans-Atlantic work should prove to be a complete failure.[101]

Now Gill wasn't so certain that wireless waves were capable of crossing the ocean at all. Light rays from the sun reflect and refract in our atmosphere, so that: "...we have light at a given point on earth long before we see the sun; in other words, the light waves to a certain extent bend to follow the curvature of the earth...approximately 1,000 miles in advance of its direct rays."[101]

The work done to date on wireless waves seemed to indicate that they would not follow the curvature of the earth indefinitely but were subject to the same limit as light, about 1,000 miles. On the subject of wireless telephony, the transmission of speech without wires, Gill seemed to be on the right track:

> There must be a variation in the intensity of the wave-train to correspond to the inflection of the voice; the wave-train must be practically continuous; and the energy imparted to the waves must be controlled primarily by the force of the voice, transmitted by sound waves through the medium of the atmosphere. If we could utilize the force of the sound waves to control an auxiliary supply of power, the problem of wireless telephony would be solved. What is wanted is a relay which will convert sound waves to ether waves and at the same time intensify the latter by the addition of power...until some device is found which will permit the use of more power, the range of wireless telephony will be very limited. The difficulty lies with the transmitter rather than with the receiver.[101]

Gill was apparently unaware of the voice transmissions of Reginald Fessenden in 1906, from Brant Rock, Mass., to ships in the Caribbean, using a heterodyne system in which voice was superimposed on a high frequency carrier oscillation.[43]

Bernard E. Norrish's M.Sc., 1909-10

Whatever the driving force, the pace of wireless activity at the Kingston campus began to pick up about this time. In the fall of 1909, one of Gill's own students, Bernard E. Norrish (B.Sc. '08) gave notice that he would be presenting himself for the degree of M.Sc. the next April, with the subject to be "Wireless Communications."[102] His proposed subject was accepted by the Board of Governors, and Gill and Clark were assigned to look after the appropriate examinations.

Bernard Esterbrook Norrish (B.Sc. '08; M.Sc. '10), Gill's graduate student.

Bernard Esterbrook Norrish (1885-1961), the son of a florist, was from Walkerton, Ontario. He had attended Walkerton High School and had registered in Science in the School of Mining in October 1904, student number 4773. He took all six available courses in E.E., and received his B.Sc. in April 1908, one of seven electrical graduates that year. Norrish then disappeared from the Kingston record for a year, perhaps working for the Topographical Survey Office in Ottawa. He turned up again in Kingston at the end of the summer of 1909, when Gill recommended him to the Faculty of Science as an assistant in the draughting room for the 1909-10 session, at a wage of $200 plus free tuition.[103] In accordance with the School of Mining regulations,[104] application for the M.Sc. was made on October 15, 1909, and on January 7, 1910, Norrish's committee recommended both the acceptance of the proposed thesis topic, "Wireless Communications," and that the subjects for examination be one paper on electric waves and two papers on modern telegraph and telephone systems.[105] Norrish was notified of the decision by letter, dated January 8, 1910, from Registrar G.Y. Chown.[106] No record exists indicating Norrish's motivation for choosing wireless as a subject.

There is a lot that is still obscure in Norrish's story, although some general notions of his qualifications for entry into M.Sc. candidacy may be surmised from the Calendar of the School of Mining of the period.[104] It stipulated that, after completing the B.Sc., the candidate must have practiced his profession for at least two years, in one of which he must have been responsible for engineering or scientific work, or he must have spent at least one session in attendance at the School of Mining after graduation with the B.Sc. In either case, the graduate must have carried on research work in that period, the results of which were to be submitted in the form of a thesis, which must be of literary as well as of scientific quality. At least one special examination was also required, on a subject related to the thesis, to be assigned by the Faculty.

Norrish's M.Sc. thesis cannot be found. It would prove a most valuable document in revealing the state of Gill's current research and might contain references to what Gill had been doing in his wireless studies since 1902. That thesis would have to make reference to the wireless

equipment then in use in Gill's laboratory, or to any such equipment constructed by Norrish or Gill for experimental use for the M.Sc.

Fortunately, the examination questions submitted by Gill and Clark
have survived, bound into the 1911 Calendar of the School of Mining,[107]
and they indicate the depth and breadth of knowledge required of
Norrish at the M.Sc. level. In the absence of any other relevant and
concrete material relating to Norrish's studies for the M.Sc., it is useful
to look at the questions relating to wireless:

Electric Waves

1. What is the period of oscillation of a circuit whose
inductance is 0.5 henrys and whose capacity is 1 micro
farad and whose resistance is 2 ohms? What effect would
increasing the resistance to 1000000 ohms have?
Explain fully how you might measure the wave length
in a Lecher system in which oscillations are occurring.
Suppose the wave length be known, how would you find
the frequency?
2. An oscillatory circuit with perfectly definite period is
acting on a receiving circuit which is out of tune. Explain
different methods of tuning the receiving circuit, and
how you could tell when it is in tune with the oscillator.
3. Explain the action of the Poulsen arc as a form of
oscillator, and state its advantages.
4. Explain the use of antennae in wireless communication
and describe the various forms in use. Describe the various forms of detectors of electric waves in common use.
5. Explain the statement: "When we are concerned with
high frequency currents, a condenser acts as a conductor
while a large inductance acts as an insulator." What
effect has the frequency upon the apparent resistance of
a circuit?
6. Upon what does the damping of an oscillation in a
circuit depend? What is the difference between closely
coupled and loosely coupled coils in an oscillation transformer?

Modern Telephone Systems

8. Describe the DeForest wireless telephone system, pointing out its limitations and possibilities.

Modern Telegraph Systems

6. Describe either the Marconi or the DeForest wireless
system, including the latest improvements. Illustrate by
diagrams.
7. What, in your opinion, are the general problems to
be solved in the wireless field?

Bernard E. Norrish received his M.Sc. degree from Queen's in 1910,
and then set out to find his life's work. He was appointed engineer on
administration in the Dominion Water Power Branch, Department of the
Interior, Ottawa. Three years later he became chief draughtsman in
charge of publication of reports and the photography branch of the

draughting room. He was put in charge of the US Essanay film crew doing contract work in areas controlled by his branch in 1916, and on his own initiative visited the parent film company in Chicago, plus a couple of other leaders in the field, to learn about the production of motion pictures. At the time, he was one of the few in Canada who knew anything about the new film technology. In May 1917, the Deputy Minister of Trade and Commerce suggested to the Minister, Sir George Foster, that Norrish be transferred to Trade and Commerce to head a new Exhibits and Publicity Bureau, sometimes called the Film Bureau, to produce, acquire and distribute motion pictures. In 1920, Norrish was recruited by Edward Beatty, President of Canadian Pacific (CPR) and later Chancellor of Queen's, to set up Associated Screen News of Canada (ASN), under the ownership of the CPR, and be General Manager. Beatty had chosen well, and ASN grew rapidly to become a leader in the development of the film industry in Canada in the educational, industrial and entertainment fields.[108]

Formal Teaching of Wireless

Some time in 1911, the various departments of the School of Mining reorganized their courses, possibly riding on an infusion of money in connection with the construction and equipping of the two new science buildings, Gordon and Nicol Halls. The number of courses offered was increased, E.E. (Course G) expanding from 7 courses and two laboratories in 1911-12 to 11 courses and four laboratories in the 1912-13 session.

One of the new courses, Electrical Engineering XI, "Telegraphy and Telephony," was to be given in the fourth year. Course XI included "The Morse System; Repeaters; Duplex and Multiplex Systems; Combination Systems; Automatic and Printing Telegraph; Railway Block Signal Systems; Modern Telephone Systems; Wireless Telephony; Simultaneous Telegraphy and Telephony." The School of Mining Calendar declared that the new Laboratory No. 4 was "supplied with a complete outfit of wireless apparatus; also telegraph and telephone receivers and transmitters." This was the first time that instruction in wireless had been officially offered at the School.[109]

In preparation for this expansion in instruction, on December 18, 1910, Gill placed an order with McGraw-Hill Book Company for the new <u>Principles of Wireless Telegraphy</u> by George Washington Pierce, Assistant Professor of Physics at Harvard. The order slip still exists,[110] but the book has long since been permanently checked out of the Queen's library system.

About a year later, on December 28, 1911, Gill ordered two portable wireless telegraph sets for E.E., "including transmitter and receiver according to specifications supplied by Prof. Gill." One wireless telegraph set was ordered from Clapp-Eastham Company, 139 Main Street, Cambridge, Mass., at an estimated cost of $240, and the second W/T was from James G. Biddle, 1211 Arch Street, Philadelphia, Pa., for $120. Both requisition slips survive,[111] as do the entries in the account book.[112]

The specifications, however, are lost, so there is no record of how the machines were equipped. All that is certain is that they were transmitter-receivers and were small enough to be carried by two husky men. The Clapp-Eastham set arrived with a bill for $120, and the Biddle set actually cost $90. Obviously, a teaching laboratory requires two W/T sets, the second to receive messages from the first and to send messages back to it. One W/T set is about as useful for teaching as a single telephone apparatus.

But the story is not so straightforward. The 1912-13 Calendar of the School of Mining says that Laboratory No. 4 was "supplied with a complete outfit of wireless apparatus," and Director Goodwin's report to the annual meeting, April 23, 1912, included in the list of new equipment in the E.E. Department "one portable wireless telegraph unit."[113] A re-examination of the record revealed a notation on the credit side of the School of Mining Cash Book, dated May 13, 1912, recording $90 received from James G. Biddle and returned to account E3, from which the money had originally been disbursed.[114] The Letter Book of the School also contains an entry of the same date, acknowledging receipt of a cheque for $90

Lester, Emma Elizabeth and Marion Rose Gill at Bath, Ontario, c. 1911.
-courtesy Mabel (Hickey) Archer

on account of unspecified apparatus returned to Biddle.[115] A search through the account book revealed no other transaction with Biddle in the amount of $90, so it is reasonable to conclude that it was the Biddle W/T set which Gill returned for a full refund. Presuming that Gill had bought in 1901, and used in the 1902 convocation demonstration, a W/T set purchased from Biddle, it is reasonable that when he came into the market for another set in 1911, he would have seriously considered the merchandise offered by Biddle. The firm did not manufacture the W/T sets that it sold, but acted as agent for the products made by De Forest and others.[116] In 1911, De Forest was embroiled in his second bankruptcy, even with the new Audion and the quenched spark gap telephonic transmitter in his hands, and production in the Newark factory was all but suspended.[117]

The Clapp-Eastham Company, incorporated in 1908, forged a solid reputation for innovative and high quality products - even its boxes and bases were beautifully finished pieces. Its slogan was "A little better than the best." After a try-out, it may have become apparent to Gill that the Biddle set was not the equal of the Clapp-Eastham, so the for-

School of Mining requisition number 4550, December 27, 1911, signed by L.W. Gill and authorizing purchase of a portable W/T set from Clapp-Eastham for the Electrical Department.

-courtesy Queen's Archives

mer was returned for a refund. The decision may have been based in part on the question of continuity of the supplier. Unfortunately, in the absence of the 1911-12 catalogues and Gill's original specifications, there is no way to compare the two purchases on any basis except cost and general reputation.

Now, Gill had only one W/T set for teaching, unless the primitive transmitter and receiver used in his 1902 convocation demonstration still survived.

Did Gill Have a Licence for W/T?

There is also a mystery revolving around the matter of government licencing of Gill's wireless. By the terms of the Wireless Telegraphy Act of 1905, which is less detailed than the Act of 1913,[118] in order to set up a wireless telegraph station anywhere in Canada, one had to obtain a licence from the Minister of the Department of Marine and Fisheries, Ottawa. Establishing a station or operating a wireless telegraph without a licence would make one liable, on summary conviction, to a fine not exceeding $50, or on conviction to a fine not exceeding $500 or to imprisonment for a term not exceeding 12 months. One would also be liable to forfeit said wireless telegraph apparatus. A Justice of the Peace was empowered by the 1905 Act to issue a search warrant if he had reasonable grounds for supposing a wireless telegraph set were operating without a licence, and the warrant would authorize the officer indicated to seize any apparatus involved. Gill's use would have been experimental, and special Experimental Licences were also available to applicants who could prove that "the sole object of obtaining the licence is to enable him to conduct experiments in wireless telegraphy" subject to special terms, conditions and restrictions "as the Minister thinks proper." From about 1910, when the level of international tensions began to increase, radiotelegraphy was regulated by the Department of the Naval Service.

The general terms and the penalties of the 1913 Radiotelegraphy Act were similar to those in the 1905 Act. The many new regulations were based on the rules adopted by the International Radiotelegraph Convention in 1912. There were now seven classes of licences, each valid for one year and expiring automatically on March 31. According to the new Act, "Experimental licences will be granted to stations intended for experimental purposes and operated with a view to the advancement of the art of radiotelegraphy. Applicants for such licences must state their technical attainments and the general lines on which they propose to pursue their investigations. It should be observed that the fact that the applicant desires to conduct experiments with his equipment does not justify or require a licence of this class, as most experiments can be conducted under the 'Amateur Experimental Licence' by the use of an artificial aerial...The power used, measured at the terminals of the transformer, must not exceed 1/2 kW...The waves emitted must be as little damped as possible," and for the first time "a distinctive call signal will be allotted to each station, commencing with the letter 'X', e.g. XAA, XAB..."[117] Authorized government commercial stations could order any interfering experimental or amateur station to temporarily cease operation.

The regulations for the two types of licence that might have applied to Gill's wireless work, under the 1913 Act, were different. The fee for an Experimental Licence was $5.00 a year. Experimental licensees were permitted to apply for the use of a particular wavelength below 200 metres, below 450 metres or above 1900 metres, and they were strictly limited to the wavelength(s) specified on the licence. If transmitting above 100 metres within the range of a commercial or coast wireless station, one needed a First-Class, Second-Class or Experimental Certificate of Proficiency in Radiotelegraphy.

Amateur Experimental Licences cost only $1.00 a year. If they were located more than 75 miles from a commercial or coast station, amateurs were permitted to use wavelengths below 200 metres. After the opening of the Marconi station, VBH, at Barriefield in 1913, if Gill had held an Amateur Experimental rather than an Experimental Licence on campus, there was a further restriction that would have applied: "Stations located within 5 miles of a commercial coast or land station or a route of navigation, shall not use a transmitting wavelength greater than 50 metres."

Careful perusal of the Department of Marine and Fisheries and the later Department of the Naval Service reports in the Sessional Papers of the House of Commons makes it plain that Gill never obtained a licence for wireless at Queen's. No Kingston addresses appear in the annual lists of wireless licence holders in Canada until 1915 and 1916. (The civil service bureaucracy must have been three years behind in its accounting. All amateur and experimental licences were suspended in 1914 for the duration of the War, but these lists were not discontinued until 1917!). Unless some sort of special arrangement existed, it must be concluded that Gill and the E.E. Department were operating illegally. It is not clear, however, whether under the 1905 Act the use of a non-transmitting dummy antenna load in place of an antenna required a licence as well. The 1913 Act is explicit in requiring a proper operating licence for sets operating with a dummy load or "artificial aerial" only.

Mrs Marion Rose (Gill) Service, the Professor's daughter, speculates that he never bothered to obtain a licence because he didn't like red tape.[119] It is equally likely that, as a wireless operator who was getting along perfectly well before any government wireless regulations existed, he couldn't see the point of registering with the government and didn't think it worth the trouble. There were very few prosecutions under the 1905 Act because there were so very few sets in the country that no one bothered anyone else and nobody really cared too much about it. One could save a dollar on a useless licence, and one's chances of getting caught were virtually zero.

Student References to Gill's Wireless Work

A professor's students should be a source of information about his research, but the students of that generation, even the engineers, had little idea of Gill's interest in things ethereal. Dr John B. Stirling (Civil '11) recalled this period at Queen's:

> I was quite unaware as a student that Prof. Gill was interested in wireless - in fact, I don't think I'd heard of wireless at that time!...The student body were conscious that Prof. Gill was working on something that we put down to research, but he was very reserved and "private" about what he was doing and was held in such high respect around the place that no one thought of prying into what he considered a secret activity.[120]

From about this time, as his work in the field use of wireless became general knowledge on the little campus, references to Gill's wireless expertise started turning up in campus publications. The references were mostly oblique, because it was something that everyone in the know on campus would twig to, and this obscurity is most frustrating. For example, in the 1912 Faculty Song, performed at the 16th Annual Science Dinner, the students made a wireless reference in their song about Gill's aeroplane:

> We are still searching in the air,
> But as yet it is quite bare
> Of a ship we wish to see
> Sailing away serenely;

We wonder if the ship's Control
Will be wireless or human skill,
For we'd hate to see our Gill
With his ship to have a spill.[121]

The Prophet of Science '14 foresaw that fifty years thence, in 1964, "everyone carries a complete wireless-telephony set."[122] One final, maddeningly obscure record of a transaction between Gill and Clapp-Eastham is partially preserved in a series of notes dating from 1913.[123] An invoice from Clapp-Eastham for $239.85, dated July 1, 1913, contains only an engmatic "Mar 21," a reference which leads nowhere. On July 15, Gill wrote to G.Y. Chown, the Treasurer, instructing him to pay Clapp-Eastham $200 on account, and to charge it to the Electrical Department. On August 11, Clapp-Eastham wrote to Chown thanking him for the remittance, which was duly credited. What goods were obtained by Gill's department in this transaction is not recorded anywhere. In general, expensive pieces of power equipment were purchased for the laboratories directly from the manufacturers, Westinghouse or General Electric, at a hefty academic discount. The companies knew that the students trained on their equipment might well be their future employees and trail-blazers in research. The surviving record suggests that most purchases from suppliers such as Biddle and Clapp-Eastham were for measuring instruments and small pieces of equipment in the $15-40 range. This was a major purchase from an electrical supply house, and therefore stands out as quite unusual.

There are reasons for supposing that this mysterious purchase in 1913 from Clapp-Eastham was a second W/T set. First of all, in 1913, Clapp-Eastham's letterhead stated: "High Potential and Other Electrical Apparatus - Wireless Telegraph Apparatus for Private and Commercial Installation." Second, Laboratory No. 4 now had only one W/T set for teaching, and one can't teach effectively with only one wireless station. Third, by mid-1915 there was a second complete portable W/T set in Gill's department, and no other plausible record of purchase exists, either in the form of a requisition or as a cash book entry. Fourth, having returned a set to Biddle in 1912, it is reasonable that Gill would go back to the last preferred supplier as soon as sufficient funds were available for a major purchase. And fifth, this purchase was made around the time the Canadian Militia was considering adding W/T to its signals training. Gill must have known Dr David Edward Mundell, Professor of Surgery at Queen's and Chief Signalling Officer of the Third Military District. Moreover, the Militia Department had promised the Queen's Fifth Field Company of Canadian Engineers a wireless for its telegraph section in 1912,[124] and Gill was the obvious man for Dr Mundell to turn to.

Donald MacClement's Recollections

The hardest evidence that Gill was conducting experiments in the field use of wireless before the beginning of the First World War is contained in the extensive reminiscences recorded informally for this project by Dr William Donald MacClement. Don MacClement was the son of Dr W.T. MacClement, the Professor of Biology at Queen's. The MacClements lived in the west wing of Summerhill, the Principal's residence, at the edge of the campus and young Donald was the typical professor's brat. The whole university was his playground. He had the run of the campus, knew every room and crawlspace in the place and had a keen interest in every exciting thing going on. From an early age he was a competent wireless amateur, one of a little group of Kingston lads who avidly read the electrical magazines, built their own sets out of homemade components, learned the Morse code on their own and "talked" to one another through the ether every chance they got.

Don MacClement: Now, I'm not sure whether this was just before the War or just after the beginning of it. The University had received a complete army field transmitting station, including the masts and aerial and the spark transmitter - the whole works. With it was a portable set for mobile operation. This military set was one of the first war-like things on campus. I remember hearing Doug Jemmett, who helped Gill with the experiments, explaining to someone that this was an army radio wireless set of the type used in England and France. They were very proud to have got hold of it. Some time before the War, Queen's had also been given all sorts of radio equipment, earphones, buzzers and keys, in order to train operators. Prof. Gill was the prime mover in getting the military wireless equipment.

Gill had a huge car, a Pierce-Arrow touring car. In those days I think it was about the biggest car on the road. Later, Queen's set up a training school for mechanics in the carpentry shop at the back end of the Mill. Gill was active in getting this school set up. Gill used to let them look at and fiddle with his big, beautiful Pierce-Arrow, but he would not let them drive it. I can remember that once some students came along when he was sitting in his car. They didn't know it was running, and one of them said, "Can we look at the engine?" and he had his hands on the hood to lift it up. I can remember Gill shouting at him, "You keep your hands off my car! Don't you kids ever touch it!"

That car carried the mobile wireless set for Gill's field wireless experiments. A pole about...oh, I guess ten feet high was fastened out on the front bumper or the radiator, and another one at the back was attached to the trunk that sat back there. Between them was stretched a four- or five-wire aerial, with the lead-in coming straight down into the back seat to the wireless set. This sat on a table that folded down from the back of the front seat. There was lots of room in the back, and whoever was operating the thing would be sitting on the back seat facing it.

Now, that set that was taken in the Pierce-Arrow, or the mobile part of the set-up, was a square, flat-sided box. It was such a nice neat little thing compared to the jumble of bits back at the base station. It had a beautifully made wood case of reddish mahogany with neat brass corners and folding brass handles. The lid was fairly deep and I remember that, when it was closed down over the top, it extended beyond the front edge of the box, and the front flap folded up and locked with it. The key was sort of set into the edge of that front flap.

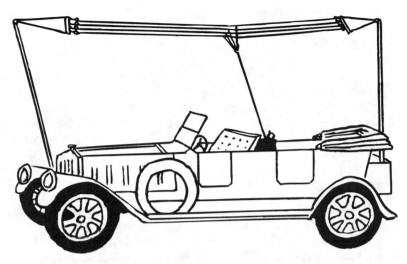

Prof. Gill's Pierce-Arrow with poles on the front bumper and rear trunk to support the aerial. The Clapp-Eastham W/T sat on a shelf that folded down from the back of the front seat.
-from a drawing by Dr W.D. MacClement

The sides of the box came up a bit higher than the surface of the top, so that the knobs, binding posts and switches were below the top edge of the sides. The larger stuff that was on the top side stuck up and could be seen when they were sitting in the car.

On the top side of this mobile set and to the left was a double-slide receiver tuning coil. I was envious because it had square rods for the sliders, and you couldn't get those. I had to use round ones. Now, I recall that on the left front face there was sort of a framework-like thing that pulled out. I have a feeling that it was a tubular tuning condenser for the receiver, with the crystal detector on the outer end of it.

In the centre of the top, near the front, there were two knife switches for changing the aerial and earth connections from the receiver to the transmitter. Behind those switches, on the right, was the transmitter spark gap. The gap was basically two rods supported on posts - not very big or impressive. In the middle of the front face of the open box was a transmitter ammeter, with the transmitter spark coil vibrator between it and the transmitter tuning helix, on the right. This transmitter tuning helix was of flat-wound copper ribbon, and there were three sort of posts coming out with notches in them to hold the coil in place. And there was no secondary coil. There was just one coil. I suppose it'd be an auto-transformer type, for both circuits. This helix was tuned by moving clips around the outer edge of the several turns. The key, as I said, was mounted on the edge of that front flap which folded down when the box was opened. I know that the reel for winding in the antenna was not fastened onto the box.

When they had that portable going in the back of the Pierce-Arrow, I can see Gill driving the car and Doug Jemmett operating the set in the back. They drove around town or out into the country on several occasions, apparently trying to see what sort of range that set had. Gill wouldn't let just anybody go in the car with him. Only Doug

1/4 kilowatt wireless telegraphy apparatus, custom manufactured by the Clapp-Eastham Company, Cambridge, Mass., c. 1911. Components similar to these can be seen in the 1916 Clapp-Eastham catalog. The adjustable spark gap is visible toward the upper right, end on. The knife-switches are at top centre and the double-slide tuning coil is at far left. The Blitzen-type transmitting helix is mounted on the front right of the box below the gap. The ammeter is to its left and the crystal detector is at far left. The key is just visible below the lip of the lid, at the lower left, and the headphones, in the foreground, are by Brandes. These components were very solidly built, in accordance with the Clapp-Eastham motto: "A Little Better than the Best."
-photograph courtesy Rev Dr W. Harold Reid

Jemmett or some other trusted assistant...I suppose because it was so dangerous with that "terrible HIGH VOLTAGE spark set" in the back seat.

Gill always tried to run the whole thing, shouting orders, bossing the bystanders and generally dashing around. He was a tightly wound up spring - completely intolerant and impatient with everything and everybody. He had an aggressive personality and relegated Jemmett to a minor position in the whole order of things. He had a way of looking at you and speaking quite sharply. He'd come close up in front of you and bark, "What's the *!#&*! all about?" Gill seemed to be nearly always telling somebody to do something, or criticizing. Gill had lots of ideas, and I can remember my father saying, "If Gill gets interested in that, he'll just run away with it, because that's the way he is." He'd go tearing into whatever it was that interested in, and just drive it hard. It takes an individual like that to get right past the ordinary hurdles that people find in front of them.

What interested me most was the other set-up, at the big base station. It was out at the back of Fleming Hall, where the ground sloped up sharply for four or five feet to a more or less level terrace running from the Mill to the rear of Ontario Hall and all the way back to

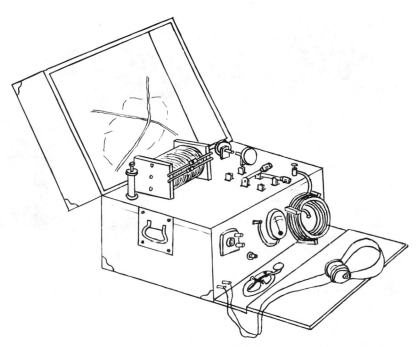

*A reconstruction of the Clapp-Eastham set drawn by Dr W. Donald
MacClement, who played with the set as a boy, based upon the two
photographs from Barriefield War Camp. On the top, clockwise from the rear
left, an unidentified ceramic component, double-slide tuning coil, adjustable
spark gap, knife switches for changing from transmitting to receiving. On the
front face, from the left, the crystal detector, hot wire ammeter (3 amperes full
scale) and flat-wound aerial tuning induction helix. A pair of early Brandes
headphones is resting on the lid. The Rev Dr W. Harold Reid confirms that the
apparatus looked as shown here.*
-from a drawing by Dr W.D. MacClement

Gordon Hall. The masts were at each end of this strip immediately
north of Fleming Hall, maybe 200 feet apart, and each mast was...oh,
pretty close to 100 feet high, with guy-wires and big insulators and so
on. One mast was behind the Mill, about where the Campus Bookstore
is now, and the other was pretty close to the back of Ontario Hall. And,
stretched in between the tops of the masts, of course, the big main
antenna. I think it was a four-wire antenna, with the wires running east-
west. The heavy wire lead-in from the antenna came straight down to
a bell tent, and came inside through a porcelain insulator taped into
the cloth. From the bell tent you sort of ran down the slope to the
back door of Fleming hall. The tent would be somewhere about where
the entrance hall is now in the Stewart-Pollock Wing of Fleming Hall.
The back door of the tent would probably be near the elevator in that
hallway.

I've been trying to remember exactly what one saw when one
looked in through the opening of that bell tent. There was a good
sized ordinary flat-top kitchen table, five feet or so long and the usual
width, and the top was pretty well covered with pieces of wireless

Typical Clapp-Eastham components, similar to those which would have been incorporated into Gill's 1/4 kilowatt portable W/T set. The "Boston" key was of nickle-plated brass, silver contacts, mounted on a polished Italian dove marble base, 3.5 x 6 inches. The adjustable spark gap had zinc electrodes and the "Blitzen" helix was of edgewise-wound copper strip held by hard rubber supports to a highly finished mahogany base.
-from Clapp-Eastham catalog S, c. 1916
McNicol Collection, Queen's University

equipment. Oh, I'd say more than half-a-dozen pieces of receiving apparatus. And about half-a-dozen pieces of sending equipment and a whole lot of dry cell batteries to operate it. This was Army Signal Corps stuff - nicely made separate boxes that you had to wire up together. All the pieces of equipment were connected together with insulated wires, which looked like ordinary 110-volt flex.

The receiver tuning coils were all inside wooden boxes at the left side. The tuning coupler was the largest, with binding posts for connecting the primary to the aerial and earth through the big two-bladed double-throw switch, and two more to connect the secondary to the rest of the receiver. On the top of the coupler box were two rotary switches, each with an arc of raised brass contacts. One was marked "primary" and the other "secondary." And, oh, there were cables coming out of the box. I know that one went up through the tent itself to the aerial, presumably, and there was another that came over along the ground. In addition to the coupler, there were two or more loading coils in the same sort of boxes, each with binding posts and a rotary tuning switch, with wavelengths marked for each contact. The tuning could be made very close to the wavelength desired. These could be connected in series with either the primary or secondary to increase the inductance when listening to longer wavelengths. Years later, I used these well-made coils when I listened to one of De Forest's first telephone broadcasts, in the Club Room in the basement of the old Arts Building.

There was more than one crystal detector, each on its own wooden base. One was a regular galena and cat's-whisker type and the other was a strongly made double-crystal type. It was a queer combination of two crystals. One of them was, I think, called "horseflesh mineral" and the other was chalcopyrites or fool's gold. The two crystals were supposed to press firmly against each other, but I could never get that type to work well. I don't see how Gill had the patience to fiddle with those detectors, for they always seemed to need adjusting. To balance the receiving circuit and help in tuning, there were at least two receiving condensers. One was a fixed capacity with merely two binding posts. The other was a fine rotary variable tuning condenser, in a small square wooden box with an engraved brass plate on top and a pointer. This was a great help in any kind of fine tuning. The students later wrecked that rotary condenser. Finally, Gill gave it to me and I made it into a wavemeter by adding a bank-wound coil. The earphones, I believe, were English-made Browns.

Between the receiving and transmitting equipment was quite a large box. As I remember, it had plain sides and maybe handles on the ends. It might have been a portable set of the kind used in artillery spotter aircraft. I know there were openings down from the top into each end of it for storing things. Out of one of these sort of pockets came a short length of flex going to the key. This key was a strongly built thing mounted on a wood block, heavy enough to prevent it from sliding around easily. The knob was grasped by the thumb and two fingers, and had large silver contacts to handle the considerable amperage in the primary circuit of the transmitter. Out of the other end of the box came the wires to the earphones. On the top of the box were several knobs and switches.

Dr W.D. MacClement standing in front of the spot, now in the hallway of the Stewart-Pollock wing of Fleming Hall, across from the old north door, where Gill set up his wireless bell tent.

-photograph by Arthur Zimmerman

At the back of the kitchen table, just a little to the right of centre, there was a big spark gap with cooling fins around each terminal where the spark jumped; that is, it was an ordinary spark gap, but with fins to cool the electrodes. The spark gap was mounted on its own wooden base and was well to the back out of harm's way. And the aerial inductance helix was...oh, a big thing. That large, loose-coupling helix was the most obvious piece on the table. It was quite an elaborate thing on a large wooden base. It sat up on end and was a good size...about a foot in diameter and at least a foot high. The primary was several horizontal turns of heavy copper wire in an open spiral supported by vertical wooden insulating posts up on the outside of it. The primary helix was connected to the rest of the transmitter's primary circuit by heavy insulated wire and spring clips. The secondary coil, of thinner wire and many more turns, was on longer insulating posts and could be moved up or down inside the primary coil. The secondary was connected to the double-throw switch. As far as I can remember, the primary wasn't flat-wound at all. Later, in trying to imitate it, I took thin brass curtain rods and bent them around and made my helix out of them. I think I can see some sort of loading coil somewhere in that secondary circuit.

In the primary of the transmitting circuit there was a condenser. I think it was variable, but I am not sure how this condenser was varied. The plates were spaced quite far apart. Also in the primary circuit was the spark gap with its cooling flanges on the rods, sitting near the back of the table. The big spark coil was in a wooden box about a foot long, with the vibrator at one end and two porcelain insulated posts on the top for the high voltage. I wonder if it was Army regulations that everything had to be inside a wooden box.

Around the outside of the tent was a big ring of iron spikes that had been driven into the ground. That was the earth connection. These spikes were all connected together with a wire, which then ran along the ground under the right bottom edge of the tent, at which point there was a knife switch on the ground. From there the wire went up to the double-throw switch on the table, and then to the set.

They ran a pair of heavy insulated wires out from the north entrance (back door) of Fleming Hall, straight across the road and up the slope to the tent to give power for the transmitter, and they put two-by-fours each side of the wires where they crossed the road. I guess just so that a passing car wouldn't damage them. This was for operation in the daytime, when the (Fleming Hall) power plant was in operation and supplying 110 volt direct current.

And, of course, when they were testing it, I'd be as close as I could get, and I remember being warned away, that I might get a big shock. One time Doug Jemmett was all alone out there in the bell tent, and he invited me in and had me help him try to operate it a bit. He knew that I had been operating my own spark transmitter for some time, and so he suggested that I work the key for a while. Gill was away driving his Pierce-Arrow and had somebody other than Jemmett in the back seat doing the wireless receiving. Jemmett had been transmitting continuous messages from that field station in the tent and trying to see how far away from town Gill could go with the mobile unit and still hear the signals. Well, I pounded away on that key sending all sorts of weird and wonderful messages, almost a continuous transmission for...oh, over an hour. I can remember sitting there on a stool and pounding away on that key - the spark zipping and ripping across the gap, making a lot of lovely noise. And I didn't notice that Gill had driven back in and had parked close by. Suddenly he was there in the door, glaring at me. He said all sorts of rude things, including, "You could have killed yourself! You don't know anything about this sort of thing. A kid like you coming in here could get seriously injured!" Jemmett had for the moment stepped away somewhere else, and I guess that Gill just thought that I was some sort of an invader into their premises. However, I was rescued by Jemmett when he came back out, and he arrived in time to hear some of the remarks. His comment to Gill was, "That boy probably knows more about wireless than you do." I left quickly. The odour of ozone from that big spark gap is still fresh in my memory.

I did not use the earphones or try to listen with the receiver. I was allowed merely to pound away on the key, which was thrill enough.[125]

Don MacClement's wonderful recollection of Gill's field wireless experiments is really the only surviving account of that work, and is the only description of the equipment used. Unfortunately, MacClement was quite young when he observed these events and could not be specific about the date, except that it was just before or early in the First World War.

Professor Gill's daughter Marion does not recall that he ever owned a Pierce-Arrow automobile: "We had trolleycars that ran from Portsmouth into Kingston, until he got tired of that and decided that he wanted a car. This was in 1912, I think, or '11. But he decided that he had to have a Ford, and you couldn't buy one in Kingston. He took the train up to Montreal and bought a Ford, and had the guy take him out and drive him around a few blocks to teach him how to drive, and he drove it back to Kingston."[119]

There may, in fact, have been large automobiles available on campus before the First War. Some time before he left for the War, Gill got a school for auto mechanics set up in the carpentry shop of the Mill, and bothered people about training wounded veterans. He may have secured an automobile for this work on loan from one of the manufacturers, as he was pushing to get this course onto the curriculum. The Department of Mechanical Engineering in the School of Mining had equipment for testing automobile engines and had the use of a 40-horsepower McLaughlin-Buick for its course in automobile engineering in 1912-13.[126] Don MacClement recalls that Gill agitated to get a Fordson tractor for the School, and remembers driving it up and down the cinder road behind Fleming Hall, and tearing up the road. So there were loaner vehicles on campus in those years, and Gill may have used one of them in his wireless field experiments.

The Date of Gill's Field Wireless Experiments

There are some benchmark circumstances that suggest several specific times at which these experiments could have taken place, times when both Gill and Jemmett were in the Kingston area and when there was a two or three week period free to do the setting up and the experiments.

In Don MacClement's story the naming of Douglas Mill Jemmett as one of Gill's assistants in his wireless field work is significant, because Jemmett was a special friend of the young MacClement. Jemmett had been a student of Gill's in E.E. in 1911-12 and 1912-13, although he did not take course XI, "Telegraphy and Telephony."[127] He may have learned his signalling in the 5th Field Company of Canadian Engineers (FCCE) at Queen's, where he was a Second Corporal in 1913. There is no question that Gill would have wanted Jemmett involved in his experiments, because he was an excellent student. It is Jemmett's presence in the Kingston area that delimits the times when the field wireless experiments could have taken place, although it may not have been necessary for Jemmett to be present all of the time.

After the conclusion of the examinations in the spring of 1912, Jemmett returned to his home in Napanee on April 16,[128] and left early in May for his summer job with Allis-Chalmers-Bullock at Lachine, Quebec,[129] where he had spent the previous summer. Thus, there is a possible window open for experiments between mid-April and early May of 1912.

From May 1913, when he graduated from Queen's with the B.Sc., until January 1914, Jemmett was foreman of electrical and hydraulic test at Allis-Chalmers in Lachine. Then he went to Fairbanks-Morse in New York for a few months, returning home to Napanee in mid-March 1914.[130] On March 21, 1914, he was Gill's guest at the Frontenac Club, Kingston.[131] It may be that the visit was to plan future experiments in wireless field work, but it is more likely that Jemmett was brought in to discuss the possibility of an appointment in the Department of Mathematics at Queen's. During that summer, Jemmett's name appeared almost every week in the Napanee Beaver. That's because he was one of the founders of the Napanee Canoe Club on May 21, 1914,[132] as well as its Captain. They held races every Thursday evening at 7.00 o'clock on the Napanee River, and Jemmett won a lot of those races. On August 9, a few days after the declaration of war, Jemmett enlisted as a sapper in the 5th FCCE at Queen's.[130] His unit, under the command of Major (Professor) Alexander Macphail, left for Valcartier on August 18 and sailed from Quebec City on October 3 as part of the First Division, Canadian Expeditionary Forces. Jemmett did not return home until July, 1916, when he was invalided to recuperate from wounds. By this time, Gill was overseas in the Great War himself.

Thus, there are a few windows of time between the fall of 1912 and the fall of 1914 in which Gill and Jemmett could have got together for a couple of weeks to set up and do experiments in wireless telegraphy on the campus. They could have done so in the fall of 1912, at the beginning of the school term, or after examinations in the spring of 1913, before Jemmett left for Lachine. The spring and summer of 1914, from late March through early August, seems to have been a time of leisure for Jemmett, except for Thursday evenings racing on the Napanee

River. A search through both Kingston newspapers and the <u>Napanee Beaver</u> in these periods revealed no evidence of mobile wireless experiments going on at Queen's. Surely the papers would have picked up on the queer sight of a huge automobile decked out with poles and wires and driven by a Queen's professor, with a student in the back listening intently through a headset. That would be as irresistable to Kingston children as a circus parade, yet it was not remarked upon in the press.

Was Gill Using Army Wireless Sets?

Could Gill have been using military issue wireless equipment, "Army Signal Corps stuff," in this work? The military had not standardized its wireless equipment in Canada by the outbreak of war, and was not at all certain that wireless was a good thing. Number 1 Wireless Detachment was set up in Hamilton, Ontario in April 1912, attached to the First Field Troop, Canadian Engineers.[133] In May 1912, the Department of Militia and Defence announced the establishment of wireless stations at Camp Petawawa, and that in future all Canadian military camps would be so equipped through the Engineers, once sufficient operators could be trained.[134] By August 1914, the Canadian Engineers had just five telegraph detachments of the nine authorized and only one of the seven wireless detachments proposed.[135] Prior to the War, there were in Canada's forces just two French Pack wireless sets and several Marconi Pack sets, under the supervision of Sergeant-Major Charles Shergold of the Royal Canadian Engineers. Only two instances were recorded of their use in Canada following August 4, 1914. The two French pack sets were used in training by the First Contingent at Valcartier Camp in September 1914, and were then returned to Ordnance. The Marconi sets saw service maintaining communication on a troop trek from Niagara Camp to the C.N.E. grounds in Toronto in the autumn of 1915.[136] Based on this information, it is apparent that the Canadian Militia really had no wireless sets to spare Gill for his experiments at Queen's. They would not have allowed wireless sets under their control to be retained by Gill after his work and then, as will be seen below, to be transferred by him to the Fifth FCCE at Barriefield Camp. Later on, in June of 1915, the Militia would not release to him the pack saddles requested for his wireless equipment. Incidentally, it will be noted that the references in the correspondence concerning the pack saddles are all to "his" wireless apparatus, not to the Militia's equipment.[137] There is no official army record of the use of wireless in Kingston or Barriefield either before or during the War.

As will be shown below, there is good evidence that the portable wireless set seen by young Don MacClement in use in Gill's field experiments was a Clapp-Eastham set owned by the E.E. Department of the School of Mining. It may have been the same portable complete set purchased in December 1911, at the same time as the Biddle set. There is, however, no way now to identify the big multi-component wireless set that was located in the bell tent behind Fleming hall. It consisted of components in beautifully made wooden boxes - a hallmark of Clapp-

Eastham, rather than of the military - suggesting that it might have been that mysterious Clapp-Eastham merchandise purchased for $200 for E.E. by Gill in 1913.

There was a second organization on campus at the time that was interested in wireless telegraphy. It was announced at Queen's that the telegraph section of the 5th Field Company of Canadian Engineers would "add to its equipment the necessary apparatus for wireless telegraphy" by September 1912.[138] The Queen's COTC documents were long ago destroyed in a fire, and there is no published record of any wireless equipment received or used by the Company. In fact, the telegraph section of the 5th FCCE did not come into existence until April 1914, when Lieut. (Professor E.W.) Henderson and 12 men laid out the wires at the Company's annual spring training session at Barriefield War Camp.[139] Furthermore, in the few notices published, there was never any indication that the Company's wireless apparatus would be supplied by the Canadian Militia. It is not impossible that Militia Signals supplied military wireless telegraphy equipment to the 5th FCCE prior to World War I and that Gill had charge of it, but there is no hard evidence to support such a supposition. It is also not unreasonable to surmise that the 1913 purchase from Clapp-Eastham may have been a big multi-component semi-portable wireless station that was to be handed over to the 5th FCCE when it was ready to begin instruction.

The Reverend Dr W. Harold Reid will have the final word on the matter:

> **Dr Harold Reid:** Now as to the reference in your letter to the W/T set that we used at our School of Signalling in Barriefield camp probably I can clear up any confusion regarding it...You mentioned that Donald (MacClement) said Professor Gill was using "an Army set on loan from the miltary."
>
> From the time the School of Signalling opened at Barriefield Camp in the spring of 1915 I was the Chief Instructor until it closed in September 1917. When Colonel Mundell discussed with me starting a W/T class he mentioned that he could get the Queen's University W/T equipment on loan. I am sure that if there had been any military W/T equipment at Barriefield Camp before this he would have told me. I am convinced that setting up the W/T class at our School was the first use of W/T at Barriefield Camp. When our S/Signalling closed this W/T equipment was returned to Queen's University.
>
> In 1916, if the Army had possessed any W/T equipment, it would not have been necessary to borrow something from Queen's. In 1917 our S/Signalling closed and the W/T equipment went back to Queen's. At that time Donald (c. 12) might have heard about this, and that it came from the military, not knowing that the Army was merely returning what we had borrowed from the University.
>
> I assure you that during the whole time our S/Signalling was in operation there was no "Army set on loan from the military" to Queen's University. The army did not own any W/T equipment at that time. I hope this will definitely clear up any confusion there may have been regarding this.[140]

NOTES

1. "Report of the Principal", QQ <u>9</u> #1, July 1901, p. 29.
2. James Cappon, "The Situation in Queen's", QQ <u>17</u>, Jan. 1910, p. 193.
3. D.D. Calvin, <u>Queen's University at Kingston: The First Century of a Scottish-Canadian Institution, 1841-1941</u>, Trustees of the University, Kingston, 1941.
4. Ibid, p. 112.
5. "Renovated Grant Hall", QJ, Feb. 22, 1935, p. 1.
6. Private letter to G.M. Grant, Nov. 8, 1901; G.M. Grant Papers, NAC, p. 4095.
7. "The Late Principal Grant's Predictions", KDS, Oct. 18, 1924, p. 16.
8. SM Minute Book, June 7, 1899, p. 121; QUA, Coll. 3695A, Box 1.
   "Dr. William Lawton Goodwin (1846-1941)", PTRSC, series 3, 39, appendix B, 1945, p. 87.
9. SM Minute Book, July 7, 1900, p. 139; QUA, Coll. 3695A, Box 1.
10. Dr. Goodwin to Principal Grant, June 8, 1900; SM Letter Book, QUA, Coll. 3701A, Vol. 1, p. 967.
11. "Louis Anthyme Herdt (1872-1926)", PTRSC, series 3, <u>20</u>, 1926, p. XXI. Dr. Herdt was appointed to the Macdonald Chair in E.E. at McGill in 1909.
    Dr. Goodwin to Dr. Bovey, August 4, 1900; SM Letter Book, QUA, Coll. 3701A, Vol. 2, p. 50.
12. Dr. W.L. Goodwin to Mr. L.W. Bill (sic.), August 15, 1900; SM Letter Book, QUA, Coll. 3701A, Vol. 2, p. 63.
13. SM Minute Book, Sept. 19, 1900, p. 141; QUA, Coll. 3695A, Box 1.
14. G.Y. Chown to L.W. Gill, Aug. 24, 1900; SM Letter Book, QUA, Coll. 3701A, Vol. 2, p. 71.
15. W.L. Goodwin to J.L.W. Gill, Sept. 12, 1900; ibid., p. 89.
16. L.W. Gill to G.M. Grant, undated; G.M. Grant Papers, QUA, <u>13</u> #4533, Reel #596, pp. 4007-8.
17. W.L. Goodwin, "The College", QQ, <u>10</u> #1, July 1902, p. 102.
18. SM Faculty Song, 1908-09, QJ, Jan. 18, 1909, p. 222.
19. Arthur Lewis Clark, <u>The First Fifty Years: A History of the Science Faculty at Queen's University, 1893-1943</u>, Queen's University, Kingston, 1943, p. 25.
20. Ibid., page 31.
    "Science", QJ, Oct. 27, 1910, p. 34.
21. SM Minute Book, Oct. 26, 1900, p. 144; QUA, Coll. 3695A, Box 1.
22. Clark, <u>The First Fifty Years</u>, p. 31.
23. Ibid.
24. Abraham Gill, Sr., "Old Charlottetown and P.E.I.: A Pioneer's Story"; manuscript lost, partial copy courtesy Dr. Pauline Hughes, New Westminster, BC.
25. "Death of Abraham Gill, Esq.", <u>Daily Examiner</u>, Charlottetown, P.E.I., Sept. 30, 1887, p. 2.
26. Article on J.L.W. Gill's older sister Elizabeth Jane (Gill) Duncan (1861-1950), by Dallas Bannister Wright; contained in an Okotoks, Alberta local history, courtesy Dr. Pauline Hughes.
27. Alexander Macphail, "The Late Lester Gill: An Appreciation", QR 13 #9, Dec. 1939, p. 263.
    Dr. Arthur E. Zimmerman, "Professor Gill's Flying Machine", <u>Canadian Aviation Historical Society Journal</u> <u>25</u> #4, Winter 1987-88, p. 103.
28. "Obituaries: Lt.-Col. J.L.W. Gill", <u>Journal of the Engineering Institute of</u>

Canada 22 #10, Oct. 1939, p. 452.
Personal communication from (Mrs) Louise Gay, Associate Registrar, University of PEI, Nov. 24, 1987.

29. "Hamilton Technical Institute Rated...Had Humble Start", Hamilton Herald 41, Dec. 14, 1929, sec. 1, p. 4.
"Our Principal's Anniversary", Hamilton Technical Institute Tech Sparks, May 1938, page 1.

30. Annual Calendar of McGill College and University, for Session 1893-94, McGill University, 1893, p. 211.

31. "Many New Engineers", Montreal Daily Star, April 28, 1896, p. 5.
"McGill College Examinations", Daily Examiner, Charlottetown, PEI, May 2, 1896.
"Certificate attesting to achievement of honours, Faculty of Applied Science, McGill University, Montreal, Session 1895-96", signed by Dean Henry T. Bovey; Archives, Dept. of Electrical Engineering, Queen's University.

32. "Hugh Longbourne Callendar (1863-1930)", in Dictionary of Scientific Biography, American Council of Learned Societies 3C.1, 1970, Scribner's Sons, New York, p. 19.

33. "Hugh Longbourne Callendar, 1863-1930", PRSL, 134, 1930, p. xviii.

34. "Royal Commission for the Exhibition of 1851: Science Research Scholarships of 1897: Opinion on the Reports submitted by Mr. JAMES LESTER WILLIS GILL, B.App.Sc., of the McGill University, Montreal, of his work during the tenure for two years of a Science Research Scholarship (1897-9)"; courtesy C.A.H. James, Secretary, Royal Commission for the Exhibition of 1851.

35. "Russell W. Stovell is Promoted", McGill News 1 #2, 1920, p. 31.

36. "Howard Turner Barnes (1873-1950)", PTRSC, Series 3, 45, appendix B, 1951, p. 77.
Dr. A. Norman Shaw, "Howard Turner Barnes (1873-1950)", McGill News 32 #5, 1950-51, p. 22

37. E.A. Corbett, Henry Marshall Tory: Beloved Canadian, Ryerson Press, Toronto, 1954.

38. John L. Heilbron, "Physics at McGill in Rutherford's Time" in Rutherford and Physics at the Turn of the Century, (ed. M. Bunge and W.R. Shea), Dawson, New York, 1979, pp. 47 and 51.
David Wilson, Rutherford: Simple Genius, Hodder and Stoughton, Toronto, 1983, p. 180.
"Frank H. Pitcher Dies in Hospital", Montreal Star, Aug. 22, 1935.

39. "Report of the Committee of Management of the Macdonald Physics Building" in Annual Report of the Governors, Principal and Fellows of McGill University, Montreal, for the year 1897, pp. 26-29.

40. J.L.W. Gill, "On a New Method of Measuring Hysteresis in Iron", Electrician 39 #22 (No. 1010), Sept. 24, 1897, p. 718.

41. Heilbron, "Physics at McGill", pp. 47-48.

42. J.L.W. Gill, "On a New Method of Measuring Hysteresis in Iron", in Report of the 67th Meeting, BAAS, Toronto, Aug. 23, 1897, Section G "Mechanical Science", p. 762.

43. Ormond Raby, Radio's First Voice: The Story of Reginald Fessenden, MacMillan, Toronto, 1970.
Helen M. Fessenden, Fessenden, Builder of Tomorrows, Coward, McGann, New York, 1940.

"Fessenden a Canadian", DBW, April 27, 1904, p.1.

44. "Robert Bowie Owens (1870-1940)", World Who's Who in Science, 1st edition, A.N. Marquis Co., Chicago, 1968, p. 1296.

45. Personal communication from Dr. Frederick S. Howes, retired, Professor of Electrical Engineering, McGill University, May 19, 1986.

46. Clark, The First Fifty Years, p. 51.

47. Minute Books, MPS, Vol. I, 1897-1904, p. 1; McGill University Archives, Acc. 1549.

   NOTE: Several years later, Ernest Rutherford presented his first results on nuclear distintegration and radiation to the members of the McGill Physical Society.

48. Ibid., p. 13.

   Nicola Tesla, "Note on an Electrical Oscillator", Report of the 67th Meeting of the BAAS, Toronto, Section A, Department I "Electricity", Aug. 24, 1897, p. 570.

49. Minute Book, MPS, Vol. 1 1897-1904; McGill University Archives, Acc. 1549, p. 19.

   J.A. Fleming, "A Method of Determining Magnetic Hysteresis Loss in straight Iron Strips", The London, Edinburgh and Dublin Philosophical Magazine and Journal of Science, 5th series, 44, Sept. 1897, p. 262.

50. "Marconi Telegraphy", Electrician 39 #21, (No. 1009), Sept. 17, 1897, p. 683.

   "Dr. Oliver Lodge's Apparatus for Wireless Telegraphy", ibid., p. 686.

51. Minute Book, MPS, Vol. 1, 1897-1904; McGill University Archives, Acc. 1549. p. 25.

52. Register of equipment available in the Macdonald Physics Laboratories, late 1890's; McGill University Archives.

53. "Marconi Wireless Telegraphy in America", Electrical Engineer 24 #500, Dec. 2, 1897, p. 532.

   H.L. Chadbourne, "William J. Clarke and the First American Radio Company", manuscript, 45pp., 1982; Special Collections, Queen's University.

54. J.L.W. Gill, "On the Distribution of Magnetic Induction in Straight Iron Rods", Philosophical Magazine and Journal of Science (London) 46, 5th series, #282, Nov. 1898, p. 478.

55. "Going to London", Montreal Gazette, April 14, 1898.

56. Stanley Brice Frost, McGill University for the Advancement of Learning, Vol. II, 1895-1971, McGill-Queen's Press, 1984, p. 37.

57. "Hugh Longbourne Callendar, 1863-1930", PRSL, 134, 1930, p. xviii.

58. Wilson, Rutherford: Simple Genius, pp. 166 et seq.

59. Ibid., p. 178.

60. Ibid., p. 132.

61. Ibid., p. 172.

62. Graduate School registration record of Lester Willis Gill, Harvard University Archives, Cambridge, Mass. UAV No. 1,272.5, DN 90 Logt R 10, orsAS RdCds, c 1895-1930, Bx6.

63. "Edwin Herbert Hall (1855-1938)", Dictionary of Scientific Biography 6, 1972, p. 51.

   Gill was enrolled in Physics 20e, a research course in electromagnetism, taught by Professor E.H. Hall; letter from Clark A. Elliott, Associate Curator, Harvard University Archives, Cambridge, Mass., July 9, 1987.

64. Lester W. Gill, East Somerville, Mass., "Fire Apparatus", Official Gazette, U.S. Patent Office 94, 1948, No. 669,492, March 5, 1901.

65. "George Washington Pierce (1872-1956)", Biographical Memoirs, National Academy of Sciences, U.S.A. 33, 1959, p. 351.

66. George Washington Pierce, <u>Principles of Wireless Telegraphy</u> McGraw-Hill, New York, 1910.
    G.W. Pierce <u>Electric Oscillations and Electric Waves</u>, McGraw-Hill, New York, 1919.
67. <u>McGill News</u> <u>4</u> #1, Dec. 1922, p. 32.
68. <u>Electric Club Journal</u> <u>1</u> #1, Feb. 1904, p. 27.
69. E. Rutherford, "A Magnetic Detector of Electrical Waves and some of its Applications", <u>Philosophical Transactions of the Royal Society</u>, series A, <u>189</u>, 1897, p. 1.
    Harriet Brooks, "Damping of Electrical Oscillations", PTRSC, second series, 5, 1899, p. 13.
    A.S. Eve, <u>Rutherford: Being the Life and Letters of the Rt. Hon. Lord Rutherford, O.M.</u>, MacMillan, New York, 1939, p. 67.
    "The Royal Society Hears an Address", <u>Ottawa Daily Free Press</u>, May 26, 1899, p. 3.
    "Wireless Telegraphy Illustrated", <u>Ottawa Evening Journal</u>, May 26, 1899, p. 8.
    E. Rutherford, "Wireless Telegraphy", <u>Canadian Engineer</u> <u>9</u> #7, July 1902, p. 190, and 9 #8, Aug. 1902, p. 207.
    "Passenger Agents Visited", <u>Montreal Daily Star</u>, Oct. 14, 1902, p. 2.
    "Invade the City", <u>The Gazette, Montreal</u>, Oct. 14, 1902, p. 5.
    "Wireless Messages to a Moving Train", <u>Scientific American</u> <u>88</u> #4, Jan. 24, 1903, p. 55.
    "Wireless Telegraphy on a Fast Express Train", <u>Electrician</u>, Feb. 27, 1903, p. 753.
70. SM Cash Book, 1893-1904; Feb. 1, 1901, page 167; QUA, Coll. 3695B, Box 1, Book 8-6.
71. "The Closing Events", DBW, April 25, 1902, p. 2.
    "Queen's University", ibid., April 26, 1902, p. 1.
72. "Shot by School Mate", ibid., April 29, 1902, p. 1.
73. "Wireless Telegraphy", <u>Kingston News</u>, April 29, 1902, p. 4.
74. "An Enjoyable Lecture on Wireless Telegraphy by Prof. Gill", DBW, April 29, 1902, p. 5.
75. "Wireless Telegraphy", QQ <u>10</u> #3, Jan. 1903, p. 268.
76. A.F. Collins, "How to Construct an Efficient Wireless Telegraph Apparatus at a Small Cost", <u>Scientific American</u> <u>85</u> #11, Sept. 14, 1901, p. 170.
77. "Science", QJ, May 9, 1902, p. 33.
78. Ibid., Jan. 3, 1903, p. 20.
79. "Report of the School of Mining", <u>Queen's Quarterly</u> <u>11</u> #1, July 1903, p. 70.
80. "Science Valedictory", QJ, May 22, 1903, p. 29.
81. "Science", ibid., Oct. 16, 1903, p. 29.
82. "Science", ibid., Dec. 1, 1904, p. 140.
83. "Science", ibid., April 3, 1903, p. 22.
84. "Science", ibid., Feb. 16, 1905, p. 333.
85. "Science", ibid., Nov. 15, 1907, p. 133.
86. Minutes, Public Utilities Commission of Kingston, Feb. 2, 1905, p. 2.
    "Visit to the General Electric", QJ, Dec. 1, 1910, p. 147.
87. <u>SM Annual Report</u>, 1907-08, p. 9; QUA.
88. "Results", <u>Montreal Daily Star</u>, April 26, 1904, p. 6.
    "McGill Degrees and Honours", <u>Canadian Engineer</u> <u>11</u> #5, May, 1904, p. 128.
    QJ, May 6, 1904, p. 32.
89. Clark, <u>The First Fifty Years</u>, p. 37.
90. "What It Proposes", DBW, April 27, 1904, p. 2.
91. SM Minute Book, Sept. 21, 1904, p. 299; QUA, Coll 3695A, Box 1.

92. L.W. Gill, "Elementary Electrical Engineering", Canadian Engineer 18 #12, March 25, 1910, p. 278, to 19 #13, Sept. 29, 1910, p. 426.
93. L.W. Gill, "Elementary Electrical Engineering", Jackson Press, Kingston, Aug. 1910.
94. A.E. Zimmerman, "Professor Gill's Flying Machine", Canadian Aviation Historical Society Journal 25 #4, Winter 1987, p. 103.
95. L.W. Gill, "Spinning Machine", Official Gazette of the U.S. Patent Office 118 #7, Oct. 17, 1905, p. 1833.
96. E.W. Henderson, "An Improvement on the Prony Brake", Electric Journal 9 #6, June 1912, p. 577.
97. 6th Annual Directory of Graduates and Students of the School of Mining, Kingston, Canada, 1912, p. 26; QUA.
International Electrotechnical Commission, Report of Turin Meeting held Sept. 1911; courtesy I.E.C., Geneva, Switzerland.
98. Journal of the American Institute of Electrical Engineering, 26B, part 2, June 27, 1907, p. 1463.
99. "Dominion of Canada Royal Commission on Industrial Training and Technical Education", Sessional Papers, House of Commons, No. 191D, 3 George V, Part IV, 1913, p. 2076.
100. "Comments on Current Events: Wireless Telegraphy", QJ, Nov. 5, 1907, p. 104.
101. L.W. Gill, "Wireless Communication", QQ, 15, Jan. 1908, p. 230.
102. SM Minute Book, 1907-17, Oct. 15, 1909, p. 9; QUA, Coll. 3695B, Series I, Box I.
103. "List of Graduates and Alumni", 4th Annual Directory of Graduates and Students in the School of Mining, Kingston, Ontario, 1910, p. 29; QUA. SM Minute Book, 1898-1909, Aug. 14, 1909, p. 378; QUA, Coll. 3695A, Box 1.
104. Calendar, School of Mining, Kingston, 17th Session, 1909-10, Article VI: Degrees and Courses of Study, p. 30; QUA.
105. SM Minute Book, Jan. 7, 1910, p. 16; QUA, Coll. 3695B, Series I, Box I.
106. SM Letter Book, Jan. 8, 1910, p. 187; QUA, Coll. 1161F.
107. Queen's University Examinations: April 1910, Faculty of Practical Science, M.Sc. Degree, Calendar Queen's College, 1910-11; QUA.
108. Bernard E. Norrish: Obituary, Montreal Gazette, Aug. 30, 1961, p. 31. Peter Morris, Embattled Shadows: A History of Canadian Cinema, 1895-1939, McGill-Queen's Press, Montreal, 1977.
109. Calendar of the School of Mining, Kingston, Ontario, 20th Session, 1912--13, p. 75; QUA.
110. SM Accounts; QUA, Coll. 3695B, Series IV, Box 8, #294, 12/18/10.
111. SM Records; QUA, Coll. 3695B, Series IV, Box 10.
112. SM Account Book, pp. 188-89; QUA, Coll. 3695B, Series III, Box 1.
113. Report of the Director of the School of Mining, Kingston, at Annual Meeting, April 23, 1912; QUA, bound with SM Calendar, 1912-13, p. 12.
114. SM Account Book, p. 247; QUA, Coll. 3695B, Series III, Box 1.
115. SM Letter Book, May 13, 1912, p. 342; QUA, Coll. 1161f, 50/112.
116. Personal communication from John James, Export Department, Biddle Instruments, Bluebell, Penna., Jan. 16, 1985.
117. Lee de Forest, Father of Radio: The Autobiography of Lee De Forest, Wilcox and Follette, Chicago, 1950, p. 275.
118. "An Act to Provide for the Regulation of Wireless Telegraphy in Canada", Statutes of Canada, 1905, 4-5 Edw. VII, Ch. 49, p. 235.
"An Act Respecting Radiotelegraphy", Statutes of Canada, 1913, 3-4 Geo. V, Ch. 43, p. 385.
"Regulations", Canada Gazette 47 part 4, June 20, 1914, p. 4456.
119. Personal communication from Marion Rose (Gill) Service, Brattleboro,

Vermont, June 1, 1986.
120. Personal communication from Dr. John B. Stirling, Montreal, July 10, 1986.
121. Faculty Song, Seventh Annual and Directory of Graduates and Students of the School of Mining. Kingston, Canada, 1913, p. 26; QUA.
122. "Science '14", PESQU 8, 1914, p. 148.
123. M Records; QUA, Coll. 3695B, Series IV, Box 10.
124. No. 5 Field Company of Canadian Engineers, Sixth Annual Directory of Graduates and Students of the School of Mining, Kingston, Canada, pp. 28-29; QUA.
125. Personal communication from Dr. William Donald MacClement, Kelowna, B.C., May 1987.
126. Department of Mechanical Engineering, Calendar of the School of Mining, Kingston, 1912-13; QUA.
     Report of W.L. Goodwin, Director, SM, April 23, 1912, Principal's Report, Queen's University, 1913, p. 12; ibid.
127. Records at Queen's Archives indicate that D.M. Jemmett took E.E. courses III and V in 1911-12 and courses VIII, IX and X in 1912-13 from Gill.
128. "Personals", Napanee Beaver, April 19, 1912, p. 4.
129. "Personals", ibid., April 26, 1912, p. 4.
130. D.G. Dewar, "Introducing the Staff: Douglas M. Jemmett", QR 27, April, 1953, p. 86.
     "Personals", Napanee Beaver, March 20, 1914, p. 4.
131. Personal communication from George Vosper, Kingston, owner of the "New Register of the Frontenac Club, Kingston", p. 1.
132. "Napanee Canoe Club Revived", Napanee Beaver, May 29, 1914, p. 5.
133. A.J. Kerry and W.A. McDill, History of the Corps of the Royal Canadian Engineers, Vol. 1, 1749-1939, Military Engineers Association of Canada, Ottawa, 1962, p. 49.
     Canadian Militia General Orders, April 1, 1912, p. 60.
134. "Wireless Station at Petawawa", DBW, May 21, 1912, p. 1.
135. Kerry and McDill, History, p. 230.
136. W. Arthur Steel, "Wireless Telegraphy in the Canadian Corps in France: The Beginning of Wireless in the Canadian Corps", Canadian Defence Quarterly 6 #4, July 1929, p. 443.
137. Correspondence between the A/Commanding Royal Canadian Engineers, 3rd Division, Kingston, and Lieut-Col. G.S. Maunsell, R.C.E., Director General of Engineering Services, Canadian Militia, June 23, 1915 and June 26, 1915; File HQ 6978-2-53, Department of Militia and Defence, State and Military Records, Federal Archives Division, NAC.
138. "No. 5 Field Company of Canadian Engineers", Sixth Annual Directory of Graduates and Students, School of Mining, Kingston, Canada, 1912, p. 28.
     No. 5 Field Company of Canadian Engineers, Principal's Report, Queen's University 1911-12, May 29, 1912, p. 29.
139. "Telegraph Section Formed", DBW, April 4, 1914, p. 1.
     "Engineer Camp Ends", ibid., April 28, 1914, p. 8.
140. Letter from the Reverend Dr. W. Harold Reid, Stoney Creek, Aug. 25, 1987.

CHAPTER 2

Prof. Gill's 46th (Queen's) Battery, CFA, CEF,
photographed in June 1916 at Bramshott, Whitley or Larkhill Camp, England.
Gill is in the second row, above the mascot's tail. Many are wearing the
extremely rare 46th Battery cap badge.
-courtesy Queen's Archives

Queen's, Wireless, and the First World War

Shortly after the First Contingent, Canadian Expeditionary Force (CEF), embarked for Europe in the autumn of 1914, a Committee on Military Training was established by the Senate of Queen's University to take steps to provide preliminary training in drill and musketry to student volunteers.[1] Professor Lester Gill was one of five members of that Committee, even though he had no previous military experience of any kind. He had tried to enlist in the engineering corps of the First Canadian Contingent in 1914, but was unable to join up at that time. He was a patriot, and going to war was probably another of those impossible challenges that he felt compelled to rise to face. In January 1915, a month after his 44th birthday, Gill enlisted in the Queen's 5th Field Company of Canadian Engineers (FCCE) with the rank of Lieutenant,[2] and became part of the 5th Company staff. He was to have succeeded to the commission of Major (Prof.) Lindsay Malcolm, who had gone overseas with one of the drafts, but that did not happen. By early summer Lieut. Gill was busy helping his two colleagues on the 5th FCCE staff, Lieut. Henderson and Capt. Wilgar, supervise construction by the Queen's Engineers at Barriefield War Camp, as they made the grounds ready for occupation by the largest force yet to train there.

The camp was about a mile and a half east of Kingston, just behind the hilltop village of Barriefield, on the 900 acres of plains between the Cataraqui and St Lawrence rivers. The area was first occupied by Indians, and then in 1758 the plain and Barriefield Hill were held by a British army sent from the Thirteen Colonies to take the French Fort Frontenac, on the Kingston side of the river, where the Tête du Pont Barracks now stand. It had been used for the drilling of local volunteers and, since the 1860s, regimental military training camps had occupied the lower commons every spring. The site was well suited for such a camp. It was high and dry, with a good water supply.

The camp at Barriefield was scheduled to take place in the summer of 1915, as usual. In March it was suddenly cancelled by the authorities - the first time in fifty years that there would be no military camp on the plains - for a saving of two million dollars to the Militia Department. Then the government acquired the greater part of the Cartwright Farm to provide more space, and the order was reversed - a five month camp would be held at Barriefield. The advance parties, including the 5th FCCE, arrived on May 22 to begin setting up facilities and, by early June, the upper commons held a real war camp, with over 3,000 troops under canvas and in training and 1,500 more expected to arrive shortly.[3] From the first, the newspapers rejoiced in the fun and conviviality of life in camp - days full of baseball, soccer, foot races, lectures and films at the YMCA tent, a camera club, singing around jolly campfires, drill and parades, route marches through Gananoque and Brockville where the citizens turned out to show their gratitude with home-made meals for the boys, and then camping out under the stars, learning about guns and horses and signals, and the thrill of being issued a uniform. Honour and arms! Oh, what a lovely war! Oh, the soldier's life is the life for me! And the volunteers streamed in from all over the district to sign up and put the kibosh on the Kaiser. It would all be over before Christmas anyway, and everyone wanted to get in on the fun before it was too late.

According to the daily reports in both Kingston newspapers, each regiment was entirely responsible for the training of its own men early that summer. Gradually things became more centralized and efficient, and provisional schools of infantry, cavalry and machine gunnery were opened. Signals training was also necessary for war, and selected instructors in these arts were brought in and courses organized by the District and Camp Signalling Officers.

Signals Training at the Barriefield War Camp in 1915

The organization of signals training at Barriefield Camp was the responsibility of the District Signalling Officer (DSO), Major David Edward Mundell (1864-1923), Professor of Surgery in the School of Medicine, Queen's University. He had served in the Canadian Army Medical Corps around 1900 and 1906, and was an old signaller besides. Dr Mundell had been the DSO of Miltary District #3, the area around Kingston, since 1906. He had once even proposed a course in signalling to the School of Mining, in 1908,[4] two years before their 5th FCCE came into existence. Mundell enlisted in the Canadian Signal Corps, Canadian Expeditionary Force (CEF), with the rank of Lieutenant-Colonel on January 15, 1915, and was gazetted DSO for the Third Military District shortly afterward.[5]

The DSO worked out the signals training in co-operation with the Camp Signalling Officer (CSO). From early June until early July, 1915, the CSO was Capt. W.S. Wood of the 38th Battalion, CEF.[6] Wood was in charge of a signalling class at Barriefield that began on June 21,[7] but he was soon replaced by Lieut. Rackham, 3rd Dragoons. Other duties occupied the new CSO and so Sergeant-Instructor W. Harold Reid did the training. Reid, now the Reverend Dr W. Harold Reid, taught signalling at the War Camp from 1915 through 1917. It was his recollections and writing about the Queen's wireless sets at Barriefield Camp[8] that led the author to the discovery of the long pre-history of wireless and radio at Queen's. Dr Reid's memories are the only accounts extant relating to the formation of the School of Signalling at Barriefield Camp and are worth recording here in detail.

Rev Dr W. Harold Reid, who was Sgt.-Instructor, Signals, Barriefield War Camp, 1915-17. Portrait, January 1983.
-courtesy
Dr W. Harold Reid

W. Harold Reid was born in 1895 in Belleville, Ontario, and became interested in signalling in the Boy Scouts, where he learned semaphore and International Morse for flags, and worked with the 34th Battery, Canadian Field Artillery. In 1912 he was in the Signal Section of the 15th Argyle Light Infantry Regiment in Belleville, and in June 1914, as Sergeant of Signals, he gave signalling instruction at Petawawa. The first winter of the War, he taught members of the CEF at a Signals School in Lindsay, Ontario, under Capt. Frank Lee, and in the spring of 1915 was instructed to report to Major Mundell at Barriefield.

W. Harold Reid: Well, you see, Mundell was a surgeon at Queen's University, but years before, for many years, he had been an officer in the military forces there. Mundell was also DSO; that is, he was responsible for signalling instruction at Barriefield Camp. He brought me down to Barriefield Camp in the spring of 1915 and he said, "We're establishing this Divisional Signal School. Now, I haven't been doing any signalling for years. I'll leave that entirely to you. I look after administration, but you'll be entirely in charge of instruction. Anything that you want, you come to me." He was wonderful. Rackham was just there as an adjutant for a while until his regiment went overseas, and from time to time another officer was appointed. So we got set up bit by bit there, and then gradually I had to have assistant instructors. Well, I told him of certain ones. I told him of a man I knew in the 109th Battalion in Lindsay. I had found this man (Louis P.) Reading, who had been an old signaller in the Imperial Army in India years ago.

He'd come back to England and then he had come out to Canada. He hadn't been in the army for a long time until I discovered him, and got him in. He was made a Sergeant, and I trained him as an instructor...While Reading was an experienced signaller, he had never done any instructing, which was quite a different thing. And then another chap by the name of Slater and others...

We started our Signals School in May of 1915. I remember the Canadian Mounted Rifles. I had some of these men in my School for visual signalling. These men were just beginning, and we had to teach them the elements first. In 1915 we spent quite a bit of time on flag signalling - semaphore, Morse flag - and Morse lamp. Those were the main things. The Begbie lamp was for night signalling, and heliographs - but they didn't use much heliograph in this country. By 1916 the semaphore was dropped completely, so we concentrated on Morse.

Our first class in Barriefield consisted of 90 men from the infantry battalions and some men from the 8th Canadian Mounted Rifles. They were divided into two groups - the beginners, and an advanced class composed of some men who had taken the preliminary training. Our second class in July was only 71, but the next one in September was larger, with 100 in attendance. Before I was promoted to Sgt. Maj. Instructor, my pay was $2.40 per day.[8]

The signalling class was conducted outdoors on the grounds directly behind the field hospital. There was no teaching of wireless or even of standard telegraphy in the School of Signalling at Barriefield in the summer of 1915. The Militia had not issued equipment, and besides, the training was mostly at the elementary level for the first few months. Dr Reid did not recall any wireless telegraphy in camp that summer. There was, in fact, some experimental wireless work going on in the lines of the Queen's 5th FCCE from about mid-summer.

Wireless at Barriefield War Camp in 1915

Just where the idea originated, to bring wireless telegraphy (W/T) to Barriefield Camp from the Department of E.E. of the School of Mining in 1915, has not been recorded, and the instigator is not known. It could have been the initiative of Prof. Gill, or Dr Mundell may have made the suggestion to Gill after the successful field tests at Queen's. Surely Mundell, as DSO and professor, had his eye on those wireless field tests. No record has survived about any transaction between Gill and Mundell in 1915. Dr Reid did not notice the wireless at Barriefield in 1915 but he recalls what happened in the next year, 1916:

From the time the School of Signalling opened at Barriefield Camp in the spring of 1915, I was Chief Instructor until it closed in September 1917. When Col. Mundell discussed with me starting a W/T class, he mentioned that he could get the Queen's University W/T equipment on loan...When our School of Signalling closed, this W/T equipment was returned to Queen's University.[8]

There is evidence that Gill was planning to take the wireless sets to

Barriefield Camp early in 1915 and was intending to take the sets out on field tests over rougher terrain than would be suitable for the automobile. A letter exists from the A/Commanding Royal Canadian Engineers, 3rd Division, Kingston, Ontario, to Lieut.-Col. G.S. Maunsell, RCE, Director General of Engineering Services, Canadian Militia, Ottawa, dated June 21, 1915:

> You may recollect me asking you some time ago if it would be possible to secure two pack saddles on loan for use of Mr. Gill, 5th Field Coy., C.E. with his wireless apparatus. Mr. Gill has his apparatus complete, wanting only the saddles, and is anxious to commence operations.
>
> Captain Wilgar has communicated with the Superintendent of the Radio telegraphs as per your instructions, and the way now seems clear.[9]

The letter implies that this was at least the second such request for the loan of two pack saddles made on behalf of Gill. The original request, made "some time ago," had been referred back to Gill's colleague in the School of Mining, Captain (Prof. of Civil Engineering) W. Percy Wilgar of the 5th FCCE, and he had cleared the way. This subsequent request was dated just a few months after Gill finally gave up on his "safety first" all-metal aeroplane, and indicates that he was resuming his interest in wireless once more. The Director General's response made reference to the two French pack sets then in Ordnance Stores in Ottawa:

> We have two of these wireless sets complete with pack-saddles, etc. here at the Engineer Training Depot, Rockcliffe, but it is considered advisable at present to keep all the different parts of the equipment together.
>
> The equipment, however, is designed to be carried either in packs or in a light spring wagon, and both have proved quite successful. It is suggested, therefore, that Professor Gill use these wagons for the present, which would probably be easily procurable in Kingston.[9]

Gill forgot about the pack-saddles and proceeded with his project using what he had. On the afternoon of Tuesday, July 20, the Engineers of the 5th FCCE put up a 60-foot multi-sectioned wooden pole, a "semi-portable" pole, on a mound of earth east of the administrative headquarters near the Gananoque Road. This pole was for holding aloft the vertical-type aerial of a wireless set, to be "constructed" by Captain Gill (he was a captain in the Queen's 5th FCCE, but a lieutenant in the CEF). A rope extended from the base of the pole to the top, for hauling up the aerial. The "stationary plant" wireless equipment, a combination transmitter-receiver brought from Queen's, was to be housed in a bell tent nearby, and the power for the transmitter would be obtained from the city of Kingston mains, tapped off the lighting wires.[10] The Daily Standard article of July 21 noted, "The military wireless outfit was built at Queen's under Professor Gill, who is superintending its erection at camp," suggesting that this stationary set may have been large and complicated enough to require assembly of components; that is, it could have been the collection of boxes from the earlier tent behind Fleming Hall. By July 31, however, the Standard correspondent had likely seen the set at camp and stated that it was "of the portable type." The extent of its portable-ness was not expanded upon by the Standard.

This camp station would be used "exclusively for military purposes," with a transmitting wavelength under 600 metres (above 500 kilocycles, and therefore in the modern AM band), so that it would not interfere with the government Marconi station on Barriefield Hill, and it would have a sending and receiving radius of about 60 miles. It would prove useful in training, particularly of the men of the 8th Canadian Mounted Rifles (CMR), in the use of portable wireless stations in the field. The other complete wireless station, still at the School of Mining and presumably a more portable one yet, would soon be part of the 8th CMR. Operators would be trained to work the sets and the cavalry would carry this more portable outfit in two pack saddles on route marches and bivouacs. They would also carry a steel pole for a vertical aerial, which could be rapidly erected or taken down, and the power would be supplied by a portable gasoline engine. This new technology would make it possible for the officers of the cavalry to keep in touch with camp headquarters in all their marches and manoeuvres.

The Standard observed: "Their latest work of installing a complete wireless outfit on the camp grounds shows that the local University is capable of great things, and its patriotism at this hour is not lacking."[11]

The stationary wireless set at the Camp was used for the first time on Wednesday afternoon, July 28, when:

> Captain Gill tuned the station with that on Barriefield Heights and messages were sent...Despite the fact that the work was done in the afternoon, the worst part of the twenty-four hours for wireless, the messages were very distinct. The portable set will be handed over to the 8th C.M.R. in the near future.[12]

These transmissions were not all that remarkable, even in daytime, for the messages had to travel only as far as the government Marconi station, VBH, on Barriefield Hill, about a mile away. It is remarkable that the operators at Barriefield were disposed to help out in the project, as they later did with many local young amateurs. The next Saturday, the Standard reported the installation complete, and that Lieut. E.W. Henderson was "turning out the coil on the machine so as to be able to get in touch with the instrument located at the University. Both instruments are of the portable type."[13]

They were still communicating with the wireless set on the Queen's University grounds two weeks later: "The wireless at the camp is working every day in connection with a station at Queen's University. The city station is erected back of the Mechanical Laboratory and the instruments like those at the camp are in a tent." [14]

This set-up is similar to the situation described by Dr W.Don MacClement, where a complete wireless telegraph set was located in a tent in the large quadrangle behind Fleming Hall, back of the Mechanical Laboratory at Queen's, and was used to communicate with a second remote and portable set carried in Gill's automobile. In MacClement's account, the wireless set in the tent at Queen's consisted of perhaps a dozen components in wooden boxes, and so cannot have been a readily portable one. This multi-component equipment was likely that taken to Barriefield in 1915 to become the "stationary set" in the tent there. In this summer of 1915, the wireless equipment in the tent at Queen's had to have been quite portable and relatively simple to set up and operate, since it was apparently the same one later turned over to the 8th CMR for its route marches. On a route march, a high-tech multi-component set requiring extensive wire-up, as described by MacClement, would have been most impractical. Therefore, the set in the tent at Queen's in mid-August 1915 must have been the smaller of the two sets owned by the School of Mining at the time, perhaps the one previously carried in the automobile.

At the beginning of August, 1915, The Whig reported: "Lieutenant Gill is very busily engaged with his wireless station which is working splendidly." [15] But suddenly there arose an unanticipated problem with the supply of power. The City of Kingston denied the camp the use of the mains specifically for the stationary wireless set. The official reason made no sense, because the small amount of current needed could be supplied by a portable generator:

> The high voltage of the current carried across the bridge is responsible for the city hesitating to furnish the camp military wireless with the current necessary to operate the newly installed apparatus. The city is sending out from its generators at the light plant about 2,300 volts, which is a very dangerous current to trifle with. At the swing bridge the City has been compelled to submerge its wires, and this makes it possible for one of the big dredges to pick up the heavily loaded wire some day, and if this is done, serious consequences might result. This is said to be the only reason for the city refusing to furnish the current asked for, for wireless use. [16]

The old wooden swing-bridge over the Cataraqui River, also known as the Old Penny Bridge, was finally being replaced that summer with a modern one. At this point, a temporary bridge was in place and the only solution for getting these first power cables across the swinging section to the War Camp and to the Royal Military College was to submerge them. The problem was solved when the plans for the New Cataraqui Bridge were modified in September and the Thunder Bay Construction Company was instructed to include an extra duct for the cables. [17] Incidentally, these first power cables from Kingston to Pittsburgh

Township were strung the mile or so from the bridge to the camp on the poles of the Great North Western Telegraph Company, and there were three lamps installed to light the way.

According to the <u>Daily Standard</u>, the true state of affairs was that the City of Kingston had enmeshed Gill and the War Camp in a power struggle. The newspaper reported that the city was trying to put pressure on the Dominion Government to reimburse Kingston for lighting the War Camp, and Gill's wireless was its pawn. There is no evidence to support this thesis in the minutes of the Kingston Public Utilities Commission.

The officers were not pleased with the City's attitude, because the situation would hamper the field work that they intended to do at Barriefield. But Gill was a fiercely determined man. He could be depended upon to lick a problem and even to convert it into an advantage. Gill immediately brought in a portable gasoline engine from Queen's,[18] and let it be known that the camp wireless station was back in business and that it could now generate enough excess power to light the camp headquarters administration building.[19] In any case, it was best for this important military wireless to be completely independent of the city. Now the military could "laugh at city plants, and supply their own juice for their station."[16]

After all of the publicity about the wireless, people were anxious to see what the portable wireless could do for cavalry on the march and bivouac, but there were no further newspaper reports of wireless activities at the camp until August 21. Exactly what caused the delay is not clear, but it may have had to do with Gill's undergoing the officers' examinations. A few days before the 21st, the engineers of the 5th FCCE were building a concrete base for the gasoline engine, and were soon going to mount a two-cylinder engine on that base, as part of the wireless plant.[20]

Ten days later, the reporter for the <u>Standard</u> declared that the wireless station was in perfect working order and that communications were still being carried on daily between the Camp and the station at Queen's. The first field test was concluded successfully on Saturday, August 28, when a portable wireless set was taken ten miles out into the country and there was no difficulty in picking up the camp station and in maintaining two-way communication.[21] There was now every expectation that they would be able to send a portable outfit with the 8th CMR on the trek to Brockville, and be assured of their keeping in touch with Barriefield Camp every mile of the trip. In anticipation of the coming test, the Camp antenna was upgraded from a single vertical pole: "Two more aerial masts have been added to the camp wireless station, with horizontal aerial wires which will add to the receiving capacity of the station."[22]

Three hundred troops and 30 officers of the 8th CMR set out from camp early on the morning of Thursday, September 2, arriving in Gananoque at 3.40 pm. The ladies of the Gananoque Red Cross Society served them supper at the Armouries that evening.[23] Somewhere on the road, the first message from the CMR was sent, from Lieut. Ross who was in charge of the transport and supplies of the trek, to Major Hamilton of the Army Service Corps at Barriefield Camp.[24] The operator at Barriefield was not credited in the story. The only other newsworthy incident relating to wireless on the trek was the following adventure:

> Lieut. W.M. Emery is in charge of the wireless outfit with the regiment, and it has proved of great service. Constant communication has been kept up with the camp and messages of different kinds have been exchanged. At Gananoque it is reported that either a wireless message or something else beyond the laws of man ordered Lieut. Emery to dismount, and he did so over the horse's head.[25]

Brockville gave the 8th CMR such a big welcome, and held so fine a recruiting meeting in its honour, that it was decided to do it all again soon, but in the other direction. The 21st of September would see 3,000 troops march from Kingston to Belleville, passing through Napanee each way.[26] Gananoque remained the wireless distance record that year, implying that the 8th CMR was unable to communicate with Barriefield from Brockville or from Belleville. They may have been foiled by strong winds on Sunday, September 25, which broke off the two 65-foot high wireless masts in Camp, and destroyed the aerial.[27] The masts were to be re-built immediately, but there was no further word of the wireless at the War Camp until late October, when it was announced that the wireless had been dismantled on the 25th and was being re-erected at Queen's.[28]

The School of Signalling continued at Barriefield until early November.

> **W. Harold Reid:** That fall I did not leave my tent in camp till November 10th, and you do not readily forget the experience of breaking the ice every morning in your water pail and then having a shave with the Army straight razor and ice-cold water. Units that had not gone overseas before camp closed were provided with quarters in Kingston for the winter...Early in January, 1916, word was received from HQ that our School of Signalling at Kingston and Barriefield would be the only one for Military District 3 apart from instruction given in Ottawa. The new arrangement for our Divisional School of Signalling was that under D.E. Mundell, DSO, who was now Colonel, there would be Capt. Frank D. Lee, who was to be CO of the School, and I was to be Chief Instructor, as I had been the previous year, along with Sgt. Reading and whatever temporary assistance might be necessary...For a few days we met in the Kingston City Hall, but on January 20, we moved to a wonderful new location. Col. Mundell...was able to get the use of Grant Hall at Queen's for our Signal School. It was a large hall with a gallery on three sides and some small rooms at the back so that it provided ample space for our separate classes and was

ideal for our work.

I had over 200 men in Grant Hall, and for flag work they'd be all spread out over the hall and up all around the galleries. If you were using one flag, you had to have at least six feet for each man. I had assistant instructors to help with that. Then, for the lectures, they all congregated in the centre, right in front of the platform. And I did all the lecturing for all classes there...to over 200 men.

Sgt. Reading was responsible for most of the practical training in semaphore and Morse flag. There was instruction at night with the Begbie lamp...there was some time given to buzzer, but this was carried on more particularly with the advanced class who had also instruction in the use of the field telephone. As it was now evident that visual signalling was giving way to the laying of lines for buzzer and telephone, some instruction was given in basic electricity.

In our Signal School I taught a class in electrical theory, to give the basic knowledge for operating Army field telephones and all this electrical equipment. The buzzer was part of your field telephone...there was one invented by Captain Lister, a signal officer in Ottawa. The field telephone had a key, so that you could use the buzzer, if you didn't want to speak. The wires were laid on the ground, and if the connection was not very good, the voice was impossible to make out, and then you could just go to the little key and buzz your message through. In fact, it was always much more satisfactory to send it by Morse instead of trying to speak. And then there was also less chance of being picked up...it could be sent in code, and that would make the enemy hesitate before they could crack the code, and then it was too late. By the Second World War they got so fast at it...(but) in my day in the First World War, it was all very slow.[8]

Already, by mid-summer 1915, word had reached camp that flag training was no longer being given in England and that visual signalling had been all but abandoned at the front. This was not turning out to be a gentleman's war, and it was proving unwise to stand up in full view of the enemy to wave one's flags. Now it was all field telephones and buzzers.[29] Next summer they would be teaching telegraph and telephone plus a new device called "the Disc," a metal wig-wag contraption that could be operated from down in the trench. The wireless would also become somewhat more important in warfare, and would be taught at Barriefield, but Gill would not be there in the summer of 1916 to further develop its use.

In the fall of 1919, after more than three years overseas, Gill returned to teaching, and very soon founded the Queen's Wireless Club, the ancestor of CFRC. It is important to try to understand what fueled his continuing enthusiasm for wireless and to know why he founded the club for his students. Gill's military experience is an important part of the story of wireless at Queen's for it influenced how things began and developed. There must have been many occasions during the War when Gill would have come into contact with practical aspects of W/T in the field - when he may have used it himself in his artillery battery - and later in teaching at the Khaki University in England. No documentation concerning Gill and W/T overseas has been found, but below is an attempt to reconstruct his military career so that future research might determine the facts and integrate them with what is known about W/T in the CEF.

Professor Gill Goes to War

After the first year of the war, the chauvinist tide was still running high in Canada and a speedy victory seemed attainable. The newspapers were not reporting the true course of the war, and no one suspected what awaited in the trenches of Flanders. The lads could hardly wait to get in on the "fun" over there. They would show those Huns that British steel is superior to German steel! Little Queen's University was all excited about sending a battery to defend the realm, and Gill was in the thick of it. Toronto Varsity students were making the 26th a superior battery, and now Queen's men were invited to join the 33rd Overseas as a Queen's Battery, "with students on all the guns and as non-commissioned officers."[30]

There was a question of whether enough men could be raised for a real Queen's Battery, because 152 were needed. After discussion, the Senate concluded that Queen's could probably not raise and sustain a permanent battery, in addition to the corps already raised. Professors Campbell and Gill were assigned to gather more information about gaining permission to raise a battery for overseas, as well as to look into the possibility of being able to raise the number of men needed.[31] Gill was also re-appointed to the Committee on Military Training for the 1915-16 session. These deliberations at the University were happening in the late summer, at the same time as Gill's involvement in the wireless work and in officer training at the War Camp.

Gill presented his report on October 1, 1915, recommending that the Senate approve unreservedly the formation of a Queen's Battery to go to the front as a unit, if permission could be obtained from the Department of Militia and Defence.[32] Negotiations were carried on with the Militia Department, and on October 29, Gill was able to advise Senate that authority had been given for the raising of a University Battery of 200 men and five officers.[33] The next day, 95 Queen's men and their friends submitted their attestation papers to join the Battery,[34] and announcement was made that Gill would be Commanding Officer.[35]

It was typical of Gill that he had to be doing several things at once, and one of these things was preparing to take charge of the Queen's Battery as soon as permission was granted. In mid-August, while the wireless equipment was being installed and tested at the camp, he was passing his test in equitation.[36] This was probably part of the course for qualification for a commission; afterward he was always referred to as Captain Gill. Around August 31, Capt. Gill reported for duty with the

32nd Battery,[37] as second-in-command to Major S.A. MacKenzie.[38] A little later, while he was recruiting and training his Queen's Battery, he was also attending the Royal School of Artillery at the Tête-du-Pont Barracks, Kingston, to qualify himself as battery commander.[39] At the same time, he was doing the administrative work of the battery and carrying on his teaching, consulting and other work at the University.[40] Although Lieutenants R.M. Elmer and G.R. Rogers were doing most of the day-to-day training of the battery, Gill's duties soon became so demanding that he had to resign his professorship.[41] He resigned outright and refused to go on half-pay, as his colleagues had done when they enlisted for overseas duty. He must have felt this to be his patriotic duty. After all, he was President of the Canadian Club of Kingston.

The Queen's Battery continued to train and grow, even accepting a contingent of 19 students from the University of British Columbia and McGill University College of BC.[42] On November 20, the men moved into winter barracks at the Old Collegiate building on Clergy Street (now called Sydenham School, across from Chalmers Church) and were issued their kit.[43] The Quartermaster Sergeant (QMS) of the 5th FCCE, F. Morris Wood (M.A. '11, B.Sc. '14; later Professor of Civil Engineering at Queen's), had joined the Battery as QMS and he issued the uniforms on November 27, making the "unit more uniform."[44]

> **F. Morris Wood**: I was in the Engineers up on Barriefield, and Lindsay Malcolm and Doug Ellis were in charge at that time. In late 1915 they told Gill, who was Major of their Battery, that I was a QM by that time, and that I would be a good QM for them, that I was ready to go overseas. So, Gill got after me and I joined up, as QM. I didn't know Gill at all until I was introduced to him as QMS. I outfitted the whole Battery! I had a hard time, because I ran out of six-foot garments. They weren't standard. All our boys were taller than usual. I had a terrible time, sending back the pygmy uniforms and their sending me back some bigger ones.[45]

Capt. Gill addressed his Battery on the 24th, outlining the work to be covered before spring - after basic obedience and physical fitness, there was foot drill, signalling, roping, knotting, and riding. QMS Wood and the other sergeants were sent by Gill to the Riding School at Tête-du-Pont Barracks, and Wood left a corporal in charge of the stores.

> **F. Morris Wood**: When we were called back suddenly to go overseas, we found that Major Gill had interchanged me with our sergeant (W.H. Brown) in the Orderly Room. So, the two of us went in and saw him, and he said, "Well, Sergeant Wood was away and he was responsible, and while he was away these six saddles were lost." You know the big wooden Mechanical Building (the famous "Tool Shed") at the entrance into the Queen's grounds, just a little further in than Gordon Hall? It's torn down now, of course. Well, this corporal that took my place, when the saddles came they went there. It wasn't my fault, so I said, "I don't know anything about the saddles." But Major Gill replied, "You were in charge, and so you're responsible." If I'd had the money, I

would have given the money. However, it worked well, because I finally left the Battery and I had a nice time flying. I'm pretty sure that Professor Gill felt sorry afterward that this had happened, because the first thing that he asked me when I came in and found it on the board and we went in and talked, was, "Do you still want to come overseas with us, Sergeant Wood?" I said, "Of course, Sir!" He was taken aback. He was upset every time he saw me, and I think he was feeling sorry that this thing had happened, because we were both likeable people. Finally, on the boat, he kept worrying about me, so he made me Sergeant in charge of Signals, and...finally he arranged that I would go to Division when they wanted a sergeant...That started me going, and that was exactly what I took up when I went into France.

Of course, his hair was against him. He was red-haired, but, apart from that, he was straightforward. Quite clear. He'd make a decision and that was it. And it was perfectly correct, in his mind, you see.[45]

The good news in 1916 was that the five Kingston Batteries (32nd, 33rd, 34th, "C" R.C.H.A. as the 45th, and Queen's; with the 34th as ammunition column) would go overseas as the 9th Brigade.[46] In mid-January it was announced that Gill's unit would be known as the 46th (Queen's) Overseas Battery, 9th Canadian Artillery Brigade, 3rd Overseas Division,[47,48] and rumours from semi-official sources were circulating that the Brigade would be ready to leave for England in a month to complete its training. The students among them would have to drop their classwork.

Overseas At Last and Into the Fray

About a thousand troops left Kingston by train for Quebec City on the morning of February 3, 1916,[49] after Gill had been promoted to temporary Major and after Major Starr had presented each man with a khaki-covered Bible. Hundreds of citizens and two military bands came to see them off, and 100 men of the 146th Battalion gave them a mighty cheer and a proper Union Jack wave. The boys had been waiting for this day and their hearts were full of gladness. They sang patriotic songs and mothers and sweethearts moistened their handkerchiefs freely. The Brigade sailed from Saint John, NB, on February 5th aboard the new CP Steamship Metagama, arriving safely in Plymouth on February 14.[50]

The 46th trained at Bramshott, Whitley and Larkhill, came under orders as a unit during the Irish uprising in May,[51] was inspected and complimented by King George on July 1st morning, and crossed over from Southampton to Le Havre on July 14 as part of the 11th Brigade, 3rd Division, under Lt.-Col. A.G.L. McNaughton.[51,52,53] The men were miffed during their final inspection, however, because the King hardly looked at them and barely acknowledged their salutes after all of their hard work to look military. He talked incessantly with Field Marshall Lord French. What the Canadians didn't know that morning was that the Battle of the Somme had been launched hours before, and the King was pre-occupied with the early reports.[54]

As McNaughton's new brigade moved into its first position, south-west of Poperinghe, Lt.-Gen. Sir Julian Byng, Commander of the Canadian Corps, embarrassed them with a surprise inspection. Very soon they were firing in the Ypres salient, and were part of the early attempts at a creeping barrage. The Queen's Battery distinguished itself on the Kemmel front as utility reinforcement artillery in support of the ANZACs and at Thiepval Ridge for the 4th Division.[55] It was also selected for hazardous duty with the 4th at Ancre Heights, Regina Trench and Desire in the ghastly Battles of the Somme.[56]

The CEF artillery was re-organized in March 1917 in front of Vimy, and Gill was put in command of the 33rd and then the 45th Battery (18-pounder field guns), 9th Brigade.[57] Andy McNaughton went up to Corps as Brigadier to take charge of special counter-battery work and Lt.-Col. H.G. Carscallen replaced him as Brigade Commander.

Although the record is not very specific, Gill must have commanded his battery in the tremendous bombardments of Vimy Ridge and in the subsequent creeping barrage in support of the 3rd Division in the Battle of Vimy Ridge on April 9.[51] His battery was also brought into the heavy artillery action at Vimy during that late spring and early summer, when the Canadians were trying to distract the enemy from Haig's build-up in the north for the advance beyond Ypres. To keep the Germans busy and concentrated in the Lens area, the Corps threatened to break into their lines along the Souchez River. They later settled on large-scale raids supported by artillery barrage, to cause as much damage as possible. In fact, the number of units holding the Ridge was greatly reduced and the few batteries in the field dodged around furiously, firing from various locations to simulate a large force. As well as a great deal of shelling in this position warfare, there was ferocious machine gunning, grenade bombing and bitter hand-to-hand fighting.

It was in a raid or skirmish on June 29, 1917, perhaps in the area of Avion Trench in preparation for the attack on Lens, that Gill got his blighty. He was gassed, probably by phosgene in a gas shell attack, and invalided to hospital in England.

Canadian Use of Wireless in Europe: Ypres and the Somme

It seemed to grow upon the British very, very slowly that wireless might be useful in directing artillery fire. There were experiments in 1914 and 1915, but commanders were reluctant to try something new. The Canadians got started in wireless in Europe rather late. They were less hidebound and had a better feel for the potential of the new technology. Canadians were soon experimenting on a small scale and building their own receivers out of cheese drums, accumulator plates and hairpin wire. Selected men of the 1st and 2nd Canadian Divisions were the first Canadians to attend army wireless schools in France. That was in November 1915, and early in January the first six Canadian Wireless Section stations were established in the Ploegsteert and Ypres areas. Little use was made of these stations, except to order rations, batteries and supplies. The 2nd Division were the first Canadians to employ wireless in action, in the re-taking of International Trench in March 1916. Unauthorized experimenting with crystal sets was done on the Bund by the 1st or 2nd Division to control the indirect machine gun fire of the Princess Patricia Canadian Light Infantry. All participants in that caper were arrested. Several operators from the 3rd Division attended the Second Army School, and they were sent into the line at Ypres in late March to man the B.A.R. receiver and Sterling spark transmitter in the 3rd Division dug-outs behind Belgian Chateau. In mid-May a 9.2-inch howitzer battery (290-pound shell) of the 3rd carried out four successful shoots by B.F. wireless and telephone relay forward of the Zillebeke Lake station.

At the Battle of Mount Sorrel in June 1916, the CEF telephone lines were knocked out and there was a long delay in turning to wireless. Canadian wireless stations were set up at the Somme in August. They eventually did useful work in the capture of the Sugar Refinery and in the attack on Courcelette. Signallers were learning slowly and by experience how to use the technology, how to exploit its possibilities and cope with its limitations. The Canadian Wireless Section left the Somme in late September to set up nine spark stations in front of Vimy Ridge. The 3rd Division had two in its sector: ZH at Battalion HQ, Neuville St Vaast, and ZQ at 3rd Division Report Centre on the Mont St Eloy-Arras Road. These were all 50-watt trench sets. None of them handled much traffic owing to lack of organization among the divisions.[58]

Canadian Use of Wireless at the Vimy Show

For the Battle of Vimy Ridge on April 9, the 3rd had a B.F. set at Brigade HQ in Territorial Trench near the Arras-Souchez Road and another at Machine Gun Fort. All stations moved up with the advance, although most of the useful work was done by the 2nd Division. In addition to these sets, GHQ gave the Canadians four new continuous wave (CW) thermionic valve (tube) sets for their first trial in the Vimy battle. They operated at 1,000 metres and were exclusively for the use of the artillery. Two each were given to the 13th Heavy Artillery Group, 4th Brigade, 2nd Division, and to the 64th HAG, 1st Division. A lot of traffic was handled by the CW sets that day, and the results were good. The forward sets advanced with the observation posts in the attack, one to Farbus Wood and the other to Le Tilleuls. Messages were carried by runners from the Forward Observing Officers to the observation posts and then wirelessed back to the guns. Wireless was also helpful in adjusting barrage fire and calling up reserves in the fighting in June around the Power Station and the Triangle.[58]

There were also air observers in the Battle of Vimy Ridge, FE2b pushers of 25 Squadron, Royal Flying Corps, in touch by wireless spark senders through an advance control wireless monitoring centre to the batteries. The pilots used little Sterling Wireless Senders, all on the same wavelength, which gave out a high pitched note. By adding an adjustable "clapper-break" to lower the pitch of the note in two steps, it became possible to sort out the messages from three spotters at once. The lowest note had the longest range.[59] The information supplied was most useful to McNaughton's Counter Battery Office in picking off the German fixed artillery.

Was L.W. Gill Involved in Wireless in the War?

Unfortunately, there is no evidence linking the wireless direction of light field artillery in these actions with Major L.W. Gill and his 46th, and later 45th, Battery, 3rd Division. His battery was in the right sector more than once and may well have acted on wirelessed information, but there is no proof of Gill's involvement. It is entirely possible that Gill had some sort of connection with wireless, since his Brigade Commander, A.G.L. McNaughton, a McGill graduate in engineering and a specialist in counter-battery, must have known about his background. Gill's practical skill in Morse may have been slight, however, so he may not have acted as a wireless operator on the battlefield. By 1917, Gill's technical knowledge of wireless was becoming outdated, as the new valve CW sets took over. Yet it is difficult to believe that a fireball like Gill could have happily directed all of his considerable energies to training and attending to his battery, when there was so much engineering work going on around him preparing the infrastructure at Vimy Ridge: laying telephone lines, installing switchboards, setting up and testing wireless and experimenting with new aerials, bringing in electricity and water, calibrating the guns, and devising counter-battery techniques. It doesn't seem like Lester Gill to stay uninvolved when there was engineering to be done.

Prof. Ebert William Henderson (Elec. '05) was Gill's assistant from 1907, and took over direction of E.E. when Gill went to war in 1916. Henderson was in charge of the Queen's wireless sets lent to the School of Signalling at Barriefield War Camp in 1916 and 1917.
-courtesy Stuart Henderson

Ebert William Henderson

While Gill was struggling in the muck of the Somme and Vimy, back at Queen's, Course XI, "Telegraphy and Telephony," continued in the E.E. curriculum, and in the summers of 1916 and 1917 Major Mundell's School of Signalling at Barriefield Camp used the Queen's equipment in the formal training of wireless operators for overseas duty.

Stepping up to replace Gill at Queen's (for the School of Mining was incorporated into Queen's in 1916 and was now the Faculty of Applied Science of the University) was his Assistant Professor of E.E. and former pupil, Ebert William Henderson (1879-1982). He was born in Almonte, Ontario, and "went down to Queen's in 1899" with his senior matriculation and the Leitch Memorial Scholarship in Classics (value $65; his total expenses for the year were $190) to join Science '03. In 1899, there were just four permanent buildings on the campus. He recalled his Physics lab in the unfinished cellar of Old Arts and first year draughting in the attic of Carruthers Hall, where the stench from the Chemistry labs "in the lower regions was something terrible.. My Professor in Mathematics was Nathan F. Dupuis, whom I always considered the best professor I ever sat under...I can remember Dr Dupuis helping me make a rosette as an ornament for a bookcase I was making at that time" in the second floor carpentry shop of the old wooden "Tool Shed."

Henderson's expenses for his second year shot up to $210, and he couldn't afford to return to Queen's in 1901-02. He dropped out to work at Canadian General Electric in Peterborough for two years and returned to school in the fall of 1903. He was one of eight graduates from E.E. under Gill in 1905, whence he and three colleagues followed Gill's trail to Westinghouse to take the Engineering Apprenticeship Course. In 1905, the apprentices were paid 16 cents an hour. The next year they got a big raise, to 18 cents. In the fall of 1907, Henderson was invited by Gill to return to Kingston to teach and was appointed Lecturer at the School of Mining in the 1908-09 session. He was promoted to Assistant Professor in 1911-12 and Associate Professor in the spring of 1920.[60] F. Morris Wood was one of his pupils before the First War.

F. Morris Wood: "Eevie?" Oh, yes, he was wonderful...That was his nickname, "Eevie" Henderson. I think it came from E.W., you see. He was young - he had just graduated in 1905. Electrical. And I liked him very much because he knew all about the mathematical theory, and that's what charmed me. He was my ideal teacher because he taught me an awful lot of electrical engineering, and that's one thing that was not my specialty. But because of the mathematical connection, I appreciated it, and I could see much more in his lectures than the average student could because I had mathematics behind me, and I just loved Eevie. In fact, I got full marks in, of all things, electrical engineering...in his course. He was a marvellous teacher. He was one of the best teachers I've ever had. Really. Oh, he was perfect. He compared to Nathan Dupuis.[45]

Henderson also seems to have been known to his students as "Volta" Henderson, according to the Faculty songs of 1913 and 1920:

Volta Henderson, you see,/ Gives us electricity,/ Gives it to us fast and hot/ Whether we want it or not./ He don't care if we attend,/ He talks on without an end/ From the time the first bell rings/ Of circuit laws and other things.

Electrify and magnetize, polarize,/ He thinks that we know what he means,/ All the time we're lookin' wise, lookin' wise,/ Thinking of far more pleasant scenes./ The permeability of our brains/ Seems to give Henderson pains,/ He says that we take the prize, take the prize,/ For the biggest bunch of noise at Queen's.

In electricity Prof. Henderson can rave/ And put old Steinmetz in the shade./ Ohms and watts all do the dance/ To his pet topic on reactance:/ He makes us think in fractions/ And "Volta" is his name.[61]

E.W. Henderson: As I remember, the apparatus in the Laboratories was unchanged by Prof. Gill during the two years I was in Pittsburgh. My memory of the details is a little hazy but I think they were as follows:

The boiler room was on the east end of the basement. Next to it was the engine room in which was a high-speed steam engine of British manufacture (manufacturer's name forgotten) driving a DC generator which supplied light to all the buildings just as heat was supplied from the boiler room to all the buildings. I think in a small gallery in the engine room was the motor generator set which was used to charge the batteries which were located, as I remember, in the second room west of the engine room. The room next to the engine room contained a DC, DC motor generator set. Opposite this, I think on a raised floor, was a DC motor set up with a brake arm and balance measuring devices such that it could be tested at different loads. This loading device was one invented by Prof. Gill and was quite unique in that, instead of a one-piece radius arm which would ordinarily be used to rest on a scale of some sort, this arm was made in two pieces, the upper one of which was hinged to the lower rod and weights were applied at the outer end from its lower arm while the upper arm was connected to a spring scale supported from a pipe arm, supported from the floor[62]...As far as I can remember, these machines, plus some loading racks made of coiled iron wire, was all that was in that room.

The next room west was the battery room, and the Lab beyond this to the west contained the following: A rotary converter driven by motor which could be used to supply either AC or DC power. I have forgotten the name of the maker of this rotary converter. A Westinghouse 2-phase induction motor arranged for loading by a brake. A test table with the necessary switches for meter connections for measuring AC power, and I think, also in this room was an AC constant current transformer of the type which had one fixed coil and one movable coil, the latter being attracted by the fixed one to maintain a constant current in the circuit in which it was connected. This room also contained a switchboard with plugs such that the current could be carried to various parts of the room, and I think for supply to run the machines there. There were also some loading racks in this room.

In the Lab nearest Ontario Hall was a cupboard containing meters and benches along two walls which contained accurate reading voltmeters and ammeters used generally for making calibration tests on the other voltmeters and ammeters in the cupboard. As I remember, this cupboard also contained shunts for DC measurement and a series transformer for AC measurements.

Upstairs in the Electrical Lecture Room hung a number of arc lamps of various makes; and on the desk was a brass disc cut radially to make a number of conductors between its hub and outside rim. I think this was similar to the one originally made by Faraday, in which he proved that an electrical voltage could be obtained by moving a conductor through a magnetic field.

As far as I remember, no other machines were bought by Prof. Gill for the Laboratories, nor did I buy any after he left, with the exception of some radio receiving equipment which I think, perhaps, Doug Jemmett put together and may have used after I retired...Naturally, no changes were made in the apparatus which we had before the war...On thinking the matter over, I think the rotary converter mentioned above was not motor driven; but in that same converter room was a CGE motor-driven alternator which supplied the AC power for this laboratory.[60]

Henderson had been a member of the 5th FCCE from its early days, as a signaller, and he was an experienced telegrapher.[63] In early 1914, he was in charge of the new telegraph section of the 5th Company. Soon afterward, he was on the FCCE Staff as Adjutant under Gill and Wilgar. Dean Clark approached Henderson some time in 1914 and requested that he not go overseas, and so he decided for that reason and also because he had three small children, that he would stay in uniform and serve in Canada. Henderson directed construction at Barriefield Camp in the spring of 1915, surveying and surfacing the roads, laying water and sewer pipe, installing electrical wiring, the telephone and telegraph wires, erecting the wireless masts and building the permanent structures. He was involved in the installation and tuning of the wireless set,[13] and was also active in recruiting volunteers for the CEF. By late January 1916, both Gill and Wilgar were making the final preparations for going overseas, and the Queen's Journal noted: "Amongst other changes that are well in accord with just deserts (sic) we notice that the popular O.C. of the Fifth Company is now signing himself Captain E.W. Henderson. The Queen's Engineers under Captain Henderson have shown themselves to be very efficient in the various jobs allotted them in this district."[64]

The 5th FCCE, led by Henderson, was scheduled to go into camp around April 19, 1916, to prepare for the re-opening on May 12.[65] On Wednesday the 19th, Henderson was "on the sick list for a few days. Lieut. (Thomas S.) Scott (Civil '97) will be in command of No. 5 Field Company, Canadian Engineers, going into Barriefield camp as an advance guard on Thursday."[66] Dr A.L. Clark, in his history of the Science Faculty, had the event out of chronological sequence when he stated, "Henderson finally broke down with the strain of double duty and was forced to leave his work for a time in order to recuperate."[67] Henderson was back in the camp working with Scott by the 24th, preparing for the arrival of 11,000 men by the end of the week.[68]

Wireless at the Barriefield War Camp in 1916

There was going to be no delay in getting the wireless going at the Barriefield War Camp in 1916. Before the end of April, "Major D.E. Mundell, commanding the School of Signalling, has applied to Queen's University for the use of the wireless equipment for the instruction of the signalling school at Barriefield during the coming season."[69] More than likely, the request was made of E.W. Henderson, who was now in charge of E.E. and the wireless part of the course. Telephone, buzzer and wireless had largely replaced flag signalling at the front, and Mundell felt himself "particularly fortunate in being able to secure the use of the Queen's outfit for his work."[70]

The initial plans were that "The permanent outfit will be installed on the camp ground in a tent, and the portable outfit will be sent out with the troops on route marches and treks. Captain Henderson is working on the condensers of the portable outfit, which got out of order last summer."[71]

Henderson was busily engaged until about the middle of May in laying out pipe lines, installing lights and making the camp ready for the huge occupation expected that summer, and so the consultation with Mundell on the transfer of the wireless equipment had to be delayed until the afternoon of May 17.[72] The Signalling School opened on June 5 "with the full signalling section of every battalion in attendance. There are eight men taking the junior course and the remainder are taking advanced work which consists of visual signalling, field telephone, trench work and wireless...Instructors Reid and Redding (sic) are conducting the school..."[73] A 60-foot aerial on a single mast was put up on the first of June in the centre of what was to be a wireless compound, composed of four marquees and one bell tent.[74] They were expecting the wireless set to be installed within a few days.

By June 10, the School of Signalling had "three tents erected on a knoll on the main parade grounds (a large tent and two small tents in front of headquarters)...the wireless apparatus is on the ground and will be set up as soon as certain adjustments can be made. The aerial has been erected and in a few days the wireless for the advanced class will be in operation and giving the instruction that is so badly needed at the front."[75] There was a delay at this point, waiting for the arrival of an experienced wireless instructor, Sergeant Lorne L. Hicks, of the 14th Regiment, who was to take that part of the course. Hicks' arrival was announced June 14.[76]

W. Harold Reid: In the summer of 1916, Col. Mundell decided to introduce a radio course because he said Queen's University would loan us their small portable radio set. I told Col. Mundell about Lorne Hicks, an electrician in Kingston, who was acquainted with wireless telegraphy equipment, procedures, etc. He was brought in as an Assistant Instructor for this special class. I had to help him with several things as he had had no army experience, and he did not know the code, but that did not matter as all students were graduates of our school. Later, a little radio shack was erected on our grounds and an antenna put up.[77]

It was Sgt. Hicks who installed the wireless equipment in 1916,[78] the same wireless equipment that had been used by the Queen's FCCE in 1915 "and which did such splendid service on treks such as the 8th CMR had;"[76] that is, Hicks' choice for teaching was the portable set in the neat reddish mahogany box with brass fittings that had once been carted around in Gill's automobile. Henderson had also supplied and erected the portable aerial which had been superceded the year before, but Hicks decided to change the aerial to a four-wire inverted L.[78] The power would come from the city mains through the camp lines, but there is photographic evidence that a portable gasoline engine was also used in 1916.

The National Archives of Canada, in Ottawa, holds a document listing the specifications of the wireless set at Barriefield Camp in 1916, an "Application for Licence to Install and Operate a Radiotelegraph Station" made to the Dominion of Canada Department of the Naval Service Radiotelegraph Branch by Major D.E. Mundell, Divisional Signalling Officer, June 13, 1916.[79]

1. Full name, address and occupation of person by whom the license is desired.	David Edward Mundell - Major District Signalling Officer MD No 3
2. The exact location of the station to which licence would apply.	Barriefield Camp for overseas units.
State distance from nearest "Coast" or "Commercial" Station	About one mile from Marconi Wireless Station Fort Henry Heights
3. Name and location of any other station or stations with which it is desired to carry on communication.	- None - Purely intercommunication in Camp grounds
4. Particulars as to ownership or occupation of each station mentioned in paragraphs 2 or 3	Marconi Wireless (Govt.
5. Purpose for installation to which the license would apply...	Experiment and instruction for overseas units
6. Particulars of installation	
a) Kind of aerial	Inverted L
b) Wavelength used for transmission (in metres)	200
c) General description of the apparatus used for transmission.	Clapeastam 25-50 M portable set
d) ...for reception	same apparatus
e) Maximum power to be taken by transmitting instruments. Current and voltage, also state whether direct or alternating with periodicity.	60 cycles - 110 volts 500 watts. primary
f) Source of power and output. State whether direct or alternating with periodicity.	From City of Kingston power - alternating
g) normal transmitting range	20 miles
7. Any other essential particulars indicating	For instructional purpose C.E.F.

the nature of the
installation, and the
circumstances in which
it is to be used.

8. State whether the persons   Sergt Instructor L.L. Hicks
who will work the station   is the one operator at present,
can send and receive   - yes - cannot state speed
signals in Continental
Code, and if so at
what speed.

Signature D.E. Mundell
Major.
Dist. Sign Off.
June 13th 1916

With the application is a sketch showing the four-wire inverted L antenna and the radio shack.

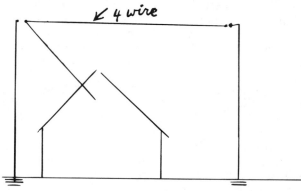

Sketch of the 4-wire inverted L-antenna and the radio shack, Barriefield War Camp, drawn by L.L. Hicks on the licence application to the Radio Telegraph Branch, Department of the Naval Service, June 13, 1916.
-courtesy National Archives of Canada
RG 97, Vol. 85, file 6202-30 Pt. 3

It was probably the experienced Hicks who recommended that they seek a licence for the Barriefield operation and Mundell must have relied upon Sgt Hicks for the technical details in the application. Gill certainly had never sought a licence.

Certain details about the wireless set used in training at Barriefield War Camp in the summer of 1916 can now be deduced from the fragmentary verbal evidence, with the aid of three remarkable photographs and the application form. The wireless telegraph set licenced for use in training at Barriefield was a portable, tunable spark transmitter-receiver manufactured by the firm of Clapp-Eastham of Cambridge, Mass. It was powered by 110-volt 60-cycle alternating mains current, and transmitted at a wavelength of 200 metres (1499 kilocycles) over a distance of about 20 miles.

The set is shown in three photographs taken at Barriefield Camp in August 1916 by W. Harold Reid. One of these photographs, plus Don MacClement's reconstruction, is shown in Chapter 1. The photographs show the reddish mahogany box with brass corners and folding brass handles. Dr Reid estimates its size as 20 x 12 inches and 10 inches high, although he recalls it as a quarter kilowatt set. It had a large induction spark coil for the transmitter (not visible), a flat-wound aerial tuning inductance, a hot wire ammeter reading 3 amperes full scale for tuning the primary of the spark, a plain unquenched adjustable spark gap as supplied by Clapp-Eastham, a double-slide tuning coil and a crystal detector. A set of what may be Brandes headphones rests on the folded down front flap of the box, and what looks like the top of a Clapp-Eastham key sits on that flap. One can also see the two knife switches on top for changing over from transmitter to receiver. This is the set seen in the back of Gill's automobile by Don MacClement at the time of the wireless field tests, and it is surely one of the sets that Gill purchased from Clapp-Eastham in 1911 or 1913. Given its relative primitiveness - a double-slide tuning coil and unquenched spark gap - it could well be the set ordered by Gill on December 27, 1911, on requisition number 4550.

What then became of the "stationary plant" of 1915, presumably the same jumble of boxes used in the tent back of Fleming Hall in the field tests? It probably remained in reserve at Barriefield, for a set like it turned up later that summer at the Kingston Fair. It is quite possible that the large set was maintained as the "stationary plant" at Camp when the troops took the portable on bivouac that summer. The small portable set would have been more suited for teaching the elements of wireless than the big one, and so it was the small one that was set up at the camp by Sgt. Hicks. Another advantage of the small set was that four or five students could "listen in" at the same time.[80]

The permit issued by the Radiotelegraph Branch of the Department of the Naval Service arrived in Camp by July 7,[80] registered to Major Mundell. The call assigned by the government was XWD, in the same series as the one given in 1914 to the Marconi test room in Montreal, XWA, which later became radio station CFCF.[81] The year before, the 79th Overseas Battalion in Brandon had held XWB, but in 1916 there was no XWB. The call XWC was held by Royal Miltary College, Kingston, in 1916. XWE was held by the DSO at Niagara Camp, and XWF by the DSO at Camp Borden, Ontario. The series "X" for experimental continued the sequence XA to XC, once issued to licenced experimental and amateur stations and now suspended for the duration of the War.

The licence specified that the apparatus could be used only between 7 am and 7 pm, so as not to intefere with the operation of the commercial Marconi station VBH on Barriefield Heights. This was no problem, since the hours of teaching fell between these limits anyway.[80]

There was a good deal of activity in the wireless section in July. On the morning of the 14th, they built a large kite "for carrying an aerial high into the air so as to try and keep in touch with the camp wireless station on each of the weekly route marches."[82] It was tried out on the 15th, but was not mentioned again. Around the 25th, the portable and a small gasoline engine were taken to Grass Creek, and messages were exchanged with the Camp wireless with no difficulty.[83]

> **W. Harold Reid:** The main thing was we had to teach them the code. The preliminary course was about six weeks. Then I would try to get as many as possible of the better men to come in for a second course, and then these were the men that would also have a chance to do wireless. Now, it all depended on how long that battalion was in the Camp. So that we had to teach them the Morse code first, for sending and receiving, and then I gave them the basic electrical technical work, you see, and then Hicks took them out on the practical wireless work. So there were really three levels, but only a few would get to the second or third. If a battalion was just formed, well they would be there for a longer time and I could get the use of some men for my second class or the third one...When they went out with Hicks, they had a little Ford car and they'd load up the stuff in this little Ford car.[8]

A photograph of that Ford Model "T" survives, snapped in 1916 by Sgt.-Inst. Reid, and showing Sgt. Hicks in the back seat and Don Robinson at the wheel. The mahogany box with the brass corners is in the back seat under Hicks' right elbow.

During an early school that he taught at Barriefield, Hicks made a startling discovery.[84] The wireless apparatus picked up messages which

Ford touring car, used to transport the Queen's portable W/T set at Barriefield War Camp, 1916. Sgt. Lorne Hicks is in the back seat, the mahogany box with the brass corners under his elbow and the portable antenna sticking out the back. Don Robinson is at the wheel. It is possible that this was the Ford car that Prof. Henderson used as a generator to run the W/T set.
-photograph by Rev Dr W. Harold Reid
courtesy National Archives of Canada, PA 110442

were clearly not wireless in origin, but were a basic lesson in Morse code sent down a wire from a buzzer. The signals were coming from a tent about 90 feet away, and Hicks attributed the reception by the wireless to absorption of electricity from the communication wires between buzzer sets. Here at last was the explanation for the Germans' being instantly privy to the details of plans sent over ground wires by low tension buzzer from Allied headquarters to the front lines. Often, when a German trench was taken, the Allied soldiers found their own orders tacked up on display. Hicks was not rewarded by a promotion for his discovery.

The first Signalling School had 90 men in the junior class, 150 in the advanced and about 20 in the special wireless class. The School closed on August 4 for examinations, administered on the 11th by Lieut. Manson, Inspector of Wireless for the Department of the Naval Service.[85,86] The officers of the School were: Col. T.D.R. Hemming, Camp Commandant; Major D.E. Mundell, DSO, Commandant of the School; Lieut. J.C.F. Munsie, Adjutant of the School; Sgt.-Instructors W.H. Reid and L.P. Reading; and three other officers: Lieut. V. Lanos of Kingston, Lieut. Hanning, Oakville, and Lieut. Smith, Vancouver. Graduates received the government certificate for operators of coast and land stations, and Lieut. Manson expressed himself well pleased with the work accomplished.[86]

The new course in signalling opened August 15, and there were hundreds of applicants.[87] Twenty-two officers and NCOs were in attendance in the wireless section.[86] Among the latter, William F. Ferguson was named Signalling Sergeant of the 156th Battalion, and reported to the wireless school for training along with Privates G.B. Fletcher and A.L. Scott.[88] Ferguson and Scott were to play minor roles in this story, while still students of the Wireless School.

On September 1, the portable wireless was taken on bivouac to Parrott's Bay, between Bayview and Millhaven, 24 kilometres west of Barriefield. Three thousand soldiers in command of Brig. Gen. Hemming marched to the pipes of the 154th and the brass and bugles of the 155th and 156th up Princess Street and out the Bath Road.[89] Munsie, Hicks, Henderson and some students were on the trek, while Reading took charge of the school at Barriefield (since Reid was ill) and Sgt Ferguson remained on duty at the Camp wireless.[90] The <u>Saturday British Whig</u> reported that on the way to Parrott's Bay, the wagon carrying the wireless set broke down, and Henderson and the others had to work into the afternoon to put the wireless back into shape. This was a gloss on the real story. A telescoped version of the truth has survived in A.L. Clark's book,[91] but here is what <u>The Whig</u> reported the following Tuesday:

> The instruments including a gasoline power plant and dynamo were in the line of march, being loaded on a van that also carried some of the men...When the wagon with the parade arrived at the camp it was driven to a point

Students of the Wireless Section of the School of Signalling, Barriefield, practicing with the Clapp-Eastham portable W/T set. Note the mahogany box with brass corners sitting on the wheelbarrow, at left. One can see the lid and body, a brass handle and brass corner, the double-slide tuning coil, an ammeter and an unknown ceramic component next to the tuning coil. The operator, wearing headphones and with pencil in hand, is sitting on one of the barrow's handles, ready to take down the message using an antenna reel for support. A portable gasoline engine supplying the power is in the foreground. This unit was probably in communication with a second Queen's wireless set located in the main bell tent at the School of Signalling (for detail of W/T box, see page 84) -photograph by Rev Dr W. Harold Reid, courtesy National Archives of Canada, PA 110469

convenient to headquarters and near the shore. The portable pole was erected and the other end of the aerial was fastened to a cross piece that was tied on a tree. The motor and power plant were erected outside on the ground while the wireless operator remained in the van with their telephone receiver.

Capt. Henderson took charge of the gasoline motor, but had some trouble in getting it to work. He however, sent for his own automobile and used that engine. The rear wheel was jacked up and a belt run from the wheel to the dynamo. The power then furnished was then reliable and regular...

"Col. Ogilvie, Barriefield Camp - Arrived 1 p.m.; troops in good condition. All settled in bivouac. (Signed) General Hemming. Time 3 p.m. Friday."

"Col. Ogilvie, Barriefield Camp - Left Camp 8 o'clock. Expect to be in by 1 p.m. (Signed) General Hemming."

These two messages were two of the official com-

> munications sent over the wireless from Parrott's Bay to
> the Camp when the troops were on the march and
> bivouac and they clearly show what a value may be put
> on the splendid equipment that the School of Signalling
> possesses in this way...The operating was carried on by
> Sergt. Inst. Hicks. At Barriefield Camp...Sergt. Ferguson
> took the messages and sent others to the bivouac.[90]

There was a thunderstorm that afternoon which interfered with
the wireless communication, "but when the air cleared the instruments
were working beautifully. Many messages could be heard from stations
in the area in which the instruments were able to receive...All day long
a large crowd of soldiers was gathered around the van and appeared
immensely interested in seeing the wireless system of communication in
actual working order."[90]

With that success behind them, Major Mundell authorized the use
of the wireless on the upcoming trek to Gananoque, with the possibility of taking it as far as Brockville, and later perhaps on the trek to
Belleville, 53 miles away. "So far, great success has been made with the
wireless outfit at Barriefield. Sergt. Inst. Hicks is in constant communication with the portable station when it is erected at some distant
point and on several occasions long tries have been made. In every case
messages could be exchanged without difficulty. In view of the way the
apparatus is working, there is no reason why it should not be used even
when the soldiers march to Belleville."[92] At that time, their record distance for this set was still Gananoque,[93] about 32 kilometres.

The Kingston Daily Standard reported on September 12 that an
Audion (De Forest's triode detector, a modified Fleming tube) had "been
installed at the wireless station which increases the receiving qualities of
the camp station to a great extent. The station at Arlington Heights
(Virginia) and other distant places can now be picked up easily by the military station at Barriefield."[94] Henderson later told A.L. Clark that he had
bought one of the new Fleming vacuum tubes and that it was installed at
the Barriefield station (i.e. in the receiver of the semi-portable collection of wooden boxes). He recalled that he then belted the generator of
the portable wireless to the motor of his old Ford car for communication
with Barriefield.[91] These stories are documented here for the first time,
although the chronology differs from Henderson's recollection.

Another improvement at the camp station was the raising of the
aerial wire by the addition of another section of pole. This added height
would improve the reliability of communication with Gananoque.[95]

The portable set went to Gananoque ahead of the troops on Friday,
September 15, and was set up in the Armouries, with Henderson's
motor car to generate the current. Hicks was in charge, but this time he
allowed his students to operate both sets. He also permitted The Whig
to send a demonstration message to Mr Britton, editor of the Gananoque
Reporter: "Give troops deserved welcome."[96] The Whig, in turn, received
a fraternal message from the Gananoque Reporter at 10.30 Monday
morning: "Troops left in return march to Barriefield Camp at 9:30 in fine

fettle. They were warmly received, well entertained, and have the best wishes and loyal support of the entire community."[97]

The <u>Daily Standard</u> had the honour of receiving a longer message from Brig.-Gen. Hemming:

> We arrived in Gananoque at 5.15 on Saturday. The troops stopped for three hours just six miles west of the place selected for bivouac, where the men had dinner and sports consisting of football and baseball. We only had one minor casualty which was taken care of by Queen's Field Ambulance. The men appear to be enjoying themselves on the trip, and are adapting themselves to existing conditions. Many citizens, including Senator Taylor, met us out from town, and led the procession into Gananoque. The citizens have provided liberally for the men, and all are being well treated.
>
> Brig.-Gen. Hemming,
> G.O.C.[95]

Henderson told the press in Gananoque that "many improvements have been made in the camp station, but the apparatus is too small and limited for a wireless school. An outfit of greater capacity - at least a one kilowatt apparatus - is needed for the kind of work being done by the school in camp. The Wireless School should also have a permanent building and a permanent aerial. This could be used all winter by the School. With another aerial on the top of the armouries in the city the school would be equipped to do good work."[98]

The next public appearance of the Queen's wireless was at the Kingston Fair in late September 1916, as part of an impressive exposition by the Militia for the encouragement of enlistment. The School of Signalling had a booth in the main building at the Fair, the old wooden Crystal Palace. The display featured a complete operational sending and receiving wireless apparatus, the first time that one had ever been shown at the Fair. Of course, Major Mundell was in overall charge, "assisted by Captain Henderson and Sergt.-Inst. Hicks, who installed the apparatus...by Sergt. W.F. Ferguson (and) Ptes. J.P. Mephan, A.L. Scott and W. Bertrand, all of the 156th Battalion, and all expert wireless operators."[99] The wireless booth attracted "no end of attention." For a silver dime you could submit a hand-written message to the officer in charge and actually watch your words being pounded out by a "khaki-clad lightning-slinger" seated at the key of the apparatus. The messages twinkled through the ether to a tent on the grounds, and as an added thrill you could go out and retrieve a copy of your own note, snatched from its furious flight and re-materialized.[100] One wag requested a keg of beer, and another asked for a date with one of those Parisian models. The <u>Daily Standard</u> observed, "The keg will not be delivered."[101]

A rich crop of dimes was handed over to the Red Cross Society. All the while, Capt. E.W. Henderson of the 5th FCCE was standing by with blank attestation papers hoping to reap a harvest of young men overcome with patriotic fervour at the sight of the Militia displays.

> **Prof. Harold Stewart:** I recall as a kid here...after the War had started, going up to the Kingston Fair, and the Signal Corps had a set-up for recruiting. And I think they had a big ten-inch spark coil there with a chap pounding away at it, and the big spark jumping across the gap, and people were very thrilled about this. Wireless was a black art then...It's become a black art again, with chips and so on.[102]

There can be little doubt that the complete sending and receiving apparatus with the ten-inch spark coil was the jumble of boxes from Gill's tent behind Fleming Hall. The Clapp-Eastham portable, station XWD, was likely used as the remote receiver at the Fair.

Between the Gananoque trek and the wireless at the Kingston Fair, Barriefield Camp experienced twin peaks of excitement. The Governor General of Canada, the Duke of Connaught, arrived for inspection on the 21st, and the Uscan Feature Photo Play Corporation of Cornwall and Messena also appeared to take moving pictures of the ceremony and of the camp's activities over the ensuing days. H. Wilmot Young, President of Uscan, and W.R. Hitchcock, Secretary-Treasurer, made the arrangements and probably shot the film. Dr Evans, of the Daily Standard, directed.[103]

The Barriefield footage was premiered before the military staff at the Strand Theatre in Kingston on November 12, 1916, when announcement was made that these 3,000 feet were to be part of a projected military feature film to be called <u>Canada in Khaki</u>. The present film showed the "boys" on parade and the white tents dotting the green hills, Queen's

The tents of the School of Signalling, summer of 1916. The men constructed a sign in whitewashed stones set into raised earthworks in front of the School. Note the main radio mast behind the wireless tent in the centre.
-photograph by Rev Dr W. Harold Reid
courtesy National Archives of Canada, PA 110416

School of Signalling Radio School tents, Barriefield War Camp, August 1917. (#99-4)

Radio Office (left) and Signalling Office (right), Barriefield War Camp, August 1917. (#100-5)
-photographs by Rev Dr W. Harold Reid

Field Ambulance, the 5th Field Company, the Bayonet Fighting School, the Army Service Corps, the Bombing School, and the Wireless School in action. In the next few days Uscan would return to shoot the Royal Canadian Horse Artillery and field artillery on manoeuvres on Barriefield Hill, convalescent homes, Ongwanada Military Hospital, headquarters staff, munitions workers leaving the Locomotive Works, and the students of Queen's. It was hoped that the film would attract large audiences in the still neutral United States.[104] The Whig reported that Canada in Khaki was being shown in mid-1917, and was a wonderful assistance to recruiting, but copies of the film have not survived. The best possible photographic record of the wireless activities at Barriefield Camp at the height of the War is lost.

The School of Signalling at Barriefield closed for examinations on Friday, October 13, 1916, and the next courses were announced for late November.[105] The wireless station was dismantled the following Monday, to be "installed" at Queen's,[106] since the signals courses during the winter were going to be held in Grant Hall. Mundell wrote to the Queen's Board of Trustees expressing his gratitude for the use of the wireless at the War Camp.[107]

Between January 1916 and late March 1917, the School of Signalling in Kingston and Barriefield turned out 780 qualified signallers for overseas service, including 50 with wireless certificates, and was famous "throughout Canadian and English training camps as being better able to train signallers than any other centre. The excellent work that has been carried on is shown by the frequent letters received from the front. English instructors have on several occasions made reference to the quality of signallers turned out of the local school...This is the only school of signalling in Canada to have a recognized wireless school running in connection with it."[108]

By the spring of 1917, the situation had changed a great deal at Barriefield Camp. Many drafts had gone overseas and several of the military schools were closing, for it was becoming increasingly difficult to find fresh men interested in going overseas to the slaughter. From the fall of 1916 the wireless classes had shrunk too. Barriefield opened in the late spring and Henderson and his 5th FCCE went into Camp around May 21, 1917, to prepare for the arrival of the troops.[109]

"School of Signalling, Kingston," card postmarked February 8, 1917, addressed
to Mrs S.W. McCullough, Wellington, Ontario, by her son Len.
Sgt.-Instr. Louis Philip Reading is seated at centre.

His company was also short of men for home service, and he was actively signing up recruits. Things were so slack that Mundell even offered
the use of the Wireless School to the vocational officer of the M.H.C.C.,
as excellent training for returned and disabled soldiers and a skill that
could be used in civilian life.[110] Headquarters returned to camp around
May 26, but the wireless masts were put up by Reid and Hicks only on
June 21.[111] The equipment was installed a few days later[112] and the
first course in buzzer, wireless and signalling began on the 27th, with
classes held from 5.15 to 6.15 pm.[113]

Right in the middle of the first course in signalling that summer,
an order was issued by the British Army closing all signalling schools in
the Dominion. All signals training would henceforth be done in
England.[114] The last school at Barriefield closed on September 7. The
wireless equipment was dismantled on that day and returned to
Queen's,[115] and the staff soon dispersed. Major Mundell was promoted to Lieutenant-Colonel and became Surgeon-in-Chief to Queen's
Military Hospital, Kingston, on October 18.[116]

Sgt.-Inst. Reid was refused permission by the Militia to go overseas,
so he left for Halifax in October to join the Navy as a radio operator.[8]
He was soon sent off to Marconi coast station VCO at North Sydney.
At most Marconi coast stations, including VBH at Barriefield, the power
for the transmitter came from a gasoline engine linked to an AC generator. When a ship called in, the operator had to run out to crank the
engine before he could transmit the signal "K" for "Go ahead." The

North Sydney station, however, had 110-volt AC from the town. The transmitter had a non-synchronous rotary spark gap, and operated at 600 metres (499.7 kHz). VCO's standard Marconi station receiver "had two carborundum detectors which were remarkably stable and very satisfactory. The carborundum point could be screwed down hard so that in bad weather at sea, when the ship was being pounded badly, it did not give out. The galena detector with its cat's whisker might produce a much stronger signal, but it was so delicate that it was useless at sea."[8] They kept a 24-hour watch and sent out regular weather and navigational reports and iceberg warnings.

Shortly after 9 am on December 6, 1917, Operator Reid began receiving coded messages from VCS at Camperdown for relay to Ottawa, with the first details of the disastrous Halifax explosion. At Halifax, all telegraph lines were down, so these messages had gone out by despatch motorcycle rider to Camperdown to be passed on to North Sydney and thence by landline to the Dominion Government.[117]

Early in 1918 Reid was transferred to VCU at Barrington Passage with its beautiful 10-kW synchronous rotary spark gap transmitter, and then to a ship on the Atlantic coast. At the end of the War, he was ordered back to the Barrington Passage station, where there was a new British 25-kW continuous wave arc transmitter, for operation to Bermuda at 4,000 and 5,000 metres. Its valve receiver often picked up signals from a similar Royal Navy transmitter on Malta. Harold Reid was demobilized in the spring of 1919 and went to Kingston to enrol in Queen's, where we shall meet him again. Sgt.-Instr. Louis Reading went to England to teach signalling at Seaford, Sussex, and later taught at RMC, Camp Borden and Vimy Barracks, Barriefield.[118]

The two Queen's wireless sets may have continued in use in the Department of E.E. at Queen's after the summer of 1917, under

Henderson's supervision. Stuart Henderson (Arts '32) was only ten years old in 1920, when his father left Queen's. He recalls his father's having a crystal receiver and teaching E.E. students, as part of the course, to wind transformers and make other components "to make the station."[119] There is no further written record detailed enough to permit positive identification of the two sets from Barriefield, but there is some evidence that the "semi-portable" multi-component set was put back into service in the Queen's Wireless Club in 1920 or 1921.

N O T E S

1. Minutes, Queen's University Senate, Oct. 14, 1914, p. 54; Senate Office, Queen's University.
2. "Principal Gill Arrives to Take Charge at Tech", Hamilton Spectator, Oct. 9, 1922, p. 1.
   5th Field Co., 3rd Divisional Area, Militia List Canada, Oct. 1915, p. 197; NAC.
   "In Military Circles", DBW, Jan. 26, 1915, p. 8.
   "Fifth Company F.C.C.E.", QJ, March 8,1915, p. 1.
   For background on the 5th Field Company of Canadian Engineers, see Kathryn M. Bindon, "Queen's Men, Canada's Men - the Military History of Queen's University", Trustees of Queen's University Contingent, Canadian Officers' Training Corps, 1978.
3. "Cancels Camp at Barriefield", KDS, March 20, 1915.
   "To Hold Big Camp at Barriefield", KDS, April 16, 1915, p. 4.
   "The Tented Cities", DBW, June 8, 1915, p. 6.
4. SM Minute Book, 1907-1917, May 5, 1908, p. 354; QUA, Coll. 3695B, Series I, Box I.
5. David Edward Mundell, Statement of Service in the Canadian Armed Forces, Personnel Records Centre, NAC.
   DBW, Jan. 27, 1915, p. 8.
6. "The Camp at Barriefield", DBW, June 1, 1915, p. 3.
   "Precautions Were Taken", ibid., July 6, 1915, p. 8.
7. "With the Overseas Troops", DBW, June 19, 1915, p. 5.
   KDS, July 17, 1915, p. 9.
8. Personal communications from Reverend Dr. W. Harold Reid, Stoney Creek, Ontario, June 3, 1986, and Aug. 25, 1987.
   W. Harold Reid, "Experiences of a Soldier Signaller-Sailor Radio Officer in World War I", Communications and Electronics Newsletter of the Royal Canadian Corps of Signals, Ottawa, 1978, p. 32.
   As a caution to accepting the newspaper stories of the period at face value, Dr Reid tells the following story:
   "There was one reporter with the British Whig, was a terrible liar, and he concocted all kinds of things. Oh, this man! When I ever saw him coming into Camp I would avoid him and never say anything. Because, no matter what you said, he would twist it. And he would make up all kinds of stories...I heard this and it's quite possible it was true, that some British official that was there at the Camp and was reviewing (some) men, and this man was there and he was wearing medals, you know...ribbons of various engagements that he had served in. And this British officer recognized that this one ribbon that he was wearing was for a battle that was fought before this man was born! But nobody else knew that. And he spotted it and pulled him out, and took it off...Oh, I knew this man and he was a terrible man, but he was a prominent writer for the British Whig. But, he wasn't to be trusted...I think eventually the Whig got wind of this and they let him go...because I didn't see him later."
9. File HQ 6978-2-53, June 23 and June 26, 1915, Department of Militia and Defence, Ottawa; State and Military Records, Federal Archives Division, NAC.
10. "A Wireless Station at Barriefield Camp", DBW, July 21, 1915, p. 5.
    "A Camp Wireless Station", KDS, July 21, 1915, p. 7.
11. "With the Soldiers at Barriefield Camp", KDS, July 22, 1915, p. 9.
12. "The Wireless at Camp Used for First Time", DBW, Whig, July 29, 1915, p. 5.

13. KDS, July 31, 1915, p. 9.
14. "Lively Time in Camp", DBW, Aug. 14, 1915, p. 5.
15. "Two Thousand Men", DBW, Aug. 3, 1915, p. 5.
16. KDS, Aug. 5, 1915, p. 7, and Aug. 7, 1915, p. 7.
17. "3. Electric Wiring for Cataraqui Bridge", in Meeting No. 23, Sept. 8, 1915, p. 57; Minutes of the Public Utlilties Commission, Kingston, Ontario.
18. KDS, Aug. 5, 1915, p. 7.
19. "Three New Batteries", DBW, Aug. 13, 1915, p. 5.
20. KDS, Aug. 21, 1915, p. 9.
21. Ibid., Aug. 31, 1915, p. 7.
22. Ibid., Sept. 3, 1915, p. 7.
23. "Gananoque", DBW, Sept. 3, 1915, p. 6.
24. KDS, Sept. 10, 1915, p. 7.
25. "Trek of the 8th C.M.R.", DBW, Sept. 7, 1915, p. 10.
26. "Troops on the March", Napanee Beaver, Sept. 15, 1915, p. 5.
27. "On Barriefield Heights", DBW, Sept. 28, 1915, p. 5.
28. "On Barriefield Heights", ibid., Oct. 26, 1915, p. 5.
    "On Barriefield Heights", ibid., Nov. 2, 1915, p. 5.
29. "Tells the Signallers", ibid., July 20, 1915, p. 8.
30. "Queen's Men May Join", ibid., Aug. 13, 1915, p. 5.
31. Minutes, Queen's University Senate, Aug. 28, 1915, p. 94; QUA.
32. Ibid., Oct. 1, 1915, p. 98, QUA.
33. Ibid., Oct. 29, 1915, p. 154; QUA.
34. "Queen's Battery", QJ, Nov. 1, 1915, p. 1.
35. "Queen's Battery", DBW, Oct. 30, 1915, p. 6.
36. "Lively Time at Camp", ibid., Aug. 14, 1915, p. 5.
37. "Men on New Batteries", ibid., Aug. 31, 1915, p. 5.
38. Ibid., Oct. 25, 1915, p. 5.
39. "News About Troops", ibid., Nov. 9, 1915, p. 5.
    "The Queen's Battery Had Its First Drill", ibid., p. 8.
    "Queen's Battery", QJ, Nov. 15, 1915, p. 1.
40. "News about Troops", DBW, Nov. 23, 1915, p. 5.
41. Annual Report of the Dean to the Shareholders of the School of Mining and Agriculture, Kingston, Canada, 1915-1916; QUA.
42. "Correspondence between Principal D.M. Gordon, Queen's University, and President F.F. Wesbrook, University of British Columbia, Dec. 7, 8 and 15, 1915; UBC Archives, courtesy Dr. Pauline Hughes.
43. "Queen's Battery", QJ, Nov. 22, 1915, p. 1.
44. Ibid., Nov. 30, p. 8.
45. Personal communication from Professor F. Morris Wood, Kingston, July 3, 1985.
46. "The Five Batteries Will Go Overseas", DBW, Jan. 4, 1916, p. 8.
47. "News about Troops", WBW, Jan. 17, 1916, p. 4.
    "News about Troops", DBW, Jan. 14, 1916, p. 8.
48. "Queen's Battery to Leave", QJ, Jan. 15, 1915, p. 5.
49. "46th Battery Leave", QJ, Feb. 4, 1916, p. 1.
    "9th Brigade Off to Front", KDS, Feb. 3, 1916.
50. "Kingston Batteries Arrive", WBW, Feb. 21, 1916, p. 1.
    "New Canadian Pacific Steamship 'Metagama'", DBW, May 17, 1915, p. 5.
    "Casualty Form - Active Service, Major Lester Willis Gill", from Statement of Service in the Canadian Armed Forces; Personnel Records Centre, NAC.
51. "46th Battery Men", QJ, March 5, 1920, p. 3.
    In Gill's record of "Promotions and Appointments" for 1916 (Statement of Service in the Canadian Armed Forces), he "proceeded o/s for 2 week

instruction 8-5-16" and returned 25-5-16. The Irish rebellion began in the last week of April, 1916, and was put down in early May. Thus, if the 46th did go to Ireland, it was likely involved in the mopping up rather than in the conflict itself.

52. "46th Battery Now in France", DBW, July 17, 1916, p. 8.
   "L.W. Gill, M.Sc., Record of War Service", sent by Gill to the Office of Harvard University War Records in 1919; QUA.

53. "Major Gill Writes", QJ, Dec. 1, 1916, p. 1.
   Letter from Sergt. Alex C. Smith, 46th Battery, QJ, Dec. 15, 1916, p. 2.
   Letter from L.W. Gill to Principal D.M. Gordon, Dec. 26, 1916; QUA, D.M. Gordon Papers, Box 2 - Correspondence 1916, Jan.-Dec.

54. Ernest G. Black, I Want One Volunteer, Ryerson, Toronto, 1965, p. 135.

55. John Swettenham, McNaughton: Vol. I, 1887-1939, Ryerson, Toronto, 1968, p. 57.

56. "Major Gill Writes", QJ, Dec. 1, 1916, p. 1.
   Letter from Sergt. Alex Smith, QJ, Dec. 15, 1916, p. 2.

57. Gerald W. Nicholson, The Gunners of Canada: The History of the Royal Regiment of Canadian Artillery, Vol. I, 1534-1919, McClelland and Stewart, Toronto, 1967, pp. 251 and 400.

58. Major Arthur W. Steel, "Wireless Telegraphy in the Canadian Corps in France", Canadian Defence Quarterly 6 #4, July 1929, p. 443; 7 #1, Oct., 1929, p. 45.

59. Wireless Sender #1, with a range of 8 miles, was developed by Fl-Lieut. B. Binyon, R.N.A.S., and manufactured by Standard Telephone and Electric Company.
   H.A. Jones, Official History of the War: The War in the Air, Vol. 2, Being the Story of the Part Played in the Great War by the R.A.F., Oxford University Press, 1928, p. 174.
   This system of wireless air observation had been perfected by Major E.R. Ludlow-Hewitt of Nos. 1 and 3 Squadrons, Royal Flying Corps, in late 1915, and the clapper-break was contributed by the Third Wing.

60. Letter from E.W. Henderson to Prof. Harold H. Stewart, Queen's University, December 2, 1974; courtesy Prof. Stewart.
   SM Minute Book, Kingston, Sept. 27, 1907, p. 297; QUA, Coll. 3695A, Box 1.
   QJ, Nov. 5, 1907, p. 86.
   Annual Meeting, Board of Governors, SM, April 25, 1911, p. 81; QUA, Coll. 3695B, Box 1.
   "Care for Yourself and Love for Others" (for E.W. Henderson's 100th birthday), St. Catharines Standard, Sept. 13, 1979.

61. Faculty Song, Seventh Annual and Directory of Graduates and Students of the School of Mining, Kingston, Canada, 1913, p. 25; QUA.
   Faculty Song, PESQU #14, 1920, p. 28; QUA.

62. E.W. Henderson, "An Improvement on the Prony Brake", Electric Journal 9 #6, June 1912, p. 577.

63. A.L. Clark, The First Fifty Years, p. 52

64. "Science", QJ, Jan. 28, 1916, p. 5.

65. "Many Military Matters", DBW, April 11, 1916, p. 12.

66. Ibid., April 19, 1916, p. 12.

67. The First Fifty Years, p. 52.

68. DBW, April 24, 1916, p. 12.

69. "Many Military Matters", DBW, April 27, 1916, p. 12.

70. "Many Military Matters", DBW, May 4, 1916, p. 12.

71. KDS, May 4, 1916, p. 7.

72. DBW May 18, 1916, p. 12.

73. "The Signalling School", ibid., June 7, 1916, p. 9.

74. KDS June 2, 1916, p. 9, and June 15, 1916, p. 9.
75. "School of Signalling", DBW, June 10, 1916, p. 9.
76. "Wireless Instructor", ibid., June 14, 1916, p. 9.
   According to the Kingston City Directory, Lorne L. Hicks arrived in
   Kingston some time in 1914 and lived in various hotels and boarding
   houses. In 1916-17 he is listed as working for Newman Electric, probably as a contractor. He was an electrician by trade, and Rev Dr Reid
   recalls him as a technician rather than an operator. This is reasonable,
   since Hicks' pupils would all have been experienced at sending and
   receiving Morse Code. Hicks' main job would have been to teach them
   the technical intricacies of the wireless equipment itself. Although he
   began instructing at the Camp in June, he did not receive his appoint
   ment to the staff, as Assistant Instructor of Signalling, until late in
   September, 1916.
77. Personal communication from Reverend Dr. W. Harold Reid, Burlington,
   Dec. 30, 1988.
78. "Installing Wireless", DBW, June 17, 1916, p. 12.
79. Dominion of Canada, Department of the Naval Service, Radiotelegraphic
   Branch, Application for Licence to Install and Operate a
   Radiotelegraph Station, dated June 13, 1916; NAC, RG 97, Vol. 85,
   #6202-30, part 3.
80. "The School of Signalling", DBW, July 7, 1916, p. 9.
81. "Report of the Department of the Naval Service for the Fiscal Year Ending
   March 31, 1917", Sessional Paper No. 38, 8 George V, A. 1918,
   Ottawa, 1917, p. 73.
82. KDS, July 15, 1916, p. 9.
83. Ibid., July 25, 1916, p. 7.
84. "An Important Discovery", DBW, July 10, 1916, p. 9.
85. KDS, June 22, 1916, p. 7.
   "Close of School", DBW, Aug. 5, 1916, p. 13.
   Ibid., Aug. 10, 1916, p. 9.
86. "School of Signalling", ibid., Aug. 16, 1916, p. 9.
87. "School of Signalling", ibid., Aug. 8, 1916, p. 9.
88. DBW, Aug. 9, 1916, p. 9.
89. Barriefield War Camp", KDS, Sept. 1, 1916, p. 7.
   "Reached Parrott's Bay", DBW, Sept. 2, 1916, p. 13.
90. "Barriefield Camp Wireless", DBW, Sept. 5, 1916, p. 9.
91. The First Fifty Years, p. 51.
92. "Wireless Is To Be Used", DBW, Sept. 13, 1916, p. 9.
93. "In Bivouac and Barracks", ibid., Sept. 16, 1916, p. 12.
94. KDS, Sept. 12, 1916, page 9.
95. "Barriefield War Camp", KDS Sept. 18, 1916, p. 7.
96. "The Wireless Messages", DBW, Sept. 18, 1916, p. 9.
97. Ibid., page 2.
   "Wireless Telegraphy", Gananoque Reporter, Sept. 23, 1916, p. 4.
98. KDS, Sept. 13, 1916, p. 9.
99. "The Big Fair Under Way", DBW, Sept. 26, 1916, p. 2.
   "Fair Week 1916", KDS, Sept. 27, 1916, p. 7.
100. "Barriefield Camp Wireless", DBW, Sept. 27, 1916, p. 9.
101. "Barriefield Camp News", KDS, Sept. 28, 1916, p. 11.
102. Personal communication from Prof. Harold H. Stewart, Kingston, Oct. 20,
   1982.
103. "Movies Are To Be Taken", DBW, Sept. 20, 1916, p. 5.
104. "In Military Circles", ibid., Nov. 13, 1916, p. 8.
105. "Barriefield Camp", ibid., Oct. 14, 1916, p. 13.
   KDS, Oct. 3, 1916, p. 9.

106. "In Bivouac and Barracks", DBW, Oct. 17, 1916, p. 8.
     "Barriefield War Camp", KDS, Oct. 20, 1916, p. 7.
107. Finance and Estate Committee, Minutes, Executive Committee, Board of
     Trustees, Queen's University, Oct. 4, 1917, p. 1; QUA.
108. "780 Signallers Passed", DBW, March 26, 1917, p. 8.
109. Ibid., May 18, 1917, p. 8.
110. Ibid., May 14, 1917, p. 8.
111. Ibid., June 21, 1917, p. 8.
112. Ibid., June 25, 1917, p. 8.
113. Ibid., June 27, 1917, p. 8.
114. Ibid., Aug. 17, 1917, p. 8.
115. Ibid., Sept. 7, 1917, p. 8.
116. "A.J. Kerry and W.A. McDill, History of the Corps of the Royal Canadian
     Engineers,Vol. 1, 1749-1939. Military Engineers Association of
     Canada, Ottawa, 1962, p. 230.
117. Rev. Dr. W. Harold Reid, "The Halifax Explosion: How They Got the
     Message Out", Atlantic Advocate, Jan. 1978, p. 44.
118. "Rare Honour Comes to Retired Signals Warrant Officer"; manuscript,
     Vimy Signals Museum Archives, Barriefield.
119. Personal communication from Stuart W. Henderson, Bedford, Ohio, Oct.
     16, 1982.

*Old Arts, now Theological Hall. The Wireless Club room was
behind the tiny barred window at ground level to the left of
the main entrance (arrow).*
-photograph by Arthur Zimmerman

The Queen's Wireless Club

Even in 1914, in the infancy of the wireless art, the Government of the
Dominion of Canada was quite aware that free use of the airwaves
was a security risk in the event of the outbreak of war. Therefore, all amateur and non-essential wireless licences in Canada were pre-emptively
cancelled and such stations ordered closed, as a war measure, by authority of Section 10 of the Radiotelegraph Act and by the Order-in-Council
of August 2, 1914 (P.C. 2030). Many violators were caught and closed
down, especially in large urban areas. The detection methods were

primitive and labour-intensive, so the government had no effective way of stopping sporadic amateur wireless activities in small centres all over Canada. The young, especially, carried on. At least three Kingston amateurs are known to have continued transmitting well into wartime, and all three were closed down at least once.

The Great War did not end officially until the signing of the peace on September 1, 1921, but by mid-April 1919 the threat of a resurgent Germany was past. The War was over for all intents and purposes, and the technical officers of the Department of the Naval Service advised that the Canadian ban on wireless amateurs could be lifted. Pre-war regulations could be resumed provided licencees complied with the "Consolidated Orders Respecting Censorship" issued under Section 6 of the War Measures Act.[1] Accordingly, the Governor General in Council ordered that P.C. 2030 be cancelled as of April 15, 1919. Licences were again made available to amateurs and experimenters by May 1, 1919,[2] but the radio frequency spectrum was now divided among the users. Amateurs, experimenters, aircraft, submarines and wireless telephones were assigned wavelengths between 0-200 metres; ship-to-shore work would be carried on at 300-600 and at 1800 metres; public correspondence between 1900-4000 metres; naval and military at 600-1,600 and 4,000-6,000 metres; and long distance naval and commercial work would be done between 6,000-17,000 metres.[3] For the first time, the amateurs were obliged to find ways to tune their wild sparks.

Wireless Research at Queen's During the Wartime Ban

All through the War years, in Lester Gill's absence, Professor E.W. Henderson acted as Head of E.E. at Queen's and had charge both of Gill's course XI, "Telegraphy and Telephony," and Laboratory 4 which contained the Queen's wireless. Licence XWD was likely not transferable from Major Mundell at Barriefield Camp to Henderson at Queen's, even though the equipment belonged to the University and had been partly in Henderson's charge at the War Camp. No evidence survives indicating whether wireless work was carried on in the student laboratory or whether it was suspended in compliance with the government ban. It is possible, though, that some sort of special permission was granted for experimental wireless work and teaching to continue at Queen's. Research into wireless communications was of some importance to the war effort and there are contemporary public references to such work going on at Queen's and/or Barriefield Camp in 1916 and 1917 in connection with government research funding.

Before the Great War, some scientific research was being done in Britain, but very little was going on in the United States or Canada in industry or in the universities. In contrast, western European countries, and especially Germany, had strong bases of industrial and scientific research, and had wonderful systems of technical schools to provide highly skilled labour to support their industry and research. Germany owed her long survival in the War to her research and technical expertise, and Gill was fully aware that pre-war Germany had a technical edge over the Allies. He was a strong advocate of similar technical schools for Canada and presented his views to the Royal Commission on Industrial Training and Technical Education in 1911.[4] Much later, however, it was brought home emphatically to Gill that the government was not serious about supporting technical training in Canada.

The Royal Commission eventually resulted in a conference at Ottawa in June 1915, called by Sir George Foster, Minister of Trade and Commerce, to which representatives were invited from the universities, the Royal Canadian Institute and the Canadian Manufacturers' Association. Dean W.L. Goodwin, rather than Gill, represented Queen's. The conference recommended the appointment of a Commission on Industrial Research to supervise and co-ordinate applied research in the universities and in individual industrial concerns, as well as basic scientific research. The implications for university funding were not lost on Principal Daniel Miner Gordon of Queen's: "Representations might be made by the Universities to the Commission regarding the equipment and staff available and the Commission would pass upon the recommendation and provide the necessary funds...Individual manufacturers having problems (might) establish scholarships... for the investigation of these problems and provide the funds for the payment of salaries and necessary apparatus."[5] He foresaw that a fund might also be established for scientific research, to which any professor could submit an application for a grant. The Commission would co-ordinate the efforts of these grantees with work already in progress under certain departments of the government. Gordon advised his Board of Trustees and they requested a report on facilities and research at Queen's.

Gordon presented the report on research in progress at Queen's and on the facilities and equipment available for research at a meeting of the Trustees on October 18, 1916. He also put forth a proposal that Queen's "co-operate" with the government in scientific and industrial research. Curiously, the news of the proposal first appeared in the Toronto Daily News,[6] edited and published by Sir John Willison, a member of Queen's Board of Trustees and a close friend of Gordon's. The Whig of October 19 did not mention the report,[7] but concentrated on the announcement of Gordon's retirement and on a new initiative to raise funds for the University by subscription. Nothing seems to have come from the proposal for co-operation in research, and there is no evidence of acknowledgement or interest by the Dominion Government. In fact, very little research work was going on at Queen's in 1916, since most of the senior professors in science were overseas. Furthermore, the government was disinclined to provide funds to set up new research, and would assist only work in progress. Later, the Finance and Estate Committee of the Queen's Trustees appointed a Committee on Scientific Research, chaired by Prof. A.L. Clark, Head of Physics.[8] Clark's committee eventually succeeded in putting research at Queen's onto a surer footing.

Notwithstanding the little work that was actually going on, the Toronto Daily News report contained a long list of research in progress at Queen's.[6] Included under "Engineering Department" was "Professor Gill has been successfully developing portable wireless equipment for field use." When that inventory was begun, perhaps in mid-summer 1915, Gill was indeed involved in wireless research work at Barriefield Camp. By the time the report was published, however, he was commanding his artillery battery in France.

A year after the Principal's proposal, Dean Goodwin reported to Senate that the Industrial Research Committee at Queen's had received a request from the Imperial Munitions Investigation Department for a list of research in progress which might be developed for use in the land war. Among the few projects submitted by Queen's was "Improvement of portable wireless outfit by Prof. E.W. Henderson,"[9] probably referring to his installation of an Audion valve in the Queen's "semi-portable" wireless set in the summer of 1916. It is quite clear that the officers of the University knew that Gill and Henderson were actively researching into the field use of wireless at Queen's in 1915 and 1916. These activities were likely based in the Department of E.E. and some of the experiments may have been done on campus, although the work was done in connection with Mundell and signalling at Barriefield. The outstanding question is whether these activities were officially sanctioned by the Department of the Naval Service as part of the war effort or whether they were technically illegal and were winked at or were undetected by the authorities.

Meanwhile, On the Battlefields of France...

During a lull in the fighting on the Somme, on November 11, 1916, Gill wrote to Principal Gordon to report the deaths of four boys of his battery two nights before, the first casualties they had suffered: "...a crushing blow...All were exceptionally good men and I feel their loss very keenly...It is now almost four months since we arrived in France, and we have been in action most of the time since...Up to the present I have been so busy that I have scarcely had time to write to anyone. I hope, however, to have more spare moments from now on, as I am getting my officers trained to take on a little more responsibility. And let me say that the responsibilities placed on a battery commander in this war, - especially on this front, - are tremendous. All operations here are primarily artillery duels."[10]

Gill answered Gordon's reply on December 26, and it provides some rare direct insight into his personality:

> I was very pleased to receive (your letter) and to learn what the University is doing for the cause. I would have been more pleased, however, if you had informed me that the academic doors of the University had been closed and the buildings all turned into Barracks and Military Schools. I have since learned that you are turning over Grant Hall and the Arts Building for Hospital purposes. Well done! The next thing is to organize a girls'

Agricultural School and utilize the Old Arts Building for
this purpose. Next, teach the medical fraternity - if you
can - that there is no scarcity of medical men at the front.
There is need however for first aid men - men who have
the courage and strength to go to the front line and give
immediate aid; also men who can assist and do the routine work at the Field Ambulances and Hospitals. Men
could be trained for this service in three months. At the
Somme the need of such men was impressed on my mind
very deeply. Men of this kind could save dozens of lives
per day. Men who are wounded and can walk go strag-
gling back to the dressing station. Sometimes they have
someone to accompany them and sometimes not. In any
case they often get lost when it is dark and that is the
last of them. I took probably a dozen of such men into my
dugout, and tied up their wounds; then sent them to the
dressing station. Some of them were lost and I heard
them crying for help. This was in our last position on
the Somme, where we were about 2500 yards from the
front line.[11]

The Queen's Battery was in the mud of the Somme for a total of
seven weeks, and lost six killed, including their cook, and three badly
wounded. Near the end of the letter Gill commented on the "Hun peace
proposals" and predicted, "If all the man-power of the Empire is mobi-
lized without delay, we will win a decisive victory before the end of
1918. If this is delayed, - well there is a question."

After the Somme, the battery took part in the Battle of Vimy Ridge.
Gill was gassed in a skirmish near Vimy on June 29, 1917, and was
declared unfit for duty. As one biographical note had it, "He was sent
to hospital in England suffering from gas,"[12] and he missed not only the
capture of Lens and Hill 70 in August, but the rest of the war. He spent
8 weeks in hospital in Oxford and further weeks on sick leave, and
was passed as again fit for duty on October 11, 1917. He continued to
suffer long bouts of what he called "bronchitis" and his body temperature remained sub-normal for years.[13] About mid-October he was
appointed to command the 1st Brigade, Canadian Reserve Artillery at
Whitley Camp.

Then, the Air Board asked for his services "on account of his splendid technical ability." At this point Gill wrote to a friend in Kingston,
"I cannot get away from the fact that I ought to be back in France,
and I would really wish to be back with my boys of the Queen's battery.
However, the authorities think I can do better work here, so I suppose
they know."[14] But, while waiting for his transfer to come through he was
picked out by the War Office in December to go to the Ministry of
Munitions to work at "expediting contracts and testing and inspecting machinery" at Whitley. When released from this duty on September
8, 1918, he assumed command of a Reserve Brigade, CFA, at Whitley
Camp and applied to return to where he wanted to be all along, with his
boys of the Queen's Battery. After some delay the release was granted and

he was on his way to France when the Armistice came into effect.[13]

On November 18, 1918, Gill was detached from the brigade and was sent to HQ Overseas Military Forces of Canada for duty with the Educational Services of the Canadian Forces Overseas, otherwise known as the Khaki University of Canada. The Director of Educational Services and the force behind the Khaki University was his old colleage from Callendar's laboratory at McGill, Col. (Dr) Henry Marshall Tory. The initial approach to transfer to Educational Services had been made by Tory that summer, while Gill was still with the Munitions Board. Lester explained to his sister Elizabeth, "I refused on the ground that I was returning to France. So when the fighting stopped, the way was open to take up this work at once. Col. Tory put me immediately in charge of all the work in England, and since the middle of December I have been working night and day. I have four stenographers in my office who are kept busy every minute."[13]

The Khaki University of Canada was based on and succeeded makeshift teaching programs operating since 1917 at Seaforth and Whitley Camps in England as well as behind the lines in France, under the sponsorship of the YMCA and the Chaplain Services. In mid-1917 there were over 300,000 men attending classes. There was even a "University of Vimy Ridge" in the battle zone until a German offensive caused its closing in April 1918.

Through the late summer of 1919, repatriation and demobilization proceeded arbitrarily and slowly, as Tory had predicted it would. The last men to arrive overseas were sent home first, morale became very low and the troops were restless and disgruntled. Riots broke out and Canadian soldiers were shot. Tory's Khaki University cleverly made use of the teaching power available in the army to occupy tens of thousands of Canadian soldiers through this dangerous period. They were kept busy in learning everything from basic reading and writing (in classroom and at bedside) to secondary school matriculation subjects, to courses in business, agriculture and teaching, to credit courses in pre-medicine, first and second year arts, first year engineering, theology and law. Some 500 soldiers who had left Canadian universities in mid-course to join up were admitted into Oxford, Cambridge, Edinburgh and University College, London, for a year and received certificates for credit.

At first, Major Gill and Major (Dr) Clarence MacKinnon of the Chaplain Services were made Assistant Directors for England. Then, in January 1919, General Currie put Gill in complete charge, appointing him Director of Educational Services for England, and "gave him permission to set up a concentration camp at Ripon, England, for university students."[12,15] From an initial staff of 100 instructors, probably inherited from the University of Vimy Ridge and other earlier attempts at continuing learning at the front, Gill built the Khaki University staff to 400 and was teaching 10,000 students at 19 centres by the end of February 1919. At the university camp at Ripon, he had 800 students and 40 instructors under his supervision.

Though desperately overworked and far from healthy that spring, Gill continued in his old work-a-holic pattern. In May he was for the second time Canadian representative at a meeting of the International Electrotechnical Commission, this time in Paris. He wrote to his former assistant at Queen's, O.J. Hickey, to boast that he "Came over from London, May 5th by aeroplane."[16]

Professor Gill Finally Returns to Queen's

The Khaki University was closed in the early summer of 1919[15] and disbanded in August. Gill embarked from London on August 13, arriving in Halifax on the 23rd[17] and in Kingston on the 25th.[18] He was demobilized and honourably discharged with the rank of Lieutenant-Colonel on August 30, in Kingston, and was immediately off to Ottawa and thence to Winnipeg and Edmonton.[19] The purpose of his Ottawa visit will soon become clear, for it was connected with the trip to Edmonton, where he surely met with Col. Tory, back at his old stand as President of the University of Alberta. On his return from the west, Gill once again took up his professorship in E.E. at Queen's.

Douglas M. Jemmett Goes to MIT

Doug Jemmett's appointment to Queen's in 1914 was originally in mathematics, but he gradually worked his way over to E.E., the subject of his B.Sc., in 1913. For 1918-19, the last year of Gill's absence, the Principal's Report states that "Mr. Jemmett gave practically all his time to the Department of Mathematics. In (the) Electrical Department he gave Electrical VII and VIII. The rest of the work was given by Professor Henderson."[20] The letterhead of the E.E. Department of that time shows three staff members, Gill, Henderson and Jemmett. With Gill back and everything returning to normal in Queen's E.E., Jemmett could at last be released from his teaching duties to go to the Massachusetts Institute of Technology (MIT) to pursue his Ph.D. studies in E.E. Jemmett left on the first leg of the trip, to see his wife and daughter off to England and to visit his mother in Napanee, around September 5. It is unlikely that he and Gill had much time to get together that autumn to talk science or to do experiments. Gill appeared in the Queen's record once more at a meeting of Faculty of Science on November 14, 1919.[21]

Gill Founds the Queen's Wireless Club

Back on civvy street, Professor Gill seems to have picked up where he had left off in wireless in 1915, and soon had things rolling again at the University. The technology had advanced rapidly during the War. Thermionic valves were now making voice transmission a practical thing and the concept of public broadcasting was just beginning to glimmer in a few minds. But Gill did not seem abashed. He probably kept up an active interest in wireless all through the War, though he likely had

Original members of the Queen's Wireless Club, founded by Prof. Gill in November 1919: Warren Alvin Marrison (President; Eng. Phys. '20), George David McLeod (Vice-President; Mining '20) and Donald O. (Red) Hepburn (Critic; Civil '24). Marrison later invented the quartz clock at Bell Labs and was given an honorary D.Sc. by Queen's.
-courtesy Queen's Archives

access to telephone and buzzer rather than wireless in the field to communicate with the spotters in his battery. On the other hand, certain batteries at the Battles of the Somme and at Vimy Ridge were in communication with one or two advance observation posts and spotter planes by wireless, and Gill's could have been one of these. He certainly would have been able to learn about the latest advances in the ethereal art from some of his engineering instructors at the Khaki University.

Now, excitement about the wonderful possibilities of the new communication technology began to build at Queen's. The first report of activity was in the city press, describing student enthusiasm for and plans to start an up-to-date wireless school at the University.[22] Without prior notice in the Queen's Journal, a meeting was held on Tuesday, November 4 at 5 pm in the New Arts Building (Kingston Hall) to form not a wireless school but a Queen's Wireless Club. "A Club of this kind is an innovation. Nothing of its kind being known amongst the Queen's Clubs previously. The object of the Club is to promote the study of Wireless and plans are being made to erect a station in Kingston."[23] The British Whig expanded on the club's plans: "It was decided to try and obtain a site in the city on which to rear an aerial, and it is very possible in the near future that there will be an interuniversity wireless club. Great interest is being shown in the project at Queen's and a large number of the students are going into it."[24] Neither of these articles named the originators of the idea.

At the November 4 meeting, Gill was appointed Honorary President of the Wireless Club, Warren Alvin Marrison (B.Sc. '20; D.Sc. '53) was made President, George D. McLeod (Mining, '20) the Vice-President, Mr Blake (probably W. Stanley Blake, Arts '24) the Secretary-Treasurer, Donald O. (Red) Hepburn (Civil, '24) the Critic and John Miller the

Reporter. Anyone interested in wireless was invited to the next meeting at 5 pm, Wednesday, November 12, 1919, in Room B13 of the New Arts Building. Further meetings were held there on November 12 and 19, but on November 26 and December 3 and 10 it met in Room 19, Fleming Hall. There is no record of what happened at the meetings nor of what plans were implemented by the Wireless Club that fall.

Officers of the Wireless Club, 1919-20

The President of the Club, Warren Marrison, was then in his final year of Engineering Physics, the first Queen's graduate in this course. He and Stanley C. Morgan (Elec. '16; who will appear later in the story) were research assistants to Dr A.L. Clark in the Department of Physics. Marrison was actually in Science '18, but like many students he put aside his studies to do his bit for the war effort. In 1917-18, he served with the Royal Flying Corps in Toronto in electrical work, mostly related to radio communication in the wireless telegraphy branch.[25] When he returned to Queen's he had become a dedicated wireless amateur.

> **Prof. Harold Stewart**: Somewhere about that time, 1919-20, when I was a radio amateur here, we had heard Marrison on the air. He had an amateur transmitter of his own, and we asked if we could go and see him. I think my friend at that time was a chap named Ralph Bunt, a son of a hardware man in town, near the corner of Princess and King. Anyway, we had a talk with Marrison and we asked him questions and he was very helpful to us. He was in his fourth year of Engineering Physics then, I think, and he knew far more than we did. But he didn't lord it over us. He was just very helpful, and I always remembered that. Later he went to Bell Labs and was very successful and he developed the quartz clock. Queen's later gave him a D.Sc. for his scientific prowess. He brought honour to Queen's, so they honoured him.[26]

John R. Rutledge (Arts '22) recalled that "in visiting Marrison's rooms in the University Avenue area, he had all kinds of transmitting equipment."[27] Marrison is listed in the 7th edition of American Men of Science,[28] wrote many articles on the measurement of time[29] and held 65 patents.

George David McLeod was an early wireless amateur in company with his friends at Kingston Collegiate Institute (KCI) in 1913. He spent the early war years as a ship's wireless operator and then returned to school. His career is outlined in Chapter 7.

"Red" Hepburn of Stratford came to Queen's in Science '23, stayed out of school in 1921-22, and was adopted by '24. He had an electrical hobby and a First Class Certificate in wireless, but he had other interests: "'Red' Hepburn, leading member of the Wireless Club, is spoiling his well developed character by attending the 'strictly private' Social Five Dances every week...We have a knight of olden days in our midst in the person of 'Red' He-b-rn. He was seen last week wearing a lady's glove pinned to his coat. It would be there yet if Hep had had his own way about it."[30]

Bill Taylor: One thing I remember in particular is honeycomb (inductance) coils they used in the tuning of receivers. Lattice wound, you know. They were supposed to be very efficient. Well, Hepburn destroyed that idea. Although they were made by Marconi, we made our own spiderweb coils which were more efficient. Wound them on pencils or nails or whatnot...Donald Hepburn was Secretary in 1922 and did all of the correspondence with Ottawa and arranged to get our licence...He took a course in commercial wireless and became a Marconi operator. He intended to go to sea on boats, but he changed his mind and started working for the Department of Highways in Long Island, NY, and stuck at that all his life. But he was a marvellous radio operator. Very good speed at telegraphy and understood the circuits...A very close friend, and always interested in amateur radio. About every ten years he'd write and ask what he should buy in the way of amateur radio, that he'd like to talk to me by radio. But he never got around to it.[31]

Nothing is known about W. Stanley Blake of Deseronto, and the Queen's Alumni Association has no record of John Miller.

Gill approached the Finance and Estates Committee of the Board of Trustees in December for funds to get the Wireless Club established, and the Committee authorized "the erection of a permanent wireless station with an aerial between Grant Hall and the Old Arts Building with an expenditure of $200."[32] In those days, the term "wireless station" could signify operation of a receiver or a transmitter, or both, so Gill's intention is not clear. Suddenly, Gill moved to Ottawa, and the activities of the Wireless Club ceased abruptly. The last meeting of the 1919-20 school year was on December 10, 1919.

Official announcement was made at the end of November that Gill had been appointed to the new post of Director of Technical Education in the Department of Labour, Ottawa. Parliament had finally passed "An Act for the Promotion of Technical Education in Canada" (9-10 George V Chap. 73, 1919) and had voted ten million dollars to encourage and support the Act.[33] H.M. Tory, President of the University of Alberta, recommended Gill most highly for the post, citing his many contributions to Queen's and his work and administrative skill while with the Khaki University.[34] The mobilization of support for this appointment was likely the real purpose of the sudden visits to Ottawa and Edmonton immediately following his discharge from the army.

An active, decisive man like Gill could not long tolerate the hierarchy and the paper-pushing life of a civil servant, and he resigned on August 16, 1921, to become the Directing Head of the Departments of Electrical and Mechanical Engineering at the University of British Columbia.[35] He was upset that his salary in Ottawa was not commensurate with his duties and status, and took his case to the Appeals Board of the Civil Service Commission. His provincial counterparts, who were dependent upon him for operating funds and advice, were better paid than he; provincial governments had even tried to lure him away from Ottawa for much higher salary. The appeal was denied, and Gill bitterly predicted doom for the Commission and the appeal process: "Duties and responsibilities are of no account. The button is pressed, the crank is turned, and the decision rolls out...The Commission has become entangled in the system like a fly in a spider's web, and, like the fly, it will probably die trying to extricate itself...The situation inside (the Commission) may well be understood by anyone who has been lost in a London fog." The more significant problem was, as had to be expected, that there were "considerable differences of opinion" between Gill and the Government about the policies to be followed in promoting technical education. He realized that he could not do a proper job on the Government terms.

After little more than a year in Vancouver, Gill resigned and headed back east, at the invitation of the Hamilton School Board, to become the new Principal of the Hamilton Technical School.[36] He was dropped into a rather messy situation by the Board,[37] but it seems that this was where he really wanted to be, where he could implement his ideas about promoting technical education in Canada and see them carried through.[38] He kept in touch with his friends at Queen's, and even proposed a scheme whereby Hamilton Tech students could be prepared for admission directly into the third year at Queen's.[39] Around 1923 he brought in a Queen's man, Ralfe J. Clench (B.Sc. '22), to be on staff at Hamilton Tech. Clench later married a woman who worked in Gill's general office.[40] Gill spent the rest of his life as Principal of Hamilton Tech, successfully building his school, creating bridges between his school and Hamilton's industry, and earning the love and respect of his students.[41] Radio was likely forgotten in these years. His daughter Marion cannot remember his even owning a radio in Hamilton.

E.W. Henderson Carries On at Queen's

Upon Gill's departure from Queen's, Ebert W. Henderson resumed the teaching load in E.E., with the assistance of Mr MacPherson, and the two carried on valiantly until the end of the 1919-20 session.[42] The Wireless Club showed no signs of life after December 10, but that's not unusual in university life where, after the shock of the Christmas examinations, students tend to drop their extra-curricular activities and concentrate on their studies. There is evidence, again supplied by the Reverend Dr W. Harold Reid, that by the spring of 1920 the money requested of the Board of Trustees by Gill had been received and spent on behalf of the Wireless Club.

Reid left the Navy and wireless work in April 1919, wanting to get back to university - he had been doing extra-mural work at Queen's in the winter of 1918-19. After demobilization he spent a year at the Ontario Agricultural College in Guelph.

Dr W. Harold Reid: When I came down to Queen's in April of 1920 to arrange for my work in the fall, I spotted this single wire antenna right away, going from Fleming Hall over to the New Arts Building. I know it went from a corner of Fleming Hall over to one of the other buildings. So I went in to investigate, to see what was going on at the end of that aerial, and met the professor who was carrying on some experiments in W/T. I recall he was a young man, of slight build. This professor had assembled some radio units which he mounted in four or five rows on a large board about four feet square. I think it was a piece of plywood, but I don't think it was exactly square. With various lengths of wire he could make many circuits, to be changed as often as required. It seems to me that he was just beginning to set up that panel. I recall distinctly that this professor had acquired a number of radio parts - tuning condensers, crystal detectors, rheostats, honeycomb coils, etc., probably twenty or more - mounted on four- or five-inch square bakelite panels, which he installed on this large board. At the bottom of each unit were two sockets - just small round receptacles - to take the simple pin at the end of a connecting lead. You didn't have to use a screwdriver or that sort of thing. With leads of various lengths he could connect the units in any way that he wanted. This made it very easy to change and test out various receiving circuits for experimental purposes. There were no connections on the back. In this way he could try out various circuits for the best reception and very easily switch around from one thing to another. Certain parts could be left connected, but if he wanted to try out some kind of a new coil, he could just plug it in and hook it up. I was interested in that because it was a very nice way of doing it. I saw this board often, discussed various circuits with him, with which I was familiar, and helped him a bit with certain circuits and things.

I have just come across my old diary of 1920, and on April 29 I had recorded, "Helped Professor Henderson with his W/T outfit." He was getting various units and putting them in, connecting them up from time to time, as he acquired them. It seems to me you go into Fleming Hall (by the rear door), and downstairs and to the right to get to the room where he had all this wireless equipment set up. That'd be to the corner of the building. Then, from that corner, the antenna went out, either to Ontario Hall or to New Arts.[43]

The De Forest Inter Panel Receiver

The De Forest Radio Telephone and Telegraph Company Catalog "B" of 1919 shows a modular panel called the "De Forest Unit Receiving Set" or "Inter Panel" set, which Dr Reid and five others have identified as the sort of thing that they saw at Queen's. It was a regenerative receiving set, the 15-panel typical set described in the catalogue as "consisting of a tuner with wave length range of 150 to 25,000 meters, a crystal and audion detector, and a one step amplifier."[44] The components for the 3 x 5 panel cost $150 to $160 in 1919, so that the 4 x 5 or 5 x 5 set recalled by Dr Reid would have cost Queen's around $200. Unfortunately, no record of purchase by Queen's has been found. A 1921 magazine advertisement for the 6 and 9 Panel Unit sets noted that the purchaser was to furnish the panel board and an "A" battery, the manufacturer to supply a detector tube, "B" battery, head phones and

A De Forest "Unit Panel" or "Inter-Panel," an early modular receiver set, similar to the one purchased for the Queen's Wireless Club in late 1919 or early 1920, with funds obtained by Prof. Gill. It was the most powerful receiver in Kingston, and people used to come to Fleming Hall to listen in.
-from De Forest Catalog B, 1919
McNicol Collection, Queen's University

a set of 11 coils.[45]

Donald C. Rose (B.Sc. '23) remembers listening to the De Forest Inter Panel set when he was a student working for Dr A.Ll. Hughes as a lab assistant:

<u>Donald Rose</u>: In 1922, I think, Professor Jemmett let me have a key to Fleming Hall, and down in the basement was a radio receiver. You go down steps to the basement floor and there's a long corridor there, with labs on both sides. At the west end of the corridor there was a small room which probably was designed as a professor's office. It was in the middle with labs on both sides, and just opened onto the corridor. It was just like the end of the corridor, with a door. And in it that summer there was nothing but a work bench along one wall and this radio receiver was on it, with a battery, I suppose, to run it. It was sort of a home-made radio receiver built on a vertical bread-board. I think most of these panels came off independently, so you could take that one off and put another one on...and it would be a different set. I used to go down there occasionally in the evenings and turn it on and listen with headphones to KDKA in Pittsburgh, the only station broadcasting that we could ever get there, at that time.

It was about two feet by three feet, and had a

Dr Donald C. Rose used to go to the basement of Fleming Hall to listen to the "Unit Panel" set in 1922, when he was summer student of Prof. A.Ll. Hughes.
-courtesy Queen's Archiv

number of switches and rheostats and two or three clear glass tubes. Valves, you know, as we called them. Vacuum tubes, and it was a super-heterodyne set and it took a lot of adjustment to get it to work. You had two dials that you had to twist, and radio frequency (honey-comb) coils that plugged into a little stand. One of the coils was on a hinge - it would turn. If you pushed the coils too close together, the set would oscillate and then squeal. It was liable to go and oscillate on its own, sort of transmit on its own if you weren't awfully careful of how you set it up. Of course, if you did let it oscillate, it interfered with everybody else, but there wasn't anybody else there for it to interfere with.[46]

The newspapers reported a large receiving set, "a Deforest set, with a receiving range that is practically unlimited and that enables any hook-up to be arranged in short order" in the new "Radio Laboratory" in February, 1922.[47] A British Whig reporter was allowed to listen in to the Queen's set in August, 1922, and reported the "home-made" set "one of the best in the city, having a splendid receiving power." The set was reputed to have received code messages from as far away as Germany, but when the reporter was there they heard Toronto, Los Angeles, Newark, Schenectady, New York City and Chicago. It seems that every night there were visitors to the station to listen to music and speeches.[48] Mrs Norman Miller heard William Jennings Bryan deliver a sermon in 1922 over a Queen's set in the basement of Fleming Hall.[49] She did not listen in for long, because it was boring. Kathleen (Whitton) Ryan (Arts '26) accompanied her future husband Frank to the basement of Fleming Hall in 1924 and saw the De Forest receiving set sitting up on a kitchen table. She said that several boys were sitting around it wearing headphones, but girls were not allowed to listen in.[50]

Honeycomb coil variable inductance module of the De Forest "Unit Panel" or "Inter-Panel" modular radio receiver. It had mountings for three coils, the outer two adjustable with gears and pinions (catalogue #ULC-100). The base is not original. Prof. Stewart recalled that when he came on staff at Queen's in 1929, he saw bits of De Forest gear in "junk" boxes in the E.E. laboratory. This variable inductance module was kept by Stewart as a souvenir of Prof. Jemmett's gear. -photograph by Arthur Zimmerman

When Harold Stewart was a student at Queen's in 1925-26, they had De Forest "gear" - a master panel with a number of components: a crystal detector fitting, vacuum tubes and honeycomb coils, one swinging and a couple stationary. When he came on staff in 1929, parts of the De Forest Interpanel set were still around:

Harold Stewart: These honeycomb coils were made by or for the De Forest Company. The coils were supposed to reduce losses in a radio circuit, and maybe they did a bit, but for most purposes you could do just as well by scramble-winding a coil on a home-made spool. These things were part of a multi-panel set-up all bolted on a big frame. One panel would be a detector, another panel would be a mounting for honeycomb coils, another panel would be a plug-in for phones, or whatever. When I came on staff there were a lot of these odds and ends around in junk boxes. I recall we rescued some of these coils and we even used them for inductances in some of the electronics labs. They were around in the early days of CFRC - part of the background, if you like.[51]

Harold Stewart inherited from Prof. Jemmett the module with the three honeycomb coils from the Queen's Interpanel set, although the base is not original.

In the 1919-20 Principal's Report, Dean Clark noted that very little had been done for E.E. for years. He urged remodelling of the store room at the west end of the Fleming Hall basement to provide needed additional laboratory space, the hiring of a teaching assistant and the updating of the E.E. equipment. He recommended immediate purchase of "wireless apparatus, rheostats and meters to the amount of $700" and a "motor-generator set costing $2500 is another absolutely essential piece of apparatus."[52] The Interpanel set may have been part of this upgrading, but it is equally likely that it was purchased with the grant secured by Gill for the Wireless Club.

Douglas M. Jemmett Returns to Queen's

E.W. Henderson, Acting Head of E.E., resigned from Queen's in early August 1920 to take up a position designing motors with Canadian Crocker-Wheeler in St Catharines, Ontario.[53] Henderson had spent his whole professional career at Queen's. He had been an Assistant Professor since 1911 and was appointed Associate Professor in the spring of 1920,[54] after Gill's departure. Henderson was not the only one besides his mentor to seek greener pastures after the War. By September 10, 1920, the entire professorial staff of Fleming Hall had resigned and left.[55] Douglas M. Jemmett was hastily summoned to return to Queen's after only a year at MIT, his thesis not completed, and he was appointed Assistant Professor of E.E. and Acting Head of the Department.[56] He brought with him a colleague from MIT, Robert Leland Davis of Grand Rapids, Michigan, who was appointed Lecturer in Mechanical and Electrical Engineering. At the same time, L.T. Rutledge was taken on as Assistant Professor in Electrical and Mechanical and Arthur Jackson as Assistant Professor in Engineering Drawing.

Harold Stewart: As I recall it, Prof. Jemmett had a radiation experiment going, even when I came here on staff in '29. He had two parallel wires strung between Fleming Hall and Gordon Hall and he fed a high frequency current around this loop. He was carrying on experiments with radiation from two parallel wires, and he was working on a Ph.D. for MIT. He ran into some integrals that nobody could solve. Now-a-days they'd just put them through the computer, but then it wasn't so easily done. His supervisor said, "Turn the damn thing in anyway. You've got a fine piece of work." But he wouldn't do it. The result is that he never got a Ph.D., which he had thoroughly earned. Too bad.[57]

Dr A.L. Clark recalled a conversation about Jemmett with Prof. Bush of the Electrical Department at MIT. Bush said, "If you have any more like Jemmett, send them to us."[58]

It is incredible that after nearly 60 years of association with Queen's and 40 years as Acting Head and Head of E.E., Prof. Jemmett should have left behind so little documentation of his life and career. Only scraps of indirect evidence remain to help determine where, when and how he became interested in wireless. There is a good chance that his interest was sparked while he was a student at Queen's or at the School of Mining. Lester Gill may have been the inspiration. Donald MacClement recalls the young Jemmett as being very helpful to him in his interest in wireless before the First World War, around the time the field experiments with Gill were going on. In 1913, Jemmett was listed as a Second Corporal in the young 5th FCCE at Queen's.[59] He may have learned his Morse signalling in the telegraphy section of the 5th FCCE before the War, although no records survive. Harry Jemmett says that his father knew Morse code and recalls his teaching him to send the code on a little box,[60] perhaps in the 1940s. Doug Jemmett's war experience, however, did not include wireless communication or electrical work.

Jemmett's Experience in the First World War

Almost immediately upon the outbreak of war, on August 9, 1914, Doug Jemmett volunteered for overseas service as a sapper in the First Contingent, Canadian Overseas Expeditionary Force. He signed his attestation papers at Valcartier Camp on September 23, 1914, and went overseas as Number 5324 with Prof. Alexander Macphail's First Field

Prof. Douglas Mill Jemmett (Arts '11; Elec. '13) returned to Queen's in 1920 from MIT, where he was working on his Ph.D., to become Acting Head of E.E. According to Don MacClement, young Doug Jemmett helped him with his fledgling wireless experiments and had assisted Prof. Gill with his W/T distance tests before WW I. In the spring of 1922, with plans sent from Pittsburgh by his friend and colleague Robert L. Davis, Jemmett and four E.E. students built a radio-telephone transmitter in Fleming Hall and obtained experimental licence 9BT. Prof. Jemmett retired as Head of E.E. in 1960 and was succeeded by his former student, Prof. Harold H. Stewart.
-courtesy Queen's Archives

Company of Canadian Engineers. They sailed from Quebec City on October 4 aboard the S.S. Zealand, in the first convoy of 33 ships to cross the Atlantic. The unit trained on Salisbury Plain, landed in France on February 11, 1915, and went into action in Belgium, in the Fleur Baix area. Jemmett received the Distinguished Conduct Medal for the night of April 22, 1915, when he and ten men detailed to guard the bridge over the Yser Canal, a mile north of Ypres, held the bridge through a gas attack and heavy shelling. Only Cpl Jemmett and three of his men survived.[61] He was wounded three times in 18 months, the last time at the Montreal Crater on Vimy Ridge in October, 1916, when the bottom of his left foot was blown off.

While convalescing in England, after spending seven months in hospital and almost losing the leg, he met Miss Maud Martineau, an English Red Cross Nurse. The next summer he was invalided home and he and Miss Martineau were married in Montreal in October 1917. Just days later he brought his bride back to Kingston and resumed teaching at Queen's.

> <u>Harold Stewart</u>: When he came back to Queen's, an official asked, looking at the D.C.M. initials, where he got his doctorate. And all Doug said was, "That's Doctor of Church Music."[57]

Evidence of Doug Jemmett's Interest in Wireless, 1917-20

After his return from the War until the time he left for MIT, there is no documented evidence that Jemmett was active in wireless either at Queen's or privately. Only two witnesses have been discovered. Don MacClement remembers that when his friend Doug Jemmett returned to Kingston he had difficulty climbing stairs and riding a bicycle because of his injured foot. He also recalled Jemmett helping him mainly with electrical theory before the War, and not with wireless itself until after his return to Queen's in late 1917. He certainly helped young MacClement with advice when he was having trouble trying to tune his spark transmitter to conform with the new regulations that came in after the War. Maud Jemmett recalled that her husband had been interested in wireless after they returned from MIT, after they had moved to the house at 61 Kensington Avenue. According to the Assessment rolls for Victoria Ward, that move was some time after September, 1920.[62]

> <u>Maud Jemmett</u>: I think it was after we'd bought and moved to a small house on Kensington Avenue that I remember his staying down there at Queen's half the night and hearing squeals which he was all excited about. And when my mother and sister came out from England to visit me, he got a squeal from somewhere or other - remote, I think it was - that he began to get really excited. So he didn't come home until about half-past one or two in the morning sometimes. It seemed to be from the other side of the world that he heard these remote sounds.[63]

The Wireless Club Is Revived

Doug Jemmett brought the Wireless Club back to life in the fall of 1920, but a good deal of preliminary work had to be done to provide proper facilities and equipment. Don MacClement remembers doing some of the spadework for him:

> Don MacClement: Doug Jemmett and I started a scheme to have a Queen's University Wireless Club, and he asked if I would put up an aerial for it and pick out a good place for the club to meet. Well, I knew all the buildings around the University inside and out and I knew just the place. And that was down in the basement of the Old Arts Building. There were some store rooms down there and underneath the tower itself there were a couple of very nice rooms. The best of these rooms had a small window, partly below ground level and right beside the base of the tower. It was full of junk and I had to help the janitor move all the pieces and bits and old broken chairs and stuff from the labs into another area. But it made a very good clubroom. Jemmett had ordered the aerial wire, which was #14 hard-drawn copper wire, and a good number of strong insulators and I proceeded to put up that aerial. One end of it was attached to the clock tower on the New Arts Building (Grant Hall tower). It ran from there right over to the top of the tower on the Old Arts Building. I don't know how many feet it was - it was many hundreds of feet anyway.
>
> Getting it up I had the help of my chum Earl Morris. We had learned how to string these long wires, because we had put up some of our own aerials the same way. You got the wire laid out, one end fastened up at the top of the clock tower, then down across the roof and along the road and doing quite a big loop out away from the building because it had to clear the eaves and any obstruction, and also a few trees. The other end was extended by a long rope up to the top of the Old Arts Tower. To get up there you went up first into the attic, then up a ladder into a kind of a loft and then out through a trap-door into the small, square, more or less flat roof inside the tower wall. One then had a loop of thin cord around the aerial wire and as one of us pulled in the rope on the Old Arts tower, the other would pull the cord around the aerial wire to one side so that it could swing up clear of the eaves and rope and other things. And then when we got that up over the New Arts Building roof, then we had to do the same thing on the other side to get it over the trees and over the roof of the Old Arts Building. After it was up in place, of course, you had to let go of one end of that cord and it came down loose. There was the aerial, right from one tower to the other. It took the two of us to pull hard enough to get it up and reasonably tight, and it was fastened to the base of the flagstaff that went up in the middle of the Old Arts tower roof. The lead-in went down vertically from the end of that wire at the Old Arts end, so that it could hang straight down to the base of the tower just outside that basement window (i.e. an L-type antenna with the drop at the Old Arts end). And there we made an anchor, and insulated the wire there very carefully so that it was fairly tight and strong, and then made the lead-in from there. We bored a hole through the frame of that window and the lead-in went through a tube insulator into that clubroom. It proved to be an excellent aerial. I know that I took my receiving

equipment over there right away and Earl and I tested it out. We could hear spark signals from much farther away than we ever had with our own aerials. My aerial at that time ran from one of the chimneys on the Principal's house (Summerhill) - we lived in the wing of it, as it were - right across a circular lawn at the back over to the top of the chimney of the Pathology Building (now Kathleen Ryan Hall). Jemmett and some others had meantime ordered parts of a very fancy receiving set that was to be the basis of the Wireless Club that was starting to form.[64]

Don MacClement was apparently unaware that E.W. Henderson had at least begun ordering what is obviously the De Forest Inter Panel modular receiver as early as the spring of 1920, and that Henderson had had W. Harold Reid's help in assembling it. Now it was going over to the clubroom in Old Arts and had to be re-assembled from scratch.

Don MacClement: Soon after that the receiving equipment that Jemmett had ordered arrived and was all accumulating on a big table over in the radio clubroom. It consisted of a vertical framework which had quite a number of square openings into which were fastened the various units of the receiving set. As I remember it, there were at least two sets of honeycomb tuning coils and two different kinds of detector units. A space for a vacuum tube - but no tube - and a tuning condenser and quite a number of switches and...oh, a lot of stuff. Apparently, some of the members of that radio club had been trying to wire it up and the wiring, of course, was on the back of the panel, and when I had looked at the back of it, it was a mess. They apparently had had some theory about how it should be wired, but almost no practical experience, because no way could it have worked the way they had put it together. Well, taking a lot on myself, I took those bits of wire connections off the back of it and put together some connections that looked at least sensible.

The tuning coils intrigued me. They were honeycomb rings which were plugged into a mount. The centre of the three honeycomb rings was fixed and on each side of it a fitting to plug in other coils, each of which could be swung towards the centre and away from it by a geared knob at the top. Well, it looked very neat and it worked reasonably well. It never did work as well as that old army tuning unit though. Well, the detectors they had in that panel - one of them was a weird looking thing - I don't know what it was supposed to be. It had, I think, a selenium chunk clamped in a mounting, and against it a rather blunt, heavy metal point that obviously was supposed to be pressed down quite hard against the selenium. Another was, of course, the usual galena crystal partly buried in white metal in a little cup and a cat's whisker that was controlled by two knobs - one for vertical movement and other for sort of lateral movement. There were several sets of earphones, and these were plugged into a special panel on that big outfit so that several people could listen at the same time.

In the meantime I brought over the parts of the mobile army set that had been sitting, I think not used at all, maybe only for demonstration, over in Fleming Hall. And I tried out the crystal detector that went with it. I wasn't very much impressed, to tell you the truth. It was a queer combination of two crystals. One of them was, I think called "horseflesh mineral" and the other was chalcopyrites, or fool's gold. The two crystals were supposed to press against each other, and this is what had been used in the army set. Well, I couldn't get nearly as good results from that as I could from one of the old galena crystal detectors.[64]

So, it would seem that the Wireless Club inherited what was left of Gill's semi-portable set. It is curious that Don MacClement makes no reference to the Audion detector, installed in 1916 by Prof. Henderson, to have it operate on continuous wave and bring it up to date.

The Wireless Club's aerial is visible in one of the so-called "Billy Bishop photographs" of the Queen's campus taken from the air in late 1919. One can see part of the wire over the open space between New Arts and Old Arts, as well as the vertical drop to the clubroom window. It is quite plain that it was an L-type antenna.

On September 25, 1987, Dr Don MacClement and the author went to look for the old clubroom in the oft-renovated basement of Old Arts. Once there, Dr MacClement quickly began to orient himself in the unfamiliar surroundings. After a few moments, we entered Room 12,

One of the so-called Billy Bishop aerial photographs of the Queen's campus. These were actually taken by S. Bonnick of the Bishop-Barker Company of Toronto, an aviator who spent 3 years as an aerial photographer for the British Government in France. This was made in the fall of 1919 from a Fokker machine flying at 90 mph. Close examination will reveal a long antenna wire extending from the north-eastern edge of the roof of Kingston Hall to the tower of Old Arts, as well as the point at which a wire drops (arrow) from the L-antenna to the basement window beside the front steps of Old Arts (arrow). The Wireless Clubroom was behind that window.

-courtesy Queen's Archives

Prof. Ann Hardcastle's tiny limestone-walled office:

Don MacClement: Underneath the front stairs here were the tanks where we kept the frogs for the Zoology lab for dissection. And the rest of it was not excavated at all. Well, you could crawl in on your hands and knees. It was a crawl space - mud and dirt and the most incredible kind of junk. And the room, before we fitted it up as a radio room, was all full of broken chairs and desks and, oh, everything left over just tucked in there. Whichever one has a little window right beside the foot of the tower...

Ah! This room has a window. This is the room! This has been re-done, but the lead-in came right in through here, through a porcelain tube insulator, and the ground stake was driven in right down there (pointing out the window to the left). And the table was over here (to the left of the window). Yeah, this was the clubroom.

The Wireless Club, 1920-21

The Wireless Club was revived in 1920, with Prof. Jemmett as Honorary President. "The first meeting was held in Fleming Hall on Friday, November 12. Mr. Hepburn occupied the chair. There was a very small attendance, but it was deemed necessary to re-organize the Club at once," with Donald O. Hepburn as President, Lawrence Wilfred Lockett as Vice-President, Douglas Geiger as Secretary-Treasurer, P.H. McAuley as Critic and Alex Ada as Reporter. "Everybody possessing any knowledge of wireless or radio, and those seriously thinking of joining, are requested to attend the next meeting at 4 o'clock, Friday, November 19th in Fleming Hall, Room 13."[65]

Lawrence Wilfred (Sport) Lockett (Mech. '23, Elec. '24), a native of Kingston now living in Ottawa, learned his wireless before coming to Queen's. His uncle was a wireless operator and his father, Captain L.C. Lockett, was Signalling Officer in "C" Company of the Prince of Wales' Own Rifles and later in the 80th Battalion, CEF. But his interest in wireless did not come through his father or uncle: "I just got interested in it, as boys did at the time I was growing up. Radio was just coming in to some prominence."[66] During his senior matriculation work at KCI he entered the Naval College.[67]

Lawrence Wilfred Lockett (Mech. '23) wa. Vice-President of the Wireless Club in 1920-2 -courtesy Queen's Archive

Lawrence W. Lockett: In the latter part of World War One, I joined the Navy and I was in a specific branch being trained as a ship's wireless operator. Actually, I was in the RCNVR, as they called it. We were training in Ottawa when the War ended, and were not demobilized until we finished the training. I never got into active service because the War ended. But I did graduate and received a First Class Marconi Wireless Certificate of Graduation as a ship's wireless operator in 1919. In the summers while I was at Queen's - four or five summers - I got work with the Marconi Company on ships on the Upper Lakes, on the St Lawrence[66,68] and actually over-seas...what they called the Canadian Merchant Marine.

So far as the Wireless Club at Queen's, I must admit that I may have been appointed Vice-President - because I had wireless experience - but I must have been a very poor one because I have no recollection of attending any meetings or anything...any active part in the club. I don't even remember being appointed. I might have attended one meeting or something. Because while I was at Queen's I really had not much interest in wireless. I was in Mechanical and I think I must have lost interest in E.E. things, other than working as a wireless operator during the summers. Although I did build a couple of little crystal sets, and while I was at Queen's I remember taking one over and giving a talk on the crystal set before the Engineering Society.[69] I used to sit at home up in the attic with my crystal set, and I could get KDKA. I had an aerial I built out back of the house, from the house to the barn. And that would be in 1922 or '23, I guess. Or '24.[66]

Lockett's graduation citation in the 1923 yearbook says, "'Sport' is an enthusiastic wireless bug, and was prominent in the formation of the Wireless Club at Queen's, of which he was president at one time."[67]

The Secretary-Treasurer, Douglas George Geiger (Elec. '22, Mech. '23) was born in Ottawa in 1900 and was educated in Brockville and at KCI. According to his wife, Goldie, he was interested in wireless in high school and around the age of 16 was listening at home and fooling around with it.

Goldie Geiger: Ken Detlor and he fiddled around with stuff. He'd tell me that he got a voice, or that he'd got Germany. Doug entered Queen's in the fall of 1918 - when we were freshmen there was a great victory parade.[70]

He never had a transmitter of his own, but he and his cousin George F. Geiger (Metall. '23) each had a one-lung receiver about 1922. Doug won scholarships in his freshman and junior years and in the summer of 1920 worked for Canadian General Electric at Peterborough, but there is otherwise no foreshadowing documented of the major role that he was soon to play in the development of radio broadcasting at Queen's University.

The Wireless Club Critic in 1920-21 was Patrick Harold McAuley (Elec. '23) of Trenton, Ontario. Not much is known about McAuley and his wireless activities, except that his graduation citation says, in part, "His foresight and consistent efforts made him a leader in radio circles, where he did much toward putting Queen's on the radio map...a leader in class, a reliable worker and an all-round good fellow."[71]

The Reporter for the Wireless Club that year was Alex Edwin William Ada (1902-1969; Arts '24; Meds '26), a Kingston boy who entered Queen's in 1919 in a combined arts-medicine course. The 1926 yearbook mentions his role as halfback on the 1924 Intermediate Dominion Championship football team.[72] No documentation remains of his wireless activities in the town or at Queen's, but he is well remembered as a local "ham" bythe Kingston old timers.

Activities of the Wireless Club in 1920-21

Late on the night of Wednesday, November 25, 1920, two unidentified Wireless Club members sat in the clubroom listening for the 10 pm time signal from Arlington, Virginia. Nothing at all was coming through, and they adjusted and re-adjusted the equipment, straining at their earphones to detect the familiar Morse signals.

> Suddenly the(ir) expression(s) changed to amazement, and they looked at each other in motionless surprise. What was the trouble? Instead of the intermittent buzzing, a customary characteristic of wireless telegraphy, the operators were amazed to hear over their rather inefficient equipment, the "Human Voice." Since the set in use was not equipped for receiving such messages, great clarity of interpretation was not possible. But several words were as distinct as on the ordinary city telephone. The message was a radio-phone message from a boat forty miles off Cape Hatteras on the Atlantic sea-board. The Club takes this opportunity of inviting all who are interested in their meetings, which are held every Friday in Fleming Hall, Room 13. Those who are conversant with the continental Morse code are very especially welcome. The Club is one of the wide-awake, up-to-date institutions of Queen's.[73]

W. Harold Reid Lectures to the Wireless Club

The next meeting of the Wireless Club took place at 4 pm on Friday, December 3 in Fleming Hall. They had the largest turn-out of students yet that year, and an important announcement was made: "Mr. Reid has consented to give the Club four lectures on the principles of wireless telegraphy. The first of these lectures will be given next Thursday evening (December 9) in Fleming Hall, Room 13. Everybody interested in the subject is cordially invited to attend."[74] It was our old friend W. Harold Reid, of Barriefield War Camp, now enrolled in Arts at Queen's. Dr Reid recalls that he was asked to deliver the series by the professor whom he helped with assembly of the Inter Panel set; that is, Prof. Henderson. Therefore, the arrangement must have been made before Henderson left, in August 1920.

A review of the first lecture appeared in the Journal the following Tuesday.[75] It was a general introduction, beginning with an historical survey, a discussion of methods of signalling used in the army and navy and a summary of the developments in wireless over its first 25 years. It concluded with a short discussion of direction-finding stations, and a prediction that wireless would not replace cable or inland telephone or telegraph. The Journal announced that lectures in the new year would be "Wireless Receiving Apparatus", "Wireless Sending Apparatus" and "Wireless Traffic and the different commercial and land stations". The second lecture was not announced or reviewed, but the third was scheduled for Monday, March 7, 1921, at 7.30 pm in Room 16, Fleming Hall.[76]

Dr Reid preserved his hand-written lecture notes, and presented the little ribbon-bound booklet to the author in 1984.[77] From these notes, the details of all the lectures in the series can be known. The first lecture was indeed delivered on December 9, 1920, for that date appears below the series title on the first page. The second lecture was given on January 27, 1921 and the fourth on March 21. The lectures were not based upon the wireless equipment at Queen's and did not refer to the club's resources:

> W. Harold Reid: The drawings are a bit crude, as they were not intend-
> ed for publication, just for my own guidance, as I put good circuit dia-
> grams on the blackboard.[43]

The topics of the second lecture were aerials, wave motion, receiving apparatus and circuits, the detector, tuning, detector types for "damped waves" (coherer, magnetic detector, rectifying crystals, Fleming valve, Audion or 3-element valve) and detector types for "continuous waves" (Poulsen Tikker, Goldshcmidt Tone- wheel and the heterodyne

Bill Taylor examining the manuscript notes of the Rev Dr Reid's 1920-21 lectures to the Wireless Club, at Queen's Homecoming, October 1985.
-photograph by Arthur Zimmerman

receiver). The notes for the third lecture, on transmitting apparatus, are the longest, some 24 pages. He described spark transmission by direct and indirect excitation, and then gave a comprehensive description of a Marconi spark transmitter and how it was operated and adjusted. He also described continuous wave transmitters, the Poulsen Arc, Alexanderson's high frequency alternator, Marconi's continuous wave transmitter and the Pliotron oscillator. He concluded with a discussion of Duplex wireless telegraphy and radio-telephony. The last lecture was on traffic management, dealing with the International Radiotelegraph Convention of 1912, the Canadian Radiotelegraph Act of 1913 and the Postmaster General's Handbook for Wireless Telegraph Operators. He explained the regulations, penalties for non-compliance, the charges for commercial messages and the procedure for transmitting messages.

In its second year, 1920-21, the Wireless Club managed to hang together in some form for the entire school year, or at least from mid-November, 1920, to late March, 1921.

Original Equipment of the Wireless Club

The members of the Queen's Wireless Club in the basement of Old Arts were given several pieces of equipment. For a year or so they had the use of the De Forest Inter Panel receiving set and they may also have inherited Prof. Gill's old multi-component wireless gear, the receiver now operating on CW with a triode valve. William A. (Bill) Taylor (Civil '24), who turned up on the executive of the Club in 1921-22 and 1922-23, recalls a wireless transmitting set:

> **Bill Taylor**: It was an old army job, and I don't remember where it came from. It was run right off the 110-volt supply. About a half-kilowatt spark coil, a later development. It was really a spark coil in effect, but there was a buzzer in it and self-contained in the apparatus - not just a plain induction coil like you use in an automobile ignition. It was spark operating the CW, meaning continuous wave. I know that it was in '21 that we were using it.[78]

"Wireless Sender #1"

Dr Donald C. Rose recalls a very different set at the Wireless Club: "It was the sort of thing an officer would carry in the field, strapped over his shoulder. He'd set it down and somebody else would be carrying a battery and they'd put it together. It was really a military unit that somebody had got out of the armed services after the War."[46]

Bill Taylor remembers that this latter wireless set used by the Wireless Club came from a World War I aeroplane housed in the early 1920s in the attic of Nicol Hall:

> **Bill Taylor**: The RE8 was an old pusher plane. A biplane. We called them pusher planes because the engine and propeller were behind the pilot. It had a wicker basket underneath the pilot, hanging down below, and that was occupied by an observer with a machine gun. He could shoot when attacked, but he couldn't defend the plane when a plane come from behind. So they were very easy targets for the enemy, and I've seen three of them shot down inside of five minutes by one German plane. They were used as observer planes, of course, and they carried a small transmitter which consisted only of a spark coil. It was called "Wireless Sender #1", and we obtained one of these little wireless sets for our Radio Club in the basement of the Old Arts Buillding at Queen's University - beneath the Registrar's Office. And we used that for some time because it wasn't so strong as to interfere with Marconi station VBH over at Barriefield. Later on we obtained a higher power set, a transmitter, which did interfere considerably with VBH, so we had to discontinue using it. The tuning of these sets was not at all critical and the signal splashed over a very wide band of frequencies. Especially this little spark coil carried in the RE8 aircraft.[78]
>
> The set was about 6" long and 4" wide and weighed approxi-

mately 12 pounds. It had a coil on the outside, a flat strip wound helically and fastened to the side of the bakelite box. To change frequency you changed the position of the clips on the helical coil. There was a separate key mounted on a piece of plywood, and it could be strapped onto the knee. There was nothing on the top side of the box except an antenna binding post and a glass aperture. This was so that you could look into the set to see if the spark on the coil was as bright as possible when sending. The set's spark was not very audible to the spotter and he had to have some way of watching the spark coil gap. It worked fine hooked up to two dry cells giving a total of 3 volts. The bottom part was the condenser of the induction coil, with sheets of tinfoil clamped into a small space.[79]

When they used several of these little transmitters at the same time for two or three adjoining artillery batteries, they used to adjust the tone of the buzzer on them. It was a spark coil with a buzzer. These sets could only send, and could not receive information back. They had one making a very high-pitched buzz, another making a medium buzz, another one making a low gruffy buzz. And that's how you distinguished between the aircraft. The transmissions were all, of course, in Morse code. They just sent in figures or letters designating the following shot, and it worked quite well...It's funny, I read in a <u>Popular Mechanics</u> magazine about 1930 or thereabouts a story about the military equipment we used in radio work during the First World War. And they mentioned this transmitter, Wireless Sender #1, that they carried in the RE8 aircraft. They told all the details of the set, exactly how it worked and how efficient it was. They knew all about it - all the details on it.[78]

oplane wireless transmitter #1, a set very similar to the one given to en's in early 1919 by the Imperial tions Board, probably mounted in a Curtiss JN-4 or in an Avro lane. This tiny set (6x2x4 inches) s made in England by the W.D. reless Factory. Top centre is the l clip, right is earth and the large nob at bottom is the spark gap uster. Below this knob is a small ow on the spark gap, for adjusting spark to optimum by eye. The less Club had one of these broad-d sets in its room in the basement of Old Arts.

otograph courtesy the Royal Air rce Museum, Hendon, London

Bill Taylor's Wireless Sender #1 may have been a Sterling transmitter of British manufacture, an interrupted continuous wave set. Normally the Sterling transmitter gave out a high note. In an experiment in mid-April 1916, the military fitted it with a device called a "clapper-break" which lowered the tone to give either a medium or a low note. Another advantage of adding the "clapper-break" was that the transmitter had a greater range on the low note and could be used for long range work. The unmodified set, producing a high note, could be used for aerial trench reconnaissance and contact patrol at close range. By adopting special techniques, including minimizing the transmission times to avoid

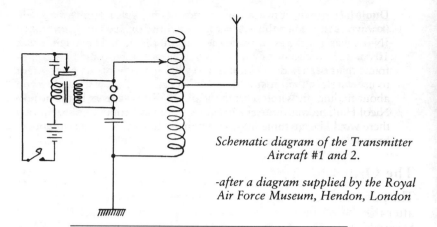

Schematic diagram of the Transmitter Aircraft #1 and 2.

-after a diagram supplied by the Royal Air Force Museum, Hendon, London

jamming, since all sets operated over a wide frequency band, they eventually learned to increase the ratio from one spotter plane to 2,000 yards of trench to one per 400 yards. Later the US Signal Corps adopted a modification, the SCR-65.[80]

As a matter of fact, the RE8 aircraft was a First World War two-seater tractor biplane; that is, the observer-gunner had a cockpit of his own behind the pilot and the propeller was mounted on the nose and pulled the craft rather than pushing from behind. But the story of the observer in a hanging basket is nicely corroborated in a story told by General A.G.L. McNaughton. He flew for the first time with Major Charles Portal, Number 16 Squadron, "in a BE2C, a pusher type of aircraft, in which he had to sit in a wicker basket out in front of the propeller."[81] The BE2C was not, however, a pusher airplane either, but an unarmed two-seater tractor reconnaissance craft. It is possible that both references were actually to the FE2b, like the RE8 and the BE2C a product of the Royal Aircraft Factory, Farnborough. The FE2b was a two-man pusher reconnaissance bomber with a kind of second cockpit, or nacelle, for a machine gunner in front of and slightly below the pilot's cockpit. Perhaps that nacelle was lined with woven wicker reinforcement. The forward gunner could not defend from the rear, and the FE2b was finally outclassed by the Germans as a daytime observer in mid-1917,[82] when Bill Taylor would have been stationed at the Vimy front in France. In the Vimy Battle, a Canadian squadron of the Royal Flying Corps in BE2C's communicated one-way with wireless stations of the Royal Flying Corps at battery positions in order to direct the fire.[83] Perhaps Lester Gill's battery made use of such aerial wireless reconnaissance at Vimy.

Dr W.D. MacClement has first-hand knowledge of how the aircraft wireless set(s) came into the possession of the Queen's Wireless Club. He corroborates Bill Taylor's story that after the First War there were indeed two aeroplanes in the attic of Nichol Hall on the Queen's campus:

<u>Don MacClement</u>: It was also Gill who wangled the gift of two JN-4's (known as the "Jenny") for Queen's at the end of the War. In one of the JN-4's was a wireless set with a key and an aerial, complete, about 8-10" x 12", with knobs and a key at one end of the top. It was in a frame thing at the side of the rear seat of the Jenny, for training observers to use wireless. And it had a wire aerial trailing. When I told Jemmett about finding the wireless set in the old JN-4 fuselage up in the attic of Nicol Hall, he was quite surprised. Apparently he hadn't thought that there would be anything like that up there. Later this set was in the clubroom.[84]

The Origin of the Aeroplanes in Nichol Hall

The published record shows where these aeroplanes came from. Shortly after the end of the War, Principal Taylor announced at a meeting of the Science Faculty that "certain aeroplane equipment was to be given to the University by the Royal Air Force," and a committee was "appointed to consider what can be done in connection with instruction in aeronautics."[85] A few months later the Department of Mechanical Engineering added a course to its fourth year dealing in part with aeroplane design and construction, and "felt that a class dealing with aerodynamics to be given by Prof. Flammer, would be a desirable addition to the instruction along this line."[86]

The "influx of aeroplanes" was completed before January 21, 1919. The <u>Queen's Journal</u> noted, "Some of the material is stored in the Nicol Hall and the remainder is in the basement of Fleming Hall" and wondered "When the two aeroplanes registered at Queen's and made Nicol Hall their home? We presume that it was during one of those heavy snowstorms last week."[87] The planes were brought into Nicol Hall through the south window of the second floor, probably not on their own power.[88]

The donation was actually from the Imperial Munitions Board, and consisted of one JN-4 Curtiss aeroplane, one Avro aeroplane (probably a 504K), four aeroplane engines of various types, mostly new and ranging from 90 to 130 horsepower, "a Vickers machine gun and a Lewis machine gun with considerable machine gun equipment...various types of bombs, carriers for the same, detonators and exploders," two aerial cameras, two sets of 200 gauges used in shell manufacture and "important wireless outfits."[89] The type of wireless sets included was not noted, nor whether from the JN-4 or the Avro, but they were presumably from some

col Hall, Queen's University, winter 1991,
vhere the JN-4 and Avro aeroplanes were
housed in 1919.
-photographed by Arthur Zimmerman

kind of spotter aircraft.

One of the planes was displayed at the Science Dance in February 1919, hanging from the ceiling of the Gymnasium (now Jackson Hall). Stewart McKercher (B.A. '19), a returned member of the Royal Flying Corps, helped set up the plane and Treffle J. (Shorty) Imbleau (Mech. '19) made the propeller go around.[90] It is a pity that there is no photograph of the 1919 Science Dance, with its oil painting of the Quebec Bridge, the artificial moon reflecting in rippling water, the crepe and the lights and the suspended aeroplane with the revolving propeller.

> Harold Stewart: I remember going over there myself when I was here as an early staffer (c. 1929) or maybe a late student (c. 1926). And I saw these things over there, and I think the aeroplane was an old Curtiss JN-4, commonly called a "Jenny" which was used for training purposes in World War I. I'm kind of an old aircraft buff and I remember these things. There was a flying school at Deseronto in World War I, and these things would fly around here. A two-seater with, I think an aluminum V-8 up front. Biplane. Two bays in the wings, and a pretty rugged old thing. Pretty slow...They also had there some old rotary engines, Gnome or Le Rhone, which were used in other World War I aircraft, like the Nieuport fighters..."[91]

Those two aeroplanes were still lying neglected in the garret of Nicol Hall in early 1934. The Queen's Journal reporter found that they had been stripped over the years for parts and souvenirs. The frames were still intact but the fabric covering was dusty, dry and brittle. Two engines, a Grove rotary (sic.) and a V-type eight cylinder, were there beside the fuselages, along with the detached wings.[88]

One old club member who wished to remain anonymous, recalled that the club had WW I army stuff: "The old aeroplane set, in a box, was the only thing we had." The University had been given a bunch of radio stuff, Lewis machine guns and two silent telegraph sets, which one connected to the ground for transmitting. They wanted to set up the silent telegraph sets and try them across the campus, but they never did. Once they got a Lewis gun out of the basement of Fleming Hall, took it out onto the ice of the Lake and fired it a few times.

Did the Wireless Club Have a Licence?

Bill Taylor remembers that Donald Hepburn, perhaps when he was President in 1922-23, did all of the club's correspondence with Ottawa and arranged to get the licence.

> Bill Taylor: Quite a cosy place. We met there frequently and we just listened. Listened mostly to KDKA, and broadcast when we weren't interfering with the Marconi station over at Barriefield, and listened around to see if we could hear other amateurs, but we never heard any other amateurs at all. None at all. Although there were quite a few on the air, I believe, in Toronto at that time. 3QU was the sign we used. Not very many people remember that there was a 3QU. We never wrote up a constitution or kept any record of the club meetings.

We did have a very good spark transmitter, and we put it on the air a couple of evenings and we got a telephone call or message from the main office or the Registrar's office saying that we were interfering with VBH, which was the radio station in Barriefield run by the Marconi Company. And we had to shut down because we were interfering with their reception.[92]

3QU would have been the ideal call for the Queen's Wireless Club - 3 for the Third District which included Ontario, and QU for Queen's University - but it appears from the Official List of Radio Stations of Canada that they never did get a licence for transmission, despite Hepburn's efforts. Like many amateurs, members just made up their own call and used it to identify themselves when they transmitted. Since the club never got a response to their CQ's, nobody except VBH knew what they were doing or that they were calling themselves 3QU. Unfortunately, the logs of VBH are lost, so it is impossible to discover when the Wireless Club, as 3QU, was on the air and interfering. The official 3QU was first granted by the Department of Marine in 1922 to E.C. Stewart of 22 Pipeline Road, London, Ontario.[93]

The Wireless Club Becomes the Radio Club, 1921-22

The first sign of life of the club in the 1921-22 session was an announcement in the The British Whig of November 4: "The Radio Club, which was organized last winter, held a meeting for re-organization purposes Thursday afternoon (November 3, 1921) when plans for the session's programme were made."[94] It is the part of the nature of the institution that there is a very short collective memory at universities, and the students were unaware that the club had been in existence since the fall of 1919. But from this point on the club is almost invariably referred to as the Radio Club. Bill Taylor remembers it only as the Radio Club. This third incarnation seemed to be a much more ambitious and forward looking outfit and for the first time advertised its assets, goals and regulations. Two articles on the Radio Club appeared in the November 15 issue of the Queen's Journal. The shorter one[95] explained that the membership fee was set at fifty cents and that, according to club by-laws, as soon as a member passed a short proficiency test, he would be allowed to use the club's equipment and would accept "joint responsibility in respect to damage to the apparatus, etc. We welcome as members those interested and full of 'pep' if they have not yet sunk their interest too deeply in other college activities...The Club has its own operating room equipped with A.C. and D.C. power, (and) the aerial is at a great height enabling it to intercept wireless waves from thousands of miles away." For the benefit of novices, the club proposed to present lectures immediately on the "Elements of Wireless" and on the "Aircraft Short Wave Tuner." This aircraft tuner was probably the first piece of club equipment that a new member would be allowed to use.

The longer article described the club meeting of November 10th, at which a new executive was elected and ambitious plans were made.[96]

Prof. Jemmett remained as Honorary President. The new President was Douglas G. Geiger, C.C. Gilbert was Vice-President, E.A. Filmer the new Secretary-Treasurer, L.R.C. McAteer was the Critic and W.A. Taylor the Reporter. "A committee of three members under the leadership of Mr. McAuley was instructed to take over the room in the basement of the Old Arts building, which was used by the Club last year, and to instal therein the apparatus which has previously been used." After the elections there was a discussion which brought out a general feeling that "this Club should set up an experimental station on a parallel with or excelling those of our neighbour Universities, and should establish some schedule of communication with these other stations." They were hoping for a larger turnout at the next meeting on the 17th (4.15 pm, Room 16, Fleming Hall) so that members could be grouped according to their wireless expertise. This notion of segregation may have been a significant change, perhaps leading to the emergence of an experienced elite which would soon go on to greater things.

Lancely R.C. McAteer (Mech. '25) started in the wireless game in 1920 or 1921 in Toronto, with a Ford spark coil transmitter. He arrived at Queen's in 1921-22 and graduated from Mechanical Engineering in 1925. He was a member of the Radio Club since at least 1921-22 and recalls that the club met in a classroom in Fleming Hall and had a room allotted in the basement, but never used it. They strung up an aerial from Fleming to Ontario Hall, and Lance put in a counterpoise. The club never really tried to build up enthusiasm, and he was at only three or four meetings. The club sort of died out, he recalls, and Prof. Jemmett's group of Electricals took over.

Lance remains active on his old "electric motor" call, VE3EM, which he has held since 1929 or earlier.[97]

William A. (Bill) Taylor came to Queen's as a war veteran in 1919. He had arrived in France in January 1917, and was stationed at the Vimy front for five months, including the Vimy Battle. He was wounded at Passchendaele in 1917-18.

Bill Taylor: I'm reminded of a small group of amateurs at Simcoe, Ontario, where I worked during 1921 as a telephone lineman. I was just putting in time for the summer, and getting experience at telephone work. I knew about their radio club and I visited their club frequently. They had an old henhouse in one of the boys' back yards, that they used as their clubhouse, and they had a regular broadcast station all set up in there. Their main objective was to broadcast music. They had some would-be musicians - two of them were one-string violinists and they made their own violins. Another fellow played a guitar, another played a mouth organ and someone had a horn of some kind. I think it was a cornet. There were five would-be musicians, but it was one of the most amusing noises I ever heard in my life, to listen to that so-called orchestra. And they actually did put out a show, but I don't think it got anywhere. I mean, the set wasn't broadcasting very far, if any more than a couple of blocks. But they were going through the motions anyhow. I remember a lot of interesting incidents about that clubhouse. The town of Simcoe has light, sandy soil. They grow tobacco around there, you know, and they had trouble getting a good ground for the transmitter. They had a sort of a ground rod that was driven into the earth alongside of the door of the ham shack. The club members were all required, as they went into the clubhouse, to urinate on the ground rod, and this was done religiously to keep a good ground on the ground rod. We'd all go in there and gather around the table and listen for a broadcast. We wouldn't hear any, then we'd go on the air and broadcast for maybe half an hour. We wouldn't have any results usually, and we'd fold up and go home again. Except to talk about their plans for future orchestras. They were all musically minded and thought they might make a good orchestra some day. I was not musically inclined and I don't think I would like to hear the orchestra.[31]

The November 17 club meeting was better attended: "The Works Committee reported that an aerial of our wireless station has been pulled up into position and that the wireless room may be fitted up with a set of instruments within a few days. This committee was instructed to instal a 'dummy' buzzer set in the room in order that, at the end of the next meeting, an initial practice in code work may be held."[98] Plans were also made for "installing one of the most up-to-date types of apparatus in the basement of the Old arts Building. The O.T.C. are to move their stores out of this building, leaving more space for the club's accommodation. At the next meeting, Mr. E.A. Filmer, the club's secretary-treasurer, is to give a lecture describing an aeroplane receiving set."[99] This would be the first in "a series of lectures, to be given by men of authority on wireless operating, on the general principles of radio engineering, and on some detail (sic.) subjects such as 'the use of vacuum tubes in oscillating circuits.'"[98] Though announced as "Aircraft Short Wave Tuner," Filmer's lecture was on "Air-craft Short Wave Tuner and Elements of Wireless," on Thursday, November 24.[100]

Prof. Arthur Ll. Hughes, the first holder of the Chown Research Chair in Physics at Queen's, addressed the club on Thursday, December 8, on "Vacuum Tubes," and Prof. Robert L. Davis of E.E. provided continuation on Monday, December 12, with a talk on "Audion Tubes." "He discussed the development of the audion, the kenetron, the triotron and the magnetron, and furnished his audience a great deal of valuable information."[101]

The new up-to-date radio equipment was still being installed in the club's wireless room in Old Arts as of mid-December. Up to this point, the club appeared to be more organized and more active than in the past, but there was still no hint of the developments that would occur in the new year. The first indication that things were changing is found in the announcement of what may have been the first meeting of the winter-spring term:

> The Club has recently added considerable new equipment of the most modern character. Three new aerials

have been added to its apparatus this winter, making a
total of four available for its work. The first aerial erected, and still the biggest, reached from the Old Arts
Building to the tower of Grant Hall. The new ones are
connected with Fleming Hall. To the receiving outfit has
recently been added a Deforest (sic.) set, with a receiving range that is practically unlimited and that enables
any hook-up to be arranged in short order. The Club is
being licenced to operate a short wave aeroplane spark
transmitter with a normal range of ten miles...In addition to practical work with the sets the members of the
Club enjoy lectures frequently, given by members of the
staff and others who are authorities in this branch of science. Prof. Jemmett of the Department of Electrical
Engineering, who is a keen wireless enthusiast, is the
speaker to the Club this week.[102]

Jemmett's subject for the meeting of February 17 was "Radiation,"
and he "dealt with the manner in which energy is radiated in the form
of electric waves from the conductors of oscillating electric currents...Professor Jemmett described some experiments which he has
performed in order to study the behavior of this radiation of energy. He
also compared his results with popular theories on the subject, and to
the great pleasure of members of the Wireless Club explained how conclusions from these technical studies might be applied in improving the
design of the antenna part of the Club's radio apparatus."[103] He stated
that he expected his results to be published during that summer. The
newspaper statement about antennas is quite vague, but Jemmett may
have been giving the first intimation that one or more of the club's four
aerials was intended to be used for transmission of wireless signals
instead of merely for reception.

The De Forest Inter Panel set was certainly not new at Queen's in the
spring of 1922, but the peculiar wording, "to the receiving outfit has been
added a Deforest set," and the "practically unlimited" range, may indicate that it was being modified, perhaps with the addition of a vacuum
tube or two. Unfortunately, Filmer's short wave aeroplane spark transmitter cannot be identified, unless it was the "Wireless Sender #1." It is
also unclear at this point, since each of the three new aerials terminated at Fleming Hall and in the light of newspaper reports of the autumn
of 1922, whether this expanded set-up was in Old Arts or in Fleming
Hall. Lancely McAteer arrived at Queen's in 1921-22, and was a member of the club executive from his first year. He did not attend many
meetings, but he doesn't remember their having a room in Old Arts at
all.[104]

A further feature newspaper report on the Radio Club attempted to
satisfy the students' curiosity about the presence of the new wireless
aerials and to expand on the club's aims and accomplishments in 1921-
22. The club had increased its membership by opening up to embrace
all faculties (including the women's faculty, Levana!). "Some of the

new members from the Freshman year are ardent enthusiasts, and are already making their presence known...Our Clubroom is now equipped with an 'Aircraft Receiving Set' and an 'Aircraft Transmitter', these sets were used in the war to control our big guns from the air by wireless; although primarily made for use in the air, they work exceedingly well on the ground. Any member of the Club is allowed to operate these sets, when he has passed a short proficiency test; also, all members accept joint responsibility in case of damage."[105]

The Royal Air Force first installed wireless receivers in aircraft in September 1918, so that there could at last be two-way communication with the ground during combat.[106] Before this time, the receivers were just too heavy and bulky and too difficult to operate. Whatever these first aircraft wireless sets were, the Radio Club probably had inherited these.

> **Lancely McAteer**: We had our own radio club and we had some old First War army equipment - little wooden boxes. It was all CW in those days. Do you remember, Bill, that we had some World War One CW transmitters? Little square boxes. I don't know the name of them.[107]...The University got a bunch of Lewis guns and radio stuff and two silent telegraph sets from the military. We wanted to set up the silent telegraphs and try them across campus, but we never did. The Radio Club had oak boxes, about 12 x 18 x 12" high - WW I Army stuff, the only wireless that there was in those days aside from what you built yourself. One box worked by itself and two of them worked together. One of the boxes had a vibrator and spark coil for sending ICW (interrupted continuous wave). But the Club didn't have the equipment set up. The old aeroplane set was the only thing we had, and it was in a box too.[104]

These are vaguely familiar descriptions, but not detailed enough to resolve whether the club had access to an updated CW version of Gill's pre-war equipment or war surplus sets of later vintage.

The Radio Laboratory is Put in Commission

The 1921-22 club prospectus in the newspaper sounds like the regular sort of Radio Club activities, although on a larger and more organized scale than before. But the article went on to describe the beginning of a new venture, voice transmission:

> It will be of interest to all the students to know that Queen's now has a "Radio Laboratory" which is being equipped with a large receiving set capable of receiving messages from a great distance, and it is hoped that soon a powerful continuous wave transmitter will be installed for direct communication between Toronto, Montreal and Ottawa, so we can obtain the results of the Rugby and Hockey matches, etc., by wireless. After all we have accomplished so far, we are not content, for "content-ment means stagnation," and our plans for the future are rather prodigious.[105]

There is very little information extant about this "Radio Laboratory." The Principal's Report for 1921-22 included a brief note on the new facility, which contained little more technical information: "The wireless room which was put into commission this winter will be ready for radiophone transmission by next September, over a 250 mile radius. The Radio Club will send out reports of all football and hockey matches and transmit any of the concerts in Grant Hall."[108] It stands to reason that the new wireless room was situated in Fleming Hall, rather than in Old Arts, though its exact location in the 1921-22 session is not known. Perhaps the installation of the room explains why three new aerials were installed terminating at Fleming Hall. The Radio Club obviously had a direct connection with the new wireless room, since it was the members' impression that they were intended to work it.

This venture into radiophone transmission, which sounds like a form of public broadcasting, is a new departure for the Radio Club. To this point in their history they had been interested only in communicating with other amateurs in code. In fact, certain of the members, such as Taylor and McAteer, were never interested in public broadcasting, and both are still active hams.

This period of transition has been the most obscure of the whole Queen's Radio study and by far the most difficult to winkle out. Little hard information is available, and the bits of story don't always fit together and make sense. Suddenly, in the spring of 1922, after two years of disorganization and faltering, the club was seized with ambition, a Radio Laboratory was "in commission" and the club took on a whole new objective. Whence came the inspiration, and who spearheaded the change? Certainly the E.E. Professors Jemmett and Davis were in the middle of things, and it seems that a student, Edward Arthur (Jock) Filmer (Metall. '25), came in for a lot of credit for his efforts in 1921-22: "Through the efforts of Mr. E.A. Filmer the Radio Club has expanded and become quite an asset to the college."[109]

Filmer was born in England and went to high school there. He was one of those people who had to know the "why" of everything, including wireless, and he had earned his British Marconi certificate. He had also served in France with the Royal Flying Corps and after the War taught radio at Camp Borden.[110] He must have entered Queen's in the fall of 1921 and immediately made his presence felt in wireless circles and elsewhere: "Jock Filmer Says 'the man who cannot speak his mind is doomed to insignificance.' Take it from us, Jock is headed for the clouds. If Hot Air was bullion Jock would be richer than the U.S. Treasury."[111] He succeeded Doug Geiger as Secretary-Treasurer of the Radio Club in 1921-22, but was not on the executive in 1922-23. Lancely McAteer remembered "a little English fellow by the name of Wolmer or some name like that was the first President when it was formed." When asked exactly what Filmer did for the club, McAteer recalled: "Filmer ran around and got people to join the club and got people to talk at meetings."[107] Only one further reference to Filmer's good works in radio exists and this will be quoted below. Patrick Harold McAuley (Elec. '23) was another student credited with important Radio Club work in this period.

The Calendar of the Faculty of Applied Science for 1923-24, listed the various E.E. laboratories in Fleming Hall: "Laboratory No. 5 is a small room occupied by the Wireless Club."[112] No. 6, which became the first home of 9BT and CFRC, was the Radio Laboratory and No. 7 was essentially Gill's old Laboratory No. 4, containing the telephone exchange and the old wireless set. Therefore, some time after the beginning of 1922 and before the fall of 1923 (when the 1923-24 Calendar was assembled) the club had apparently moved to Fleming Hall from Old Arts. Or else it had split into two parts, one remaining in Old Arts and interested only in trying to talk to other amateurs in code, and the other, newly installed in Fleming Hall, on the verge of embarking into public broadcasting.

What Became of the Radio Club?

There is some indirect evidence that the club had indeed split in the 1922-23 session. Bill Taylor recalls that a small group of Electricals under Prof. Jemmett had control of the new broadcasting station in Fleming Hall. He and other non-Electricals in the old Radio Club were not allowed access to the new facility. Taylor denies that the radio station derived from the Radio Club, but Doug Geiger has written that this was indeed the case.[113]

Lancely McAteer recalls that there was a split, and the underlying reason was that the Radio Club was disorganized:

> **Lancely McAteer**: The club met in a classroom in Fleming Hall. They had a room in the basement allotted, but they never used it. I just went to the meetings - only three or four. At the meetings there were lectures and talks by different people, and they tried to build up the enthusiasm, but the Radio Club never really got rolling. They had no licence...Then, in the spring of 1923, Dr. Jaffrey, a Queen's graduate living in Hamilton, donated a lot of radio gear to Queen's...mostly tubes, condensers and coils. The Radio Club was excluded from the radio station because they weren't Electricals. Doug Geiger took a dim view toward them and excluded them from the laboratory after Dr. Jaffrey's gift arrived from Hamilton. So, Prof. Jemmett's group sort of superceded the club. Electrical took charge of the stuff donated by Jaffrey, and the club sort of died out and Jemmett's group took over...The radio stuff was in the basement of Fleming Hall, west end, to the right (north) side.[104]

Sidney Victor Perry (Elec. '23) arrived at Queen's in the fall of 1919. He was a charter member of the Wireless/Radio Club, although his name never appeared on the record because he never held office. In an oral history recorded in 1982, he could not remember when he joined the Club:

> **Sidney Perry**: Prof. Jemmett was not there that year. He was taking extra work at MIT, as I understand it. And so we just heard about Prof.

Jemmett. But there wasn't much done. I don't remember whether anything was done or not.

However for our second year, our sophomore year, the club was
drawn together. Dr. Jemmett was...teaching some of the classes and we
got in on that. And somehow or other he started the thing off. I don't
remember just what we did about it, but there wasn't much of an
organization, if any. We had very little equipment...We had one or
two of those English (Mullard) vacuum tubes. His contact was mostly with the English companies, I think, and I should tell you somewhere in here who else was involved. There was P.H. McAuley, and
there was a guy from a former year. Geiger - there were two Geiger
boys. I think they were in different years, and they really didn't belong
to Science '23. As far as I'm concerned, it wasn't any club...because it
wasn't organized, see? It was just a few students that were interested
that got together and did something or other. Not very much, either.

We got a little bit more active as the years went along. That's the
second and third years...third and fourth....We got together one
Saturday morning. This should be in our second year, now (1920-21).
I'll simply say that under Jemmett's leading we built a squirrel-cage type
of antenna from Fleming Hall (was that where the Electricals were?)
over to the Grant Hall. It cut across a fairly open section there, and we
put that up. That's one of the things we did, and several people probably were in on that. The purpose of the antenna was to pick up other
stations and so on. It wasn't a broadcasting antenna, although we
used it for broadcasting...We broadcast nothing in particular. We got
a comment from the radio operator (at VBH, Barriefield). He called up
one of our people one night and said that we were interfering with
them. He didn't report it or anything. He just told us about it and told
us we'd better not do it any more.[114]

When asked if he recalled the first broadcast with the new antenna:

Sidney Perry: I think I was behind the microphone. We apparently
were building a new transmitter at that time. And it was late at night.
Yeah... it was around 10 o'clock at night maybe, when everything
seemed to click all right and I had to sign off. Prof. Jemmett wasn't there
at the time. I was all alone, I believe, and that's why I had to sign off the
station...It could have been 1922, not 1921.[114]

Perry's memory was very patchy and he could not be specific about
details nor certain about chronology. He could not remember very
much about the club, and he did not distinguish between what may
have been different phases of the activity under Jemmett's direction.
Perry's impression was that the erection of the antenna between Fleming
Hall and Grant Hall took place in the fall of 1920 or in the spring of
1921, perhaps a year before the fact appeared in print.[102] At this point,
the club's room was down in the basement of Old Arts and they were
still sending out CQ's in code trying to raise some response. There was,
however, something a little different happening in the basement of
Fleming Hall. A small group of E.E. students under Jemmett was working apart from the Wireless/Radio Club:

<u>Sidney Perry</u>: I suppose we could say that the Radio Club was interested in more or less a public type of broadcasting, rather than just ham broadcasting. First they wanted to cover the Kingston area, because that's where the biggest audience would be. The rest of it's not clear to me at all, as to what went on...The outside connections of what we were doing (i.e. the licence), there wasn't much said about it and Jemmett took care of it himself, as far as I know. Whatever connections he had to make with the government and so on.[114]

The Radio Club Carries On

The first meeting of the Radio Club in the new school year, 1922-23, took place in Room 13, Fleming Hall, on the afternoon of Thursday, October 19, 1922. The outgoing President, Douglas G. Geiger, B.Sc., was in the chair, and a new executive was elected. Lancely McAteer succeeded to the presidency, Bill Taylor became his Vice-President, Joseph Taylor Thwaites (Physics '25) was the new Secretary-Treasurer, Donald O. Hepburn (Civil '24) the Critic, Wilfred Ernest Patterson (Chem. Eng. '24) the Reporter, and the Work Committee consisted of Samuel Welberne Small (Chem. Eng. '25), Gerald Stanley Lyons (Elec. '24) and Sheldon Julius Cohen (Arts '27). Prof. Jemmett was again the Honorary President.

> After the election of officers, the meeting adjourned to the Club Room and examined the equipment. There are now two aerials and two receiving sets at the disposal of the Club so that members may listen in every night they please. It is hoped that the membership will increase now that we have equipment to work with.[115]

Of the new names on the executive, documentation on radio activities exists for only three. Joe Thwaites later became involved with CFRC, Sheldon Cohen was a local ham and co-founded a commercial radio station in Kingston in 1925, and of Gerald S. Lyons, the 1924 Yearbook said, "After spending three years with the Bell Telephone Company and many hours upon brain-racking radio diagrams, Jerry decided that a course in Electrical Engineering was his only salvation."[196]

It is still not clear what class of equipment the club had access to, nor in which building the clubroom was located in 1922-23. Furthermore, this <u>Queen's Journal</u> report[115] is the very last that appeared in the Queen's or Kingston papers on the Radio Club. Lancely McAteer remembers the club continuing until around 1925, but there is no supporting documentation after October 24, 1922.

Bill Taylor summarized the activities of the Queen's Radio Club as he remembers it:

> <u>Bill Taylor</u>: Well, we were just interested in listening to radio and doing the odd broadcasting. But as an amateur, to talk to other people. We never did make any contacts with others except VBH, accidental-

ly, over in Barriefield. They, of course, could read our telegraphy and knew who we were by our call sign. And they promptly gave us the devil for interfering with them...We used to delight in listening to the programs of KDKA, because they came in loud and clear on our little set at Queen's. Usually music and brief talk shows too, and they tried to interject a lot of comedy - telling jokes and whatnot, and it came through clearly. Oh, we were not very active at all, actually. We met about every two weeks, not on a scheduled day basis, and we'd talk about providing some music if we ever learned how to transmit by phone, but we only talked about it. The same rules applied in those days as now. When you get an amateur radio licence, it's only for code for the first year. You must be active for that whole year before you can apply for a phone licence. We were never active for a whole year. But we did operate two years in succession. I remember that quite clearly - at least two years. I think it was '21 and '22, and possibly part of '23. We never did apply.

I don't know about Jemmett (as Honorary Chairman) - I don't think he was ever connected with our amateur Radio Club. Of course, he was a Professor for some of the students there and they probably put his name down because of that. Oh, he had nothing to do with the amateur Radio Club, as such. Nothing to do...at all. Just students, that's all. Very green students at that. I think I was the only one of the group that had overseas experience - I'd been in the First World War and had some signalling experiences, and that helped me. I was all ready to transmit by CW if they ever got the thing set up. And we did send out CQ's quite often - that's "anybody answer who hears" - but we never got any answers.[79]

G.S. Lyons   D.G. Geiger, B.Sc.   W.A. Taylor

Members of the Wireless Club,
G.S. Lyons, Doug Geiger, W.A. Taylor, E.A. Filmer.
-courtesy Queen's Archives

NOTES

1. Department of the Naval Service: Radiotelegraph Reulations; Order from the Governor General in Council, April 29, 1919; Dominion of Canada, Sessional Paper No. 52f, May 6, 1919.
2. Letter from Deputy Minister, Department of the Naval Service, Ottawa, May 3, 1919; NAC, RG 97, Vol. 320, file 6270-1.
3. Memo to Deputy Minister, Department of the Naval Service, March 19, 1919; NAC, RG 97, Volume 132, 6208-1, part 1.
4. Chapter LII, "Queen's University and the School of Mining at Kingston", Section 5: "Information Obtained from Mr. Lester W. Gill, Professor of Electrical Engineering", in The Dominion of Canada Royal Commission on Industrial Training and Technical Education; House of Commons Sessional Paper No. 191D, p. 2076, 3 George V, Part IV, 1913.
5. "Research Work" in Report of the Principal, Session 1915-1916, Queen's University, May 1, 1916, p. 6.
6. "Queen's University Aids in Industrial Research", Toronto Daily News, Oct. 21, 1916, p. 20; reprinted, DBW, Oct. 24, 1916, p. 9.
7. "Queen's Desires $50,000", DBW, Oct. 19, 1916, p. 2.
8. "Scientific and Industrial Research" in Principal's Report, 1916-17, Queen's University, April 28, 1917, p. 30.
9. "Dean Goodwin...on behalf of the Industrial Research Committee", Minutes, Queen's University Senate, Feb. 23, 1917, p. 144; QUA.
10. "Major Gill Writes", QJ, Dec. 1, 1916, p. 1.
11. L.W. Gill to Principal D.M. Gordon, Dec. 26, 1916; Gordon Papers, QUA.
12. "Hamilton Branch", Journal of the Engineering Institute of Canada 6 #1, Jan. 1923, p. 28.
   John B. Nayler (Elec. '23) was involved in all of the action at Vimy Ridge and recalled that by the middle of 1916 the Germans were no longer releasing chlorine gas, but were firing gas shells containing phosgene (chloroformyl chloride or carbonyl chloride), which is ten times as toxic as chlorine. Phosgene hydrolyzes in the lungs to release hydrochloric acid and may cause severe pulmonary damage and edema (i.e. pneumonia). Gill likely got a dose of phosgene gas at Vimy Ridge. Mr. Nayler also recalled that at Vimy the Canadians used Iroquois Indians to man the field telephones, so that there was no way for the Germans to crack their messages. Personal communication, Oct. 15, 1988.
13. L.W. Gill, at the Royal Automobile Club, Pall Mall, London, to his sister Elizabeth Jane Duncan, Okotoks, Alta., March 16, 1919; courtesy Dr. Pauline (Duncan) Hughes.
14. QJ, Jan. 11, 1918, p. 1 (report taken from the KDS).
   "L.W. Gill, M.Sc., Record of War Service" (typed on letterhead of the Queen's E.E. Dept. and hand corrected by Gill for submission to the Harvard University War Records, Cambridge, Mass., c. 1919); Harvard University Archives.
   "Lester Willis Gill: Statement of Service in the Canadian Armed Forces"; Personnel Records Centre, NAC.
   "Prof. Gill Comes East Again", The McGill News 4 #1, Dec. 1922.
15. E.A. Corbett, Henry Marshall Tory: Beloved Canadian, Ryerson, Toronto, 1954, pp. 138-56.
16. Postcard from L.W. Gill to O.J. Hickey, Esq., dated May 9, 1919, Paris; courtesy Mrs. Mabel (Hickey) Archer.

17. "Proceedings of an Officer or Nursing Sister Struck Off Strength of the Canadian Expeditionary Force", in Gill's "Statement of Service in the Canadian Armed Forces"; Personnel Records Centre, NAC.
18. "Major L.W. Gill Returns", DBW, Aug. 25, 1919, p. 1.
19. "Told in Twilight", ibid., Aug. 30, 1919, p. 3.
20. "Department of Electrical Engineering", Principal's Report, Queen's University, 1918-19, p. 28.
21. Minutes, Faculty of Science, Queen's University, Nov. 14, 1919; QUA.
22. "Wireless School for Queen's", DBW, Nov. 4, 1919, p. 18.
23. "Wireless Club", QJ, Nov. 11, 1919, p. 4.
24. "Queen's Wireless Club", DBW, Nov. 13, 1919, p. 16.
25. "Marrison, Warren A.", The Canadian Who's Who 10, 1964-66, Trans Canada Press, Toronto, 1966.
    W.R. Topham, "Warren A. Marrison - Pioneer of the Quartz Revolution", National Association of Watch and Clock Collectors Bulletin, April 1989, p. 126.
26. Personal communication from Prof. Harold H. Stewart, Kingston, Oct. 20, 1982.
27. Personal communication from John R. Rutledge (Arts '22), Oct. 16, 1982.
28. American Men of Science: A Biographical Dictionary (ed. Jaques Cattell - 7th edition), Science Press, Lancaster, Pa., 1944, p. 1170.
    Around 1926, when Western Electric was first recording sound on film, a new department was set up for this work and Marrison was put in charge of it (QR 1, April 1927, p. 61). In 1947, he received the Gold Medal of the British Horological Institute for pioneer researches in the development of the quartz clock (Queen's Alumni Record, Nov. 3, 1954).
29. W.A. Marrison, "The First Electric Clock", PESQU 29, 1940, p. 15.
    W.A. Marrison, "The Crystal Clock", Proceedings of the National Academy of Sciences, U.S.A. 16, 1930, p. 496.
    Bell System Technical Journal 8, 1929, p. 368; 27, 1948, p. 510.
30. QJ,, Nov. 28, 1919, p. 4.
    "Science '23", QJ, Dec. 12, 1919, p. 8.
31. Personal communication from William A. "Bill" Taylor (B.Sc. '24), Bath, Aug. 21 and Oct. 19, 1982.
32. Minutes, Executive Committee, Board of Trustees of Queen's University, Dec. 10, 1919, p. 74; QUA.
33. "Prof. L.W. Gill Appointed", DBW, Nov. 24, 1919, p. 3.
    "Professor Gill's Appointment", McGill News 1 #2, March 1920, p. 30.
    "Technical Education", Report of the Deputy Minister, Department of Labour, Sessional Papers, House of Commons, Paper #37 57 #9, 1921, p. 16.
34. Letter from Col. Tory, President, University of Alberta, to the Civil Service Commission, Ottawa, Sept. 6, 1919; H.M. Tory Collection, Archives of the U. of Alta. at Edmonton, acc. 68-9-418.
35. Gives Some Reasons", Ottawa Journal, Aug. 17, 1921, p. 4.
    "For British Columbia", DBW, Aug. 18, 1921, p. 7.
    Ibid., p. 1.
36. "Principal Gill", HS, Oct. 9, 1922, p. 1.
    "Local News", DBW, Oct. 10, 1922, p. 16.
37. "Seeks New Principal", HS, Sept. 8, 1922, p. 1.
    "Principal Sprague to be Retired", ibid., p. 1.
    "Dismissed! Sprague Scrapped", ibid., Sept. 15, p. 1.
38. "Principal Gill Cites", HS, April 15, 1931, p. 20.
    "Apprentice Training", HS, April 18, 1931, p. 6.
    "Obituaries: Lt.-Col. James Lester Willis Gill", Journal of the Engineering Institute of Canada 22 #10, Oct. 1939, p. 452.

39. A.L. Clark, The First Fifty Years, p. 78.
40. "You Can Actually See the Difference", Hamilton Techical Institute Tech Sparks 12 #48, May 1938, p. 6.
    Personal communication from Ralfe J. Clench, Jr., Kingston, Jan. 28, 1987.
41. "Lester W. Gill Succumbs", HS, Aug. 25, 1939, p. 7.
    "Gill", QR 13 #8, Nov. 1939, p. 249.
42. "Changes in Staff", Principal's Report, Queen's University, 1919-20, p. 22; QUA.
43. Personal communications from Dr. W. Harold Reid, Aug. 9, 1984; Sept. 5, 1984; May 15, 1985; May 26, 1986; June 3, 1986.
44. De Forest Post War Models of Radio Apparatus, Catalogue B, De Forest Radio Telephone and Telegraph Co., 1919, p. 23.
45. "DeForest: Newest Ideas in DeForest Unit-System Receiving Apparatus", Pacific Radio News 2 #7, Feb. 1921, p. 232.
46. Personal communication from Dr. Donald C. Rose, May 9, 1986.
47. "Happenings in the Wireless World", KDS, Feb. 15, 1922, p. 7.
    "Wireless at Queen's", QJ, Feb. 21, 1922, p. 5.
48. "Radio at Queen's", DBW, Aug. 11, 1922, p. 2.
    "Everybody..."Listening In", ibid., Aug. 23, 1922, p. 12.
49. Personal communication from Mrs. Norman Miller, Kingston, Sept. 28, 1982.
50. Personal communication from Kathleen (Whitton) Ryan, Ottawa, Nov. 7, 1987.
51. Personal communication from Professor Harold H. Stewart, Kingston, Oct. 20, 1982.
52. Report of the Dean, Faculty of Applied Science, Principal's Report, Queen's University, 1919-20, May 1920, p. 27.
53. "Prof. Henderson Resigns", KDS, Aug. 5, 1920, p. 12.
54. Letter of appointment from G.Y. Chown to Prof. E.W. Henderson, April 28, 1911; SM Letter Book, p. 908; QUA, Coll. 1161f 41/121.
    Letter to Prof. H.H. Stewart from E.W. Henderson, Dec. 2, 1974; H.H. Stewart's papers.
    "Care for Yourself", St. Catharines Standard, Sept. 13, 1979.
55. "Fleming Hall Staff", KDS, Sept. 10, 1920, p. 2.
56. Minutes, Finance and Estates Committee, Board of Trustees of Queen's University, Sept. 13, 1920, p. 96; QUA.
    "New Appointments 1920", QJ, Oct. 20, 1920, p. 7.
    "Further Additions to Queen's", KDS, Oct. 21, 1920, p. 12.
57. Personal communication from Prof. Harold H. Stewart, Kingston, Aug. 24, 1982.
58. A.L. Clark, The First Fifty Years, p. 65.
59. "No. 5 Field Company of Canadian Engineers: Non-Commissioned Officers", 7th Annual Directory of Graduates and Students in the SM, Kingston, 1913, p. 33; QUA.
60. Personal communication from Harry Jemmett, Kingston, June 8, 1988.
61. D.G. Dewar, "Introducing the Staff: Douglas M. Jemmett", QR, 28 #4, April 1953, p. 86.
    "Douglas Mill Jemmett: Statement of Service in the Canadian Armed Forces"; Personnel Records Centre, NAC.
62. "Assessment Roll for 1921, made in 1922 for Victoria Ward, in the City of Kingston, Ontario, #436"; QUA.
63. Personal communication from Maud Martineau Jemmett, Kingston, Oct. 5, 1982.
64. Personal communication from Dr. W. Donald MacClement, Kelowna, B.C., May, 1987.

65. "Wireless Club", QJ, Nov. 19, 1920, p. 4.
66. Personal communication from Lawrence Wilfred Lockett, Ottawa, April 24, 1985, and Oct. 15, 1988.
67. "L.W. Lockett", Science '23, Queen's University Yearbook, 1923, p. 153, QUA.
68. "To Resume His Studies", DBW, Sept. 18, 1922, p. 5.
69. "The Bulletin", QJ, Dec. 15, 1922, p. 1.
70. Personal communication from Annie Lenore "Goldie" (Bartels) Geiger (B.A. '22), Edmonton, June, 1986.
71. "P.H. McAuley", Science '23, Queen's University Yearbook, 1923, p. 158; QUA.
72. "Alexander Edwin William Ada", Faculty of Medicine, Queen's University Yearbook, 1926, p. 110; QUA.
73. "Wireless Club", QJ, Nov. 30, 1920, p. 7.
74. "Wireless Club", QJ, Dec. 10, 1920, p. 7.
    This announcement was published a week later than it should have appeared. The meeting announced actually took place on Dec. 9.
75. "Wireless Club", Queen's Journal 47 #18, 1, Tuesday, December 14, 1920.
76. "Wireless Club", QJ, March 1, 1921, p. 8.
77. "A Course of Lectures on Radiotelegraphy Delivered to the `Wireless Club', Queen's University", 28 double-sided hand-written pages, 137 x 213 mm., presented by Dr. Reid to the author on Aug. 9, 1984.
78. Personal communication from William Arthur "Bill" Taylor, Bath, Nov. 30, 1982.
79. Ibid., November 23, 1988.
80. H.A. Jones, Official History of the War: The War in the Air, Vol. 2: Being the Story of the Part Played in the Great War by the Royal Air Force, Oxford University Press, London, 1928.
    Bill Gould, "Airplane Radio Telegraph Transmitting Set Type SCR-65" Old Timer's Bulletin: Official Journal of the Antique Wireless Association 9 #1, May 1969, p. 11.
81. John Swettenham, McNaughton: Vol. 1, 1887-1939, Ryerson, Toronto, 1968, p. 96 footnote.
82. Chaz Bowyer, The Encyclopedia of British Military Aircraft, Arms and Armour Press, London, 1982, p. 40.
83. Sidney F. Wise, Canadian Airmen and the First World War: The Official History of the Royal Canadian Air Force, Vol. 1, University of Toronto Press, Toronto, 1980, pp. 401-2.
84. Personal communication from Dr. W. Donald MacClement, Kelowna, B.C., July 26 and Aug. 1987.
85. Minute Book, Faculty of Science, Queen's University, Dec. 13, 1918, page 20,; QUA.
86. Ibid., Feb. 14, 1919, p. 30; and April 10, 1919.
87. "Science", QJ, Jan. 21, 1919, p. 2.
88. "Two Aeroplanes", ibid., Jan. 23, 1934, p. 1.
89. "Aerodynamics To Be Added", ibid., Jan. 21, 1919, p. 5.
    "Department of Mechanical Engineering" in Annual Report of the Dean of Applied Science (Dr. W.L. Goodwin) Principal's Report, Queen's University, April 30, 1919.
90. "Science '19", QJ, Feb. 21, 1919, p. 6.
91. Personal communication from Professor Harold Huton Stewart, Kingston, Oct. 20, 1982.
92. Personal communication from William Arthur Taylor, Bath, Aug. 21, 1982.
93. Official List of Radio Stations of Canada, Department of Marine and Fisheries, Ottawa, Aug. 1, 1922.

94. "In the Halls of Queen's", DBW, Nov. 4, 1921, p. 2.
95. "Radio Club Notes", QJ, Nov. 15, 1921, p. 6.
96. "Radio", ibid., p. 8.
97. Personal communication from Lancely R.C. McAteer, Havelock, Aug. 21, 1982 and July 5, 1985.
98. "Radio", QJ, Nov. 25, 1921, p. 6.
99. "Notes of Queen's", KDS, Nov. 18, 1921, p. 16.
100. "The Bulletin", QJ, Nov. 22, 1921, p. 1.
     "In the Halls of Queen's", DBW, Nov. 25, 1921, p. 2.
101. "Notes of Queen's", KDS, Dec. 13, 1921, p. 5.
     "The Bulletin", QJ, Dec. 9, 1921, p. 1.
102. "Happenings in the Wireless World", KDS, Feb. 15, 1922, p. 7.
103. "Radio", QJ, Feb. 21, 1922, p. 5.
104. Personal communication from Lancely R.C. McAteer, Havelock, Dec. 3, 1988.
105. Wireless at Queen's", QJ, Feb. 21, 1922, p. 5.
106. S.F. Wise, Canadian Airmen, p. 246.
107. Personal communications from Lancely C.R. McAteer, Havelock, Aug. 21, 1982 and June 5, 1985.
108. Principal's Report, Queen's University, May 1922, p. 19.
109. PESQU, #16, 1922, p. 107.
110. "Edward Arthur Filmer", Science '25, Queen's University Yearbook, 1925, p. 98; QUA.
111. "Queen's Radio Club", QJ, Oct. 31, 1922, p. 6.
112. "Electrical Engineering Laboratories", Calendar, Faculty of Applied Science, Queen's University, 31st session, 1923-24, p. 89; QUA.
113. "D.G. Geiger, "Radio Broadcasting at Queen's" , QR 2, 1928, p. 9.
114. "Personal communication from Sidney Victor Perry (1901-1986), Cherry Hill, N.J., Oct. 12, 1982.
115. "Queen's Radio Club", QJ, Oct. 24, 1922, p. 6.
116. "Gerald S. Lyons", Science '24, Queen's Yearbook, 1924, p. 212.

CHAPTER 4

*The west end of Fleming Hall. The Queen's radio station was
in the basement from 1922-27, on the first floor from 1927-
33, and on the second floor from 1933-58. The corner room,
second floor (left), housed the transmitters from 1933-90.
-photograph by Arthur Zimmerman*

Radio Station 9BT

The Radio Club's most promising year yet was winding down by the
middle of March 1922. The beginning of the examination schedule for
Science was announced for April 19, 1922, and the <u>Daily Standard</u>
reported, "The approach of exams has dampened the activity of the
Radio Club and its apparatus is being dismantled for the summer. The
private sets about the college are still in operation."[1]

Things were progressing in exciting directions by the spring. The
club's equipment was now in a small laboratory in the basement of

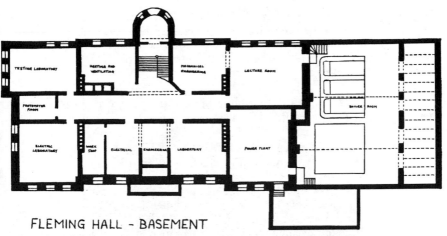

FLEMING HALL - BASEMENT

The original floor plan of Fleming Hall basement, circa 1902. The original 9BT was probably set up in the photometer room at the west end of the basement hall. The space may have been used as a professor's office before the Radio Room was installed there. Some time later, possibly in 1923, the station was moved to the Testing Laboratory in the north-west corner of the basement, and when music was to be broadcast, a wooden frame studio was built out in the hallway and draped with the Drama Club's old curtains.

-from Queen's Calendar

Fleming Hall, next to the new Radio Laboratory of the Department of Electrical Engineering. Plans to do public broadcasting had been formulated, probably in the fall of 1921, and the by the end of the winter the Radio Laboratory equipment was in readiness for the club to do real radio-telephone broadcasting from the campus, just like the big new commercial stations in Toronto, Montreal, New York and Pittsburgh. The Principal's Report, published in May, had even announced the plan officially: "The wireless room which was put in commission this winter will be ready for radiophone transmission by next September, over a 250 mile radius. The Radio Club will send out reports of all football and hockey matches and transmit any of the concerts in Grant Hall."[2] So, it appeared that the Radio Club, without ever having had a licence to transmit code, with few experiencd operators and no experienced announcers, was now going to embark on the transmission of voice and music programs to half the province of Ontario. It would be the first university broadcasting station in the Dominion.

The only extant description of the new wireless room in the basement of Fleming Hall comes from the Calendar of the Science Faculty for the 1923-24 session, prepared for publication in the spring of 1923; that is, a full year after its installation:

> Laboratory No. 6 contains the experimental transmitting station 9 B.T. There is also a receiving set of very flexible design (i.e. the De Forest Inter Panel set). The laboratory in addition to being a Radio laboratory is used for the study of the characteristics of electron tubes

as generators, oscillators and amplifiers. A number of tubes with the necessary variable condensers, reactors, A and B batteries and wavemeters are available. Direct current up to 2500 volts and 0.25 amperes is provided for plate voltages.[3]

That summer The Daily British Whig published two reports about the receiving set "of very flexible design" at Queen's:

RADIO AT QUEEN'S
The University Home-Made Set a Good Receiver

Toronto, Los Angeles, Cal.; Newark, N.J.; Schenectady, N.Y.; Chicago, Ill.; New York city, and many other places were picked up at the radio station at Queen's University on Thursday evening when a representative of The Whig had the opportunity of listening in to concerts in many parts of the United States. The atmosphere was almost devoid of static disturbances and it is expected from now on there will be fewer interruptions to clear listening. The set at the University is home-made, but is one of the best in the city, having a splendid receiving power. Code messages from as far away as Germany have been heard, although nothing has ever been picked up from the Eiffel Tower in Paris. The dance music played by the best orchestras in the United States came through clearly and it is possible that at the science dance next winter several of the numbers will be danced to radio music.[4]

Two weeks later, The Whig printed the second installment on the De Forest receiver in the basement of Fleming Hall:

At Queen's University there is an excellent set which has a great receiving range and as there are few disturbances in the neighbourhood the concerts come in very clearly. Every night there are visitors at the station who hear the music from Newark, Schenectady, Toronto and many other places quite plainly. There is also the possibility that at some of the dances next year a set will be installed in Grant Hall and that the latest fox trots from distant stations played by the foremost American orchestras will be heard through the wireless. By means of amplifiers the sound could be made loud enough to be heard throughout the hall. It would be a most interesting innovation and quite within the range of possibility.[5]

From its opening, July 2, 1922, one could also go to listen in at the Canada Radio Stores on Princess Street, where Orton H. Donnelly, Merle E. Ward and Gordon A. Thompson carried high-grade equipment at reasonable prices. Most nights they could easily pick up as many as 12 stations, including WHAM (Rochester), WJZ (Newark), CFCA (Toronto Star Station), WHAS (Louisville), KDKA (Westinghouse in Pittsburgh), WGY (the General Electric station in Schenectady, the most powerful in the world), WWJ (Detroit) and WOH (Indianapolis).[6] Sometimes reception was broken up by interference from the government-owned, Marconi-operated station, VBH, at Barriefield. And as early as 1921, one might occasionally tune in programs of gramophone records broadcast, illegally, by a few Kingston amateurs.

The radio phone craze peaked in early 1922 and was showing signs of faltering by late summer. A Marconi official in London, England, thought the American boom rather premature. Marconi was cautious and was planning to rent receiving instruments to subscribers if the government lifted the restrictions. The company would not provide transmitters to the public. "We do not want to see a radio epidemic here," said Marconi's Managing Director.[7] In one year, by mid-1922, the number of receiving sets in the United States had leapt from 60,000 to 600,000, and in one week $5,000,000 in new capital had been invested in the radio business. One manufacturer of radio parts reported a 60,000% increase in production per month since 1919, and another, who had sold 40,000 vacuum tubes in all of 1920, sold 200,000 in March 1922.[8] Some 80 stations were already operating in the United States by May 1922, and there were complaints that "the ether has become congested with commercial, political, amusement, religious and amatory communications."[9]

Listeners in could hear grand opera and church services, business reports, theatre parties featuring the latest Broadway stars, symphony concerts, dance music, jokes, news, sports actualities, lectures and almost anything else that the genius of man could conceive of broadcasting. There was an aeroplane wedding broadcast, where listeners in could hear the nuptial kiss.[10] "Bringing Up Father" in the comics had radio references. Someone had found a way to send photographs by radio.[11] Paul Whiteman played a radio concert for his mother's birthday party 2,000 miles away, and it was estimated that there were three million radio guests.[12] Police and taxis were beginning to adopt radio receivers now that they were portable and cheap, and farmers were already relying upon the new service for weather and market reports.[9] The Whig predicted in late August that Kingston would have 200 new radio sets installed by winter, bringing the total to over 500, and "Everybody will soon be 'listening in'."[5] Moreover, if a broadcasting station were established in Kingston, there would quickly be over 1000 sets in the city.

There were predictions of doom too. All the radio waves clogging the ether had some scientists worried that strange and powerful atmospheric forces might be aroused and unleashed.[13] In the United States, millions were listening in to church services by wireless, and there were fears that church attendance would drop calamitously, with consequent deterioration of the general moral fibre. A fellow in New Jersey "complained to the police that he had a hat full of radio waves and did not know what to do with them." A veritable radio bug! People were also starting to complain of hearing radio messages in their heads when local stations were sending, probably through the fillings in their teeth.[14]

By mid-summer the novelty seemed to be wearing off, and the market for radio was in a slump. The manufacturers thought that the slackening was due only to the summer doldrums, compounded by the relatively poor reception characteristic of that season,[15] and they were right. Radio took off again in the fall and kept right on going.

The Beginnings of 9BT

Before the start of the Queen's Radio Oral History Project in mid-1982, only a very small stock of facts had been assembled about the facilities and activities of the first licenced Queen's radio station, 9BT. The plan for the first transmitter is lost. An intensive investigation was made to try to discover any facts that could be checked against the meagre published record and against the few memories of this period gleaned from the oral histories recorded for the 60th anniversary radio series. During six years of research, many lost facts about 9BT were unearthed, as well as documentation establishing the date and content of the first licenced public broadcast from the Queen's campus. A good many more particulars are now known about 9BT, but a lot of vital information has not been discovered. The origins of the idea to venture into public broadcasting from the Queen's campus are still unclear, and the events of the spring and summer of 1922, leading up to the first broadcasts the new radio station, remain obscure.

Some very basic information is contained in the only almost contemporary sketch of the beginning of 9BT, published early in 1924. The station "occupies one of the Electrical Laboratories in the basement of Fleming Hall, the room being entirely fitted for and devoted to radio work. During the summer of 1922, Prof. Jemmett collected sufficient apparatus for an experimental set and also secured an Experimental Radio Licence, 9BT, which allowed the use of wavelengths 175, 275 and 1050 metres, but for experimental purposes only. This set was assembled in September, Electrical students doing much of the work, and was put into operation during October."[16]

Written accounts of the beginnings of 9BT were left by two original participants, Douglas Jemmett and Douglas Geiger, but these were composed years after the events and are not as rich in detail as could be wished. Jemmett read the following statement in a fully scripted radio interview over CFRC on October 27, 1962, as part of the celebration of forty years of public broadcasting by 9BT and CFRC:

> A group of us decided in the spring of 1922 to investigate the possibilities of radio broadcasting. R.L. Davis was acquainted with some of the work that had been done at KDKA in Pittsburgh. He drew up circuits and plans. Professor Hughes, in Physics, made two rectifier tubes for a 60 cycle, single phase, double wave rectifier. I designed and had built a three section filter to smooth the output wave of the rectifier - and a large tuning coil for coupling the set to the antenna. Two final year Electrical Engineering students worked with us - and we

worked for long hours.

The Radio Branch at Ottawa issued us an experimental licence and assigned the call letters 9BT. By the fall we were ready to try a broadcast. A number of friends and members of the Queen's Radio Club tuned their receiving sets to our signal. The first sounds of Queen's University on the air were of poor quality compared to present day standards but they were very exciting for us then. We did a lot of experimenting that session - enough to convince us that we would try regular broadcasting the following session. One thing that helped to convince us was a grant of money from Dr. W.R. Jaffrey of Hamilton - we bought a motor generator with that.[17]

In another CFRC production, called "Personality" and broadcast that same day, Jemmett was again interviewed on the origins of radio at Queen's, and revealed a few more facts:

Well, it was purely experimental, that first set-up. We worked through the summer of 1922, making most of the station equipment, although we did buy a few parts. I designed a three-section filter to smooth the output wave of the rectifier, which Professor Hughes of the Physics Department had built. A number of vacuum tubes, ranging in size from the smallest up to those rated at 125 watts plate dissipation and 2,000 to 3,000 plate volts, were obtained from the Mullard Wireless Company of London, England. We erected a cage aerial between masts mounted on the roofs of Fleming and Ontario Halls, and made a large tuning coil for coupling the set to the antenna. The set was connected up and we did a lot of experimenting. (It was) successful enough that we were sure that with the addition of some pieces of equipment and a few meters, particularly a 3,000 range voltmeter, very good results might be obtained. (The first equipment) didn't look like very much at all. It wasn't fancy according to present day standards. No pale blue cases with chrome trim. It sat on a table in a corner of a downstairs lab. We were working on improving it all the time. In view of those first experimental broadcasts, we decided to attempt regular broadcasting during the following session...Quality didn't enter into the picture the way it does to-day. We didn't have equipment to bring out the quality. That was true of the whole process, from microphone to receiver.[18]

Point form notes exist from an earlier Jemmett interview, in 1952 for the program "Meet Your Prof." Here he revealed that the radio station was to be a second term course for fourth year Electricals, a voluntary session for two hours a week. They thought that a radio station would be good laboratory experience for the students, and decided to build a transmitter. The transmitter was in the basement of Fleming

Hall, scattered about on a table. It had a power output of approximately one-quarter kilowatt and a range of 100 miles. In "October of 1922, 9BT broadcast a football game, years ahead of any athletic event in the country."[19]

Much of the information in Jemmett's scripts, indeed much of the phraseology, comes from an article by Douglas George Geiger (Elec. '22; Mech. '23) published in the Queen's Review in 1928.[20] Geiger's is the most comprehensive account existing of the early days of radio broadcasting at Queen's. Ten years later, Harold Stewart updated Geiger's article on CFRC, adding some new facts which he obtained from Geiger.[21] What follows is an amalgam of the two articles, spiced with comments made by Prof. Stewart during a 1982 oral history taping:

In the spring of 1922, as an outgrowth of the activities of the then existing Radio Club of Queen's University, Prof. D.M. Jemmett, Head of the Department of Electrical Engineering, decided to investigate the possibilities of radio broadcasting. Jemmett had brought in Mr R.L. Davis from M.I.T., where they both had been pursuing graduate studies. Davis was a specialist on communication and Jemmett brought him up to Queen's to be Assistant in the Department and to develop that phase of their curriculum. Davis, who was acquainted with some of the work that had been done at KDKA in Pittsburgh and who later joined the engineering staff there, drew up circuits and plans for an experimental transmitter for Queen's. From the Radio Branch at Ottawa an experimental licence was secured authorizing transmission of spark on 175 metres and continuous waves on 275 and 1050 metres, and assigning the call letters 9BT. During the summer of 1922 the required equipment was drawn together and the first station assembled. Dr Arthur Ll. Hughes, Chown Research Professor in the Physics Department, made the first two rectifier tubes, which he rated at 2000 volts and 125 watts each. It was a single-phase double-wave rectifier operating from 60-cycle city power. Prof. Jemmett designed and had built in the machine shop of the University the power supply filter, a three-section filter to smooth the output wave of the rectifier. He also had constructed the antenna-coupling coil, a large tuning coil for coupling the set to the antenna. The transmitter tubes, ranging in size from the smallest up to those rated at 125 watts plate dissipation and 3,000 plate volts, were obtained from the Mullard Wireless Company of London, England. All had tungsten filaments and required considerable filament power. The transmitter was of the modulated-oscillator type and the aerial was a cage slung between masts mounted on the roofs of Fleming and Ontario Halls.

During the session of 1922-23 the set was connected up and a very considerable amount of experimenting was carried out. Two final year Electrical students, Sidney Victor Perry and Patrick Harold McAuley, acting under the direction of Prof. Jemmett, did a great deal of the work. The first program went on the air from the basement of Fleming Hall when Mr George Mitchell Parsons (Elec. '23), gave a cornet solo. He used a microphone of the telephone type, which was supposed to have a flat-frequency characteristic over a "very wide range." While granting that it would not live up to a modern (1938) interpretation of this term, it probably had better characteristics than the receivers used to listen to it. The experiments showed that with the addition of some pieces of equipment and a few meters, particularly a 3,000 range voltmeter, very good results might be expected. Also, in view of the results obtained, it was decided to attempt regular broadcasting during the following session and to have one of the staff of the Electrical Engineering Department engineer and operate the station.

It is interesting to note that people in general were very skeptical about radio telephony in those days, and in spite of the previous activities of KDKA and others many felt that the whole thing was an out-and-out fake, with a concealed phonograph under the table.[21]

Professor Arthur Llewelyn Hughes' Role

Dr A.Ll. Hughes was the first Chown Research Professor at Queen's, appointed in 1919. There were two candidates for that position, Hughes and Arthur H. Compton. Hughes was selected because it was thought that, as a British subject, he was more likely to remain in Canada than the American Compton. Hughes had worked in the laboratory of Charles Barkla, ultimately a Nobel laureate for his work in X-rays, and then with J.J. Thomson at Cambridge. He came to Queen's from the Rice Institute in Texas, and worked on photoelectric phenomena, experiments involving blown glass tubes containing metal electrodes. Hughes was expert at glassblowing and at making glass-to-metal seals, and it was only natural that Doug Jemmett ask him if he could make the rectifier tubes for 9BT.[22] Dr Donald C. Rose was a lab assistant for Hughes during the summer vacation of his third year in Engineering Physics, the summer of 1922.

Donald C. Rose: He was doing thermoelectricity work or conduction of electricity in gases, and most of the work was high-vacuum work. That was my first experience in learning how to do high-vacuum, a bit of glass blowing and things like that. I remember hearing about those rectifier tubes. I think I remember once he told me about them, but I don't remember ever seeing them. I would say that they might have been fairly long, but wouldn't have to be more than two or three inches in diameter, which is about the maximum size of glass blowing we could do then. I don't know when he did them...but it was not while I was working with him and I had no part in it. It was probably in 1921 or early 1922 that he did it.[23]

Don MacClement: Now that side of Physics was very early in the whole business of X-ray tubes, and I gather that Hughes was doing quite a lot of experiments. I know one of his assistants got badly burned or whatever by the radiation from the tubes they were using. I imagine they didn't consider any shielding necessary. But they had the equipment for making those tubes, because they made their own as far as I know. That's the first time I ever saw a person working with and shaping large glassware on a turning lathe and a flame. That was fascinating. And it was over to their lab that I used to take my small receiving tubes, UV 200's and 201's, and also the old UV 199's that had to be pumped out harder - a higher vacuum so that they'd stand a heavier plate voltage. At that time they were working on radio tubes, as I remember them, big bulbous ones. The reason that I had gone over there was that Jemmett had referred me to Hughes, and my contact was right after the First War.[24]

The Other Equipment of 9BT

Prof. Jemmett's requisition books, from June 1922 through the 1940s, were preserved by Harold Stewart. These books contain the carbon copy records of the particulars of his purchases for the Radio Laboratory, as well as the contracting of additional construction work and important incidentals. The first requisition was dated June 7, 1922, when Jemmett ordered three 0/250, two 0/20 and two 0/30 transmitting valves from Mullard, as well as two Type K.A. and two Type R receiving valves plus six valve bases. Jemmett was likely just beginning to order the parts at that time. The total cost of the Mullard order was £34-14-10.

The 0/250 transmitting valve was a baseless spherical bulb with vertical glass arms, top and bottom. The bulb itself was 5 inches in diameter and 11 inches long overall, including the arms, and had wires protruding from the arms. Its anode dissipation was 250 watts, anode voltage 1500-2000 volts, filament voltage 10.5, filament current "A" 3.6 amps and filament current "B" 4.7 amps. The 0/20 and 0/30 valves were 2.25 inch diameter spherical bulbs with 4-prong bases, and were only 4.25 inches tall, including the pins. Their specifications were slightly different, the anode dissipation of the 0/20 being 20 watts and of the 0/30 being 30 watts. The anode volatges were, respectively, 200-400 and 600-1200 volts, the filament voltages 5.5 and 6.0 volts, and the filament currents 1.5 and 1.2 amps. The Type R receiving valve looked the same as the 0/20 and 0/30, a French or 4-prong type. It was "specially designed to give robust service with good amplification and can also be used for detection...Best operation is obtained with about 4 volts on the filament and 60-80 volts between anode and filament." The K.A. receiving valve looked like a conventional cylindrical 4-prong radio tube, one inch in diameter and 3.75 inches tall: "The Type K is a very compact valve and is recommended where especially good amplification is required, being particularly suitable for use in high frequency amplifiers. It requires about 3.5 volts on the filament and only 20-30 volts between the anode and filament."[25]

Extrapolating from the design of the 1927 transmitter, also a modulated oscillator type, one may conclude that the receiving Type R tubes were to be used as microphone amplifiers in the 9BT transmitter.

Jemmett ordered a Brown loud speaker, 2,000 ohms resistance with horn, and a pair of Brown headphones, 8,000 ohms, on July 6 through Mr F.W. Pye, the instrument maker in Ontario Hall. Requisition No. 2736 states "Accepted cheque or Money Order payable at par in Ottawa to 'Department of Naval Service' for $5.00 to cover Radio Experimental Licence fee to Mar 31-1923." That was to cover the cost of the first application for a radio licence, which turned out to be experimental licence 9BT. On the 19th, Jemmett filled out a requisition for small parts - condensers, honeycomb coils, grid leaks and transformers - from the Manhattan Electrical Supply Company. The first microphones, a Western Electric 284-W and a 323-W, then known as "transmitters," were ordered from Northern Electric in Montreal on August 17, along with #22 and #26 magnet wire. An order was put in to the General Radio Company of Cambridge, Mass., on August 28, for a Type 174 direct reading wavemeter from page 974 of the General Radio catalogue. James Bews, in charge of buildings and grounds, was sent a work requisition dated August 31 stating, "An order to Mr. Bews for labor in erecting wireless poles on Fleming and Ontario Halls, stringing aerial and installing a ground." Sixteen Burgess No. 2156 B.P. 22.5 volt "B" batteries were orderd from the Canada Radio Stores, Kingston, on September 15, and those are all of the requisitions that survive from 1922.

Now, with some idea of what transmitter equipment Jemmett purchased in 1922, it is instructive to compare these known bits and their specifications with the components shown on the surviving blueprint of the 1927 CFRC transmitter. While it is not possible to draw hard conclusions about the vanished plan of the 1922 transmitter, the 1922 tubes, at least, appear to be earlier versions of the tubes in the 1927 set. Therefore, it can be conjectured that the fourth transmitter of 1927 was an up-dated version of the original modulated oscillator 9BT transmitter, but with modern tubes and a better power supply. (see blueprint in Chapter 10)

The Experimental 9-Call

The Radiotelegraph Regulations of 1922 laid out the conditions for obtaining a licence wth an experimental 9-call:

> Experimental licences will be granted to stations intended for purely experimental purposes and operated with a view to the advancement of the art of radio. Applicants for such licences must state their technical attainments and the general lines on which they propose to pursue their investigations. It should be observed that the fact that the applicant desires to conduct experiments with his equipment frequently does not justify or require a licence of this class, as most experiments can be conducted within the limitations of an "Amateur Experimental Licence" or by the use of an artificial aerial.[26]

Applicants for an experimental licence had to specify the wavelength(s) that they desired to use, the normal ones being 175 metres on spark and 275 metres on continuous wave (CW) or radiophone. The station had to be worked by a person holding at least an Amateur Experimental Certificate of Proficiency if operating below 275 metres, and by a person holding a First Class, Second Class or Experimental Certificate of Proficiency if operating above 275 metres. Normally the power used, measured at the terminals of the transformer or generator, was limited to half a kilowatt. The signal was to be damped as little as possible, and "a distinctive call signal will be allotted to each station, commencing with the figure 9, e.g. 9AA, etc. This signal is to be transmitted twice at the termination of every transmission." There were other more detailed regulations which applied to experimental stations.

Thus, someone with a higher certificate of proficiency had to be in charge of the new experimental radio station at Queen's. A few members of the Radio Club may have had their own amateur broadcasting licences, but the only students likely to have held higher certificates would have been those who had been Marconi operators. E.A. (Jack) Filmer had a British Marconi Certificate, so he would have been qualified to operate 9BT. There is no evidence that any of the other active Club members whose names are known, including Prof. Jemmett, had the proper operator's qualifications.

In 1922, the only other universities in Canada to hold 9-calls were Acadia University in Wolfville, NS (9AT), and the Physics Department of the University of Alberta at Edmonton (9AU).[27] Other Canadian universities had broadcasting stations with different classes of licence, such as the University of New Brunswick, which held 1DO.[28] Incidentally, there was another 9BT, held by an amateur in the Chicago area, District 9 of the United States, and this made for some confusion.

George Parsons and the First Broadcast

George Parsons was uncertain, after 60 years, whether his broadcast over 9BT took place in February or March 1922, or in October of that year:

G.M. Parsons

It could have been October or it could have been March. My memory doesn't go back clearly that far. February or March was my impression. The only thing that mattered in those days was my classwork. So far as I recall it, it was Prof. Jemmett who asked me. Why? Well, simply because I was the bandmaster of the college band there. We had organized a band in Queen's University that was the first college band in Canada. We weren't very good but we had a lot of fun. And we had no money. We had no uniforms. We just simply were a group of about 18 or 20 fellows (including some veteran members of the 14th PWOR band) who had had some band experience. I was asked to go over to the Fleming Hall, which was the headquarters for E.E., and play my trumpet - cornet as we called it in those days - on the Queen's radio station. (Question: Did they ask you because the cornet was particularly loud or clear?) I don't think they were discriminating that way. They were looking for somebody that would put on some sort of a number. Well, I wasn't any great shakes of a player, but I acceded and went over. As I say, I was working very hard then on classwork, and as I recall I went down from my room - I used to room up on Princess Street - and went down there and I wasn't away from my room more than an hour altogether. There was no studio - no formal studio. I played in the basement of Fleming Hall where some of the radio equipment was set up. My recollection is that I played entirely alone, and I played "The Bluebells of Scotland" with my trusty cornet over the air at that time. My understanding was that it was the very first time that they broadcast. Now, my memory is very foggy. I don't remember very much of who was there in the studio or what else went on, but that was the one and only time that I played on the radio.[29]

The newspapers did not pick up the story of George Parsons' pioneering solo flight through the ether. The only thing in print linking Parsons to the Radio Laboratory was a note in the Queen's Journal after the November 30th Medical At-Home, under "Things We'd Like to Know": "If disclosing the mysteries of the wireless room to the fair sex at 11 p.m. was more fascinating than listening to Jardine's Orchestra. Ask G. P-rs-ns."[30]

Evidence That 9BT Was On the Air

The first public indication that something was about to happen in the radio line at Queen's was a squib in The Daily British Whig on September 11, 1922: "Queen's is to have a broadcasting outfit for its radio service. Kingston news will be distributed."[31] The 82nd session of Queen's began on September 27, with registration and classes.[32] The first issue of the Queen's Journal did not appear until October 10, so only the Kingston dailies were able to publish the press release about the first scheduled broadcast of 9BT.

The first scheduled broadcast over 9BT was not of a football game, as Prof. Jemmett recalled.[18] It was a broadcast of the summary of a football game, and not of a regular league game either. This made it difficult to track down in the papers, because it was before the opening of the season. It was the exhibition game of Saturday, October 7, 1922, against the Hamilton Rowing Club of the Senior ORFU. The notices of the broadcast appeared in the city papers on the very day it was to take place. The Daily Standard gave it more space than The British Whig:[33]

WILL BROADCAST FOOTBALL STORY
Queen's Radio Station Will Send Out Story Tonight at
8.30 - First Attempt in City

Kingston will be placed on the Radio broadcasting map tonight, when Professor Jemmett and his staff of the Queen's Broadcasting station will send out the story of the Hamilton Rowing Club - Queen's football game of this afternoon.

Last night the apparatus was tried out and it is believed that it was a success, and that the messages reached a radius of from 150 to 200 miles. The wave length of the station is 200 metres.

The Queen's broadcasting station is the only one in the city and one of the few in Eastern Ontario.

The British Whig was the only one to publish the call letters, but they called the station "Nine BP." It may be argued that George Parsons' broadcast occurred on the night of October 6 as an unscheduled transmission in the testing of the apparatus, but there is no proof possible now.

Both newspaper accounts of the great football summary broadcast deserve to be quoted in full. Neither paper mentioned in its coverage of the broadcast that Queen's had defeated Hamilton by 25 to 1. This time the Standard had the shorter report:

> The broadcasting station at Queen's University was used on Saturday evening for the broadcasting of news in connection with the football match, and this, its initial try out, is reported as having been very successful. The broadcasting station has not as yet attempted anything in the way of amplification, but in the very near future amplifiers are to be installed when a far greater radius of usefulness will be the result. Prof. Jemmett, when spoken to in this connection by The Standard today, forecasted great things for Queen's in radio activities.[34]

> The first complete broadcasting programme of the Queen's University radio station was sent out on Saturday night, and, although the fog and weather conditions made receiving difficult from all stations, the modulation of the set was excellent. Radio receivers stated that everything came through very clearly and they are looking forward to listening in on more of Queen's output.
>
> George P. Awrey, athletic director, had the honour of broadcasting and gave a complete summary of the football game between Queen's and Hamilton at the George Richardson stadium on Saturday afternoon. The broadcasting set has a power of one-half kilowatt which is one half the power of the Pittsburgh station frequently heard in Kingston. The whole power, according to listeners, was not used on Saturday night and even better results are in prospect for the winter season.
>
> The set has a wave length of 200 metres and when it is working at full power will have a radius of from 150 to 200 miles. It is believed that Saturday night's programme was heard within a radius of over 125 miles.[35]

The newspaper reports of the broadcast were not entirely accurate. Coach Awrey was away in Toronto on October 7, scouting the McGill-Varsity game for Queen's[36] for their impending intercollegiate clash on October 28 in Toronto. Neither did he return to Kingston that night to take part in the broadcast. George Awrey went home to Hamilton after the game in Toronto,[37] and didn't return to Kingston until Wednesday.[38] The Captain, John L. (Red) McKelvey, handled the Queen's Tricolor that day,[39] and probably did the radio broadcast from Fleming Hall in the evening. The simplest explanation is that the reporters were told that the broadcast was done by the coach of the game, and they assumed that it was George Awrey.

Incidentally, Queen's Tricolor was on its way to the Dominion Championship in 1922, and defeated Toronto for its second consecutive win on the 28th. The team stayed at the King Edward Hotel: "Hal Salton on arriving at Union Depot, Toronto, was standing waiting for the gang, when a porter stepped up and said, 'King Edward, Sir?' 'No! Just Hal Salton of Queen's'."[40]

The new radio station was welcomed by the new radio equipment supplier in town, the Canada Radio Stores, in advertisements in the Queen's Journal on October 13 and 17:

THANKS TO QUEEN'S

Kingston now has an up-to-date Radio Broadcasting Station. To hear this costs little. You are sure of getting the best at Kingston's ONLY EXCLUSIVELY RADIO Supply House.

CANADA RADIO STORES
269 1/2 Princess St. Phone 1207j [41]

One of the owners of the Canada Radio Stores, Orton Donnelly, was also the District #3 (Eastern Ontario) correspondent, Ontario Division, to "The Operating Department" section of the amateur radio magazine QST, the official organ of the American Radio Relay League (ARRL). Through the late autumn of 1922, Donnelly reported to QST on the progress of 9BT, although from the viewpoint of an amateur interested only in Morse relay work. In October: "...report Queen's University has lots of power available there (C.W.) if they can be prevailed upon to use it for relay work,"[42] and in December: "9BT is going now, but mostly phone broadcast."[43] Donnelly's report in the December issue is probably of the October 7 football summary broadcast.

In April 1923, 9BT (Canadian) was reported in the "Calls Heard" section of QST,[44] as having been picked up by station 6BEK, Puente, California! 6BEK belonged that year to Harry R. Green, 1814 South Vermont, Los Angeles, California. "Calls Heard" listed only the call and did not indicate date of reception or contents of the broadcast. Correspondents to "Calls Heard" usually distinguished among spark, CW and "fone" (radio telephone or AM). Green indicated that all stations he picked up were CW, which may indicate that 9BT was then broadcasting in code rather than by voice. Supposing that it took four months from the date of reception by 6BEK for the report to pass through the hands of the Pacific District Manager and be forwarded to QST, Green could have heard 9BT as late as December 1922. None of the old Kingston hams contacted in this study remembered hearing the broadcasts of 9BT or noted its reception in their logs. Harold Stewart, for one, was interested only in DXing; that is, in receiving distant broadcasts and getting confirmation through QSL cards.

Jemmett maintained, in a fragment of a radio script from 1957,[19] that 9BT broadcast a Queen's-Varsity football game from Richardson Stadium in October 1922. Varsity's only appearance in Kingston that year was on November 11, although the Toronto Argonauts played an exhibition game at Queen's on November 6. There are, however, no reports in the Kingston papers of a radio broadcast from Queen's 9BT after October 7. The Canada Radio Stores would have continued plugging the new Queen's service had it continued, but they remained silent on the matter too.

Jemmett's radio boys must have felt encouraged by their initial broadcasting successes. There is a piece of oblique evidence from mid-October 1922, suggesting that they were emboldened enough to contemplate applying for a commercial broadcasting licence the next year. Oblique evidence, because it is just a squib in the Queen's Journal, and reflects sophomoric contrariness as much as anything: "Friend Jack F., who is eminent in wireless circles wishes us to broadcast the information that he is authorized to hire about twenty men to act as linemen for the P.D.Q. wireless station duties to commence at the closing of the spring term."[45] Taken at face value, it might mean that Jemmett, Filmer and the radio boys had plans to string telephone lines from Richardson Stadium, Grant Hall, the rink and the gym to the radio transmitter in Fleming Hall, so that they could try experimental remote broadcasting of sports, music and lectures.

The foregoing few references are the only reliable evidence of 9BT actually going on the air, except as a piece of apparatus in Jemmett's research on radiation from parallel wires. 9BT may not have undertaken many public broadcasts in the 1922-23 session, because there were serious problems with the quality of the transmitted signal:

> ...it did not work well because the wave-length, 275 metres, was too near the natural period of the aerial, and also because of a disagreeable hum from the A-C voltage supply which could be heard in receiving sets. As no one in the Electrical Department had sufficient time to devote to improving the operation of the set, no experimenting was done.[16]

9BT's modulation was likely very poor as well and, combined with the primitive microphones that they had, caused the speaking voices to sound thin and buzzy.

Two Anomalous Recollections of Early Broadcasts

Among the stories collected for the Oral History Project in 1982 were two interviews which placed the beginning of football broadcasting from Queen's before 1922. These two anomalous recollections have now been checked against the available facts, and neither story seems to be supported. They should not, however, be totally discounted. Bill Taylor says that he heard a Queen's-Varsity rugby game over Queen's Radio in 1921. This would have had to be the game of October 8,

1921, the inaugural game played at the new Richardson Stadium.[46] He recalled that he was living at Mrs English's boarding house that year and that the quality of the broadcast was very professional:

> **William Taylor**: 1921 is the year I had the little hand-made radio set made with a Northern Electric peanut tube - a one-tube receiver, and I remember distinctly listening to the broadcast from the Richardson Stadium that year, the game between Toronto University and Queen's. And it was a good transmission. I copied it very well. I had no antenna, but just a short piece of wire in the room, and the sound came in very well, in a house at the corner of Johnson and University Avenue. I remember that very distinctly, because I had regretted that I had missed seeing the game. I was confined to stay in my room. I had a bad cold and I had to do some studying, so I listened to the game until it was almost over, and some friends of mine bust in just before the game was over and sort of spoiled the end of it.

The record shows that Bill lived at Mrs English's at 336 Johnson Street in 1923-24.[47]

Mr W.J. Hurn,[48] who lived in Kingston as a boy, recalled hearing a Queen's football broadcast the year before he left Kingston permanently for Timiskaming, a year before the great Haileybury fire of October 1922.[49] His friend, Harry Coventry, had bought a crystal for fifteen cents and with some brass tacks built a receiver in a de Milo cigar box, and they listened to the Queen's games. The kids also used to climb the trees on Union Street to watch the rugby games for free. Hurn still remembered the Queen's yell and recalled that after a Queen's victory, hundreds of students used to snake dance down Princess Street[50] and "rush" the theatres. The boys of the town swiped baskets of fruit from the stalls of the local merchants during the confusion, and the Queen's Alma Mater Society had to pay the damages. He also remembered Alfie Pierce at the games, which would place his memories at least into the fall of 1922, when Alfie Pierce returned to Queen's as mascot.[51] There is additional evidence that Mr Hurn cannot have left Kingston the year before the Haileybury fire, for he remembered that Queen's Radio had the call letters CFRC, standing for "Canada's Famous Rugby Club." The call CFRC was not granted to Queen's until the spring of 1923.

A Wireless Broadcast of a Sort from Kingston in 1921

The inaugural game in Richardson Stadium may not have been broadcast from the campus over Queen's Radio, but reports of that game's progress were sent out by wireless. One of the Associate Editors of the Toronto Varsity newspaper, Eric Druce, was in Kingston for the game and arranged for the government station at Barriefield, VBH, to transmit the score at the end of each quarter. The code messages were picked up in Toronto by Gordon W. McClain of 342 Brunswick Avenue (who held the licence for 3GE) and by VBG, the government wireless station on Centre Island. The Varsity kept in telephone contact with both of these stations and posted the notices as quickly as possible on a bulletin board in Hart House.[52] It turned out to be the first time in 13 years that Queen's had defeated Toronto. The <u>Varsity</u> was going to get fuller reports from the Montreal game the next week, but did not boast of a second wireless success.

The Contribution of Robert Leland Davis

In the Queen's record, R.L. Davis' role in the birth of Queen's Radio was summarized by Prof. Jemmett in a few lines in one of the 1962 anniversary broadcasts,[17] as well as in a fragment of a 1957 radio script where he said of Davis, "He was a big help in the preliminary work of constructing the station at Queen's." Douglas Geiger[20] and Harold Stewart[21] also credited Davis with designing the original 9BT transmitter. What was not clear was whether Davis helped in the actual building of the transmitter.

A few of the alumni interviewed in the Oral History Project remembered R.L. Davis as a teacher in Science at Queen's. None recalled his radio work, his personality or his Christian names. One fact proved to be the key to finding out about him - that Davis had been brought to Queen's from the Massachusetts Institute of Technology (MIT) by Jemmett when he returned to head the Electrical Department in the fall of 1920. The Reference Archivist at MIT was able to supply Davis' Christian names and date of graduation, but had little other information, except that Davis' last known address, in 1975, was the same as that given when he entered MIT in 1919. A telephone call to the operator in Grand Rapids, Michigan, revealed that there was still a Davis listed at the old address. This turned out to be Henry B.

Robert Leland Davis, who was brought to Queen's from MIT as a lecturer in E.E. by his friend Douglas Mill Jemmett in the fall of 1920. Davis had been a wireless amateur since around 1913, and his expertise may have been the inspiration for the installation of a Radio Room in Fleming Hall in the winter of 1921-22. In the summer of 1922, Davis worked in Dr Frank Conrad's radio laboratory at Westinghouse station KDKA, Pittsburgh, and never returned to Queen's.
-courtesy Henry B. Davis

Davis, a nephew, and the executor of his uncle's estate. Henry Davis has most of R.L.'s papers and letters, though at the time of the initial contact they were still unsorted in hundreds of boxes.

A good deal of Davis' background has now been clarified, and a little more can be deduced about his contribution to the beginnings of public radio broadcasting from the Queen's campus.

Robert Leland Davis was born in Grand Rapids, Michigan, May 4, 1894,[53] the son of George A. and Alice Glover Barnard Davis, both from New England colonial stock. George A. Davis settled in Grand Rapids during the furniture boom in the 1880's and co-founded Stow-Davis, a manufacturer of office furniture. Robert graduated from Central High School in 1912, specializing in science, and entered Oberlin College.

According to Henry B. Davis, Robert began building radios in high school or while he was attending Oberlin, around 1912-14. Henry Davis still has some components of these spark sets: coils, capacitors, spark gaps and other bits.

Robert transferred to the University of Michigan Engineering School, Ann Arbor, in 1914 and kept up his interest in radio while at U. of M., although there is now no way to determine whether he was part of their radio station, 8XA. This was an experimental (denoted by the leter "X") private or limited commercial service with no special hours, and it operated at 450, 600 and 1,800 metres.[54] In 1916, the magazine QST cited the U. of M. Ann Arbor station, which it called 8XN, for its "striking efficiency."[55] After graduating from E.E. in 1917, instead of going into graduate school as he had planned, Robert was taken up by the Cutler-Hammer Company in Milwaukee to work with the US Navy devising electrical controls for submarines as part of the war effort. After the War, in 1919, he enrolled in the Master of Science program in E.E. at MIT, and it was here that he met Doug Jemmett. According to Maud Jemmett,[56] Davis and Jemmett lived in the same boarding house in Cambridge. Davis' thesis, submitted to MIT in 1920, was not in the radio field at all. It was titled "A Comparison of the Diesel Electric System of Ship Propulsion with the Direct Drive by Steam and Diesel Engines for Trawlers."[57]

There was opportunity for both Davis and Jemmett to be involved in wireless at MIT, because it had a 500-watt amateur radio station in 1919-20, with the call 1AN.[58] No evidence survives that they were members of the MIT Wireless Society. Davis seems never to have gone as far as to obtain a broadcasting licence for himself, either in the US or in Canada. He is not listed as a licence holder in Radio Stations of the United States in the 1913-16 or in the 1920-25 editions in districts 1 or 2.

When Jemmett was hurriedly summoned back to Queen's in the fall of 1920, because all of the faculty in the Electrical Department had left,[59] he invited Davis to come back with him to take up a position in his department. Davis confirmed the story in a general way in a letter to Prof. Arthur Jackson:

> In 1920, while at M.I.T., I became associated with
> Professor D.M. Jemmett of Queen's who was there doing
> graduate work before becoming Head of Queen's E.E.
> Department. In September, 1920, he asked me to come to
> Queen's as a lecturer in E.E. I stayed at Queen's a sec-
> ond year when they appointed me as Assistant Professor.[60]

On the evidence contained in Davis' letters, they knew each other
before the Christmas holidays in 1919 and may have been classmates at
MIT. When Davis wrote home to his mother on returning to Cambridge
from Michigan in early January 1920, he told her, "I went over to see
Jemmett this evening a few minutes. He has been here all this week."[61]
Davis wrote to his mother on September 1, 1920, asking her advice:

> Jemmett has recommended me for a job up at Queen's, as
> one of his assistants with a pay of not less than 1500 and
> more likely 1600 or 1700 for 7 months work. He has
> not heard definitely which, but not less than 1500. Would
> you consider it advisable for me to consider such a job?
> I would probably have to go directly there from here and
> would not be able to get home before Xmas. It would be
> better in every way except the name of it than taking
> Jackson's job at Tech. It looks pretty good in some ways...I
> saw Jemmett last night and he said that the dean at
> Queen's wanted an answer by Sat of this week at latest.[62]

He wrote to an aunt on the 14th, saying that he was "in a terrible
mess trying to figure out what to do." The job offered by Jackson was
only a "dish washing job" paying just $1200 for nine months. "However
the school (Queen's) does not have the reputation and is in a sort of out
of the way corner. Also the term begins in about ten days...Neither one
would really be much in line toward being a consulting engineer, and you
know Parker said I ought to be in engineering before I started to teach
if I ever wanted to be much at it." He was also considering applying to
a ship engine firm in Connecticut or returning to Michigan to use the
family home as a base for consulting work.[63]
Jemmett departed MIT for Kingston on the evening of September 15,
leaving Davis in a perplexed state,[64] but cabled on the 20th offering
$1500 a year.[65] In the meantime Davis had turned down Jackson's job
and a prospect at Hartford did not look good, so he jumped at the
Queen's offer and cabled his reply that night. He expected to reach
Kingston at noon on the 24th, and classes began on the 29th:

> I'll probably get more out of teaching at Queen's than I
> would at Tech and learn a lot more than I would from
> teaching just one "canned" course, but I'll probably have
> to study quite a bit to keep a head of the game. Of course
> the reputation of Queen's isn't up to that of Tech but I
> should worry and I'm blamed glad to have something to
> do...I don't know very well yet what I'll have to teach
> but I'll take a chance on it...Funny Jemmett telegraphed

as I had turned him down only Wed. and hadn't a chance
to let him know.[66]

Davis was appointed Lecturer in Mechanical and Electrical
Engineering at Queen's in the fall of 1920.[67] Jemmett found his friend
a room in a clean, quiet house at 5 Wellington Street, an eight minute
walk to campus across Macdonald Park. Davis was made responsible
for an elementary course in electrical theory, for one on DC machinery,
for a third in electrical apparatus for non-Electricals and for the corresponding laboratories. He thought that he might have to teach a telephone course after Christmas - Gill's old Course XI, "Telegraphy and
Telephony." Davis was worried that it would be a lot of work keeping
ahead of the class - he and Jemmett had been left no course outlines to
follow in preparing the work.

Davis wrote his mother his impressions of his new home. Queen's
had a little over 2,000 students and was

> a full University with Lit, Medicine, Divinity and
> Engineering Schools. The Engineering School has between
> 400 and 500 in it...There are about a dozen buildings in
> all (no dorms)...There are 3 teaching mech. engr. Jemmett
> and I in elec. and a couple of men who teach drawing
> etc....Besides doing the teaching the E.E. dept is expected
> to look after and do any electrical engineering the univ.
> may need done. At present they are figuring on a new
> generator set in the power house and Jemmett and I are
> suppose to make recommendations on what to put
> in...Kingston is a city about as large as Muskeegan...Most
> of the houses are older than in G(rand) R(apids) but kept
> up better than in Boston and Cambridge so the place
> does not look as ancient. Practically all are brick and
> stone. There are lots of trees and parks...There are three
> street car lines in the city but the cars don't run on Sunday.
> 5c fares also, however I am near enough to everything, not
> to need to ride.[68]
>
> They don't heat the (university) buildings at night
> or on Sundays, but if I can't get it one way I'll get anoth-
> er so I rigged up a big electric heater out of a lab rheostat
> so now I have plenty of it any time, steam or no steam.[69]

Davis gained interesting impressions of and insights into his new
home and relayed them to Grand Rapids in his Sunday letters. In late
1920, a group of students calling themselves the "Bolsheviki" issued a
manifesto calling for the revival of the old customs which nurtured the
Queen's spirit, such as the annual parade, the rush, the old form of
election and dramatic club nights at the Grand Theatre.[70] The Bolsheviki
had some effect:

> Friday night the students put on a play and I went down
> to see that. It was the night before the campus elections
> and between the acts the students had a general rough

house in the theatre and more or less filled the pit with bales of waste paper, corn, beans, etc., which they through down from the gallery. Last night there was a big parade with floats of the different classes and departments. I worked the most of yesterday afternoon helping the senior engineers rig up theirs, which was an engine and dynamo which ran some electric signs about the Science dept.[71]

This last week has been quite a busy one with the Science dance and Annual banquet...While the University is officially dry nevertheless quite a bit was brought in by the returned graduates and some of the undergraduates had it so that things were far from dry.[72]

Evidence Linking R.L. Davis and Radio at Queen's

From that critical period when the idea of creating a public broadcasting station at Queen's was germinating, virtually no hard information survives to indicate the genesis of the idea, the exact timing, nor the roles of individuals such as Professors Jemmett and Davis. Fortunately, Davis and his family were all meticulous keepers of letters, and the entire story of the development of Queen's Radio may eventually emerge from these letters. Henry B. Davis is still mining the huge legacy of his uncle's documents. Some new information, below, has already been unearthed in Robert's letters from Kingston to his mother in Grand Rapids.

Davis was worked very hard that first year at Queen's and had little time to follow any of his personal interests: "I have got two tons of lab reports to mark. Since I had to take on the D.C. design I haven't had time to mark any and they have accumulated in vast quantities. I rather imagine I'll have to grade quite a number of them by dumping them down the stairs and grading them inversely as the distance from the top that they come to rest."[73] Then, early in 1921 he wrote,

> I wish you would look around at the house and see if you can find a papercovered (gray) book on Radio Telegraphy by the U.S. Army Signal Corps and send it to me as soon as you can as I'm afraid I've got to give a lecture for the Wireless Club on that stuff before long.[74]

That term, W. Harold Reid was keeping the Wireless Club busy with his series of lectures, so Davis did not address them until December 12, 1921, when he talked about Audion tubes.[75] The Queen's calendars for 1920-21 and 1921-22 show Davis as responsible for Gill's old Course XI and Laboratory #4, which included wireless telegraphy and telephony.[76]

Davis may have spent the summer of 1921 at home in Grand Rapids, for he commented in his letter of October 9 that he was quite rusty from having done little thinking over the summer. The third year classes in the new term were so large that he figured he would have to run two night labs. That autumn he was promoted to Assistant Professor of

E.E.,[77,78] and he moved to 120 Wellington Street, next door to Orton Donnelly, the young radio amateur. Before leaving Grand Rapids he had asked his mother to forward "the box of Radio stuff,"[78] but by November 27 he hadn't "had time to do much yet with my radio but am getting some wire and having some things made for it so I hope to have it going before long."[79] That day he had had dinner with the Jemmetts and was invited by Dr Hughes to supper. Davis seemed to socialize with the Jemmetts quite a bit.

Old timers remember Davis as a nice man, but their memories are too vague to convey a sense of his personality. He was included in the annual Faculty Song in both his years at Queen's, but the lyrics are pretty silly and lack radio references:

> Jemmett and Davis from Boston Tech.
> Are our Electrical Profs.
> They lecture the third and final years,
> But they haven't tried the sophs.
> Davis is a fusser bold,
> He loves the ladies fair,
> But Jemmett being a married man,
> He finds no solace there.
> We'll have to teach them
> Electricity is no joke.[80]

> "Dropsy" Davis, from the States,
> Is Prof. in Electrical III;
> He has a soothing line of talk,
> And just between you and me
> He's got a hunch that every one
> Thinks that he is mighty nice,
> For he's discovered a method new
> For electrocuting mice.[81]

On the afternoon of Saturday, January 7, 1922:

> Jemmett and I strung up a wireless aerial between this building and the Chem bldg (Fleming to Gordon Hall). It has a span of about 400 feet long and it is more than 40 feet high. It ought to (be) good for receiving anything for a thousand miles or more around here. I hope to have spare time this week and get my radio apparatus working.[82]

More than half a century later, in 1974, Harold Stewart wrote to R.L. Davis: "Arthur Jackson pinned a note on the copy of your letter asking me if I knew who took down the aerial that you put up between Gordon and Fleming Halls. As I recall that aerial, Doug Jemmett was using it for measurements in high frequency skin effect as part of his M.I.T. doctoral research work...I do not recall that aerial being used for any other purpose such as broadcasting. The AM broadcasting aerial has been between masts on Fleming and Ontario Halls as long as I have been here."[83]

Some of Robert Leland Davis' wireless apparatus, photographed at the Davis family home in Grand Rapids, Michigan, June 1988. A home-made adjustable spark gap with zinc electrodes mounted on a wooden base and an enclosed induction coil for producing discharges across a spark gap. The binding posts for the gap electrodes are on top of the box. The collection included an antenna tuning helix, a very large induction coil of novel design, and a home-made glass plate capacitor.

-photographed by Arthur Zimmerman

Davis wrote home on January 15, 1922:

There have been examinations in second and third year classes but in as much as I didn't have any it gave me some time to get my radio working. I got it to gether Thursday afternoon and that night got the concert from Pittsburgh and time signals from Arlington. The music and talk from Pittsburgh came through as clear and loud as if only over a telephone line between two rooms in a house. The weather was clear and cold, just right for radio work. Friday I could just make some of it out. Sat night Jackson and his wife came over and we got parts of it fairly good. This afternoon I got parts of the church service at 3 o'clock that is sent out by the Westinghouse Co. They had a double quartet which sang and the last hymn came in very good. The minister took for his subject "The quest for truth" and his text was "Ye shall know the truth and it shall make you free". Soon after the sermon started a big snow storm came up and there was a lot of static in the air for awhile which mussed up the talk, but he had something about Esau selling his birthright and toward the end when the air cleared up a bit he was talking about the truthfulness of the early men in the U.S. and their part in the Revolution. There will also be another radio service tonight at 7.30. It ought to come in better as transmission is always better at night.[84]

On the 22nd,

I have put in what time I could on my radio, but wish I had a lot more for such things. During the week I fixed up a form of loud speaking receiver and horn, and when the transmission is good I can use that as the music etc. can be heard all over my office. I can also get Arlington so loud that it can be heard clear down in the library room from my office. One night during this week the music came out as loud as a phonograph using fibre needles.[85]

He continued through February and March to tune his radio in to KDKA in Pittsburgh to hear Sunday sermons and music. About this time he wrote that he was "all covered up with work" and was too busy to eat or sleep. On April 3 he told his mother about the burning down of the curling rink on campus, and how the gymnasium and Carruthers Hall were nearly set on fire. (The new Jock Harty Arena, to the east, was also spared.) Don Nickle and some students were on the roof of Carruthers Hall with water buckets, keeping the roof fire under control. "However, aside from a few cracked windows no damage was done except to the rink, and most everyone except the curling club was glad to see that go."[86]

That summer Davis worked in the Radio Department at Westinghouse in Pittsburgh, through some helpful pull from his friend Don G. Little, a classmate from the University of Michigan, who worked under Dr Frank Conrad. Little had helped Conrad design and build the first 100-watt KDKA transmitter on the roof of a Westinghouse factory building, and Little was the engineer in the first KDKA broadcast, the returns of the Harding-Cox Presidential election, November 2, 1920. In August 1921, Little helped design the 500-watt power modulation replacement transmitter for KDKA, a machine that was the prototype for Newark (WJZ), Chicago (KYW) and East Springfield, Mass. (WBZ).[87] Davis must have had something special to contribute to Westinghouse radio, for he was asked to stay. But he was uncertain of his value to them. Davis did not like Pittsburgh, its climate or the job, although he appreciated that others thought that working at Westinghouse was more prestigious than teaching. He wondered whether there might be more for him there than in Kingston. Certainly more money. The work was easier and the hours shorter. And he wondered whether he was better suited for this work than for the career he had wanted in general engineering. He liked the atmosphere and the healthy living at Queen's, and there was another attraction:

This next year, I could have a good time with radio there as the University is going to get the stuff for the E.E. department to put up a sending set. During the last week I have written Jemmett a lot of stuff about what to get. Aside from that, things will be the same as before and nothing much ahead as to the future."[88]

All through July, Davis agonized about his future:

I wish I knew what to do about next year. I ought to do something about that soon, or it will leave Jemmett in a bad hole if I don't go back. I don't like the idea of living in Pittsburgh, but somehow I can't help but feel it is a more permanent job than going to Kingston for year to year, and about the only thing I am fitted for, to say nothing of having a lot more money in it. Jemmett is going ahead planning on a lot of new stuff in the radio line expecting me back...I certainly wish I knew what to do.[89]

He wrote to his friend Jemmett on July 13 explaining his dilemna, and stating that he might not come back. The agreement among Westinghouse, General Electric and Western Electric with respect to radio patents was unsettled at that point, and if it collapsed, Westinghouse might expand. After thinking about it for a week Jemmett replied, in part, "In the first place, from purely selfish motives, I would be very sorry to see you leave Queen's, because I could not want a better man for your place, but I think if I had your prospects at the W.E. & Mfg. Co., I would throw up my own job here and leave. It seems to me that you have more than an ordinary chance to reach the top of the heap. There is a lot of money in wireless stuff and for the next 4 - 5 years, before things are quite standardized, it would seem that there should be a wild scramble to get men like yourself for the job. There are also vast possibilities that may become realities at any time in future electron tube history." Jemmett held out the hope of Davis' getting much better pay at Queen's and assured him that "The colleges are always looking for a 'practical man' and if he has some 'theory' as you have he can get a good offer at any time...However the matter is entirely in your hands and I will be tickled to death to have you here next year."[90]

Robert Davis mused about radio in a letter to his mother in late July:

With the sort of training I have had, radio work etc. is about the only thing I am fitted for, and of course that means working with W.E. & M. or some similar place...The Radio dept here certainly does have the best of things, and gets paid a lot more than else where in the works...I really can't tell whether I like radio enough to stay with it indefinitely. In some ways it seems more like an amusement with me but I like the general layout here about 5 times as well as I did the work at C.H...
Today we worked on starting a train by radio which

was featured as a big event at the Company, as it was used to start a train of 30 cars of stuff for Chilli. A lot of big bugs from there came out to look it over and see the radio station. We had quite an exciting time up in the radio station; just before the thing was to come off the set quit and there was a lot of chasing around to get it going again.[91]

In early August Jemmett was "pulling hard with the registrar" to get Davis to come back to Queen's,[92] while Don Little was counselling him to stay with Westinghouse. The last letter in the series from Davis to his mother told of Queen's turning down Jemmett's request: "I don't know as there is any use of considering returning to Kingston considering the attitude taken by the officials there...If they had merely said they were shy of money I would have been inclined to forget the matter, but they seem to be inclined to argue about it and be a little disagreeable. I feel like telling them to go to the deuce."[93] And so Robert Leland Davis did not return to Queen's in the fall. He stayed in Pittsburgh to make his career in the newly formed Radio Development Program of Dr Frank Conrad's Radio Engineering Department. Robert's letters suggest that his brother may have passed through Kingston to pick up his belongings.

Could Davis' letters to Jemmett in late June 1922, containing "a lot of stuff about what to get"[88] have included circuit designs for the 9BT transmitter? Jemmett's and Geiger's recollections of Davis' contribution of circuits and plans for 9BT[17,20,21] include the statement that Davis was acquainted with some of the work done at the Westinghouse pioneer radio station KDKA. Certainly at this point, in the summer of 1922 after a month or two with Donald Little at Westinghouse, Davis would have been acquainted with the workings of KDKA. The 9BT project, however, was clearly conceived in the spring of 1922 or perhaps in the late winter of 1921-22. Jemmett had been collecting materials for the transmitter all spring, probably including Hughes' rectifier tubes, and in early June, before Davis wrote his letter "about what to get," he ordered eleven expensive valves from Mullard in England. In mid-July, he made out the requisition for an experimental licence. Thus, whether Davis' letters contained "circuits and plans,"[17,20] modifications of some earlier plans, or specifications, the radio transmitter project was well under way in some form before Davis left Queen's at the end of the spring term.

The 9BT transmitter was put together in September, when Davis was not in Kingston and so he could not have been involved in that work. On the other hand, Davis had the know-how and the equipment to be able to do pre-licenced broadcasting or testing while he was living in Kingston. Evidence to support this latter possibility is lacking.

Was the 9BT Transmitter a Clone of KDKA?

An intriguing possibility is that Davis may have sent Jemmett transmitter plans based upon the KDKA set. Don Little published a simplified schematic of KDKA in 1922.[87] The schematic of 9BT no longer exists but, as will be shown later, Mark IV of 1927 was probably just a modification of the original, and a schematic for that exists.

John Fraser: Basically, the 1927 CFRC transmitter is a modulated oscillator with shunt modulation - Heising modulation. The oscillator is a shunt-fed Hartley. It's very, very basic technology, about as early as you can get with a tube-type transmitter. The modulator consists of two triode tubes in parallel, so they would have to be class "A" modulators. Antenna coupling is by induction, with a simple coupling coil.

The KDKA transmitter is much the same type of circuit except that the oscilltor consists of four tubes in parallel instead of a single tube, and the modulator has five tubes in parallel. So, it's obviously a higher power transmitter. They use the same type of oscillator circuit, the Hartley circuit. The only difference is the size and type of antenna coupling. Instead of having a separate coupling coil to couple the antenna to the oscillator tank, this one merely has a moving tap on the oscillator coil, which is connected to the antenna. Direct coupling. The circuitry design is similar, but you couldn't say that one is derived from the other. There were probably dozens of transmitters using the same basic design.[94]

R.L. Davis' Career at Westinghouse and After

Robert Davis is credited in a biographical sketch with

pioneer work in radio broadcasting, then in its infancy and about to burst upon the American scene in earnest. He was soon immersed in the exciting details of building station KDKA, the world's pioneer broadcasting station and experimental transmitter set up by Westinghouse. This station was equipped to broadcast both at the normal frequency of 980 kC, but also as a short wave station. Mr. Davis and Mr. Little were chiefly responsible for the design, construction and early experimental operation of this pioneer station.[95]

In fact, KDKA had then been on the air on a regularly scheduled basis from the roof of Westinghouse's East Pittsburgh Works since December 23, 1920. After Davis joined Little as #2 man in Radio Engineering, they became involved in designing and constructing modern self-contained broadcasting facilities for KDKA at a site on high ground two miles from the Westinghouse factory. The new transmitter was of the modulated oscillator type, the same general type as installed at 9BT. The new plant went into operation in March 1925, and later that year Davis and Little presented a technical paper on these new facilities.[96] "Even Mrs. Davis became involved. Mr. Davis was attempting to locate a suitably high hill in the Pittsburg area. He obtained topographic maps of the area, and asked Mrs. Davis to locate the high point. She did, and her selection at Saxonburg turned out to be available, where the station transmitter was constructed."[53,97]

Robert L. Davis was involved with experiments in television at Westinghouse as early as 1927. He was listed in the Westinghouse directory of management employees in 1930 as "Engineer in Charge, Radio Development, Engineering Department, East Pittsburgh."[98] That same year the Radio Engineering Department moved to Camden, NJ, to become part of RCA and Davis was transferred to the Chicopee Falls plant near Springfield, Mass. Here he was Manager of Radio Engineering, in charge of 150 employees. That move broke up the first office intercom system in all industry, an ingenious arrangement. Every row of desks in the long engineering and draughting room, every private office, laboratory and dark-room had a miniature telegraph sending and receiving set. Every employee in the department, including the stenographers and office boy, had to know Morse code and had his or her own call letters. Extended conversations were forbidden, because everybody would stop work to listen, but at lunch time witty conversations in code provided lots of amusement.[99]

The illness of his father necessitated Davis' leaving Westinghouse in 1934 to return to Grand Rapids. After four years in engineering consulting there, he re-enrolled in the U. of M. Ph.D. program in E.E. His thesis topic was to be the electrical measurement of sound, but the writing was interrupted by the war, and in 1940 he went to work for the American Seating Company, Grand Rapids, developing safety equipment for fighter pilots. He retired in 1952, the holder of 36 patents, 29 of them from Westinghouse and four from American Seating. In 1975, four years before his death, Davis was awarded an Honorary Doctor of Science by Olivet College in Olivet, Michigan.[100]

Robert L. Davis wrote to Prof. Arthur Jackson in 1974, in response to a letter from Harold Stewart:

> I think the students in E.E., as I remember them, were very good and should have made good engineers. I would like to feel that I had some small part in getting them interested. Later, some came to Westinghouse in Pittsburgh, where I was. In particular, A.C. Monteith, who distinguished himself and became a Vice-President of the company. Something Queen's can be proud of...There must have been others too that gained prominence, and for them I also like to feel I may have had a little to do with their growth in engineering. Particularly high frequency technology or radio. Incidentally, I like to think that I encouraged Queen's E.E. Department to develop further in that line.[101]

Prof. Jemmett's Radio Boys

Not very much is known about Jemmett's "Radio Boys," Sidney Perry and P.H. McAuley, nor about their specific contributions to the development and success of the Queen's radio station, 9BT. Their involvement is recorded only in Douglas Geiger's 1928 article on radio at Queen's.[20]

W.K. Detlor B.Sc.   D. GEIGER   P.H. McAuley.   S.V. Perry.

The original 9BT radio boys who worked with Prof. Douglas M. Jemmett in
building and testing the station in 1922:
W. Kenneth Detlor (Elec. '22; Mech. '23)
Douglas George Geiger (Elec. '22; Mech. '23)
Patrick Harold McAuley (Elec. '23)
Sidney Victor Perry (Elec. '23)
There is some evidence that Donald O. (Red) Hepburn (Civil '24) and E.A.
Filmer were close to the 9BT radio group.
-courtesy Queen's Archives

Since they both left Queen's in the spring of 1923, neither had anything to do with the second radio station, CFRC. Two others, Douglas G. Geiger and W. Kenneth Detlor, were probably also part of the original group, though data are lacking.

Sidney Victor Perry (Elec. '23) was born in London, England, arrived in Canada in 1911 around the age of 10, and graduated from Trenton High School. He entered Queen's with Science '23 in the fall of 1919, was a scholarship winner, a Douglas Tutor and runner-up in the Dominion Wrestling Finals in 1922.

There are a few tantalizing squibs in the Queen's Journal about Perry, but none reveal anything useful about his radio work: "When questioned as to his phenomenal success in Physics 1A, Mr. Perry answered, 'Well, I don't do much fussing like a lot of the other fellows, like Cameron for instance, and besides I just love to play with my sliding rule'."[102] "Perry the physicist was engaged on work of so profound a character that the Dean confessed himself completely flabbergasted."[103] His graduation caption in the 1923 Tricolor says, "In radio circles, perpetual difficulties only served to increase his interest."[104]

Perry was one of the students who did a lot of the work of building the 9BT transmitter in the 1922-23 session. In an interview in 1982, he recalled putting up an aerial from Fleming Hall to Grant Hall in his sophomore year, 1920-21, and that it was used for broadcasting as well as for receiving. He remembered that they were building a new transmitter around 1922. Sid was sitting behind the microphone and had to sign off the station late one night when everything finally worked and it got on the air. But, when questioned as to what they did, he replied, "Nothing much. I should tell you somewhere in here who else was involved. There was P.H. McAuley. And there was a guy from...I don't know how he got into the Queen's thing, but he was from a former year. Geiger. There were two Geiger boys, and I think they were in earlier years and they really didn't belong to Science '23. I don't remember anybody else."[105] He added, "George Parsons and I lived in the same boarding house. You know he had a twin brother? He came in and lived with his brother George. He was Gerald, I think. George and Gerald, and that was about, I think our third and fourth year." Perhaps it was Perry who suggested to Jemmett that his housemate, the cornetist and bandmaster George Parsons, would be a good choice for a musical test of the new 9BT transmitter.

> **Bill Taylor**: Sidney Perry, who lived in New Jersey, I understand, was the first President of our amateur Radio Club at Queen's University in 1922. I think he was replaced the following year by someone else, probably L.R.C. McAteer, at which time W.A. Taylor, myself, was Vice-President. But Sidney was actively engaged in designing and building 5-metre transceivers, which was very unusual in those days. And he was quite successful. He had very good capability in radio work and developed a good transceiver which operated on 5 metres. And I was surprised when I met him about ten years ago at a convention at Queen's University of the old timers, Sidney told me that he had not had a screwdriver or a pair of pliers or a soldering iron in his hand since he left Queen's. He gave up the ham radio activity completely at the time of graduation. Apparently he went into executive work with the Westinghouse Company in Pittsburgh.[106]

The published record does not show Perry holding any executive office in the Queen's Radio Club. He worked for more than 40 years in the Home Instruments Division of RCA in Camden, NJ, and helped develop their "Golden Throat Sound" after 1945.[107]

Even less is known about Patrick Harold McAuley (Elec. '23). He too graduated from Trenton High School in 1919, and entered Science '23 at Queen's. The 1923 Tricolor said, "An unusually bright student, he found time to take an active interest in many phases of student life. His foresight and consistent efforts made him a leader in radio circles, where he did much toward putting Queen's on the radio map."[108] Nothing specific is known about his contributions to the building of the 9BT transmitter, except that he and Sidney Perry did a lot of the work under Jemmett's guidance. "He retired in 1962 as Director of Engineering Laboratories at Westinghouse Corp., after 38 years of service in the fields of power transmission, high-voltage insulation and testing, and industrial control."[109]

Douglas George Geiger (Elec. '22; Mech. '23) was on the executive of the Wireless/Radio Club in 1920-21 and 1921-22, and he may have been one of Jemmett's original radio boys, along with Perry and McAuley.[20] As with the other two, it is no longer possible to determine his role in 9BT. Sidney Perry graduated and left Queen's in the spring of

1923 and remembered Geiger as part of the group, so there can be little doubt that he was in it. His cousin, George F. Geiger (Metall. '23), "the other Geiger boy," may have been involved too, but only peripherally. In an interview in 1983, he denied being part of the radio group and he remembered very little of his cousin's activities.[110]

W. Kenneth Detlor (Elec. '22; Mech. '23), another of Jemmett's radio boys, was born in Deseronto in 1903 and entered Queen's in the fall of 1918. He was a scholarship student, a boxing champion, an excellent hockeyist and a Douglas tutor.[111] In 1923 he joined the Bell Telephone Company at Kingston, but was scheduled to co-teach two laboratories in E.E. at Queen's in 1923-24.[112] He was later transferred to Montreal and then to Toronto, where he was transmission engineer involved in telephone and radio communications. He died after a very brief illness in 1930, leaving a wife and an infant child.[113] Nothing survives of his role in 9BT, but Mrs Douglas (Goldie) Geiger recalled him as one of the original radio boys.[114]

Radio at the 1923 Science Dance

In the second Whig report, in August 1922,[5] on the wonderful wireless receiver at Queen's, a prediction was made that radio would be installed at some of the dances in Grant Hall that school year to bring in "the foremost American orchestras" and "the latest fox trots...By means of amplifiers the sound could be made loud enough to be heard throughout the hall."

A radio dance was attempted at the Science Dance in early 1923: "The radio numbers, although not a brilliant success, were a real novelty and demonstrated the ability of the Electricals to pick up music anywhere. To hear the Radio announce that the next number is to be played by a New York orchestra makes one feel that New York is getting closer to Kingston every day."[115] Della Munro (Arts '24) recalled the event for her 60th re-union: "We danced for the first time to radio. The engineers brought it (in), and told us it was from New York. We were so excited. But later we found it had come from Fleming Hall."[116]

Sidney Perry was involved in that "broadcast" from Fleming Hall: "We played music for the senior year dance. We had our source of sound in Fleming Hall, and the dance, of course, was in Grant Hall. It was not broadcasted - it was reproduced in Grant Hall. By wire."[105]

Bill Taylor has a slightly different twist on that historic non-broadcast, because the people of the Radio Laboratory needed the help of the Radio Club to bring it off:

> Bill Taylor: I remember that in my second last year at Queen's, when we went to our Queen's Dance, we expected to have a broadcast from Fleming Hall to Grant Hall and that we'd be able to dance to that music. But it didn't work very well, so we ran a wire from Fleming to Grant Hall and relayed the music of a phonograph by means of this wire so that we could dance to it. It wasn't very good even at that because we had no amplifying equipment. Oh, yes that was CFRC. We, the

Radio Club, weren't actually involved with that as radio people at all. Except that we ran the wire to help them out.[106]

The members of the Radio Club were not normally welcome in the radio precincts of Fleming Hall, especially if they weren't Electricals. Bill Taylor remembers Doug Geiger guarding a room on the second floor of Fleming:

> **Bill Taylor**: I do remember that in '23 they had a lot of radio equipment in Fleming Hall and I went in several times and had a peek at it, you know, just because I was interested in wireless junk. Geiger had all kinds of amateur equipment in Fleming Hall, that he had seen advertised in radio magazines, and bought by the University. One thing I remember in particular is honeycomb coils, and they had a stack of them. I think I swiped a couple of them. It was a laboratory, that's all it was, on the second floor. It was full of junk, like a modern radio amateur that's not very tidy. He had bits of wire and coils and tubes kicking all over different desks and whatnot in the room that I visited there. Oh, they had a lot of equipment there, and some big tubes for broadcasting too. Some fair sized tubes, so they must have had a fair amount of power. A half a kilowatt anyway.[106]

The First College Stations in Canada

The only college station licenced in Canada before the First World War had the call XBT. Dr Augustin Frigon, Professor of E.E., held the licence on behalf of the Ecole Polytechnique in Montreal.[117] At that time there were three series, XA-, XB- and XC-, and 47 licenced amateurs in the country.

Dr Frigon was a member of the 1929 Aird Commission and became Assistant General Manager of the new CBC in 1936. He succeeded Gladstone Murray as General Manager in 1942.

When C-calls were first published in Canada, in August 1922, there were no college stations on the list. The licencees were mainly electric companies, telephone companies, radio stores, newspapers and automobile dealers.[118] There were, however, four all-Canada experimental licences (9-calls) issued to educational institutions: Acadia University, Wolfville, Nova Scotia, had 9AT; the Physics Department at the University of Alberta, Edmonton, was 9AU; Sprott-Shaw School in Vancouver held 9AX; and Queen's University got 9BT.

In 1923, there were two college stations with C-calls: CFRC at Queen's, and the University of Montreal, 185 St. Denis Street, with CFUC. The latter was not renewed in 1924. That year there was a fourth college 9-call, 9BZ, belonging to the E.E. Department, University of Toronto.[119]

CKUA, at the University of Alberta, Edmonton, came into existence in 1927, and later, around 1933, there was CKIC at Acadia University.

NOTES

1. "Notes of Queen's", KDS, March 23, 1922, p. 7.
2. "Report of the Dean of the Faculty of Applied Science", Principal's Report, May 1922, p. 18.
3. "Electrical Engineering Laboratories", Calendar of the Faculty of Applied Science, Queen's University, 31st Session, 1923-24, p. 89; QUA.
4. "Radio at Queen's", DBW, Aug. 11, 1922, p. 2.
5. "Everybody", ibid., Aug. 23, 1922, p. 12.
6. "Heard Via Radio", ibid., Aug. 24, 1922, p. 13.
7. "Radio Phone Craze", ibid., April 21, 1922, p. 1.
8. "The Radio Experimenter", KDS, Oct. 10, 1922, p. 15.
9. "The New Rage", DBW, May 1, 1922, p. 6.
10. "An Aerial Wedding", ibid., April 29, 1922, p. 1.
11. "Photographs by Radio", ibid., May 16, 1922, p. 6.
12. "Three Million Guests", ibid., Aug. 11, 1922, p. 11.
13. "What May Happen", ibid., April 27, 1922, p. 12.
14. "Short Flashes", KDS, Oct. 7, 1922, p. 19.
15. "Radio's Future", DBW, Aug. 25, 1922, p. 6.
16. "Radio", QJ, Feb. 22, 1924, p. 1.
17. Script of "Documentary for Saturday, October 27, 8:30", 1962; CFRC Archives.
18. "Early Days at Queen's Radio, with Dr. Jemmett and Professor H. Stewart", script of CFRC programme "Personality" of Saturday, Oct. 27, 1962; CFRC Archives.
19. Interview with Prof. Jemmett, from a point-form script for the CFRC programme "Meet Your Prof", broadcast on a Thursday night at 7:40 some time in 1952.
20. D.G. Geiger, "Radio Broadcasting at Queen's", QR, Jan. 1928, p. 9.
21. H.H. Stewart, "The Origin and Development of CFRC", PESQU 28, Sept. 1938, p. 30.
    Personal communication from Prof. Harold H. Stewart, Kingston, Aug. 10, 1982.
22. B.W. Sargent, "Arthur Llewelyn Hughes, 1883-1978", PRSC, Series 4, 23, 1985, p. 109.
23. Personal communication from Dr. Donald C. Rose, Brockville, May 9, 1986.
    Letter from Dr. Donald C. Rose, Brockville, Dec. 4, 1983.
24. Personal communication from Dr. W. Donald MacClement, Kelowna, B.C., Aug. 1987 and Sept. 25, 1987.
25. Specifications from Lists V.R.1, 7, 8, 12 and 16 of the Mullard Radio Valve Co., Ltd., London; courtesy Mr. Ian Nicholson (formerly of Mullard Co. Ltd.), North Littleton, Worcs.
26. "Radiotelegraph Regulations", Supplement to the Canada Gazette, Sept. 23, 1922.
27. "Experimental Stations", Official List of Radio Stations of Canada, Dept. of the Naval Service, Ottawa, Aug. 1, 1927, p. 77.
28. "Wireless Station for University of New Brunswick", Electrical News 31 #18, Sept. 15, 1922, p. 60.
29. Personal communication from George Mitchell Parsons, St. Lambert, Quebec, Aug. 14 and Oct. 13, 1982.
30. "Science: Things We'd Like to Know", Queen's Journal 49 #17, 6, Friday, December 8, 1922.
31. DBW, Sept. 11, 1922, p. 2.

32. "Queen's University Session", ibid., Sept. 27, 1922, p. 12.
33. "Will Broadcast Football Story", KDS, Oct. 7, 1922, p. 1.
"Queen's to Broadcast", DBW, Oct. 7, 1922, p. 18.
34. "Results Broadcasted", KDS, Oct. 9, 1922, p. 12.
35. "Rugby Report Sent Out by Radio", DBW, Oct. 9, 1922, p. 1.
36. "Local News", ibid., p. 12.
37. "Awrey Surprised with Varsity", KDS, Oct. 11, 1922, p. 8.
38. "Mr. Awrey's Opinion", ibid., p. 12.
39. "Local News", DBW, Oct. 9, 1922, p. 9.
40. "Queen's Alumni Banquet Team", QJ, Oct. 31, 1922, p. 1.
41. "Thanks to Queen's", QJ, Oct. 13, 1922, p. 3, and Oct. 17, 1922, p. 3.
42. "The Operating Department: District #3, Ontario Division, A.H.K. Russell, Mgr.", QST 6 #3, Oct. 1922, p. 50.
43. "The Operating Department", ibid 6 #5, Dec. 1922, p. 54.
44. "Calls Heard", ibid., 6 #9, April, 1923, p. 69.
45. "Science '24", QJ, Oct. 20, 1922, p. 6.
46. "Queen's Win Senior Game", DBW, Oct. 10, 1921, p. 3.
47. Personal communication from William A. Taylor, Bath, Ontario, August 21, 1982.
Student Directory, Queen's University, Kingston, 1920-21 to 1923-24; QUA.
48. Personal communication from W.J. Hurn, by telephone from Noranda, Quebec, Oct. 19, 1982.
49. "Forest Fires Sweep", DBW, Oct. 5, 1922, p. 1.
50. They would also collect empty furnitre boxes from behind James Reid's store, drag them out into the middle of Princess Street and make colosal bonfires. The snake dances often ended in rushes of the local theatres, with hundreds of students crashing through the turnstiles and rushing down the aisles.
51. Fire Wipes Out", DBW, July 5, 1922, p. 1.
"'Alfie' Pierce", ibid., Aug. 18, 1922, p. 10.
52. "Wireless Service", The Varsity, Oct. 10, 1921, p. 1.
53. "Robert L. Davis: Biographical Data"; courtesy Mr. Henry B. Davis, Grand Rapids, Michigan, June 1988.
54. "Special Land Radio Stations", Radio Stations of the United States, Department of Commerce, Bureau of Navigation, Radio Service, Washington, D.C., July 1, 1915, p. 62.
55. "Last Minute Changes", QST 1 #5, April, 1916, p. 66.
56. "Personal communication from Maud Martineau Jemmett, Kingston, Oct. 5, 1982.
57. "A Comparison of the Diesel Electric System of Ship Propulsion with the Direct Drive by Steam and Diesel Engines for Trawlers", thesis submitted to M.I.T. in the fall of 1920; courtesy Henry B. Davis.
58. Radio Stations of the United States, June 30, 1920.
59. "Fleming Hall Staff", KDS, Sept. 10, 1922, p. 2.
60. R.L. Davis, Ann Arbor, Mich., to Prof. Arthur Jackson, Kingston, July 10, 1974.
61. Robert L. Davis, Cambridge, to his mother, Ann Arbor, Jan. 4, 1920; courtesy Henry B. Davis.
62. R.L. Davis to his mother, written on the Boston-Gloucester train, Sept. 1, 1920.
63. R.L. Davis to his Aunt in Springfield, Mass., Sept. 14, 1920.
64. R.L. Davis to his Aunt, Sept. 16, 1920.
65. Western Union Telegram, Jemmett to R.L. Davis, 335 Harvard St., Canbridge, Mass., Sept. 20, 1920.
66. R.L. Davis to his mother, dated at Chester, Vt. en route to Montreal and Kingston, Sept. 22, 1920.

67. Minutes, Board of Trustees, Queen's University, Oct. 20, 1920; QUA. "New Appointments 1920", QJ, Oct. 20, 1920, p. 7.
68. R.L. Davis to his mother, dated at Kingston, Sept. 26, 1920.
69. Ibid., Nov. 28, 1920.
70. "Queen's and Elections", DBW, Dec. 5, 1921, p. 3.
71. R.L. Davis to his mother, Dec. 20, 1920.
72. Ibid., Feb. 20, 1921.
73. Ibid., April 10, 1921.
74. Ibid., Jan. 30, 1921.
75. "Notes of Queen's", KDS, Dec. 13, 1921, p. 5.
76. Calendar of the Faculty of Arts and Science, Queen's University, 1921-22; 1922-23.
    E.W. Henderson was listed in the calendar as teaching Course XI-Laboratory #4 in 1920-21; he left Queen's in late summer of 1920, after the calendars had gone to press. Similarly, Davis was shown teaching this combination in 1922-23, though he had decided not to return to Queen's by late August, 1922.
77. "Notes of Queen's", KDS, Sept. 26, 1921, p. 2.
78. R.L. Davis to his mother, dated at Kingston, Oct. 2, 1921.
79. Ibid., Nov. 27, 1921.
80. "Faculty Song", PESQU #15, 1921, p. 49.
81. "Faculty Song", PESQU #16, 1922, p. 25.
82. "R.L. Davis to his mother, dated at Kingston, Jan. 8, 1922.
83. Harold H. Stewart to R.L. Davis, July 18, 1974.
84. R.L. Davis to his mother, January 15, 1922.
85. Ibid., Jan. 22, 1922.
86. Ibid., April 3, 1922.
    "The Curling Rink Burned", KDS, March 28, 1922, p. 1.
87. Donald G. Little (B.Sc., Michigan; D.Sc., Kalamazoo College) was involved in radio development work during Word War I, and continued afterward at Westinghouse in Radio Engineering and at KDKA under Dr. Frank Conrad.
    Erik Barnouw, A Tower in Babel: A History of Broadcasting in the United States - to 1933, Oxford University Press, New York, 1966.
    Gleason L. Archer, History of Radio to 1927, American Historical Society, New York, 1938.
    "Frank Conrad", National Cyclopaedia of American Biography 35, p. 179.
    Hugh G.J. Aitken, The Continuous Wave: Technology and American Radio, 1900-1922, Princeton University Press, Princeton, N.J., 1985, p. 471.
    D.G. Little, "Radio Equipment at KDKA", Electrical Journal 19 #6, June 1922, p. 245.
88. R.L. Davis, Pittsburgh, to his mother, July 2, 1922.
89. Ibid., July 11, 1922.
90. D.M. Jemmett to R.L. Davis, July 25, 1922.
91. R.L. Davis to his mother; dated at Pittsburgh, July 31, 1922.
92. Ibid., Aug. 7, 1922.
93. Ibid., Aug. 24, 1922.
94. Personal communication from John Fraser, Department of Electrical Engineering, Queen's University, Nov. 2, 1989.
95. "Robert L. Davis: Biographical Data", courtesy Henry B. Davis, 1988.
96. "D.G. Little and R.L. Davis, "KDKA", Proceedings of the Institute of Radio Engineers 14, 1926, p. 479.
97. R.L. Davis and V.E. Trouant, Westinghouse Radio Station at Saxonburg, Pa. 20 #6, June 1932, p. 921.
98. From a directory of management employees, Westinghouse Electric and Manufacturing Co., 1930; courtesy Charles A. Ruch, Westinghouse

Electric, Pittsburgh, June 17, 1985.
99. "Two Minutes", feature in Westinghouse employee magazine, Jan. 1930, p. 3; courtesy Charles A. Ruch.
100. "Bro. Robert Davis Honored", Scottish Rite News, Jan.-Feb. 1976, p. 5. "Broadcast Pioneer", Grand Rapids Press, Dec. 22, 1979.
101. Personal communication from Prof. Harold H. Stewart, Kingston, Oct. 20, 1982.
102. "Science '23", QJ, Feb. 10, 1920, p. 4.
103. PESQU #15, 1921, p. 185.
104. "Sidney V. Perry", Science '23, Queen's University Yearbook, 1923, p. 162; QUA.
105. Personal communication from Sidney Victor Perry, Cherry Hill, NJ, Oct. 12, 1982.
106. Personal communication from William A. Taylor, Bath, Aug. 21, 1982.
107. "Obituaries: Sidney Perry, Former RCA Engineer", Courier-Post, Feb. 26, 1986, p. 4B; courtesy Richard A. Perry, Collingswood, N.J.
108. "P.H. McAuley", Science '23, Queen's University Yearbook, "Tricolor", 1923, p. 158.
109. "Patrick Harold McAuley", QR, 52 #5, Sept.-Oct. 1978, p. 25.
110. Personal communications from George F. Geiger, Southbury, Ct., 1983 and March 12, 1985.
111. "W.K. Detlor" Science '23, Queen's University Yearbook, 1923, p. 143.
112 Calendar of the Faculty of Applied Science, Queen's University, 31st Session, 1923-24, p. 89.
113. "Deaths: Detlor", QR 4 #6, Aug. 1930, p. 220.
114. Personal communication from Goldie Geiger, Edmonton, 1982.
115. "Science Dance", QJ, Feb. 9, 1923, p. 1.
116. "Alumni to Find Same Fun", QJ, Oct. 12, 1984, p. 5.
117. "Annual Report of the Radiotelegraphic Branch", June 15, 1914, in Report of the Department of the Naval Service, Sessional Papers, House of Commons, Paper No. 38, 50 #26, 1915, p. 72.
"Experimental Licence to use Radiotelegraphy", No. 44, Ottawa, April 1, 1914, issued to Dr. A. Frigon, on behalf of the Ecole Polytechnique, Montreal; NAC, RG 97, Vol. 85, File 6202-30, Part 1.
118. Official List of Radio Stations of Canada, Department of the Naval Service, Ottawa, August 1, 1922.
119. Official List of Radio Stations of Canada, Department of Marine and Fisheries, Ottawa, July 31, 1923.

CHAPTER 5

Mary Isabella Hickey at the family radio, circa 1924.
-courtesy Mabel (Hickey) Archer

The Beginnings of CFRC

Building experimental radio station 9BT from Robert Davis' plans was a very exciting project for Prof. Jemmett and the boys. 9BT worked well enough to fulfill some of their immediate expectations. It was also a good piece of apparatus for laboratory work, but there were severe problems with signal quality that made it virtually useless for public broadcasting. Jemmett and the boys did some experimenting with the 9BT transmitter, probably in late 1922, and were convinced that, with a motor generator to provide smoother and more even high voltage and better filtering to make the signal hum-free, they could do public broadcasting from the Queen's campus in the next session. "One thing that helped to convince us," said Prof. Jemmett, "was a grant of money from Dr. W.R. Jaffrey of Hamilton - we bought a motor generator with that."[1]

The groundwork for the new Queen's station may have been laid in Hamilton a short time after the first 9BT broadcast. In late November 1922, Principal R. Bruce Taylor and W.F. Nickle, MPP, were guests at the annual dinner of the Queen's Alumni Association of Hamilton and Wentworth, Ontario. Dr W.R. Jaffrey (Meds '13) was Secretary-Treasurer of that branch that year.[2] Jaffrey was also a frequent visitor to Queen's, addressing the Medical Faculty and students on his specialty, dermatology.[3] He had been very interested in wireless since his teens, and was a keen listener and DXer in the early 1920s. Therefore, it would not be surprising that around the time of the Alumni dinner, Dr Jaffrey heard about the problems that 9BT was having with its AC rectifiers and decided to come to the rescue.

Dr Jaffrey's Gift to Queen's

William Reginald Jaffrey was born in Fredericton, NB, in 1887. After graduating from Queen's in Medicine in 1913, he became a fellow in Pathology at Queen's and Assistant Bacteriologist to the Ontario Board of Health. The following year he became Assistant Bacteriologist to the New York Post-Graduate Medical School, and in 1915 moved to Hamilton to begin a practice in dermatology. He was City Pathologist for Hamilton from 1915 to 1919 and a Captain in the Army Medical Corps during the First War.[4] He was one of four founding members of the Inter-Urban Dermatology Association in 1926, an organization which became the Canadian Dermatological Association 20 years later.[5] It is not clear when Jaffrey became a wireless operator, but from at least 1926-27 until the end of his life in 1950, he held the licence for 3DC and VE3DC.[6] Not only did he build his own fixed transmitting and receiving equipment, he even had a portable transmitter in his car in the 1940s. His name on air was "DOC" and his home base in the Dundas Valley was shown as "OIDAR" on his QSL cards.[4]

The initial approach by Dr Jaffrey concerning the gift to the Queen's radio project was made to Dr J.R. Currie of Queen's on April 6, 1923,

Dr William Reginald Jaffrey (Meds '13) of Hamilton had been a wireless enthusiast since his teens, and gave Queen's a gift of $500 in early 1923 "In memoriam to a very dear friend of mine..." to "give the nucleus to a broadcasting station...and hope it will do more good than any other sort of memory I could contribute and be as lasting...it will give Queen's some honest advertising and at the same time the public something worth while." Prof. Jemmett used the money to buy a motor generator set for the station, and this enabled him to upgrade the operation and apply for the call CFRC.

at a dinner of the Hamilton Branch of the Queen's Alumni Association.[7] Dr Currie relayed the information about the offer to Principal Taylor, who wrote to Prof. Jemmett on the 26th:

> Would you be good enough to let me have a statement saying what you can do in the way of broad-casting with the Five Hundred Dollars which the Hamilton Alumni proposes to give for this purpose. They are so keen upon their gift that I am anxious to let them feel that I appreciate it.[8]

Prof. Jemmett took a few days to think the matter over, and replied to the Principal on May 2. The letter deserves to be quoted in large part, as it gives the only thorough account of his aims in seeking a commercial broadcasting licence for Queen's that summer.

> This summer I have planned the erection of a 60 ft. pole on top of the Physics building and with more power in the oscillating and modulating tubes I can see no reason why we cannot be heard all over Old Ontario, and perhaps as far north as Haileybury and Timmins where there are large numbers of Queen's men. As we have a great deal of the necessary equipment I believe that this could be done with the Five Hundred Dollars the Hamilton Alumni Association are prepared to give.
>
> In order to make a success of the scheme it would be necessary to leave our set always in readiness and not use parts of it for general laboratory purposes. This means a slight duplication of equipment. Also some scheme to outline and carry through a satisfactory radio programme should be put under way at once. Finally a fourth year student or demonstrator who is properly qualified should be paid, say $200 a year to keep in order and operate the outfit.
>
> This is the very cheapest form of advertising, for in addition to a few Alumni Associations who are interested in Queen's there is a vast audience who perhaps have never even heard of us but who will hear our station and if we can put on worth-while lectures and concerts will judge the University by its radio. The very fact of its existence and their knowledge of it is a good advertisement.
>
> No Canadian University has a broadcasting station as yet, but I believe that the University of Toronto is installing one this summer.
>
> I believe that the Trustees would be well advised to spend $300-$400 a year on the maintenance of this station. It would save many times this sum in ordinary newspaper publicity and be a form of advertisement in keeping with the spirit of the times.[9]

The Principal wrote to Dr Jaffrey the next day, enclosing a copy of Jemmett's letter, and endorsing Jemmett's vision: "This is indeed a most timely and generous suggestion...It seems to me that within the University we have to organize this broad-casting idea and so make the University more widely known throughout the country. There is no reason why a good deal of educational work might not be given in this way."[10] Dr Taylor had recently gained some first-hand experience with the power of radio. His speech at a Sunday Evening Club dinner in Chicago on April 6, 1923, was broadcast all over the mid-west, and he was quite impressed with the vast potential radio audience. Perhaps Queen's might benefit from its own wireless mouthpiece.

The mails must have been efficient in those days, for Jaffrey's reply arrived dated May 4, 1923. In this letter he made clear the source of the gift:

> I am glad you are interested in an idea which will give Queen's some honest advertising and at the same time the public something worth while. Prof. Currie misunderstood, this is a personal gift in memoriam to a very dear friend of mine who has gone to the great beyond, but I trust this will not deter your acceptance of the offer. Let this be the nucleus of the Queen's Broadcasting Station. For the upkeep I should think the City of Kingston would be willing to help some and have certain uses given them.[11]

Dr Jaffrey also mentioned that the Hamilton radio station, CKOC, with a range of 400-500 miles, was going to install a more powerful transmitter and would let the present one go at a reasonable price. He would also send specifications on some transmitters made for the US Army, which were for sale in Buffalo. One of these might be imported duty-free as scientific equipment. Jaffrey continued:

> Perhaps Prof. Jemmett would rather build his own, I leave that to your option but will give the nucleus to a broadcasting station in memory of my dear friend and hope it will do more good than any other sort of memory I could contribute and be as lasting.[11]

As an afterthought, he noted at the bottom of the page that CKOC had just bought a compact motor generator for its set for $185.

Jemmett responded through the Principal that the Buffalo sets were too expensive, but "The money that he is willing to give would go toward the antenna and counterpoise, Dr. Hughes' high voltage M(otor) G(enerator) set or its equivalent, and a suitable loading coil. We have most if not all of the other equipment and a very few dollars would go toward small details."[12] The Principal forwarded a copy of Jemmett's letter on the 30th, assuring Jaffrey that: "During the summer we shall have this broadcasting arrangement erected and I shall keep you informed of the progress of the work."[13]

Dr Jaffrey sent in his cheque for $500 on June 5:

> Herewith my promised donation to a memorial which I
> feel will be more useful and lasting than the same sum in
> a monument to one who has left this life and would be
> pleased with the spirit of the gift. It is not that I am over-
> loaded with this world's gifts that I am doing this, even
> more a privation on my part.
>
> I hope the use this will be put to will spread the good
> work of Queen's and when we personally meet I will
> explain the circumstances of the gift.[14]

The circumstances of the gift and the identity of the "dear friend"
may have been confided to the Principal, but he did not preserve this
information for the record. Dr Jaffrey's son, Bill, speculates that the
friend may have been a buddy killed in the Great War.[4]

Purchases for the New Broadcasting Set

In the new year, before Jaffrey's gift was offered, Prof. Jemmett purchased a few things which could have been for the radio station or
might have been for his own research: an RCA radio frequency transformer and a potentiometer, from the Canada Radio Stores on January
12, and a few 10" Electrose radio insulators on March 21. He also
requisitioned a cheque for five dollars to be sent to the Department of
Marine and Fisheries, Ottawa, for the renewal of Experimental Radio
Licence No. 45, call sign 9BT, for the fiscal year 1923-24. The last was
on April 2, 1923.[15]

Jemmett began to buy equipment for the new transmitter on June
28, ordering three 0/250B, two 0/30A, two 0/30B and four 0/5 Mullard
transmitting valves and two O.R.A. "A", three O.R.A. "B" and four
R.A. receiving valves, all from Mullard.[16] The total projected bill was
£49-16-6, or about $250. The 0/250B was very similar to the 0/250A
which was used in the 9BT transmitter, and it was likely the oscillator.
The 0/30A and 0/30B were likely just modifications of the Type 0/20 and
0/30 from the old set, and were part of the amplifier. No information was
found on the 0/5 valve, the only novel type of valve ordered for the
new set. It may have been a modulator. The O.R.A. "A" would Oscillate,
Rectify and Amplify and is a descendant of the Type KA and of the
French "R" (reception) valve used in the receiving set in 1922-23. It
could be used as a local oscillator to heterodyne incoming waves and
extract the audible signal in a receiver.[17] The O.R.A. "A" was probably
used in the receiving set in 1923-24. The same day he spent about $150
on capacitors and inductances.

Prof. Jemmett decided not to use Dr A.Ll. Hughes' motor generator set, perhaps because Hughes was then on the point of accepting an
appointment at Washington University in St Louis. An Esco "Dynamotor-
Generator set, to deliver 0.4 amps at 3000 volts and to run from the campus 110 volt D.C. power supply, as per second proposition your letter
July 12-1923" was ordered on July 30 from the Electric Specialty
Company, Stanford, Conn., at a cost of $450, less 15%.[18] It had three

commutators, producing 1,000 volts DC each. In August, Jemmett purchased some transformers, variable condensers, a grid leak condenser, rheostats, honeycomb coils and mounts, three Radiotron tubes (one UV 200 detector and two UV 201A amplifiers) and some "B" batteries. The purchases made until this point were charged to Electrical Engineering, but after the beginning of September the money came from a "Radio Special" account.

The material for the aerial was ordered on September 4 on requisition No. 2840: 80 feet of 3" galvanized pipe for the mast, a 3" cap and a 3" pipe flange, 600 feet of #10 galvanized wire, two pulleys, and 100 feet of 0.5" manilla rope.[19] The order to Mr Bews to erect a radio mast on Ontario Hall was signed on September 18, on requisition No. 2847.[20]

The new aerial was likely a cage aerial, similar to the 1922 model. This horizontal array, known as a squirrel cage, was a hexagonal cylinder of six wires kept open and supported by several sets of three wooden spreaders. Prof. Harold Stewart explained the theory of this kind of array:

> The squirrel cage antenna was designed to increase the capacitance, to draw up a lot of current to be radiated by the vertical portion. The useful radiation came from the vertical portion. The cross-wires provided an incentive for the current to flow upward, to charge the horizontal sections, so that the vertical portion could radiate waves.[21]

It was also an inverted L antenna, for there was a vertical wire rising almost straight up from the transmitter to connect to one end of the horizontal array.

No new microphones were ordered for the first year of the Queen's broadcasting station. The Western Electric 284-W and/or the 323-W, bought for 9BT in 1922, were used.

The New Private Commercial Broadcasting Licence

There must also have been an order that spring or summer for a cheque for $50 to the Department of Marine and Fisheries to pay for the new Private Commercial Broadcasting License for Queen's, but neither this nor the licence can be found. The number of that first CFRC licence, #33, is recorded on requisition No. 2460, dated April 5, 1924, for the first renewal of CFRC. The first edition of The Official List of Radio Stations of Canada to include CFRC was published July 31, 1923.[22] The wavelength assigned was 450 metres (666.3 kilocycles), with 1,500 watts anode power input and a range of 500 miles.

The Private Commercial was a new class of broadcasting licence, introduced under the revised Radiotelegraph Regulations published in September 1922.[23] The annual fee was $50, and licences were granted for one year, from April 1 until March 31 of the following year, "for the broadcasting by radiotelegraph or radiotelephone of news, information, entertainment or other service." The licence form for a Private Commercial station still referred to it as "a radio Land Station."[24]

The regulations were few in those days, and here are some of the more interesting and relevant:

> The working of the station must be strictly limited to the hours prescribed in the license and the station must use such wavelength as is specified therein...The licensee shall so work the licensed apparatus as not to interfere with the working of any other Radio station in Canada, or with marine signalling on the waters or territory of Canada or neighbouring waters or territory...The licensed station must be provided with an accurate wavemeter of approved type...must be provided with a connection with the local wire telephone system...The licensed apparatus shall not, without the consent of the Minister, be altered or modified in respect of any of the particulars mentioned in the schedule hereto...The station must be operated by a person who is the holder of a "First Class" or "Radiotelephone" Certificate of Proficiency in Radio...All operators at the said station shall be British subjects, and must be of such number and the holders of such Certificate of Proficiency as are specified in the schedule annexed hereto.

The licencee was also forbidden to collect fees or tolls for any service performed, thus precluding advertising on the US model. This rule was changed in 1928 to permit indirect advertising - mention of the sponsor's name as opposed to a direct, extended pitch on the product.

It seems strange by to-day's commercial licencing practices to include the following statement in the regulations: "The allotment of the wavelength or wavelengths specified...does not confer a monopoly of the use of such wavelength." In 1922 the AM broadcast band was defined as lying between 200 and 545 metres (1,499 to 550 kilocycles), giving 95 channels with ten kilocycles separating adjacent channels. There was normally one frequency allocated to a city. Up to three stations in each city had to share that one frequency on a mutually negotiated schedule. Station A might sign on at 10 am, broadcast for an hour or two and sign off. Station B might sign on the same frequency as soon as station A's transmitter was shut down. Station A might sign on again for a time later in the day, or it might not go on the air again for a few days. Radio stations depended on batteries for their operation. These would run down after a couple of hours on the air, and the station would have to suspend operations to re-charge, unless it were wealthy enough to own a spare set.

The wavelengths allocated to the first group of broadcasting stations all across Canada were 400, 410, 420, 430, 440 and 450 metres (749.6 to 666.3 kilocycles), toward the lower end of the present AM dial.[25] In 1923 the US, with 600 radio stations, appropriated practically every one of the 95 channels available in North America. The 40 or so Canadian stations actually in operation then got severe interference from the high-powered stations south of the border. In October 1924, the US

Department of Commerce agreed to reserve 6 channels in the upper band as exclusive to Canada and allowed Canada to share some 12 channels used in the southern States. By 1924-25 there were 19 channels in use in Canada,[26] but the authority of the US Department of Commerce was weakened when it lost an action to keep the Zenith Radio Corporation of Chicago off 329.5 metres (910 kilocycles), a band exclusive to Canada.[27] In 1927, Canada wanted to increase the number of its exclusive channels to make possible a chain of high-powered stations coast to coast. The US insisted on 77 channels exclusive to itself, 6 to Canada and 12 shared. The negotiations broke down,[28] until US law was changed in 1927.

There was now a strict requirement in the Dominion Regulations regarding the keeping of proper radio logs: "When using a wavelength greater than 275 metres a proces verbal of all signals transmitted, giving date, time and nature of such signals shall be kept by the licencee...The licencee shall preserve all proces verbaux for such period as is from time to time prescribed by the Minister..."[24] Incidentally, the licence forms of this period for amateur, experimental and private commercial stations all had space for the rating of a motor generator, as if it were understood that this was now becoming a requirement for broadcasting.

Prof. Jemmett ordered "1 - Note Book as log book for Radio Station" to be sent to him from Tech Supplies, Fleming Hall, on requisition No. 2417, dated October 29, 1923. This black-covered laboratory note book is now in the Queen's Archives.[29]

CFRC: Do the Call Letters Have Some Special Significance?

The first round of applications for the new "C" calls took place in the spring of 1922. By the time of the publication of the first Official List of Radio Stations of Canada in August 1922,[25] there were 55 broadcasting stations in Canada. Many of them never made it to air and some did not survive their first year. Several were phantom stations with their own "C" call but using other stations' facilities for their broadcasts. When CFRC came on the air in the fall of 1923, there were perhaps 34 Canadian private commercial stations broadcasting.[22]

When asked, Prof. Jemmett always replied that CFRC stood for "Canada's Famous Rugby Champions," and Prof. Stewart confirmed this.[30] Mr W.J. Hurn, who said that he left Kingston for good in 1922 (more likely in 1923), insisted that CFRC meant "Canada's Famous Rugby Club."[31] By the time of the granting of the call CFRC in 1923, the Queen's Tricolor had won its first Intercollegiate Championship of four in a row and its first Grey Cup of three in a row, and the team was hardly as famous as it would become by 1925. Yet the Queen's Journal, in December 1924, said that the letters CFRC "mean Canada's Famous Rugby Champions, or they may mean Crazy Fellows Raising Cain...or more melodious epithets which could be unprintable in this

Journal."[32] The last was a reference to the severe interference that CFRC was causing right across the dial for radio fans without selective sets.

There is nothing in the regulations to indicate that an applicant could or could not suggest the call letters desired. Canada was assigned the series CF— and CH— to CK— under the 1912 International Radio Convention. In 1922, the Department of Marine and Fisheries in Ottawa issued call letters in four main series, CF—, CH—, CJ— and CK—, without any apparent rationale. There was certainly no regional significance to the distribution of calls. In the first round of granting C-calls, in the spring of 1922, they probably took pains to assure that nothing could be construed from the call letters, and the same spirit seemed to pervade the 1923 and 1924 series. All four Canadian series showed two general patterns of call letters; for example, the CF series for 1924 had the two sub-series CFC- and CF-C. There were CFCA, CFCF, CFCH, CFCK, CFCL, CFCN, CFCQ, CFCR, CFCT and CFCU in the first series and CFAC, CFDC, CFHC, CFLC, CFQC, CFRC, CFXC and CFYC in the second. It is not impossible that Jemmett may have been permitted to select his own set of letters, but if that were so, the letters had to fit within one of the existing patterns. If he had had a free choice, he might well have selected CKQU for Canada-Kingston-Queen's University.

It wasn't until about 1926 that Canadian stations were able to apply for call letters that had special meaning for them. Examples are Toronto stations which got their calls changed in 1927, CKGW (Gooderham and Worts; began as CKCW in 1925) and CFRB (Rogers Batteryless; began as CJCQ in 1925), and a third, CKNC (National Carbon), which began operations in 1927.[33]

Building the First CFRC Transmitter

What happened in Fleming Hall during the rest of that summer and early autumn of 1923 is known only sketchily. Prof. Stewart consulted Douglas Geiger for his 1938 article, and stated there that the transmitter was rebuilt in the fall of 1923 by Geiger.[30] D.G. Geiger himself wrote in 1928:

> During the summer the experimental connections (for the 9BT transmitter) had been dismantled, so that the entire set had to be rebuilt. In doing this a few changes were made, the main one being in the antenna coupling coil...The (new) 3000 volt, 1200 watt motor generator set (replacing) the rectifier tubes... delivers a much smoother and more even voltage wave, and is more easily controllable than the rectified apparatus...At the opening of the fall session of 1923 the writer, then instructor in electrical engineering, accepted the offer to engineer and operate the station...[34]

A 1924 <u>Queen's Journal</u> article is more specific about the date:

> Work on the reconstruction of the set was begun about the first of September, but there were delays in the arrival of some of the new apparatus so that it was not until just a few days before the Queen's-McGill game, October 27, that the set was ready for testing.[35]

Prof. Jemmett's requisition book shows rush orders for magnet wire, condensers and other small parts from October 10th through the 15th.

Prof. Stewart believed that the 1923 CFRC transmitter also had the Heising constant current system of modulation, or plate modulation, in which the plate circuit of the oscillator and the plate circuit of the modulator were in common.

The experimental radio broadcasting station was housed in Laboratory No. 6, in the north-west corner of the basement of Fleming Hall, where 9BT had been. There was also in that room "a receiving set of very flexible design (i.e. the De Forest Inter Panel set). This laboratory in addition to being a Radio laboratory is used for the study of the characteristics of electron tubes as generators, oscillators and amplifiers. A number of tubes with the necessary variable condensers, reactors, A and B batteries, and wavemeters are available. Direct current up to 3,000 volts and 0.4 amperes is provided (by the Esco set) for plate voltages."[36] Laboratory No. 6 was described in the calendar until 1926-27, when CFRC moved upstairs in Fleming Hall. The complete outfit of portable wireless apparatus was still in use, and was in Laboratory No. 7, along with the old telephone exchange, lamps, a photometer and a Duddell oscillograph.

Plans for Radio Broadcasting

The Registrar and Vice-Principal, Dr W.E. McNeill, informed the Principal of the plans and questions in hand concerning the radio station on October 9, 1922:

> As you know, Professor Jemmett is very anxious to have material for broadcasting. He wishes to arrange to have someone report the two Inter-University Matches played in Kingston, and thinks that Professor Jolliffe, if he can be induced to do this, would be the best person. The plan is to have the game reported as it is played. Professor Jemmett, in addition, would like to broadcast weekly talks of about twenty minutes' duration. I suggested to him this morning that a small Committee representing the three Faculties might be appointed to confer with him and decide on topics and speakers. I suggest that you appoint such a Committee now in order that plans may be made promptly. Professor Jemmett's present plan is to broadcast on Wednesday nights.[37]

The Principal replied next day:

> I shall ask Professor Jolliffe to report the Queen's matches.
>
> As for a Committee to arrange for broadcasting, Dean Clark, Professor Matheson, Professor James Miller together with Professor Jemmett could, if they would act, handle this matter admirably.[38]

Professor Richard Orlando Jolliffe

R.O. Jolliffe was a Kingston native, born in 1876, the son of a Wesleyan Methodist minister. He was an Arts graduate of the University of Toronto in 1897 and received his Ph.D. from the University of Chicago in 1916. He first taught classics at Picton and Owen Sound, was then Professor of Latin at Wesley College, Winnipeg, from 1903-14, and became Professor of Greek at the University of Manitoba in 1914. In 1920 he came to Queen's to be Professor of Latin. He was an outstanding layman of the United Church,[39] and an inveterate speaker, in church and after dinner. There are newspaper reports of Dr Jolliffe's delivering Sunday sermons all over Ontario in the 1920s.

Prof. Alfred W. Jolliffe remembered his father as quite an athlete:

> He taught out at Winnipeg. In those days, apparently, a school could drag in some of the professors to play on their inter-collegiate teams, and Dad was a goalie in the Wesley College. One of my earliest recollections is Dad being brought home on a stretcher with a broken leg which he got in his goal tending.
>
> Now, as far as I know, my father was the first broadcaster of the Queen's football games. I think he was more or less semi-selected by Prof. Doug Jemmett. They were both members of an athletic group from the faculty that used to meet in the old Queen's gym and play basketball and so on. No question about it, Doug Jemmett was the star basketballer of this professors' athletic group, and I remember one professor of English who was probably the dirtiest basketball player I ever came up against. I was just a kid at the time but I filled in occasionally with the professors. At any rate, it was a nice little group and I think it was from that association that Jemmett

Dr Richard Orlando Jolliffe, Professor of Classics, was the announcer on the first public broadcast of CFRC, a football game against McGill, October 27, 1923. He sat at a little table in front of the stands at the old Richardson Stadium and called the play-by-play into a telephone handset.
-courtesy Catherine Jolliffe

suggested that my father broadcast the football games at Queen's University. He certainly wouldn't have anything to contribute to the electrical or the technical side, but for several years he broadcast the Queen's games. And this was very pleasant, because we used to go with him - a bunch of kids would go with him - all the way to Toronto for the big games including the Grey Cup Final and so on, back in the early '20s.[40]

R.O. Jolliffe was also an ardent football fan,[41] and even served as timekeeper at a Queen's-Argo exhibition game in Toronto.[42] And then there was the quality of Prof. Jolliffe's voice. The first year co-eds were always "startled when six feet and a voice of thunder walked into the room."[43] He addressed the students at a mass rugby rally in Grant Hall: "'Broadcast,' said the representative of the Athletic Board of Control. 'The best cure for a six-footer with a pee-wee voice is the yell - and yell, again - broadcast.'"[44]

All of these qualities, his interest in sports, his knowledge of the game of rugby, the excellence of his voice and his great oratorical powers, combined to make Dr Jolliffe a natural choice as the rugby voice of Queen's on CFRC. His students, however, suspected that he would call the games in Latin.

Preparations for the Big Broadcast

Now that the transmitter was re-assmbled and modified, it had to be tested and the operating characteristics of the tubes determined. Doug Geiger kept a detailed record in the back of the CFRC log book[29] of the tubes used, in both the transmitter and receiver. Each tube was assigned a number, and its accumulated hours and filament, plate and grid voltages were entered. The record is obscured, unfortunately, because he referred to the tubes by wattage rather than by manufacturer's model number. The earliest testing date in the book was October 22, 1923.

Geiger also left a brief account of the process of testing the new radio transmitter prior to the first broadcast:

> The day before this broadcast an interesting test had been made. Through the co-operation of Mr. O. Donnelly, one of the amateur radio operators in the City, contact was made with another radio amateur near Rochester. The Rochester amateur listened to announcements from C.F.R.C., each one consisting of a series of numbers of which each number corresponded to a different setting of some piece of the equipment. He then reported in code to Mr. Donnelly which numbers had sounded best to him, and Mr. Donnelly passed the information to the writer, who was making the adjustments. In this manner it was possible to determine the best adjustments of the apparatus, and in practically every case these were found to agree with those previously determined from the known characteristics of the vacuum tubes.[34]

One of Orton Donnelly's younger brothers, Robert, was only about seven years old at the time, but he was very interested in Orton's radio work with his amateur station 3HE, and he knew about the testing with CFRC:

There was a ham operator in Charlotte (pronounced "Shallot"), New York, near Rochester, that talked to Ort quite often, practically every night, and I believe that there was a chap from Queen's which also was in on this...and they were carrying out some kind of test or something. There used to be a three-way hook-up at times between Queen's, Ort and the fellow in Charlotte. But I have no idea now what his name is.[45]

Arthur John Wehman, a young Kingston amateur living at 251 Division Street, listened in on the earliest series of pre-broadcast tests with Orton Donnelly's 3HE and CFRC, and recorded what he heard in his log book, as entries 149 to 152.[46] In Art Wehman's shorthand, "Ex." meant that the reception was excellent, and the A.P. tube was in Art's receiving set. "Doug" referred to Doug Geiger rather than to Doug Jemmett.

> 149. 3.H.E. calling & testing with C.F.R.C. Test no. one, two, etc. (Ex.) 10.30 p.m., Oct. 23. A.P.
> 150. C.F.R.C. Queen's University (Kingston) testing with 3.H.E. test no. 1, two, three, four, etc. (Ex.) 10.30 p.m., Oct. 23. A.P. tube.
> 151. 3.H.E. calling C.F.R.C. Hello there "Doug". I didn't get that. The other numbers were all right. 11.55 p.m. Oct. 24. A.P. tube (Ex.)
> 152. C.F.R.C. testing with 3.H.E. (Ex.) A.P. tube. 11.55 p.m. Oct. 24.

Therefore, the testing of the CFRC transmitter actually took place as early as the Monday night before the big game. Art Wehman's logged notes suggest that in the tests Donnelly communicated with CFRC by wireless rather than by telephone, and this is corroborated in a 1924 Journal article:

Thanks to the co-operation of Mr. O.H. Donnelly of this city, who listened to the tests, and reported back by radio-phone, final adjustments were made. The final test on the afternoon of the day before the game was the broadcasting of speech direct from the stadium using a telephone line rented from the Bell Telephone Company between the stadium and the transmitting room.[35]

Early in this study, since no requisition for a Bell Telephone line could be found, there was speculation that the radio boys may have strung their own telephone wire from Fleming Hall over to Richardson Stadium. This evidence that they rented a Bell line shows that the radio boys did not have their own means of amplifying line signals, just as they

had had none for the Science Dance "broadcast" seven months before.
Lancely R.C. McAteer of the Radio Club also had a job to do for the
new station, possibly before the first CFRC broadcast: "They had an
antenna from the Physics Building (Ontario Hall) over to (Fleming
Hall), and I put up a counter-poise for them, oh about eight feet above
ground, and it had two cross-arms and about six wires ran across for the
counter-poise."[47] A counter-poise was sort of a second aerial, known as
an "earth screen," generally hung below the transmitting aerial close to
the ground, and used in those days in places where an efficient ground
connection could not be made. In a theory propounded in Germany, it
was the opposite of the elevated antenna.

The First CFRC Broadcast, October 27, 1923: Technical

Doug Geiger left a brief acccount of the first broadcast to use the call letters CFRC:

> On October 27, 1923, the first broadcast, a play-by-play
> account of the Queen's-McGill senior rugby game direct
> from the Stadium, was transmitted. Professor R.O. Jolliffe,
> sitting at a table in front of the grand stand, spoke into an
> ordinary telephone. The voice currents from the telephone
> travelled over a telephone circuit from the Stadium through
> the telephone exchange to the station at Fleming Hall,
> where, instead of going to a telephone set, they were
> applied to the first of the amplifier tubes of the transmit-
> ter. Thus Professor Jolliffe's voice was put directly on the
> air.[34]

Two Queen's Journal articles, published in 1924, tell the story of the
first broadcast in wonderful detail. One of them has a thorough contemporary description of the path of the signal from the microphone
through the new transmitter to the aerial. Combined with our knowledge of the tubes purchased for the new set, this information might
help in a future general reconstruction of the first CFRC transmitter:

> The best way to describe the set would, perhaps, be to fol-
> low a wave from the time it reaches the microphone as a
> sound wave till it leaves the aerial as an electromagnetic
> wave. Sound waves reaching the microphone, which in the
> transmission from the stadium was an ordinary desk
> phone, are changed by it to electric waves of exactly the
> same frequency and form. These electric waves or oscil-
> lations go from the microphone through a pair of wires to
> the primary of a transformer, known as a modulation
> transformer. In this transformer they are very weak, so
> that they must be amplified many times before they are
> strong enough.
>   The secondary of the modulation transformer is con-
> nected to the input of the first tube of the speech ampli-
> fier, which is a Mullard amplifying tube. The output of

this tube is coupled to the input of a second tube by another transformer. This second tube is a 5 watt Mullard. The output of this second tube is led to two 20 watt tubes operating in parallel through a reactance coupling. The output of these two tubes in turn feeding the grids of the two 250-watt modulator output of these two tubes, totalling 500 watts, is an exact reproduction of the input to the modulation transformer, but is increased in volume a million or more times.

Two other 250-watt tubes, the Oscillators, are so connected to a circuit which includes the primary of the aerial inductance that they generate and supply to this circuit an oscillating current at a frequency corresponding to a wave-length of 450 metres. These are the radio frequency oscillations. Also the two modulators are so connected to the two oscillators that the speech frequency waves from them are superimposed on the radio frequency oscillators in the primary of the aerial inductance. These composite oscillations are then impressed on the aerial through the secondary of the aerial inductance. Part of these oscillations are radiated from the aerial as electromagnetic waves.[35]

The CFRC log book,[29] a black-covered laboratory notebook, records the proces verbal of the broadcast of that Queen's-McGill game. There wasn't much to that proces verbal - the broadcast began at 2.00 pm and ended at 4.30 pm, Saturday, October 27, antenna current 6 amps. CFRC received letters reporting clear reception of the broadcast from nine listeners:

S.H. Hubbard, 162 East 2nd St., Oswego, N.Y. (40 miles)
C.H. North, Picton, Ont. (35 miles)
John Burkitt, Madoc, Ont. (50 miles)
J.A. Brown, M.D., Colborne, Ont. (75 miles)
R.J. Traill, 97 Belmont Ave., Ottawa, Ont. (90 miles)
E.T. Johnson, Antwerp, N.Y. (40 miles)
Percy Dorner, Seeley's Bay, Ont. (25 miles)
Lt. Col. D.W. Cameron, Ottawa, Ont. (90 miles)
F. Riley, Canfield, Ont. (190 miles)

and indirect reports from Montreal and Ste-Agathe, Quebec, and Brampton, Ontario, probably through Orton Donnelly of the Canada Radio Stores on Princess Street.[48,49] Dr Douglas W. Geiger, D.G. Geiger's son, confirmed that the first pages in the CFRC log book are in his father's hand.[50]

The First CFRC Broadcast, October 27, 1923: On the Field

A sports actuality going "live" to air today requires as basic equipment, in addition to the broadcast-quality telephone line to the transmitter, a separate ordinary telephone line for off-air communication between the announcer, or his producer, and the station. The station needs to be able to contact the remote site at any time to arrange cues, station breaks, commercial messages, late-breaking news and scores and, in general, to iron out any glitches.

In 1923, the techniques and practices for live remote radio broadcasting were still being developed, and people had to improvise as they worked. Whether Prof. Jemmett had obtained a second, off-air telephone line for Jolliffe in Richardson Stadium in 1923 has not been recorded. If there were no second line, the station would not have been able to communicate with the announcer once the broadcast started, except by runner, for example in case of telephone line or transmitter failure. The announcer, provided with only a one-way means of communication, would have had to keep talking through the entire period of the game, from the kick-off to the wrap-up. There were no such things as commercial breaks or interludes of pre-recorded interviews or music. This is why Prof. Jolliffe was admirably suited for the role of announcer in these early broadcasts.

No one has left an account of how the rugby broadcasts were produced in 1923. The two long Queen's Journal articles of 1924, however, tell a good deal about how the broadcasts of rugby games by CFRC worked in 1924, after J.W. Bain took over the operation of CFRC from D.G. Geiger. In the second article, a good deal of what went on behind the scenes was explained, as if by someone who was with the announcer at the stadium:

> Professor Jolliffe of the Arts Faculty has been the voice of C.F.R.C. on those great occasions when we have shown Varsity and McGill how to play football. He has been the eyes for the thousands of rugby enthusiasts scattered all over the province and even outside of it. From the time the ball is put into play till it is dead he is telling them what is happening or what he thinks is going to happen and then when the yelling has died down he tells them what actually did happen. It is no easy job we can assure you and Professor Jolliffe deserves great credit for the splendid way in which he has handled it. Then there is Professor Bain of the Electrical Dept. in the basement of Fleming Hall jiggling rheostats and wiggling condensers and otherwise nursing our bag of tricks (and a very good bag of tricks it is too, except for a few imperfections which will be remedied as soon as we get the cash). Finally, we must not forget Professor Jemmett who is the guiding genius of us all.
>
> We have a special line which runs through the local exchange and thence back into the Stadium. At the Stadium end we have a special long distance transmitter hooked up to the regular telephone apparatus so that we can call Professor Bain as soon as we are ready to start. As soon as he is ready he gives us a ring, then throws over our circuit and we are on the air. Running parallel with the

> aerial line inside the radio room, and spaced about two
> feet from it, is a line to which is attached Professor Bain's
> receiving apparatus so that he can correct any defects
> either at the Stadium or in the apparatus.[51]

The description of the telephone connections, above, defines the situation quite clearly. First of all, the "long distance transmitter" was a special microphone, perhaps putting out an augmented signal. It is most probable that there was just one ordinary telephone line installed by Bell, providing two-way communication between the stadium and the station. Before the broadcast they could call back and forth, but once the circuit was thrown over to the CFRC transmitter, the line became the unidirectional microphone line and Jolliffe was entirely on his own until the game ended. Back in the radio room, Bain was monitoring the quality of the signal from the aerial lead-in by induction.

The kick-off was at 2.30 pm. An important detail left out of the 1924 accounts, but implied in the log book, was that the broadcast did not start with the kick-off, if the plan for the second broadcast followed that of the first: "Those who tune in on our 450 metre wave length at 2 p.m. Eastern standard time will hear the first reports, line-up of the teams, weather conditions, and the like, to be followed at 2:30 by the running account of the game."[49] So Prof. Jolliffe had to begin each rugby broadcast with a half hour free-form monologue.

At half-time of the game, a topical demonstration was put on to entertain the fans and complement the efforts of the Queen's Band. Jolliffe was obliged to remain on the air throughout, informing his audience. And what a magnificently rich and ripe description it must have been! Had the stentorian professor been interested in an announcing career, he would have given heavy competition to Foster Hewitt's nasal tenor.

> At half-time a boxing ring was quickly fitted up in mid-
> field and Mrs. Dempsee and Mrs. Furpow gave a bur-
> lesque of the recent heavyweight championship battle.
> Both young "ladies" possessed splendid physiques and one
> looked very much like a prominent Queen's hockey play-
> er and sprinter. Mrs. Dempsee won easily. A free-for-all,
> with the contestants blindfolded, then went into the ring.
> All were knocked out in short order. The referee won.
> Spark Plug and Barney Google took a holiday from The
> Whig's Saturday comic section and paraded around for
> the inspection of the crowd.
>
> Another interesting little sidelight of the half-time
> session was the introduction to the crowd of "Alfie"
> Pierce, famous coloured mascot of other days, who still
> follows the team in all its ups and downs. "Alfie" lost
> his voice cheering for Queen's but his enthusiasm will
> never die. The crowd gave him a good Queen's yell.[52]

Queen's won the game, defeating McGill 19 to 3, and "Pep" Leadlay was sensational, scoring all of Queen's points by himself.[52] A professional photographer was there taking pictures of the game. One is a famous action shot of the legendary Leadlay-Batstone end run, but the camera was never turned to show Jolliffe at his little table in front of the grandstand.

Prof. Jolliffe had other problems than keeping track of the play, trying to maintain a continuous monologue and inventing sportscasting and colour commentary as he went along:

> **Alfred Jolliffe**: Dad told me after one broadcast, he got quite a lot of criticism and he said he was quite innocent. He broadcast from a seat in the Richardson Stadium, and he said right close by was a rather inebriated fan who was very loud-mouthed and coarse and so on, and he'd blast off a number of these obscenities and they would go out over CFRC and drown out Dad's voice. He claimed he was quite innocent but this other broadcaster, unofficial, was stealing his thunder.[40]

The Canada Radio Stores knew something about promotion. It provided a radio, and probably an operator, to The British Whig so that The Whig could follow the play on CFRC and post the results on the bulletin board in its front window. It was traditional for citizens, usually of the Liberal persuasion, to gather in front of The Whig's window to get the latest election and baseball results. People also telephoned The Whig for up-to-the-minute sporting results.[53] This tradition came to an end only around 1929, when most people had access to radios. Right from the first CFRC game broadcast, the Canada Radio Stores also piped the play-by-play onto Princess Street through a loudspeaker in front of the store.

Bill Taylor heard some of Jolliffe's rugby broadcasts over his homemade receiver. Bill is that unique and happy combination of radio amateur and avid football fan with a good memory:

> **Bill Taylor**: In the fall of 1923 they were broadcasting the Queen's rugby game. I did hear that in my room (Johnson Street at the corner of University) when I was going to Queen's. And that's the year before I graduated. I do remember listening to a football game from CFRC in the fall of 1923, and it was a very good broadcast. I could hear everything that was described about the game and I listened to it right through the whole game. It was quite well done. A good announcer and a good broadcast. A good transmission and it was a very clear reception. It was a Queen's game against McGill, which I wasn't able to go to because I was pressed for studying for exams and I listened to the full game...Yes, I think Queen's won the game.
>
> We could only find out by verbal enquiry when they were going to broadcast. Oh, it was mentioned very briefly in the newspapers, but there was no set means of telling when they were going to broadcast any day.
>
> One fault of CFRC is that they haven't been listed in a number of publications. And they're still not listed. And that had always applied

to CFRC.[54]

There was, in fact, no announcement in any newspaper prior to the broadcast of the October 27th game, but from the next week on there were sporadic notices in the "Coming Events" column on the front page of the Queen's Journal and also in The Whig.[55]

D.G. Geiger in Charge of CFRC

Prof. Jemmett proposed to appoint a fourth year student or a demonstrator to run CFRC when it got going.[9] In the fall of 1923, Doug Geiger had just completed an additional year of study to add Mechanical to his degree in Electrical. He had been a half-time demonstrator in E.E. in 1922-23, was appointed a full-time demonstrator for 1923-24, and was put in charge of CFRC in the fall of 1923: "...this set is now very complete. Arrangements are being made for broadcasting of weekly lectures to be given by members of the staff. It is hoped with the help of Mr. Geiger, who has been appointed to superintend this work, to make wireless of use both to the public and to the University."[56] Doug Geiger was probably the operator for all of the first year's broadcasts, as those CFRC logs are all in his hand.[50]

Bill Taylor and Lance McAteer enjoy remembering that Doug Geiger didn't know as much about the new technology as he should have. Keep in mind, though, that they were peeved that Geiger wouldn't let them into the Radio Room or the laboratory:

> I remember that Dr Jaffrey from Hamilton, he donated the vacuum tubes, great big things like basketballs, and the equipment for the first station. I remember the fellows in the Engineering Department, they didn't know that they had to put a bias on the grid of the tubes. They were trying to operate them without any but zero bias and they burnt up the first couple of tubes. Do you remember that, Lance?
>
> Yeah, I remember Prof. Jemmett and that other fellow who was his assistant, but I forget his name. He took a dim view of the Radio Club and excluded us from the radio station after Dr Jaffrey's gift arrived, because we weren't Electrical. He was a mean son-of-a-gun, Bill.
>
> Well, you're referring to Geiger, probably. Doug Geiger.
>
> Yeah, that's the guy! He wouldn't put any bias on the grid of those big Belgian lanterns, and he burnt up the first couple. But they finally got it going, and it was on voice.[57]

The story of the burning out of the large tubes is reflected in Jemmett's requisition book. On November 10, he ordered eight more 0/250B tubes from Mullard.[58] The 0/250B bulb was about five inches in diameter and was probably the oscillator (the final), ancestor of the VO/350 in the 1927 transmitter.

The Second CFRC Broadcast, November 3, 1923

The second CFRC broadcast was announced for 2.00 pm and the kick-off was scheduled for 2.30.[49] There was a student parade from the gymnasium to the stadium, beginning at 1.15, but the log shows that the transmitter was put on at 1.00 rather than at 2.00. Perhaps there was some testing done before the game, as there had been the day before, when CFRC tested with Harold O. Quick (8QL; though the log says 8COI) of Syracuse, NY.[29]

Ambitious plans were quickly announced for CFRC, likely emanating from the Broadcasting Committee which was now in place and made up of Professors Matheson, McNeill and Jemmett and chaired by Dean Clark. The Queen's radio services were directed at grads and extra-murals all over Ontario:

> ...another strong link will be forged in the bonds which bind the graduate to his Alma Mater; and the extra-mural, within reasonable proximity of the college will no longer need to consider himself "beyond the walls".
>
> Besides the daily broadcasting of sports news, elections, important speeches and so forth, the Journal has arranged to give a summary of the "Week's doings" each Wednesday evening.
>
> Word was sent to Miss A.E. Marty, B.A., LL.D. asking her view on the broadcasting of her address when she lays the cornerstone of the new Women's Residence, on Saturday next. Miss Marty has accepted, and her words will hence in double fashion express the growth and advance of Queen's.
>
> It is understood that within three weeks, arrangements will be made to give Radio connection with Convocation Hall, so that Queen's enthusiasts not within the precincts, will be able to obtain the speeches of visiting celebrities, and university lectures red hot off the wires.
>
> Connection with Grant Hall and the Jock Harty Arena will also allow for the sending out of concerts, convocation services and hockey games. It is also planned to broadcast the College Frolic to those unfortunates unable to be present in person.
>
> It is estimated that at least thirty per cent. of our former students possess receiving sets, while the number of schoolboy amateurs who will in this way hear about Queen's is numberless.
>
> The wave length is 450 metres. The call is CFRC - Ubiquitous Queen's.[59]

The Third and Fourth CFRC Broadcasts, November 10, 1923

Dr Marty's address and the laying of the corner stone of the new women's residence, Ban Righ Hall, were broadcast by CFRC from 4.30 to 5.30 pm on November 10, immediately following the broadcast of the Varsity-Queen's III rugby game from Richardson Stadium.[29] Miss Aletta

Marty was handed the silver trowel by Miss Charlotte Whitton, M.A., of Ottawa, President of Queen's Alumnae Association, and Principal Taylor spoke briefly afterward. The ceremony and speeches were reported in the papers,[60] but there is no reference there to the CFRC broadcast and no photograph of the proceedings. Neither is there any credit in the log book to the producers nor any account of the techniques employed in their mounting back-to-back remote broadcasts from different locations.

Ban Righ Hall began going up in the spring of 1924. Kathleen (Whitton) Ryan was at Queen's all through the construction and lived in one of the old residences: "The Hen Coop was the original women's residence, at Clergy and Earl streets, and even in 1923 they were using coal oil lamps because it was in the spirit of the early days of Queen's, as long as the lamps didn't upset. And then there was the Avonmore, and the girls at the Avonmore were a bit more sophisticated. They had electric lights. I'm glad to say that and get back at the Avonmore."[61]

The Journal radio news service was begun at 7.00 pm on Wednesday, November 14, with John Campbell McGillivray and Mr Lyght speaking.[29] The program was thirty minutes in length and presented material from the Queen's Journal of the preceding Friday and Tuesday: "Members of the Journal staff spoke into the transmitting apparatus in Fleming Hall and told the radio public what was going on around the University. In the future this will be a regular department of the Journal, supplementing the publication, with a digest of all the current events taking place at Queen's."[62] As of November 21, the time of the Journal broadcast was changed to 9.00 pm, and it continued irregularly on Wednesdays until the end of the spring term.

John Campbell McGillivray (Arts '23 & '24) was the Journal's Radio Director for the 1923-24 session, and Kathleen Ryan knew him: "McGillivray was, I think, head of the Officers' Training Corps. Very tall, and very intelligent chap. His father was Professor of German there."[61]

Those interested in radio were anxious to hear CFRC's first program with music and variety entertainment, just like the big stations south of the border. The Journal correspondent for Science '25 submitted a proposal for a radio schedule "for the first Broadcasting Programme of the Queen's Radio Club."[63] This gives a sample of current college humour but, more importantly, reflects the common form of radio scheduling of the day:

> 8:00 p.m. - Introductory announcement, "Station Q.B.C.
>   speaking, we beg to announce.." -
> 8:10-8:10:30 - Queen's yell to clear the air of static.
> 8:15-8:30 - "Why Hamilton boys always make good
>   (liars)" by one of them.
> 9:00-9:25 - Marshall Reid's Orchestra will render "Soup
>   of the evening, beautiful soup".
> 9:30-9:45 - A soprano solo "Ain't Nature Grand" by
>   the Levana Choir.
> 9:50-10:15 - Talk "The Effect of Dancing on Morals"

by Bert. Quance.
10:20-10:25 - Song by Skinner (E.W.) "Here's to Good
    Old Queen's, (Drink Her Down)".
10:30-10:45 - Choir recital by entire Staff, "How Dry
    We Are".
10:50-11:05 - Time Signals by the Grant Hall Clock.

The Fifth CFRC Broadcast, November 24, 1923

The fourth and last rugby game broadcast by CFRC in its first year
was the Eastern Canada Finals between the Queen's Tricolor and the
Hamilton Tigers, played in Kingston on November 24. The radio transmission began at 2 and finished at 4 pm, according to the log. Listeners
wrote in to tell CFRC that "the cheering and shouting can be distinctly heard while such details as the blast of the referee's whistle and the
shouts of encouragement to the various individual players also come in.
One fan writing from a point in Prince Edward County said he could distinctly hear the band but that it sounded like a dinner orchestra."[64]
The listeners surely heard the band playing Oscar Telgmann's new composition "The Mascot - Boo-Hoo's March,"[65] dedicated to Queen's
bear mascot. Once more, the Canada Radio Stores had a receiving set
in The Whig offices to supply bulletins to the wireless-less public. "The
radio account of the game was exceptionally clear and interesting..."[66]
One of the reception reports sent in to CFRC was from Mrs E.A. Geiger
of Brockville, Doug Geiger's aunt. Doug had built the receiver for her.[50]

    One can hardly imagine Prof. Jolliffe's florid description of the
spectacle that took place over and on the Richardson field before kick-
off:

> Just before the game started a Curtiss Plane from the
> Hamilton Daily Spectator flew over the Stadium and
> dropped several parachutes made up of cloth of black
> and yellow, the Hamilton colours. On its last trip over it
> dropped a big floral horse-shoe made up of yellow 'mums.
> This was presented to Captain Tuck, of the Tigers, by
> Controller Treleaven, of Hamilton. After Queen's had
> emerged with a victory (13 to 5), Capt. Tuck passed the
> emblem of luck on to Capt. "Doc" Campbell, with the
> best wishes of the Ambitious City Team.[67]

    This action by the Hamilton Captain "was about the most graceful
thing that has ever been done in Richardson Stadium, and the Queen's
players appreciated it."[68] The next Saturday, the Queen's Tricolor was
going to Varsity Stadium in Toronto to play Regina for the Dominion
Championship and the Grey Cup.
    A couple of days before the final, trainer Billy Hughes turned his
back on the team mascot, Queen Boo-Hoo the bear cub, and she ate the
good luck floral wreath.[69] Perhaps this was for the best, because the
wreath hadn't brought much luck to the Tigers. Then, very early on
Saturday morning, someone let Boo-Hoo out of her suite in the King

Edward Hotel, and she ran around gawking at the bright lights of Toronto until the pajama-clad team rounded her up.[70] Queen's was favoured - by some to double the score on Regina. But the Regina team was big and strong, and confident because recent heavy rains had made the field sloppy and treacherous. The Toronto Telegram and the fans at Varsity Stadium were for Regina, but the west had nothing to counter the Tricolor machine and Queen's routed them 54 to 0.

CFRC did not carry the game from Toronto, and there is no evidence of a local Toronto broadcast either. CFCA, the Toronto Star Station, aired a summary only, at 4.30 pm. Besides, 600 fans went from Kingston, and probably the Queen's radio people were among them. The Whig had the scores telegraphed from Toronto, and these were chalked up on the score-board at the RMC v/s Grand Trunk final at Richardson Stadium. The score-keeper at the Stadium ran out of chalk that day.

According to the CFRC log,[29] Doug Geiger et al. tested with Mr H.B. McKenzie on November 28, with Mr J.D. Murray (Elec. '25?) on the 29th and with Prof. Jemmett on December 5, the afternoon of a Wednesday evening news broadcast. The tests used between 5.0 and 6.0 amps in the aerial. The next Wednesday, from 9.00 to 9.30 pm, CFRC logged a broadcast of music for the first time, although no details survive. The CFRC microphones were incapable of transmitting any sort of frequency range, so it probably sounded worse than music over the telephone.

On November 26, Jemmett wrote up two reqisitions. The first was "An order on F. Salisbury, 21 - 6th Street, Kingston, for two oak cabinets for Radio broadcasting," estimated cost $8.00,[71] and the second was for five sheet steel boxes, 6"x6"x4" deep. Don't imagine that the CFRC transmitter was completely enclosed in a protective cabinet, because it wasn't. An open design was rather typical of amateur designs of the 1920s. One visitor to Fleming Hall knows that there was no safety enclosure:

> Dr Donald Rose: One silly thing I do remember is...they had this transmitter out on a table in one of those rooms in the basement of Fleming Hall. It must have been the summer of '24 or thereabouts. I was poking around and I put my finger on a hot...or on one of the high frequency things. Well, it was all right as long as I held my finger there. You don't get a shock at high frequencies, you know, but when I tried to take my finger away it went into an arc about that long. Burned a hole in my finger. Well, I knew when I was doing it. I knew once I put my finger on it, how am I going to get that off? And I was scared to leave it there long, because, if it was transmitting, I was upsetting the frequency, you see. So, I just picked my finger off quick, but it still burned a hole in my finger.[72]

> G.R.M. Garratt: A commercial transmitter, even in the early '20s, had to be safe for untrained operators, whereas the 1927 CFRC transmitter has got unprotected wires and terminals carrying high voltages, perhaps 3,000 to 5,000 volts, immediately accessible to impart a lethal shock to any careless operator! Extremely dangerous for any-

one who didn't know **exactly** what he was doing and even then, a moment's carelessness and...![73]

The next job was the stringing of non-Bell telephone wires around campus for the relaying of remote broadcasts from various spots to Fleming Hall. Jemmett submitted a requisition on November 28 for the "Services of Electrician to install wires from Fleming Hall to Grant, Convocation Halls, Gym & Rink. This should be done at once before bad weather starts."[74] They did a very sensible thing, running four wires to each of the four locations: "Two of these wires are used for transmitting the speech and the other two for signalling between the transmitting room and the place from which the broadcasting is done."[35] A long and ambitious series of sports broadcasts followed throughout January and February 1924, hockey from the first Jock Harty Arena and basketball from the old gymnasium, now Jackson Hall. The technical details of the transmissions are recorded in the CFRC log book,[29] but the announcer is not identified.

The first venture with the new lines was a match between Queen's All-Stars and the Canadian Olympic Hockey Team, 7.40 to 10.20 pm, January 7. Queen's was trounced 7 to 0, and just one reception report was received, from 600 miles away, in Chicago. A week later it was Queen's against the University of Montreal in a Senior Intercollegiate game, and people wrote that they had tuned it in at Gananoque, Sea Cliff, NY, and Milwaukee. The Wednesday evening Journal News transmissions on the 9th, 16th and 23rd got responses from as far away as Hyde Park, Vt, and Boston, estimated at 200 and 300 miles, respectively.

The next game broadcast was the Intercollegiate hockey match against Varsity, on February 8, when CFRC was heard in Napanee, Trenton, and Marlboro, Mass. The following evening they aired a basketball game against Varsity, reports coming in from Trenton, Worcester, Mass., and Plymouth, NH. Ambition got to them on the 15th and 16th, and they really stretched themselves. At 8.00 on the first night it was a hockey game against McGill, and they received reports from Windsor, NS (600 miles), Philadelphia (320 miles), and Wollaston and Brookline, Mass. (275 miles). They also got a telegram from Dr Jaffrey in Hamilton, indicating that they had indeed reached an alumnus "beyond the walls."

The following night was a real battery bruiser, when they broadcast in succession Belleville-Queen's II basketball, McGill-Queen's Senior basketball, and the West Point-RMC hockey game! CFRC signed on at 7.15 and stayed on the air until 10.30 pm. Jaffrey tuned in to 450 metres again that night, and so did the fellow from Worcester. CFRC also got mail from Worthington, Minn. (950 miles), Rennselaer, NY, and Wallingford, Pa.

Broadcasts of Women's Sports

There followed a series of broadcasts of the Girls' Intercollegiate Basketball Meet at Queen's, February 21-23. There were two games transmitted on the first night, KCI versus Queen's and then McGill against Varsity. These were picked up in Greenville, SC (750 miles), and Shelton, Conn. (250 miles). Varsity played Queen's on the 22nd, and McGill played Queen's on the afternoon of the 23rd. This last was heard in Rennselaer and Syracuse. The Journal did not report the games or the scores.

The story of the final broadcast of the first season needs a little backgrounding. Queen's co-eds were genuine pioneers in ice hockey. There was a women's hockey team at Queen's years before the Toronto and McGill distaff ever started playing the game. The Queen's teams even endowed a cot in the Queen's Overseas Hospital in the First World War.[75] The women's faculty at Queen's, Levana, wanted to revive hockey in 1923. So, in December 1923, the Queen's women challenged Varsity and McGill to admit them to the Intercollegiate Hockey League, and they began to advertise for players.[76] McGill opted out, and Western, Ontario Agricultural College, Aura Lee and North Toronto declined, so the series was Queen's against Varsity for the Beattie Ramsay Cup, currently held by Varsity.[77] Varsity won the first match 4 to 1. The second, broadcast on February 25, saw Varsity defeat Queen's by 2 to 0 in a hard-fought match and retain the Ramsay Cup.[78] One listener wrote in to CFRC, from Boston, Mass.

The Queen's Journal was on the air only once in February, at 9 pm on the 20th. There was an announcement of a short studio program for 9 o'clock on the 22nd,[79] but this does not appear in the CFRC log book.

Jock Harty Arena Burns Down

The new Jock Harty Arena, built "to perpetuate the memory of the late Dr. J.J. Harty, one of Canada's greatest athletes and a star member of the famous Queen's hockey team of `the days of Captain Curtis'," ran east-west behind the New Medical Building (renamed Kathleen Ryan Hall) from Arch Street through what is now Humphrey Hall and a parking lot. A week after the last CFRC broadcast from the arena, the second Queen's-McGill women's hockey match, the structure was destroyed by fire.[80] The radio station surely lost its lines and jacks. It comes as no surprise that young Don MacClement was on the spot that night.

Don MacClement: The whole thing was steel frame - covered - and all the seats were oiled wood. It was so hot it melted all those big girders. They just crumpled up like doughnuts. And our club's stuff - we had a Circle Six hockey team - our stuff was down in the basement in one of the little dressing rooms. There was deep snow between the end of the Jock Harty Arena and the Curling Rink, and the big pile of snow that was in there saved the Rink. It was steam - great clouds of steam running out of that melting snow. We got into the window of our dressing room and we were collecting uniforms and the rest and stuffing them out the window. I remember a fireman putting his head through the window and he says, "Do you want to come out now, or are you going to be buried in there?" We got all of our stuff out and eventually the whole thing collapsed.[81]

Fred Simpson, Jr, Radio Editor of the <u>Daily Standard Junior</u>, announced in late May that "CFRC, Queen's University, will not be on the air until next fall."[82] It is much more likely that he heard this news through the amateur grapevine than through the University.

Altogether, in its first year of existence, CFRC was on the air 23 times, totalling 60 hours:

> Full running reports were broadcasted of all Senior Intercollegiate games, and, as the numerous replies (about sixty) showed, were much appreciated. The <u>Journal</u> gave a radio edition on Wednesday evenings and is planning to enlarge this feature next year. The station was heard as far east as Windsor, N.S., 600 miles; as far south as Greenville, S.C., 750 miles, as far west as Worthington, Minn., 950 miles, and Oregon, 2300 miles. There is no question of the range or efficiency of modulation of C.F.R.C. local opinion to the contrary notwithstanding. Very little more is needed except a special microphone for music.[83]

Notwitstanding the optimism of the Principal's Report, there continued to be problems with CFRC interfering with local reception of stations on adjacent frequencies. Non-selective tuners were still quite common and, when it was transmitting, CFRC would swamp the whole dial of these primitive sets. CFRC just couldn't be tuned out. This was long before the days of transmitters with master oscillators and crystal-controlled frequency. The frequency stability of the first CFRC transmitters was determined solely by the circuitry and by the constants surrounding the final, the transmitting tube itself. The frequency would even vary with temperature and with loading.[84]

In the spring of 1924, Doug Geiger got a research job with Bell Telephone in Montreal and left Queen's.[85] He was an engineer with the Transmission Division, General Engineering, and designed, built and installed a single-channel telegraph with a capacity for 400 words per minute. J.W. Bain (B.Sc., McGill, '14) was assigned to look after CFRC in 1924-25.

What Became of 9BT?

Prof. Jemmett renewed the licence for 9BT in the spring of 1923 for use in his own continuing research into electromagnetic radiation from parallel wires. The publication of his results, expected for the summer of 1922, did not materialize. His progress was reviewed by the Science Research Committee:

Professor Jemmett's work, in more popular language, is an attempt to get precise quantitative information as to what goes on when wireless messages are being sent out from the aerial of a sending station. There was some not clearly understood discrepancy between the amount of electrical energy given to the aerial in the form of wireless energy. There can be no doubt that the more accurate information we have on such matters, the better for the advance of wireless telegraphy and telephony. It is almost needless, in view of the well known widespread interest in wireless at the present day, to point out the opportunities for making it more perfect. Professor Jemmett's work has an indirect bearing on the possibilities of transmission of power by wireless, a feat, if ever realized, destined to outshine all the other triumphs of wireless.[86]

The last news published on this research project appeared in the 1924 Principal's Report: "Professor Jemmett has made considerable progress in his research on the radiation of energy from wireless antennae. He is planning to make a brief investigation of the accuracy of induction type watt-hour meters with rectifier loads."[87] The intractable integral had still not yielded, and Jemmett was switching his interests over to another problem.

Experimental radio licence 9BT was not renewed by Jemmett for 1924-25. It was assigned to W.E. Beattie, 287 King Street, London, Ontario, in 1926-27,[88] and from 1929 to 1932, as VE9BT, belonged to the Ontario Forest Branch, Port Hope.[89] VE9BT was not listed again until 1937, when it was held for one year by the Borealis Co., Ltd., Toronto.

NOTES

1. CFRC script, "Documentary for Saturday, Oct. 27, 8:30" (1962).
2. W.F. Nickle Papers; QUA, Coll. 1027.
3. "The Halls of Queen's", DBW, Feb. 14, 1922, p. 2.
4. Personal communication from William Jaffrey, Hamilton, Oct. 3, 1985.
5. R.R. Forsey, "History of the Canadian Dermatological Association",
   Archives of Dermatology 91, May 1965, p. 486.
   A.R. Birt, "The Canadian Dermatological Association: First 50 Years",
   International Journal of Dermatology 16 #4, May 1977, p. 289.
6. W.R. Jaffrey to Prof. Harold Stewart, Jan. 17, 1939.
   Personal communication from Dr. John Card, Hamilton, March 17, 1985.
   "Amateur Experimental Stations, District No. 3, Licenced Fiscal Year
   1926-27" in From Spark to Space: The Story of Amateur Radio in
   Canada, Saskatoon Amateur Radio Club VE 5AA, Saskatoon, 1968,
   p. 135.
7. Secretary to the Principal, Queen's University, to J.S. Macdonnell, May 3,
   1926; QUA, Coll. 1250, Box 19, Principal's Office, Series #1, Subject
   Files R-S, Radio Station CFRC, 1923-52.
8. R.B. Taylor to D.M. Jemmett, April 26, 1923; ibid.
9. D.M. Jemmett to R.B. Taylor, May 2, 1923; ibid.
10. R.B. Taylor to W.R. Jaffrey, May 3, 1923; ibid.
11. W.R. Jaffrey to R.B. Taylor, May 4, 1923; ibid.
12. D.M. Jemmett to R.B. Taylor, May 25, 1923; ibid.
13. R/B. Taylor to W.R. Jaffrey, May 30, 1923; ibid.
14. W.R. Jaffrey to R.B. Taylor, June 5, 1923; ibid.
15. Requisition No. 2810, April 2, 1923; from Prof. Jemmett's first requisition
    book, courtesy Prof. H.H. Stewart.
16. Requisition No. 2826, June 28, 1923; ibid.
17. Personal communication from Ian Nicholson, North Littleton, Worcs.,
    (formerly of the Mullard Radio Valve Co., Ltd.), Sept. 26, 1987.
18. Requisition No. 2831, signed by Prof. D.M. Jemmett, July 30, 1923.
19. Requisition No. 2840, signed by Prof. Jemmett, Sept. 4, 1923.
20. Requisition No. 2847, signed by Prof. Jemmett, Sept. 18, 1923.
21. Personal communication from Prof. Harold H. Stewart, Kingston, Sept. 23,
    1987.
22. Official List of Radio Stations of Canada, Department of Marine and
    Fisheries, Ottawa, July 31, 1923.
23. "Radiotelegraph Regulations", Supplement to the Canada Gazette,
    Ottawa, Sept. 23, 1922.
24. "Radiotelegraph Regulations": Form W69, "Private Commercial
    Broadcasting Licence", p. 17; ibid.
25. Official List of Radio Stations of Canada, Ottawa, Aug. 1, 1922.
26. "58th Annual Report, MF: Radio Service", Annual Departmental Reports
    6, Ottawa, 1924-1925, p. 136.
27. Ibid., 3, 1926, p. 144.
28. "60th Annual Report of the Deputy Minister, MF", Annual Departmental
    Reports 4, Ottawa, 1926-27, p. 142.
29. The original CFRC log book, now kept at Queen's Archives, begins with
    he first CFRC broadcast, Oct. 27, 1923, and ends with that of Feb. 22,
    1932.
30. H.H. Stewart, "The Origin and Development of CFRC", PESQU 28, Sept.
    1938, p. 30.
31. Personal communication from W.J. Hurn, Noranda, Quebec, Oct. 19, 1982.

Mr. Hurn says that his family moved away from Kingston the year before the Haileybury fire, which was in October, 1922. Since he remembers the call CFRC, it is more probable that he left the year after the fire.

32. "Radio Forms Interesting Journal Dept.", QJ, Dec. 2, 1924, p. 1.
33. T.J. Allard, Straight Up: Private Broadcasting in Canada: 1918-1958, Canadian Communications Foundation, Ottawa, 1979, p, 38. The International Radiotelegraph Convention, London, 1912, also assigned VAA to VGZ to Canada. The 1927 Convention in Washington assigned CFA- to CKZ-, as well as VAA to VGZ, to Canada.
34. D.G. Geiger, "Radio Broadcasting at Queen's", QR 2, Jan. 1928, p. 9.
35. "Radio", QJ, Feb. 22, 1924, p. 1.
36. Calendar of the Faculty of Science, 32nd Session, Queen's University, Kingston, 1924-25, p. 89.
37. Dr. W.E. McNeill to R.B. Taylor, Oct. 9, 1923; QUA, Coll. 1250, Box 19, Principal's Office, Series #1.
38. R.B. Taylor to Dr. W.E. McNeill, Oct. 10, 1923; ibid.
39. "Deaths: R.O. Jolliffe", QR, 7 #1, Jan. 1933, p. 26.
40. "Personal communication from Prof. Alfred W. "Fred" Jolliffe, Kingston, Aug. 27, 1982.
41. "Queen's Alumni Banquet Team", QJ, Oct. 31, 1922, p. 1. "Staging Great Comeback", KDS, Nov. 20, 1922, p. 10.
42. "Local News", DBW, Nov. 24, 1922, p. 20.
43. "The Bunk", QJ, Oct. 18, 1927, p. 8.
44. "Mass Meeting", ibid., Oct. 7, 1924, p. 8.
45. Personal communication from Robert Donnelly, Brantford, June 30, 1986.
46. "Radio Stations: Personal Log of Arthur Wehman", 1922-23; courtesy Arthur Wehman.
47. "Personal communication from L.R.C. McAteer, Havelock, Aug. 21, 1982.
48. "Queen's Broadcasting Report", DBW, Nov. 3, 1923, p. 7.
49. "Broadcasting", QJ, Nov. 2, 1923, p. 8.
50. Personal communication from Dr. Douglas W. Geiger, Kingston, Sept. 6, 1982.
51. "Radio Forms", QJ, Dec. 2, 1924, p. 1.
52. "Leadley's Sensational", DBW, Oct. 29, 1923, p. 9.
53. "Whig Bulletins", DBW, Nov. 5, 1923, p. 10.
54. Personal communication from William Arthur Taylor, Bath, Aug. 21, 1982.
55. "Queen's Broadcasting of Rugby", DBW, Nov. 3, 1923, p. 7.
56. Minutes, Board of Trustees, Queen's University, Oct. 19, 1923, p. 171; QUA.
57. Personal communication from William A. Taylor and Lancely R.C. McAteer, Aug. 21, 1982.
    Personal communication from L.R.C. McAteer, Dec. 3, 1988.
58. Requisition No. 2425, signed by Prof. Jemmett, Nov. 10, 1923.
59. "Queen's Broadens Educationally", QJ, Nov. 9, 1923, p. 1.
60. "Corner Stone", DBW, Nov. 13, 1923, p. 9. "Concerning Milestones", Alumni News 8, Dec. 1923, p. 26.
61. Personal communication from Mrs. Kathleen Ryan, Ottawa, Nov. 7, 1982.
62. "Coming Events", QJ, Nov. 9, 1923, p. 8. "Broadcasting", ibid., Nov. 16, 1923, p. 8.
63. "Science '25", ibid., Nov. 16, 1923, p. 6.
64. "Broadcasting Welcomed Afar", ibid., Nov. 23, 1923, p. 7.
65. "Prof. Telgmann's Spirited", ibid., Nov. 30, 1923, p. 1.
66. "Whig's Scores Were Received", DBW, Nov. 26, 1923, p. 10.
67. "Eastern Rivals Vanquished", QJ, Nov. 27, 1923, p. 1.
68. "Terrible Tigers Sent Home", DBW, Nov. 26, 1923, p. 7.
69. "Boo-Hoo Eats Floral", ibid., Nov. 30, 1923, p. 3.

70. "Queen's Bear Breaks Away", ibid., Dec. 1, 1923, p. 1.
71. Requisition No. 2431, signed by Prof. Jemmett, Nov. 26, 1923.
72. Personal communication from Dr. Donald C. Rose, Brockville, May 9, 1986.
73. Personal communication from G.R.M. Garratt, retired Keeper, Telecommunications and Aeronautics Section, Science Museum, London, UK, Feb. 20, 1988.
74. Requisition No. 2435, signed by Prof. Jemmett, Nov. 28, 1923.
75. "World of Sport: Queen's Girls", DBW, Feb. 4, 1922, p. 15.
76. "Levana: Girls' Ice Hockey", QJ, Dec. 14, 1923, p. 5.
77. "Levana: Few Women Turn Out", ibid., Jan. 18, 1924, p. 5.
78. "Varsity Co-eds Win", ibid., Feb. 29, 1924, p. 1.
79. "Radio", ibid., Feb. 22, 1924, p. 1.
80. "Jock Harty Arena Destroyed", ibid., March 4, 1924, p. 1.
81. Personal communication from Dr. W. Donald MacClement, Kelowna, BC, Sept. 25, 1987.
82. "Daily Standard Junior", KDS, May 31, 1924, p. 9.
83. "Electrical Engineering", Report of the Dean, Faculty of Applied Science, in Principal's Report, Queen's University, May 7, 1924, p. 27; QUA.
84. G.R.M. Garratt, Esher, Surrey, to Ian Nicholson, North Littleton, Worcs., Nov. 21, 1987; courtesy Ian Nicholson.
85. "Additions to Arts Faculty", QJ, Sept. 28, 1926, p. 4.
86. "Science Research Committee Report to Finance and Estates Committee", p. 1; W.F. Nickle Papers, QUA, Coll. 1027, Box 2, file folder 55.
87. "Report of the Science Research Committee", Principal's Report, Queen's University, May 7, 1924, p. 52.
88. From Spark to Space, p. 143.
89. Official List of Radio Stations of Canada, Jan. 1, 1932.

CHAPTER 6

VBH, *the government-owned Marconi-operated ship-to-shore station on Barriefield Hill, 1914-1977.*
The single storey operating house is visible to the right of the trees, and the semi-detached dwelling houses are behind the trees. One of the antenna poles, red and white striped, is at top centre, and the cross-braced tower at the right is part of a small weather station.
-photograph by Harry Marlatt

The Marconi Station at Barriefield, VBH

The presence of the Marconi coast stations all around Canada's Atlantic coast, up the St Lawrence and through the Great Lakes, and along the Pacific coast, had a profound effect upon the birth and development of amateur radio in this country. The signals from these stations were almost the first that our amateurs could detect, and in many cases the young Marconi operators generously assisted boy experimenters in their building and research into the new medium. The Marconi shore wireless station on Barriefield Hill, VBH, was certainly a vital component in the development of amateur wireless in Kingston.

The historic memorandum of agreement between the Dominion of Canada and Marconi, which gave rise to the Dominion Wireless Service, one of the earliest chains of Marconi-operated shore-to-ship stations in the world, came about because of an accumulation of accidents of history and opportunities seized. It is useful to outline how the great inventor came to make one of his earliest contracts for the construction of a wireless chain with the Dominion of Canada.

Marconi Comes to Newfoundland

Signor Guglielmo Marconi's major project of 1901 was to demonstrate that he could send wireless telegraph messages 2,000 miles across the Atlantic Ocean. Many prominent scientists maintained stoutly that electromagnetic radio waves, like light waves, travelled in straight lines only. Accepted wisdom was that the waves would not bend to follow the curvature of the earth but would shoot off tangentially to pass unavailable, miles above the intended point of reception. Marconi was not aware of Heaviside's theory of the existence of the ionosphere, the layer which would reflect certain wavelengths back down to earth, but his practical experience had shown him that the waves could be picked up at long distance and therefore must bend to get there.

The original plan was to send from his new station near South Wellfleet, Cape Cod, Massachusetts, and receive at his station at Poldhu, near Lizard Point, Cornwall. On both sides of the ocean Marconi's antennae were huge circular arrays of twenty wooden masts, 200 feet high and rather fragile,[1] with an inverted cone of wires inside. High winds brought down the Poldhu masts in September 1901, and the next month the Cape Cod station was destroyed in a storm. Marconi replaced the Poldhu antenna with a simpler structure, a huge fan of nearly vertical bare wires 200 feet across at the top and hung between two 150-foot masts.[1,2] This new antenna was used in November when he achieved his best distance to date, 225 miles between Poldhu and Crookhaven on the west coast of Ireland.[1] As for his trans-oceanic signalling experiments, however, Marconi figured that the three months required to rebuild Cape Cod was too long to wait, so he decided to try transmitting from his powerful station at Poldhu to a "purely temporary installation"[2] across the Atlantic at St John's in the old British colony of Newfoundland.

T.J. Murphy, Newfoundland's Minister of Marine and Fisheries, had recently sent Marconi a newly published chart showing the locations of shipwrecks in Canadian coastal waters.[3] It showed 80 wrecks off the Newfoundland coast, which had cost 600 lives and $20,000,000. Marconi was attracted by such dangerous spots because they were obviously the most useful places to put wireless ship stations. If a ship foundered just out of sight of port in those days, it was beyond contact and might as well have been 1,000 miles at sea. This is partly why Cape Cod had been his first trans-Atlantic choice, and partly why, after correspondence with Murphy, Newfoundland became his second. It was also important to Marconi that there be nothing but open water between his stations. Experience told him that wireless waves travelled most efficiently over water, and that intervening islands would slow the signals or derange the experiments. Newfoundland was, however, a curious choice for a finicky wireless experiment in the month of December, when the weather there is frightful - stormy, with high winds, rain and sleet.

Marconi arrived in St John's aboard the RMS <u>Sardinian</u> on December 6 with two assistants, his manservant George S. Kemp and technician P.W. Paget, piles of wooden crates and a large wicker hamper containing two 14-foot balloons and six Baden-Powell kites.[1,3,4] There was no time to put up an array of masts here - besides the weather was too bad - but by means of a balloon or kite a single-wire temporary aerial could be rapidly raised hundreds of feet. From the time his ship came into the Narrows at the entrance to St John's harbour, Marconi had his eye on Signal Hill, on the ship's right and overlooking the open ocean straight across to Cornwall.

At that time, there was a local marine wireless station going up at Cape Race, Newfoundland,[3] as well as a pair working across the Strait of Belle Isle between Chateau Bay, Quebec, and the Belle Isle lighthouse. The submarine cable across the Strait had been put out of commission by ice in October 1901, and all through 1902 Bell Isle and Chateau Bay maintained contact by wireless.[5] Emergency "aerial telegraphy" across the Strait may have been the idea of the Society of Lloyd's, London, which would soon be giving a discounted rate to ships using Marconi's wireless.[6] The Dominion signal station at Cape Race was even then a Lloyd's reporting station, before it had wireless capability, and the Society was interested in including Belle Isle in its system. The Government of Canada appreciated the potential of the new signalling technology too, and was awaiting news of Marconi's work at St John's, since it was thinking of installing a chain of wireless stations around the Gulf of St Lawrence.

Marconi talked to the press on Friday the 6th, the day of his arrival, but he kept his true plans secret. He told them that he hoped to achieve a transmission distance of 400 miles during his 3-week stay in the Ancient Colony, to a ship in mid-Atlantic. His plan after leaving Newfoundland was to visit Canada and the United States.[3]

By the next day, Saturday, December 7, Marconi had already selected his site. That day he visited the Governor, Sir Cavendish Boyle, and also the Premier, Sir Robert Bond, and his ministers, who promised him every assistance. Later that day the contract was let to P.F. Moore and Company to cover the ground in the area of the apparatus with large sheets of zinc, and the 25 hydrogen cylinders and other pieces of apparatus were carted up to Signal Hill. In the evening Marconi, his two assistants, T.J. Murphy and a couple of officials visited the hill to settle on the site for the wireless experiments. Marconi was pleased with the two acre plateau on top of the hill, but had to check first to see that there were no iron or other deposits locally to interfere with his waves. They chose to locate their laboratory in a second floor room in an old barracks, known as the Old Fever Hospital, not far from the Cabot Tower. Mr Murphy arranged for the use of a room in the Tower as well.[7]

The wireless receiver - popularly called the electric eye - which Marconi brought to Newfoundland consisted of "a single earphone (actually an ordinary telephone receiver)...a few coils and condensers and a coherer, no valves, no amplifier, not even a crystal."[1] Marconi claimed that at Poldhu he had a syntonic or tuned transmitter and that he also had a syntonic receiver with him in Newfoundland. Since the length of the receiving aerial and therefore the capacitance of the system varied as the supporting kite or balloon rose and fell, he could not maintain his tuning. Therefore, his regular syntonic receiver would not work well and had to be modified. He replaced his usual coherer with a self-restoring microphonic coherer, which was not good for syntonic effects and could not activate an event recorder.[2]

Marconi maintained contact by trans-Atlantic cable with his director at Poldhu, Dr J. Ambrose Fleming. Poldhu had instructions to send, at specified times and at a pre-arranged speed, a series of S's followed by a short message every ten minutes, alternating with a five-minute rest. Marconi told the press on Monday that he was awaiting cable messages informing him of the sailing of ocean liners with which he was going to conduct his experiments.[7] This ruse of short distance communication was to ensure against bad press in case of failure and was intended to put off the competition. The deception continued through the following Saturday.

The Trans-Atlantic Reception of Signals

Some experimenting began on Tuesday evening, December 10, when Marconi told the Evening Telegram that he would soon try to pick up the Cunard liner Lucania out of Liverpool, as soon as it arrived within 300 miles of the coast.[8] In fact, that day he cabled Poldhu to begin their daily sending on Wednesday, between 11.30 am and 2.30 pm, Newfoundland time. One of the 30-guinea balloons was lost in a gale on Wednesday,[9] and the next morning they lost one kite and then got another up to 400 feet. On Thursday and Friday Marconi told the reporters that he would try to receive from the boat on Saturday, and the Saturday papers told of the failure to get the second balloon aloft in the foul weather after a dozen tries, and of a kite blown into the sea. He announced that he intended to make another attempt that night if the wind subsided, but it did not.[10] It seemed to the world that Marconi had failed so far, but all the while the long distance experimentation was proceeding secretly. At week's end Marconi cabled his business associates in England with news of his trans-Atlantic success and late on Saturday afternoon he broke the story to the press.[11] They had received the letter "S" in Morse - three dots - from the Poldhu transmitter on the other side of the Atlantic at a specific speed and at pre-arranged times, although they could not pick out the appended message. At some point in the interview, when asked how he knew that he had not been hearing up some ship's transmissions, he replied, "It is impossible. I was tuned only for Cornwall."[12]

The world's press descended on Marconi on Monday the 16th, and then he revealed that he had first heard Poldhu, 2,170 miles away, on Wednesday just before the balloon broke away, and that they had arranged to repeat the experiment on Thursday. It was at half past twelve noon on Thursday, Marconi said, using their second kite with a 500-foot trailing antenna, that he heard the three dots in the low-pitched whine of the Poldhu transmitter through the telephone receiver. He then passed it to Kemp, who heard it too. Paget was slightly deaf and could not make it out through the buzzing and crashing interference. The signals faded out when the height of the kite, and consequently the length and capacitance of the antenna, changed, but they were heard again at 1.10 and 2.20. George Kemp noted the event that night in his diary, now in the possession of the Canadian Marconi Company.[13] Faint signals may have been received on Friday as well.[14] No one but Marconi and Kemp heard any of those signals, and the times of receipt were not corroborated by anyone else, giving rise to a persistent skepticism about the success of the transmission. Many scientists at the time did not believe the report. Edison at first called it a "newspaper fake," but later said that he believed since Marconi had published a signed statement.[15] Marconi could have proven reception by recording the signals on paper tape with a Morse "inker," if he hadn't had to use a microphonic coherer in his receiver. In long distance shipboard wireless experiments in the Atlantic in February, Marconi used a syntonic receiver with an antenna of fixed length attached to the ship's superstructure, and so he was able to use an "inker" to record the incoming signals. He had the ship's captain observe the reception and then sign the tape.[14]

The day following the news of the trans-Atlantic success, cable operators and owners were in shock. The submarine cable company stocks plummetted, and remained low all week. Marconi was in great spirits. He told the New York Herald that he was going to build a $50,000 permanent station in Newfoundland, and on Monday the 16th he went to Cape Spear to look around for a likely spot.[1,16,17] At his hotel that night Marconi was presented with a lawyer letter from the Anglo-American Cable Telegraph Company maintaining that he was in violation of the company's monopoly of all telegraphic construction and communication and thus of all experiments in any way connected with telegraphy in the Colony.[12,18] The letter was sent on instructions from the head office in England, but the local Anglo-American manager had no complaints. Marconi had cabled over 50,000 words through his office the previous week, at $0.25 a word!

Alexander Graham Bell cabled Marconi on Wednesday offering temporary use of his estate near Baddeck in Cape Breton.[19] Governor Boyle was most displeased with the Anglo-American position and did what he could to show it. He cabled King Edward about the successful experiments, threw a grand luncheon for Marconi and the Newfoundland cabinet on Thursday[20] and then made an official visit to Signal Hill to show his support. Unfortunately, Marconi could not now legally demonstrate his apparatus to Sir Cavendish Boyle's party. Anglo-American protested Boyle's partisanship, asserting that the government should not take the side of either party. But Anglo-American had won. Newfoundland was legally out of bounds for Marconi, and he had to find still another site for his permanent station.

It is not clear exactly what Marconi meant on December 6, when he told the St John's Evening Telegram reporter that he would visit Canada, as well as the United States, before returning to England. A stop in

Cape Breton was part of the Newfoundland to New York journey in those days. He did in fact go from Cape Breton to Montreal and Ottawa toward the end of December, but it was never made explicit that these two particular visits were part of his original plan.

There are two detailed stories of how Canada "kept" Marconi after Anglo-American made him stop his work in Newfoundland, each with its teller as the hero.

Canada Determines to Keep Marconi: Story #1

Degna Marconi[1] accepted the story of Alexander Johnston,[21] editor of the Sydney Daily Record and Member of Parliament for Cape Breton, of how he single-handedly persuaded the great inventor to bring his experiments to Canada. Johnston entered the story just after Christmas when a deeply disappointed Marconi arrived by steamer in North Sydney, Cape Breton, on his way to New York and thence to London. According to Johnston, he rushed to the pier at North Sydney to meet and interview Marconi before he boarded the train for New York. Johnston, "juggling charts against an east wind," persuaded Marconi to consider Cape Breton as his North American centre of operations. The next day Johnston, the Hon G.H. Murray, Premier of Nova Scotia, and Cornelius Shields, the General Manager of the Dominion Coal Company, escorted Marconi on a private train tour of the coast between Sydney and Louisburg. Marconi was taken with Table Head, between Bridgeport and Glace Bay, and Shields, on behalf of Dominion Coal, offered to give him the land. Then Johnston accompanied Marconi and his assistants to Ottawa to meet the Minister of Finance, Prime Minister Laurier and Sir Sandford Fleming. When Marconi left Ottawa for New York on January 9, a draft contract had been signed to construct a station at Table Head, and the Dominion Government had agreed to assist him to the tune of $75,000.[22]

Canada Determines to Keep Marconi: Story #2

The other version was written by William Smith, who was Secretary of the Federal Post Office Department in 1901.[17] Smith was in St John's on departmental assignment and was staying at the Cochrane House, where Marconi stopped. They soon became acquainted, and Smith was discussing the Cape Spear trip with the inventor in the hotel dining room on the evening of December 16 when the lawyer letter from Anglo-American was handed to Marconi, threatening legal action if he did not stop his work in Newfoundland. Marconi made his reply that night, agreeing to discontinue the experiments. Smith, however, went to work on Marconi immediately that the Anglo-American letter was read, to convince him to transfer his North American operations to Canada. Marconi was determined to leave for England by the first boat, but Smith persuaded him to delay leaving St John's for a few more days, while he set off on the next evening train for Ottawa to raise some official interest and to secure an invitation for the inventor to come to Canada.

Arriving in Ottawa on Saturday night, Smith was taken to see the Hon William Stevens Fielding, Minister of Finance and former Premier of Nova Scotia, who had by then cabled Marconi twice inviting him to visit Canada. Marconi had accepted and Smith was advised to "hasten back to North Sydney to receive Marconi on behalf of the Government." On the Sunday train between Ottawa and Montreal, Smith was lucky enough to meet and talk with Prime Minister Laurier, and later he spoke to the Hon B.F. Pearson of Halifax, who gave him letters of introduction to Mr A.J. Moxham, General Manager of the Dominion Iron and Steel Company, which controlled the railroad between Sydney and Louisburg, and to Mr Cornelius Shields, General Manager of Dominion Coal. He reached North Sydney in the wee hours of Tuesday, the 24th. The Hon Mr Murray, Premier of Nova Scotia, had already been notified and he called upon Smith in the morning. Later in the day Smith visited Messrs Moxham and Shields to obtain their assistance.

On the morning of Marconi's arrival in Cape Breton, the 26th, he and Kemp were met by Smith, Premier Murray "and a number of the leading public men of Nova Scotia." According to the Sydney Daily Record, the SS Bruce was met at North Sydney by Premier Murray, Mayor MacKenzie, Hon John Armstrong, General Manager Shields, William Smith of the Post Office Department, Alex Johnston, MP, Captain Carlin and others. "Mr. Marconi said that his object in coming to Canada was to confer with the Government with regard to the establishment of an experimental station on the Atlantic seaboard. He was seizing the opportunity, while in Cape Breton, to visit Louisburg, and spoke of that point, Mira Bay or South Head (Port Morien), as being advantageous locations for the work."[23] The party arrived at Sydney aboard the SS Peerless, courtesy of the Cape Breton Electric Company, and partook of luncheon at the Sydney Hotel.

In the afternoon Marconi was accompanied on the famous train trip to Louisburg by Smith, Murray and the rest of the party, again including Alex Johnston, MP.[24] At Glace Bay, "we proceeded on foot to Table Head, an upland overlooking the ocean. The site delighted Marconi, and he would have decided upon it then, if the deputation from Louisburg had not invited him to look over the sites they had to offer. We went there the next morning, taking the (S&L 9.15 train for Louisburg and the) Dominion Coal Company's steamer from Cow Bay. Our trip, unfortunately, was made in a heavy snow storm, which shut out the shore and prevented inspection after we reached Louisburg. But Marconi was not disappointed. Table Head met all his needs."[17] With the Table Head site selected, Smith and Marconi started for Montreal in a Dominion Government private car on the 28th, the inventor to be the St James Club luncheon guest of the Elder-Dempster Steamship Company, which had been using the Marconi system on its boats for the past season.[25] They left for Ottawa on the 30th, where Marconi met Fielding and Laurier on the 31st.[26] Smith recalled telling the frugal Fielding that Marconi needed $80,000 to build his station and that Canada should go into a kind of partnership with Marconi to keep him in Canada. "Two or three days later an arrangement on that basis was discussed between Mr. Fielding and Marconi. Marconi left Ottawa for New York on his way to London on January 9, 1902, expressing satisfaction with his welcome, and with the result of his negotiations with the Government."[17]

The Discoverer: Johnston or Smith?

Smith's role in bringing Marconi to Canada is confirmed in an article of December 30, 1901, in the Halifax Herald, detailing his efforts and saying that he "discovered the discoverer."[27] Johnston's newspaper, The Daily Record, certainly supported Marconi editorially from the first,[28] and as early as the 18th cordially invited him to continue his work on Cape Breton.[29] The paper's editor, Alex Johnston, MP, was indeed at the North Sydney pier on the morning of the 26th to welcome Marconi to Canada, but he was one of a large welcoming party, and by then Marconi needed no further persuading to consider Cape Breton as a base for his operations. Johnston was also on both inspection tours of the coast, on the 26th and 27th, but there is only his word that he accompanied Smith and Marconi to Ottawa on the 28th.

At the end of 1901, C.A. Hutchins, Superintendent of Lights, proposed to the Halifax Board of Trade that Marconi stations be erected along the western landfall, at Cape Race, Newfoundland, on Sable Island and at Camperdown at the entrance to Halifax harbour.[30] History shows that Mr Hutchins chose well.

Further Honours to Marconi

The American Institute of Electrical Engineers gave a sumptuous banquet in Marconi's honour in the Astor Gallery of the Waldorf Astoria, New York, on the 10th of January. There were 300 guests, including Charles Steinmetz, Michael Pupin, Alexander Graham Bell and Elihu Thomson, and letters of congratulation were read from Edison and Tesla. Marconi announced that he was going to have two wireless stations on each side of the Atlantic: in the Old Country at Poldhu, Cornwall, and at Nieuport in Belgium, and on the American side in Nova Scotia and at Cape Cod. The Dominion "Government was very sympathetic and assisted me in every way possible. Sir Cavendish Boyle and Lord Minto of Canada were exceedingly kind to me in every way. I received through Sir Cavendish on behalf of certain Canadian capitalists offers of financial backing, but have not as yet arrived at a conclusion on the subject."[31] Prof. Bell again offered the use his estate in New Brunswick, but Marconi declined as it was too far inland. He was still uncertain whether the ether waves could pass over land masses as easily as over water. Earlier, R. Lemieux, MP for Gaspé, had offered Marconi a mountain site in his riding.[32] Bell was skeptical about the possibilities of wireless, because the many land stations needed would interfere with one another's transmissions: "Wireless telegraphy will never be made practicable for service on land. I believe the invention will supplant the oceanic cables when it is sufficiently developed to be reliable for the transmission of messages, but it can never take the place of the telegraph or the telephone."[33]

The Historic Memorandum of Agreement

Marconi returned to Ottawa on March 17, 1902, when he, on behalf of Marconi's Wireless Telegraph Co., Ltd., and the Marconi International Marine Communication Co., Ltd., and Prime Minister Laurier, President of the King's Privy Council for Canada and representing His Majesty King Edward, signed a memorandum of agreement[34] based on the draft contract of two months before. A week later, Marconi announced that the machinery for Table Head had been ordered, and that work on the site would begin at once.[35] Four cross-braced towers were quickly constructed at Table Head, Cape Breton, by a contractor from Amherst, Nova Scotia, and the station was almost complete by the end of July.[36]

The Memorandum of Agreement[34] contained a number of points that would develop significantly. Marconi's companies would set up a station in Nova Scotia and one in the United Kingdom to carry on a trans-oceanic commercial communication service. The Dominion Government agreed to pay a maximum of $80,000 toward the erection of the Canadian station, the companies to supply any excess cost. As far as possible, Canadian machinery, materials and labour had to be used in the construction. The station was required to use only government land telegraph lines in Canada, and to charge no more than government rates. The trans-Atlantic wireless rates were to be at least 60% lower than charged by the commercial cable companies; that is, no more than ten cents a word. The government was free to put up any additional seacoast or inland waterway stations it thought necessary to navigation, and the companies were bound to provide all necessary apparatus to these stations at fair prices. The companies would have the option of erecting, maintaining and operating any such government stations at their own expense. Furthermore, any station maintained by the Dominion of Canada on the coast of Newfoundland was deemed to be a part of Canada, so that Anglo-American's monopoly of telegraphy in the "Old Colony" was now effectively broken.

The first official Marconigram from Glace Bay to Poldhu was sent on December 22, 1902, from Governor General Lord Minto to the King. The Dominion Minister of Commerce, Sir Richard John Cartwright, originally of Kingston but now MP for Oxford, sent the second Marconigram, addressed through the London Times, "to congratulate the British people on the accomplishment, by Marconi, of the greatest feat modern science has yet achieved."[37] The Glace Bay station was scheduled to open for commercial business in four months, and Marconi was turning his attention to perfecting the Cape Cod station. Meanwhile, cable shareholders were frightened and cable stocks were still plentiful on the exchanges. Marconi contributed to the panic by predicting that his wireless would become so cheap, perhaps a cent a word, that it would eventually replace the postal service.[38,39] Marconi thought of his wireless as for interpersonal communication only, and not in the modern sense of public "broadcasting." In March, 1903, the Times of London published its first overseas news sent "By Marconigraph" on a contract basis from New York, and the editor chided British business for ignoring this useful advancement in science.[40]

The headquarters of the Canadian Marconi Company was established in Montreal early in 1903, and plans were announced for the building of a transcontinental chain of stations across the Dominion and into Alaska.[41] Things did not go smoothly for Marconi. The desperate cable companies continued attacking his trans-Atlantic wireless as fraudulent, and when the public tired of that they contended that wireless signals could be diverted from their intended target and stolen away, or blocked by mountains or by stormy weather. About the same time other wireless telegraphy systems emerged to compete with Marconi for government contracts, those of Fessenden, Lodge-Muirhead, de Forest, Armstrong-Orling and of the psychist Ernest Rhymer. There were nasty patent litigations too. The Dominion Government was severely criticized for giving a monopoly to the Marconi group, and also for the size of its grant to Marconi. In essence, the contract provided a huge research stipend to a private firm in aid of technological development, as well as a retaining fee. $80,000 seems a small sum to-day, but in 1902 the entire budget of the Dominion of Canada was only $51,000,000. McGill University was upset too. McGill was worried because a Marconi wireless plant going up on Mount Royal, not 700 yards from the Department of Physics, might interfere with research. The biggest fear was that the radiated energy might cause fires on the campus. A Marconi official called their ojections "absolute nonsense."[42]

On top of all these problems, five years passed and Marconi's promised trans-Atlantic commercial wireless service was still being promised. In late 1907 our old friend Prof. Gill of Queen's, who had been an enthusiastic supporter, was wondering whether Marconi had ever bridged the ocean with his waves.[43] The Marconi Company's excuse since 1903 was that the British government telegraph service refused to relay the Marconigram messages inland to their final destinations.[43,44] But Marconi's image and troubles with bureaucracy are not our concern here. This chapter is about the development of the Marconi wireless service to ships in Canada, and in this matter Marconi fulfilled his committments.

The Beginning of the Dominion Wireless Service

The Annual Reports of the Dominion Department of Marine and Fisheries for 1902 and 1903, in the section titled "Signal Service," mention the wireless telegraph stations "established by the Marconi Wireless Telegraph Company (Limited) at Belle Isle and Chateau Bay,"[45] possibly through the mediation of Lloyd's. The report for 1904 lists 6 new Marconi stations built that season on the River and Gulf of St Lawrence, all working months before the close of navigation in late 1904.[46] The stations were at Belle Isle (which was re-built) and Fame Point, Quebec; Heath Point, at the eastern tip of Anticosti Island; Point Amour, Labrador, on the Strait; Cape Ray near Port aux Basques; and Cape Race on the south-eastern tip of Newfoundland. The effective ranges of these stations were from 100 to 130 miles. Lord Minto, the departing Governor General, made use of the two Newfoundland stations to send farewell messages to the government and people of Canada, and the government greeted the new Governor General through the Cape Race facility.

Ship owners praised the wireless as a great supplement to the existing fog signal service; the etheric weather and news reports were invaluable to both crew and passengers. Besides, the weather forecasts, navigation aids and distress traffic were transmitted free of charge. Owners soon discovered that there was great advantage to them in having the government wireless stations communicate with one another, a practice which was not intended in the first place. The government decided for 1905,[46] therefore, to boost the power of the existing stations to increase their range and to build two new stations under contract with Marconi. As soon as the weather permitted, a station would be erected at Sable Island and another on the mainland in the Canso area. Three government steamers, the Minto, the Stanley and the Canada were outfitted with Marconi apparatus in 1904, and the first two helped in the transportation of materials to the new Marconi station sites as well as in the testing for distance. Rather than two, seven wireless stations were built on the St Lawrence route in 1905, with new ones at Sable Island; Point Rich, Newfoundland; Point au Maurier, Quebec; Camperdown (Halifax harbour) and Cape Sable, Nova Scotia: Partridge Island (St John), New Brunswick; and Cape Bear, PEI. They were especially proud of the installation on Sable Island, the "Graveyard of the Atlantic." Marine and Fisheries had, to date, spent less on wireless telegraphy than the cost of laying a cable from the island to the mainland. Furthermore, with wireless, the station could aid with navigation on the North Atlantic as well as communicate with the mainland.[47] Despite this crowing, the Sable Island station soon disappeared from the record, apparently replaced by one at Sydney, Cape Breton. In 1906, two stations on the St Lawrence River proper were on the drawing board, at Father Point and Clarke City (Sept-Iles),[48] and were completed by December.

Several significant changes were announced in the 1907 report. The government had taken absolute control of the wireless plants on its ships, and had replaced the Marconi operators with its own. Company operators would not submit to ship discipline nor would they maintain the apparatus properly. The new Wireless Telegraphy Act had also come into force,[49] and gave rise to a displute with the Marconi Company which refused the issued licences, contending that the form of the licence infringed on its contract rights. The government had issued an experimental licence to the Dominion DeForest Wireless Telegraph Company to operate on Grindstone Island (Magdalen Islands, Quebec). They had also decided to install wireless stations along the west coast to serve Vancouver and Victoria. These would be working by January 1, 1908. Many boats on the Pacific coast were using the Massie wireless system, and since "the principle of intercommunication has never been accepted by the Marconi Company," the new stations would operate with the Shoemaker system. Besides, a study had shown that the Shoemaker system was far cheaper, easier to maintain, more up-to-date and better.[50]

By early 1910, there were 29 stations in the radiotelegraphic service. Eight on the west coast and 15 on the east coast were owned by the government and operated under contract by Marconi. North Sydney and Pictou were owned by Marconi but operated under government contract, and 4 on the east coast were owned and operated by the Marconi Wireless Telegraph Company: at Montreal, Trois-Rivieres, Camperdown and Sable Island.[51]

Wireless Time Signals

Time signals are required by ships' navigators for correct determination of longitude - war ships, the merchant marine and cable ships - and the correct time is also needed by railways, watchmakers, caretakers of public clocks and by the general public. The Canadian Meteorological Service made its own sidereal observations at the Dominion Observatory at St John, New Brunswick, maintained the standard clock for the Maritime Provinces and disseminated the information as best it could. The St John Observatory was connected by land telegraph lines to relay the 10 am signal from its mean time transmitting clock to every Western Union office in the Maritimes. A local telephone line in St John carried the beats of a sounder connected to the standard clock, and there was also an official clock in the St John Post Office. The St John and Halifax harbours each had a tower with a falling ball device, a "time ball" with an electrical release, much like the New Year's falling ball in Times Square, New York City. The falling ball signalled 1 pm, 60th meridian time, every week day. They also had systems of electric lights in harbours for night signalling, but that was largely for weather reports. Therefore, in order to receive the Dominion Observatory's official time signals, a ship had to be within sight of certain harbours or else an officer had to get to the end of a telegraph or telephone land line.

D.L. Hutchinson, Director of the St John Observatory suggested in 1905 that the new Marconi wireless station at Camperdown near Halifax be equipped with machinery to permit it to transmit the daily time signals from the Western Union lines to ships at sea. The Director of the Canadian Meteorological Service approved the scheme, and by May of 1907 an automatic key connected to the transmitting clock at St John was sending the time signal instantaneously down the telegraph lines to the Camperdown station each week day at 10 am, Atlantic time.[52] There, a special wireless apparatus automatically transmitted the signal - all of the relaying was automatic. Hutchinson concluded in his 1907 report, "Thus the daily signals from the transmitting clock at St. John will be available to ships at sea, equipped with the wireless apparatus, within the wireless zone of the above station."[53]

Canada was the first country in the world to broadcast a daily time signal, although the word "broadcast" was not yet used for wireless. The best known daily time signal, from the famous high-powered US Navy station, NAA, at Arlington, Virginia, did not come on until February, 1913. The US Navy opened bids for equipment for NAA in 1909, specifying a year-round radius of reception of 3,000 miles. The contract went to the National Electric Signalling Company, Reginald Fessenden's company, which supplied a 100-kW synchronous rotary spark set.[54]

The Great Lakes Coast Stations

Even before Edward VII died, the sabres started rattling in Europe. Soon after, Canadians rallied to the support of the Empire and militia groups formed and drilled, including the Queen's 5th FCCE, and armouries were built. General Hamilton declared that war is not unthinkable and that we must be prepared. It was admirable to work for peace, he thought, but a wave of moral indignation could arise at any time and sweep away peace ideas.[55] At the Imperial Conference in London in 1909, talk was of forming colonial navies, and Canada decided to establish her own. Even before the passing of the Naval Bill, a Department of the Naval Service was set up to administer the Canadian Navy. Wireless telegraphy, hydrographic survey, tidal survey, naval militia and fisheries protection were taken away from the Dominion Department of Marine and Fisheries and attached to the Naval Service.[56] As war approached, the government moved to complete its wireless coast stations, with a chain of nine at 180-mile intervals the length of the Great Lakes, plus a pair in Hudson's Bay. There was additional pressure from the steamship companies operating on the Lakes to build at once. They wanted to protect their precious boats with wireless. The Ontario Hydro-Electric Commission, through Sir Adam Beck, offered to supply the power for the Lakes stations.[57]

A small station at Port Arthur was built by the Marconi Wireless Telegraph Company in November 1910, and assisted its first vessel in distress on December 7. The Port Arthur station, later known as VBA, was purchased by the government and opened under contract in the spring of 1911.[58] On April 5, 1911, the Dominion Government and Marconi settled some of their differences and signed a contract which set out the guidelines for Marconi's operating of all Canadian coastal stations.[59] The way was now open to complete the Great Lakes wireless chain, and sites were secured and building begun at Sault Ste. Marie, Tobermory, Midland and Point Edward.

G.J. Desbarats, Deputy Minister of Marine and Fisheries, sent a draft contract to the Marconi Wireless Telegraph Company of Canada on February 23, 1912, proposing that Marconi operate for the government the three newest wireless stations along the Ontario shoreline, VBB at Sault Ste. Marie, VBC at Midland and VBD at Tobermory, as well as five more which His Majesty wished to build.[60] With respect to the three existing stations, His Majesty agreed to pay the sum of $3,000 each per annum for their operation, maintenance and repair. The terms were the same for the five proposed stations, at Point Edward (Sarnia, Lake Huron), Port Colborne, Port Stanley, Toronto Island and Kingston. Furthermore, any technical upgrading desired by the Minister would be carried out by Marconi with expenses to be borne by the government. There was some negotiation on the basic fee, and it was raised to $3,500.

Call Letters

Referring to the stations by call letters at this stage is an anachronism. Letters were assigned only in accordance with the Radiotelegraphy Act of 1913,[61] which in turn followed from the International Radio-Telegraphic Convention, signed in London, England in 1912.[62] The calls of the east coast stations were generally VC- (except Halifax Dockyard, owned by the Dominion Government, which was VAA), the west coast had VA- and the Great Lakes and Hudson's Bay were VB-. The calls for the new Great Lakes stations became VBE, Point Edward near Sarnia on Lake Huron; VBF, Port Burwell on Lake Erie (which replaced Port Colbourne and Port Stanley); VBG, Toronto Islands; and VBH, Barriefield Common near Kingston. Each had a theoretical range of 350 nautical miles. Port Nelson on Hudson's Bay was VBN, and a land station at Le Pas, Manitoba, was VBM. Incidentally, the call of the original Marconi trans-Atlantic station at Glace Bay was GB.

The ocean stations worked all the year round, and indeed their services were more valuable in the winter when ships were more likely to get into trouble on the stormy seas. The Lakes stations closed down in early December each year, when the waters froze, and they re-opened with the beginning of the shipping season around the first of April.

The Kingston Coast Station

News of the plans for building a Great Lakes wireless chain, the longest in the world, reached the Kingston public in February 1912.[63] The station to be built at Kingston would have sufficient power to reach Montreal, to link up with the Atlantic coast system, so that a passenger in mid-Atlantic would have through communication by relay as far inland as Sault Ste. Marie.

The Kingston station was to be erected at 76° 27' 30" W, 44° 14' 05" N, high on the Barriefield Common just back of Fort Henry, and it would be a duplicate of VBD, Tobermory. Apparatus for the Marconi marine station at Barriefield arrived at the end of November 1913, and the engines were expected from Toronto any day.[64] The two wooden masts, the operating house and the two dwellings had been put up by government contract the previous summer and fall, and the Marconi men expected the station to be in commission by spring. Edwin F. Robinson, an Englishman on the Marconi technical engineering staff, and his assistant, H. Dashwood, arrived in Kingston at the end of the first week of December to supervise the installation. Robinson had been in charge of the work at Midland and at Point Edward, both opened earlier that year. He told The Whig that Sarnia had received a freak wireless message from Australia, and that news raised Kingston's excitement still higher. Local boys interested in wireless came up to Barriefield Hill every chance they had to watch the installation of the complex grounding system, to see the masts and buildings going up and finally to witness the fabulous machinery being put into place.[65]

Technical Specifications

The Daily British Whig provided technical details to the public:

> The two masts at Barriefield are 450 feet apart, and 185 feet high. The cross-arms are 20 feet wide and are used to space the aerial wires. The aerial, in this case, will take the form of the letter "T" because it will radiate and receive equally in all directions. With a properly shaped aerial, it is possible to send out messages in any direction. The most common is the inverted "L". There will be two gasoline engines, of eight horsepower each, and two special high frequency generators. These travel at a speed of 200 revolutions per minute.
> It appears that this station will give out a musical note when messages are being transmitted, probably about middle "C" on the piano. This is accomplished by a disc which has a number of points passing very close to two fixed points at a high rate of speed. There will be in the neighbourhood of 100,000 volts on the aerial. Around the outside of the operating house is buried a quantity of copper netting and zinc plates. This is to act as a return circuit to another station, the air being used in place of wire. A chief operator and two assistants will be permanently at the station when completed.[65]

The report of the Department of the Naval Service, drawn up in 1914,[66] gave further information:

> A suitable site for this station was secured on the Militia and Defence reserve on Barriefield Common, and a complete new station consisting of a type No. 2 operating house, 40 feet by 20 feet, two 185-foot housing masts and a duplicate 10-horsepower 5 1/2 kW radiotelegraph equipment, was established during the year, at a total cost of $21,534.28.
> Public tenders were invited for the erection of two masts and the operating and dwelling-houses, and the contract was awarded to Messrs. McFarlane, Pratt & Hanley, of Toronto, who submitted the lowest tender of $12,650 for the work.

The radiotelegraph equipment consists of two 5 1/2 kW 240-cycle synchronous disc transmitters, with all necessary auxiliary apparatus, each generator being belt connected to a 10 horsepower Canadian Fairbanks Morse gasoline engine.

The receiving equipment consists of a tuner capable of receiving wave lengths between 200 and 3,000 metres, and a carborundum crystal detector.

The contract for the installation of the above apparatus was awarded to the Marconi Wireless Telegraph Company of Canada, Limited, for the sum of $7,106.

The station has a normal range of 350 nautical miles over water.

Work was completed and the station placed in commission in January, 1914.

The Schedule of Apparatus for the Kingston Station

There is no list of specifications for the Kingston Marconi station. The National Archives, however, holds the list for an identical plant at Tobermory,[67] and it is instructive to look at some of the major pieces. Everything in the operating house was supplied and installed in duplicate by Marconi.

The earth connection consisted of a continuous circle of No. 9 gauge, 8 x 3 foot zinc plates, set on edge in a 4-foot-deep circular trench 200 feet in circumference (65 feet in diameter) around the centre of the operating house. No. 12 bare copper leads, running underground from the centre of the operating house, were soldered to the zinc plate every 4 feet, and 1" mesh No. 22 gauge copper netting ran radially from the centre in four sections, ending soldered to the bare copper wire. All of the earth wires were soldered together and run up into the engine room through a porcelain tube for connection to the two transmitters and to the earth lightning arrestor plates.

The aerial was a standard double T, wave length 600 metres, stretched between the yards on the wooden masts (steel masts each cost $3,000 more). The government specified two No. 42/23 silicon-bronze horizontal wires, with a down-connection of similar wire soldered or clipped to the middle of the horizontal wires. The vertical wires were separated near the ground by a 12-foot-long white-painted spreader, and thence soldered to a brass rod running through the lead-in insulator.

Neither Tobermory nor Barriefield had electric power before the First War.[68] Therefore, to power the transmitter, each station was supplied with duplicate 8-horsepower gasoline-fuelled Canadian Fairbanks Morse Horizontal Special Electric Engines with throttling governor. These were installed in the operating house on a block on top of 8 x 6 foot concrete beds, 3.5 feet high, so that the fly wheels were at least 6 inches above the floor. There is no indication of the diameter of the fly wheels, but they could have been quite large. The engines were each connected by 25-foot x 4-inch double-ply leather belting to one of duplicate belt-driven 5.5-kW, 240-cycle, single-phase motor generator sets with a Marconi Disc Discharger attachment in a ventilated sound-deadening metal case. Each case was fitted with a small window to enable the spark to be observed. The fixed electrodes, which were attached to that metal casing enclosing the rotating disc gap, were themselves capable of being rotated, not less than 3 inches, for adjustment to the phasing of the generator.

There may have been about 24 projecting studs or points on the VBH rotating spark gap disc. When the transmitter was in operation, the spark would leap across the gap from the fixed electrode on one side of the enclosing case to the nearest stud on the moving disc, flow across the wheel and jump from a passing stud to the fixed electrode on the other side. They could get 24 bangs per rotation if each stud on the spinning disc came into timed alignment once with the fixed electrode on each side of the casing, to conduct a maximum of two discharges per rotation. If it were a synchronous rotary gap, operating with alternating current, there would be a discharge at the maximum positive and negative of each cycle. 240 cycles AC would produce 480 sparks per second and a pure musical tone, so the disc would have to spin at 20 rotations per second or 1200 rpm. The amateurs believed, however, that the VBH transmitter produced a 240 cycle per second note, just shy of middle C, in which case the disc would have spun at 10 rotations per second or 600 rpm.

The motor generators or alternators were mounted on slide rails to facilitate tightening of the belts. The duplicate transmitting sets were also in the engine room and they each had a high-frequency primary variable inductance solenoid, a 0.06 microfarad condenser in an oil-filled tank, transmitting jiggers, 5.5-kW 50 or 300 mH tapped aerial tuning inductance helix in an open hardwood case and a 5.5-kW 240-cycle 0.06 M.F. oil-insulated transformer in a cast iron case.

In the instrument room there was one main switchboard, with a panel of blue Vermont marble mounted on an angle iron frame, complete with AC and DC voltmeters, AC ammeter, generator and exciting rheostats, pilot lamp and single- and double-pole throw switches to control all changes and instruments. It was designed so that either transmitter could be put into circuit and operated from the instrument room. The two transmitting keys were of the standard Marconi type, complete with telephone break attachment and one pair of platinum contacts. The receiver was to be either the silicon (and cat-whisker) type-1-p 175 or a standard Perikon type (zinc-chalcopyrite combination crystal rectifier detector) supplied by the Wireless Specialty Company, New York, but a standard double magnetic receiver complete with telephone condenser was also supplied. The tuner for use with the crystal detector was of the valve type, and equipped with a safety spark gap. Three pairs of headphones with adjustable head band, called double head telephones, were supplied, one with low resistance for the magnetic detector and two of high resistance with adjustable magnet.

Inspection Report of January 19-21, 1919

A brief inspection report for VBH, on Form W. 14, turned up in the

National Archives. The two sheets give some details of the operation and physical plant. At the time, Ernest Cashell was the only operator, and the hours of service were not yet fixed. VBH was not connected with any telegraph lines, but the inspector was assured that telephone connction would be made in the spring of 1919. VBH had two 185-foot masts, supporting a horizontal "T" aerial, 291 feet long, with a 174-foot vertical wire. The licenced wave lengths were 600 and 1,600 metres, but the actual waves emitted were 620 and 1558 metres. Even though the range of VBH was 250 miles, signals were exchanged on both long and short wave with the Midland and Sarnia stations only. In 1919 there were still "No public mains run in or available" on Barriefield Hill and no batteries. Power for transmitting was supplied by a Canadian Fairbanks Morse 8-horsepower horizontal gasoline engine, special electric type. The dynamo or motor generator was a "Standard Motor - Robbins & My", supplying 5.5 kW and producing 240-cycle AC, 440 volts single phase. The transmitting apparatus took 425 volts, 12 amperes.[69]

Recollections of VBH

Mr N.A. Good was in charge of VBH when it opened on January 17, 1914. In 1926, Ernest Cashell who came direct from Ireland, was Chief Operator and N.A. Good and George H. Gurney assisted him.[70] Cashell was Chief Operator when Harold Stewart was a boy and was still there in 1933. While Cashell was aloof, George Gurney was helpful to the young amateurs, and even helped them to learn the code:

> **Harold Stewart**: The buzzers that we had were not tuned. We just stuck them on the air and the wave was very broad, which means that it was highly damped, and it would smear a receiver right across any kind of a tuning band. Hard to tune it out. It may be that we raised hell over at station VBH. Keith MacKinnon and I and others made friends with the operators over there, at least with one of them...George Gurney. He helped us in the early days, making tuners, and letting us try them out on his big antenna over there.[71]

Bill Taylor remembers George Gurney too: "My wife and I - the girl who is now my wife - used to walk over there quite regularly to visit Gurney. And they used to put on quite a show even as late as 1924, of the transmitter, which was out-of-date at that time. A big spark job that gave a 500 cycle note, and they used to crank it up just for our benefit and you could hear the darn thing about a mile away." Lancely McAteer adds: "When they cut the grass over there under the aerial, they had to re-tune the transmitter."[72]

Harry Marlatt started working for Marconi in 1920 at VBE, Point Edward, Ontario. VBE had a 5-kW open spark gap, and it could be heard for miles - one could hear the spark jumping the gap and read the code. VBE had electric power, but not Kingston:

The operating station is a one storey building in the foreground, and the semi-detached pair of residences is centre left. One antenna pole, red and white striped, is at top centre and one of the cement pylons, removed in the summer of 1986, can be seen just above the right end of the clothesline.

-photograph by Harry Marlatt

<u>Harry Marlatt:</u> She was an open rotary spark set with a gasoline engine driving it. The wheel was about a foot in diameter, and she kicked out quite a spark. You could hear it, I imagine, downtown some days. It was a nice note to read. They had a nice note, but it'd light up the whole engine room when you were on the air and working. We had no 550 power up there, and we had to go out and wind up this big brute every time anybody called us. And if you had to change frequencies you had to go out and shut it off and wind up the other one. We had 600 metres and around 1,700 for the working channel. 600 was the stand-by channel. All ships all stood on that and, when they called in and you had traffic, you switched them to what we used to call the long wave. We ran on that until, I guess, around about the early 20s when they got the power in, and then it was just a matter of throwing a switch. But every time you wanted to change frequencies, you had to change transmitters too. I came here in '37 and we had the rotary spark in then, and then gradually they put in the valve sets, and that's where she ended up. And I walked out in '67 on pension and I've never been in a wireless or radio station since.

We had two houses up there (actually a semi-detached pair), for a first and second operator, practically adjoining the station. Made it very handy in the wintertime, slipping across to work - no driving to do. There was somebody on duty 24 hours a day, 7 days a week, in three shifts. We worked 8 hours a day, 7 days a week. I'd come off at 8 o'clock this morning and go back on 4 o'clock to-morrow afternoon, and work 4 to 12 and come back at midnight and work til 8 and come back at 8 in the morning and work until 4, and then I was off for 32 hours. There were only three operators for the steady shift. And we had to cut the grass too, but I don't recall our having to re-tune afterward.

You had to make an entry in your log every three or four minutes. You had to protect yourself, and you got accustomed to it. Otherwise you could walk into a station to relieve one operator, and when he goes out of sight you could go home, if you haven't got a log covering you. The amateurs never bothered us - to me they were no trouble. We had enough interference without amateurs - a lot of stations and ships. You see, the freaks is what'd get you. A freak signal'd come in and pretty near take your head off, and it might be some station in California. Yet you couldn't work a ship 50 mile away, but you could get some fellow 3,000 miles away, and he'd knock you off your point, of course, your carborundum or your galena. If you got a good point and were getting good signals, you left her there. But if you were on board ship and then some other ship come up real close to you and started up with a spark set - they were pretty wide - they'd knock you off your point and you'd have to tune it again. We had an "A" battery through carborundum at the station when I went there, and finally we got 201A tubes, and we were right at the top of the world then. Oh yes, we were on code - I think it was '39 or '40, roughly, that they put the voice in, and now code is gone out like high boots. You don't hear it at all.

The station was a lovely little place, just a small kind of a 3-room bungalow, sitting out by itself, with the engine room in the back to the north, and the operating room in the middle. And I forget what the heck was in the other room. Cupboards and stationery, I think. There wasn't much to it. Just a nice little - all the room you needed for your equipment, comfortable and warm. They were all the same - if you were in one station, you were in them all. They were lovely homes, easy to heat, about 75 to 100 feet along a cement sidewalk from the station, two single houses with 6 rooms. Very high rent - $9 a month. And the houses were all the same. The station was beautifully situated - it had a beautiful view. You could sit up there and look all over Kingston at night from your big front verandah.

We handled marine messages, ships' arrivals, trouble aboard and where to go and where to go to get a cargo. Just ordinary telegraph messages. We never handled any personal messages. On the passenger boats we handled personal messages. In those days we had to relay messages. We'd get one from Toronto and we'd relay it to Montreal, or a message coming down from the Upper Lakes would come down from Port Arthur to the Soo. In good conditions you could work the Soo, but on average you worked it through Midland and re-transmitted it on down to Montreal. You'd get "Msg," which is the prefix, then the steamer whatever it was - Keewaitin - the number and the amount of words to check, the file time and the station via, which is the original station. That was all the pre-amble of the message. And you got it addressed to Can Steam, Montreal, and the body of the message. Just an ordinary telegraph message.

They were very good and very considerate - as considerate as they could be. You had to be ready to move at all times, from one ship to another. I was running up to Newfoundland in a coal boat, and coming back down to North Sydney we got a terrible beating crossing the Gulf, and they had to put her in dry-dock in Halifax. And I was taken off that coal boat in the morning and put on a CN passenger ship in the afternoon. Can you imagine a change from a coal boat to a passenger ship? I had to get out the white uniforms for South America. And I came here from New Zealand. I was up in the Upper Lakes all summer and I went down below, and I come back to Toronto at 8 o'clock in the morning from New Zealand and 4 o'clock in the afternoon I was on my way down here to Kingston. So they give you lots of time to think of where you're going. Then I got an extra year's extension, and when that was up they wanted me to be on stand-by. Well, that'd mean standing out on the verandah with a blue uniform and a white uniform and waiting for them to tell you to go some place. Because their operating covered it all, and if an operator got sick here or there, why they just had to replace him right away. We used to get some quick jaunts, but thank heaven that was before they used aeroplanes. But they were a good outfit and I liked them.

I don't think George Gurney ever was aboard ship. He was on coast stations all the time. He retired before me - he was senior to me. And Rain, the other fellow, he was from North Sydney, he's gone; and Bartlett, the original OIC, he's gone. He was from Port Arthur. Oh, they're all gone.[73]

Keith MacKinnon and his pals Harold Stewart and Ralph Bunt used to bicycle up to VBH often:

Keith MacKinnon: He and I and Stew knew the second operator. The first operator was Cashell, and he was an Englishman that came over when they brought the Marconi stuff in, and he had a lovely fist. Oh, it was perfect. A lovely fist! But the second operator was George Gurney, and his fist was no hell. We always knew who was on the air, you see, and we wouldn't go over when Cashell was there. He wasn't interested in seeing us - didn't want anybody bothering him - but Gurney always made things open for us. As a matter of fact, he helped me get my commercial licence. He helped me with all the plans and everything. When you do your licence you have to draw the whole damn thing out, by memory, and he'd loan me the official prints of all the equipment and I'd take them home and copy them. And then I could study them that way, and he showed me everything that was going on. He'd let me send the probs out - they'd send the weather out all the time in the daytime. Those damn husky Marconi keys are great, big bastards, and how Cashell could handle them the way he did, I don't know. It was amazing. He had a lovely fist, but he was brought up on those things. VBH was very good to me and I went to Ottawa and passed the exam and I got the certificate. So I have a commercial licence for the old VBH transmitter, which was the 5 1/2-kilowatt, 240-cycle Marconi coast station equipment, and I still have it framed up at home. I was going to go one summer on the lake boats, but it was a bad season for the grain with no ships going and nothing to do, so I played in a dance band all summer.

It was a big 240-cycle rotary spark set, with about two dozen prongs on the wheel. It was going like hell, 240 explosions a second so you got to go at a high speed. You couldn't see them.

I didn't know the night operators - they were young chaps who just came in for the season. Cashell and Gurney did all the day work and lived there all the time.

VBH had nice equipment. It was in duplicate, you see, in case one failed, and they used to use them alternately. They ran by an old gasoline engine. Bang, bang, bang, bang...it ran like that, you see, and it had a big fly-wheel, and a belt driving the 240-cycle generator. You got the engine started by climbing on the spoke of the big wheel and you'd get enough weight on it just to get it over after setting everything up. They used to fire the engine by a contact - not a spark plug, but two things that got together and made a spark. And Gurney was a light chap, you know, and he'd just climb on the wheel and get the thing going. You see, when they'd get a call none of the equipment would be going. So, he'd have to go into the engine room and get the thing going. And after it was running he'd come back into the operating room and see that the board was right. He had a big switchboard there with the ammeters on it and everything, telling that all the generators were going. It was a high class installation that Marconi put in for those days. And then he'd go ahead and play and there was a hell of a racket when those sparks went. It lit up the room. Oh, yes, it was tremendous.[74]

Harold H. Stewart: My recollection of VBH goes back into the '20s when I was building my own ham gear and my friend Keith MacKinnon and I made so bold as to go over there to see the operators, and we got acquainted with George Gurney, who was number two man. We met Cashell, the Chief operator at one time, but he was rather austere and we didn't feel so free with him as we did with George. George was very kind to youngsters who were breaking into the game, as we were, and one time when we made our own tuners we took them over and tried them out on his big aerial. We brought in all kinds of stations just as loud and clear as the Marconi gear, which maybe wasn't saying too much. But George's favourite detector was a galena one, with a cat-whisker, although they had others. A piece of carborundum clamped between two electrodes was another one which was used. The only virtue it had was that it wouldn't be knocked out of commission by the local transmitter. And they also had a magnetic detector, which was not very sensitive, and I gathered from George Gurney that really it was just a fixture there and they didn't make any use of it at all. It had a couple of reels on it and a piece of metal or iron tape that went around over these two reels and through some kind of a pickup head. When George said that it wasn't up to much, we really didn't bother much more about it. It was better than a coherer. Almost anything was better than that. I think they had one of those over there too. I never saw it work, but I recall we had one in high school physics class one time, and the teacher set out to demonstrate the thing. I was never sure whether the bell was set off by the radio signal or whether he kicked something under the table. We didn't really trust it.

The spark transmitter at VBH was a fascinating thing because it moved fast and it made lots of noise. Oh, it rotated a few hundred rpm. The frequency of the alternator that fed it was, I think, 240 cycles and the output was about 5 kilowatts. I recall when we were over there at different times we'd be in the receiving room listening, and George'd get a call and he'd have to down everything and run out to the transmitter room and climb up on the belt that belted the gasoline engine to the generator, run up it and pull it back way down and then chug-a-chug chug-a-chug chug chug bang bang bang! The thing would fire up and you'd hear the roar of the alternator. Then he'd come in and we stayed in the transmitter room while he sent messages. If you were too close to it the spark was almost deafening. It was just the sheer spark discharge between the electrodes that made the noise. I know you could hear the thing half a mile away on a quiet night.

It gave a note because the frequency of the current flowing through the arc was 240 cycles a second, and no doubt that that would make the arc vibrate in and out at that frequency as the current rose and fell. The arc intensity would rise and fall and therefore the volume of the arc fire would shrink and enlarge, so that made the note. And, yeah, what a note![75]

Don MacClement: During the last few winter months of 1919, particularly, I'd been in contact with the staff over at the Barriefield station, VBH, and it was to quite an extent due to their encouragement that I think I cleaned up my act in terms of receiving and sending and my whole attitude toward operating a wireless. It needed a certain amount of determination on my part, for it was a case of biking across town from the University grounds out across the causeway over to Barriefield, then pumping all the way up that quite steep slope to the top of Barriefield Hill, where the wireless station was. It was right up on the crest of the hill and you went off to the right down a dirt road to the rather isolated location of that little shack there underneath the big towers of their antennae. The fence around the outside had the obvious signs of "BEWARE OF DANGEROUS HIGH FREQUENCY CURRENTS". However, inside it was usually pretty cosy. It was a comfortable little shack and there were usually two operators there - one busy and the other sitting around doing not very much. Of course, we talked wireless endlessly, and watching their beautiful transmitter with the powerful spark you could feel the tension of that transmission. I don't know just what-all reacted, but I know that one felt almost as though one's hair was sort of standing out a little bit more than usual when that spark was ripping across the gap. Late in the afternoon when it was getting dark, one could see a glow out from the lead-in that went up to the aerial and this was quite powerful...certainly to me.

When there was no other business for them to work on, we'd listen to all sorts of things, particularly they were interested in some of the ham stations around town which were, most of them, struggling to get down below 200 metres. I felt very proud when they said that I was the best-tuned station in the city, and they were trying to encourage me to go on and train to be a commercial wireless operator. Well, I was tempted, because it was fascinating. They held out the idea of getting a job on a freighter or a ship out on the Atlantic and going around the world and having a wonderful adventure. Or, if I wanted to stay closer to home, there were ships on the Great Lakes. The regular ones that I heard were mostly the old Car Ferry #1 and #2 going across the lake to Rochester. Gradually I was allowed a little more access to the equipment and I found that I could improve my sending ability without too much effort because they were critical, and it had to come up to their standards. They sent out news bulletins - quite long things - two or three pages of news. They were apparently intended for general use of the ships on the lower Great Lakes. Well, I was pretty soon given the job of pounding away, hammering out that news bulletin. Quite often I'd get a little written comment, like "space that r wider apart" and "don't crowd your dots so close together" or "send words, not letters separately" and "make your sentence sound as though it comes to a full stop."

By the time school finished that spring, I had read and studied practically all the manuals and textbooks that I could get my hands on. Quite a lot of it supplied by the VBH operators.

Then I went up to Ottawa to apply for my Limited Commercial Licence. Well, it was easy. I could answer all the questions without even trying to think about it because I'd been living and thinking those circuits and the stipulations and regulations for weeks and weeks. During some of the short spells that spring when I'd been allowed to do some transmitting from VBH, I'd identified a very distinctive fist. It had a sort of swoop to it. It'd start off a little bit slowly at the beginning of a sentence and then speed up and rush through it and then slow up at the end. It was distinctive enough that I knew at once that it was that same person that was sending, and it turned out to be the Junior Operator on Car Ferry #1. Of course, I made myself known, that I had my own private tramnsmitting set, 3MC, and immediately was told, "Yes, I've heard that queer note of yours on the air once in a while." Well, here was my own distinctive identification, and that turned into a sort of a correspondence.[76]

The Junior Operator on Car Ferry #1 wanted to get out of the ship wireless business, and asked if Don MacClement would like to take his place on the ferry. One of the things he needed was a letter of reference from a top class wireless operator, and young Lawrence W. Lockett (Mech. '23) of Kingston, who had served on fruit boats down the Atlantic coast to Central America, supplied the recommendation.

Don MacClement: When I was hunting around for advice about becoming a professional wireless operator, I called on one of the Lockett brothers and on Howard Folger. From Folger I received a brush off with something like, "I've outgrown all that, and have no interest in it now." From Lockett I received a rather pessimistic description of life on a banana boat going between Cost Rica and the eastern USA. He was not interested in ham radio and had found the operator's work monotonous.[77]

Don MacClement got the job with no trouble, and when he reported to the wharf, his friend the Junior Operator was all packed and ready to disembark:

A fair bit of (wireless) business transaction was carried on by the people on the car ferry...and I had to sign a declaration that under no circumstances would I divulge any contents of any message that I had to handle. I made myself known to the operators at VBH, and passed along a few comments which I thought were off the record, but later found that they put them all down in their log and they used to tease me about it afterward...Their receiver was a combination of the older crystal detector and a Marconi peanut tube detector...It was a good one. The receiving part of the business was my interest, really, for one had a good condition when one was out on the lake for receiving from almost all over North America. I can remember picking up the American Navy somewhere on an island out in the middle of the Pacific. All I could get was the call sign - the rest of it seemed to be all in groups of numbers, I suppose in code. And there was a great deal of commercial traffic, I think quite a bit of it was related to the newspapers, and sometimes there were long transmissions of news...So, the summer passed very gently and easily.[76]

VBH in the News

Early in the spring of 1914 groups of young boys were caught "swinging on the guard wires leading into the wireless station at Barriefield. It is a very dangerous practice as they are liable to get a shock."[78] They were probably swinging on the guy wires of the antenna rather than on the surrounding chain-link fence. There was at least one young boy who was paid to have fun on the guy wires. Each summer for several years in the late 1930s, Ernest Muller of Kingston used to help a man grease the wires on the two red and white wooden towers. The towers were made of poles, one on top of the other, each spliced onto the side of the one below, and there were 12 guy wires supporting each tower. Ernest had a horse and the man had a boatswain's chair with a hook on it. Under the boy's direction, the horse would pull the man up on each wire, and then the horse would be backed up and the man slid down the wire and oiled it. Ernest got a quarter each time he pulled the man up.[79]

Government regulations requiring wireless on all passenger steamers on the Great Lakes came into effect on January 1, 1914 and the Naval Service had a busy time inspecting for compliance.[80] The service proved useful in all sorts of emergencies. One night in the mid-summer of Barriefield's first year of operation, young Herbert Miller had his arms and legs scalded by steam aboard the steamer Syracuse out of Rochester. Wireless instructions were sent to VBH at Barriefield to telephone Mr E.E. Horsey, who then arranged for a physician and ambulance to meet the boat at the Kingston wharf.[81] By August, 1914, Barriefield's record distance was 300 miles.[82]

Wartime Restrictions

Two days before Germany declared war on the United Kingdom, there were reports that Canada's Parliament would be called into special session, that the Naval Service would take over the wireless service and that the Militia Department would set up censorship over the land telegraph.[83] In fact, the order-in-council cancelling all non-essential wireless licences and transferring control of the wireless service had already come into effect as a war measure on August 2, 1914. On August 5, the day Britain went to war, an armed guard was posted at the Barriefield station to keep trespassers away, and all messages were censored.[84] The next day there were ten soldiers of Kingston's 14th guarding VBH and

Fort Henry, and no one was allowed within 300 yards of the station.[85] By Saturday, a 24-hour guard was extended to the canal, dry dock, waterworks and Cataraqui Bridge and an incident occurred at VBH that night. Two inebriates drove up to the station to look around. They had alighted and were toddling toward the installation when a sentry confronted them and ordered them to leave. One sozzle ignored the command, the guard pointed his rifle, whereupon the two explorers retreated in disarray without their conveyance. A son of one of the gents came up next morning to get the car.[86] A week later, the station at Port Arthur was attacked by two armed agents of the Kaiser who cut the supporting cables and brought down the aerial but not the masts.[87] Probably not the two men from Kingston.

The Post-War Regulations and Interference

After the War ended, the government lifted most restrictions on wireless operation, and licencing was resumed. There was, however, inadequate inspection and so the amateurs were free to fill the ether with undamped sparks from their homemade buzzers and their Ford spark coil transmitters. Most of the boys' transmissions were practice sessions, chatter or just gibberish. Harold Stewart was one of the unlicenced boys in the early 1920s: "We had spark transmitters when we first started. I'm sure if anybody had a spark transmitter in the city, it would probably interfere with VBH. We had spark transmitters when we first started, made of Ford coils, and when you tuned it in with a receiver, it would be as broad as a barn door, so to speak. And that's the trouble with the spark. It occupied too much of the spectrum. Anyone who went on the air hogged the whole thing."[88] So, VBH received a lot of interference from the Kingston boys, as well as from the Queen's Wireless Club and its Wireless Sender #1, unofficially 3QU. Before they got licences, most of the boys used their initials as call signs. Everybody in town probably knew who they were anyway because everybody knew everybody else. If Orton Donnelly's unlicenced transmissions were interfering with the government station, they would send out "OD STP VBH," and Orton would have to get off the air. The Marconi operators soon learned the patterns of some of the worst offenders and used to telephone directly to the culprits to warn them to get off the air.

One or more of the Kingston amateur stations actually blocked out the reception of messages by VBH in May 1920, and a Dominion Government inspector was called in from Ottawa. The British Whig editorialized, "Privately owned wireless installations in the city put the Barriefield station out of business. Cannot the government station on the hill compete with small private plants?"[89] Three amateur wireless sets were seized and taken to the police station, and the owners, who will be identified in a later chapter, were required to apply for licences and to reduce their power.[90] In the early years after the War, especially, VBH inhibited to some extent the activities of the Kingston amateurs. Had the government station not been here, complaints and inspections would have been less frequent and the boys would have enjoyed more freedom to fiddle and to experiment.

The Kingston Radio Association[91] was a short-lived attempt by local members of the American Radio Relay League in early 1921 to relieve the congestion of the Kingston airwaves by policing the amateurs, by teaching them the code and by involving them in the discipline of relay work. In mid-1921, Orton Donnelly (licenced with the call 3HE, President of the Kingston Radio Association and Superintendent of Ontario Division District #5 for the magazine QST) was reported as saying "that most of the transmitting carried on there now is done by means of low power buzzer and spark coil sets to avoid interference with the local Marconi station."[92] That statement cannot have been true, because a couple of months later, when the amateurs were switching over en masse from spark to continuous wave tube transmitters, Orton told QST: "Owing to spark interference with the government station in Kingston, C.W. looks to be the one best bet in Kingston City with spark sets in Belleville for further assistance" with message relaying.[93] The Kingston Radio Association is discussed in detail in Chapter 8.

The new Radiotelegraph Regulations of 1922[94] made no distinction between the telegraphers and the radiophone broadcast listeners. The authorities wanted to promote good will and co-operation between the two groups, and at the same time were very much in favour of the amateurs and their pioneering work. In the new Regulations, the operators of radiophones and CW code transmitters were encouraged by being permitted to use 150-225 metres, and in special cases to use 275 metres, because they were more likely to be properly tuned to the frequency. Amateur spark stations were pushed down to 175 metres, whereas previously they could go as high as 200 metres. Kingston amateurs were still restricted to 50 metres, since they were within five miles of a Marconi station. Maximum power was now to be 500 watts, which was not as much as was needed by spark for good distance. So spark began to lose its appeal and was rapidly abandoned in Canada, giving way to CW. The authorities hoped that the new rules would hasten spark on its way out. QST congratulated the Dominion Naval Service on its policies and opined, "That sounds like ding dong bell for the poor old pebble squashers."[95] Spark operators were variously known as brass pounders, stone crushers or pebble squashers. Interference with the work of VBH, and the consequences, will be discussed in a chapter on the Kingston amateurs, below.

Interference Caused by the VBH Spark

There is another side to that interference story, which is now long forgotten. It so happened that VBH, in its turn, interfered with the playing of the boy wireless bugs in Kingston. The Radiotelegraph Act specified that commercial service always took priority over amateur activities, so the boys had to try to fit in and be content with their lot. After the failure of the Kingston Radio Association, the local boys were unorganized and had no lobbying power. Soon, however, VBH ran up against the general public, and this was a different matter entirely. As the new commercial radiophone stations rapidly expanded their services in the US and Canada in 1922 and after, the number of radio receivers in Kingston grew into the hundreds and the proud owners found that VBH often ruined their enjoyment of this new companion and entertainer. The Canada Radio Stores sponsored a daily radio listing and review in The British Whig: "Due to interference from the Barriefield station, receiving was broken up somewhat during the evening, but several good selections were received from Newark, Rochester, Pittsburgh and the usual Tuesday evening concert of Schenectady, N.Y.".[96] Some of the broadcasts were lavish affairs. At 7.45 pm, October 3, 1922, WGY scheduled a studio production of Gilbert and Sullivan's H.M.S. Pinafore.[97]

The Wireless Association of Ontario was sensitive to the interference problems that its members were causing. The Association curtailed its members' activities to allow for the nightly radiophone programs, ordering them to refrain from transmitting from 7.00 to 10.30 every evening over the winter of 1922-23.[98] The independent young bugs probably followed suit by example, and they may have been kept from playing with their transmitters on Sundays as well, since radio church services quickly became a regular feature of the day. Sundays in the fall of 1922 Kingstonians could hear an entire religious service from a Schenectady church at 10.30 am, and then at 4.30 in the afternoon a vesper service conducted in the studio of WGY, Schenectady. You could also tune in a Sunday service broadcast from the Eastman School of Music in Rochester where the Rev T. Byron Caldwell, a former pastor of the First Baptist Church, Clayton, NY, presided.[99] In the US, ten million people on land and sea listened in to a New York City church service one Sunday, and the authorities worried that people would not come to church any more.[100]

Marine and Fisheries reacted to complaints of interference from the listening public, and in 1922 started to remove its ship and shore stations from the public broadcasting channels, which soon spread from 400-450 metres to 295-510 metres. One of the wave lengths, 300 metres, was still an international ship wave, and in its 1924 report the Department stated:

> One of the notable developments of the year, and from the public standpoint the most interesting, is the application of the radiotelephone to broadcasting. This interest is reflected in the number of broadcasting licences issued, a total of 62...The interference from ship stations...using 300 and 450 metres, is an international matter and the question of an international reservation of a band of wavelengths for broadcasting would appear to be the only satisfactory solution, and until this is done good programmes will continue to be spoiled by spark signals, to the intense annoyance of the broadcast listener. In the meantime the Department cannot do very much more than see that Canadian ships and stations keep off the

broadcasting wavelengths during broadcasting hours and eliminate all possible interference from this source.[101]

The next year the Department began to deal with the rest of the problem:

> Having dealt with interference from ships and stations using spark transmitters on waves actually within the limits of the broadcast band, we found considerable interference from spark stations using other waves, these spark transmissions being so broad as to splash over into the other waves. The only remedy for this trouble is to replace the existing apparatus with apparatus of a non-interfering type. With this end in view, the Department has, since the inauguration of broadcasting, actually equipped five of its stations with this new apparatus, viz: Prince Rupert, B.C.; Victoria, B.C.; Toronto, Ont.; Montreal, P.Q., and Quebec, P.Q., and now has in course of installation similar sets at four further stations: Port Arthur, Ont.; Sault Ste. Marie, Ont.; Point Edward, Ont., and St. John, N.B. Most of these new equipments will be actually in operation shortly after the opening of navigation in 1925, and the balance a few weeks later. These replacements represent an expenditure of approximately $60,000, all of which is being undertaken in the interests of the broadcast listeners.
>
> The interference from the spark equipment in ships is not as serious as that from the coast stations in that ships rarely use their apparatus until they are twenty or thirty miles away from the broadcast listener centres.[102]

Canada and the US took steps to remove the bands used on the waters of the Great lakes further from those used for public broadcasting, reasoning that since only those two countries would be affected, they need not worry about adhering to the international wavelength of 600 metres. At a conference in Detroit on May 4, 1925, they adopted a new wavelength of 715 metres for use on the Great Lakes, with an alternative wave at 875 metres for traffic, to become effective at midnight, July 15, 1925.[102] The "pleasure broadcast" bands in North America continued to be 200 to 545 metres, and the interference from commercial ship traffic was greatly reduced by the move.

While the interference problems in Toronto and several small spots were thus solved by Marine and Fisheries, the Barriefield station continued to operate with the old broad wave spark transmitter and to spoil radio reception in Kingston and in northern New York.[103] The reaction came to a head in early December 1926, when there was a flurry of letters and articles in the newly amalgamated Kingston Whig-Standard, prompted by a story from Cape Vincent, NY, in the Watertown Standard. VBH was increasingly encroaching on wavelengths reserved for US "pleasure broadcasting" stations, and when VBH started up it could not be tuned out. There was nothing to do but put the headphones away for the night. Listeners as far west as Toronto were disturbed by the Barriefield spark.

The Whig-Standard interviewed the VBH operator by telephone, and he believed that the trouble was not as bad as painted by the Americans: "During the past few days on account of the ice conditions in the upper lakes, a number of messages were handled, but the operator said that he did not think for one minute that the public was much inconvenienced. He said that it being a government station and being used for government work, it was necessary for the people who are on the air for entertainment to stand by for about two minutes each time while messages were being received and sent to steamers on the lakes."[104] One day that week VBH was not on the air at all! Radio dealers said that interference from VBH was affecting their sales, and they thought that replacing Barriefield's 1913 spark apparatus with more modern CW equipment would solve the problem.

Dr W.A. Jones, radiologist at the Kingston General, did not agree with the operator of VBH: "...the constantly recurring staccato volleys of sparks are an unmitigated nuisance...The protest of the Watertown residents may afford amusement to the officials of the local station, but our friends across the line are merely expressing the same sentiments that Kingston people have been for several months expressing verbally. For several of us the amusement stage has passed and we experience now only a feeling of intense annoyance. At one time the people of Toronto were subjected to a similar nuisance. More modern equipment was later installed and around Toronto this interruption is now abolished."[105] Dr Jones' letter caused considerable local discussion and letters to the editor. Val Sharp, a Kingston ham, defended VBH and thought that installing a quenched-spark set there would help minimally, while ships on the lake with crystal receivers could not pick up CW transmissions.[106] R. Baldwin suggested in his letter the same day that the local MP, Dr A.E. Ross, could look into the matter, and that Kingston fans might get up a petition to send to Ottawa: "With oscillating (regenerative receiver) sets and Barriefield, which during the World Series and the football games of Queen's always managed to volley the air with sparks, just at the crucial times of the games, and on election night, it is fierce."[107]

In his Annual Report for 1925-26, the Deputy Minister of Marine and Fisheries announced a program to gradually replace the spark sets of all of the Great Lakes stations, particularly those near large cities, with non-interfering CW apparatus. VBH was neither mentioned in that part of the report nor included in the first wave of spark set replacements, but that did not mean that Barriefield was given a low priority. Solving the interference problem had to be delayed because the station at Barriefield was still without a source of hydroelectric power. Special arrangements with the local utility commission had to be made before the CW installation could go ahead. The job was completed within fifteen months, some time before the next Annual Report, dated March 31, 1928: "Kingston, Ont. - A 1,600 watt CW and ICW valve transmitter was installed. The operating house and dwelling at the station which were wired for electric lighting when built, were connected to a power line constructed for the Department by the Kingston Public Utilities Commission."[108]

Keith MacKinnon had one of those strange circle-of-life experiences when he worked at the National Research Council in Ottawa in the 1940s:

> It was all replaced, you know, the spark equipment with all tube stuff, at VBH. When the War started, I was working on radar with the Research Council, and what showed up one day was the VBH 240-cycle generator, sitting out in the field at our field station. Some of our radar worked on 1,000 cycles, and this was something that would be good - a big generator, 240 cycles, and some of the vacuum tube boys were going to use this for seeing what the hell it worked like. And what surprised me - THERE was the thing I got my licence on. The VERY generator, sitting out in the field with all the rest of the junk. What a surprise, you know. The VERY generator. Not the same type, but IT! There it was. I don't know what happened to those things, but the Council had them, and they tried them, but I don't know how much they used them. I was working on antennas and I didn't bother with all this stuff, what they called tube-and-socket boys.[74]

The Last Days of VBH

VBH continued to operate for many years, without exciting much further interest from the Kingston public. Somewhere along the way, the Department of Transport took over responsibility for the commercial ship radio service, and in the 1950s, when air traffic and air navigation aids became much more alluring and prestigious than the work of the old ship stations, the section of Radio Aids to Marine Navigation was given to Air Services. George Gurney, the Officer in Charge of VBH, retired in late 1958,[109] and in 1970 the VBH co-ordinates were moved to the area around Odessa, Ontario, about ten miles to the west of Kingston.[110] In 1977, under the Canadian Coast Guard, Transport Canada, the old station at Barriefield was closed down and the facilities were operated remotely from Cardinal, Ontario, on the St Lawrence, east of Brockville.[111] The towers, fence and houses were removed soon afterward, and the last traces of the old station, the sidewalk around the houses and the great concrete pylons that anchored the guy wires, were removed in May 1986.

The research on VBH, Barriefield, would have been greatly facilitated by access to the station logs. These records were thought by some at the National Archives to be at Tunney's Pasture in Ottawa, by others to be at the head office of the Lakes Coast Guard in Toronto, while others were of the opinion that the logs no longer existed. If these logs do survive, they will require a long and difficult chase through the catacombs of officialdom to unearth. The search in connection with the present work was abandoned after running into several stone walls.

NOTES

1. Degna Marconi, My Father Marconi, (2nd edition, revised), Balmuir, Ottawa, 1982, p. 85 ff.
2. G. Marconi, "The Progress of Electric Space Telegraphy", Electrician 49 #10, June 27, 1902, p. 388.
3. "Marconi Here", Evening Telegram, St. John's, Dec. 6, 1901, p. 4.
4. "Local Happenings", ibid., Dec. 6, 1901, p. 4.
   "Sardinian Here", Daily News, St. John's, Dec. 7, 1901, p. 4.
5. T.J. Allard, Straight Up: Private Broadcasting in Canada: 1918-58, Canadian Communications Foundation, Ottawa, 1979, p. 4.
   "Marconi System Success in the Gulf", Family Herald and Weekly Star, Montreal, Oct. 15, 1902.
   "34th Annual Report, Department of Marine and Fisheries", Sessional Papers, House of Commons, Paper No. 21, 36 #9, 1902, p. 22.
   "35th Annual Report, MF", ibid., Paper No. 21, 37 #8, 1903, p. 23.
6. "Wireless Telegraph", New York Times, March 16, 1902, p. 13.
7. "A Site Chosen", Evening Telegram, Dec. 9, 1901, p. 4.
   "Notes in Brief", Daily News, Dec. 9, 1901, p. 4.
8. Local Happenings", Evening Telegram, Dec. 11, 1901, p. 4.
   "Mr. Marconi's Tests", ibid.
9. "Notes in Brief', Daily News, Dec. 12, 1901, p. 4.
10. "Mr. Marconi's Tests", Evening Telegram, Dec. 14, 1901, p. 4.
11. "Marconi Triumph", Daily News, Dec. 16, 1901, p. 4.
    "The Marconi Transoceanic Experiments", Scientific American 86 #1, January 4, 1902, p. 4.
12. C.H. Sewall, Wireless Telegraphy: Its Origins, Development, Inventions and Apparatus, Van Nostrand, New York, 1903, p. 32.
13. "Robert Collins, A Voice from Afar; The History of Telecommunications in Canada, McGraw-Hill Ryerson, Toronto, 1977, p. 297.
14. W.P. Jolly, Marconi, Constable, London, 1972, p. 106.
    "Marconi's Latest Feat", Scientific American 86 #11, March 15, 1902, p. 186.
15. "Marconi's Success", Evening Telegram, Dec. 16, 1901, p. 4.
    This account said that Marconi had received the letter "S" twenty-five imes on Wednesday, beginning at 11.30 am, which was at the beginning of the sched-uled time of transmission. Other reports say that the balloon broke away at 3 pm, and that would make sense if it were 3 pm British time, which is 11.30 am in Newfoundland.
    "Edison Now Believes", Daily Record, Sydney, NS, Dec. 18, 1901, p. 1.
16. "Big Monopoly", Halifax Herald, Dec. 18, 1901, p. 1.
17. William Smith, How Marconi Came to Canada, QQ 38 #3, Summer, 1931, p. 405.
18. "To Restrain Marconi?", Daily Record, Dec. 18, 1901, p. 1.
    New York Times, Jan. 14, 1902, p. 1.
19. "Graham Bell's Offer", Daily News, Dec. 19, 1901, p. 4.
20. "The Governor's Luncheon", Daily News, Dec. 20, 1901, p. 4.
    "Marconi Banquet", Daily Record, Dec. 20, 1901, p. 1.
21. "How the Canadian Government Staked Marconi", Seaports and the Transport World, June 1964, p. 16.
    This article is based on a radio broadcast made by Alexander Johnston in Dec. 1929, when he was Dominion Deputy Minister of Marine and Fisheries.
22. "Marconi", Ottawa Evening Journal, Jan. 9, 1902, p. 3.

"Marconi", New York Times, Jan. 10, 1902, p. 1.

23. "Coming to Sydney?", Daily Record, Dec. 23, 1901, p. 1.
"Marconi Arrives", ibid., Dec. 26, 1901, p. 1.

24. "Still Investigates", Daily Record, Dec. 27, 1901, p. 1.
In this interview, Marconi revealed that in 1897 the Prince of Wales had an apparatus installed on his yacht, the Osborne, and another at Queen Victoria's residence, Osborne House. They exchanged several messages by Marconi's wireless.

25. "Signor Marconi", Halifax Herald, Dec. 28, 1901, p. 1.
"Marconi", Daily Record, Dec. 28, 1901, p. 5.

26. "Treating Marconi", Halifax Herald, Dec. 31, 1901, p. 1.

27. "With Marconi in Cape Breton", ibid., Dec. 30, 1901, p. 1.

28. "Marconi's Wireless", Daily Record, Dec. 17, 1901, p. 4.

29. "Where Marconi Can Experiment", ibid., Dec. 18, 1901, p. 4.

30. "Marconi", Halifax Herald, Dec. 31, 1901, p. 1.

31. "Marconi Here", New York Times, Jan. 13, 1902, p. 1.

32. "Marconi's Chance", Ottawa Evening Journal, Jan. 8, 1902, p. 9.

33. "No Wireless System on Land", DBW, July 8, 1902, p. 3.

34. "Memorandum of Agreement", Sessional Papers, House of Commons, Paper No. 51a, 36 #13, Labour Miscellaneous, 1902, p. 1.

35. "Cape Breton Station", New York Times, March 24, 1902, p. 2.

36. "Marconi's Triumph", DBW, July 30, 1902, p. 4.

37. "Sir Richard Cartwright", ibid., Dec. 22, 1902, p. 5.

38. "Editorial Briefs", ibid., Jan. 2, 1903, p. 4.
"Marconi's Speech", ibid., Jan. 3, 1903, p. 6.

39. Thomas J. Curren, "The Wireless Telegraph Station at Glace Bay: Including an Interview with Mr. Marconi's Chief of Staff", Canadian Magazine 20 #3, Jan. 1903, p. 246.

40. "New Statue", DBW, March 30, 1903, p. 1.

41. "Canadian Marconi Company", ibid., Jan. 5, 1903, p. 4.

42. "It's Nonsense", ibid., Feb. 3, 1903, p. 1.

43. L.W. Gill, "Wireless Communication", QQ 15, Jan. 1908, p. 230.

44. "Editorial Notes", DBW, March 12, 1903, p. 4.

45. "35th Annual Report, MF", Sessional Papers, House of Commons, Paper No. 21, 37 #8, 1903, p. 22.
"36th Annual Report, MF", ibid., Paper No. 21, 38 #8, 1904, p. 26.

46. "37th Annual Report, MF: Appendix 11: Marconi Wireless Telegraph System", ibid., Paper No. 21, 39 #9, 1905, p. 114.

47. "38th Annual Report, MF: Appendix 6: Wireless Telegraphy Marconi Stations", ibid, Paper No. 21, 40 #9, 1906, p. 118.

48. "39th Annual Report, MF: Wireless Telegraphy", ibid., Paper No. 21, 41 #9, 1907. p. 11.

49. "An Act to Provide for the Regulation of Wireless Telegraphy in Canada", Statutes of Canada, Ch. 49, 1905, p. 235.

50. "40th Annual Report, MF: Appendix No. 8: Wireless Telegraph Stations", Sessional Papers, House of Commons, Paper No. 21, 42 #10, 1907-08, p. 95.

51. "43rd Annual Report, MF: Appendix 12: Radiotelegraphic Service", ibid., Paper No. 21 45 #13, 1911, p. 200.

52. D.L. Hutchinson, "Wireless Time Signals from the St. John Observatory of the Canadian Meteorological Service", Transactions of the Royal Society of Canada, 3rd Series, Sect. III, 2, 1908, p. 153.

53. "40th Annual Report, MF: Appendix A", Sessional Papers, House of Commons, Paper No. 21, 42 #10, 1908, p. 107.

54. Hugh G.J. Aitken, The Continuous Wave: Technology and American Radio, 1900-1932, Princeton University Press, Princeton, N.J., 1985, p. 87.

55. "Gen. Hamilton", DBW, June 14, 1913, p. 1.
56. "43rd Annual Report, MF: Naval Department", Sessional Papers, House of Commons, Paper No, 21, 45 #13, 1911, p. 60.
57. "Wireless Stations", DBW, May 1, 1912, p. 2.
58. "Annual Report of the Department of the Naval Service: Construction", Sessional Papers, House of Commons, Paper No. 38, 48 #27, 113, 4 George V, A. 1914.
59. Richard Brown, "The Big Storm - Navigation on the Great Lakes; Wireless Telegraphic Communication, 1911-1914", Archivist 13 #3, May-June 1986, p. 6.
60. G.J. Desbarats, Deputy Minister, Ottawa, to the Marconi Wireless Telegraph Company of Canada, Montreal, Feb. 23, 1912; NAC, RG 12 E 4, Vol. 36, File 5604-14, vol. 1.
61. "An Act Respecting Radiotelegraphy", Statutes of Canada 1-25, Ch. 43, 3-4 George V, 1913, p. 385.
    "Annual Report of the Radiotelegraphy Branch, Department of the Naval Service", Sessional Papers, House of Commons, Paper No. 38, 50 #26, 1915, p. 75.
62. "Annual Report of the Department of the Naval Service: The International Radiotelegraph Conference", Sessional Papers, House of Commons, Paper No. 38, 48 #27, 1914, p. 119.
63. "Vagaries of Wireless", DBW, Feb. 16, 1912, p. 8.
    "Wireless at Kingston", ibid., May 10, 1912, p. 1.
64. "Wireless Apparatus Arrived", ibid., Nov. 27, 1913, p. 2.
65. "Wireless Equipment", ibid., Dec. 13, 1913, p. 36.
66. "Annual Report of the Department of the Naval Service", Sessional Papers, House of Commons, Paper No. 38, 50 #26, 1915, p. 88.
67. "Schedule of Apparatus for 5 1/2 K.W. Installation with Duplicate Engines and Transmitters: Schedule A, Tobermory", NAC, RG 12, Vol. 397, File 5604-14, part 2.
68. KDS, Aug. 7, 1915, p. 7 (see Chapter 2).
69. Inspection Report, on Form W.14, for Marconi Wireless Station at Kingston, Ontario, Jan. 19-21, 1919; NAC, RG 42, Vol. 1037.
70. "Radio Station Does Big Work", KDS, May 15, 1926, p. 2.
71. Personal communications from Prof. Harold H. Stewart, Kingston, Oct. 20, 1982; June 23, 1986.
72. Personal communication from William A. Taylor, Bath, Aug. 21, 1982.
73. Personal communication from Harry Marlatt, Glenburnie, June 16, 1986.
74. Personal communication from Keith MacKinnon, Ottawa, Aug. 15, 1987.
75. Personal communication from Prof. Harold H. Stewart, Kingston, June 23, 1986.
76. Personal communication from Dr. W. Donald MacClement, Kelowna, BC, May, 1987.
77. Ibid., Sept. 17, 1988.
78. "Boys Swinging on Wires", DBW, March 30, 1914, p. 5.
79. Personal communication from Ernest Muller, Kingston, June, 1986.
80. "Wireless on Canadian Boats", KDS, July 25, 1914, p. 7.
81. "Procures Aid by Wireless", ibid., July 22, 1914, p. 6.
82. "Wireless Has No Orders", ibid., Aug. 4, 1914, p. 4.
83. "Special Session", ibid., Aug. 3, 1914, page 5.
84. "The Kingston Corps", DBW, Aug. 5, 1914, p. 8.
85. "Two Companies of 14th", ibid., Aug. 6, 1914, p. 5.
86. Men of 14th Guard Drydock", ibid., Aug. 8, 1914, p. 8.
87. "Wireless Station Attacked", ibid., Aug. 15, 1914, p. 1.
88. Personal communication from Prof. Harold H. Stewart, Kingston, Aug. 10, 1982.

89. Editorial, DBW, May 27, 1920, p. 6.
90. "Must Reduce Voltage", ibid., May 26, 1920, p. 16.
91. "Kingston Radio Association", KDS, March 24, 1921, p. 3.
   "Club Gossip", Radio News (USA), May 1921, p. 795.
92. "The Operating Department", QST 5 #1, Aug. 1921, p. 34.
93. Ibid., 5 #3, 39, Oct. 1921.
94. "Radiotelegraph Regulations", Supplement to the Canada Gazette, Ottawa, Sept. 23, 1922.
95. "The Operating Department: Ontario Division", QST 5 #10, May 1922, p. 43.
   Ibid., 5 #11, June 1922.
96. "Heard Via Radio", DBW, Aug. 9, 1922, p. 11.
97. Canada Radio Stores adv't., ibid., Oct. 2, 1922, p. 4
98. "Amateur Stations", Toronto Daily Star, Oct. 3, 1922, p. 7.
99. "Religious Services from WGY", DBW, Oct. 21, 1922, p. 5.
   "Held Service by Radio", ibid., Nov. 8, 1922, p. 5.
100. "Notes and Comments", KDS, Dec. 5, 1922, p. 4.
101. "Annual Report, MF: Radiotelegraph Service", Sessional Papers, House of Commons, Paper No. 28, 60 # , 1924, p. 140.
102. "58th Annual Report, MF: Report of Deputy Minister: Radio Service", Annual Departmental Reports 6, 1924-25, p. 136.
103. "Barriefield Station", KWS, Dec. 10, 1926, p. 15.
104. "Complaints", ibid., Dec. 9, 1926, p. 1.
105. "Dr. W.A. Jones Declares", ibid., Dec. 13, 1926, p. 1.
106. "Defends Barriefield Wireless", letter to the Editor of KWS from Valentine Sharp, Dec. 14, 1926, p. 9.
107. "In Accord with Dr. Jones", letter to the Editor of KWS from R. Baldwin, ibid.
108. "61st Annual Report, MF: Report of Deputy Minister: Radio Service", Annual Department Reports 3, 1927-28, p. 161.
109. Radio Aids to Marine Navigation: Atlantic and Great Lakes, Air Services, Department of Transport, Ottawa 3 #2E, Aug. 1, 1958, p. 7.
110. Ibid., 15 #2E, June 1, 1970, p. 67.
111. Ibid., 21 #4E, Dec. 1, 1976, p. 110.

CHAPTER 7

The Armouries on Montreal Street, where Kingston boys
received signals training and took rifle practice after WW I.

Kingston Radio Amateurs, Part 1

The Wireless Telegraphy Act of 1905 set out the general terms for
the operation of wireless telegraph equipment in Canada and made
specific provision for licences to be issued by the Minister of the
Department of Marine and Fisheries to private individuals for experimental purposes: "Where the applicant for a licence proves to the satisfaction of the Minister that the sole object of obtaining the licence is
to enable him to conduct experiments in wireless telegraphy, a licence
for that purpose shall be granted, subject to such special terms, conditions, and restrictions as the Minister thinks proper."[1]

Canada's first experimental wireless licence granted under the terms
of the 1905 Act was likely that to the Dominion DeForest Wireless
Telegraph Company for its station on Grindstone Island in 1907.[2] For
the next few years only the coast stations in the public service and those
licenced on government steamers were listed in the annual report of

the ministry, until the new Naval Service took over responsibility for the radio-telegraphic service. Then, some time between April 1, 1910, and March 31, 1911, the Department of the Naval Service issued an experimental licence "to Mr. Frank Vaughan, St. John, N.B., for the erection and operation of an experimental wireless station."[3] By the time of the next annual report on the Radio-Telegraphic Service, there were 11 amateur and experimental wireless stations in Canada, 8 east of the Ontario-Quebec border and 3 in British Columbia. Call letters were assigned in that year and Frank P. Vaughan got XAO,[4] ninth in alphabetical sequence. XAB and XAC were also assigned but, curiously, XAA was not. In the year ending March 31, 1913, Canada had 28 licenced amateur and experimental stations, mostly XA-'s plus XBA to XBG. There were 5 in Prince Edward Island, 4 in Nova Scotia, 2 in New Brunswick, 7 in Quebec, 9 in British Columbia and just one in Ontario.[5] Why Canada's coasts had amateur wireless stations so long before the inland locations is a puzzle, but it may have to do with local exposure to the delights and mysteries of the coast wireless stations.

The 1915 report listed 47 licenced amateurs, 5 of them in Ontario, including H.P. Folger of Kingston, who had call XCD.[6] The next year's report, the last before the closing down of the amateurs for the duration of the War, showed two more licenced amateurs in Kingston. The new ones were J.C. Giles, with the call XDW, and K. Coward, who was licenced for reception only.[7]

Howard Price Folger (XCD, 3GV)

Kingston's first licenced amateur was the son of Howard S. Folger (Arts '87), investment broker, District Manager of the Equitable Life Assurance Society, Deputy US Consul in Kingston, and former General Manager of the St Lawrence River and Thousand Islands Steamboat Company. The Folgers lived in a huge house, Edgewater, at 1 Emily Street in Kingston, right beside the lake. Howard Price Folger was born in Kingston in 1896, graduated from Kingston Collegiate Institute in 1913 and entered Arts '17 at Queen's at the age of 17. At Queen's he was Secretary of his year, member of the Year Book Committee, Senior Prosecuting Attorney of the Arts Concursus and a participant in intramural hockey, rugby and soccer. In December 1916, he became Second Vice-President of the Queen's Alma Mater Society, at the same time as Miss Charlotte Whitton became Assistant Secretary. "Howard's personality, and all-round qualifications make him representative of the highest type of Queen's men."[8] After graduating from medicine at Queen's in 1922, Howard interned in New York, did post-graduate work in Vienna and opened an eye, ear, nose and throat practice in Kingston around 1926. He became Professor of Ophthalmology at Queen's and a famous specialist in diseases of the eye.[9]

Many people contacted in this study knew or remembered Howard Folger, but only one, Don MacClement, knew that he had been a wireless amateur in his youth:

The first licenced wireless amateur in Kingston was Howard Price Folger (Arts '17; Meds '22). He exchanged messages with the wireless station built by four students at KCI in 1913. In 1922 he obtained the call 3GV, in the new series issued by the Department of the Naval Service.
-Queen's Journal clipping courtesy Dorothy (Folger) McDiarmid

Don MacClement: Howie Folger...oh boy...he was a very capable guy. The Folgers, of course, they were big shots in Kingston here. I think there was at one time an attempt to set up some kind of a city broadcasting set. Now, I just vaguely think of Folger being associated with it somehow. Howard Folger, I guess, was one of the most popular young men around town at that time, and lived right down by the park, and I believe he did have a radio set. I can remember asking him about it at one time, but he was much too dignified for a little kid asking questions. I knew him more down at the Yacht Club because they were keen yachting people, but also in connection with radio.[10]

That was all that was salvageable regarding the wireless activities of the young Howard Folger until the author made an unexpected discovery in 1986 at the National Archives of Canada in an old file on inspection reports and prosecutions under the Wireless Telegraphy Act. Preserved there is the report of an inspection of his apparatus carried out on January 20, 1914.[11] Later, the NAC turned up a copy of his 1913-14 licence and his "Declaration of Secrecy in the Operation of Radiotelegraphic Apparatus," signed November 22, 1913.[12]

The Documentation

Howard Folger's licence to establish and operate an experimental wireless telegraph station at 1 Emily Street, Kingston, was "issued in accordance with the provisions of the Telegraph Act, Revised Statutes of Canada, 1906, Chapter 126." The licence was dated April 1, 1913, signed by the Deputy Minister of the Naval Service of Canada, G.J. Desbarats, and it cost one dollar. Howard got Licence No. 38 on Form W. 20, assigning call letters XCD. His wavelength was not to exceed 50 metres, his aerial could be no longer than 30 feet, and the power "absorbed by the primary of the transformer or induction coil shall not exceed 1/2 K.W."

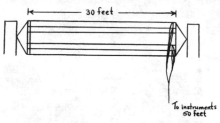

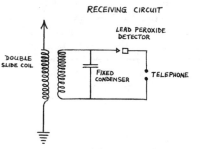

The inspection report contains a great deal more information than the licence. Folger had a double 3-wire flat-top inverted L type of antenna, 30 feet long and stretched between two chimneys on the lake side of the house. The lead-in from the antenna to the instruments was 50 feet long and his earth connection was made with the heater coils in the house. Power was obtained from the 110-volt 60-cycle Kingston power mains, stepped down by transformer to 6 volts. The transmitter was based on an ignition-type coil, powered with the 6-volt output of his step-down transformer. He had a pair of glass- and zinc-plate fixed condensers in a 8x6x1" box in series in the secondary and a 1/8" spark gap in shunt (parallel). The antenna coil was an open helix of 8 turns, 9" in diameter and 9" high, but it is drawn as tapped by both sides of the secondary. The range was "not five miles."

Folger's receiver had a double slide tuning coil in the antenna, but it is drawn like a loose coupler. There was a fixed condenser in shunt in the secondary circuit of the receiver, a lead peroxide (galena) cat-whisker detector and a telephone ear-piece. Under question 16, "Stations with which signals are exchanged," was entered "Collegiate Institute, Kingston, Ont."[11] Folger retained his licence through the early 1920s, when his call was 3GV.[13]

Howard Folger's antenna configuration, transmitting circuit and receiving circuit, from a drawing on a Canadian Government Radio-Telegraphic Service inspection report of his first licenced station, XCD, January 20, 1914. His 6-wire L-antenna was mounted between the two large chimneys on the south side of the family home at 1 Emily Street, and his earth connection was heater coils in the house.
- National Archives of Canada, RG 42, Vol. 1037

What Else Can Be Deduced About Folger's Station?

The section on wireless telegraphy in the Revised Statutes relating to the Telegraph Act of 1905 is written in quite general terms. There are almost no specific regulations except those giving the Minister powers and the requirement for a licence.[1] Since Folger probably got his first licence when the new act of 1913 was in the works, he may have come under the accumulation of regulations to the 1905 Act. Thus, it might be instructive to look at the new legislation and its statutes.

By the terms of the new Radiotelegraphy Act of 1913, an amateur experimental licence was issued "to small stations for instruction, experimental purposes or amusement by persons relatively inexperienced in operating." Nevertheless, such a station could be worked only by the holder of an Amateur Experimental Certificate of Proficiency. Folger must have possessed such a certificate to qualify as a licencee under the new Act, but it is not known when he earned it. The certification examination was held in Ottawa or at a convenient radiotelegraph station, such as VBH, and it was both practical and oral. A candidate had to demonstrate proficiency in adjusting and operating his own apparatus, had to know the appropriate departmental and international regulations and be able to send and receive International Morse Code at five words per minute. He also had to be able to distinguish from other signals, "SOS," "STP" and his own call repeated at ten words a minute. Finally, the licence fee was $1.00. Licences were granted for a period of one year, expiring on March 31. Therefore, Folger must have received his first licence some time during the year ending March 31, 1914. He would then have been 17 or 18 years old. Since his station was located within five miles of the new commercial coast station at Barriefield, he could not transmit at a wavelength longer than 50 metres, nor could he interfere with commercial business. The signal emitted had to be as little damped as possible. The logarithmic decrement of a complete oscillation could not exceed 0.2. Power for any kind of experimental station was limited to 500 watts.

Jack Clifford Giles (XDW)

The National Archives of Canada holds the application for a radiotelegraph station, submitted on June 8, 1914, by Jack Clifford Giles of Kingston[14] and made in accordance with the Radiotelegraph Act of 1913. The form is stamped "Answered July 15, 1914, Radiotelegraphic Branch." Giles described himself on the form as a schoolboy living at Calderwood, which was a large house at 186-189 Union Street West, next to the present Donald Gordon Centre. Since the boy was still a minor, the form required that his parent or guardian's full name be included. Accordingly the name Lieut.-Col. G.M. Giles, I.M.S. (retired), possibly the boy's grandfather, was also entered. Don MacClement remembered Jack Giles: "Jack Giles and Howie Folger were fairly well known. They both belonged to very wealthy families - I'd call them high society people in Kingston."[10]

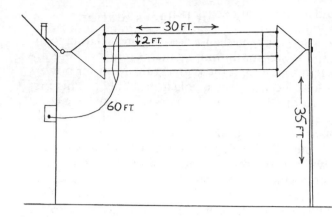

*Jack Clifford Giles'
antenna
configuration, taken
from his licence
application to the
Radiotelegraphic
Branch, Department
of the Naval Service,
June 8, 1914.
-National Archives
of Canada, RG 97,
Vol. 85*

Interestingly, "Howard Folger, King Street" was entered under question 3, "Name and location of any other station or stations with which it is desired to carry on communication." Additional evidence required in that section, that the second party be willing to be involved with the applicant, was not supplied. Giles added that he was "wishing to learn code" and could "not at present" send or receive in Continental Code. The purpose of the installation was "solely for experiments," so it seems reasonable to assume that Folger was going to be involved in training young Jack.

The applicant was required to include the particulars of his installation, including a general description of his transmitting equipment. He wrote "1 inch coil, Leyden Jar, Zinc spark gap, Helix, Key." The receiver comprised "Two Loose Coupling Receiving Transformers, Two Loading Coils, one Variable Condenser, 2 Fixed Condensers, Either Silicon, Galena or Ferron Detector, Two 1,000 ohm receivers." His sketch of the aerial shows a flat-top of four parallel wires, spaced two feet apart, 30 feet long, and looking very much like an inverted L, although he called it a T. The antenna was stretched between a chimney near the edge of the sloping roof of the house and the top of a 35-foot pole. The lead-in, shown going to a second floor window, was 60 feet long.

Giles' licence was suspended in the general clampdown at the outbreak of the Great War, so he may never actually have made it to air. Col. G.M. Giles, of the Indian Army, died about three years later. His name disappeared from the <u>Kingston City Directory</u> after August 1917, and the government obtained the house in 1919 and converted it into Calderwood Hospital for incurable soldiers.[15] No one has been found who knows what became of Jack Giles.

K. Coward

Prof. Harold Stewart was the only person interviewed in this study who remembered K. Coward, the holder of a "Reception Only" licence in 1914.[7] Keith (Kirk?) Coward of 386 Albert Street got the young Stewart (404 Albert Street) started in wireless telegraphy, and he remembered that Coward had a receiver in a shack in his back yard. According to the Official List of Radio Stations of Canada, Coward never obtained a licence to transmit. There were, however, half a dozen or more boys in town who were quite proficient in flashing their electric waves into the ether and in snaring distant message-laden oscillations therefrom, without benefit of official licence.

The KCI Connection

Four boys at Kingston Collegiate Institute (KCI) built a complete wireless set, transmitter and receiver, in November 1913. They could send up to 50 miles and receive from 1,500 to 2,000 miles:

> At midnight they are capable of receiving messages from Arlington, Virginia, from the wireless station there. They can also make out signals from the Marconi station at Montreal and Cape Cod. The whole apparatus has been made by the boys, with the exception of a few coils and ear-trumpets which are too delicate to be made satisfactory. The boys who have rigged the wireless up are George McLeod, a KCI student, Arthur Cruse, Harold Farrar, and Willard Roche. The wireless is placed in a small room off the assembly hall on the upper flat of the Kingston Collegiate Institute.[16]

When Howard Folger's station was inspected in January 1914, he indicated on the form that he was exchanging signals with the Collegiate Institute, Kingston;[11] that is, with the four KCI wireless boys. It is a puzzle why he made this admission to an inspector, since there is no evidence that the wireless set at KCI was ever licenced, either for sending or for receiving.

The KCI boys and Howard Folger may have been inspired, encouraged and helped in wireless by the physics teacher, Captain James Wilfred (Cap) Kelly (Arts '10). Don MacClement recalled, "You ask about the four boys at KCI. I can remember something about Mr Kelly, who was the teacher of science at KCI then and, I believe, helped some of the students - and probably this'd be the group - to make some wireless equipment. I don't remember ever seeing it."[17] Tall and mustachioed, standing straight as a ramrod, Kelly was a Captain in the Princess of Wales Own Regiment. He was appointed to KCI in 1912 and retired in 1946.

In the fall of 1913, Howard Folger was newly graduated from KCI and a freshman at Queen's. Cap Kelly could have been the person who started him on his wireless adventures while at KCI. This connection will bear more research.

It has proven difficult, at a distance of 75 years, both to track down details of the KCI wireless story and to secure information on the participants. Family members of two of the boys were located, and Ronald Ede, Guidance Counsellor at KCVI, provided valuable assistance and information.[18]

Arthur Cruse was the son of James G. Cruse, head caretaker at KCI from 1896 to 1948. There were five children in the Cruse family, all living in the tiny house on the school property at 136 Alfred Street. Arthur Cruse entered KCI in September 1913, when he was 16. He spent most of his life in West Orange, NJ.

Harold John Farrar lived at 119 Alfred Street, the son of an assistant superintendent at the Canadian Locomotive Company. The family had come to Kingston from Pittsburgh, Pa., and Harold enrolled at KCI at age 15. He wrote on the Honour Matriculation Scholarship in 1916, won the McLaughlin Scholarship[19] and entered science at McGill University.

Farrar wrote a fine article on wireless for the K.C.I. Journal in June 1914, when he was in grade 11. The article makes use of a water analogy to explain electrical phenomena, but gives no certain clues as to the wireless activities of the KCI boys. It does, however, contain some details which may be based upon experience:

> This alternating (electrical) current is caused to change its condition in the "transformer" from a current of large size and small pressure to one of small size and tremendous pressure, although the actual amount of electricity is unaltered. In the latter state it is very violent and intent upon causing a disturbance. However, it is restrained in the "condenser" which absorbs this violent energy gradually and then suddenly discharges it in a concentrated form. This discharge takes the form of a crashing blue flame between two stationary zinc points and the teeth of a zinc wheel moving at a high rate of speed, which apparatus is called a "spark gap."[20]

The rectifying detector was a "mineral crystal" and for hearing the signals one uses:

> delicate watch-case receivers which fit over the operator's head. These consist of small permanent magnets wound with fine wire. The pulsating current, aiding the magnets attracts and releases a thin diaphragm very rapidly, causing a humming sound. Communication is carried on in the regular telegraphic code of dots and dashes. These are made by interrupting the flow of the current in the sending set so as to produce short or long "hums" in the receiving set. Each operator can change from "sending" to "receiving" at will by a switch. Both sending and receiving sets are equipped with "tuning" apparatus, i.e. instruments by means of which all undesired stations may be "cut out."[20]

As a cautionary note, this description could be based partly on observations of the apparatus of the new Marconi station at Barriefield, which was opened in January 1914.

Willard Roche was shown living at a non-existent address in his only listing in the <u>Kingston City Directory</u> in 1914-15. He may have been the son of William Roche, a machinist at the locomotive works, but the surname Roche, previously shown at two addresses in town, disappeared from the <u>City Directory</u> after the 1914-15 edition. No record exists at KCVI of Willard Roche having attended that school.

George David McLeod was the only one of the four boys known to have tried wireless telegraphy as a career. He was born in Hamilton, Ontario, in 1896, the son of Peter C. McLeod, a carpenter from a New Brunswick seagoing family. The family moved from 17 St Lawrence Avenue, Kingston, to 204 Alfred Street about 1913. The school records say that George entered KCI in September 1912, when he was 15, and that he left in December 1913 because he did not want to be put back to Form I.[18] There is other information in the press that he left KCI in April 1914, to accept the position of wireless operator on an ocean-going boat.[21] At some time along the way he had earned a Marconi wireless operator's certificate.

> McLeod has been studying the wireless apparatus for the past couple of years. Some time ago he, in company with a couple of boys from the Institute, erected a wireless apparatus on the roof of the Kingston Collegiate Institute. After considerable work they were able to get in communication with some of the wireless stations in this vicinity. After perfecting the apparatus, G. McLeod decided to enter the profession. He left for the city of Three Rivers, Quebec, where he will sail from that port.[21]

He was stationed by the Marconi Wireless Telegraph Company at Trois-Rivieres, at Quebec City, and on the salvage boat SS <u>Lord Strathcona</u> when she went to the aid of the grounded CPR liner <u>Montford</u> below Quebec City. In early May, promoted to O.I.C. for wireless, he left Quebec City aboard the steamer <u>Wacousta</u> of Sandefjord, bound for Norway.[22] On his return he was reposted to the <u>Strathcona</u>, working on the

George David McLeod (1896-1973) was one of 4 boys who built a wireless transceiver at KCI in 1913. He went to sea as a wireless operator in April 1914 for the Marconi Company. He returned to Kingston to complete a degree in mining engineering in 1920, and was one of the founding members and the first Vice-President of the Queen's Wireless Club in 1919.
-courtesy Peter C. McLeod

wreck of the CPR liner <u>Empress of Ireland</u>, which had been rammed and sunk off Rimouski with the loss of 954 lives.[23] He had signed onto the German steamer <u>Berkintilds</u> in July 1914 and with war about to erupt, his family was very worried.[24] George wrote from the <u>Wacousta</u>, bound for Scotland, to calm their fears. The German boat had been ordered home and he had refused to go with her, while the <u>Strathcona</u> and all Canadian government steamer wirelesses were closed and guarded: "I have orders not to leave any instrument unattended at any time on the trip over, which I take to mean that we may possibly have some fun."[25] Later, he was wireless operator on the SS <u>Clutha</u>, a captured Austrian liner serving as a British Admiralty transport, and wrote "of thrilling times dodging German submarines off the English coast. They were near the SS <u>Iberian</u> when that boat was torpedoed",[26] and that summer he got as far east as the Dardanelles. In August 1916, on one of H.M.'s transport ships, <u>English Monarch</u>, "He was slightly wounded while participating in a little exploit under naval supervision near La Basse and Armentiers. Later his ship was wrecked and he lost all his belongings, including some souvenirs he picked up near the firing line in France."[27]

McLeod was granted a special permit from the British government to attend Queen's University in the winter of 1915-16, and in the spring he reported for duty again with the rank of sub-Lieutenant and wireless operator in the Imperial Service.[28] The last date of discharge in his "Continuous Certificate of Discharge" was September 11, 1917, at which point he may have returned to Queen's. Apparently he went back to sea several times in all, and graduated in Mining Engineering in 1920.

McLeod was a charter member of the Queen's University Wireless Club in the fall of 1919, and was its first Vice-President while Warren A. Marrison was President and Prof. L.W. Gill the Honorary President.[29] George used to do the wiring and lighting for the Queen's science dances because he couldn't afford the price of admission to take his future wife.[30] Peter C. McLeod remembers his father being crazy about radio, and that he built radios during the Great Depression and afterwards: crystal sets, including one that could sit on a finger tip, vacuum-tube sets and short-wave radios. He was always interested in astronomy and navigation too, and after his retirement, he used to go out to shoot the noon-day sun with his old sextant.[31]

W.D. MacClement Tells His Own Story (3MC)

Dr William Donald MacClement is the son of Dr William T. MacClement, Professor of Biology at Queen's and first Director of the Summer School. The family lived on campus, and young Donald's playground was the whole of the university grounds. He knew the few campus buildings of his youth from cellar to attic, and oversaw all of the happenings in his domain, including the building of Lester Gill's aeroplane and his mobile wireless experiments. Donald says that he started building wireless receivers from the age of nine or ten, and that he was pretty well established in the Ford spark coil game before World War I. He has retained in his incredible memory the most amazing detail, and this wonderful narrative about the hobby of his youth points up the industry, inventiveness and perseverance of the wireless amateur.

<u>Don MacClement</u>: Summerhill was our home - just the west wing. It was a separate structure then. Originally, when the University moved into the building, administration was the ground floor and Theology the top floor of the centre part. Every part of Arts was over there. On the far side it was Medicine, and when Dr Scott moved in and started cleaning up the basement, which was only partly excavated then, he found parts of old coffins, because the subjects for dissection somehow got here in the coffins. And I can remember Mrs Scott...oh, she was horrified to find these. She said, "I'm going to find a skeleton down there some of these days."

There was a wooden shingle roof up there. They've put a metal roof on now. Near the west end at the back was my fort - a wonderful place. The wall was wide enough at the top that you could stand there and about half a dozen of us professors' kids, Don Matheson (Arts '26; Meds '34), "Snag" Skelton (Arts '27) and several others, we manned this fort and challenged all the kids in the neighbourhood. There used to be a lot of horse chestnut trees in the front, and in the fall you'd collect great bags of those for ammunition. We had wooden swords and defended the walls - they even built scaling ladders. I had an almost perpetual job of replacing the glass in those windows.

My bedroom was on the second floor at the back, and it still has the French windows in it. Mother got those put in so that the whole thing could be opened. She was a fresh air fiend. To the east of my window was a chain-link ladder up the wall to the roof - where the drainpipe is now. I almost wore out that ladder, going up and down there. The roof was a playground.

For a youngster living in the University here, it was a never-ending place to explore. Everything was fascinating! In the chemistry building I watched them doing experiments, and tried to do some myself, or made things out of the glass tubes that had been thrown away and wonderful chances like that. And over in the Mining Engineering building, the smelting in the big furnace. The Biology Museum. Where the

Dr W. Donald MacClement standing below the French windows of his former bedroom at the rear of Summerhill Annex, Queen's. This area was his boyhood playground, including the roof which was accessible up a chain ladder hanging to the left of the window.

The rear of Summerhill, showing Donald's bedroom with the French windows, centre, and the drainpipe, which is roughly where the chain-link ladder led to the roof.
                    -photographs by Arthur Zimmerman

metal grill is, over the basement window on the east side of Old Arts, used to be a great place for catching frogs in the spring. It was a sort of pit, about three feet deep, and the frogs would come up the hill for some reason and they'd get into that pit. I've caught a couple of dozen frogs at a time, hopping around in the bottom of that. They couldn't hop high enough to get out.

During the First War, the top floor of Old Arts was used as a shooting gallery for target practice. The target was made of wood with a piece of metal on the back, and it was almost chewed to pieces with the thousands of .22 calibre bullets. On the floor was a sort of big pit underneath the target, and the chewed up wood and bullets all fell into that. So, I got pails of water and dumped the wood and the rest in. The wood floated and the bullets went to the bottom. I got bags of them. It took me two trips carrying those spent lead bullets downtown to sell them.

About Friend Douglas Jemmett

I believe the first time that I met Doug Jemmett was in the electric generating room in Fleming Hall, where they made the 110-volt DC. I'd made a little one-pole electric motor, and had noticed the distinctive odour created by the spark from the commutator. I'd noticed the same odour around the generator there in the power station. I'd been shown through the power station by Mr Hickey, who ran that part of the University, and I'd gone back to find out more about the source of that very distinctive odour. While I was there, one of the people in engineering came along and asked me what I was doing. That was Doug Jemmett. I said that I was trying to find out more about the source of that odour, and he explained that it was ozone, a gas produced by the spark. We apparently discussed it and I told him about my motor. That was the first time I met him, I think, and I was fairly frequently over in the engineering building because that's what fascinated me. Somewhere along about then I'd been disturbed because I couldn't understand the voltage difference with one or three dry cell batteries. Doug Jemmett, when asked about that voltage difference, showed me through the electrical labs where the students worked on rows of lightbulbs on bases and a big panel with a whole lot of little brass places where you plugged in cables and made up various kinds of voltages and so on.

Now, that was well before the First World War and wireless hadn't come into it at all then. This was just the beginning of my interest in electricity, and I think it was Jemmett's patience in explaining the mysteries of electricity and voltages and amperages to me...I can remember his trying to teach me the difference between things connected in series and in parallel, the difference it made and resistance. Certainly, I didn't have his help during the first stages of putting together any wireless equipment. During that period my friendship with Jemmett was connected only in the realm of electricity generally.

He was sort of a quiet, steady type. He didn't shout and dash around like Gill did. The two of them, I think, sort of balanced each other. Gill was the tremendous enthusiasm that went helter skelter for something - if there was something he didn't understand, he'd go at it head first. He just drove for it. I was a bit scared of him, because he was always a bullet whizzing somewhere. Impatient and demanding and authoritative, but he must have had a tremendous brain. Jemmett was very much the junior, the one who made sure that this was done and that was set up and that they moved from one stage to the next - progressively and according to some sensible plan. He was in a sort of student-teacher relationship with Gill. Jemmett was changed quite a bit when he came back from the army. It might have been partly the responsibility that made him seem so much more solemn and serious about things - and he married when he came back.

As soon as there was news that Jemmett had come back from the War, I hurried over to Fleming Hall and hung around until I finally met him. I poured out all my enthusiasm about wireless and building receiving and transmitting sets. I think he had a limp at that time. He took quite a while getting up the stairs in Fleming Hall. There was something the matter with his foot, and for some time he couldn't ride a bicycle.

Jemmett was definitely interested in wireless when he came back to Queen's from the War. But it wasn't until Jemmett came back from MIT that he was much of an authority on wireless. He set up in one of the rooms at the west end of Fleming Hall some of that equipment that must have been from the experiments in the bell tent. The aerial ran across on a slant upwards - a single wire across to something on the top of Ontario Hall. I don't know what was on the Fleming Hall end of it, but it came down into one of the windows of that classroom. I can remember asking him how to balance the tuning of a set, and he referred me to textbooks, which he loaned me, full of how to calculate capacity, inductance and the shape of radiation waves. I remember that I had been disappointed at how few books the Queen's library had on it. I could comfortably fit them all under one arm.

He used to take a bunch of us kids out on hikes and bicycle rides through the country, and he knew a tremendous lot of all sorts of other things. Not just electrical information, but almost anything, and he was a good teacher in that he got a lot of pleasure out of simply getting his ideas or the facts across so that they were understood. His personality was, to me, delightful. I lived just across the lawn from Fleming Hall and if something came up I'd head for Jemmett rather than go to my own father, because he was a person that would be patient, listen, give advice. He was that sort of person, and eventually became our substitute Sunday School teacher at Chalmers Church. Most of us professors' kids had been kicked out of Sunday School because we were such a nuisance. But he took us on, and we agreed that if we finished the lesson first so he could tell that to the Superintendent, for the rest of the time we could talk about or do whatever we liked. That was tremendous. We'd hurry through the lesson so that we could get into questions about science and everything under the sun, and he was always helpful.

The Beginnings of Interest in Wireless

About the first thing I can remember in connection with radio was one day when I was visiting my school chum Lewis Donovan Clark (Mining '29; the son of Dr Arthur L. Clark, Dean of Science). It'd be about 1910 or 1911 and I'd gone out to his house on Albert Street to see some of the things he'd received for Christmas. Well, one of the presents intrigued me. It was a wireless set, made by the Gilbert Company who made various kinds of toys. Now, I knew what the word wireless meant - it was some sort of weird communication device, and the name of Marconi, of course, was somehow associated with it. But this was a set that he could operate! It was a flat breadboard layout, on which were mounted some coils of wire and various gadgets and knobs to turn. There was a key at one end, a place to connect some dry cell batteries and binding posts at the back where it said "aerial" on one and "earth" on the other. Donovan told me he could communicate right across town to our house, which was in the University grounds, without any wires or connection in between. He warned me to stay back because there were some very powerful electric currents in it, and he connected on the batteries with quite a flourish, of course, and moved a switch or two or turned something and then he pressed a sort of strap key at the side of the board and a bright, sizzling spark jumped between two points in the machine and he proceeded to tap out a whole lot of dots and dashes and things. Then he disconnected it and said, "There, I've sent a message." Well, to me that was pretty close to a kind of magic. I wanted to know more about it and when I went home I asked father if I could have a wireless set like Donovan Clark had. He said, "Oh, it costs too much, and you'll probably have to make one for yourself."

Well, that was the way it started. I can remember digging up magazines over in the Staff Reading Room,[32] which was over in the Old Arts Building then, trying to find out what the score was on how to make a wireless set. I think <u>Popular Mechanics</u> and <u>Scientific American</u>. I had some knowledge of electricity for I had played with dry cell batteries and little lights and even made a small electric motor that ran very well. I'd learned the hard way of how 110 volts feels, because I'd tried to light my little 6-volt lamp off the 110-volt direct current which was then supplied to all of the buildings in the University grounds from the power plant attached to the end of Fleming Hall with its tall stack. The resulting...I was going to say energization of my muscles by that 110 volts threw me right across the room against the wall on the other side. I was lucky. It hurt a bit but no further damage except a few burn marks on my fingers. Queen's ran everything during the day-time, during class hours anyway, on 110 volts DC. The wires were very simple...almost crude...insulated more or less, and mounted on porcelain cleats in pairs that ran through the attics and under the floors and up the walls and so on. Those porcelain cleats made excellent insulators for aerials and things like that. I know our house, up in the attic, later lost some of the cleats when I needed some for my aerial.

One of the very first things I saw in the magazines was a coil of wire. Well, it had to be wound around something, and the size and shape of it was very similar to that of a round Quaker Oats box. The wire I got at the Engineering Lab, from an old burnt-out motor, the field coils of which had somehow shorted. It had been taken apart, part-ly for demonstration, and I asked if I could get some of the coils of wire from the field. It did very well for making coils - single, cotton-covered wire, about size 18, I think. The piece of equipment I was attempting to make was called a loose coupler. It consisted of an outer coil, which was that Quaker Oats cannister, and inside it and free to slide along a couple of rods there was a smaller inner coil wound around a cylin-drical cardboard salt box. It was just about the right size. I proceed-ed to wind wire around the boxes very tightly, the turns close togeth-er until each whole surface was covered. To support the inner tube on the rods, one cut out circular plugs of wood and fitted them into the ends so the rods going through the end plugs could let it slide along in and out of that outer coil, the outer one being mounted on two square blocks of wood which were then fastened to a baseboard. The wood-working was very crude. I had very few tools and almost no money to buy any with.

Another instrument which seemed to be essential was a coherer. This looked fairly simple - a bunch of iron filings in a glass tube, a fixed metal plug at one end and an adjustable metal plug at the other. The whole thing was vibrated by a hammer of an old electric doorbell. The alternative to the coherer was a detector or rectifier using a gale-na crystal and a cat's whisker. Well, after the receiving instrument had been put together, and the two types of rectifiers tried over and over again, I soon found that no way could I get that darn coherer to function anything other than just momentarily and not satisfactorily. The galena detector was more successful. It required infinite patience to find the right spot on the surface of that galena lump, but I seemed to have lots of patience then.

I could get galena crystals very easily because the Mineralogy Department had great bins of lumps of galena, and they let me have any parts I wanted. So, I always had a few bits of galena lying around and could whack up a little detector with a cat's whisker. Eventually

I got a spool of bronze-phosphor wire. Now, that was the grand stuff for making cat's whiskers because it was springy enough, a good conductor and didn't curl very much. You were supposed to mount the lump of galena in White's metal, a very soft kind of solder. I didn't have means of melting solder, so mine were mounted in a little brass or copper cup with a screw through the side of it to clamp the galena crystal in. It worked pretty well. Some time later I learned how to melt White's metal and half bury the crystal in that.

Making and putting up the aerial was probably the most fun. Up on the roof, at either end of our wing of the house, were two matching chimneys, spaced about 40 or 50 feet apart, and between these I fastened up what looked quite like the pictures of an antenna, with wooden spreader and about four wires, each one insulated at each end with a couple of those porcelain cleats. And a drop-wire connected to each wire and all connected together in a "Y" to the main lead-in which came down across the roof and out around the eaves-trough and into the window of my bedroom. I wonder whether the mark is still on that window frame, where I made a hole through it. Keeping the lead-in wire up from the roof was quite a problem. I finally made little tripods of wooden sticks and had a series of those coming along down across the shingles of the sloping part of the roof. A long stick, fastened into the top of the eaves-trough with an insulator on the end of it, held the lead-in wire out over the metal eaves-trough. The wire came in behind the hinge-line of my bedroom window.

My first serious purchase was a single-blade double-throw switch on a porcelain base. I bought that at an electric store downtown, and fastened it on the outside of the window frame of my room. And to the centre post of that was attached the aerial wire. From the top terminal of that switch, the lead-in wire came in behind the hinge-line of the French window of my room, and from the bottom one a wire went straight down to the ground where I'd driven an iron stake, four or five feet down into the ground outside the wall of our house. To make sure that the earth, as it was called, was properly functioning I read that one was supposed to pour some water on it every once in a while to make sure that the earth was damp.

Now I had the loose coupler, the tuning system, both a coherer and a galena detector and all I needed really was a set of headphones. They were going to cost a lot more money than I knew how to find. I had tried to complete my home-made wireless receiver by disconnecting the earphone from our wall telephone - by the way, our telephone number was Number 5 in the city - and had used it, not very satisfactorily, until one day father came home and found no earpiece on the set on the wall. Of course, I got hell. So, I had to put it back and promised never to take it again. I then went to the telephone office downtown trying to get a telephone receiver. They said they had given some telephone equipment to Queen's and they didn't have any more spares. Well, the obvious thing to do was to go back to E.E. and ask if they had any earphones that I could use. And there, the person I asked was Gill and he gave me, not a long tubular sort of earphone that hung on our wall phone, but a short squat thing fastened on the end of a black handle. Gill said I could use it as long as I promised to bring it back just as soon as I was able to get an earphone of my own.

Well, I took it home and connected it onto the circuit of my would-be receiving set. First I tried the coherer and set the buzzer going, tapped it, adjusted the plug at one end and fiddled with it and got nothing but sort of scratching sounds, a little humming and didn't seem to be getting anywhere. I tried moving the connection along the outer coil of the loose-coupler and it didn't seem to make any difference. After a while I connected in the galena detector and tried that. Well, the sound all at once was different. I heard hissing noises, scratching noises and a bit of hum - I had changed the connections on the outer and inner coils. And as I wiggled around with that cat's whisker on the galena lump, suddenly I heard "beep, beep, beep" signals. This was radio! Wireless signals - and I was fascinated. It was my first success in the radio field. The slightest movement of that cat's whisker and the signal disappeared. I hunted and hunted everywhere on the surface of the crystal where I could hear a bit, and finally realized that it was no good holding the cat's whisker in my fingers. They weren't steady enough. So, I devised a system by which I could turn a screw on the post that held the cat's whisker and bend it over one way or another way, and finally got a thing that would work. The signal I was listening to was quite strong. I could hear it very clearly indeed, and at that time I didn't know it was VBH, the Marconi transmitter at Barriefield doing its stuff as it did off and on all during the day and quite often at night too. I was mesmerized by it and listened for a long time, pretendng that I could understand what the dots and dashes meant.

Well, that meant right away that I had to learn the Morse Code. In a magazine I had seen a list of the alphabet and numbers in Morse Code and a diagram and description of how to make a practice set so that one could learn it. It required a buzzer, a key and a battery. I had the buzzer from the coherer set-up, and I made a key out of a piece of springy brass stuff. Batteries were easy enough to find - a couple of #6 dry cells - and soon I was buzzing away, going through the alphabet, memorizing it and practicing making those dots and dashes into some sort of a logical sequence.

It was a kind of fascination. During the day, even at school, I was saying to myself dit-da, dit-da, dit and da-dit-dit-dit, until some of my chums thought I was a little bit odd. Odder than usual. Anyway, it didn't take too long to learn that code. Then, of course, one had to apply it. I can remember the first serious attempt to write down the signal from VBH. I could get a letter and by the time I'd written it down, another letter had gone by and I'd missed it. I'd get a letter here and a letter there and I'd put them on a line across a page as fast as I could scribble. It took quite a bit of practice, and first of all I could read the call sign VBH. It was repeated over and over again, and it helped to learn how to put those letters together.

Just about then there was an ad in the newspaper about training in signals, given free by Sgt. Tom Brown, instructor of the local signal corps. And the lessons were to be held down in the Armouries on Montreal Street and a chum and I, Earl Morris, went down to see what it was all about. Well, they immediately enrolled us in a course and told us that when we finished the course, not only would we get the $5 that we were promised, but we'd also get a limited commercial operator's licence if we went far enough in it. This was an army training system and they set up a timetable in which groups of 10 or 12 would meet so many times a week down at the Armouries and practice signals. The first thing you were taught, of course, was dots and dashes. Well, I had already learned a fair bit of that and it didn't take long to get speed enough in sending and receiving on the buzzer sets that they had. I could get up to almost 10 words a minute before very long. Then they switched us over to Aldis lamp, and you strained your eyes to watch the flashes away down at the other end of the Armouries hall and shout out the letters to somebody who was writing it down. Then they switched from that, after you got fairly proficient, to flags...much more difficult. I'm sure it was during the winter, and of a group of 20 of us that started, at the end of it I think there were about a dozen of us. Besides myself and my chum Earl Morris there was Keith MacKinnon, a school chum, and Ralph Bunt, known as "Rulla" Bunt, and there was one of the Cohen boys, and Norry McLeod. Oh, there were others. Just a bunch of the guys that I'd known most of my life from around town. Well, as soon as we thought we were pretty smart sending and receiving Continental Code, we learned that the thing to do was to have a transmitting set, so that we could send to one another across town.

That was not difficult nor expensive in those days. A ready-made high frequency coil was available in almost any garage because the old Model T Ford spark coil provided a dandy spark. Oh, anywhere from a quarter inch to half an inch across a gap if you hooked it up to a battery. Pretty well everybody in this gang - there must have been about a dozen - got hold of spark coils and connected one side of the gap to the aerial and the other to the earth and the battery to the low tension circuit with a key in it. And we proceeded to torture the air with our signals - untuned and just wide open oscillations. At noon, we'd run home from public school and go up to our room and one of us was designated that day to send out a complete message. We could send anything we liked. Most just sent out a passage from a story or a poem or something like that...oh, maybe 100 or 200 words. Not too long because we had to eat lunch and then get back to school again. Everybody else in that group was supposed to be listening, writing it down. And then we'd meet at school and criticize whoever'd been sending, the mistakes they'd made or how they made their signal. Well, it was really good fun and certainly gave us practice. One, of course, was competing with the other to send as fast as he could. Accurately, if possible, but it didn't take long to get up to 12 or 15 words a minute. Receiving was a little bit more difficult if you had to scribble it down on paper. I can remember Rulla Bunt said that the way to do it, if you really wanted to do it properly, was to get a typewriter, and then it all came out nice and neat and quickly. Far faster than you could write it down with a pencil. Well, I didn't have a typewriter, and so it was just pencil and paper.

Anyone who had particular skills, mine was in winding coils and making the tuners and the rest...and I used to wind coils. I think I made receiving set tuning coils for half a dozen anyway, around town, and finally got a few tools to make them more accurately than I'd made that first one of mine. Let's see, Alex Ada, Don Matheson, Norry McLeod, Sheldon Cohen and somebody else all used my tuning coils.

A tuning condenser was a great help in accurately tuning in a signal and it was a rather expensive item. Some of the gang sent away and got these - mostly from Montreal. William B. Duck put out a catalogue that we treasured and memorized, and when we could get a little cash I think we bought one thing or another out of that; half a dozen of us ordering on one order form. Well, I couldn't afford the rather expensive rotary multiple-plate tuning condensers, so I went back to Doug Jemmett to ask if there was some way I could make my own condenser. He took out a textbook in which some of the early wireless people had designed and made their own tuning condensers consisting of three squarish metal plates held vertically in a block of wood and two other metal plates in another block which could slide in grooves in between those three plates. I got hold of some of Father's old photographic plates, stuck lead foil on each side of them and connected them up in a wooden frame. To make the two sliding blocks run smoothly was a little beyond my skill, and it seemed that nearly always the thing stuck when I was close to getting it properly tuned. Trying to get the loudest signal out of a weak station, I soon found that there was an optimum combination of inductance on the coil and capacity in the condenser to get the best signal.

I think the tuning of the primary was almost incidental. Quite soon we found that you could have a very good set with an auto-transformer type, in which there was just one coil of wire. One end of it was connected to the aerial and connections tapped off it at intervals on the way down to the other end. You formed a secondary by connecting the aerial lead not only to the coil but also to the rest of the receiving set, and then made a connection back to the rest of the tuning circuit somewhere less than the amount of turns that were on the primary, which was of course leading back down to the ground. This was easier to use and make and much less bulky, and became more or less a normal tuning system. That and some kind of variable condenser was about all one needed.

Earl Morris was the chum that I worked with for most of my...let's say escapades or activities. He lived down on Barrie Street and was one of the sons of Johnnie Morris, who ran the Imperial Oil outfit in the Kingston area. Earl was one of the first ones to get infected with wireless, because he and I were together on all sorts of activities. Pretty soon he had an aerial from the chimney on his house down to the ridge of a barn in their back yard, where his father always kept a high-stepping stallion for his driving. That aerial was according to a picture in one of the magazines - a fan aerial. I think it had 20 wires in it. A row of double-cleat insulators along the ridge of that barn and a big, lumpy insulator - I don't know where we got it - up on the chimney on his roof. The lead-in went into his bedroom which was one of the upper rooms in that house. The tranmsmitter was a very fierce one. We had wound a transformer of quite large size - somewhere we had found that you could wind the coils by driving a shaft through a spool and could wind on any amount of wire. It was a huge thing...oh, several inches in diameter, and it was slipped over a primary and a big iron core went through it. The spark was terrific and it ripped and snarled at you when you pressed the key. It was probably one of the ones that bothered Barriefield,[33] because it splattered all over the whole world of ether that was available.

Then, Earl was the muscle man in the setting up of a long antenna. The first one that we put up ran all the way from a mast on the Principal's house right across Arch Street and across the block to the front chimney of Earl's house on Barrie Street. It took quite a bit of doing, because it had to go over not only houses and trees, but high power wires on Arch Street. But we got it up, eventually, and he hooked on to one end of it and I hooked onto the other end of it. Now, how in the world the single antenna could work quite well for both of us, I don't know. But it was quite good for a while until a storm took it down. It was after that that I put up the aerial from our house across over to the Pathology Building, and it lasted quite a bit longer.

I guess I should relate an incident...more a confession than anything else. Just about that time, when we were putting up that long aerial from our house down to Earl's house, we decided that my end of it, at the west end, wasn't high enough because it had got touching the trees on the other side of the house. So, Earl and I and some others built a mast out of 20-foot 2x4s. Now, it was quite a mast. The bottom piece was three 2x4s, the centre one a little shorter than the outer two. The next one was two 2x4s and the top was one 2x4. It'd be a good fifty feet high anyway, by the time it was all together. It was made with guy wires in four directions, with a pulley at the top to haul up the antenna wire. Somehow or other when it was all put together, we got it up that chain ladder that went up the back of our house to the roof, then dragged it up the roof and fastened the base of it firmly - bolted it or fastened by wire - to the west side of the main chimney at the west end of the Principal's house; that is, Summerhill. Then the guy wires were attached to the edges of the roof so that it held up quite firmly. When the antenna was connected and hauled up to the top, it was nicely clear of all trees and power wires and everything. That was wonderful, and we admired it a great deal. Until one day, the Principal, Dr R. Bruce Taylor, coming across from the Old Arts Building, suddenly saw this huge, great monstrosity sticking up out of the top of his house...his house, not ours...with an aerial on it. Since it looked like something of my wireless stuff, he came charging into our house and I can remember he came right up to my bedroom and told me exactly what he thought of

Dr W. Donald MacClement pointing to the main chimney on Summerhill, the Principal's residence, to which he attached a 50-foot wooden mast in his youth. This mast anchored one end of an enormously long antenna that stretched over trees and wires across Arch Street to the front chimney on Earl Morris' house on Barrie Street.
-photograph by Arthur Zimmerman

people who put all sorts of crazy, dangerous and unsightly structures on top of his house, and without permission. And I was to get it down right away. We got it down that night, and didn't dare put up another one.

Bruce Taylor, the Principal's son, had started in wireless but hadn't the patience to persevere. But, it was a hobby that carried a youngster along almost anywhere, and there wasn't another like radio, where one could get so much pleasure out of it and which would grow as your skill and knowledge grew. If you got interested in radio, it was all there and it was so fascinating. All you had to do was dig hard enough and keep on plugging away at it. Of course, the code was a barrier that made it almost unintelligible in the ordinary sense of communicating. But, once you got past the barrier of the code, when that was familiar and you could handle it easily, it was a wonderful outlet, because you reached out. I can remember Principal Taylor asking me why I was so interested in it. I said, "I can listen to ships in the middle of the Atlantic Ocean," and he said, "Oh? Can you hear them out there? Right here in Kingston?" And I tuned up the old set and he listened and said, "Can you read that? Can you?" So, I wrote down some of the messages - just ordinary, routine ship-to-shore messages and things like that, and he got quite interested in it.

I had a lot of fun out of the set and was quite proud of what I could do with it. I'd been receiving signals and copying them down, all the way out into the Atlantic and the middle west and regularly got time signals from (NAA) Arlington, Virginia. Transmitting was limited, of course, locally to the gang around town. Then, at the beginning of the War we had a visitor - the Minister of Transport and Communications, I believe it was. He came to Queen's to get an honorary degree or something like that, and he stayed at our house because he knew Father. They had been on some committee together or something. Anyway, Father boasted about my wonderful wireless set that I operated and this dignitary from the government came up to my bedroom and looked at it. I gave a little demonstration of transmitting, and before I'd tapped out more than a few letters, he called a halt to it and told me in very solemn terms that there was a war on and any unlicenced wireless transmitter was against the law. This was a shock, and to make it even more definite, he disconnected the lead-in from the aerial and with a blob of wax...I guess it was sealing wax because father used to use it...he sealed up the end of that lead-in wire and said that under no circumstances was I to re-connect it until the war was over. Well, I kind of thought he was a pompous ass. That was a real nuisance, and I remember that very soon after he left to go back to Ottawa, I had that wireless going full speed. I didn't need the aerial outside. I soon found that the bedsprings or the radiator were quite good enough to transmit around the city.

Some of the others had received a note warning them that it was illegal to transmit, but I know that for some time during that war we went on. We didn't give our call signs or our names, so that we thought we were being anonymous. I can remember the operator at VBH used to copy down some of our signals and, if they could identify the person, you got a telephone call saying that your transmission had been monitored by an official and warned that you would be subject to legal prosecution and so on.

About that time there was news of vacuum tube receiving devices - there was the Fleming valve and the De Forest "Audion." During the War I vowed that I'd get an Audion tube just as soon as they were available anywhere. They came on sale, but it wasn't until after the War was over. By then, there were a couple of other vacuum tubes available. There was the little Marconi peanut tube, a little cylindrical one with a metal contact at each end and one on each side, and there was another which didn't have a base or anything. It just had wires sticking out of the ends of this larger glass tube - about an inch in diameter and about four inches long. It had two filaments in it, carbon filaments, which didn't last very long. The second filament was sort of a back-up to the first one when it burnt out. The first one of those that I saw was one that Rulla Bunt had bought, just after the War.

During the winter of 1917-18, I'd gone south with Mother for her health and had access to a very good public library in St Petersburg, Florida, that had pretty up-to-date textbooks on wireless and the American Radio Relay League periodical QST. In it were all sorts of excellent suggestions on how to tune a spark transmitter. At the time there was a lot of difficulty convincing everyone that it was practical to have amateur sets operating accurately below 200 metres. QST suggested that one of the very important things was to have the antenna system tuned accurately to the desired wavelength, with a condenser and inductance coil through which the transmission was being fed into the antenna. I came home that spring with all kinds of sketches and ideas of how to build that tuning system for my spark transmitter and, of course, it meant going back to Doug Jemmett. He hadn't a very great knowledge of construction and referred me to textbooks on how to measure wavelengths and how to make a useful wavemeter for monitoring my output signal. This consisted of a simple circuit of a coil and a condenser. The length of wires on the coil and the number of turns and the tappings off it and at last a really good rotary condenser that I could calibrate were all part of this simple little unit. I wrote away and got lists of stations and their wavelengths so that I could spot the identified signals on that little coil and condenser. Setting this little wavemeter beside my regular receiver made it quite easy to know when I had the meter tuned onto the same signal I was listening to. By identifying the tone and the code signal from that transmitter, I got each wavelength on a fairly complete chart. I was more interested in the wavelengths down towards that 200 metre hurdle that had to be got behind. I made three sets of coils for it so that I could go on up to the long wave lengths into the couple of thousand metres anyway.

Measuring my own transmissions was not nearly so easy. Even though I had a crude condenser system made of flattened tin cans and a coupler to the primary and secondary, connected to the aerial, I still was splattering the signal all over the place and I had to get that little wavemeter of mine quite a distance away. The usual place was to take it down to Earl Morris' house on Barrie Street, and get him to listen in and find out where the maximum was and how far you could hear it up to a certain audibility on each side of it, while I made adjustments in the tuning.

I had taken down my main aerial, which originally was between the two identical chimneys on our house, and had strung a wire from the nearest of the big chimneys on the Principal's house, next door, right across to the centre chimney on the Pathology Building, now Kathleen Ryan Hall. That was very good for receiving, but it was too long to contain my signals for transmitting to a reasonable tuning with the rest of the set. So, I had to cut the ends off it...actually I took a lot off the far end at the Pathology Building and shortened the lead-in coming into my window. I got it accurately measured down to about 175 metres length. For some reason I started experimenting with decoupling the primary and secondary of that transmitting tuner and almost at once found that with the secondary, the small coils attached to the antenna-ground, moving it some distance away from the primary, I could get the signal down quite sharply, down into the 200 metre range. After a little experimenting I found out that it was even possible to get lower than that.

I should explain some of the development of that wave meter of mine. It started with a very crude sort of little receiving set - a coil, a condenser, a detector and earphones. That was one of my earliest attempts to make a little portable set, mounted on a breadboard. It was awkward because the coil was large - a Quaker Oat box coil to begin with - and it needed reducing in size to make it truly portable. Well, I read about making bank-wound coils; that means you pile a whole series of windings on top of each other in a sort of a little pyramid in groups, and pack into an inch or two of space. Once I had that inductance coil wound - and I sprayed it with a bit of shellac - it was in good, tough shape and I put it together with the condenser in a small wooden box.

Now, that box has a bit of history. It was a square wooden box with a large knob on top and a big pointer with engraved numbers all around a ring on the lid of it. This was the same receiver rotary variable tuning condenser in the square wooden box with the engraved brass plate and brass pointer on top, that had belonged to the big wireless set on the kitchen table in the bell tent out behind Fleming Hall. The condenser inside was tin, maybe, and had been taken apart. It had multiple series of flat plates - one set fixed and the rest of them moving in between - beautifully made and the sort of thing I couldn't afford. I can remember admiring that, sitting dismantled on a bench over in one of the E.E. Lab rooms, and I asked Gill what it was used for. And he said, "Oh, the students take it apart just to see what it's like. They're destroying the thing because they're so clumsy they can't put it together properly." I offered to put it together properly and I set to work and slowly straightened out the plates of that condenser, smoothed the edges of it, flattened them back accurately, mounted them on the rotating centre stem very, very carefully and got the thing into reasonable shape. I remember Gill came along when I was sighting through the plates, and he said, "Why don't you just take that thing? It's no good in the hands of the students. They'd just destroy it."

So, I became the proud possessor of a variable condenser, and it was a nice little one. It had been in this nice little wooden box and I mounted it back in there again, took it home and was very proud of it because none of the other hams around had as beautiful a condenser as that. Inside that box and to one side of the condenser I was able to mount the bank-wound coil. Then, to make it useful when someone was listening to my transmission, say down at Earl's house, I fixed on the top of that surface a small, neat little detector made up of what was left over from an old transmitting key. It served very well, very stabilized and strong, and simply it was cut into the circuit of the coil and the condenser. The key had, as part of it, a shorting switch which closed the circuit and cut out the detector if you ever needed to. Then, in series with that again, I mounted a couple of little binding posts with a bridge across, which could be removed. On those posts one could then connect the earphones, so that it became a complete small receiving set. The coil itself had two connections out at the edge and two more binding posts - one for the earth and one for the aerial when one wanted to tune it to a distant station. There were very few stations operating below 200 metres in those days, but one got a few of them. I was delighted to hear something quite sharp and clear...about 170 or 175 metres...coming from New York City or somewhere down that way, and another one from Norfolk, Virginia. There seemed to be mostly army signals down in that range. I'm not sure why they got down there. I had expected that there'd be more from ships, but only one or two seemed to creep down toward 200 metres. Most of them kept well up in the thousands of metres. Gradually the metal dial of that wavemeter accumulated enough scratches on it to identify quite a number of stations 200 metres and below.

That humble Ford spark coil was still the basis of my transmitter, although now it had moved up into a much more elaborate transmitting set. The biggest change was introducing a rotary spark gap. Over at the machine shop I had picked up a discarded buffing wheel, about two or three inches in diameter. Around the edge of it were tufts of fine brass wires that had been worn down until they were about a quarter of an inch long. I saw potential in this, took it home and straightened out those little tufts of wires. The large hole in the centre fitted onto a bakelite knob from an old variable resistance. This was then clamped onto the shaft of a little Meccano motor. The whole thing, with two large binding posts, one on each side of the disc, was mounted on a piece of bakelite. Each binding post held a small piece of flat copper for an electrode which could be adjusted pretty accurately up to the ends of those little bristles on the disc. To make sure that they were even, a thin file could be pressed down while the motor was running, and the tips of those wires were evened up quite accurately.

The primary of my tuning coupler consisted of a spiral made from brass curtain rods, about a quarter of an inch in diameter, and curved into quite a neat, smooth spiral held in position by a number of small bakelite strips that were fastened onto a base. This was in imitation of the big loose coupling aerial inductance helix of heavy copper wire that Prof. Gill had in the tent out behind Fleming Hall for his distance tests with the Pierce-Arrow. The secondary consisted of a larger number of turns of heavy copper wire, smaller in diameter, and arranged on a glass rod that I'd mounted right in the middle of the primary. The secondary could slide up and down on it to vary the amount of coupling between the two coils. An alligator clip allowed the connection onto each coil to be adjusted when tuning it. The combination of shortening the aerial and also reducing the number of turns on the secondary got the signal reasonably down below 200. It turned out to be just below 175 metres.

Calling "CQ" in those days was a sort of mark of graduation from local contacts to an attempt to reach out longer distance. I don't think I called more than half an hour or so, CQ and my call sign and then listening, tuning up and down below the 200 range, and there was an answer! Somebody was calling me! This was a tremendous thrill. I think one of the greatest I had in radio. Actually, it was a clear, sharp signal not very far from mine. A little below it actually, and a spark. It wasn't a rotary. Straight gap. The keying was neat and clean and I, of course in great excitement, answered and we started a conversation. That ham was in Oswego, across the lake down on the New York shore, and apparently he'd been listening to quite a lot of my transmissions as I was tuning and fiddling with the various variables in that transmitting set. His first comment was, and I think I can quote him, "Your note is sing-ey." Now, at first I didn't know just what he meant. Gradually I realized that my rotary gap had a slight tremolo in its note and made quite a distinctive sound. Well, that was a high point in wireless as far as I was concerned. That contact was kept up for several years afterward. Gradually I spread my contacts or CQ calls across over to Watertown and another half-dozen places in New York State, up to Oshawa, Brockville, and once I got an answer from somebody in Ottawa. The connection was poor and I don't think I ever quite identified it. All this tuning and developing of that little spark transmitter was during the winter of 1918-19 and a bit into the summer of 1919.

At the end of World War One, there was an announcement that things were going to change drastically in that whole free-for-all set-up of spark transmitters. Anyway, I was told by Jemmett that all amateur wireless transmissions had to be kept below 200 metres. Well, I can remember expostulating with him about it because I had checked the transmissions of any of the wireless spark transmitters and some of them had a range of hundreds of metres. You could get to the loudest place in the middle of the band, but it had a wide spread. And I can remember Doug Jemmett saying, "Well, all you have to do is tune your transmitter accurately. You have to have a transformer, tuning coil with a secondary and primary through which you send out your signal and you have to have it tuned accurately with a tuning condenser, and all of this has to be made so that it will stand the high voltage of that spark." Well, that started another game - making a tuned transmitter system. One wound a big cage of wires around wooden posts on a base and had variable clips that you could clamp onto it for tuning, and I devised a system of tuning the transmitter by having sheets of thin metal...pieces of tin cans which I could move back and forth on a rack, and they were far enough apart that they didn't arc over when the transmitter was working. The Ford spark coil was still the main means of transmission.

The transmitting condenser began with squared pieces of an old broken plate glass mirror. The plates were to slide in grooves on a board, except that I could never be sure of a good contact with the darn mercury film on the back of the glass. And the adjustment was very hit-or-miss and the whole thing was distorted. Pictures of variable condensers in which the one set of plates rotated forward or away from the fixed plates made me want to have something a little more business-like. I found some nice, neat, carefully shaped squares of aluminum - oh, they must have been four or five inches square, maybe a sixteenth of an inch thick, and very smooth and flat, with rounded corners and a rounded edge - somewhere up in the woodworking shop above where Prof. Gill stored bits and pieces of his old aeroplane. Now, it may have been some parts left over from making his aeroplane. They were simply lying on the floor there, and, of course, I automatically assumed that they could be used and proceeded to make them into my condenser. I think it was five sheets spaced something like a quarter of an inch apart and these were fastened on edge with a bolt through holes in two corners of it, to hold them rigid. The moveable group would probably be four. These were also bolted together but only at one corner, by a long bolt with washers in between the plates to keep them spaced properly, and tightened up quite tight and with a knob at the one end of it and the whole thing turned in a rather crude bearing. One could turn the knob and swing those plates in between the other fixed plates or away from them. I was very impressed by the continued repeat instructions, "You must keep the resistance low." I simply soldered heavy copper wire onto the back end of each of those bolts - wire large enough to carry many times the EMF that I was handling with that little transmitter.

It wasn't until I'd re-tuned the primary circuit, but mostly by the alligator clip on the coil of wire, that I got a fairly sharp, clean signal down in the range that I wanted. Then began a bit of juggling, and I was never sure just what one should do, because as one reduced the number of turns on that primary coil, one could compensate by changing that variable condenser in the circuit. It seems as though you could come back to that same tuned wavelength that matched the aerial and secondary at any number of points just by adjusting those two factors. Well, after a while I was given a hint as to how to know when I was doing it right. In one of the QST magazines there was a suggestion that a very neat and inexpensive way to find out how much energy you were putting into your antenna was to put a little flashlight bulb in one of the leads. This turned out to be a winner, and I put that tiny bulb down inside a fountain pen barrel so that I could see the light even in the daylight. I could then adjust the two tuning positions until the brightness of that light told me that I was putting out the most power that I could. Actually, if I could get it to light up at all, even dimly, it was a success. Connected into the circuit right at the clamp where the earth connection was made onto the hot water radiator and down in a dark corner of the room, I did a lot of peering over the end of the table watching that little light as I slid the clamps around and turned that condenser knob. It was soon pretty evident that if I increased the capacity of that condenser in the primary circuit and reduced the number of turns on the primary of the coil, I could get a brighter light, and therefore I presumed there was more current going out. And this was quite independent of how I juggled the clip on the secondary coil. When I got those two circuits tuned fairly accurately together, I noticed almost at once that there seemed to be a much decreased brightness and sharpness of the spark at the rotary gap. I felt that this probably was also a sign that the antenna circuit of the secondary was absorbing effectively the current that was being produced in the primary, and it seemed to draw quite heavily on that circuit from the spark coil. One had to adjust the gap and the rotary wheel right down to a very small distance to be sure that it was a smoooth and steady spark. These many adjustments meant endless long series of V's, of course broken now and again giving my call sign, 3MC, and then going back to tuning and more V's and so on. It was weeks before I finally came to the conclusion that all the adjustments I could make were at an optimum. And then I started seriously to start calling CQ.

Using some of my pay from the car ferry, I very soon headed over to the States to buy a radio tube - it was a UV 200 - and proudly brought it back, of course, in my pocket. That tube changed the range and type of listening I did. The circuits had changed also, and with the tube I set up a regenerative circuit with a tickler coil in the plate circuit which fed part of the energy back around to the primary, and built up a very much stronger amplification than I could otherwise get. It could be built up until, of course, the tube would start oscillating, and as I got used to the fine adjustment of that tickler coil, I could bring that set up to a peak of amplifying that brought in extremely weak signals.

Some of those stations were broadcasting continuous wave. Continuous wave was not so easy to receive with a crystal, but with that you got a V-shaped whistling tone - the side bands on each side were quite audible - and to listen to code, of course, you had to tune it to one or other of those side bands.

KDKA was the very first wireless or radio transmitter that regularly sent out music and voice and quite a range of stuff to listen to. As I can recall some of the very early broadcasts, the operator used to chat very casually, informally, describing the set and his location. Apparently the thing was a temporary arrangement - a tent or a shed up on the roof of a building - and you could hear the traffic around the building faintly as a background to most of what he was saying. You could hear conversations in the room that were not supposed to be broadcast, I think. You'd hear someone coaching the singer and the pianist, when a little concert was supposed to be broadcast. Someone would be telling the singer, "Now don't get the microphone too close to your lips because you'll get all sorts of noises that you won't want them to hear." And to the pianist, "Now, just keep that loud pedal easy. We don't want to destroy anything with the volume of noise you can make with that," and so on. Well, I can even remember when the operator got bored with the whole affair. Had nothing further to say at the moment, and he said, "Here, I'll let you hear the traffic." I remember he took the microphone and lowered it on a cable down over the side of the building so that we could hear the cars tooting, people shouting and noise from the street below. It wasn't long before there were several broadcasting stations in the United States...WLW, WJZ, and several others that were farther away and not quite so easily listened to. Listening to voice and music over the radio was still quite a novelty in those days, and I can remember inviting the Principal, who lived next door, to come in and hear part of a concert that was being broadcast from, I suppose, New York City.

Pretty soon superheterodyne got to be THE thing to have and, if you had enough money, you would buy one of the superheterodyne receivers. It was tricky. You had to tune each of the radio frequency circuits, one after another, until you got them all tuned just right before the thing'd work right. Once it was tuned right, it certainly brought in long-distance stations and lots of them.

My interest then was in vacuum tube transmitters, for it was quite a different thing tuning down to the shorter wavelengths. Getting under 200 metres was the easiest thing possible. All you had to do was to measure the coils and your aerial and get them balanced together and tuned accurately. It was no trouble at all making a little transmitter tune accurately and quite effectively down on the 40- or 20-metre band. My experience in tuning a spark transmitter paid off because soon I had a little transmitting set using a UV 201 tube, which I had pumped out to a higher vacuum over at our Physics Department, and putting 200 volts on the plate of that I could get quite a good signal out. It was, of course, necessary to have a different aerial, and I soon set up an aerial and a matching counterpoise between my bedroom window and the Anatomy Building, which is only a very short distance from the corner of our house. The plate voltage was 200, made up of the 110 volts DC in our lighting system in the house plus a couple of 45-volt "B" batteries. So, a nice steady and even plate current was obtainable without much difficulty. The trouble of that transmitting tube, or rather the plate of it, getting too hot, was one of the first things one ran into. It was quite easy to operate that tube with the plate glowing slightly cherry red. I soon had a busy schedule of contacts, some of them quite good long distances. One very regular one was with an amateur out in Denver, Colorado. Another one was down in a town in Mississippi, near the Gulf, and another one was in Tampa, Florida. A Boston ham, who had been one of my first contacts with the spark, was still on tap, and of course, the group over in New York State, which by then were all old friends. So that 3MC got hammered out across the airwaves quite regularly. Daylight was the best time if I wanted to get something a long distance away and I used to sometimes sneak in some transmission at noon on 20 metres and pretty soon was experimenting with shorter ones - I think it was 9 metres. That was not so easy to handle, because making the tuning equipment for that meant some very, very careful measuring and adjusting. However, all that was great fun amd that was part of the background that led me into making small, portable transmitters and receivers.

The UV 199 tube, which was almost a miniature tube not much bigger than a peanut tube, proved to be excellent for many uses. As a receiving tube it was fine, and one could make a very neat little compact set. The whole thing could fit into a cigar box - all except the batteries, of course. Even after the tube had been pumped out to a much higher vacuum, it acted as a good transmitting tube. I soon had a small transceiver, using UV 199s and had quite a respectable performance from it with two tubes in push-pull application in the transmitter beside the oscillator. And then two tubes, one just an ordinary detector and the other an audio amplifier, for the receiver. The whole thing was about as big as two good sized cigar boxes. To get enough power for that transmitter though, you had to lug around big, heavy "B" batteries, and that was a nuisance.[10,34]

Alexander A.E. Ada (3HN)

Alexander Edwin William Ada (B.A. '24; Meds '26) entered Queen's in 1919 in the combined arts-medicine course. He was the Reporter, on the executive of the second incarnation of the Queen's Wireless Club, in 1920-21. Also holding office that year were Donald Hepburn, L.W. Lockett, Doug Geiger and P.H. McAuley. Ada held the call 3HN in 1921, and was mentioned that year in QST, the magazine of the ARRL, as active in relay work.[35] He was also a memeber of the Kingston Radio Association, in the Donnelly-Thompson wireless circle.

Alex Ada was certainly active in 1920-21, but may have gradually lost interest in wireless work. He was halfback on the 1924 Dominion Championship intermediate rugby team and was the captain in 1925. He led his class in medicine and surgery. While a house surgeon at Kingston General Hospital in 1927 he won the Hoffman Surgical Fellowship and went to St Luke's Hospital in New York City. He spent his career in the United States and was a founder and president of the American Thoracic Society.[36]

> **Don MacClement:** Alex Ada, who later went into medicine, was interested in wireless very slightly, and I built a receiving set for him and taught him the code. He was with me on several occasions over in the Radio Club Room in the Old Arts Building, and I believe he was there the evening when I heard Lee De Forest's broadcast.[10]

Robert M. Davis (3IL)

Robert Davis was active in wireless as early as the spring of 1920, when his equipment was seized by a government inspector for interfering with reception by VBH.[33] Davis was Orton Donnelly's Vice-President in the short-lived Kingston Radio Association in 1921.[37] In 1922 Davis was noted as quite an active and successful DXer, with a 125-foot mast and a 10-watt CW set.[38] He succeeded his friend Donnelly as the head of the radio department at the Newman Electric Store when Orton left to start up the Canada Radio Stores in 1922.[39] The next year Davis was appointed Radio Inspector for Kingston and District and seems to have stopped his relay work.[40] He moved to Belleville in 1925 to open the Ontario Radio Stores and Frank J.C. Dunn succeeded him as Radio Inspector.[41]

Davis sent his list of calls heard to QST early in 1922, boosting his own Kingston circle by including them: 3HE (Orton Donnelly), 3HF (Gordon Thompson), 3HN (Alex Ada) and 3NE (K.L. McAlpine).[42]

> **Don MacClement:** Bob Davis was of the Davis Drydock family. They lived out on Union Street and he got to be a very enthusiastic ham. Just as soon as the radio tubes started to be available, especially the transmitting ones, he had a very grandiose transmitter put up. His father's company set up a hundred foot tall mast on their property - it was on the corner. I know he had several 50-watt transmitting tubes in his transmitter, and I don't think he did an awful lot with it, but he had

a whole lot of stuff there. I know that later on, when Chalmers Church got infected with that...at that time very popular idea of broadcasting the sermons, I was asked if I would put together and set up a broadcasting station for them, or at least see if it was possible. Bob Davis, who had gone on to other interests, offered to donate all his radio equipment for that purpose. I looked it over, and came to the conclusion that the Church was getting into an awful lot more complications than would be warranted. And I advised against it.[10]

Keith A. MacKinnon: They were wealthy. His father had the Davis Shipyard Company - kind of in the ship salvage business - and he built him a real high class metal pole - a good high one and very obvious - just set up in the back yard at Union and Collingwood. We used to see Bob Davis often. I remember once Stew and I went to his family's cottage away out the Front Road right on the lake. He thought if we were there at night and had a big antenna, we'd hear a lot of things. And we went and stayed there the whole damn night...half asleep, you know. We weren't used to staying up. We had crystal detectors, and there was lots of stuff on the air, but it wasn't any better than if you were in town. Maybe the antenna was a little better.[43]

Fred Jolliffe recalled listening to KDKA around 1921 at the home of his friend, Bob Davis, at the corner of Union and Collingwood streets. He remarked that his friend built radios at KCI.[44]

Ralph L. (Rulla) Bunt (3AT; 3MX)

Ralph (Rulla) Bunt, 3AT and 3MX, photographed on Nottingham Island at the west end of Hudson Strait. He wrote on the back, "Me and me trusty plane! Spring 1928, not bad, eh?"
-courtesy Keith MacKinnon

Keith MacKinnon: You know, "Rulla" Bunt lived within a half a block of me, near Johnson Street, and we played together a lot. He called me "Kye" - that was my nickname. He had a lot of brothers, but they weren't in the radio business. There was Albert and Charlie was the youngest, and "Fordie" - Fordham was the third one - and Ralph. Their father worked in the post office. Rulla and Stew and I went through public school and high school together, and the three of us worked together all the time, listening. We'd call somebody and see whether they'd answer or not. Way back in the First World War they wanted to train kids to shoot and everything, and we used to go in the evenings after supper to the basement of Victoria School and shoot with the .22 rifles for a couple of hours. Rulla was a great shot. Well, he and I and Stew, we went to VBH often because we knew the second operator, George Gurney, who always made things open for us.

When he left high school, Rulla was working in a machine shop. Then the next I heard he'd moved to Ottawa and in 1925 got his commercial ticket and went away for ships[45] and I never saw him much after that - I remember an amazing number of people that were radio operators. He worked ships down on the east coast,[46] and then he was on the government ice-breakers and all that sort of stuff. I have an excellent picture of Ralph Bunt standing on the float of one of the planes on Nottingham Island in the spring of 1928. As I recall it, he was there for a year as part of the aircraft expedition organized in connection with the use of the port of Churchill on Hudson Bay for grain export. Ottawa wanted to learn how soon the ice broke up in Hudson Strait. To this end, the RCAF set up three points on the strait for carrying out air photography of the ice: Nottingham Island on the west end, Wakeham Bay in the centre and Cape Hope's Advance on the east end. The radio communication between the sites was a Department of Transport operation, and Ralph was the operator on Nottingham Island. They used to go up every day and photograph. He was great on hunting, and he came home with silver fox and stuff from away up north, and wanted to know if I wanted it or something. I worked him up there on Nottingham Island one Sunday in code on 52.5 metres, the Canadian frequency, but I never could get him again. He used to write me lots of letters. The way they spoke English in Newfoundland - that really got him - and he wrote me a poem that he'd picked up somewhere there. All I remember is the first two lines. It's a model of inexactitude, and it goes like this: "'Twas on a lorn Juvember morn/ Come last Octobruary/ But maybe might was Tuesday night/ I can't remember very." Finally, in the 1940s, he came back to Ottawa as a radio inspector of some sort - on interference.[43,47]

Don MacClement: A tube which I looked at very seriously the summer I worked on Car Ferry #1, was about four inches long and more than an inch in diameter and had wires coming out of both ends of it, but it didn't have a socket. Three wires out of the top end and two wires out of the bottom end I'm pretty sure that's the way it was. The uniqueness of this tube was that it had two filaments, each one starting at the sort of centre wire at the top and then running down to each one of the two wires coming out of the bottom. The idea was that if one filament burned out, you still had another one to work with. This was the tube that I saw in Ralph Bunt's radio equipment that summer. I think he was the first person in the city to move across to the vacuum tube type of equipment. Several of us used to come along to his house in the evening and sit in and listen to some of the stations he could hear. He made his own equipment and it was very well made - very neat and efficient. He had, as I remember, basket-wound or honeycomb coils or something like that, very neat and they worked very well.

Now, Rulla Bunt was the most active and skilful of all the hams in the city. He was always ahead of everybody else - the first to get a radio tube to help his receiving. When I was working up in the Arctic, between 1949 and 1960-61, first for the Defence Research Board and then for Northern Affairs, I had to travel around pretty well all over the Northwest Territories and at some of the wireless stations there they talked about "Old Man Bunt," who had been the pioneer in setting up and designing those communication stations. Now, could that have been Ralph Bunt? Whoever it was came from Kingston, so maybe it was.[10]

The Official List of Radio Stations of Canada shows R.L. Bunt, VE3MX, living in Ottawa beginning with the September 1, 1930 supplemental issue.[48] His old friends in Kingston heard that he retired to the east coast, around 1970, but he has not been located.

Donald I. Matheson (3NW)

Don Matheson (Arts '26; Meds '34) was the son of a Queen's professor and he lived in a grey frame house on Alice Street, now known as Grey House, 51 Queen's Crescent. His interest in wireless was a "fleeting high school boy's fancy," and he maintains that his expertise was not great and that radio was forgotten when he entered Queen's in the autumn of 1923. Howard Folger taught him in medicine at Queen's.

His wireless post was his attic bedroom in the south-west corner of the house, and he put up a single wire as an antenna from the main part of the roof over to a tree in the back yard. His receiving set had a silica crystal, a cat-whisker and a loose coupler wound on a round box, with a switch on the inner coil of wires. He recalled building one himself and lacquering it. Wireless communications were exciting to hear, even though they were too fast for him to decode. His sending set consisted of a key, a Ford spark coil, a spark gap and two batteries. He remembers coming home at noon, rushing up to his bedroom and pounding out simple messages at a pre-arranged time to some of his friends, probably "Turk" MacClement. He also remembers Orton Donnelly, Alex Ada, Robert Davis, Sheldon Cohen and Ralph Bunt. They went a few times in the evening to the Kingston Wireless Association, in the Golden Lion Building diagonally across from the old Post Office at Wellington and Clarence streets. There they practiced sending and receiving in Morse code, but he doesn't remember who or how many went. He also listened and thrilled to hear KDKA in Pittsburgh and the Arlington noon time signal in code.

Don Matheson knows nothing of Doug Jemmett's wireless work at Queen's, but Jemmett was his Sunday School teacher at Chalmers Church for a short time, and he knew him quite well. He remembers Jemmett telling the class about his experiences in the War, about being wounded in the leg and having to walk with blood swishing inside his boot. Matheson later sailed against him in dinghy racing at the Kingston Yacht Club.[49]

Sheldon and Harold Cohen (3ACF)

Sheldon Cohen (Phys. '27) was born in 1905, and his brother Harold Arthur (Dick) (Eng. Phys. '30) in 1908. Their father, Isaac, was in the scrap metal business, and got involved in municipal politics. He served as city alderman for a term, which was most unusual for an immigrant. He acquired the Monarch Battery Manufacturing Company at auction in Guelph in 1921, brought the business to Kingston,[50] and developed it successfully. The first battery factory was at the corner of King and Queen, and the office was at 254 Ontario Street. The second factory was at Montreal and Joseph streets.[51] At one time, in the mid 1920s, they made batteries for the "Eatonia" brand, sold by Eaton's of Canada.

The family lived in a big house at 209 Queen Street, and it was there that Sheldon had a big, powerful rotary spark gap transmitter. After 1924 they moved to 165 University.

Dick was a radio "bug" from the age of 11 or 12. He still has Amateur Experimental Licence #430 for 3ACF, dated August 4, 1922, on Form W. 44 of the Naval Service, surcharged "Marine and Fisheries, Ottawa." This lists an inverted L antenna and a 1/2 kW spark oscillation transformer-condenser transmitter, powered by the 60-cycle city mains. Wavelength was not to exceed 180 metres.

> Dick Cohen: I really got interested in radio through my brother Sheldon, a wireless pioneer, because he was older than I was and going to Queen's. He was a member of the Wireless Club - they built their own radios with double-crystal Perikon detectors. Even in high school my brother was interested in radio, and he'd tell me about these things. I was also interested in electric light. We thought it was a great achievement to buy a couple of #6 dry cells, hook them up to a little 2-volt light bulb and have a light of our own. I remember even in public school I took a medical coil to class - simply a high-voltage spark hooked up to a couple of handles, and you just grab hold of them and feel a shock. There was no danger. You didn't get any powerful shock, but you could feel a tingle. That would be maybe 1916, when I was 8 or 10.
>
> Well, I was in radio long before 1922. I got my official licence for 3ACF in 1922, but I'd been playing around with radio many years

The Cohen family residence at 209 Queen Street. Dick had a 62-foot high mast holding one end of 3ACF's antenna. (inset Harold A. Cohen)
-photograph by Arthur Zimmerman

before that. I had a rotary spark gap before KDKA came on, but never used it because I didn't have a synchronous motor. There was a little radio club and we met in Dr Mylks' home, at Bob Davis' and then at the Golden Lion. I think the first thing my brother and I did was to rig up a spark transmitter in the attic in our house on Queen Street, and we had another one in the basement. One of us would be upstairs and one downstairs, and we had crystal detectors and earphones and everything. We were so close together, actually, that when you went to adjust the crystal you could see a spark on it. A spark on the crystal! And we would send code to each other. That's the earliest I can remember back. I guess you were supposed to have a licence, but all we used was initials to call someone. If you wanted Orton Donnelly, you called "OD," you know, da-da-da da-dit-dit.

We used to talk to each other around town - Orton Donnelly, Wally Reid, Gordon Mylks, Bob Davis - he had the highest mast in town. I think he had about a hundred foot pole. Wally had a very long tuning coil and he could pick up the long-wave station POZ in Nauen, Germany. Most of the boys went to Queen's eventually.

My father put up a lean-to radio shack in the space between our house and the garage. On the Sydenham Street side. I had a mast that was about 62 feet high, but that's not so high. I just ran the antenna from the pole to a high tree that we had, and I got out quite a bit with 5 watts CW. Oh, I never got out of the city with spark. You'd send out "CQ" or you'd answer a CQ and once I contacted somebody on the US coastguard cutter Modock off the coast of Charleston, South Carolina, and I still have his QSL card. I got an STP order from VBH only once and was never caught by the inspector.

We all made condensers, just for the spark transmitters. We used to take just ordinary window glass plates - and then put what we called tin foil in between, but it was actually lead foil. The glass acted as a dielectric, and in that way you built up a condenser that could stand a pretty good high voltage. My quarter kilowatt transformer was rated at 8,000 volts, but this condenser stood up very well with it. Of course, once we went into CW, I don't think very many built their own condensers.

Most of the amateurs used transformers and rectifiers. They used to refer to chemically rectified AC and they probably used a bank of aluminum-lead electrodes in a borax solution. That converted the AC to DC, and it did a good job too. But it was very bulky with all these glass jars. The one I had was about 30 inches long and 18 inches wide, with anywhere from 12 to 24 cells. You were rectifying maybe 1,500 volts and the rectifiers were good for only about 30 to 50 volts per cell. So that you'd need about 30 cells. But it did a good job, and then you had to filter it, which meant that you needed condensers and choke coils to get rid of the hum. Otherwise you'd get quite a hum from the 60-cycle. I bought from Newman Electric what we called an oscillation transformer, which was used as a spark transmitter. I guess they can also be used in CW for that matter.

If you broke the ground wire of a regenerative receiver, then by making and breaking the ground you could interrupt the regenerative squeal and send out dots and dashes.

There were a lot of problems at that time, and you realize how little we knew about it then when you see what's been accomplished today. It's progressed not geometrically but logarithmically. Actually, crystal detectors were solid state diodes, but I don't know whether we realized it or not. Of course, they were handling only microamperes, whereas the solid state rectifiers of today can handle enormous currents.[52]

Don MacClement: Well, it got pretty quiet during the latter part of the War. Listening to signals became more important because the number of transmitters around town got to be very few. I think the last one to go off the air was Sheldon Cohen's. His father, who was in the scrap metal business, somehow picked up a half-kilowatt Thordarson high frequency transformer. And I can remember going out to his house one time, warned of secrecy, and shown...Oh, my gosh! The spark - it was an inch and a half long and made a noise like a gun going off. A terrific racket! When I went over to see Sheldon's huge, big half-kilowatt transmitter, he'd already had threats that it would be confiscated and made me keep very quiet about having seen it and knowing that it was there. I guess that soon after that it was put out of commission anyway. He'd been warned a couple of times by VBH to keep off the air, and I can remember later on, when I got to know the operators over at VBH, I was told that the reason he was warned to get off the air was that he was drowning out the signals that the Marconi station were trying to listen to.[10]

Someone told the Oral History Project a story of the Dominion Radio Inspector searching for the Cohens' powerful transmitter with a special automobile equipped with a directional antenna on the roof, trying to get a fix on the source of the signal by triangulation. The year was not known.[53]

Art Wehman recalls that some time later, Dick was working with radiotelephone, and called him on the telephone one evening: "I wish you'd tune me in. I'm experimenting and want to try something. Call me on the phone if you receive me." Art tuned in and heard, "How do you receive me now, Art? I made a little change in my microphone position. I'm going to go off the air for about five minutes and I'm going to call you back." Dick called back on the phone and said, "Okay. Will you tune me in again? I'm going to test again." Art soon heard through the headset, "How do you receive me now?" Art said that it was darned good and Dick laughed, "I'm talking to you from down in the coal bin."[54]

Harold Huton Stewart (3NF)

Harold Stewart: I was born in 1903, and I graduated from Queen's in E.E. in 1926. I can't say too much about Doug Jemmett, my Prof. He was my ideal for as long as I knew him. As I recall it, he had an experiment going with radiation from two parallel wires, even when I came here on the staff in '29. He had two parallel wires strung between Fleming Hall and Gordon Hall and he fed a high-frequency current around this loop. His lectures opened up a vista - the whole electrical field - to me that I didn't understand before. I was an amateur radio operator, I built my own transformers and I went by the book. They worked, but I didn't know why they worked. For example, when you calculate the number of turns on the primary of a transformer, you come out with a number that will fit the formula, but if you figure out the resistance of that amount of wire, and divide it into the applied voltage, you get a current that would blow the thing sky high. The answer is that it's the reactance of the thing that limits the current. I didn't know that. Doug taught us that. After three years in industry, with Westinghouse in Hamilton, helping to build motors, transformers and that sort of stuff, I was offered a job on staff at Queen's. This was a joint appointment, half in Electrical and half in Mechanical - a lecturer at the magnificent sum of $1,500 per annum.

I recall as a kid here, after the War had started, going up to the Kingston Fair, and the Signal Corps had a set-up for recruiting. I think they had a big ten-inch spark coil there with a chap pounding away at it, and the big spark jumping across the gap, and people were very thrilled about this.

Before I went to Queen's I was an amateur operator. In 1919-20 I just had a receiving set, and in 1921 I had a little transmitter. People used buzzers - a door bell with the bell off. You could put a key in the battery circuit and take leads from the opening and closing contacts, and run one to ground and one to aerial and you had a transmitter. It would get all over town and probably farther than that too. The buzzers that we had were not tuned. We just stuck them on the air and the wave was very broad, which means that it was highly damped, and it would smear a receiver right across any type of tuning band. Hard to tune it out, but there wasn't any broadcasting at the time, so there wasn't anyone on the air in this area except the amateurs themselves. Nobody to raise a fuss, you know. We were clobbering one another, that's all. The next stage was a Model T Ford spark coil - four coils in a box on the dashboard. They were more powerful than a buzzer. You'd put a key in the primary circuit and make a little spark gap, take the leads from the spark gap to the aerial and ground and you were in business. Better than a buzzer. And, if you got smart, you'd put in a tuning coil, maybe, but nobody bothered with that much. Again, there were no licences. It may be that we raised hell over at VBH. Keith MacKinnon and I and a few others made friends with the operators over there - at least with one of them, George Gurney. He helped us in the early days, making tuners and letting us try them out on his big antenna system over there.

Looking through my log book, I found a log of stations received over a couple of years. Mostly American. I don't find, unfortunately, any reference to 9BT, although here I am in the same town, a few blocks away. I just didn't connect - I just wasn't on at the same time on the same wavelength.[55]

Prof. Stewart recalled that he and some friends went to the Armouries on Montreal Street in the early 1920s to take instruction in military signalling: line telegraphy, heliograph (Aldis lamp) and flags. The instructors there were Fowler and Sgt Reading, an old soldier from the northwest India frontier, the same man who had worked with Harold Reid in the School of Signalling at the Barriefield War Camp, 1915-17. The Captain of the Signal Company at the Armouries was S.A.

Lee, DSO, who was Honorary President of the Kingston Radio Association in 1921. Capt. Lee's signature is on Stewart's signalling certificate, issued in 1922.[56]

When the Kingston Radio Association was formed in the spring of 1921, Harold Stewart and Gordon A. Thompson were the two traffic managers.[37] In 1922-23 his name appeared often in the "Operations Department" section of QST as one of the three or four involved in relay work in Kingston. Asked about his own gear, Prof. Stewart said that he had a receiver sprawled out on a table, a so-called breadboard arrangement, with a loose coupler wound on a Quaker Oats box, a rather large primary with a smaller one, a tin can, inside. It was arranged to slide on rods inside. Taps on the secondary fed into the detector. He also had a De Forest tube, an Audion, that he had carried across the bridge at Niagara Falls past Customs. Later he carried a big tube across that way under his coat.

From mid-1924 through May 1926, when Harold Stewart got his B.Sc. and left the city, Kingston radio amateur Fred Simpson, Jr, was editor of a radio column in the Kingston Daily Standard's weekly Daily Standard Junior page. This series gives fascinating glimpses of many individual ham activities in Kingston, and documents a good deal of Stewart's work in this period. All the locals hams knew of that quiet fellow's prowess in wireless and that "when Stew sets out to do a thing he usually does it." During the winter of 1923-24 he had been received by WNP, the Macmillan Arctic Expedition station "at the North Pole." In October, 1924, he was blowing 5-watt tubes, and decided to suspend operations until he got his new radiating system up and could use his new 50-watt tube.[57] A few weeks later, he and some friends began putting up an 80-foot high lattice-work mast and counterpoise, based on an article in QST.[58] The plan specified splicing 1x2" 16-foot lumber and cross-bracing with plaster laths every two feet, the whole thing to be pre-coated with hot tar. The base was two feet square. The pessimists had a pool on how long it would stay up,[59] and Stewart had to guard it to make certain that it survived Hallowe'en.[60] On November 5, a guy wire "egg" insulator broke during a heavy gale and the mast snapped into four pieces and collapsed. He managed to salvage a 72-foot tower out of the wreckage, and within a few days was expected to be "pounding the brass" again, using his new tube.[61] But his luck was not improving:

> On Friday (November 14) 3NF, with the help of several friends, succeeded in breaking his 72-foot lattice work mast in two pieces. One section is forty-four feet long, the other is twenty-eight. When the mast was being raised the main rope which was being used was fastened down too low and the weight of the upper part of the mast broke itself away from the other part. On Saturday 3NF and the friends raised the 44-foot section in his own backyard and was on the air Saturday night, and part of Sunday. In the heavy gale on Sunday a guy wire which had been greatly strained by the previous disaster broke loose

The rear of 404 Albert Street, showing the attic dormer of young Harold Stewart's bedroom.

-photogrph by Arthur Zimmerman

about two feet from the top of the mast. The structure wobbled around for a while and while 3NF was looking at it through a window, he saw it make a final effort and come whistling through the air. The end of the mast missed a window by only a very few feet. By now it is expected that the 44-foot section will be in position where it originally stood. When 3NF gets his mast up to stay he is going to write an article on expert mast wrecking.[62]

During the next couple of weeks, Stewart was on the air with his new 50-watter and causing no interference to Bill Cameron, 3AFZ, half a block away, or to local listeners. Then he shut down for his examinations at Queen's.[63] In January he was getting good results at 75 metres, with a high message total, and was doing good DX work, talking regularly to amateurs all over North America.[64] Then he replaced his single-wire aerial with an inverted L and was able to work 2NM in England for half an hour on February 1,[65] and California afterward. Stewart was said to be one of the neatest and fastest senders in Canada. No local "pounder" could send too fast for him to read. Later he worked Puerto Rico[66] and was heard in Morocco,[67] the first in Kingston to achieve such distance. A great dream was realized on the morning of February 18th, when he heard Australia, and later he may have "raised" 2CM of Sydney, Australia, 10,300 miles away.[68] Because of his achievements, he was invited to a convention of Toronto amateurs by the ARRL Canadian General Manager to be inducted into the "Royal Order of Trans-Atlantic

Brass Pounders."[69]

He awoke on March 15 to find that the little mast attached to his house had snapped in two spots and his aerial was on the ground again. It took a few hours to untangle the aerial and put it up again in a temporary location. He told 3AEL that he was going to put up another 80-foot lattice tower in the fall, and that it would have a nearly vertical antenna.[70] He was having rectifier troubles and was talking of "S" tube rectifiers and a good filter being on order for use in the fall. He would build a "Super BCL Hades" that would be "the berries" (BCL was a slightly derogatory term for Broadcast Concert Listeners), to operate on 40 metres, when he returned from his summer job as concrete inspector in St Catharines.[71] He remained active there over the summer on 3KP, a friend's set, and often contacted Kingston to get the latest news.

> 3NF and the gang raised up his mast in his back yard on Thursday afternoon (October 8, 1925) and now a new aerial is in place and 3NF is on the air with 1,000 volts AC on the plate of his 50 (He also has a new receiver for low wave bands). He will soon have his new (4,000-volt) transformer completed and then he will have his (50-watter with) "S" tubes (rectification) in operation. This year he expects to work on the 40-metre band and will make another try for Australia and New Zealand.[72]

Over the next few weeks he worked Cuba and California with his "ether buster" and 4,000-volt transformer (a "nice job"), and announced that he was thinking of trying a Hertz (single-wire) antenna. This Hertz fad called for a small light bulb to be installed at the null point of the antenna. The rectifying system was installed on October 23 and helped in his communication with Cuba, but one of his lovely "S" tubes turned out to be a dud.[73] In late 1925, the wireless art had advanced to the point where the goal was to achieve daylight communication with Europe at low power and short wavelength. It was certain that Harold Stewart would soon put Kingston on the world map. Bill Choat, 3CO of Toronto, who had been radio operator of VDM, the C.G.S. Arctic polar ship, visited 3NF in November, and Stewart, Donnelly, Simpson and Acton went to Montreal to represent Kingston at the ARRL All-America Convention.[74]

Early in November he erected his Hertz antenna to replace his other aerial, and reported that it worked quite well but had not yet helped him connect with the antipodes;[75] "3NF slapped 1,600 volts AC on the plate of his 50 and received flattering reports from all over the continent. He had better watch out or he'll wreck those New Zealanders' phones yet."[76] The next week he blew his plate transformer, and had to operate with low input,[77] but soon after that he announced that he would be leaving the city after his final examinations, and would soon have to dismantle his apparatus.[78] It seemed possible for a while that his brother John would take over 3NF with a De Forest "H" tube in the fall, but Harold announced that he would sell his apparatus and not return to the air until he could get a motor generator set for his power supply.[79] On May 14, 1926, armed with his B.Sc., "Mr. Harold Stewart, Kingston's oldest and one of Canada's most popular amateurs, left the city to take up a position with the Westinghouse Electric and Manufacturing Company, located in Hamilton."[80] He returned in the autumn of 1929 to replace H.J. Hartman as Lecturer in E.E. at Queen's.[81]

Keith Abbott MacKinnon (3HO)

<u>Keith MacKinnon:</u> I was always interested in signalling. My father was a soldier and I remember he was teaching my brother and me the code. My uncle was a Morse line operator, and it's a different code. Incidentally, I'm a dual operator - I know both codes. I have a sounder at home, all working, and I keep my hand in, just for the hell of it.

Stew lived just on the next street (404 Albert St), around the corner from me (76 Mack St). I was on Mack Street and Albert, near Victoria Park. Our yards were side by side, so I guess we both started together. He always called me "Mac," you see, though my nickname was "Kye." We bought the crystal detectors and the headphones, which were the most expensive things - they cost five or six bucks. We made everything else. We'd make our own cat's whiskers and loose couplers, and we had antennas up. We did it all ourselves. There was no one there to help us. We were originals. There was no one before us that we knew of, so we'd get our ideas in books or magazines mainly, to do with the radio amateur. We'd build the thing up, and then we'd have to get the right number of turns and everything. I remember I made my own variable condenser, with plates and bringing them close together - pretty elementary, but it worked. And I made wave meters, in a coffee can, that I had all calibrated. It was quite a job to calibrate a thing like that. 600 metres was where all the ships were, so I could get that one, and then they used to jump to 300 to send messages, so you could get that on the dial. After you were using the set for a while you'd get to know.

The antenna I had first, when I lived on Mack Street beside Stew, was strung up inside the attic. My father had a lot of electrical stuff. He was in the artillery business and he made spotting machines. Anyway, he had lots of wire and so I fixed that one-wire L-type antenna up there. My father was interested - I inherited some of his skills and stuff. My mother didn't know what was going on, but she knew not to touch the crystal detector. I had a little corner there which I had canvas around, and with a heater in the winter - it was cold there! And the ground wire going away down. It's amazing the stuff you could pick up on the crystal detector with just a little antenna there. Then we moved down about a half a block on Albert Street, and I had it in my bedroom. My uncle ran a lead mine out the Perth Road, and he brought in two poles and we put them up. Had a damn good antenna then, up about 40 feet and a long stretch between the house and the barn.

I noticed through my log Stew was saying, "Can you help me hang this up on the roof?" He was putting one up - it used to be Doyle's Bakery down there (48 Mack Street, the next door neighbour of the MacKinnon residence, on the Nelson Street side). It was a single L antenna, and he built the mast himself out of wood. It looked just like a steel mast. It was very good. It's a hard thing to get up. Well, he put this in the neighbour's back yard - he did a nice job there - and that was later on.

When the First World War ended, the French government had built a whole bunch of small cargo ships with radio, and a Great Lakes outfit bought a whole bunch of them and put them on the Lakes. Anyway, they didn't want all the damn radio stuff and a friend of mine, an electrician, had to take all this junk out, you see. He gave me an old French key and an ammeter. I still have the key, a real fancy job - real well-made solid brass - and the ammeter is hanging up in my workshop. The French wouldn't have a rotary gap, but they'd have a big spark gap with a tremendous fan blowing on it like hell - a hell of a fan blowing on the gap, so that would blow the spark out. It would break it or quench it, and take the place of the rotary gap. Stew and I both had "cootie" keys also - side-swipers, with two contacts - oh, some time early in the War. It doesn't vibrate like a Vibroplex key. People call them side-swipers too, but that's a "bug" - their trademark was a bug. That's why some people called this a "cootie" key - same as a "bug."

Originally I had a spark coil that my father had. They used to make those things - Ruhmkorff induction coils - and the biggest trouble was to get a decent note - a tone. Without the vibrator on it you got just one spark, while with the vibrator you got a bunch of sparks, and so you had to adjust it. That's what gave you the note, and it'd be a terrible note. If you could get them in a certain place, you could get a really good tone out of them, but that was quite hard. I remember I tried to make a rotary contact thing, but I didn't have the facilities to make a good one. I should have had a brush on and a commutator thing. And then, we didn't have it matched up or anything, so we never could get much distance.

Later on, when we got vacuum tubes and stuff, we really got going then because we could get pretty good signals out. That was on CW and with the sky wave you could work hundreds of miles. The 201 was RCA's amplifier tube, but the 202 was their five- or eight-watt transmitting tube. I guess we had some of those. They were pretty big - even the 201 amplifier was a big tube, about four inches high. I don't remember much about the circuitry. The other thing was the power supply, and we needed thousands of volts DC, so we made rectifiers with aluminum and zinc in some kind of a mixture - a whole lot of little jars, you see. We could get up to a high voltage then - it wasn't pure DC, but it was pretty good.

We had no regular message relay routine, and I don't remember working any Canadian stations. In Kingston it's not so good east and west. North and south work better. There was Ralph and Stew and myself, and Stew had a chap in Napanee - that's 25 miles, pretty good distance. His name was Bill Caton - he became a Marconi operator on the coast and he ended up as a radio inspector in Ottawa. Stew worked him, but I never did for some reason. We'd just talk to anybody we could, by key - I never did anything on radiophone at all - you'd hear somebody on the air down in the States and you'd call them. If they happened to be there, they'd answer. It's amazing what we got with the little thing, and when the sky wave was good at night you'd get the whole coast. I think I got KPH once, in San Francisco, when the conditions were right. Of course you could easily do that when you had a tube set, but on a crystal that was quite a feat. They're not very sensitive, you know.

We were good operators. By that time we could read anything they had on the air. They had no high speed transmitters then. Everything was by hand. I remember the father of one of my playmates thought that it was somebody tapping on the thing. Didn't believe it at all.

We never had anything to do with Don MacClement and the boys who used to meet at school, as far as I know. I can't remember ever working any of those chaps at all. But I knew, because I'd heard them on the air. Prof. Matheson's son was one of them and Earl Morris. I never had anything to do with the Wireless Club at Queen's, and Stew had nothing to do with them either. The same with MacClement, you see. He went to school with me at Victoria and I knew him very well - a nice chap.

Keith MacKinnon's Second Log Book

I've still got some old logs I had of 1920 - actually, this is my second log. I started and made things up the winter before, and in the spring I was starting to work. On the first page it says "Started November the first, 1920," and it ends February 5, 1921. The logs won't mean anything to you - it's all mixed up - straight copying. It's all as I copied them out, and all telegraphy.

But, at this time, in 1920, we didn't have licences, and Stew had his own call, XDA, and Rulla Bunt had XRB. This log is all filled up with XDA and XRB. The real XDA, incidentally, was the Mexican main transmitter during the First War. POZ, the Nauen station in Germany worked XDA, and this was how they handled all their stuff - all to do with spies. I don't know why Stew called himself XDA. He just did. He may have thought it sounded all right, but XDA was the Mexican station and I used to hear it occasionally. Well, there was nobody to check on us. No one could hear us for more than a few miles. We might have trouble if we interfered with VBH, you know, but that was quite a way away, and we just had little spark coils. I remember George Gurney saying he could hear us operating. They thought we were professional ship operators who had something going here. I mean, we were pretty good operators. I remember George saying, "We heard these people there...," and they were us.

About Alex Ada because I happened to see this here in my log. We didn't know who 3HN was at the time, and I didn't know who I was copying. I imagined it might be Stew or Ralph...Rulla Bunt, you know. And here's "What's 3HN doing on 400 metres?" and "Who is this 3HN? He's interfering with the radiophones on now. 350 metres. Do you hear the rotten 3HN radiophone on now?" He was all over. We didn't know who 3HN was, and that was just the call before mine - 3HO. I don't know when we found out who 3HN was. Stew and I went over to see him once - he wanted to show us the equipment - but I never knew him. I knew his brother George better. We never worked 3HN - I don't know whether he worked code or not, because he had this (radio)phone apparently...And I don't know who I was copying here, but I imagine it was Stew or Rulla Bunt.[82]

Keith MacKinnon's log is a sheaf of a hundred small sheets torn from a pad of merchant's delivery forms and strung on a loop of the same #24 or #28 cotton-covered wire that the boys used for winding their coils. At a casual glance, the log looks like strings of single letters and gibberish in a youthful round hand. On closer examination it becomes apparent that these are snatches of messages jumbled together, copied letter by faltering letter from the crackling earphones. Some was transmitted by friends at low speed for practice and some came in at a ferocious clip from the commercial operations. A few examples of the more coherent messages reveal something of the boys' daily lives, their friends, and their utter fascination with wireless:

> throw your da mned old leyden jar and sedrx out of with o you dam ned take the head off note pure but not bad like wlc try buzzer now, co nnect lact to aerial and armature. now jaz z it off.
>
> eastern glen advise promp tly ex pected arrival norfolk amount ball ast and repars anti cipated is ge cracio cracio master ack no wred ge to sh ip board nor folk atd wash ington 1610 end - kotn de NAA
>
> very loud tho all up to end of programme guess ill have asleep now and seeif any american amateurs are on yet
>
> I dont know. have to take ashes thas and sift too may afternight do ne got to go to cullens now, worse luck. your note is ok, same ash rk can you get naa loud enough d?en to copy? which note do you like best this or this/ naa - g ok - you will have to keep speed down or i cant read you. it just is a acy wiggles all over the table oa as ouch noise as an old sounder come on to school now, eh!
>
> no qsa? - tune to 200 naeters, ab out two hundred on at on cc worst than 600 meters is the note pure - amuhing primary of cou pler to tune am send ing on now
>
> no, lots on, wo ts the matter so long - call me when you come home will you? how many volts are you using now can hear you all over room and hall, did you take helix out of c ircuit?
>
> QST de XDA - any one want to practise?/ send something snappy for practise nr qu you want somethingersal by dist i lled and bottled by h limited at a lke cus le ontario, canada oa tured in cask inaware souses avsi ch are during be cold reason and bott led by the distiller
>
> n bond, undergo ve rnment exc ise on ascendfe d by anofficia l over be capsu le of bottle all our coa ks and capsules bear our nam e, this brand wa c arc and label are regi stered, end -
>
> did yu get em ce meon the big aerial yu said ut, dot n dj sold his bet XRB is up on his hind legs howl ing at us eh wh at? Sunday Jan 30 1921

how did you get that? - how did you get that? dont get in a rage. cam yoself

XRB de XDA - nr1-hr- qrm-is bad but you cut thru i m y boy no serr mon yet - XDA - no latter/ who is ndm?/ ndm is a ham ham in this ci ty qrm ing sim no sermon yet, eh. are you hearing any phone?

where did you get that patheti c twist to your "k", oa de VBH - XDA clock stopped. thin g wkc

how the sedl do i know was thatn old VBH - Jan 27, 6.30 p.m. QSD - 6.57 now. cape cod or herring or some darn fish

got tube oscil lating, buzzer just screeches at me, you roar some thing aw ful. qrm is strong

5HT-ok-3HE-k. 5HT de 3HE will retry c.w. put k ey in series wit h ground. make the bulb oscill ate send /// then change to spark. 3 HE do 5HT ok...

glad to hear you receive me good, will you please inform my uncle mr. S. Knapp staying at the paisley house that you receive me. we ll send again on friday for own six to seven 3 HT de 3HE

what time did you get home last nite l - x note hi hi but b- guess I'd better stop stop the qrm, cassel will be up in the air cursing...

about what about kingston radio club tob e formed ok. I don't know what you mean, I never hearda anything about it. plai. before supper you send....but the idea is to get after ch arteer tem us to start the ball rolling - end - did you get it? - end - XDA, k

qrm - is old your motor speed now the note is dest when she is low signals are much str onger too on low note.- ican hear you with phones on table now. jake note should get yun wlis very strong all

Jan 15th/21 storm o r VBG de VBH=r1 the end i p icdg start agn ok. did VBC send a to se

XRB you have a rotten note XPB....XRB. you r note ery strong. XRB-k. XRB de igot you that time. XPB

XDA de 3IR. XRB de 3IR.

vin doyle wants me to take aerial off chimney hisand icould put it on corner of barn, will you aideze moi demain?/ solong, tatanks! will you assist me now? I know its a holiday.

QST de XDA - msg mer ry christmas and happy new year to all...how do you get me on 200 metres? pretty broad tuned, eh? I bet old cashel is cursing me now to with him! have you heard wst to night?

I can hear the sound of my spark coming out of the gas heater in cellar

QST de VBH - probs - re lower lakes esgeongi mode rateve c inu cloudy sunday, fres gut and sou th s modtry cloudy ind n local i of or ra - .....Sunday QST de VBH probs - lower lakes fresh south winds nor cloudy and mil d today for lowed tonight and on Monday kj incaes g winds and showers in life is comparab fr so is progr ess of the arts and science. what appears fantastic al wholly visi onary today ha s a knack of co ming true on the morrow. co nse-quen on l ook ing back on ly over the micr oscopic all y sm all stretch of tim e of twenty-five years, we find the are of radio...VBG de VBH - to do vo gper slack for toron-to oese c y af train gpre - however wehave still an enor-mous mountain to cli mb asfar as the radio art is con-cerned compared to the ultimate goal, our present ach ievements are like the smallest peaks in comparison with himalayas...standing on the little pebbles we dont get much of avan tage point, but we can at least lo ok back upwar road we travel and we can als o...

QST de XDA = msg - probs for lower lakes and geor bay mild er, with local showers to night and on saturday

ears tired. had them on for two hours, will call you in half hour (4.20) amiain - am hearing some on 1500 metres = heard nau san juan porto r ico. did you get (jabbed) - dam ithink I heard key west, got his QST. NA miss ed his last letter while you were sending heaso about 1000 metres too, so hi'em going to have a holiday de o a / QST de XDA - XRB de XDA - msg - come on up and over 3HO de XDA - come on over - were the re many away pho - I am going to tmlk this afternoon, I heard wsg wro gug on 1500 and 2500

...of wreck age and three derelicts ?? to most ten-felt out of water over dangerous to navy at - end - QST de NAA ast a n my wrist is scramp d gagak ga - my ears are tired - amateur is going to du d stew is it hows your vaccination? have you got your lesson yet. come on down. please send reply via XDA and signed Charlie Bunt

...night I got 16. - yes. mlo metetiar many u on i - sure never noticed any difference. from ahw gh beggar then. li ste while donner lly called you thompson it was 3HF did you hear hv - 6.10 VSZ VCA de VBH...

got new cat whisker and you come in twice as loud and mor prtee of no

lakes country, said something about amurder last night, I wrote five pages last night, he sent from 10 till 11.40 last night. darn good practice. was just listening to nnz, now was chear and lad with secondary 3/4 out rpt...

XRB de XDA - come on up and bring your home work and ears, home work and phones. I saw the edmon-ton on swifts w arf she has aerial on her. come on up, XDA

MacKinnon was included, with the call 3HO, in the first <u>Official List of Radio Stations of Canada</u> in August 1922.[83] His call pops up in the <u>Daily Standard Junior</u> in the fall of 1924, when he was inactive as an amateur. The editor expected him to return to the ether with 5 or 10 watts, but soon realized that he would not return that winter. 3HO was busy working on his first class Commercial Operator's Certificate, with assistance from George Gurney at VBH. He went to Ottawa to take the examination some time in November, was successful,[84] and got word from the Marconi Company to "stand by" to embark in early May, 1925.[85] But the grain shipping business was poor that summer, and the call did not come. The next winter he did not operate as an amateur[86] but played the saxophone in a local theatre orchestra.[87] After graduation, in the spring of 1926, he stayed at Queen's to do post-graduate work in engineering.[88] His call, 3HO and then VE3HO, remained active in Kingston and at Queen's into the early 1930s.

Keith MacKinnon worked for the old Canadian Radio Broadcasting Commission between 1932 and 1936, under Colonel W. Arthur Steel, and later was with the National Research Council working on the secret RDF project, which turned out to be radar. Later, he was an Ottawa-based free-lance consulting radio engineer, dealing with all aspects of commercial radio antennas and broadcasting contours. He was the consultant for CFRC for many years, and prepared the first brief for CFRC-FM in 1952.

Kenneth L. McAlpine (3NE)

Kenneth Langrill McAlpine (B.A. '23) was born in Ostrander, Ontario, in 1898. He transferred to Queen's from the University of Toronto, where he had completed the first year of a course in chemistry. In the Official List, August 1, 1922, he was shown as living at 67 Gore Street and holding the call 3NE.[83] Don MacClement, Harold Stewart and

Orton Donnelly associated the name McAlpine with a Kingston radio amateur, but could not recall the person. While Arthur Wehman did not recognize the name either, he remembered hearing 3NE on the air in 1923. One oral source connected McAlpine with early CFRC. Documentation is lacking, although he could have belonged to the Queen's Wireless Club.

McAlpine became a research assistant to Professor Goodwin at Queen's in April 1922,[89] went to the Connaught Laboratories in Toronto to do his M.A. in biochemistry, and may have been associated there with Banting and Best.[90] A letter surviving from 1924 mentioned building a loud talker, listening-in until 1.45 am, visiting the local Tillsonburg station and constructing a 9-tube superheterodyne receiver.[91] He later established McAlpine Pharmaceuticals in Toronto and spent the rest of his life there.[92]

Kenneth Langrill McAlpine (Arts '23), a Kingston ham circa 1923, had the call 3NE. -courtesy Lynn McAlpine

Abbreviations Used in Radio Communication

QRM - being interfered with
QRN - strong atmospherics
QSA - signals strong
QSD - my time is...
QST - general call to all stations

Kingston Amateurs in the 1920s

Call	Name	Address
	George David McLeod	204 Alfred Street
	Arthur Cruse	136 Alfred Street
	Willard Roche	379 Alfred Street?
	Harold John Farrar	119 Alfred Street
XCD/3GV	Howard Price Folger	1 Emily Street, Edgewater
XDW	Jack Clifford Giles	186-189 Union West, Calderwood
3AO	Newman Electric Co.	167 Princess Street
3HE	Orton H. Donnelly	118 Wellington Street
3HF	Gordon A. Thompson	247 Earl Street
3HN	Alex E.W. Ada	51 Arch Street
3HO	Keith A. MacKinnon	76 Mack Street
3IL	Robert M. Davis	210 Union Street
3LO	R.D. Travers*	(Walkerville)
3MC	W. Don MacClement	Queen's campus, Summerhill
3AT/3MX	Ralph L. Bunt	5 Birch Avenue
	and R.E. Hunt	
3NE	K.L. McAlpine	67 Gore Street
3NF	Harold H. Stewart	404 Albert Street
3NT	J. Earl Morris	114 Barrie Street
3NW	Don I. Matheson	Alice Street
3OQ/3JG	Jack M. Campbell, Jr	5 Emily Street
3VS	Valentine Sharp*	269 Princess Street
		ex-2EM of Montreal
3AAY	Canada Radio Stores	269 1/2 Princess Street
3ACF	Harold Cohen	209 Queen Street
3ACX	Wally Reid	361 Division Street[93]
3AEL	Fred H. Simpson, Jr	85 6th Street
3ADR	Frank C. Dunn and	119 William Street
	R.B. Grimm	
3AFZ	William M. Cameron	189 Pine Street/35 Stephen Street
	Sheldon Cohen	209 Queen Street
	L. Donovan Clark	Albert Street
	Norry McLeod	?
	Sergt. Tom G. Brown	Armouries, Montreal Street
	Dr W. Gordon Mylks	122 Wellington/79 William Street
	J.R. Mitchell	11 Ontario S.
	W.A. Acton	?
(R)	K. Coward	386 Albert Street
(R)	Arthur Wehman	251 Division Street
	Warren A. Marrison	(Queen's University)

3DF	J.G. Shearer	12 Hewitt Avenue, Toronto or John L. Shearer (Civil '28) of Ottawa (?) attended Queen's from 1922

(R) = Reception only
* arrived January 1926

Kingston and Area Calls, 1938[94]

VE3ABA	O.H. Donnelly	560 Princess Street
VE3ACH	R.E. Freeman	532 Albert Street
VE3ADG	W.K. Salisbury	45 Sixth Street
VE3AQY	P. Joron	132 Wellington Street
VE3BU	J.V. Wiskin	32 Nelson Street
VE3GO	C.A. Millar	Ontario Street East (c/o Knapp's Boat House)
VE3IV	A. Gillingham	c/o R.C.C.S., Kingston
VE3JG	J.M. Campbell	c/o Gananoque Electric, Light and Water Supply Co., Ltd.
VE3MC	J.J. McWatters	86 Princess Street
VE3NT	H.H. Stewart	435 Frontenac Street
VE3TL	L.A. Milton	109 Alfred Street
VE3VX	L.G. Askwith	Queen's University Radio Club, c/o Students' Union, Queen's
VE3MX	R.L. Bunt	251 Fifth Avenue, Ottawa
VE3WZ	R.D. Travers	1 Duke Street, Hamilton

Amateurs Attending Queen's or RMC in the 1920s

3BT	?	Cobalt (Queen's 1925-26)
3DB	?	Hamilton (Queen's 1924)
3DF	?	Ingersoll, operating at 3NF and 3AEL (Queen's)
3FT	?	? (RMC 1924-26)
3IS	?	Brockville (Queen's 1925-)
3JY	?	Ottawa, operating at 3NF and 3AEL (Queen's 1925-)
3KJ	?	Ingersoll (Queen's 1924)
3OB	?	? (Queen's)
	W.A. Taylor (VE3AN)	? (Queen's 1919-24)
3ME	L.R.C. McAteer	Toronto (Queen's 1920-25)
3TY	?	Kitchener (cadet, RMC 1924)
1OL	?	United States (Queen's 1924)

Other Amateurs Mentioned in the Text

3BP	E.S. (Ted) Rogers	49 Nanton Avenue, Toronto (1922)
3CO	W.F. Choat	241 Robert Street, Toronto (1922)
3ES	H.D. Marlatt	77 McCaul Street, Toronto (1921)
3VZ	L.L. Hicks	Kemptville, Ontario (1922)
3DB	J.R. Bain	65 Garfield Avenue, Hamilton (1921)
3FE	William A. Caton	Napanee (1921); Ottawa (1922); Toronto (1929-)
3FG	G.H. Daly	Napanee (1922)
3DC	Dr W.R. Jaffrey	64 Hughson Street South, Hamilton (1927-)
9AL	A.H. Keith Russell	11 Pinewood Avenue, Toronto (1922)
VE2EM	L.R.C. McAteer	2052 Victoria Street, Montreal (1929-)

NOTES

1. "An Act to Provide for the Regulation of Wireless Telegraphy in Canada", Statutes of Canada, Ch. 49, 1905, p. 235.
   "An Act Respecting Telegraphs: Wireless Telegraphy", Revised Statutes of Canada, Vol. 3, Ch. 126, 1906, p. 2229.
2. "40th Annual Report, MF: Appendix 8: Wireless Telegraph Stations", Sessional Papers, House of Commons, Paper No. 21, 42 #10, 1907-08, p. 95.
3. "Annual Report of the Department of the Naval Service: Report on the Radio-Telegraphic Service", ibid., Paper No. 38 46 #23, 1912, p. 42.
4. "Experimental and Amateur Stations", ibid., Paper No. 38, 47 #25, 1913, p. 58.
5. "Licenced Experimental and Amateur Stations", ibid., Paper No. 38, 48 #27, 1914, p. 99.
6. "Licenced Experimental and Amateur Stations", ibid., Paper No. 38, 50 #26, 1915, p. 72.
7. "Licensed Amateur Stations", ibid., Paper No. 38, 51 #27, 1916, p. 120.
8. "Alma Mater Candidates", QJ, Dec. 5 , 1916, p. 3.
9. "Heart Attack Proves Fatal", KWS, Nov. 23, 1946, p. 1.
10. Personal communications from Dr. W. Donald MacClement, Kelowna, BC, May, Aug. and Sept. 25, 1987.
11. Unsigned inspection report on Form W. 14, Coast and Land Stations, Canadian Government Radio Telegraphic Service, Proces Verbal, January 20, 1914; NAC, RG 42, Vol. 1037.
12. "Licence to Use Wireless Telegraphy", Experimental Licence No. 38, Call Letters XCD, on Form No. 20, Dominion of Canada, April 1, 1913; NAC, RG 97, Vol. 85, File 6202-30, part 1.
    "Declaration of Secrecy in the Operation of Radiotelegraphic Apparatus", on Form W.12, signed by H.P. Folger, Nov. 22, 1913; ibid.
13. "Amateur Experimental Stations: District No. 3", OLRSC, Department of the Naval Service, Ottawa, Aug. 1, 1922, p. 41.
14. "Dominion of Canada, Department of the Naval Service, Radiotelegraph Branch, Application for License to Instal and Operate a Radiotelegraph Station", Form W. 4, signed June 8, 1914; NAC, RG 91, Vol. 85, File 202-30, part 2.
15. "Kingston's New Hospital", DBW, Oct. 2, 1919, p. 4.
16. "Wireless Amateurs", ibid., Nov. 20, 1913, p. 5.
    "Boys at Wireless", ibid., Nov. 22, 1913, p. 5.
17. Personal communication from Dr. W.D. MacClement, Kelowna, B.C., Aug. 1987.
18. Personal communication from Ronald Ede, teacher and Guidance Counsellor at KCVI, March 26, 1986.
19. SM Letterbook, June 12, 1918, p. 159; QUA, Coll. 3701A, Vol. 7.
20. Harold John Farrar, IIIb, "Wireless Telegraphy", Kingston Collegiate Institute Journal, June 1914, p. 12.
21. "Kingston Boy Leaves", DBW, April 3, 1914, p. 6.
22. "George D. McLeod", ibid., May 6, 1914, p. 5.
23. Kingston's Young Wireless Operator", ibid., June 5, 1914, p. 2.
24. "A Wireless Operator", ibid., Aug. 5, 1914, p.2.
25. "A Letter Received", ibid., Aug. 6, 1914, p. 8.
26. "George McLeod", ibid., Aug. 18, 1915, p. 8.
27. "Wireless Man's Experience", ibid., Aug. 11, 1916, p. 2.
    George D. McLeod's "Continuous Certificate of Discharge", No. 834759;

courtesy Peter C. McLeod, Manitouwadge, Ont.

28. "Was Wireless Operator", DBW, Nov. 19, 1915, p. 2.
    "George McLeod", WBW, April 20, 1916, p. 4.
29. "Wireless Club", QJ, Nov. 11, 1919, p. 4.
30. Personal communication from Mrs. Margaret McKay-Clements, daughter
    of George D. McLeod, Haileybury, Ont., Sept. 23, 1987.
31. Personal communication from Peter C. McLeod, Manitouwadge, Ont., Sept.
    23, 1987.
32. Old Arts is now known as Theological Hall and the old Staff Reading
    Room is today Room 118, Rotunda Lobby.
33. "Wireless Seized", KDS May 26, 1920, p. 5.
    "Three Wireless Apparatuses Seized", DBW, May 25, 1920, p. 1.
    J. Earl Morris, 114 Barrie Street, Robert M. Davis, 210 Union Street and
    Orton H. Donnelly, 118 Wellington Street, had their sets seized by the
    Dominion Government Radio Inspector for interfering with reception
    by the Barriefield Marconi station.
34. Shortly before vacuum tube amplifiers appeared on the market, Don
    MacClement invented a silicon crystal amplifier which impressed Prof.
    Jemmett. He could not afford to pay the various fees to get it patented,
    and dropped the matter, but now muses that "it probably was a little
    touch of the first bit of magic of the silicon world". For MacClement's
    story of his crystal amplifier, see the unedited manuscript of this book;
    QUA.
35. "The Operating Department: Ontario Division", QST 5 #1, Aug. 1921,
    p. 34.
36. "Alexander Edwin William Ada, Faculty of Medicine", Queen's University
    Yearbook, 1926, p 110; QUA.
    "Medical Graduates", DBW, May 25, 1926, p. 2.
    "Dr. Ada Gets Hoffman", KWS, April 2, 1927, p. 1.
    "Dr. Alexander Ada", New York Times, March 3, 1969, p. 35.
37. "Club Gossip: Kingston Radio Association", Radio News (U.S.A.), May
    1921, p. 795.
    "Old Time Club", KDS, Sept. 6, 1924, p. 11.
38. KDS, April 29, 1922, p. 2.
    "The Operating Department: Ontario Division", QST 6 #5, Dec. 1922,
    p. 54.
39. DBW, June 14, 1924.
    "Brief Notes", KDS, June 14, 1922, p. 5.
40. "Radio Inspectors Appointed", Electrical News 32 #16, Aug. 15, 1923,
    p. 58.
    "The Operating Department: Ontario Division", QST 6 #11, June 1923,
    p. 47.
41. "New Radio Inspector", KDS, Sept. 12, 1925, p. 9.
42. "Calls Heard: Communication from 3IL, Kingston", QST 5 #9, April
    1922, p. 68.
43. Personal communication from Keith A. MacKinnon, Ottawa, Aug. 15,
    1987.
44. Personal communication from Dr. Alfred W. Jolliffe, Kingston, Aug. 27,
    1982.
45. "Daily Standard Junior", KDS, June 13, 1925, p. 9.
    "Daily Standard Junior", ibid., Aug. 1, 1925, p. 5.
46. "City and District", KWS, Jan. 17, 1927, p. 12.
47. Personal communication from Keith A. MacKinnon, Ottawa, Sept. 19,
    1987.
48. Supplement No. 1 OLRSC, Dept. of Marine, Radio Branch, Ottawa, Sept.
    1, 1930.

OLRSC, Dept. of Transport, Ottawa, March 31, 1938.
49. Personal communication from Donald I. Matheson, Brockville, July 31 and Aug. 30, 1987.
50. "A Letter", KWS, Dec. 2, 1926, p. 13.
51. "Storage Batteries", DBW, Jan. 25, 1922, p. 10.
"Second Fire Breaks Out", ibid., July 10, 1922, p. 7.
Ibid., Sept. 15, 1925, p. 10.
52. Personal communication from Harold A. "Dick" Cohen, Kingston, Sept. 29, 1982.
53. Frank J.C. Dunn, Kingston Radio Inspector, to C.P. Edwards, Director of the Radio Service, Radiotelegraphic Branch, Dept. of Marine and Fisheries, concerning an amateur spark coil, (probably not the Cohens'), Jan. 1926; NAC, RG 97, Vol. 130, file 6270-1, part 1, Prosecutions - Radio, Amateur Stations.
54. Personal communication from Arthur Wehman, Kingston, March, 1987 and June 14, 1990.
55. Personal communication from Prof. Harold H. Stewart, Kingston, Oct. 20, 1982.
56. Ibid., June 1, 1987.
57. "Listening In: With the Amateurs", KDS/DSJ, Sept. 13, 1924, p. 8; Oct. 4, 1924, p. 7.
58. Gordon Hammond (3CEL), "An Eighty-Foot Latticed Mast", QST 7 #11, June 1924, p. 39.
59. "Listening In", KDS/DSJ, Oct. 18, 1924, p. 9.
60. "Listening In", ibid., Nov. 8, 1924, p. 7.
61. "Listening In", ibid., Nov. 15, 1924, p. 11.
62. "Listening In", ibid., Nov. 22, 1924, p. 11.
63. "Listening In", ibid., Nov. 29, 1924, p. 11; Dec. 6, 1924, p. 7; Dec. 13, 1924, p. 8.
64. Ibid., Jan. 17, 1925, p. 11; Jan. 31, 1925, p. 9.
65. "With the Amateurs", ibid., Feb. 7, 1925, p. 9.
66. "With the Amateurs", ibid., Feb. 14, 1925, p. 7.
67. "Listening In", ibid., June 6, 1925, p. 9.
68. "With the Amateurs", ibid., Feb. 21, 1925, p. 9.
"All Records Smashed", ibid., May 30, 1925, p. 11.
69. "With the Amateurs", ibid., Feb. 28, 1925, p. 10.
70. "With the Amateurs", ibid., March 22, 1925, p. 10.
71. "With our Amateurs", ibid., April 11, 1925, p. 9.
72. "Listening In", ibid., Oct. 10, 1925, p. 8; Oct. 17, 1925, p. 9.
73. Ibid., Oct. 24, 1925, p. 10.
H.M. Williams, "The Hertz Antenna at 20 and 40 Meters", QST 9 #7, July 1925, p. 24.
74. "Radio Operator", KDS, Nov. 17, 1925, p. 17.
"Big Radio Convention", ibid., Nov. 11, 1925, p. 12.
75. KDS/DSJ, Nov. 14, 1925, p. 8; Nov. 28, 1925, p. 9.
76. "Listening In", ibid., Jan. 23, 1926, p. 7.
77. "Listening In", ibid., Jan. 30, 1926, p. 8.
78. "What the Radio Amateurs are Doing", ibid., Feb. 13, 1926, p. 7; Feb. 20, 1926, p. 7; Feb. 27, 1926, p. 7.
79. "What the Radio Amateurs", ibid., April 17, 1926, p. 8.
80. "Kingston Loses Popular Amateur", ibid., May 15, 1926, p. 7.
81. "Influx of Frosh", QJ, Oct. 1, 1929, p. 5.
82. Personal communication from Keith A. MacKinnon, Ottawa, Aug. 15, 1987.
83. OLRSC, DMF, Ottawa, Aug. 1, 1922.

84. "Radio Notes", KDS/DSJ, Nov. 22, 1924, p. 11.
85. "Listening In", ibid., May 9, 1925, p. 7.
86. Ibid., June 20, 1925, p. 10.
87. "What the Radio Amateurs", ibid., Feb. 20, 1926, p. 7.
88. "What the Local Radio Amateurs", ibid., May 8, 1926, p. 7.
89. Science Research Committee Minutes, Queen's University, April 17, 1922; QUA, Coll. 1196, Vol. 2.
90. Personal communication from Mrs. Lynne McAlpine, Saskatoon, June 20, 1990.

    K.L. McAlpine, "The Action of Insulin in vitro," University of Toronto Archives, MA 1925, file T79-0081 .[59].

    "Chemist Backs Florist", Globe and Mail, April 14, 1949, p. 25.
91. K.L. McAlpine to Miss Hazel Lorimer, July 20, 1924; courtesy Mrs. Lynne McAlpine, Saskatoon.
92. "Deaths, Memorials, Births: K.L. McAlpine", Toronto Star, Nov. 26, 1983, p. B7.
93. This may be Wallace Middleton Reid (Chem. Eng. '26).
94. OLRSC, Radio Division, Department of Transport, Ottawa, March 31, 1938.

CHAPTER 8

The Donnelly home at 118 Wellington Street (centre), with the Young Irishmen's Hall next door, to the left.
-photograph by Arthur Zimmerman

Orton Donnelly, Kingston's Principal Amateur

This chapter is largely about Orton Donnelly, the most serious wireless amateur in Kingston. He was an innovator, a builder, an experimenter, and the best known, most adventurous and widest ranging amateur in the city. In the vernacular of the day, he was a full-fledged wireless "bug." Orton was involved in radio in some way or another throughout most of his life. Unfortunately, he did not retain his logs for verification and documentation of his activities, but his siblings, Edward, Robert, Josephine and Douglas, helped to reconstruct his story.

Orton Hepburn Donnelly was born in Kingston in 1904, the son of Foster and Florence (Hepburn) Donnelly and eldest grandson of Captain John C. and Elizabeth Ewart Donnelly. Capt. John was the wreckmaster of the Calvin and Breck Company from around 1850 and he founded Donnelly Salvage and Wrecking Co., Ltd., in 1884. The offices were at 193 Ontario Street and their yards were located at the eastern side of West and Ontario streets and later beside the prison in Portsmouth, where the Olympic Harbour site is now. Capt. John was one of the original subscribers to the Kingston School of Mining in 1892, and a member of its board. When he died in 1900, the company was taken over by two sons, John (Mining '98) and Thomas. John Jr's son Miller ran the company later, and it was sold around 1929.

> Bob Donnelly: Dad worked on the wrecking company in the naviga-tion season and, when that closed up, his winter job was at the Grand Opera House. I know when I was a kid, we used to get taken in the backstage door every once in a while to watch a show. We'd sneak in, then sneak through the door and up into a seat. The older ones of us were quite familiar there at the time. I think it is quite possible that Orton met the Dumbells and so on through the Grand. There was a chap by the name of Marty Gorman who had been with the Dumbells overseas, but he lived next door on Wellington Street. He operated the spotlights at the Grand, as a part-time job in the evenings. And I'd go up and sit with him when he was switching the spotlights off and on. Of course, I got in free through my father, and I had to get out of the way to let the paying customers in.[1]

According to family legend, Orton was around 12 years old when he became interested in radio. This was about the time he was pro-moted to Senior Fourth at Central Public School, in 1916, when the family was living at 204 Montreal Street. Orton was the apple of his paternal grandmother Elizabeth's eye, and it was she who gave him the money to build or buy his first radio set. This was probably a crystal receiver, as Orton recollected: "In our younger days, just before there was broadcast, one or two of us, we had crystal sets and listened to code, like Toronto (VBG) and Kingston (VBH) communicating with the boats on the lake."[2]

Grandmother Donnelly had suffered a broken hip in a street acci-dent and couldn't walk, so her son Foster and his family moved in with her at 118 Wellington Street around 1917 or 1918, about the time Orton started at KCI. There, he could been influenced to expand his interest in wireless by "Cap" Kelly. Apparently, Orton was so wrapped up in radio while at KCI that he wouldn't even look at girls. His grand-mother turned the two large rooms in the attic over to Orton, the front one for his bedroom and the back one for him to tinker in. Family leg-end has it that he was interfering with VBH at Barriefield or violating the wartime ban on amateur radio and was caught. He got notice to appear in court and was fined $100. Elizabeth Donnelly cheerfully paid the fine for her "golden" grandson.

Orton's Rotary Spark Transmitter

Orton must have had a buzzer or Ford spark coil transmitter in the early days, but within a few years he somehow acquired a powerful

Orton Donnelly's sideswiper Vibroplex key.
-photograph by Douglas Donnelly

commercial rotary spark gap apparatus.

Bob Donnelly: I believe that rotary spark transmitter came off a French battleship that was being scrapped at Hamilton or Detroit. It was just at the War's end, things were expanding and they needed steel. Because of the Rush-Bagot Treaty, the boats couldn't be driven through the Great Lakes under their own power, so they were towed up. Now, either my father, who was associated with the wrecking company, or my uncle John saw this spark transmitter and got it for Ort. Or Ort was down there and wanted it right off the bat.

I never saw the spark transmitter itself, as a transmitter. The front panel was made out of inch and a half blackish-grey quarry slate. It'd be about three feet-plus by four feet. It was drilled and there were a few meters and knife switches bolted onto the front. I saw the slate front a lot, sitting in the woodshed in Wellington Street. My brother Ed and I took the remaining nuts and bolts off it and laid it outside the woodshed door. We used it as a doorstep out the back woodshed on Wellington Street for a long time after. Everyone used to go and slide on it as they walked out there. The spark gap I don't think I ever saw. I've probably seen it, but I don't remember. I started Louise School in the 1921-22 year, and about that time he was using the tube - just started on the tube - so it was prior to that. But the year he got the ship's transmitter, I don't know. He may have got it in 1918 or 1919. That was the first transmitter that I remember he had.

I recall that one of his transmitters, probably the rotary spark apparatus, was using so much current that they had to have an electrician put in a separate circuit just for his transmitter room. I heard the neighbours say that he was drawing so much power off one particular transformer, feeding maybe five or six houses, that he was blinking their lights on them. And he was also putting out a beautiful RF signal that was wild, and things were hot all over the house every once in a while. Touch a hot air register or something and you were liable to get a shock when he was transmitting. RF all over the place![1]

Ed Donnelly recalled that after dark the kids on the street could read Orton's code from the blinking lights of the neighbours' houses. Orton remembered that the light bulb in the basement of the house across the road, even though it was switched off, would glow intermittently as he keyed the big rotary rig.

During renovations of 118 Wellington Street in the mid-1980's, Peter McMillan, the current owner, found a large slate slab on the rafters in the storage area of the attic, under the eaves. He gave it away to a Queen's student who broke it.

One seizure of Orton's rotary spark gap equipment by the authorities is documented in the press. This may be the same incident noted above. On May 25, 1920, his aerial - "the wires on the top of the house" - or his whole wireless apparatus according to another report, was confiscated by a Dominion Government inspector and handed over to the police. Orton and two other city amateurs, Robert M. Davis and J. Earl Morris, had been interfering with reception of messages by VBH, and they were told that their equipment would be handed back as soon as they reduced their high voltage and took out licences.[3] The Whig complained editorially, "Privately owned wireless installations in the city put the Barriefield station out of business. Cannot the government station on the hill compete with small private plants?" A search through the newspapers did not turn up any court appearances or penalties for these offences. One version of the family legend has it that Constable Vernon Campbell came to the house to make Orton dismantle his set. Orton wouldn't come home for supper because his mother, the disciplinarian of the family, and Officer Campbell were sitting and waiting for him.

> **Douglas Donnelly:** I knew Vern Campbell through his daughter Margaret. We kids used to play at the Kiwanis Playground in the old Swamp Ward, around Bagot and Raglan Road. I seem to have been attracted to Vern's daughter and we used to sit on the verandah in the evening. Vern had left the force by then and was driving a McCaul-Frontenac Red Indian Gasoline truck,[4] and he would sit out there and tell me about when he was on the force. He had to come up and serve notice on Orton that he had to take down his transmitting equipment and stop sending and receiving because he didn't have a licence to operate. And he expressed very serious regret that he had to do that, because he thought Ort was a very intelligent young man with a great future in that area. He hated to do anything that would impede or discourage him.
>
> Well, according to Vern, and later told to me again by my aunt, Ort knew he was in trouble. He had heard that the police were coming and he didn't know what was going to happen. He did not go back to school after lunch that day, torn between protecting his equipment and making himself scarce because he didn't want a confrontation. The problem was to be there but not appear to be there, so he resolved this by going behind the Quebec heater set up in the hall for keeping that part of the house warm. Apparently he sat in behind that most of the afternoon trying to outwait Vern, and it got pretty

warm back there, but he toughed it out. Eventually he was confronted by Vern and was instructed to tear down the equipment. The aerial was part of it, and was installed from our rooftop to the rooftop of
the Convent, which abutted our back yard - where the library is now.
After Vern left, they were debating about whether to dismantle the aerial from the Convent side and reel it in or what. Meanwhile, my brother Ed got a .22 rifle and is purported to have shot out the insulator at
the Convent end and solved the problem while they were standing
there discussing it. "That's your aerial down" sort of thing. That's
just what Ed would do - he had that spirit of adventure.[5]

It was probably just after this date that Orton applied for a licence
and got 3HE, while Bob Davis got 3IL and Earl Morris was given 3NT.
Less than a year later Orton was involved with the American Radio
Relay League and was trying to organize and regulate amateur wireless
traffic in Kingston through the Kingston Radio Association.

Orton Donnelly's Radio Gang

The boys in Orton's group in the very early 1920s were probably Alex
Ada, Robert M. Davis, Gordon Thompson, Ralph Bunt, Harold Stewart
and Keith MacKinnon.

> **Bob Donnelly**: Ort started out with the "boy experimenter" magazines.
> Ralph Bunt, Ort and another chap - they were all chums together.
> This was the initial radio gang. I think the man that really got them
> going on radio, the one fellow that really got them inspired, was old
> George Gurney at VBH. George and the chap down at the CN tele-
> graph office, or CP. George was the type of guy, if you asked him
> anything about the code, he'd teach it to you. He was a very pleasant
> sort of chap. Saturday or Sunday afternoon you walked over to VBH
> and looked in the windows, if you could get through the fence, and see
> them operating there. I knew George in later years, and he taught
> dozens of kids the code. There was a place on the north side of Princess
> Street, near the newspaper or in the next block, and you went up a set
> of stairs and they had rooms there and they used to teach the code with
> a bunch of buzzers. And the kids'd all go up there - it might have
> been Sea Cadets. I believe George was part of that. He was the one that
> really got them interested. I would imagine it would be before Ort
> went to Schenectady, because when he came back he was perfectly
> familiar with code.[1]

Union College, Schenectady, New York

Orton left KCI in October 1920, at the beginning of grade 12, to attend
Union College in Schenectady. He did not remain long, for he was back
in Kingston by mid-March, 1921.

Union College had a spark station from 1915 or 1916. Two professors next built a radiophone station in the back of the E.E. laboratory
in the winter of 1916-17 and did experimental broadcasts, including the
transmitting of music from an Edison phonograph. A radio club was
formed around that transmitter before the United States entered the

First War, and they obtained the call 2YU. In 1919 they had three transmitters in the E.E. labs and a fan-type antenna 70 feet high, 90 feet wide, with seven down-leads. They broadcast a Victrola concert on October 14, 1920, the first of a schedule, and were soon granted the call 2XQ. Every Sunday evening 2XQ gave a faculty lecture. They also experimented transmitting live remote dance bands: "A horn, to which was fastened a telephone transmitter, was hung over the orchestra and wires run to the phone set in the Electrical Laboratory." Union College obtained the call WRL in 1922, but could not meet the commercial competition and did not renew the licence in 1923. Later they broadcast over over WGY, the General Electric station,[6] and now hold WRUC-FM.

> <u>Bob Donnelly</u>: I think that they let him out of high school to go to Schenectady provided he got a job with General Electric and would attend Union College in his spare time. The whole idea, and his ambition, was that when he came back he would go into the electrical trade, which was distinct at that time from the radio trade. He switched when he got to Schenectady - I think it was a bug that sort of hit everybody at the same time. But I think he kind of went wrong and he went to work and got into radio in Schenectady. When he came back, after a brief sojourn, he carried it on. He was back from Schenectady, actually, when I started becoming conscious of all this stuff. My memory starts really with the old radio room upstairs - fooling around up in there - being a kid and told not to go in there or you'd get a shock, holding wires for them to wind coils and so on.
>
> I think that the time at Schenectady was how he really got in with the Newman Electric Company. He was supposed to be electrical. But he'd already become interested in radio in Schenectady, and this was his natural branch-off. I think he got in with Newman on the electrical end, and the rest of it sort of just grew out of that. But, he wasn't interested in electrical at all - after Schenectady it was strictly radio, radio, radio.[1]

Orton Donnelly's Activities in ARRL

The American Radio Relay League (ARRL) of Hartford, Conn., still exists and has its Canadian headquarters in London, Ontario. It was originally a sort of continent-wide boys' amateur radio club. In the early days of spark, when transmission distances were short, and radio was conceived of as a private telephone, the idea was to establish affiliated stations all over North America for the relaying of messages in the public service. The League organized relay competitions to improve equipment, speed and distance, to bridge the continent, and then the oceans. Its policy statement in 1923, published in its journal, <u>QST</u>, read as follows: "The American Radio Relay League, Inc., is a national non-commercial association of radio amateurs, bonded for the more effective relaying of friendly messages between their stations, for legislative protection, for orderly operating, and for the practical improvement of short-wave two-way radiotelegraphic communication."[7]

By the mid-suumer of 1921, US amateurs were discarding their spark sets for CW sets. The August 1921 issue of <u>QST</u>, shows Orton

Donnelly as Superintendent of the Kingston District of the League's Ontario Division, having just succeeded Bill Caton of Napanee (3FE), who had resigned and left for Ottawa:

> Mr. Donnelly, Supt. of Kingston District, reports that most of the transmitting carried on there now is done by means of low power buzzer and spark coil sets to avoid interference with the local Marconi station, but the Kingston Radio Association is installing a C.W. set and expects to be handling messages in a short time. Total msgs - 10; 3HF - 3 (Gordon Thompson); 3HE - 3 (Orton Donnelly); 3HN - 1 (Alex Ada). A test was carried out to try to establish communication direct between 3GE, Toronto, and 3HE, but on account of QRM (interference) and QRN (atmospherics) it was a failure. Mr. Woodley, Supt. of Belleville District, reports that a C.W. station is being erected there which will be a very important link in our chain from Windsor to Montreal. The distance between Toronto and Kingston is rather great for steady daylight transmission, and this station midway between the two cities will act as a stepping-stone.[8]

In his November report, Orton concluded that Kingston's only hope for achieving distance was in continuous wave, "which he hopes to supply."[9]

> Bob Donnelly: Gananoque, Kingston and Belleville or Napanee - three groups were experimenting back and forth in these early days reporting each others' signals and there were plans to form a relay net through there.[1]

In October, the Ontario Division, under Manager A.H. Keith Russell,[10] was reorganized into 7 new subdivisions: "District 1 is under W.J. Carter, 3DH of Windsor; 2 controlled by Gowan, 3DS in Kitchener; 3 controlled by Russell in Toronto, with 3GE as assistant; 4 by (Ted) Rogers in Newmarket, 3BP; 5 by Donnelly, 3HE in Kingston, and 6 by Major Steel in Ottawa; the 7th District is not yet organized."[11] Through the next year or two, Orton was the chief relayer in the Kingston district, but he was having trouble getting his signals out to the north, east or west, even as close as Ottawa, Montreal or Toronto, since Kingston, surrounded by escarpment, is in a radio hole. East-west relays had to go around Kingston, to the north or to the south. In Kingston, 68 years later, it is still very difficult to pick up AM radio from Ottawa, Montreal or Toronto. The amateurs were experiencing new problems too, because the "broadcasting programmes have cast rather a damper over the relaying of messages."[12] By June 1922, Kingston was regularly copying Toronto and Montreal, but a good station was still needed here to forge the link between the two metropoli: "Tests are now being run between Brockville, Kingston and Gananoque on voice, which is another good omen for the fall."[13]

Orton did not yet have a radio telephone in the summer of 1921. It was probably late in 1921 when he converted to CW transmission with tubes and was able to get better distance. With CW, perhaps for the first time, he was able to to get his messages consistently out of Kingston to the north, east and west. Previously, the main traffic from Kingston had been to the south only, across the lake.

The Kingston Radio Association

Kingston wireless telegraphy and telephony amateurs had apparently long been hoping to form their own radio club, and the impetus to finally get one going may have come from successful radio clubs in Ottawa and Peterborough. Twenty-five prospective members of the Kingston Radio Association got together on March 23, 1921, probably in the home of Dr W.G. Mylks, 122 Wellington Street, whose son was interested in wireless. That evening they made the club rules, talked about traffic rules, drew up the constitution and by-laws and elected the officers. Membership was $1.00 per year, and each member got a "snappy" certificate. The Honorary President was Capt. S.A. Lee, MC, District Signalling Officer. Orton Donnelly was President, Robert M. Davis the Vice-President, Staff Sgt. T.G. Brown of the Armouries was the Secretary-Treasurer. The Traffic Managers, to look after the traffic rules, were Harold Stewart and Gordon A. Thompson.[14]

Meetings were appointed for every Friday evening at 7.30, where

The Kingston Radio Association first met at the home of Dr W. Gordon Mylks,
122 Wellington Street,(left) in 1921.
-photograph by Arthur Zimmerman

reports would be received and "prominent radio men in Kingston" would lecture. They would hold code practice every Tuesday evening and after the weekly meetings, Class A for beginners and B for those working on their speed. The Association was also going to erect an aerial and install and operate a transmitting and receiving set: "Altogether, this will be a live wire organization, as all the members are doing their best to make this one of the live wire radio clubs of the country."

At the second meeting, on April 1, the traffic rules for the amateurs were drawn up, and four new members were admitted. Members with their own transmitting sets were assigned to take turns sending out the QST, a general call to all stations, every evening at 7.00, with special club news, first at 15 then at 8 words per minute, at the compulsory wavelenth of 50 metres. And a committee was appointed to look into acquiring the club wireless equipment to link Kingston into the chain with the rest of the wireless world.[15] The next week they decided to test all amateur sets in the city with a wave meter to make sure that none was tuned to more than 50 metres, in accordance with government regulations: "Steps to do away with the congestion of the air were also taken. Amateurs will be allocated an interrupted (sic) period of fifteen minutes each evening to send, unless no other 'ham' is requiring to send."[16]

There was a meeting on April 29 and the one of May 6 was the first held in the club's new rooms on the top floor of the Golden Lion Grocery Building, located in the triangle bounded by Brock, Clarence and Wellington streets. Membership cards were distributed and new little blue

Golden

Lion

Grocery

The Kingston Radio Association had its rooms on the top floor of the Golden Lion Grocery block at the corner of Brock and Wellington streets in 1921.
-from the Daily British Whig

and white KRA pins promised. Reports by members continued, of reception of a Texas amateur, and of the concerts of KDKA and Marconi station CHCB in Toronto. Code practice was extended to 3 days a week, Monday, Wednesday and Friday.[17] Through May they continued to discuss wavemeters and their use. On the 20th the Association discussed holding a moonlight excursion in July.[18] At the next meeting, the KRA discussed putting up the aerial, the material for which had arrived.[19] At the meeting of June 10, they announced that a local electrical dealer was going to stock wireless goods, and about that time Newman Electric began to advertise: "We are stocking a complete line of wireless apparatus."[20] Amateurs would no longer have to send away to Toronto, Montreal and New York for parts. Before the last reported meeting of the KRA, June 17, 1921, the ground wire had been put in, and the members were anticipating that their aerial, on top of the Golden Lion Block, would be visible all over the downtown.[21]

The KRA became inactive during the height of the summer, because in the hot days reception of wireless signals was plagued by terrible static. The KRA was supposed to re-form in the fall, when the new receiving set was to be installed "thanks to the generosity of a few interested citizens,"[22] but the Association never met again. Of the 35 or so members of the KRA in 1921, none had yet a wireless telephone transmitting set, although there were at least 15 who could receive voice transmissions from great distances, implying that they did have tube sets.

> Dick Cohen: Well, there was a little radio club, the Kingston Radio Club. We were meeting in Dr Mylks' home, and there was Gordon Mylks, and the group that I mentioned were all there - Bob Davis, Orton Donnelly and Wally Reid, but my brother Sheldon wasn't."[23]

Dick Cohen also remembered a meeting at the home of Bob Davis, and a meeting on the top floor of the Golden Lion Building, above McRae's store. He thought that maybe he was really too young to have been there himself, at age 13.

> Don MacClement: I think there was at one time an attempt to set up some kind of a city broadcasting set. I just vaguely think of Howard Folger being associated with it somehow. It was a sort of get-together of the hams that had been chattering away with our little Ford spark coils - a very casual sort of affair. We used to have discussions as to how we could get everything down below 200 metres. It was essentially an attempt by Orton Donnelly and I suppose his partners in that radio station at Newman Electric Company to get the spark transmitters of the gang around town out of the way. I can remember threats that we'd have legal procedures taken against us if we didn't get our transmissions below 200 metres. It had, so far as I know, no popular support in town, and was dominated by the newer vacuum tube broadcasting side of things almost entirely.[24]

Some Kingston amateurs were among the 300,000 who heard the

Dempsey-Carpentier heavyweight championship fight of July 2, 1921, from Jersey City, broadcast at 1600 metres from the Lackawana Railway yards in Hoboken over RCA's makeshift WJY.[22] Major J. Andrew White, editor of Wireless Age, called the fight into a telephone at ringside. The listeners-in heard another voice, however, probably technician J.O. Smith at Hoboken, reading a transcript of White's words typed from the telephone. David Sarnoff of RCA had borrowed the transmitter, just completed by General Electric for the US Navy, through his friend Franklin D. Roosevelt. The transmitter was not designed for continuous operation at full power, and at the end of the broadcast it burned up.[25]

Orton Donnelly at Newman Electric

Henry W. Newman owned the H.W. Newman Electric Company, 167 Princess Street, the "pioneer radio supply house of Kingston." The company had been in business since about 1907 or 1908, boasting a staff of about 20 in 1919, selling the standard electrical supplies and appliances, and doing electrical contract work as well. Newman was a Kingston alderman in 1914, when he brought daylight saving time to the city. In 1919, he was elected mayor, calling himself "the father of day light saving" during his campaign. Just before H.W. Newman died in 1926, he sold the business to J. Bruce Saunders (Elec. '22), and it became Saunders Electric.[26]

Newman's stocking of wireless apparatus was announced in advertisements in the Daily Standard, June 9-11, 1921. It is possible that Orton went to work for Newman as early as June 1921, although no proof has been found.

> Orton Donnelly: I was going to school at the time and I had a job there. I was a sort of an electrical apprentice and it was during the holidays I worked for the H.W. Newman Electric Company. We knew the boss at that time as "Daylight" Newman. I believe he was one of the proponents of daylight saving time.
>
> He was interested in radio and I agreed at the time to put in a radio stock - the first radio stock in Kingston. We got the few existing catalogues - there weren't too many at that time - I made a list of what to order and he put in a line of crystal detectors, loose couplers, variometers, tuners - no complete sets. Mainly the components to build a complete set, and we were quite successful. The stock moved very well and he was quite happy with the stock. I wasn't there very long, but we did launch a successful selling item. People were beginning to take an interest.[27]

Orton was asked if he did any broadcasting from Newman Electric, and he replied: "At the time we put on some phonograph records and made a few announcements. Just a little bit to stir up some excitement around town. Our call was 3AO. We had no programs as such at the time."[2]

Orton's engagement by Newman Electric is first indicated in two newspaper announcements in mid-November, 1921. On the 11th: "A local dealer is now putting in a line of wireless goods which surely ought to boost radio along in Kingston."[28] On Saturday, November 12, 1921, between 8.30 and 9.30 pm, crowds of customers at Newman Electric were privileged to hear "the famous prima donna" Amelita Galli-Curci and a tenor singing over KDKA in Pittsburgh:

> Orton Donnelly, an expert electrician in the employ-ment of the company, has established a receiving sta-tion. Mr. Donnelly has received messages from many distant points in the Atlantic and Pacific, but the concert in the Pittsburgh opera house may be heard several evenings every week. The opera house is equipped with a wireless transmitter, and the whole concert may be heard wherever there may be a receiving station...The receiving apparatus is located near a show window and several persons may be accommodated at once...The music sounded very much like that from a phonographic record or a voice on a long distance telephone, or more like the telephone perhaps.[29]

The next "Saturday evening a large number gathered at the office of the H.W. Newman Electric Co., where they 'listened in' over the wireless plant of the firm to a splendid concert given by the Davis Concert Co., in Pittsburgh, Pa. The concert could be heard distinctly in the earlier part of the evening, but later a storm interfered with the cur-rents and broke up the service somewhat... The wireless was operated by Mr. Orton Donnelly, who explained many of the mysteries to those who gathered to listen."[30] Newman's advertising exhorted the radio amateurs to be up-to-date: "Wireless Concerts: Amateurs! Would you like to hear them? Discard your crystal detectors and get one of our new Bulb Detectors complete for $15.00."[31] A very good receiver with the latest Audion bulb detector cost $20 to $40 in Canada, batteries not included.

By the new year, Orton had installed a radio transmitter at Newman Electric to give his customers something to tune in:

> Thanks to a local store, who have a wireless telephone transmitting set, other wireless enthusiasts are able to enjoy local wireless concerts practically every evening. This transmitting station will now be supplied with the latest and best records every concert night, thus enabling persons who cannot go down town to hear them a chance to visit some friend with a wireless receiving set and be entertained via wireless telephone. The call of this station is 3AO, wave length 200 metres. A 5 watt transmitting bulb is used. These concerts will no doubt please every amateur within range and interest other people into installing a set of their own.[32]

The Newman wireless concerts continued into February.[33] There was a break in late February, and afterward the concerts were no longer attributed to Newman, but were "local amateur radio telephone broadcastings" which "still continue every evening from 6:45 to 7:30 and are meeting with a great reception by every Kingston amateur."[34] It could be that the Newman set really belonged to Orton and that, after a break to permit him to move and rebuild it, he began to do the broadcasting from his attic studio on Wellington Street. Orton was still a very active amateur during this period. On February 18, he was credited by the Standard with establishing a long-distance transmitter that finally succeeded in being heard in Toronto: "This enterprising wireless man will fill a long-felt need of a good relay station in Kingston to correspond with Montreal and Toronto. May his radiations never cease!"[35]

Dick Cohen remembers hearing Orton broadcasting from Newman Electric, and that he played records and announced the Newman Electric Company. Dick says that it was good quality because Orton had battery power - at the store he had access to a lot of 45-volt "B" batteries.[23]

> **Bob Donnelly**: He told me one time that they wired up a hand-cranked generator off an old telephone to a thumb tack in the toilet seat in the back at Newman's. The boss' wife came in instead of the boss, and they got fired.[1]

Before he left Newman Electric to start the Canada Radio Stores, Orton, with the help of another wireless amateur and on behalf of Newman, produced what may have been a first in Canada. In March 1922, he brought off the remote broadcast of a Queen's revue "live" from the stage of the Grand Opera House. Nobody remembers the identity of the second amateur, but he may have been Orton's successor in charge of Newman's radio department, Robert M. Davis (3IL), or his future business partner Gordon A. Thompson (3HF). Newman was then featuring the new Marconi set, "the latest idea in wireless receiving apparatus."[36]

The Queen's College Frolic Broadcast

Student life at Queen's in the 1920's must have been harrowing in the extreme at times. It began with the freshmen, who were painted green, dunked in axle grease and sawdust or were anointed with flour and shellac. Then there were the rushes, when students charged the theatre turnstiles and snake danced down the aisles. At "Students' Night" at the Grand, Denis P. (Dinny) Brannigan, the Manager, probably locked himself in the cellar and wept. The evening always started with the cacaphony of competing faculty yells, augmented by horns and tin pans. Between the acts the occupants of the galleries deluged the lower regions with beans, pennies, eggs, flour, bran, confetti, toilet paper and even live chickens and pigeons.

The "first ever" college theatre night was planned by the Alma Mater Society for February 1922, and each faculty was asked to contribute a skit to the program, proceeds to go to the War Memorial and to the Women's Residence Fund.[37] "The College Frolic of 1922" under the direction of Charles E. Gates, was set for 8.15 pm Tuesday, February

28, one night only, and prices were 50c, 75c and $1.00.[38]

There were ten acts, over 100 participants, and it was a sold-out house and a "Rollicking Success."[39] Some had tried to keep the "Honourable Bolsheviki" out of the Frolic, but they were the scream sensation of the evening with their court scene, "The Kingdom of Upsidownia." The execution of J.B.C. Trottings, "leader of those who say Huzza," drew great cheers from the audience. The Women's Society, Levana, came a close second with their startling and evocative production of <u>Bluebeard</u>, songs composed by Miss Honora Dyde and Miss Georgie Ettinger:

> Margaret Porteous, as Bluebeard's wife, Fatima, gave evidence once more of her versatility as an actress. Her song with the chorus, "I've Got Those Dog-gone Bluebeard Blues," was one of the best of the whole evening, while the ease with which she inspired in the audience a horror of the secret chamber, suggested the training of a professional. Miss (Marion) McArthur gave to the part of Sister Anne a truly oriental interpretation...The peak of perfection was reached when the dancers appeared. One might have thought that they were the leads in "Mecca" specially imported for the evening at a fabulous salary. No greater tribute could have been given them from such an audience than the absolute silence - nay, breathlessness - which pervaded the entire house. Miss (Anita) Martin and Miss (Helen) Tofield syncronized perfectly and their impeccable involutions gave testimony to a very assiduous rehearsal. Miss (Clara) Farrell floated airily before the audience in a remarkably well interpreted incense dance...The entire number was faultless in its poetry.[37]
>
> The identity of "Bluebeard" is shrouded in mystery, but he was indeed a fearsome person and quite able to decapitate any number of wives.[40]

Messrs Lansbury and Lyght, as the Brothers Halbaldi in "A Pair of Drawers," threw about rapid fire campus quips and puns as the former tossed off cartoon caricatures, ending up with a life-size portrait of Principal Taylor. For music there was the large Queen's Orchestra, the Hawaiian String Trio and "Slim" Monture's Melody Minstrels. The delightful Metropolitan Trio of Misses Goldie Bartels, violin, Lois Taylor, piano, and Anna Corrigan, voice, sang and played "Mighty Lak' a Rose" - straightaway unjazzed music.

The Frolic got such good notices and drew so much attention from the general public that it was decided to repeat it on March 13 as a "Citizen's Night," with the patronage of the Rotary and Kiwanis Clubs, for the benefit of the Queen's Athletic Board of Control.[41] The show was trimmed to seven acts and the <u>Queen's Journal</u> "hoped that the students will behave themselves and not throw rice, cabbages and live chickens at the stage, since the good people of Kingston will be in attendance."

The day of the repeat performance of the Frolic, both Kingston papers announced that the production would be broadcast. Apparently Orton brought a receiver and a loud speaker with him when he set up on the evening before the performance:

> The first show in Kingston and possibly the first in the whole of Canada to be broadcasted by radio telephone will be "The College Frolic for 1922", which is being presented this evening in the Grand Opera House. A wireless telephone transmitting set has been installed by two wireless men of the city and a two-wire aerial, fan type, stretching from the tower of the Kingston Auto Sales on Brock Street to the roof of the Grand Opera House, has been erected. Kingston amateurs who wish to listen in, will be able to hear all that goes on just as good as the audience.
>
> Music, speeches, etc., from Pittsburgh (KDKA) and Newark (WJZ), were received on Sunday evening for the benefit of some of the Grand Opera House staff and the Rex (Snelgrove Repertory Theatre) Company. William Jennings Bryan was in the pulpit of the Pittsburgh, Pa., (First Episcopal) church, and his lecture was plainly heard by all who were there, although his speech was said to be a "dry one."[42]

The broadcast of the three-hour show from the stage of the Grand was a success. Amateur wireless stations all over the district received it clearly and distinctly. No details of the broadcast itself survive, except a note that between the second and third acts it was announced in the theatre that the show was being transmitted by wireless telephone, and "the attention of the audience was drawn to the wireless receiver (sic) on the stage."[40] "Considering the short time that was given to install the radiotelephone set at the Grand, the results last evening were excellent and it is hoped to improve them much more."[43]

Toronto wireless men were just then arranging to broadcast church services, as was already being done in the United States.[43] The British Whig editorial page glowed:

> For the first time in the history of Kingston, the music, the songs and the dialogue of an entertainment were broadcasted by radio to many city residences and to homes in Napanee, Gananoque and other places on Monday evening...Few people will fully realize the far-reaching importance of this innovation. It will perhaps be only a matter of a short time before one can remain comfortably at home and listen to a concert, a sermon, a lecture, a band concert, etc. If the programme, at any point, fails to interest, one can "ring off" and await the next turn..."The night shall be filled with music", is literally true for those who "listen in."[44]

Orton Donnelly's name did not appear in print in connection with the broadcast from the Frolic, but Newman Electric assumed the credit:

> A few more successful broadcasting of shows from the local theatres will add to the family of wireless fans in Kingston. Amateurs report the broadcasting of Queen's show last Monday evening as very good. There will be more under the same direction if all is well. These are being given by a local electrical dealer and two wireless amateurs.[45]

> Through the courtesy of the Newman Electric Company, arrangements have been made to broadcast on the Radio Telephone, concerts and musical plays which are to be given in the city. It has been arranged to have a set installed in the Grand Opera House to broadcast the performance when Fiske O'Hara appears at the Grand, and other musical plays will also be broadcasted. The Newman Electric Company experimented with the radio at the Queen's play at the Grand and met with a great success.[46]

Fiske O'Hara, everybody's favourite actor-singer, scored a triumph in Anna Nichols' fascinating romantic story and comedy of class, The Happy Cavalier, on March 22. There were no newspaper reports of a radio telephone broadcast of the show and, in fact, Newman Electric was next reported putting on a wireless concert at the YMCA in the hope of promoting installation of a permanent set there.[47] On March 25, the Daily Standard assured Kingstonians that the local radiotelephone transmitting set would soon be in operation again.[48] The implication is that Orton Donnelly was re-assembling his transmitter in his attic and would continue his broadcasts from there. Whether there was a falling out at this point is not known. Within a couple of months, however, Orton and two partners opened the Canada Radio Stores in competition with Newman Electric's radio business.

Unfortunately, Orton remembered very little of his broadcast of the Frolic. His brother Bob, however, recalled his role in the broadcast in great detail:

> Bob Donnelly: He used his first tube transmitter at the Grand Opera House. I would imagine that he built it around '20, '21, in that area. The transmitter itself, as I remember it, wasn't exactly portable, but it wasn't too hard to move. There used to be a chap by the name of (John) Sinnott, who handled all the baggage coming in to the two inner railway stations, at the foot of Johnson Street and by the City Hall. He had an old horse and cart and I recall that Ort got his services and they loaded this thing - I helped them carry all these darn pieces down from the attic and put them on the wagon - this is how I remember it so well. Up to Montreal Street and in the stage door, and Ort assembled it. He used the same transformer that he'd got off the boat

- we had to lug that thing. That thing weighed a ton. He'd had it rewired for household current, because the voltages weren't the same. I start talking about it and I laugh because I think about the first O.B. in Kingston - the first out-board broadcast - and then I think about this old horse and wagon.

We came up to the stage from the stage door, on the loading side of the stage. We didn't cross the stage to put the stuff in, so it would be just inside. There was a little office that was back there, almost on the stage and a ladder going up to the overhead ropes and sandbags and the scenery, and the set was right in that corner. It'd be right stage, on the Brock Street side. I didn't help set it up. I just carried it in and hung around for a while.

I don't know what the play was - I have no memory for that. But they had erected an antenna over to the old church that was on the corner of Montreal and Brock Street. It was later a garage (now a bowling alley), but it had been a church. They'd slapped up an antenna on there to a piece of 2x4 on the Opera House, and dropped the antenna down to the stage.

I'd only have to speculate about the microphone, but I imagine it'd be the standard double-button carbon mike, and they were Northern Electric type transmitters or Western Electric. It would be a rather wide range. The mouthpiece had perforations in the bottom, knocked out or drilled out, and then you just fitted on the hard rubber mouthpiece that screwed into it, like a little megaphone. Actually, I heard several broadcasts with these things. They worked![1]

The rear loading door of the Grand Theatre, 1991, with the chimney and tower of the former Kingston Auto Sales, Brock Street, on the right. It was here that Orton Donnelly set up a two-wire fan-type aerial for his broadcast of the Queen's College Frolic on March 13, 1922.
-photograph by Arthur Zimmerman

Orton Donnelly: There is one incident that I remember quite well in the rather hectic broadcast of the "Queen's Frolic" on the stage of the Grand Theatre. The scene was a lonely cabin up in the north country, and during the scene the hero was wrapped up like an Egyptian mummy - he wasn't gagged, which I think was a mistake. In the stage was a trap door, used mainly in magic acts for disappearing. They took hold of the bound up hero and lowered him through the trap door to the cellar below. Pre-arranged there was a soapbox below, and one or two members of the College were to catch this chap as he came sliding down. However, they had taken time off to go to the corner to have a smoke, and the chap came sliding down through the opening and hit the soapbox with quite a crash. Now the villain in the piece took the cue up promptly and said, "Ah, hear him splash!" Incidentally there were a few oaths that came from down in the cellar.[2]

More Recollections of the College Frolic

Three young women who appeared in the College Frolic of 1922 remembered the show quite clearly after 60 years, Goldie (Bartels) Geiger, Georgina (Ettinger) Logan and Clara (Farrell) Brooke, but none recalled that the repeat had been broadcast.

Goldie Geiger (B.A. '22) was the violinist-leader of the five-piece pit orchestra at the Grand in 1922, playing for the travelling plays, revues, films and for the Rex on weekends.

<u>Goldie Geiger</u>: The Grand was very well patronized. The Royal Canadian Horse Artillery always took the first two or three rows, and I always **knew** it was the Horse Artillery that were in there. I had a piano, cornet, drums and a clarinet. Mignon Telgmann was in one of the movie houses - she and a pianist did the music - and she told me how to do things when I got started. The shows didn't have much music, really. We always had to play before people came in, and then we always had to play a chaser, and between acts in the Rex Snelgrove thing. So I had to have an awful lot of music. Oh, I had to provide the music!

I think the Grand orchestra did the Frolic. I can't remember having anybody else there. I did play a violin solo as a separate act - I was one of the "stars" of the thing. It was a great excitement for me, because they got up and yelled, "More! More! More!" Science '23 were wild people, and they had a club called the "Bolsheviki," and they did a hospital skit that was pretty vulgar. They didn't leave anything to the imagination. The Medics were even worse. A fellow called D.O. Robinson did a black-face thing - "Nobody." It was a good show -it was really great!

I got the music all settled for <u>Bluebeard</u>. I didn't compose it - we used the music from an opera thing that was very, very good, and it just fitted it. What I remember most was the sight of these girls with their faces all white, and all you could see was the supposedly cut-off heads of all the wives staring out at the audience. And Sister Anne being terrified. It was quite exciting and lots of fun. And the students were terrible. They marched up and down the aisles, and threw toilet paper down from the balcony. They did everything crazy - these awful Science men. Poor old Dinny Brannigan. He used to get so mad, but couldn't do very much.

I have no recollection of that repeat. Isn't that funny? I remember the great first performance because I had a lot to do with it, and I did the great solo and got all the applause and the cheers. I felt like a great artist, you know.[49]

<u>Georgina Logan</u>: This is the first I'd ever heard of it having been broadcast. "Red" McKelvey and two or three of his friends were very original, and they did very clever skits for the Bolsheviki side of that thing. I remember they had a skit where it was supposed to be Grecian, and one of the Greeks died so they had him on a stretcher and took him up to heaven. And he went up and he fell out...fell from grace.

I played the part of Bluebeard, and we had to find a costume, so I got some rope and dyed the rope blue and made a beard and a moustache, and then we had to make oriental trousers, an old oriental kimona and a thing around my head. The whole performance, in our eyes, was ridiculous, and we had a lot of fun. It was a composite thing. No person did any one thing - everybody got together and put in out of their ideas. To give it a little oriental flavour, Honora decided we'd say "one beer sot" instead of a quarter of an hour. It was quite a long time ago, and most of us didn't go around in the streets in pants, and I was reluctant to go on the stage wearing trousers. I said I'd play Bluebeard if they'd leave my name off the program and nobody knew who it was. So they did, and I discovered I'd created more furore for not having my name on the program than if I'd quietly left it on and gone about it.[50]

The Women's Editor of <u>The British Whig</u> unbearded "the murderous tyrant Bluebeard...a dramatic figure...in his oriental draperies and astonishingly long blue beard" a few days after the repeat performance.[51]

Receiving "Those Radiophones" in Kingston in 1922

The Department of the Naval Service granted Canadian amateurs the use of all wave lengths below 200 metres for the winter of 1921-22, until April 15, when the ship stations reopened. By late February, the Naval Service had recognized the superior tunability of the continuous wave and radiotelephone sets and temporarily granted them the use of 200 metres, while keeping spark down between 50 and 150 metres. Because of the proximity to Barriefield, this change did not apply to Kingston amateurs, who were stuck with 50 metres.

KDKA, the Westinghouse station in Pittsburgh, was the first public broadcasting service received in Kingston, and the first to broadcast church services. Many Kingston amateurs heard the initial one, from the Calvary Episcopal Church on January 2, 1921.[52] There were then only a few radiophone transmitting sets in Canada. The Marconi Company, Montreal and Toronto, broadcast concerts every Tuesday and Friday evenings at eight o'clock on 1,200 metres. There was also the Canadian Independent Telephone Company, Toronto, which aired a concert every Monday and Thursday evening, at 450 metres. Many citizens had heard the newfangled wireless at Newman's and were "now satisfied that talking without wires is a reality, not a dream."[52] That year, for the first time, Christmas stories were broadcast especially for children on Christmas Day, over KDKA. The whole family and their friends could listen in at once if they could afford one of the new loud talkers or loud speakers. The best one was the Magnavox.

By early 1922, other Westinghouse stations were on the air: KYW, Chicago; WBY, Roselle Park, NJ; WJZ, Newark and WBZ, Springfield, Mass., all at 360 metres. The Department of the Naval Service, Ottawa, wrote all Canadian radio organizations

...outlining suggestions for the improvement of conditions in the ether...Among the suggestions is that no amateur be allowed to transmit during radio telephone concerts, between certain hours in the evenings. As the radio telephone concerts are increasing to a vast number now and free musical services are rendered to the receiving amateurs, the advisability of this rule is easily seen. In Kingston quite a lot of interference is noticed during these concerts by some unlicenced young amateurs who, not knowing the code, do not send anything but dots and dashes. The note of the average spark coil is quite enough interference, but when the transmitter does not know the code it is far worse. If this suggestion becomes law, it will be a fine thing for radio telephone reception, especially in Kingston.[53]

These sorts of problems, caused by quite young boys hacking away on home-made spark coil senders, are what likely prompted the formation of the Kingston Radio Association in the spring of 1921, to try to teach them the code, to regulate their transmissions and to bring some kind of order to wireless traffic. While the Radio Association disintegrated after a few months, the problems continued. By late 1921 there were 12 licenced transmitters in Kingston and many more unlicenced. Only two had a licence for reception.[52]

While a local radiotelephone concert is in progress, there is nothing more annoying to hear than interruptions by spark coils. The local radiotelephone transmitting set is powerful enough to be heard by even the most unsensitive of crystal tuners so why not have from 7 till 10, free hours for those who really do want to hear these concerts? These unnecessary interruptions are generally caused by some unlicenced amateur with false call letters, although the familiar 3's are even heard at times interfering. These amateurs, if they need code practice, should take off their aerial lead and leave the air free for the concerts and those who want to hear them.

There are...the usual few who do not want to hear the music and seemingly delight in interfering with their spark coils. The majority of these are not licenced by the Government and wish to keep their identity a mystery. The transmissions are nothing more than a lot of useless dashes and dots which go to show that the senders do not know the Continental Code. One young boy, although repeatedly warned, is still continuing to interfere every night. Every amateur who receives radiotelephone music in Kingston knows who this is and if this useless "hamming" does not stop, it will surely come to the ears of the Licence Inspector, who is reported at the present time to be in Kingston. For the benefit of the interferers, it might be said that the majority of real amateurs are "fed up" with the Q.R.M. and the boys who cause this interference are advised not to jam the air when concerts are going on. This can be done all day, leaving 7 to 10 every evening free for the reception of concerts. Another practice in the city is for a licenced amateur, to avoid being recognized by his call, to sign off with an X call. It is not playing the game with the rest of the fellows to send out this kind of call, especially during radiotelephone broadcasting hours.[34]

Orton Donnelly's Radio Room

<u>Bob Donnelly</u>: I can't remember a time on Wellington Street when there wasn't a radio in it. There was always the radio room, and when I got heavy enough to carry stuff, I was always the guy that got dragooned into carrying it up there. The top floor was two rooms, and very large - approximately half the house. The radio room was to the rear, towards the Convent.

His radio room was rather long - oblong shape - and under the eave there was a storage area, where the roof slanted down. A carpenter built Orton's bench along there, and he had his equipment on there. He got the high voltage for the transmitter through the Hydro lines, from the mains, 1,200 to 2,200 volts through a chemical rectifier, depending on where he tapped the transformer. Everything else was battery. Under the bench were batteries, the Tungar-Apwaite battery charger, and behind the partition were the slop-jar chemical rectifiers. The transformer was in there, and I believe he had a great big reactor - a great big son-of-a-gun - looked like a transformer. It was a reactor choke for the glass plate type capacitors. I took them apart to get the glass out of them. The special wiring came through the wall, through these custard cup insulators, to feed the finals on the transmitter.

The chemical rectifier was in a drawer that he swiped from a dresser, and in it he had these imperial pint mason jars. There'd be maybe forty or fifty all lined up one alongside the other. They were filled with a solution of borax...boric acid in water. In the first jar there'd be an aluminum electrode, and then there'd be a piece of lead and a piece of aluminum bolted together, shaped like a horse-shoe, so that it would go into two jars together. Thus, you had aluminum and lead in each jar. You'd get hydrogen peroxide actually, and this was a rectifier for high voltage rectification. The slop jars, they were called, were all in series. They were connected up to either 1,200, 1,500 or 2,200 volts from the big transformer.

And there were two types of condensers. One was just heavy waxed paper wrapped around sheets of tin foil wrapped around plates of glass. And they were connected up in piles. They were filter condensers. And later on there were some for lower voltages - the same thing, glass and tin foil and they were full of molten sulphur let cool. They were a higher dielectric type, but they were better for lower voltages.

He built most of his first tube transmitter himself, and it was laid out breadboard style. My uncle was a wood-worker and every once in a while he'd build a nice hardwood frame with four 2x2 posts up it, shellacked, and they put a piece of hard rubber on the front. Or bakelite. And switches and an ammeter on the front to measure aerial current. This framework held the power unit of the transmitter. The transmitter itself was a MOPA rig: Master Oscillator Power Amplifier. He had an old German tube, a Mueller - a 50-watter which he got in Schenectady - a Florence flask with the piece on the bottom, where you have a heavily plated cap on top for the plate current and the connections went through the glass and directly into the tube. And the oscillator - I forget whether it was like the old UX models with a federal base - the oscillator was quite a little ways beside the big tube. The tubes were drawn glass, with tips like the old tungsten light bulbs. And the modulator fed in from outside. The tank coils on it were wound with quarter-inch copper tubing out of an old automobile. The tank coils sat on glass rods, the sort Woolworth's used to sell for bath towel racks. The antenna coil was coupled by sliding it up and down the glass rods - a simple arrangement. And the modulator was a sort of breadboard panel sitting beside her.

When that big Mueller tube lit up you didn't need a light in the room. Oh, it was BRIGHT! It had a big rheostat on it, a real heavy thing to control the voltage. The heaters on it - the filaments at that time were battery-driven.

The antenna fed out through the attic window through two custard cups and holes cut in the window. The custard cups had a little gasket between them and they were held together with a piece of threaded brass rod. This gave him insulation for the winter. Then, the antenna was on the roof of the house and it ran over to the Sisters of Notre Dame Convent, which was where the Public Library is now. I imagine the antenna was maybe fifty feet high by the time it hit the Convent. And the antenna was tuned - in those days they put a 6-volt car headlight in the middle of the antenna, right in the null. If you were tuned up, you were in the null, and the light would light. There was a piece in the newspaper because a bunch of people gathered on Johnson Street one night and they thought they were seeing a comet or something. Ort was transmitting, and this light was blinking - the antenna wire was invisible at night, and they were sitting there watching this thing. The guy from the paper came up and they finally discovered it was Ort. [This may have been the "Hertz" antenna of late 1925, mentioned below.]

He had several receivers. The one I remember was a regenerative receiver, a blooper type, battery driven on the heaters, "B" batteries driving the plate supplies. And big, monster honeycomb coils - one fixed coil in the centre and two tuners on hinges that you could swing back and forth to increase or decrease the coupling. I believe it had a real big old vernier knob on the front of it and a little hole cut through the face plate so you could see if the tubes were lighting - to estimate your heater current. How bright your tube was, that's how much current you were using. There weren't all that many meters in those days. A blooper was a regenerative circuit where you took the RF off the air and fed back a portion through it, and it'd break into oscillations. Very, very sensitive, but they'd also raise hell with anybody else in the area because they'd send out a bunch of squeals. If they were mistuned they were a little transmitter, so they weren't too darned popular.

You could use a regenerative receiver circuit or an ordinary oscillator and a power amplifier, and modulate the antenna rather than feeding it in through an ordinary demodulator system. This was primitive. In fact, the first regenerative sets some of the kids used to do this with a microphone, and broadcast all over town before they got kicked off. It was very common at one time. They could tune very roughly. It would just splash right across the whole works. Quality was poor. Music? Music was terrible! You'd have a little oscillation out through the music on the higher and lower notes. They wouldn't sound too much like music - more or less like singing through a barrel.

He used to go up there right after supper, maybe about seven o'clock, play records for 15, 20 minutes, half an hour, and ask for QSL cards, as a check-up on his signal. He had the one thing that he used to say all the time. It was, whether you were on the same street or around the corner or miles away, please send me a letter and let me know that you're receiving. He received letters - I think he got one letter from Vancouver one time, on a night broadcast. And he'd get letters all the way around Lake Ontario - Gananoque, Clayton, Watertown - so he was putting out a fair signal.

There was always somebody going up there to the attic to broadcast. There was no piano or accompaniment up there, but they'd be talking. Always had somebody sitting there. In fact, some nights it'd be crowded up there. They sat around and talked radio and then go on the air and play a few records, tell a few jokes or ask for somebody to send them QSL cards. Most of them were fellows around town interested in radio. C.W. Lindsay was a store on the north side of Princess Street above Wellington that sold phonographs, and Ort used to get records on loan from them. They'd play them and announce who made the record and so on. He'd play them a couple of times and take them back. I used to carry some records over from there, and take the old ones back all wrapped and packed in a parcel. The old record player was a hand-powered outfit. Weighed about 90 pounds with a

Rear of 118 Wellington Street facing the former convent. Orton's radio room was in the attic. The Irishmen's Hall is the large building to the right.
-photograph by Arthur Zimmerman

diaphragm-type pick up on it. They used to put the mike in front of the horn and let the records go. We had two telephones, an extension and a main phone in the house. None of them ever had a mouthpiece on them. He was always stealing them to make those little megaphones out of them. One time he had a set they'd built hooked up downstairs in the parlour, and I heard one of his broadcasts. Of course, we didn't need an aerial or anything because it was just a couple of floors below. And I was listening to two of them playing out and picking up some stuff around on the mike. You could hear very plainly. The carbon mike was real noisy. You had to graduate the size of this megaphone thing or you'd get too much background noise.[1]

Douglas Donnelly, nineteen years younger than Orton, dates his first memories from around 1927:

I remember vaguely the transmitter as a sort of square metal frame - four posts of pieces of angled metal at the corners, maybe about 18-20 inches high and maybe about 14 inches square. I remember seeing this enormous tube in the centre of the frame, maybe 4 inches in diameter and about 10 inches in height. Orton had all of his equipment set up in the attic, and there were very definite instructions to the rest of us to keep away. He used to have little devices for keeping us at our distance, like a copper plate wired up at the inside of the door that if you stepped on it you'd get a little charge. Enough to attract your attention and let you know that you were where you weren't supposed to be.

That attic room in the back part was big enough. I think it was a good 15 feet in one direction and at least that much or more in the other. With a peaked roof, so you'd have to stoop if you got over too far to one wall. It would hold ten people - no problem.[5]

The Canada Radio Stores

The announcement of the opening of the Canada Radio Stores appeared in the <u>Daily Standard</u> on June 24, 1922: "Kingston's first and only exclusively radio supply house at 269 1/2 Princess Street, under the management of Merle E. Ward, Gordon A. Thompson and Orton H. Donnelly."[54] They offered "complete, ready to use sets installed by thoroughly efficient

The first advertisement for the Canada Radio Stores, announcing the opening, appeared in the Kingston Daily Standard on Saturday, June 24, 1922.
-photographed by Arthur Zimmerman

men," and their motto was "Right Goods at Right Prices." They offered only the best makes of commercial receivers, kits for the adventurous, high grade parts and helpful advice from experienced radio men: "If you have been 'stung' on radio elsewhere, come to us and be cured." The partners must have been forward looking and organized too, for right from the beginning the store had its own radio licence, 3AAY.[55]

Radio was the ultimate in high technology in 1922, so the partners made every effort to be friendly and helpful to the would-be radio "bugs" who were brave enough to step in. Their earliest advertisements solicited enquiries, invited people in at any time for an explanation of how to build their own crystal or bulb set, for advice on operation and handling, invited them to come in for the posting of the baseball scores every evening at 7.30, or to listen to the latest news, or concerts such as the WGY production of HMS Pinafore on October 3, 1922.[56] A few days later, the Stores was the first to publicly congratulate Queen's on the birth of 9BT.[57] The sidewalk in front of 269 1/2 Princess became a popular spot, a place where things were happening. A few months later the Queen's Journal wondered, "Who was the goof who took his lady friend down to hear the concert outside the radio store when it was 20 below zero?"[58]

> Jack M. Campbell, Jr: I used to go to Orton's store. The Canada Radio Stores was a very small and narrow place, just about where the House of Sounds is now, and Price's Dairy was just above it on Princess Street. It was run by Gordon Thompson, who was the instigator of the set-up, Ort and another chap, an elderly chap. Sort of a senior partner. When I say old, he was old to me at the time. Merle Ward. He was a big man and had trouble geting in and out of the narrow store. Any amateurs or anybody interested in radio used to go over there to get their bits and pieces.[59]

Kingston, and the rest of North America, went radio nuts that summer. The demand for sets exceeded the supply by three-fold. In June The British Whig began a "dandy" new serial called "On Wings of Wireless,"[60] and in August the Allen Theatre featured the first all-radio film: "How it Works - Radio - Revealed Pictorially."[61] There were radio marriages, radio advertisements, radio sermons, photographs by radio, artillery fire directed by radio, radio in autos, on bicycles and on horseback, political speeches, circus broadcasts, Amundsen promising broadcasts from the North Pole, baseball, football and prize fight broadcasts. Wires were being rigged up at a frantic pace wherever there was a tree or a chimney, and Orton Donnelly continued his adventures in broadcasting. There were the skeptics too: "I do not think the radio so fine; I get more news on our old party line."[62]

Early Canada Radio Stores advertisements gave out an invitation to drop in any night for baseball scores by radio, and they brought the first World Series broadcasts to the Kingston public. On October 5, 1922, they attracted hundreds of people to the front of the store to follow their posted bulletins of the first game of the Giants-Yankees series.

The lucky ones inside the store got to wear one of six sets of receivers and could hear the play-by-play on WGY, Schenectady. Listeners were flabbergasted by the immediacy of the new medium. The Whig reporter

> could hear the voice as that of the umpire calling 'Ball'.[63]... the announcer was broadcasting the game from the press box at the Polo Grounds, and as every play was made he announced it in the broad-caster. At one time the announcer called out over the phone, "Ruth just hit a two-bagger to centre," and those in charge of the phone at the Canada Radio Stores prepared to put a bulletin in the window to this effect. Scarcely had they set down their receiving sets, when the announcer called out, "Stengel got under the ball." This only went to show the rapidity with which the account of the game was received in Kingston, and baseball fans at the Canada Radio Stores simply had to sit back, with the receiving sets to their ears and imagine they were in the Polo Grounds."[64]

It is quite likely that the source was WJZ, Newark, at 360 metres, which had secured the silence of the other New York area stations for the duration. Grantland Rice, of the New York Tribune, called the game into a telephone at the Polo Grounds, and Tommy Cowan repeated his words into the microphone in the WJZ studio.[65] CFCA, the Toronto Star Station, picked up the series from the second game.

Bob Donnelly: We had quite a few little ventures from the Store, mostly in the form of advertising, and we were given a fair amount of publicity. We used to broadcast the World Series, and we used to block traffic on Princess Street. We'd put out a large Magnavox horn loud speaker, and we had quite a crowd. The police co-operated pretty good. I know we did it several years.[1]

Orton Donnelly: After graduation from high school, I took an interest in radio and decided for the time being to discontinue my formal education, and through the interest of another chap, Gordon Thompson, we opened the Canada Radio Stores. We put a large structure on top of the building and attached an antenna, and got busy in the business of selling radio parts, helping out a dedicated group of people around town that were interested in radios - making radios...We obtained an amateur licence - 3AAY. We actually never used the station for broadcasting to any point because we were not licenced at the time for public broadcasting. We used it on an experimental basis to try out equipment which we had on hand, for receiving, to try out the sensitivity of different products which we sold. However, through my own licence at the time, 3HE, I was able to use the station for transmitting and a certain amount of receiving. I had a very strong leaning toward amateur radio. In the early days we did a certain amount of advertising and we really had more than we could handle, and we were busy. We had a couple of chaps winding couplers. Actually we had a small manufacturing plant at the back of the shop, and what we weren't able to get ahold of we more or less assembled ourselves. As a matter of fact we built and sold our own neutrodyne sets with batteries - come to think of it, we didn't have a licence for manufacturing. Things seemed to be more or less wide open then. There wasn't the restrictions that we have today. Radio was more or less wide open and something new. In making up these sets I remember myself and a couple of employees sitting at the back of the store winding numerous coils to be used as inductances, and also trying out different crystals which were used as a detector or rectifier of radio signals - galena, chalcopyrites, silcon and carborundum. Silicon happened to be the most sensitive, and it was much easier to find a point.

We went along for a time with crystal radios. Then one day a salesman arrived for a new radio maufacturing plant in Toronto, called Rogers Radio. I had previously met the owner of the Rogers Radio plant - he was also a ham - and we had quite a long conversation, which wound up in us taking a contract for handling Rogers Radio. Later we handled Fada Radio and Zenith Radio, quite a nice set and well designed and built. I also happened to contact their radio station in Chicago - 9ZN - they were one of the top amateur stations. I remember that one chap, an old farmer, came in and said to me, "Can I listen on the headphones?" I let him listen for an hour to KDKA. He gets up and hands me a dollar and walks out. I didn't want the dollar, but it was there and he was out the door before I could say anything. It just shows you the novelty of it.[2]

Bob Donnelly: Merle Ward was in sales, and Gord and Ort were running it. Merle was sort of the head salesman - the outside man. He'd be on the counter, but he'd also be out selling and demonstrating. These were sold more or less like vacuum cleaners are now. Put one in the customer's home and hook it up and let him listen. If he liked it you would try to sell it to him. Gord, of course, was on the counter most of the time and on sales, and Ort was working in the back. They had some other chaps working there making up pieces. In those days they used to have home-built sets. They'd build them themselves and sell them. I believe they sold Fada for a while and then the Rogers batteryless, about 1926-27.[1]

Arthur Wehman: I did know Gord Thompson very personal. He was a few years older than I, and my older brother and he were very good friends. So Gord used to come up to the house quite often. Gord was avery qiet chap, but a good business man, and very well liked.

What was his call? 3HF? I've heard him. I've heard him on the air. Why should I remember all this stuff, 70 - 75 years ago? Well, that's amazing!

The Jack Elder Broadcast

Jack F. Elder ran a cigar store and also sold books and musical instruments at 260 Princess Street, one door below the old Strand Theatre. Around 1926 he ran the Elder Aerated Water Works and manufactured a pure fruit drink called "Smile." In the 1930s Elder was the local Pepsi-Cola bottler. Bob Donnelly recalled that his whole family knew

Jack, a Harry Lauder type of singer with a Scottish repertoire.[1] In fact, he was known in 1922 as "The Harry Lauder of Kingston." Orton put him on the air on November 22, 1922, from the Canada Radio Stores.

> The distinction of being the first Kingstonian to have his voice broadcasted goes to Jack Elder, the well-known local tobacconist, who is better known outside the city and in some circles within it as a very able singer and Scottish entertainer. On Wednesday evening the Canada Radio Stores tried out their broadcasting apparatus, much to the delight of many amateurs in the district, and the first voice to carry over the air was that of Mr. Elder, singing some of his favourite Scotch songs.[66]

Nothing is known about the circumstances of that broadcast. Jack Elder must have been a decent performer, because in 1925 he was invited by the broadcasting department of the Canadian National Railway to sing over their station CNRO in Ottawa.[67] He was still an active entertainer in Kingston in 1951.

The Dumbells Broadcasts

The Dumbells, the Canadian Third Division Concert Party, was born in the mud of World War I near Vimy Ridge in France. The troupe was led by Honorary Captain Merton W. Plunkett, and was named after the divisional insignia. The Dumbells played Kingston at least once a year after the War, Biff, Bing, Bang in 1920, The Maple Leaf Overseas Revue Camouflage in 1921, The Dumbell Revue of 1922 and their Fifth Annual Canadian Revue, Cheerio, in 1923. They continued to stage shows in Canada and abroad until 1929.

Cheerio, starring Al Plunkett and Ross Hamilton, "the wonderful impersonator known as 'Marjorie'," played to packed houses at the Grand Opera House for three nights, beginning on Thursday, November 29, 1923. The hits of that show included "Winter Will Come," "Down By the Old Apple Tree," "Swinging Down the Lane" whistled by Jack Lougheed, "The Adventures of Barney Google," and "Old Granny O'Mine" sung by Al Plunkett assisted by T.J. Lilly as Granny and brother Morley Plunkett as the soldier boy. Ross Hamilton, in several changes of gown, gave "The Life of a Rose," "What I Think of You" and "Building a Bungalow." Pat Rafferty sang "Tony, the Gay Mountaineer" and Stan Bennett sang "Archibald." They also included a near-opera called "Disorderly Room" and the tabloid musical skit "O'Brien Entertains" closed the show. Fred Treneer of Kingston had recently left Treneer's Orchestra to join the troupe, and he played the saxophone, sang in the chorus and accompanied "Marjorie" at the piano.[68]

Orton Donnelly remembered broadcasting The Dumbells once, from the Young Irishmen's Christian Benevolent Association Hall at 116 Wellington Street. In fact, he aired them two evenings in a row. On Friday, November 30, the troupe came up to his attic studio at 118

Wellington to broadcast over 3HE, and then on Saturday night they put on a public performance at the Irishmen's Hall, which was also sent out over 3HE.

The British Whig said:

> The Dumbells, who played at the Grand this past week, are certainly bitten by the radio bug very badly. Every evening after the show was over they listened in to Chicago, Davenport, Fort Worth, Texas and other points, and they became so enthused Friday night that they staged a little concert of their own which was put on the air by O.H. Donnelly, of the Canada Radio Stores, who is the owner of one of the finest transmitting sets in the province. So many good reports of this broadcasting was heard that the company got together on Saturday and from 6:10 till shortly after 7 o'clock, the air around Kingston and many miles further was sweet with "Cheerio", "Granny", "I Passed By Your Window" and other hits of the Dumbell show. All the company took part and the whistling solos by Jack Lougheed and the orchestral selections by the company's players were much appreciated.
>
> The second broadcasting took place from the Young Irishmen's Hall on Wellington Street and this was connected to Mr. Donnelly's transmitting outfit. The hall was packed and this concert, although short, was much appreciated by the many who were unable to buy tickets to see the whole show. The Dumbells are to be congratulated on being the first company to broadcast in Kingston by radio.
>
> One report from Leighton, Pa., says that this broadcast came in very loud and clear. No doubt other equally as far distant points will be heard from.[69]

Recollections of the Second Dumbells Broadcast

Orton Donnelly: Another one of our illegal adventures occurred - I happened to live in Wellington Street next to a fairly large wooden building known as the Young Irishmen's Hall. It was an association for the promotion of young Irishmen - more or less a club. They held dances to raise finances, and I was right next door, and my radio station was on the top floor within spitting distance of the Irishmen's Hall. My next door neighbour, Mr Fowlie at 120, had friends from up around Orillia, where some of the cast of The Dumbells originated. Several members happened to drop in on the Fowlies - they had two sons attending Queen's at the time, and we were very good friends - and they acquainted me with the members of The Dumbells. The boys knew I had a station next door and the idea came up that possibly we could arrange a broadcast from the Irishmen's Hall. The officers there were quite agreeable, and let us use the hall as a space for broadcasting. It was just a matter of running microphone wires out my window and across about fifty feet over to the building. On an off-night from their show at the Grand Theatre we arranged a sort of program, that ran for about an hour and a half, probably two hours, and we put the show on ham radio, which I think probably was a first in Kingston, even if it was illegal. It didn't do any harm to the gate receipts at the Grand Opera House.[2]

Freeman Waugh: When I was in my early teens, The Dumbells came to Kingston. The backbone of the troupe were brothers by the name of Plunkett, who were brought up in the Orillia-Barrie area where my father spent his early youth. When they got to town, the Plunketts met Orton Donnelly, a young scientifically inclined man who had started up a small amateur radio station. The Plunketts asked him if he could put on a radio show to help them advertise their stay at the Grand Opera House. It was arranged that the show would be put together at the Irishmen's Hall on Wellington Street, right next door to our house. Orton Donnelly's house was immediately on the other side of it. One of the snags in putting the show together was the fact that the piano in the Irishmen's Hall hadn't been tuned since 1898, and sounded like it. Merton Plunkett, who was the oldest and the boss of the group, saw my father's dentist sign next door to the hall at 116 Wellington, and the name rang a bell. He went to see whether it was the man he thought it was, and sure enough, they were old friends from away back. Mr Plunkett explained to my father the dilemna they were in because they didn't have a piano, so father said, "Borrow ours." Without wasting any time at all, Mr Plunkett assembled the entire cast of The Dumbells and they proceeded to take the piano from our place to the Irishmen's Hall. This was no easy feat, because it was a big piano and there were seven steps from the front porch to the sidewalk, and a good many more steps to get the piano upstairs in the Irishmen's Hall to the place where the show was to be put on. One of the smallest people in the troupe was a man named Pat Rafferty, who immediately grabbed the piano stool as his contribution to the combined effort. They got the piano down the stairs and up the stairs, and the show went on.

I remember when I followed the piano up the stairs I noticed that the only microphone visible was hanging from the ceiling, with Orton Donnelly fussing about it. I have always thought, after watching that performance, that it was the greatest show that they ever did in their lives. It was very funny. Some of the songs that they sang during the show, or during their shows in Kingston, were such numbers as "Oh! It's a Lovely War," "Dear Old Granny Mine," "I Know Where the Flies Go in the Wintertime" and another one that was, I would suppose, considered rather risque in those days, about man's infidelity to woman. I can't remember the title of it, but the last few lines went something like this: "Though your wife's name is Maud/ And you love her a heap/ There'll come a time you'll call her/ Suzie in your sleep."[70]

Bob Donnelly: Plunkett and The Dumbells. Yes, I recall that very well. I did see a part of The Dumbells program, but just a very small part. Then I got thrown out. I was too small to really know what happened - in the six-seven-eight range. I got thrown out because I was the kid brother getting in the way. "You get the hell out of here" sort of thing, if you got spotted. They were using a carbon mike, and they had about fifty, sixty feet of mike wire running to the transmitter next door in the house. The mike started getting awful damn hot - they put two batteries on it and it was really sizzling. So I got called back. I had to dip some towels in cold water, wring them out and carry them over for them to wrap the microphone and keep it from burning up. It was an old Northern Electric telephone transmitter (microphone), like an old candlestick telephone - pedestal type, with the hook missing - except that this one had been nickel plated.

In The Dumbells there was a chap by the name of Marty Gorman that had been with them overseas, and he lived next door to us. There were several other chaps around Kingston that had been with them.[1]

<u>Ian Baines</u>: In those days, single- or double-button carbon micro-phones were connected into the cathode lead of the first amplifier, which meant they were hot. Those microphones could have 100 volts on them. More than one announcer put his lips too close, and had the time of his life. There was a lot of power dissipation in those microphones.[71]

<u>Ed Donnelly</u>: From the Irishmen's Hall they broadcast Jack Lougheed, artistic whistler, and the Plunketts. After The Dumbells broadcast Orton said, "We just got away with murder." The Radio Inspector (Bob Davis) got after him for the broadcast, but did not carry it far.[72]

Rex Snelgrove ran a repertory company, the Rex Theatre Company, which rotated performances through Kingston, Brockville and several other eastern Ontario towns. Brother Bob thought that Orton could have broadcast a performance by the Rex: "I know that Ort had some-thing to do with them. I recall that I was thrilled when the great actor came around to the house one night to talk about this, but I don't know if he broadcast or not."[1]

Sid Fox Broadcasts from the Irishmen's Hall

Sid Fox started out in music in Kingston around 1918, playing trombone or saxophone with the PWOR Band and with the Queen's Band at football games. He played violin in the pit for silent pictures at the old Strand Theatre on Princess Street and at the Grand Opera House for Al Jolson, Blackstone, Thurston, and The Dumbells. When the talkies came in and the Grand closed, he started up his own dance band. Sid Fox had the biggest and best dance band in Canada, at the Chateau Laurier in Ottawa. He also played hundreds of dances at Queen's in the 1920s and 1930s. Some time in the mid-1920s, Orton invited Sid and a small orchestra to broadcast a program from the Irishmen's Hall:

<u>Sid Fox</u>: Orton Donnelly's father was stage manager at the Grand for around 25 or 30 years. He would be in charge of the stage for The Dumbells or any show that came through. He was a real show man. He looked something like Archie Bunker - a rough sort. You had to be rough to be stage manager, but I know he didn't take no interest on radio. Well, Ort went into the radio business. I remem-

ber Orton inviting us down to Wellington Street, and he had quite a job hooking it up - I don't know what he had to do, but he had to get some wires and connect it up - I remember it just vaguely. There was four or five in the orchestra - Bob Warmington was one of them - and no audience. I don't know whether it was an experimemt for him or not, but I just vaguely remember going over to the hall, and he was explaining what was going to happen and we took it more or less as a caper. He was serious in that business. That's the only thing that he was concerned with.[73]

A Circus Parade Broadcast from the Canada Radio Stores

<u>Bob Donnelly</u>: I went to Louise School - started in '21-'22, and this is what I'm taking my point of reference from, because there was another broadcast. Of the Sparks' Circus parade. This was the first time I'd ever seen a circus, so I remember that like yesterday. We were let out of school about 9 o'clock to go see the parade. It was a big day, and I high-tailed it down to the Canada Radio Stores to look for Ort. He was up on the roof setting up a mike. And the parade was coming, and I held up these megaphones, I call them, I held these towards the parade for the sound, and while we were up there a chap from the circus came on and he started doing a little bit of the broadcasting. After the thing was all over, he asked me if I wanted to go to see the circus. So he handed me a pass, and I had to go home, of course, to get permission to skip school, and back up to the Fair Grounds. Circuses always came around in May or June, so it'd be in '22 or '23. A few years later, about '26, the Canada Radio Stores started with the first Bell and Howell home movies. That year they broadcast the circus parade and also took movies. They had to send them to Hollywood to have them developed, and they used to show them in the store window.[1]

Norman Brokenshire Broadcasts on 3AAY

Perhaps the exploit that made Orton Donnelly the proudest was his broadcast of the famous radio announcer Norman Ernest Brokenshire of WJZ, New York. He was born in Murcheson, Ontario, June 10, 1898, the third of five children of the Rev William Henry (B.A. '01) and Georgina (Jones) Brokenshire. The Rev Mr Brokenshire was a Kingston native, but served in mission work for many years in western Canada, Japan and New England. Mrs Brokenshire's father, Cornelius Ansley Jones, had been a backwoods itinerant minister and rapids runner, known locally as "Moccasin" Jones. After the turn of the century he operated a locksmith, bicycle and revolver repair shop at 293 Princess Street.[74] Norman may not have spent much of his life in Kingston, but his family had its roots in the city and the people regarded him as a native son who made good.

On a visit to New York City in 1924, after graduation from Syracuse University, he answered a newspaper advertisement for "a college man with a knowledge of musical terminology." He was one of four selected from 400 applicants by Mr Popenoe, Manager of Broadcast Central, to be an announcer on WJZ. His colleagues were Tommy Cowan

(ACN), Milton J. Cross (AJN; later of the Metropolitan Opera broadcasts) and Lewis Reid. In those days, announcers were not permitted to use their names on the air, and Norman was known as AON. He should have been ABN, for Announcer-Brokenshire-New York, but someone else was ABN. He rapidly rose to fame, broadcasting the donnybrook from the 1924 Democratic National Convention at Madison Square Garden with Major J. Andrew White (who broadcast the first prize fight), the first horse race broadcast, again with Major White, from Belmont on Labour Day, and the reception at Mitchell Field for the US Army Airforce pilots on their way to completing their round the world flight.[75] On March 4, 1925, Calvin Coolidge was inaugurated and Norman broadcast it for the little RCA network (WJZ, WJY, WRC, KDKA and KFKX) in a fierce competition with Graham McNamee of WEAF and the AT&T transcontinental network.[76] "Broke" was the broadcast host at every presidential inauguration for the next three decades. His distinctive honey-throated voice and trademark, "How DO you DO!", helped him win the title "King of Announcers" in the New York Mirror Contest in 1932.[77]

Later, his voice was familiar in every household on the continent as the announcer for "The Chesterfield Hour," "The Eddie Cantor Radio Follies," "Major Bowes' Amateur Hour," "The Good Gulf Show" and "The Theatre Guild of the Air." He also appeared in one of "The Big Broadcast" films, playing himself.

> Orton Donnelly: Along with the activities which were all quite legal, I'm going to mention at my own risk a few illegal activities, which we had quite a bit of fun out of. Norman Brokenshire was a well known announcer at WJZ, New York, and he came from Kingston. His family lived in Kingston - his brother Bill (Wilbur) Brokenshire was a reporter for The Whig-Standard, and Norman was home visiting the family when we happened to run into him and mentioned the fact that we had the capability of broadcasting from our little station in Canada Radio Stores, 3AAY. He said, "I would like to do a little program so that my mother could actually hear my voice here. She has never heard my voice by radio." So, we set up the station, and we had a little studio rigged up at the back of the store - a little desk and a microphone on it - and we arranged for a local orchestra, which unfortunately did not show up. However, Norman filled in very nicely with the aid of a little phonograph music, and talked for approximately an hour of his experiences in the States, and we were very glad to have him. He was very pleased to be able to transmit from Kingston and let the family hear him.[2]

The date of Norman Brokenshire's broadcast from the Canada Radio Stores has not been established. The most likely time is the late summer of 1926. His father died in Kingston on March 21, 1926,[78] but Norman did not visit Kingston until after the funeral. In his biography, he recalled that he visited Canada with his scrapbooks before he quit WJZ, September 25, 1926. At that point he had just broadcast his first Miss America Pageant, around the 11th, and the implication was that he came to Canada between the end of the pageant and the time he quit WJZ. A search through the Kingston papers for the whole summer of 1926 did not turn up any information on the visit.

Orton Donnelly: We lived near the Notre Dame Convent, a stone building where the Public Library is now, and I got along pretty well with the Mother Superior. I used to broadcast symphony music on Sunday afternoons, on a loud speaker out the attic window so they could hear it. The girls used to gather around to listen to the music and applaud. It sounds kind of corny now, but we thought it was quite a thing in those days.[2]

Bob Donnelly: Ort had the first Rogers batteryless radio set hooked up in 1926 for the first Wrigley Marathon Swim at the Canadian National Exhibition. He had one in the parlour at home on Wellington Street, sounding out through the front door, and another one at the Canada Radio Stores. We had crowds in both places. The people thought it was an auction sale on Wellington Street. It was reported in the paper.

KDKA was the pioneer, but in this area it wasn't really THE station. The station here was WLW in Cincinnati. If you could get WLW on Saturday night for the Crosley Follies, you were listening to **something**. It was a variety show, and ran all night. And there was "Billy Jones and Ernest Hare/ We're the Inter-woven pair/ Call each other heel and toe/ Happy-go-lucky wherever we go." For Inter-woven socks. Other early commercials were for beer companies, and it was prohibition. So you had Fleishman's yeast, Pabst Blue Ribbon yeast. All the beer companies put out a yeast, you know. They'd tell you all about their yeast, but you knew darn well the beer companies were putting these shows on.[1]

The MacMillan Expedition

Bob Donnelly: He still had the other transmitter, but he built up a small breadboard type of rig on 40 metres, which was very experimental at that time. I'd heard that he had MacMillan, and England knew that he'd had that. But this wasn't the same as his broadcasts. I believe he got a QSL card from them.[1]

Rear-Admiral Donald Baxter MacMillan mounted the Crocker Land Expedition of 1913-17, searching for a predicted land mass in the Beaufort Sea, and later the Arctic Expedition of June-October, 1925, with Lt-Commander Byrd. For the latter they took along Zenith radio equipment operating from 16 to 40 metres, to make possible low power daylight transmission. The expedition kept in daily touch with civilization and the National Geographic Society through the 1,200 volunteers of the ARRL, and messages were picked up all over North America, England, Australia and New Zealand.[79] Orton Donnelly talked with the MacMillan Expedition at Etah, Greenland, 11.5 degrees from the pole, but the The British Whig thought that contacting England was a much bigger deal:

> Mr. Orton Donnelly...who operates an amateur radio
> station 3HE, recently established double communica-
> tion with station 2KZ, Middlesex, England, operated
> by B. Clapp...(He) did not use the official Canadian wave
> length of 52.51 metres, but resorted to the 40 metre
> wave-length. He has also been very successful in long
> distance work, having communicated with MacMillan
> when the explorer was situated at Etah, Greenland. In the
> English communication, two five watt power tubes were
> used with success. When he connected up with
> MacMillan, he was using a ten-watt transmitter with
> 300 volts on the plate.[80]

Orton was still addicted to the big German tubes: "The use of a
German tube, using 4,000 volts on the plate, is being tried out by the
Kingston amateurs...So far this tube has met with a great measure of success in Kingston, and is especially adapted for long distance work, it is
being installed by Mr. Donnelly."[80]

Reports in the Daily Standard Junior, 1924-26

Orton Donnelly was far more active than the other Kingston amateurs
in this period, not only because most of them were still students and
unable to afford expensive "junk," but because he had access to literature, catalogues and the latest gizmos as a partner in the Canada Radio
Stores. He was also able to travel a good deal in his "Lizzie," attend
meetings and make personal contact with other amateurs. Orton experimented with new equipment constantly, building and replacing at a
furious rate. When the Daily Standard Junior reports began in late
1924, Orton was getting good results operating at 15 or 20 watts, but
talking of going to 50 and changing his antenna radiating system. He was
also having trouble getting his transmitter down to 80 metres, but was
soon on and doing good work at 80 and 120 metres.[81]

Early in 1925 he increased his signal strength by replacing his single-wire counterpoise with a 16-wire array, then went to 5 and finally
down to one wire. He obtained a "bug" key (a high-speed side-swiper,
which 3NF and 3AEL already had) and was nominated to be Ontario
Division Manager of ARRL.[82] He was experimenting with Heising
modulation for his 'phone work [83] and was able to communicate with
9AD, Selkirk Mine, Manitoba, the most northerly station in Canada,
with the Atlantic coast and with England using his new counterpoise.[84]
An amateur relay using voice was set up in April, consisting of Orton's
3HE in Kingston (180 m.), WVAT (a US Army School near Watertown;
"on a wavelength a little higher than WEAF"), 8CPZ at Evans Mills, NY
(225 m.), and 8CFV at Gouverneur, NY (135 m.). The relay operated
every Monday, Thursday, Saturday and Sunday at 12.30 and 5.30 pm,
and the BCLs could listen in.[85]

In May, Orton built a new rectifier, re-arranged his equipment and
finished a short wave antenna: "At present he has so many counter-
poises and aerials around his home there isn't room for any more. He

is contemplating building a 5 metre transmitter. The other day he accidentally put 2,000 volts on the plate of a lone five-watt tube and burned a hole in the plate about the size of a small 5 cent piece."[86] Then Orton left the air for a while. One reason was that there was now a YL (young lady) in his life, and another was that he was building a wavemeter and a portable 40- to 80-metre CW and ICW transmitter using two 201-A tubes and a "plate supply from a specially treated spark coil."[87] He took the portable station, and perhaps the YL, in his Lizzie to Kingston Mills on May 31, and communicated with 3AFZ on 150 metres,[88] sounding as strong as if he were in Kingston. Orton also built a low power 20- to 40-metre breadboard transmitter, with a baseless 5-watt tube (later two 5-watters) as oscillator and a coupled Hartley circuit,[89] with 1,000 watts on the plates. Although his signal strength was good, he wanted to put a 50-watter into the socket and then a 1-kW Mullard MT6.[90] He took his portable set in his Lizzie to the Radio World's Fair in New York,[91] "with a sign bearing his call and the name Kingston." In July his large antenna was replaced by a new one for 40, 80 amd 120 metres, and then in October his 5-wire aerial was replaced by a single-wire "Hertz" antenna with a light bulb at the null point.[92] At one point the bulb blew, but it was too icy to go up onto the roof and haul in the antenna.[93] Eventually the Hertz antenna came down (perhaps Ed shot it down) and a new aerial and counterpoise went up. He got a French tube, and was able to work England and all US districts with it for a few weeks until something happened and it wouldn't light. After that he started using a 150-watt Mueller tube from Germany.[94]

Next he tried remote control, like his correspondent 2FO in Montreal, with his transmitter in the attic and his apparatus downstairs,[95] but that didn't last very long, and the transmitter went back to the attic. He converted to the "All Canada" frequency of 5713 kilocycles (52.5 m), for exclusive use within the British Empire. The final report in the series shows Orton thinking about putting up a gutter pipe mast, used successfully by the Rochester amateurs.[96]

> **Keith A. MacKinnon:** I don't remember especially about Donnelly, but I remember he was quite active. POZ was the old call of the big Nauen station in Germany in the First War. And I remember Donnelly went out in the country somewhere and strung up a great big antenna somewhere to pick up POZ. Whenever I see POZ car, I think of Donnelly. It was all long wave, do you see, 15 kilocycles (19,900 m). It was their key station, practically world-wide on these very low frequencies, and they sent to all their submarines and stuff.[97]

Epilogue

Some time in 1926 or 1927, Orton sold out his interest in the Canada Radio Stores and took up a contract to instal wireless systems in northern Ontario. Val Sharp may have gone to Red Lake as well about that time.

Douglas Donnelly: Ort went up to Red Lake (north of Kenora), and stayed about a year in the Howie Mines, installing the communications equipment. I was about 4 when he came back and that would put the year about 1927.[5]

Orton's 3HE was not in the Official List of Radio Stations of Canada in 1927, but returned in the first supplement in November.[98] The 1938 edition shows him holding the call VE3ABA, and living at 560 Princess Street. In the 1930's he was trying to re-establish himself in the radio business.

The Kingston City Directory and newspaper advertisements show Donnelly's Radio Service at 139 Clergy Street, near the corner of Queen, in 1935; at 155 Brock Street in 1939 and at 420 King Street East in 1946. After retirement and right to the end of his life[99] he still loved to play with radios and other electrical gadgets.

Some time in 1926 or 1927, Orton Donnelly sold out his interest in the Cana Radio Stores and went to northern Ontar to install and operate a radio station on government contract at Red Lake. Both t station and Orton's living quarters were i small log cabin.
-courtesy Josephine (Donnelly) C

Douglas Donnelly: I remember how Ort used to operate from his shop on Clegy Street, and he had an old white panel truck that he used to make his pick-ups and deliveries. He used to take me on the calls with him. The truck had no heater, and I'd be sitting there while he was inside chit-chatting with clients.

During the War, he was in Remy and overseas. He worked in the Base Ordnance Depot at Borden, in England, preparing the wireless equipment in tanks for the invasion. When Ort came back from overseas, he was thinking about going back into the radio repair busines, and possibly include sales. I guess that didn't work out as well as he had thought. He had gotten out of the army, and he decided that he would go over and work as a civil servant in the workshops at Barriefield - 207, I believe it was. He worked there for a number of years, in radio, until he reached retirement age, and that was where he stopped working professionally.[5]

Bob Donnelly: He enjoyed having a radio or something like that go on the blink. Oh, he loved it! He'd be into it up to his elbows. He was playing with tape recorders, taking them apart and trying to change the speeds on them and things like that. We were there in Florida for his birthday one year, and his wife, May, had bought him a tape recorder. In five minutes he had that all apart. She said, "Orton Donnelly! It's new. I just bought it." And he said, "But there may be some dust in it. I have to dust the insides of it." You wouldn't have believed. It was all picked apart. He had to put it back together again. And he had the lawn mower picked apart. He'd never come into the house that he didn't tune the radio if it was playing. Or he'd be in front of the TV, twirling the dials, "Well, maybe I can get a better picture." He just couldn't leave things alone.[1]

NOTES

1. Personal communication from Robert J. Donnelly, Brantford, June 30, 1986.
2. Personal communication from Orton Donnelly, Hay Bay, Ont. Sept. 29, 1982.
3. "Wireless Seized", KDS, May 26, 1920, p. 5.
   "Three Wireless Apparatuses", DBW, #133, May 25, 1920, p. 1.
   "Must Reduce Voltage", ibid., May 26, 1920, p. 16.
   Editorial, ibid., May 27, 1920, p. 6.
4. Vernon Campbell was taken on as Constable on the Kingston Police Force on Dec. 1, 1919, and left on Nov. 31, 1924; courtesy Kingston Chief of Police Rice, April 3, 1989.
5. Personal communication from Douglas Donnelly, Toronto, March 31, 1989.
6. S.E. Frost, Education's Own Stations: The History of Broadcast Licences Issued to Educational Institutions, University of Chicago, Chicago, 1937.
7. "The American Radio Relay League", QST 6 #10, May 1923, p. 6.
8. "The Operating Department", ibid., 5 #1, Aug. 1921, p. 34.
9. "The Operating Department", ibid., 5 #4, Nov. 1921, p. 36.
10. "Who's Who in Amateur Wireless: A.H.K. Russell", ibid., 6 #8, March 1923, p. 50.
    Keith Russell held pre-War licence XRE for an old Clapp-Eastham Hytone set (Certificate #98, April 21, 1915, in Sessional Papers, House of Commons, Paper No. 38, 52 #21, 1917, p. 97) and later 9AL.
11. "The Operating Department", ibid., 5 #3, Oct. 1921, p. 39.
12. "The Operating Department; Ontario Division" ibid., 5 #9, April 1922, p. 45.
13. Ibid., 6 #1, Aug. 1922, p. 37.
14. "Kingston Radio Association", KDS, March 24, 1921, p. 3.
    "Club Gossip: Kingston Radio Association", Radio News, May 1921, p. 795.
    "Old Time Club", KDS/DSJ Sept. 6, 1924, p. 11.
15. "Radio Organization", KDS, April 2, 1921, p. 5.
16. "Radio Assoc. is Growing", ibid., April 11, 1921, p. 12.
17. "With the Kingston Radio Association", ibid., May 3, 1921, p. 2.
    Ibid., May 10, 1921, p. 5.
18. "Radio Assoc.", ibid., May 24, 1921, p. 5.
19. "With the Kingston Radio Association", ibid., May 31, 1921, p. 5.
20. Ibid., June 14, 1921, p. 2.
    "Radio Amateurs", advertisement by the H.W. Newman Electric Company, ibid., June 9, 1921, p. 7.
21. "With the Kingston Radio Amateurs", ibid., June 21, 1921, p. 5.
22. QM-Sgt Tom Brown, Secretary, KRA, to Editor re Amateur Wireless Telephone, ibid., July 14, 1921, p. 8.
23. Personal communications from Harold "Dick" Cohen, Kingston, Sept. 29, 1982; Oct. 31, 1985.
24. Personal communications from Dr. W. Donald MacClement, Kelowna, May and Aug. 1987.
25. Gleason L. Archer, History of Radio to 1926, American Historical Society, New York, 1938, p. 213.
26. "H.W. Newman, Ex-Mayor", DBW, May 3, 1926, p. 1.
    "Originated in Kingston", ibid., Sept. 21, 1926, p. 2.
    (re Saunders Electric) ibid., Sept. 15, 1926, p. 6.

27. Personal communications from Orton Donnelly, Hay Bay, Aug. 21 and Sept. 29, 1982.
28. "Happenings in the Wireless World", KDS, Nov. 11, 1921, p. 10.
29. "Kingston People Hear Concert", DBW, Nov. 14, 1921, p. 7.
30. "Enjoyed Wireless Concert", KDS, Nov. 21, 1921, p. 5.
31. "Wireless Concerts", adv't., DBW, Dec. 1, 1921, p. 10.
32. "Happenings", KDS, Jan. 10, 1922, p. 10.
33. "Happenings", ibid., Jan. 21, 1922, p. 8; Feb. 1, 1922, p. 3.
34. "Happenings", ibid., March 4, 1922, p. 10; March 15, 1922, p. 3.
35. "Happenings", ibid., Feb. 18, 1922, p. 7.
36. "Brief Notes of City News", ibid., June 14, 1922, p. 5.
37. "In the Halls of Queen's", DBW, Dec. 15, 1921, p. 2.
38. Grand Theatre adv't, ibid., Feb. 24, 1922, p. 14.
      QJ, Feb. 14, 1922, p. 1.
39. "College Frolic of 1922", DBW, March 1, 1922, p. 7.
      "Theatre Night Performance", QJ, March 3, 1922, p. 1.
40. "College Frolic Repeated", DBW, March 14, 1922, p. 15.
41. "Citizens' Night", QJ, March 10, 1922, p. 1.
      "Monday, March 13", DBW, March 9, 1922, p. 2.
      "Theatrical", ibid., March 11, 1922, p. 19.
42. "To Broadcast Local Show", KDS, March 13, 1922, p. 5.
      "William Jennings Bryan", DBW, March 13, 1922, p. 2.
43. "Heard Play at Wireless", KDS, March 14, 1922, p. 5.
44. "All May Listen In", DBW, March 16, 1922, p. 6.
45. KDS, March 18, 1922, page 5.
46. "To Broadcast Music", ibid., March 15, 1922, page 5.
47. "Wireless at the Y.M.C.A.", DBW, May 1, 1922, p. 2.
48 "Happenings", KDS, March 25, 1922, p. 5.
49. Personal communication from Annie Lenore "Goldie" (Bartels) Geiger, Toronto/Edmonton, Sept. 6, 1982.
50. Personal communication from Georgina (Ettinger) Logan (Arts '22), West Hill, Oct. 16, 1982.
51. "News and Views for Women", DBW, March 16, 1922, p. 8.
52. "Happenings", KDS, Dec. 16, 1921, p. 13.
53. "Happenings", ibid., Feb. 4, 1922, p. 5.
54. "Radio: Announcing..." ibid., June 24, 1922, p. 14.
55. "Now Open" (adv't), DBW, July 4, 1922, p. 4.
56. "Oh! Say" (adv't), ibid., Oct. 2, 1922, p. 4.
57. "Thanks to Queen's" (adv't), QJ, Oct. 13, 1922, p. 3.
58. "Science", ibid., March 23, 1923, p. 6.
59 Personal communications from Jack M. Campbell, Jr., Gananoque, Nov. 11, 1982; Oct. 7, 1988.
60. "Kingston and Vicinity", DBW, June 14, 1922, p. 5.
61. "Don't Miss..." (adv't), ibid., Aug. 28, 1922, pp. 4, 10.
62. "Along Life's Detour", ibid., Sept. 16, 1922, p. 6.
63. "Game by Radio", ibid., Oct. 6, 1922, p. 18.
64. "Radio Stores are Up to Date", KDS, Oct. 7, 1922, p. 9.
65. "World's Series", Toronto Daily Star, Oct. 3, 1922, p. 7.
66. "First Kingston Voice", DBW, Nov. 24, 1922, p. 2.
67. "Jack Elder Sang", KDS, Nov. 2, 1925, p. 2.
68. "Plunkett Company", DBW, Nov. 30, 1923, p. 2.
69. "Give Impromptu Programme", ibid., Dec. 3, 1923, p. 14.
70. Personal communication from Freeman Waugh, Kingston, Dec. 8, 1982.
71. Personal communication from Ian Baines (Elec. '74), Burlington, Ont., Aug. 23, 1986.
72. Personal communication from Edward Donnelly, Kingston, Aug. 20, 1986.

73. Personal communication from Sid Fox, Kingston, Oct. 26, 1982.
74. Kingston City Directory, July 1907-July 1908; July 1913-July 1914, Leman
    A. Guild and George Hanson, Kingston, Ontario.
75. "City Welcomes", New York Times, Sept. 9, 1924, p. 1.
76. "National Audience", ibid., March 8, 1925, p. 8:15.
77. Norman Brokenshire, This Is Norman Brokenshire: An Unvarnished Self-
    Portrait, David McKay, New York, 1954.
    "Norman Brokenshire," New York Times, May 5, 1965, p. 47.
78. "Obituary: Rev. William H. Brokenshire", DBW, March 22, 1926, p. 3.
    "Rev. Brokenshire Died", KDS, March 22, 1926, p. 12.
79. "MacMillan in the Field", National Geographic 48 #4, Oct. 1925, p. 473.
    D.B. MacMillan, "The MacMillan Arctic Expedition Returns," ibid., 48
    #5, Nov. 1925, p. 477.
80. "Radio Communication", DBW, Feb. 9, 1926, p. 11.
81. "Listening In", KDS/DSJ, Dec. 20, 1924; January 10, 17, 24, 31, 1925.
82. "With the Amateurs", ibid., Feb. 7, 1925, p. 7; Feb. 14, 1925, p. 7.
83. "With the Amateurs", ibid., March 28, 1925;
    "Our Amateurs", ibid., April 4, 1925.
84. "Listening In: Heard in Manitoba", ibid., April 11, 1925.
85. "Listening In: An Amateur Relay", ibid., April 25, 1925.
86. "Listening In", ibid., May 9, 1925, p. 7.
87. "Listening In", ibid., May 23, 1925; May 30, 1925.
88. "Listening In", ibid., June 6, 1925, p. 9.
89. Ibid., June 20, 1925, p. 7.
90. Ibid., June 27, 1925, p. 9; July 4, 1925, p. 8.
    The MT6 vacuum tube was used in Canadian commercial broadcasting
    stations. CFRC was mentioned as using it, but the record does not
    show that CFRC ever had an MT6.
91. "Listening In", ibid., Sept. 19, 1925, p. 7.
92. "Listening In", ibid., Oct. 3, 1925, p. 9.
    H.M. Williams, "The Hertz Antenna at 20 and 40 Meters", QST 9 #7,
    July 1925, p. 24.
93. KDS/DSJ, Dec. 12, 1925, p. 11.
94. "Listening In", ibid., Jan. 23, 1926, p. 7.
95. Ibid., Nov. 7, 1925, p. 7; Nov. 14, 1925, p. 8.
96. "What the Radio Amateurs", ibid., April 17, 1926, p. 8.
97. Personal communication from Keith A. MacKinnon, Ottawa, Aug. 15,
    1987.
98. OLRSC, DMF, Marine Branch, Ottawa, June 30, 1927.
    Supplement No. 1 to OLRSC, ibid., Nov. 30, 1927.
99. "Births, Deaths: Orton Donnelly", KWS, Feb. 13, 1984, p. 19.

CHAPTER 9

The first edition of Jack Campbell's voice transmitter (right) and receiver (left) in his radio room, the 3rd floor den of 5 Emily Street. The potential transformer was on the floor, his licence on the wall at far right. A copy of QST on the shelf, above, is dated March 1923.
-courtesy Jack M. Campbell

Kingston Radio Amateurs, Part 2

By 1922-23, all of the most competent amateurs in Kingston, those who were active experimenters and builders, were members of the American Radio Relay League (ARRL) and devoured QST every month. Through the magazine and widespread contacts with amateurs, they had access to the latest ideas, equipment, schematics and suggestions for improvement in their art. It should be apparent from the detailed accounts of amateur activities in the Daily Standard Junior, that innovations and fads moved in waves through the amateur world - the counterpoise as the "earth screen," striving for high power and then the discovery of the magic of low power on low wavelengths, the move down below 200 and then to 40 metres and lower, the lattice-work mast, abandoning the cage antenna for a single wire and then flirtation with the Hertz antenna, indoor antennas, the gutter-pipe mast, remote control and receiving sets as transmitters. Most of the best of the amateurs tried these new ideas in the hope that their distance would be improved. The bit of record that exists shows that they were trying pretty much the same things at the same time. They all tried to get involved in the trans-continental relays and the trans-Atlantic tests, listened for the MacMillan Expeditions, for the <u>Shenandoah</u>, DXed all night and sent out their QSL cards.

Jack M. Campbell, Jr (3OQ/3JG)

John M. Campbell, Jr, is the son of John M. Campbell, Sr, owner and later manager of Kingston Light, Heat and Power, forerunner of the Public Utilities Commission, and unsuccessful Liberal candidate for Kingston in the Dominion elections of 1921 and 1925. Jack, Jr, was an active radio amateur in Kingston from the end of the First War and a member of the ARRL from at least 1922. He is now Chairman of the Board of Gananoque Light and Power.

<u>Jack M. Campbell, Jr</u>: I know that Orton Donnelly built a transmitter about 1918, because he worked with me. I got the idea of building

The second edition of Jack Campbell's voice transmitter (right) and one-tube (Mullard) regenerative receiver (left). *Note the telephone transmitter (microphone) on the table in front of the transmitter. His old 110-2400 volt powerhouse potential transformer is at centre. The features on the face of the transmitter are, from left: dial on the tuning condenser, hot-wire antenna RF ammeter from Canada Radio Stores, filament voltmeter.
-courtesy Jack M. Campbell

one and so I used to have a close association with him, and we used to talk back and forth a bit. We were on before KDKA in Pittsburgh came on commercially because we heard it come on the first time they were testing. So it must have been 1918 or 1919 that we were active on the air. In those days you'd just crank the thing up to see if you could make it work, and then you applied for a licence. Because you couldn't interfere with anybody, at that time.

I have a photograph of my first set-up. Just a very simple receiver, a three-circuit one-step audio receiver. It was a regenerative receiver with one Mullard tube. The transmitter was a coupled Hartley circuit, and the one oscillator tube did the whole thing. Pretty crude - just on a piece of plywood. The transmitter worked by absorption loop modulation. We modulated it by wrapping about a turn and a half of wire around the tank coil (an open wound helix I made myself - I don't think they were on the market), which was the oscillating coil. The one tube was tuned to the wavelength of the tank coil and the condenser with which it was regulated, and you'd talk into the microphone to modulate it. A very queer way to modulate, but it worked. If you were to put two or three loops around the tank coil, it would melt the carbon in the microphone. Oh, it was voice. I think I never got into the code part at all - I liked the voice part.

We used to use an electrolytic rectifier - a combination of borax, lead and aluminum - a whole bunch of jars, a jar each for so many volts. And I had a 110- to 2,400-volt potential transformer, one of the original, typical power house types. It had been superceded, and was given to me by my father. The transformer was on the floor, hooked up so that we could get 1,100 volts out of it. Orton Donnelly suggested putting 50 watts into the plate of the 5-watt tube, because they were always very conservatively rated. We used to run the plate at white heat. If you were running for a long time you would cut her back to dull

Jack Campbell, Jr, had a 4-wire cage antenna on wooden bicycle hoops strung from the roof of 5 Emily Street to a maple tree of the same height, and held taut by pulley and weight. Two bicycle hoops are visible, one just above the dormer and the second just below the centre of the photograph.
-courtesy Jack M. Campbell

Jack Campbell's radio room, on the top floor of 5 Emily Street. Another radio amateur owns the house now, and his antenna can be seen on the roof.
-photograph by Arthur Zimmerman

red, but if you wanted to really get out, you'd push it up to the white. I used to get down pretty near to New York City with it, though usually we used to talk to Troy and places over just across the border. They were quite active over there. When Barriefield came on, you just closed shop and forgot it! It blanked everything out - must have covered the whole wave band from one end to the other. I don't know what frequency they were on at that time. I don't think there was much restriction actually, because there were virtually no broadcast stations operating. The second edition of my transmitter was vastly improved and modernized and I have a photo of it too.

There's a little room up there (on the third floor at 5 Emily Street, facing south, overlooking the garden), a mahogany-panelled den up in the top, instead of the bottom, and that was the radio room. It had a closed stairway access and a hardwood floor - I must have ruined that floor, spilling borax solutions all over it. The feed came out and hooked onto the cage aerial on the roof - a four wire cage. We used wooden bicycle rims to spread the wires. The theory was then that the cage acted as a piece of wire of the diameter of the cage, rather than as 4 thin wires. And the bigger the diameter of the cage the more you radiated. Soaked the ground, I guess. The 4 wires came together at the end and attached to an insulator, which was connected to a wire over a pulley and was weighted to keep the aerial taut. The aerial came from the house over to a maple tree of the same height. We used a counterpoise under the aerial, rather than a ground, and found it worked much better than just putting it into the ground. Away better. You got a better coverage. I used to get up on the roof fixing the aerial and the lady next door would watch me and then phone up my mother and say I was going to fall off the roof.

I know I had the thing operating before 1921. My chum, Don MacLean lived around the corner at 53 King Street East, corner of

Maitland. His father ran MacLean Packing Company, a boat outfitter. In those days most people had gas for lighting, but I think they used candles. So I ran a pair of twisted wires over from 5 Emily Street into Don's bedroom window - I had a little generator, a motor driving a DC generator - I ran this DC over and he had a little flashlight bulb that I lit up from the generator. So that he could have a 6-volt light to read by. When I was away at school in Port Hope he died. There was a sort of miniature tornado that lifted the roof off the Murney Tower in Macdonald Park, carried it a block over to O'Kill Street, which we always used to take going to school, and came down on his head and killed him deader than a doornail.[1] It was before that that I had a radio transmitter. That's the only way I could tie this together for dates.

I wanted to get an RF meter to measure the radio frequency antenna current, and the Canada Radio Stores got me one. A Jewel meter - I remember it was quite a thing. When it arrived, they put it out in the window, and I went down there and picked it up. That was quite an innovation in those days to have an RF meter in your antenna circuit.

In 1923, I ran a wire down from the transmitter to a lower floor in the house, where the piano was. I put the microphone down in where the piano wires were, got my mother to play and I went down to the Canada Radio Stores and listened to it. And boy! The sound came through perfectly good, for the type of equipment. She played away. I don't know what she played - I don't think she knew more than four or five tunes - but she kept playing them and playing them until I came back...She was a good sport. It'd be as early a music broadcast as any of them, but that's the name of the game. The beauty of this amateur radio business is, no matter how far back you go, an awful lot of the development that's occurred commercially has resulted from amateurs experimenting.[2]

The Daily Standard Junior noted ex-3OQ planning to return to the air in October 1925, with a phone set on 125 metres, an illegal wave for phone, using "S" tubes for rectification. Later he was quite successful with short wave, 40 metres, on phone.[3] Jack Campbell held 3OQ in 1927, and disappeared from the Official List until 1932, when he lived in Toronto and obtained the call VE3JG.[4] In 1937, he was listed with VE3JG c/o Gananoque Electric, Light and Water Supply Co., Ltd., and he has retained his interest in amateur radio to this day.

Kingston's Second Radio Station, CFMC

Very few people know that Kingston had a second commercial radio station in the 1920s, long before The Whig-Standard ever considered getting into the broadcasting business. The story of the origins and accomplishments of radio station CFMC are still a little hazy, but there are three threads to follow. The Cohen brothers, Sheldon and Harold (3ACF), and William Cameron (3AFZ) were the prime actors, and the Monarch Battery Company of Kingston perhaps brought them together and acted as the catalyst. Two of the threads, 3ACF and 3AFZ, can be winkled out of the weekly "ham" reports in the Daily Standard Junior.

Jack M. Campbell, Jr: CFMC, the Monarch Battery Company. They were way up on Princess Street, and for a while the factory was down near where S&R is now. I went there with my father and met Isaac Cohen. He was a great guy, and he said, "Here," and he gave my father a battery in a wooden case, and he said, "Take it and see if it works all right. Try it out." They were a great pair. They used to get together a lot.[2]

Harold A. (Dick) Cohen (3ACF)

In September 1924, Dick Cohen held 3ACF in his name. He was interested in DXing, and was likely operating mostly with CW in code. He was using two 5-watt tubes in his transmitter, and was dreaming of a 50-watt Dutch tube.[5] Instead of increasing power, he built both a motor generator to rectify AC for high-voltage plate supply and a high-voltage storage battery. He rejected the latter.[6] Later it was said that he had so many sources of power supply that he couldn't make up his mind.[7] Like his fellow "bugs", he regularly blew 5-watt tubes and couldn't decide what wavelength or power to settle on, but actually "got out pretty good" over the winter of 1924-25. Dick tried remote control, the transmitter outside the house, and rejected it as N.D.G. months ahead of Orton Donnelly. Before shutting down for examinations at Queen's, his phone transmitter was reported interfering with the broadcast concert listeners (BCLs). In May, he "threw the junk together and it worked so I'm going to leave it that way."[8] He talked of building a "Super-Het" and then an efficient short wave receiver described in QST.[9] The latter was actually built for him by Bill Cameron.[10] In the fall of 1925, after the debut of CFMC, Dick acquired a 150-watt Mullard tube for 3ACF, and after trouble getting started, he transmitted on 80 metres and answered CQ's on 40.[11] In the early winter he dismantled his set, mast and all, intending to stay off until the summer. But "the radio bug is a hard one to kill," and Dick was assembling 3ACF again in February 1926.[12] There the Daily Standard Junior reports trickle off.

William M. Cameron (3AFZ)

William Cameron was listed in the 1927 City Directory as an electrician with the Monarch Battery Company. He turned up in the Daily Standard Junior in October 1924, having moved to a new location within the city and preparing to put a new transmitter on the air.[13] He put up a 64-foot fragile-looking lattice work mast,[14] and he got on the air with 5 watts in late November.[15] His mast soon blew down, as predicted,[16] and he intended to get it back up, double his power and reach England. He got involved with AYL and didn't get the mast up until February.[17,18] Between the YL, illness and business, he didn't get much air time. For the next few months he was continually blowing his 5-watt tubes, building a transformer and planning to build a new 30-metre transmitter and a portable transmitter for his motorcycle, but he was also experimenting successfully with low plate voltage on phone at 80 metres.[19] He built a Tesla coil and demonstrated it at the KKK (Kingston Kiwanis

Karnival),[20] joined the ARRL and worked on getting his first-grade commercial certificate in the fall.

It is not clear just when and how 3AFZ and 3ACF became associated in their radio work, and whether Bill's employment at Monarch was a factor. Dick Cohen recalls that he heard 3AFZ on the air one night, and made contact with him. There is some record of their working each other in late 1924, probably in CW code.[21] The next hard evidence of a connection was the report that Bill was building 3ACF a low loss, short wave "Schnell type" receiver from QST.[22] This set turned out to be one of the best in the city and it inspired 3ACF to get back to working with his "Concert Buster" once more. About this time: "3AFZ is going to use the storage battery plate supply of 3ACF for about a month while the latter is going to try 3AFZ's transformer."[22]

Radio at the 1925 Kingston Industrial Exhibition

On August 15 the Daily Standard Junior announced that 3AFZ was very busy evenings and was building a phone station to be used for demonstration purposes during the Kingston Industrial Exhibition.[23] No other details were given. The 1925 Kingston Industrial Agricultural Exhibition ran from Tuesday, September 15, until Saturday, September 19, and, of course, the newspapers reviewed the major exhibitors on opening day. The Monarch Battery Company was in a large tent outside the Crystal Palace, Mr I. Cohen in charge.[24] Dick recalls that Bill Cameron built the transmitter not specifically for use at the fair, and he did a good job with the materials he had to work with. They had a radio receiver at the fair for demonstrating Monarch batteries, although it was difficult to pick up KDKA in the daytime. As amateurs, however, they knew that one could do wonders with a local transmitter. They did, in fact, broadcast to the receiver in the tent at the fair from a temporary studio located in the Monarch Battery showroom at 290 Princess Street.[25]

The First Broadcast of CFMC

A little confusion about the call letters surrounded the first broadcast by Kingston's new radio station, CFMC.

> A new local radio station made its first bow to the unseen audience of the country last night when station CFMB of the Monarch Storage Battery Company broadcast a very good programme. As it was the first occasion on which this station took the air, some experimental work had to be done, but the results obtained were on the whole very satisfactory. Mr H.A. Cohen was in charge of the broadcasting.
>
> Many local people heard CFMB last night, and at the request of the announcer telephoned him to say that his programme was being received very well. Violin solos, phonograph records and a talk by Mr Cohen were the features of the programme.[26]

They went on the air again on Friday: "Station CFM (sic) of the Monarch Storage Battery Co., Limited broadcasted an excellent programme last night from their temporary studio on Princess Street. The programme was rendered by local talent, consisting of piano, violin and vocal selections." Obviously the Daily Standard was still confused about their call. It was also announced that the station would be on the air about once a week and hoped to have Mayor Angrove, Hon Dr Ross, MP, and Hon W.F. Nickle, MP, address the radio audience.[27,28]

The Cohen brothers had applied for the call "C-MB" for "Monarch Battery," and at first thought that they had been granted CFMB. They announced that the call was CFMB on the first broadcast. A few days later they learned that they had been given CFMC by the Radio Branch,[27] though Dick thinks that perhaps the letter "M" did stand for Monarch. They had pretty good transmission and no hum from the studio because they used a high-voltage "B" storage battery and didn't have to rectify AC power.[29]

Mayor Angrove delivered a "most interesting" Chamber of Commerce-type address on Kingston over CFMC on September 22, speaking "feelingly of the splendid city of which he is Chief Magistrate. He referred to the beauty spots of the city and took occasion to mention the free shipping facilities provided here...Mr Sheldon Cohen is studio director of the station; H.A. Cohen is assistant operator and Mr W.M. Cameron, operator of station 3AFZ, is chief operator and announcer."[30] The station received several appreciative telephone calls following the program.

A few days later, on the 25th, CFMC broadcast the speeches from a big Conservative party election rally at a packed Grand Theatre. The Daily Standard was ecstatic about the event, The Whig a good deal less so. Brig-Gen Dr A.E. Ross, MP and Conservative candidate in the October 29 Dominion election, Senator Gideon D. Robertson and ex-Prime Minister Arthur Meighen made speeches, the last for a full 50 minutes![31] The same night, John M. Campbell, Sr, was nominated to bear the Liberal standard at a meeting in Memorial Hall.[32]

CFMC's equipment and expertise were not yet well suited for the broadcasting of remote speeches. They had used a microphone designed for studio musical pick-up, and the quality of the remote broadcast was not very good. To remedy the situation, CFMC promised to instal a new microphone for the broadcasting of speeches and make other improvements to the station's equipment.[31] Their next project, "a series of weekly talks on the subject of radio interference, as compiled by the Radio Branch, Department of Marine and Fisheries and its inspectors and given by the Radio Inspector of Kingston and District," began on September 25 from the Princess Street studio.[33] The title of the first talk was "How Interference is Caused."

CFMC Broadcasts "Blossom Time" from the Grand

J.J. Shubert, the Broadway impresario, liked producing high-toned romantic musicals. These quasi-operettas were big "box office" in the

1920s. He got the inspiration for <u>Blossom Time</u> after seeing a popular German operetta loosely based on the life of the composer Franz Schubert (no relation!). Jake had the libretto re-worked for the American audience, so that the story bore absolutely no resemblance to anybody's life, and he got Sigmund Romberg to prepare the score. The show, a fiction about Schubert's desperate love for his pupil Mitzi, was a smash. It played Broadway for three years, budding off road companies all over the place. Jake used to hire a bunch of no-names and has-beens for next to nothing, let them play one night at New York's Ambassador Theatre, and then send them on the road billed as "Direct from Broadway."[34] <u>Blossom Time</u> was announced for one performance in Kingston, Thursday, October 1, 1925, billed as "The Great Musical Hit of the Generation" with "the ideal New York singing cast," and a "Golden Girl Chorus."[35] The Grand was packed from pit to gods for this high class historical and cultural entertainment. And the Kingston performance was graced by the presence of Milton Shubert, nephew of the famous Shubert brothers.[36] In all of the excitement, the attempt by CFMC to broadcast <u>Blossom Time</u> went totally unnoticed by the Kingston press.

> <u>Dick Cohen:</u> We tried to broadcast the musical <u>Blossom Time</u> over our station from the Grand Opera House, but I don't think it was very successful. The people that had charge of the concert gave us permission to do it. They were quite happy to have it broadcast, but we weren't really competent to handle it. We should have run an amplified line to get the sound over to our studio, but we didn't have one. Bill Cameron and I climbed over blocks of roofs, stringing a plain double wire from the Grand to our transmitter. I don't remember just how successful we were in getting the sound from <u>Blossom Time</u> to our station.[37]

A fire broke out in the basement of the store that housed the Monarch Battery showroom and CFMC on November 2, 1925, but aside from thick smoke from the overheated furnace, little damage was done to the building or to the radio station. The program announced for later that week would go on as scheduled,[38] but there were few newspaper reports on CFMC's activities until the next September. CFMC announced that it would broadcast the September 14, 1926, Dominion election returns, at 267.7 metres (1120.7 kilocycles), the first broadcast since station CFMC was re-built.[39] No report was published. It tested on September 20, 1926, in preparation for "a busy season," and boasted a new microphone which had been used by Lloyd George, Governor General Byng and Lieutenant Governor Cockshutt.[40] After that, CFMC's activities did not make the press, and the station disappeared from the record. CFMC made its last appearance in the <u>Official List</u> of June 30, 1927.[41]

Dick Cohen brought out a few more points in musing on the brief history of CFMC. Originally the radio activities were on Montreal Street, but there was no studio there so they moved up to Princess Street for all

The former site of the Monarch Battery Company showroom at 290 Princess Street, where the studios of CFMC were located in 1925.
-photograph by Arthur Zimmerman

of the broadcasts. CFMC was at 290 Princess below Clergy, where the A&P used to be, but the store was half the width it is now. They didn't stay there for long either. The studio was in the store, in a room with cloth drapes around it and a minimal microphone which sounded all right. The single-wire antenna was on the roof of the building, about 20 or 30 feet high and 40 to 50 feet long. They aired mostly records lent them by the C.W. Lindsay store down the street. Harold and Sheldon announced and so did Bill Cameron, who had a good voice for radio. Dick also remembers broadcasting a special Kiwanis meeting, possibly from the British-American Hotel. Isaac Cohen was a Kiwanian. Their greatest success was when Sheldon played the violin a few times, and people called them up to say that it sounded beautiful and that they had enjoyed the music.

Dr Irwin Sugarman remembers Sheldon playing the violin over the air from a little dark room on the small second floor above and at the front of the Monarch factory on Montreal Street.[42] George Ketiladze (Elec. '29) played the piano for the CFMC audience several times and got some favourable calls. Dick recalls that they probably put on broadcasts only a couple of dozen times in all, with very little advertising in advance. In all, 8 broadcasts by CFMC have been documented in this project.

CFMC was a low power station, putting out about 20 watts,[43] and it was difficult to get the signal out of the Kingston area. CFMC shared 267.7 metres with CFRC in 1926. CFMC did have good quality with no hum, however, because it did not have to rectify AC. It was pure DC, using high voltage storage "B" batteries. The Cohens wanted a power of 500 watts for CFMC, but that would have meant getting a commercial transmitter, which cost $20,000 then. They knew the person who developed CKOC, Hamilton, but turned down his offer of a 50-watt transmitter from his failed station in St Catharines. They thought that if they could get some distance, and cover a radius of 150-200 miles, CFMC would have been commercially viable. Their particular interest was in going east and west over a large area, as far as Montreal and Toronto, to advertise their Monarch batteries for radios. They did not appreciate then that a strictly local radio station would be worthwhile and that it would have local commercial possibilities for them, or they would have stuck with it. They operated on and off for a few years and then gradually petered out.[44]

> Keith MacKinnon: Bill Cameron, a radio operator, worked in the storage battery company run by the Cohens. He was a kind of a funny-looking chap with a long neck and everything, and thin. In the War, you see, I went to the National Research Council and they were after all the radio men for the radar, and guess who showed up? This chap, and I recognized him. He was a good chap too. See, radar was very secret at this time. We called it RDF then, and this chap had written in to the Council and proposed the scheme which was the radar scheme. He was a smart chap, and Colonel Wallace, the boss, said, "You know, guys like this, we either hire them or shoot 'em." So he hired him. And he worked there, but he didn't recognize me when I went down to say hello.[45]

Frederick H. Simpson, Jr (3 AEL)

Fred Simpson was born in New York State around 1908, the son of Frederick H., Sr, and Helena C. (Hogan) Simpson. Frederick, Sr, was the proprietor of a hat cleaning company at 163 Princess Street. The family also lived there, above Treadgold's Sporting Goods, not far from the Daily Standard offices. By his own account, Fred got into the radio game around 1920, held licence 3AEL from around 1923 and was a member of ARRL. In 1924, the Simpsons moved from the shop to a house at 85 Sixth Street, now Hamilton Street.

The boy entered KCI in September 1922, and matriculated after Fourth Form in June 1926. While at high school, he was on the staff of the K.C.I. Times and wrote at least one major article for them about amateur radio, including blurbs on the radio alumni of KCI.[46] Perhaps through his talent for writing, he became Radio Editor of the Daily Standard Junior in 1924, succeeding to the post of Editor-in-Chief in 1925. Fred was an excellent publicist for amateur radio, explaining to Kingston BCLs the aims of the amateurs and the value to radio science of their relay work. Most of the detailed information known about the activities of the Kingston "hams" between May 1924 and May 1926 comes from Fred's "Listening In" columns. A good deal of the information published is about Fred and his station, 3AEL.

Fred was a very active DXer in 1924, and for a time was one of only two in Kingston who could tune down low enough to pick up VDM, the station of the Canadian government steamer Arctic in her annual cruise to Baffin Bay. VDM's radio operator was Bill Choat, 3CO of Toronto, who relayed the "all daylight" reports from WNP of the MacMillan Arctic Expedition at the North Pole to his fellow amateurs of North America.[47] The 24-hour Arctic daylight made it impossible for MacMillan's WNP to communicate directly by wireless with the amateurs, so WNP relayed the messages through VDM. At specified times, amateurs were permitted to use 120 metres to contact VDM. In August 1924, Fred's new set-up on Sixth Street was a honeycomb coil three-circuit tuner and one stage of AF amplification, with Baldwin phones. His antenna was "1 wire 38 and 15 feet high, 80 feet long." He reported hearing California, Texas, Louisiana, Cuba, France, Germany, including POZ at Nauen near Berlin, IHT in Italy and UCT (UFT?) in Paris[48] and several English stations on about 2,000 metres. He announced that when his new transmitter was in operation, he would soon be sending out musical programs for the testing of new tubes and microphones. Broadcasting music would ensure listener co-operation in reporting reception. He might also feature requests played by the Daily Standard Junior Orchestra, which was "too jazzy to play anything but popular music."[49]

> **Neighbour:** I see your wife hangs the wash on your aerial.
> **Radio Bug:** Yes, very convenient. When it has been out long enough the loudspeaker starts singing "How Dry I Am" and she takes it in.

3AEL had a 26-foot pole, the top spar of a ship's mast, mounted on a 12-foot high garage behind 85 Sixth Street, but he wanted to add three 10-foot lengths of 3-inch gutter drain pipe to make 65 feet.[50] Then he put up a 40-foot high four-wire flat top on 12-foot spreaders, with a fan cage lead-in. It got twisted up in the same storm that brought

Garage behind Fred Simpson's home at 85 6th Street (house at right), now Hamilton Street. Fred mounted the 26-foot top spar of a ship's mast on the 12-foot high garage to make an antenna for 3AEL.
-photograph by Arthur Zimmerman

down AFZ's 64-foot lattice mast.[51] In the winter of 1924-25, he picked up LPX (LPZ?), Argentina, XOX and POZ in Germany, and 8AB in France, every night. 3AEL was an ARRL observer during the eclipse of the sun in January, noting that at totality the local stations faded and LPZ, POZ, OA (Ottawa) and KDKA were very loud on 80 metres.[52] He was getting out well with 5 watts and he, 3NF and 3HE were the only stations in the city using a sideswiper, a "bug" key, to increase sending speed to 30-40 words a minute.[53] Wednesday nights at 12.30 were Canadian Nights on the exclusive 125 metres, when Canadian "bugs" could make contacts. He even came up with a scheme whereby the BCLs could send applause to their favourite radio stations with a free radiogram through 3AEL, 85 Sixth Street, telephone 1488.

Fred enjoyed the same triumphs and suffered the same disasters as his peers. He successively installed a new 40-jar rectifier, blew his plate transformer, built a new one plus a filament transformer, and then his antenna fell down, twice.[54] He applied for and received the appointment as second local official ARRL relay station, in addition to the much more experienced 3HE.[55] In May 1925, he and 3AFZ were experimenting on very low power input, and got good results sending "weird noises which he called music" with phone on 80 metres.[56] His phone was unstable at 80 metres, but contrary to "theory" it modulated better at 40-50, and still better at his "own" wave, near his fundamental of 38 metres.[57] This dropping of the wave soon became a trend, down to 40 and 20 metres. 80 metres quickly became very crowded with a lot of interference, since all the amateurs went to the same things about the same time, usually soon after each was described in QST. Thus, each new notion got a good deal of thorough, if uncontrolled and imaginative, experimenting so that the bad ideas were rapidly sifted out.

With the new low wave, 40 metres, he could pick up WNP of the MacMillan Expedition's schooner Bowdoin regularly on a three-circuit receiver with one stage of audio frequency amplification.[58] On 40 metres he could hear 7EC in Denmark, NRRL and KDUH in Hawaii, WNP on its way north, and several west coast stations. His 2,000-volt plate transformer was almost completed and approaching 60 pounds, and with a re-built rectifier he would get out in fine style.[59] He was beginning to raise his numbers in the handling of relay work, and to add to their totals he and 3HE offered to send messages free to any part of the world. He piled up 46 messages in one 7-day period in July.[60] At the same time he took 2 wires out of his 4-wire inverted L antenna and one of the two in his counterpoise to get his fundamental of 39 metres. With this arrangement he received 2YI in Australia and NRRL in Samoa in daylight and other places from New Zealand to Brazil to Switzerland and Italy at night, and was received in daylight in St Louis, Missouri, and in Florida.

Another great adventure for the amateurs was when the giant US Navy Air Corps blimp Shenandoah made a transcontinental flight from Fort Worth to Seattle via San Diego in October, 1924. The amateurs were alerted to track her and handle her press dispatches, since they were skilled at handling short wave and the government stations weren't. 3AEL heard the Shenandoah during that trip and again a few months later.[61] During the late summer of 1925, 3AEL was once more in contact with the airship's radio, NERK-1, on her midwest tour. That was just 3 days before she broke apart in a wind storm and crashed near Caldwell, Ohio, with 13 killed, including the radio officer.[62]

In pursuit of better distance, Fred took down his old flat-top inverted L and replaced it with a far superior single-wire "Hertz" antenna 45 feet long plus a 65-foot counterpoise, and did experiments to prove to the BCLs that there was absolutely no interference from his radiating system.[63] The neighbours claimed, however, that his "Hertz" aerial with the light bulb was starting fires. Fred was now able to listen to the world. At 40 metres he picked up the antipodes, Hawaii, 2NM at Caterham in Surrey broadcasting records to the United States, France, WNP, WAS and VDM in the Arctic and Mexico all in one week-end, but had trouble under 1,000 miles after 10 pm: "3AEL is getting out well and says multi-wire antennae are the 'bunk'. AEL wishes he had tried a one-wire sooner. Perhaps after he breaks a few records, a few more stations will use a one-wire aerial. All his receiving is done with a twenty-foot indoor antenna."[64] He was also experimenting with receiving sets as transmitters, and building a Schnell-type tuner.

During the summer and fall of 1925, 3AEL led all Canada for four months running in the total of messages handled. He boldly announced that he had a plan to get him the lead in North America in 1926 by boosting his message total to 1,000 a month.[65] Now that he was an "Official Broadcast Station" of the ARRL, with the job of broadcasting the latest news bulletins for local amateurs, Fred decided to go for a 9-call and higher power.[66] He played around with a 100-watt Muller tube that had a "wicked punch," but he never announced that he had got his 9-call.

His high school friend Norman Asselstine dared Fred to send a relay message to New Zealand, 10,000 miles away. It was sent on January 21, 1926, via Denver, reaching Dunedin in 5 days, and the letter of acknowledgement took 6 weeks.[67] Fred's activities had been slowing down since late autumn, probably because of school, and in January he shut down.[68] He said that he hoped to operate with a United States call that summer.[69] Just before the Daily Standard Junior terminated publication and regular Kingston amateur radio news stopped, Simpson was appointed City Manager by the ARRL.[70] Inspired by his new office, Fred attempted to get the Kingston Amateur Wireless Association going again to check interference.[71]

After this, there were sporadic reports of Fred's talking in code with Kingston in Jamaica, Uruguay[72] and MO-2, a ship in the harbour at Castagena, Columbia:[73] "while Mr. Simpson was sitting with his feet over a hot air radiator the other man was carrying out his work in the sweltering heat and with a high-speed fan going overhead." In January 1927, Fred was going again. He announced that he had handled 80 messages in 20 days, and that if he reached 100 messages before the end of the month, he would become the first Canadian ever to get into the upper category of ARRL operators.[74]

His ambition on leaving KCI had been to work for a radio station back in his native United States, but he remained in Kingston for more than a year after graduating. In 1927, the City Directory showed Fred, Jr, employed as a mechanic at the Canada Radio Stores. For a short while that spring, Fred and Mr W.J. Marshall wrote a daily "Radio Reception" feature for the new <u>Kingston Whig-Standard</u>.[75] Fred did not renew his licence for 3AEL in 1927, but by late autumn he had the call 1AH, based in Liverpool, Nova Scotia.[76] Then, in January of 1928, the press reported that he was having great success in his own contract business in Boston, installing radio equipment in boats and on land.[77] His old friends in Kingston received a few letters from him and, when these stopped, they never heard from him again. Years later they learned that he was Chief Engineer for General Electric in Oakland, California.

Richard D. Travers (3LO)

Richard Travers (RMC '31; Mech. '34) was born in Napanee in 1909. His first memory of radio was sitting in with George H. Daly (3FG), who owned a one-tube regenerative receiver, listening to the voice broadcast of the Dempsey-Carpentier fight. That was July 2, 1921.[78] He recalls that the announcements were just the bare bones, like a computer voice, with no dramatics or colour, and he and Daly were terribly disappointed because Carpentier lost. That experience got him started in radio.

The Travers family arrived in Kingston via Walkerville in 1925 and around January 1926 his call first appeared in the <u>Daily Standard Junior</u>. According to these reports, Travers met the Kingston gang and told them that he was going on with a "fiver" on 40 metres. He worked the local hams first, using his receiver.[79] He got his junk perking late in January, with the help of 3AFZ, and operated locally on 40 metres with his 5-watter.[80] He must have impressed as quite a competent fellow, since he was soon reported winding a 700-volt transformer for his plate supply and building a rectifier to use with it.[81] His antenna, suspended between two trees, fell down and he too opted for a Hertz antenna.[82] One of the last reports told of his rebuilding his transmitter for work on 7,500 Kc. (40 metres).[83]

> <u>Richard Travers</u>: We lived in the outskirts, one house in on the southeast corner of Bath Road and the Pen Road, a little west of where the bus station was. Actually, I borrowed that 700-volt transformer from another ham. The only way I could think of to break the circuit was to key the whole 700 volts, and my hand slipped. I nearly killed myself. My father made me return it.
>
> It wasn't a Hertz antenna, but a Windham, an off-centre-fed dipole with a single-wire feed line. It's an all-band antenna, if you cut it to the right length and put the feed point in the right spot. They're still being used.[84]

Jack M. Campbell, Jr: I wish I could locate this fellow Dick Travers. He was a great fellow on the code. He used to come down and run my outfit on code all the time. He was really fast.[85]

Travers left Kingston around 1934-6, and was listed as holding VE3WZ in Hamilton, Ontario, in 1938.[86] After retiring, he took up the radio game again in 1980, and now holds VE3MFJ in Picton.

Valentine Sharp (ex-2EM/3VS)

Valentine Sharp moved to Kingston from Montreal early in 1926. He had held the call 2EM in Quebec, but nothing is known of the extent of his wireless experience there. The call 2EM first showed up in print in Kingston in October 1925,[87] with the news that Sharp expected to be operating in Kingston in the winter. He told Fred Simpson that he was given special permission to use the call 3EM when he renewed his licence for use in Ontario in the spring.[88] After he arrived in Kingston to stay, however, he tried his Amateur Experimental Proficiency Certificate test[89] and was granted the call 3VS instead.[90] Sharp started work with his junk in the spring, as predicted,[91] and announced that he would go on air as soon as his licence was renewed for 1926. Sharp must have gone on the air with 3VS just as Harold Stewart was shutting down 3NF.

Sharp got embroiled in the controversy about VBH interfering with the BCLs when he defended the Marconi station in a letter to the editor of The Whig-Standard. He pointed out that the "staccato volleys of sparks" from VBH that disrupted listening could be the response to an emergency. A quenched spark transmitter would still cause interference and a continuous wave set could not be received by ships having only crystal receivers. Besides, BCLs had the whole winter free, so why complain about a few minutes of noise at a time during the summer? He suggested that BCLs tune down to 100 metres, where there was no interference and better results besides.[92] Another correspondent pointed out that no one disputed the necessity of VBH, but Kingston could do with a modern wireless station, like the one in Toronto, which did not interfere with radio broadcasts.[93]

For a while, from mid-1929 to mid-1930, Sharp was living in Aurora, Ontario,[94] the site of Ted Rogers' original CFRB transmitter poles. Sharp's mailing address was in care of radio station CFRB, Toronto, so he may have had a job with CFRB to do with transmitter engineering. In September 1930, he was back in Kingston,[95] and appeared in the Official List at various rooming house addresses every year through 1936. There was a V. Sharp, with the call VE2CR, in Montreal from late in 1930,[96] shortly after Sharp returned to Kingston from Toronto. It could have been the same person. He turned up again in Toronto in May 1961, as VE3LJ and President of the Queen's Park Radio Club.[97]

Keith MacKinnon: Valentine Sharp...oh, I knew him pretty well. He was a nice chap. The main thing I remember about him, he used to often see me whenever he came to Ottawa for something or other, he'd call me up. Way back in those days, I think it was Red Lake was opening a gold mine and it wasn't on any highway or anything, and they had no communications, so the Ontario Government got Val Sharp and some other guy (possibly Orton Donnelly) to set up, and they worked the whole damn thing for the winter.[45]

Arthur Wehman

Art Wehman lived at 251 Division Street in his youth and was part of the Stewart-MacKinnon-Cohen lunch-time wireless group, though he never sought a licence. Like all his fellow young "bugs," he was constantly experimenting to improve his gear. He is perhaps still best known in Kingston as the local CPR telegrapher.

<u>Art Wehman</u>: I started way back around 1915 or '16. A pal of mine, Norm Veale, and I were kind of inclined to be inventive sort of kids with electricity. We got a spark coil out of my Dad's old Ford truck and a Hot-Shot 6-volt battery and we began fooling around with shocking machines. Hallowe'en night we rigged up our doorbells with this spark coil, and as the kids come along to ring our metal doorbell, they got a little shock. We had a lot of fun and the kids did too, and then we called them in and give them their treats. Unfortunately, I left that on all night. The milkman come along in the morning and he got a shock and he didn't like it, so he promptly told my mother. His name was Leslie and he turned out to be a pretty good scout.

We saw an ad in an American newspaper for a small crystal set, so we went over to Clayton, NY, on a moonlight excursion on the old steamer <u>Waubic</u> and picked up a chunk of galena. Then we got some #22 cotton-covered wire and a couple of cardboard "pickle jars" - you'd get a quart of pickles at a grocery store and they dumped them in this little cylindrical wax-covered cardboard container - for winding our coils and we built a small crystal set. We picked up experimental stations and the Marconi stations on the lake, VBG in Toronto, VBD at Tobermory and VBH over here in Barriefield. And we got to know those wireless operators very well. George Gurney was one of the operators and he was actually one of the gang too, and we'd congregate in the Canada Radio Stores when it opened up on Princess Street, with Ort and Gordon, the Radio Inspector and a few others and exchange ideas. One big happy family: "Why don't you try this? Oh, this is better than that. I'd do this if I were you."

Now we needed an aerial, so I went down behind the Bell Telephone plant on Brock Street and, delving into the garbage, I picked up perhaps two-foot lengths of copper wire, painstakingly scraped the ends and soldered them together, and that was our aerial. It went from my bedroom window to a clothesline at the end of the yard, about 50 feet away. We would sit up there and yack away and listen to KDKA or WBZ. I would hustle home from my paper route, put the old earphones on and pick up CFCA, the Toronto Star station. I was the only one who could get it, since my aerial happened to be running in that direction. For ground wires, most used a rod driven into the ground, but I got a brain wave. I got a 5-gallon tank, soldered a wire

onto that, filled it with water, dug a trench, thinking that it would be
an improvement. It was, a little. You went along experimenting...

We used to go home from school at lunchtime and talk around
town on our radios...Dick Cohen and Sheldon and Harold Stewart
and anyone, to improve his set or try things. Every kid experimented
with this and that, crazy things, but a lot of them turned out to be
components of even TV's of to-day.

As time went on I saved up enough money to buy a 3-element
tube, and it cost me $8.50 - seems to me it was a Cunningham. It was
a dandy! So Norm and I branched out into a one-tube set - receiving
only, but we could transmit with it too. We had a nice piece of board
for a panel and we made a vario-coupler and a variometer. Our variable condenser was two test tubes, about 6" long, one inserted in the
other, and the outside of each was covered with tinfoil. We fastened that
to the panel with flexible wires. To operate the condenser, we put a cork
in the centre and drew it in and out with a matchstick very, very
patiently and we would get our station. That was really brilliant and
it was our idea. A year or so later we bought a variable condenser
where the rotor plates interleaved with the stator plates. For a rheostat
we picked up a German flexible metal armband, cut it in two, nailed
it to a flat board and with a flexible piece of copper we managed to control the filament voltage for the tube.

Every Saturday night I would sit up all night long listening. This
was a regenerative circuit and by turning the condenser, the vario-
coupler and the variometer I could bring in stations. At first I'd hear just
a faint whistle but with lots of patience I could bring in very faint
voice. I had amazing results because at that time I had lots of patience,
which I haven't got now. One morning, possibly two or three o'clock,
I heard Winnipeg and Tacoma, Washington, and Portland, Oregon, testing - talking back and forth and making tests of their own. I heard
the big ship <u>Shenandoah</u> testing in mid-Atlantic; and Los Angeles and
Mexico City. One night I almost fell off the chair when I heard this chap
say he was Cuba. He was an amateur in a small town near Havana, and
he said he was playing records. I frantically wrote down the names
and the times. He didn't have any call, but asked if anyone on the
mainland - and he meant Florida - should hear him would kindly write
to him. I wrote to him next day and I was the first one in Kingston to
have heard that distance because my little set was fantastic.

You could use that receiver as a transmitter. You'd open up the
ground circuit, put a wireless key in that circuit, turn up the volume until
the tube began to screech, turn your condenser up until you got a high
frequency note, and you were in business with a CW transmitter.
Possibly a range of 5 or 10 miles. Now, if you wanted voice, you took
the key out, made your ground regular again, took possibly 8 or 10
turns of #22 cotton-covered wire attached around the outside turns of
the vario-coupler, made a loop and in that loop you put a microphone
- an old carbon grain microphone discarded from a Bell telephone.
Crude, but workable and you again turned your volume up and your
condenser and got the best you could between whistles, and you were
on the air with a little voice transmitter. I didn't have a licence - in
those days you could get away with murder, as long as you didn't
interfere with VBH or VBG. Besides, the inspector was one of us in
those days. Bob Davis. Lived in a big brick house out on the corner of

Collingwood and Union, with a beautiful great big mast sticking up from his roof.

My Dad and my older brother kicked in and we got a re-chargeable 45-volt car battery and a battery charger and I added two more tubes to the set. We didn't get any greater distance but we were able to operate a small loudspeaker.

I knew Earl Morris very well and Don MacClement. Wally Reid (3ACX), holy cow! His parents ran a little gocery store on the corner of Pine and Division Street. He was a little older than I was, and had a brother "Dooley." We were neighbourhood friends, played a little shinny together or ball. Two fine boys. Alec Ada's dad had a small garage at 24 Division Street. I knew him slightly from getting together in the Canada Radio Stores. Never heard of Kenneth McAlpine...but 3NE? Yes! I **heard** him on the air! Yes! I would say around 1922. Hah, hah, hah! And Dr Mylks' son Gordon - I think he was phone.

Norm's uncle George was manager of the CN in Toronto and, knowing that I wanted to become a Morse land line operator, suggested I go to his friend the manager here in Kingston, put my name in, become a telegraph messenger boy and learn it. By fate I finally got hooked up with the CPR. I had to push the wireless to one side because the codes were a little bit conflicting. Later, when war broke out, I enlisted and went overseas as a wireless operator. So, that meant I brought out the wireless code from between the ears, and was put on some big stations to copy German frequencies. Let me tell you, they had some real smart operators and fantastic equipment.[98]

Other Regional Radio Clubs in 1924

The Island City Radio Association, 10BB, in Brockville had 136 members in June 1924, and a six-wire cage aerial 75 feet long and 25 feet off the top of a four-storey building, the Code and Tetts Electric Store on King Street West. The ICRA had a 10-watt phone transmitter and broadcast on 250 metres. Meetings were held on Wednesdays at 8 pm.[99]

The Belleville Radio Association was formed January 28, 1924, with 51 members, and met every second Friday at the Chamber of Commerce rooms.[100] There was also a radio club meeting in Napanee in 1924.

Interference Reports

Radio interference was a constant problem in the mid 1920s, especially in the warm weather, the "QRN season" (i.e. strong atmospherics). Broadcast Concert Listeners blamed the amateurs, but their numbers had fallen off in Kingston from over 50 in pre-broadcast days to a solid and responsible half-dozen or so by 1925. The worst and most unpredictable interference was from atmospherics and storms. There was terrible machine-gun interference from VBH and from boys with spark coils, but it also came from leaks and sparks from street cars, neighbourhood power transformers, telephone generators and automobile spark coils. Household appliances such as heating pads and ultraviolet ray instruments caused radio interference. So did machines containing motors with commutators - oil furnaces, vacuum cleaners, washing machines and sewing machines - and door bells and buzzers too. In October 1924, the Public Utilities Commission began removing the old arc lamps from the street standards, installing high-current electric light bulbs in their place. The old arc lamps used to interfere with radio reception, and people were happy to see them disappear.[101]

Later, the Government Radio Interference Squad returned to discover that a major cause of local interference and buzzes was power leaks in the generators at the telephone exchange.[102]

F.J.C. Dunn, the Kingston area Radio Inspector, was plagued by someone flooding the city with spark coil interference late in 1925, and tried to locate him with a selective receiver and loop. The Director of the Radio Service informed Dunn that the loop was unsuited for direction finding with spark. He advised that Dunn rely on signal strength to find the district and then make a house-to-house search for aerials. Three law-abiding amateurs were enlisted to help, including Orton Donnelly,[103] and newspaper articles appeared warning of the arrival of the RCMP and inspectors from Ottawa.[104] Dunn and Donnelly discovered and eliminated the source of a buzz in the area of Kingston Junction.

On top of all this were the omnipresent squeals from poorly tuned local regenerative receivers. Even CFRC was known to cause some interference to Kingston's BCLs. But the amateurs and kids with spark coils were constantly under the gun from the BCLs, and editor Fred Simpson rose to their defence:

> Interference: For the past four years the amateur has been called all names imaginable by the Broadcast Concert Listener who thinks every commercial station, street car noise, power leak, telephone generator, etc., is an amateur. They even went so far as to have a local amateur's licence cancelled...All single circuit tuner listeners and unlicenced B.C.L.'s have no cause for complaint at all concerning amateur interference as the single circuit tuner is one of the least selective tuners on the market.[105]

Many BCLs didn't understand the fine art of tuning their single-triode regenerative receivers. The tickler or feed-back coil in the output or plate circuit is coupled inductively to the input or grid circuit coil, the secondary of the tuner, to return a portion of the radio frequency current from the plate to the grid circuit and so obtain regeneration or feed-back. This produces more current in the plate circuit and acts as an amplifier. If it isn't set just right, at the zero beat point, feeding back the right amount of energy in the correct phase, the tube oscillates and sends out squealing radio signals of its own. In New York, the practice was that a squealer would first be sent a card by the victim(s). If things did not improve, the squealer would open his door one morning to find a package containing his aerial insulators pounded into powder, his aerial wire cut into 6-inch lengths and another warning. Canadians were much more civilized. In Canada the complainant telephoned the main local broadcasting station, which then announced that a person in such a block was interfering by not tuning his set properly. The next step was to announce the perpetrator's name, address and telephone number, the result being "that he is usually kept so busy answering the phone that he cannot tune his set."[106]

Paul Painleve, French Minister of War and a distinguished scientist, was of the opinion that the dull, rainy weather that covered the northern hemisphere in June 1926 was the fault of the strong worldwide radio activity of the wireless:

> If you introduce radiotelephonic emissions into a tightly enclosed room where the air is absolutely transparent, in a moment little drops of water will begin to form on your face. What has happened? The Hertzian fog has become ionized and electricized, and the invisible fog, found even in every home has turned into water. You have made rain in your own home. One, therefore, can conclude that the multitude of radiotelephonic waves existing in the ether just now produce the same effect upon the atmosphere, the result being the very beautiful month of June with rain each day.[107]

In 1927, when there were 624 licenced receiving sets in Kingston, local inspector Frank Dunn wrote a series of articles for The Whig-Standard on radio interference.[108] There were then 200,000 receivers in all of Canada, although only half of all owners paid the one dollar annual licence fee.[109] Every little while the Radio Inspector let it be known through the press that there was a drive on to round up the unlicenced delinquents, or that officials from Ottawa and the police were snooping around the city with a direction-finding automobile looking for unlicenced receivers. Big fines were promised.

Other Kingston Amateurs, To 1938

VE3GO	C.A. Miller	104 Mimico Avenue, Mimico c/o Knapp's Boat Livery, Ontario East (from Nov. 1/33)
VE3JC	R.J. Chambers	94 Division Street (from Sept. 1/30)
VE3FG	O. Shaw	C.N.T. Repeater Station, Montreal Street, Kingston Junction (from Sept. 1/30 to Jan. 1/32)
VE3KR	F.J. McDiarmid	205 Stuart Street (from Jan. 1/32)
VE3NT	H.H. Stewart	62 Livingston Avenue (from Nov. 1/33)
VE3AI	A.W. Richmond	78 Clergy Street West (from March 31/34)
VE3ABA	O. Donnelly	560 Princess Street (from Nov. 1/34)
VE3BU	J.V. Wiskin	32 Nelson Street (from March 31/36)
VE3BY	G.W. Motherwell	Old Arts Building, Queen's University (from March 31/36)
VE3MC	J.J. McWatters	86 Princess Street (from March 31/36)
VE3ACH	R.E. Freeman	532 Albert Street (from March 31/37)
VE3JG	J.M. Campbell	Gananoque (from March 31/37)
VE3VX	G.W. Motherwell	Queen's University Radio Club (from March 31/37)
VE3ADG	W.K. Salisbury	45 Sixth Avenue (from March 31/38)
VE3AQY	P. Joron	132 Wellington Street (from March 31/38)
VE3IV	A. Gillingham	c/o R.C.C.S. (from March 31/38)
VE3TL	L.A. Milton	109 Alfred Street (from March 31/38)
VE3VX	L.G. Askwith	Queen's University Radio Club (from March 31/38)

NOTES

1. "Cyclone Strikes", DBW, Sept. 30, 1921, pp. 1 and 20.
2. Personal communication from Jack M. Campbell, Jr., Gananoque, Oct. 7, 1988.
3. "What the Radio Amateurs", KDS/DSJ, Oct. 24, 1925, p. 10; Feb. 20, 1926, p. 7; April 17, 1926, p. 8.
4. OLRSC, DMF, Marine Branch, Ottawa, June 30, 1927.
   Ibid., Department of Marine, Radio Branch, Ottawa, Jan. 1, 1932.
5. "Listening In", KDS/DSJ Sept. 13, 1924, p. 8.
6. "Listening In", ibid., November 1, 1924, p. 11.
7. "With the Amateurs", ibid., June 27, 1925, p. 9.
8. "Listening In", ibid., March 22, 1925, p. 10; May 2, 1925, p. 7.
9. "Listening In", ibid., May 9, 1924, p. 7; June 6, 1925, p. 9.
10. "With the Amateurs", ibid., June 20, 1925; July 4, 1925, p. 8.
11. Ibid., Oct. 24, 1925, p. 10; Oct. 31, 1925, p. 10; Nov. 7, 1925, p. 7.
12. "What the Radio Amateurs", ibid., Dec. 5, 1925, p. 9; Feb. 27, 1926, p. 7.
13. "Listening In", ibid., Oct. 4, 1924, p. 7.
14. "Listening In", ibid., Oct. 18, 1924.
15. "Listening In", ibid., Nov. 29, 1924, p. 11.
16. "Listening In", ibid., Dec. 13, 1924, p. 8.
17. "The Amateur", ibid., Jan. 31, 1925.
18. "With the Amateurs", ibid., Feb. 7, 1925.
19. "Listening In", ibid., April 25, 1925; May 2, 1925, p. 7.
20. "Listening In", ibid., June 6, 1925, p. 9.
21. "Listening In", ibid., Dec. 13, 1924, p. 8.
22. "With the Amateurs", ibid., July 4, 1925, p. 8.
    "Radio Notes", ibid., July 11, 1925, p. 7.
23. "What Local Radio", ibid., 5, Aug. 15, 1925, p. 5; Aug. 27, 1925, p. 7.
24. "Kingston Fair is Opened", DBW, Sept. 15, 1925, p. 1.
    "Kingston Fair", KDS, Sept. 15, 1925, pp. 1 and 5.
25. Personal communication from Harold A. "Dick" Cohen, Kingston, Feb. 15, 1988.
26. "Local Radio Station Opened", KDS, Sept. 16, 1925, p. 2.
27. "Station CFM on the Air", ibid., Sept. 19, 1925, p. 5.
28. "Radio Station CFMC", ibid., Sept. 21, 1925, p. 12.
29. Personal communication from Harold Cohen, Kingston, Oct. 31, 1985.
30. "Gave Address Over Radio", KDS, Sept. 23, 1925., p. 5.
31. "Fine Broadcasting", ibid., Sept. 26, 1925, pp. 1 and 2.
32. "John M. Campbell Chosen", DBW, Sept. 26, 1925, p. 1.
33. "To Broadcast", KDS/DSJ, Sept. 26, 1925, p. 8.
34. Jerry Stagg, The Brothers Shubert, Random House, New York, 1968, p. 187.
35. "Amusements: At the Grand", KWS, Sept. 26, 1925, p. 13.
36. "Shubert Official", KDS, Oct. 2, 1925, p. 2.
37. Personal communication from Harold Cohen, Kingston, Sept. 29, 1982.
38. "Fire on Princess Street", KDS, Nov. 2, 1925, p. 12;   "Radio Not Damaged", ibid., Nov. 3, p. 12.
    "Fire This Morning", DBW, Nov. 2, 1925, p. 2.
39. "Broadcasting Returns", KDS, Sept. 14, 1926, p. 14.
40. "CFMC Preparing for Season", ibid., Sept. 21, 1926, p. 12.
41. OLRSC, DMF, Marine Branch, Ottawa, June 30, 1927.
    Ibid., Jan. 1, 1929.
42. Personal communication from Dr. Irwin Sugarman, Kingston, April 11, 1989.

43. "Foreign Radiobroadcasting Stations by Countries: Canada", World Radio Markets in 1926, United States Department of Commerce, Washington, 1926, p. 15.
44. Personal communications from Harold Cohen, Kingston, Oct. 31, 1985; Feb. 15, 1988.
45. Personal communication from Keith A. Mackinnon, Ottawa, August 15, 1987.
46. Radio 3AEL, "The Radio Amateur" Times of K.C.I., 1926, p. 61.
47. "Work in the Arctic", KDS/DSJ, July 12, 1924, p. 11. \
    "Arctic Reports", ibid., July 19, 1924, p. 11.
48. "Listening In", ibid., Sept. 6, 1924, p. 11.
49. "Listening In", ibid., Oct. 4, 1924, p. 7.
50. "Listening In", ibid., Oct. 18, 1924, p. 8.
51. "Listening In", ibid., Dec. 6, 1924, p. 7; Dec. 13, 1924, p. 8.
52. "The Eclipse", ibid., Jan. 31, 1925.
53. "With the Amateurs", ibid., Feb. 14, 1925, p. 7.
54. "Listening In", ibid., Feb. 28, 1925, p. 10.
    "With the Amateurs", ibid., March 22, 1925.
    "With the Amateurs", ibid., June 27, 1925, p. 9.
55. "Listening In", ibid., March 28, 1925.
    "Listening In", ibid., April 25, 1925.
56. "With the Amateurs", ibid., May 2, 1925, p. 7.
    "Listening In", ibid., May 16, 1925.
    "Listening In", ibid., May 23, 1925.
57. "With the Amateurs", ibid.,, May 30, 1925.
58. "With the Amateurs", ibid., June 27, 1925, p. 9.
    "With the Amateurs", ibid., July 4, 1925, p. 8.
59. "Radio Notes", ibid., July 11, 1925, p. 7.
60. "Radio Notes", ibid., July 18, 1925, p. 8.
61. "Listening In: Shenandoah", Oct. 18, 1924, p. 8.
    "Listening In", ibid., Jan. 17, 1925.
62. "Giant Airship Shenandoah", DBW, Sept. 3, 1925, p. 1.
    "Twice Widow by Disasters", ibid., Sept. 4, 1925, p. 1.
    "Listening In", KDS/DSJ, Sept. 19, 1925, p. 7.
63. "Listening In", KDS/DSJ, Sept. 28, 1925, p. 8.
64. "Listening In", ibid., Oct. 3, 1925, p. 9.
65. "3AEL Led Canada", ibid., Feb. 13, 1926, p. 7.
66. "Listening In", ibid., Jan. 16, 1926, p. 7.
    "Listening In", ibid., Feb. 6, 1926, p. 8.
67. "Sent Radio Messages", KDS, March 13, 1926, p. 8.
68. "What the Radio Amateurs", KDS/DSJ, April 3, 1926, p. 8.
69. "What the Radio Amateurs", ibid., April 17, 1926, p. 8.
70. "What the Local", ibid., May 8, 1926, p. 7.
71. "Radio Amateurs to Re-organize", KDS, Sept. 28, 1926, p. 5.
72. "Local Amateur Reaches Jamaica", KWS, Dec. 13, 1926, p. 16.
73. "Long Distance Conversations", ibid., Dec. 15, 1926, p. 2.
74. "Kingston Amateur", ibid., Jan. 18, 1927, p. 7.
75. "Radio Reception", KWS, March 28, 1927, p. 12.
76. OLRSC, DMF, Marine Branch, Ottawa, June 30, 1927.
    "Supplement No. 1 to OLRSC, ibid., November 30, 1927.
77. Great Success Met With", KWS, Jan. 11, 1928, p. 2.
78. Gleason L. Archer, History of Radio to 1926, American Historical Society, New York, 1938, p. 213.
79. "Listening In", KDS/DSJ, Jan. 23, 1926, p, 7.
80. "Listening In", ibid., Jan. 30, 1926, p. 8.
    "Listening In", ibid., Feb. 6, 1926, p. 8.

81. "What the Radio Amateurs", ibid., Feb. 27, 1926, p. 7.
"What the Radio Amateurs", ibid., March 6, 1926, p. 7.
82. "What the Radio Amateurs", ibid., Feb. 13, 1926, p. 7.
"What the Radio Amateurs", ibid., April 17, 1926, p. 8.
83. "What the Radio Amateurs", ibid., April 3, 1926, p. 8.
84. Personal communication from Richard Deming Travers, Picton, Feb. 24, 1991.
85. Personal communication from Jack M. Campbell, Jr., Gananoque, Nov. 11, 1982.
86. OLRSC, Department of Transport, Radio Division, Ottawa, March 31, 1938.
87. "Listening In", KDS/DSJ, Oct. 3, 1925, p. 9.
88. "Listening In", ibid., Jan. 16, 1926, p. 7.
89. "Listening In", ibid., Feb. 6, 1926, p. 8.
"What the Amateurs are Doing", ibid., Feb. 20, 1926, p. 7.
90. "Granted Licence", ibid., March 20, 1926, p. 7.
Just after this time, the Department of Marine and Fisheries changed its policy and began to grant personalized call letters. Valentine Sharp's old call, 2EM, was given to Lancely R.C. McAteer (Mech, '25) by Marine and Fisheries in 1929.
91. "What the Radio Amateurs", ibid., April 3, 1926, p. 8.
92. Valentine Sharp to the Editor, KWS, Dec. 14, 1926, p. 9.
93. A. Williams to the Editor, ibid., Dec. 15, 1926, p. 2.
94. "Supplement No. 2 to OLRSC, 1929 Edition", DMF, Marine Branch, Ottawa, July 1, 1929.
95. Supplement No. 1 to OLRSC, 1930 Edition, Department of Marine, Radio Branch, Ottawa, Sept. 1, 1930.
96. "Supplement No. 2 to OLRSC, 1930 Edition", Dept. of Marine, Radio Branch, Ottawa, Nov. 1, 1930.
97. Saskatoon Amateur Radio Club VE5AA, From Spark to Space: The Story of Amateur Radio in Canada, Saskatoon, 1968, p. 91.
98. Personal communication from Arthur Wehman, Kingston, June 14, 1990.
99 KDS/DSJ, June 28, 1924; ibid., July 5, 1924.
100. "Radio Clubs: Belleville", ibid., July 12, 1924, p. 11.
101. "Listening In: Editorial Chat", ibid., Oct. 4, 1924, p. 7.
102. "Have Located Interference", KDS, Nov. 4, 1925, p. 12.
103. "Listening In", KDS/DSJ, Feb. 6, 1926, p. 8.
104. "Radio Inspectors Expected", DBW, Feb. 3, 1926, p. 12.
105. "Editorial Notes", KDS/DSJ, Feb. 7, 1925.
106. "Editorial Notes", ibid., Feb. 21, 1925.
107. "Radio Is the Cause?", DBW, June 17, 1926, p. 4.
108. "Local Inspector", KWS, Nov. 26, 1927, p. 13.
"Sources of Radio Interference", ibid., Dec. 3, 1927, p. 11.
109. "200,000 Radio Sets", ibid., Dec. 28, 1927, p. 12.
"City and District: Radio Licences", ibid., p. 7.

CHAPTER 10

The early CFRC "squirrel cage" L-antenna, with three sets of three 4-foot spreaders visible, between Ontario Hall and Fleming Hall in the late 1920s. Instead of a ground, this system was operated with a counterpoise - a second wire slung under the antenna close to the ground, not visible - since the earth in that spot is shallow, sandy and poorly conducting. The pole on Ontario Hall is on the same spot as the pole which supported that end of the recently removed CFRC-AM T-antenna. Note the sloping roof of the pre-fire Fleming Hall.
-courtesy Queen's Archives

CFRC, 1924-1936

Professor Douglas M. Jemmett signed requisition number 2460 on April 5, 1924, to authorize "A certified cheque for $50.00 to Renew Radio License #33 Call - CFRC payable to Dept. of Marine and Fisheries, Ottawa."[1] The Department of Electrical Engineering was granted "License to use Radio" #22, valid from April 1, 1924, to March 31, 1925.[2] The licence states that the daytime range of CFRC was 500 miles, that the licencee had an inverted L aerial, a choke control type of transmitter taking a maximum power at the anode of 1,500 watts and

3,000 volts, and the decrement per complete oscillation was not to exceed 0.08. The wavelength, 450 metres, was noted as subject to amendment. With respect to the operator, the licence states: "If the operators are not holders of Canadian Certificate of Proficiency, they must be competent to work the apparatus and will be subject to the approval of the Minister." Doug Geiger did not hold a Certificate of Proficiency, and neither did his successor.

J.W. Bain Succeeds D.G. Geiger at CFRC

When Doug Geiger left Queen's in the spring of 1924 to work for Bell in Montreal, James William L. Bain was put in charge of CFRC. Bain was born in St-Polycarpe, Quebec, in 1883. He matriculated from Abingdon School in 1901, and must have spent some time working because he joined Science '14 at McGill for the 1913-14 session and graduated in 1914. Bain was for a while Assistant Engineer to Montreal's Electrical Commission. He went overseas with Princess Patricia's Light Infantry in September 1914, in the 1st Division, and from mid-1917 he was in intelligence work with the 1st Army. On his return to McGill in 1919, he was appointed Senior Demonstrator in Electrical. He came to Queen's for the 1922 session as Lecturer in E.E., replacing Robert Leland Davis. Bain taught courses III and VII and the laboratory for course XI, Lester Gill's old "Telegraphy and Telephony," through the 1925-26 session. His trail is lost after 1926, but in 1946 he was working for the Radio Branch of the Department of Transport in Ottawa.[3]

Professor Bain started into his new charge at the radio station at Queen's with a real zest and very ambitious plans.

Early in the 1924-25 term, an article about radio at Queen's appeared in the Journal, articulating the aims and hopes of CFRC. The ideas could have been Jemmett's, but it seems rather that they reflect Bain's enthusiasm.

> First, it is our purpose to keep our Alumni as closely in touch with our doings here as our radio equipment will permit. Second, it pays to advertise. In the Industrial World it is a recognized fact that a good broadcasting station does more to win friends than any other agent. We want to use our station to boost Queen's not only for the benefit of the University itself, but also for the benefit of the graduates and undergraduates. At present our rugby team is by far our best advertisement, and yet those of us who cross the line are often asked if Queen's has a football team! It is up to C.F.R.C. to tell them about it.
>
> In these days you are being fervently adjured to get behind so many things that it is no wonder that your lights are completely obscured. Therefore we are not going to ask you to get behind us; we want you to get in front of us and criticize for all you are worth. We hope that everyone from the most learned senior to the most timid Freshette will give us their ideas as to how we can

James William Bain was hired in September 1922 to replace R.L. Davis as Lecturer in E.E. at Queen's. When Doug Geiger left Queen's for Bell in the spring of 1924, Bain was put in charge of CFRC. He carried out many ambitious broadcasting projects in his two years with CFRC, including the re-building of the Mark II to create the Mark III transmitter in 1925-26.
-courtesy Queen's Archives

be of more and better service. Send in your suggestions to the Radio Director and leave it to us to sift the wheat from the chaff. Another thing - please tell all your relatives, friends and acquaintances about it and ask them to tell us how we come in. Professors Jemmett and Bain need this information in order that they may improve the operation of the set. We have been picked up as far South as South Carolina, but reports from all points will be welcomed.[4]

In preparation for the new season, Jemmett invested in a new microphone, a Western Electric R-373 costing $115, plus a mounting for the microphone, for an extra $60.[5] He also had Marine and Fisheries calibrate the CFRC wavemeter, for $3,[6] and instructed the "Bell Telephone Company to install a pair of wires from Stadium to Fleming Hall with local battery phones as last year."[7]

CFRC Broadcasts in 1924-25

CFRC tested on October 1, 1924, with Mr Mager of Massena, NY, and on the morning of the exhibition football broadcast of the Queen's-Royal Military College game, October 4, tested from the stadium and announced the afternoon's radio program.[8] That was the first of eight games broadcast from the stadium that autumn, four senior (exhibition matches October 4, 11, and the usual two regular league games in Kingston, October 18 and November 1) and four intermediate (October 11, November 10 and 29, and December 6). The October 11 broadcast consisted of two games, back to back, beginning at 1.15 and concluding at 5.00. Presumably, Professor Jolliffe continued to handle all of the on-air duties, and the program format remained unchanged: "We go on the air at least fifteen minutes before the games are scheduled to start and we endeavor to keep talking from then until the final whistle goes. The hockey games, etc., are also to be broadcast in the same manner."[4] The amateurs heard that the Queen's-Varsity match on October 18 was received in Toronto and by alumni all over the province,[9] although the CFRC log did not record any reports.[8] The Queen's Tricolor finished the season with six wins and no losses, the Intercollegiate Championship, the Eastern Championship and, by default, the Dominion Championship and the Grey Cup. The Senior finals were not played in Kingston, so CFRC could not carry them. CFRC did, however, air Queen's II defeating the Sarnia Hard Oils for the Intermediate Championship on December 6.

Donald Ridgeway McLeod (Elec. '26) was elected for 1924-25 to continue the work begun by John C. McGillivray, and the office of Journal Radio Director was created for him.[4] The Journal News, broadcasting "news of interest to graduates of the University,"[9] began on October 15 and continued faithfully every Wednesday from around 9.00 to 9.45 pm, until it was pre-empted by the Queen's I-RMC hockey broadcast on December 10. That was the day of the formal opening of the new Jock Harty Arena. On November 19, a "Post Office Broadcast" was added to the Journal News, probably a notice about mailing early for Christmas or advice on wrapping parcels. The Journal News resumed on January 14, 1925, but was knocked out by transmitter failure on the 28th and by a special remote dance broadcast a week later. The Journal crew re-grouped on March 4 and 11 to finish off the broadcasting season.

The winter fare on CFRC was mostly hockey from the re-built Jock Harty Arena and basketball from the gymnasium, but not just Queen's games:

> The Whig has received many good reports regarding the broadcasting over the radio on Friday evening last, of the hockey match between Brockville and Kingston (Intermediates) at the Jock Harty Arena.
> "I was listening-in on the game," said a well-known resident of the Bath Road, "and I heard everything that went on. It was certainly great to stay at home and get all the particulars regarding the game, in my own home. The service was excellent. I got everything, play by play, and could hear the cheering of the fans, and it was one of the finest treats of the season over the radio."
> The Whig heard several other radio fans talking about the wonderful service on Friday night's game.[10]

In late November, CFRC had been reported causing bad interference to the BCLs, and listeners were wondering why CFRC did not broadcast any concerts.[11] The breakdown on January 28 could have been caused by a faulty tube but, unfortunately, the record of tubes was abandoned in November, so we can't know for certain. In any event, the interference was cured at about that time.

CFRC was back on the air at noon on January 29th with some harmonica solos,[12] but the log contains no reference. This music was a test which was part of a larger plan, explained by CFRC announcer and Journal Radio Director D.R. McLeod in a letter to the Queen's Journal on the 29th:

> I would like to use this means of announcing that the experiments which Professor Bain and myself have been conducting with a view to broadcasting the Arts and Science Dances, for the benefit of students and residents of Kingston, have so far been successful. We will be making some more tests this morning (Thursday), and after midnight to-night, which will show us whether we can expect success. We are still having trouble but from the results of the experiments which we have already made we feel safe in advising radio fans, who want to hear Jardine's Orchestra, to look out for us at 8 p.m. on Friday. We will be operating at reduced volume to give Kingstonians the best possible conditions for reception. We should, however, have a range of 100 miles (about one-eighth of normal), and we will increase it if we can. I may say that the Canada Radio Stores, 269 1/2 Princess St., usually puts us on a loud speaker in their store.[13]

The next evening they attempted to broadcast dance music played by J. Wilson Jardine's Orchestra from the Arts "At Home" at Grant Hall. This went to air half an hour after the completion of the Kingston-Brockville Intermediate hockey match at the Arena.[14] The CFRC log says that the transmitter was very unstable, but "CFRC met with fair success when they broadcasted the concert on Friday evening. Although the music was not very loud, the results were considered successful by those in charge of the station...CFRC does not cause so much interference as formerly."[15]

The CFRC transmitter was fired up on Saturday night at 8.00 for the Senior Basketball game, when the Queen's hoopsters defeated McGill 31 to 29. At half-time, Queen's women trounced KCI in an exhibition match.[16] On February 4, Bain broadcast the music from the annual Science "At Home" in Grant Hall, again provided by Jardine's Orchestra of Toronto. This time the transmitter was on from 9.30 pm until 2.30 am.[17]

Bain installed Magnavox loudspeakers in the second Jock Harty Arena and, with the new CFRC microphone, the broadcast quality was improved: "CFRC is getting a little better results since he installed the new microphone. While CFRC was broadcasting a hockey game recently the musical affair was playing in the Arena, and of course, was broadcast too. A local listener, not knowing this, spent some time trying to separate the speaker from the music as the latter was interfering a little. He gave up after a few minutes of unsatisfactory and unnecessary trouble."[18]

J.E. Macpherson, Executive Assistant, Bell Telephone Company of Canada in Ottawa, wrote to Principal R. Bruce Taylor late in December 1924, enquiring whether CFRC was giving lectures to extramural students, as he had been told the year before. The Principal had to reply that they had "not yet done anything in the way of giving Radio lectures to extra-mural students."[19] There was already a radio university going in Berlin (Hans Bredow School) and Mercer University in Atlanta had begun transmitting four full credit courses in November.[20] The last CFRC broadcast of the session on March 25, however, made a beginning in this direction. It was the "first lecture in connection with University Extension work." Professor B.K. Sandwell, Head of the English Department, Queen's University, spoke on "Canadian Poetry."[21] In all the long history of Queen's Radio, though extension lectures have been aired from time to time, CFRC has never been used for transmitting university credit courses to extramural students.

According to Jemmett's requisition book, Bain was paid $300 for looking after the radio broadcasting and D.R. McLeod received $150.[22] Although Doug Geiger wrote[23] that H.J.D. Minter was Bain's assistant in 1924-25, there is no record that Minter was yet working for CFRC or was paid that year.

That session the logs show that CFRC received reports from Toronto, Rochester, Syracuse, Ottawa, Millbrook and Oswego. The best distance in the fall was Moncton, NB,[24] but the best all year was about 700 miles. A report that CFRC was heard in Seattle in April turned out to be incorrect, since CFRC was then off the air.[25]

In May, the Radio Branch of the Department of Marine and Fisheries changed CFRC's wavelength from 450 metres to 267.7 (from 666.3 to 1120.7 Kc)[23]: "This was a change in the wrong direction from the station's point of view because the lower the frequency or the longer the wavelength, the less absorption there would be on the ground wave and the better the travel - the longer the range you'd have. However, more powerful, full-time stations were coming into operation and it was necessary for the smaller, part-time stations to give way to them."[26] The CFRC licence was renewed with a cheque for $50 to Marine and Fisheries, issued on requisition number 6811 on May 28.[27]

Fashion Note: The fad among college girls in the spring of 1925 was to roll their stockings down to their ankles and to paint pictures on their knees.

Renovations to the Station: Mark III Transmitter

Beginning in May 1925, the basement of Fleming Hall was extensively remodelled for the installation of new electrical apparatus, and during the summer the radio station was also thoroughly overhauled and re-built.[23,28] These changes were based upon extensive testing that had gone on during the year. No additional purchases of equipment were made that summer for the radio station, however, according to Jemmett's requisition book. In November 1926, after a lot more work, Jemmett, the Director of the station, pronounced the Mark III apparatus 90% perfect.

CFRC announced in March 1925 that it was going to broadcast regular musical programs beginning in the fall, replacing the Journal News, and that a regular studio would be installed for the purpose.[29] Jemmett requested Mr Bews to make some changes to Room 9 in Fleming Hall, the Radio Room, on September 4.[30] A week later Jemmett ordered 34" x 105 yards of flannelette at $0.37 a yard and 325 linear feet of dressed lumber, 3 x 7/8", for building some sort of temporary framework studio in Fleming Hall.[31] The Radio Room itself could have been draped with cloth, but since it was quite narrow, the studio may actually have been built out in the corridor. Unfortunately, no one remembers what the Radio Room looked like. One or two have a vague recollection that it was just a bare room with lots of equipment inside. Some time later, Jemmett ordered four 5 x 1 yard strips of "cocoa matting" for the studio floor,[32] and then rented an upright piano from the C.W. Lindsay Company for $6 a month.[33] With the draperies hung, things were now all set to begin experimenting with studio broadcasting from the basement of Fleming Hall.

CFRC Broadcasts in 1925-26

Harold John Duncan Minter (Elec. '25), a Douglas Tutor and demonstrator in E.E., was Bain's assistant as well as the announcer in 1925-26.[34] Minter was born in Suffolk, England, in 1901, but grew up in Ottawa. There must have been other members of the radio station who helped anonymously and did a lot of listening in the basement, but were neither qualified as operators nor gifted as announcers. "The radio association" was often mentioned in the Journal, especially in 1925-26, but its composition was never defined or published. Its members have to be deduced from the CFRC log and from anecdote. Some of the students who were associated with broadcasting between 1924 and 1929 were, besides H.J.D. Minter, Donald Ridgeway McLeod (Elec. '26), Joseph Gerald Burley (Elec. '26), Joseph Taylor Thwaites (Physics '25), Donald John McDonald (Elec. '26), William Alex Richards (Mech. '26), William Gordon Richardson (Elec. '26), Sylvester Frank Ryan (Arts '27), George V. Ketiladze (Elec. '29) and possibly John Reid (Radio) Bain (Elec. '28).

Tests were conducted on October 2, 7 and 9,[8] and the first broadcast, the Queen's-Varsity football game, was fed over the re-installed Bell line from the Stadium at 2.15 pm on October 10. Transmitter trouble developed after the first quarter and the broadcast was stopped at 3.00. On October 17 the complete Queen's-RMC Intermediate contest was broadcast, and the next Saturday the Queen's-McGill Senior game.

CFRC's first studio musical program went to air on Friday, October 30, using the rented piano in the draped studio, from 11 pm to midnight. The five numbers were listed in the log, but further details were supplied by the Journal:

> Queen's radio broadcasting station CFRC, delivered the first of a series of studio programmes last Friday evening. With the completion of the new studio in Fleming Hall, the radio association feel that they can now broadcast a

concert which will be appreciated by the fans. The talent is secured by the student body and is voluntary.

The concert was somewhat delayed owing to trouble experienced with the plate generator, but they were able to go on the air at 10:45 on a wave length of 286.7 metres with the following programme:

1. Piano Solo - Fantasie Impromptu, Composition by Chopin; - George Ketiladze
2. Baritone Solo - Three for Jack, Composition by Squire; - Verne Zufelt
3. Piano Solo - Prelude in c# minor, Composition by Rachmaninoff; - Stewart Harper
4. Baritone Solo - Jest Her Way, Composition by Aitken; -Verne Zufelt
5. Piano Solo - Waltz Caprice, Composition by Newland; - George Ketiladze

The accompaniments were played by Stewart Harper. Announcer, Mr. H.J.D. Minter.

The association are very grateful to those who assisted.

The operating control room was in charge of Prof. Bain, assisted by Mr. G. Burley. Mr. Minter looked after the programme and studio arrangements.

After midnight Prof. Bain and Mr. Minter conducted a series of tests, using special and carbon microphones.

Reports from various individuals in the city were received. All spoke favourably of the programme, but complained of a slight buzzing. It is hoped that this will be eliminated and that the second of the series of studio concerts will prove as delightful as any received from outside centres.[35]

A letter was received in early November from Minnesota, over 1,000 miles away, reporting reception of CFRC, and in the testing local BCLs seemed to think that, since its overhauling, the station was working far better than it had for some time. "Mr. J.W. Bain, Lecturer in Electrical Engineering, is in charge of the station. Prof. D.M. Jemmett has had a great deal to do with the broadcasting and Mr. McLeod has assisted in announcing."[28]

Bain operated for the Queen's-RMC Intermediate football game on November 2 and H.J.D. Minter was the operator when CFRC broadcast the Science '26 Social Evening from Grant Hall, 8.30 to 11.00 pm on November 2. The Science Dance featured the first broadcast over CFRC of a local orchestra, Sid Fox and his Screnaders. The whole program[36] was sent out and "Those who heard the musical numbers report upon them most favourably."[37]

CFRC's second studio program was transmitted on November 6 from 9.30 to 10.00 pm, Joseph Taylor Thwaites operating. George Ketiladze was to be featured playing the piano, with assisting violinist and vocalist.[38] The program was changed before air time and, fortunately, the CFRC log listed the numbers and artists:

> 1. Petite Waltz. Op. 28 #1 - Henselt; Piano - C.Y. Hopkins
> 2. Piano Solo, Selections from Il Trovatore - Verdi;
>   - K.A. MacKinnon
> 3. Violin Solo - Medley of Popular Airs; - W.W. Ashford
> 4. Piano Solo - Hungarian Rhapsody #12 - Liszt;
>   - Stewart Mathews
> God Save the King.[8]

> Many reports from local and outside listeners were received congratulating CFRC on the high quality of the programme, and on the excellent manner in which it was received. Messrs. Minter and Thwaites, to whose efforts this success is due, are well pleased to know that they can now produce a concert which "is appreciated as much as any concert from the American side," as one of the listeners reported. This week-end CFRC will broadcast on Friday afternoon the game, McGill vs. Queen's II from the stadium; the Medical dance programme this evening; also the exhibition game, M.A.A.A. vs. Queen's I, tomorrow afternoon.[39]

Dr Clarence Yardley Hopkins (B.A. '24; M.A. '26; Ph.D., New York) does not remember much detail of his short broadcast, nor does Keith A. MacKinnon (Elec. '26):

> **C.Y. Hopkins**: I recall the incident itself quite clearly. I believe it was in...Fleming Hall and possibly in the basement...It seems to me another man and I played a duet on an ordinary upright piano. Was the man's name Peacock? If so, he was the one who made the arrangements with me...I knew Joe Thwaites at the time...I was not much interested in the project (was more interested in girls then!) and just went in, played the short number and left. There seems to have been only one other present, the announcer. I have no idea who he was. My memory for most things is still pretty good but, after all, that was 64 years ago.
>
> The station was either a small rather bare room or a bare corner of a large room. I focussed on the piano and the music, and recall nothing of the radio equipment... No spectators. No one told me that they had heard the broadcast. It impressed me as possibly an experimental broadcast. I did not see the other performers. Perhaps my effort was a sort of rehearsal or dry run for the operators, although we understood it was going out over the air.[40]

> **Keith A. MacKinnon**: I had nothing directly to do with the technical operation of the station, but you told me about playing the piano, which I'd forgotten all about. You mentioned the names of the people, and one was W.W. Ashworth (Phys.'29), and he used to play violin with me in the Queen's Collegians. We had a dance band; and he ended up as General Manger of the cable wire works down in Brockville. And you mentioned about playing "Il Trovatore." I knew that. I can play it now for you, because it was in my mother's old music book - it

must have been printed about 1870 or something. It was a lovely piano part, and I knew it in my head, so that's why I played it. I was just playing it the other day, seeing if I could remember the thing off-hand.

I was a flute and piccolo player in the theatres right after the War. My father was a musician too, you see, and I played in the theatres. The talkies came in and that killed the theatre, and I played in dance bands on the saxophone then for about ten years. But not piano. I never played piano except when the piano man had to leave for some instant or something.[41]

Some sort of problem was developing with the transmitter in early November, for beginning with the broadcast of the Loyola-Queen's Junior Intercollegiate game from the stadium on the afternoon of November 7, they reduced the antenna current. Joe Thwaites was the operator and the log states "This was purposely broadcasted on low power (4.6 amps instead of 5.4) to avoid modulator oscillation."[8] After that, CFRC began using only 1.4 to 1.6 amps antenna current.

There were five more studio programs broadcast from the cloth and frame studio in Fleming Hall. From 9.20 to 10.25 pm, November 11, the soloists were Mr M. Thurling (piano), Mr J. Laflair (vocal) and Mr W. Orr (violin), with Mr Thurling acting as "accompaniest" as well as playing solos. Joe Thwaites and H.J.D. Minter operated. The concert of November 20 featured Miss Dorothy Dowsley (piano and accompaniment), Mr H. Fairbairn (violin) and Mr B. Philips (piano), with Minter operating. Minter and Bain operated the January 8, 1926, studio concert for which no program survives, and Minter alone worked the February 3 program in which Miss Sanders, Miss Pearl Seal and Mr Madron played the "3 B's." On February 24, Joe Thwaites operated and J.W. Bain announced, but the program for that is lost too.[42]

Bliss Carman came to Queen's Convocation Hall to speak to a large crowd on November 12, when he was introduced by Principal Taylor and thanked by Dean A.L. Clark. The Daily Standard noted that the address had been "broadcast from station CFRC, the University station, and was heard with interest by many local people in that way."[43] The CFRC log shows that Joe Thwaites had the station on the air that night at 9.30, but only for a "short announcement re week-end broadcast."[8] The Daily Standard had other insider information: "Convocation Hall is excellent for broadcasting, experiment has shown. It is understood that the next concert that is held there is to be broadcasted by Queen's station, CFRC." The next concert, at 8.15 pm on the 19th, featuring violinist Geza de Kresz accompanied by his wife Nora Drewitt de Kresz, was not broadcast. The Standard further predicted that a vocal and piano concert would be aired from Convocation Hall on Saturday, November 20.[44] The log shows, however, that the Saturday broadcast was a studio concert.[8]

On the day of the November 20 concert, the Radio Association appealed through the Journal for talented volunteers "in order to continue these concerts and to keep CFRC strictly Queen's own... Tonight you will have an opportunity of performing before the public over the air. Kindly get in touch with Mr. J.D. Minter at Fleming Hall and have your name added to the programme."[45]

Ernie Bruce and His Million Dollar Orchestra (violin, banjo, saxophone, xylophone and piano) supplied the waltzes and fox trots on November 13 for the "Medical At Home" in Grant Hall,[46] and "they had more pep than a barrel of monkeys." Number 16, "My Sugar," turned into an old-fashioned snowball fight, using for ammunition the artificial snowballs sitting on the decor fenceposts. Minter operated for CFRC.

The keepers of the CFRC log were not terribly fastidious about keeping a complete record, so perhaps the Daily Standard was not mistaken about the Bliss Carman and music broadcasts.

That year, Keith MacKinnon announced a football game impromptu over CFRC. It was a Queen's II-Loyala game, and was on either November 7 or 21, during the transmitter trouble. He recalls that Minter was the operator, so it was likely the 21st.

> Keith MacKinnon: Once I announced a game, probably in '25. They were having trouble with the transmitter, I think. Well, this Minter was running the station and there was this football game on. I think it was Loyola or something - the second teams, you know. I happened to be around in the hall and Minter said, "We're kinda working on this now and we think we got it going. Why don't you go out and announce the football game?" I'd never done anything like that before. Well, okay. So I went out to where we had the game, and I had a table, a chair and a small microphone on a stand and the wires, sitting right out in the field just beyond where the deadlines were (on the Union Street side). And that was all. I couldn't speak back or anything. We had no cuing system at all. And there wasn't a soul around except me and the players on the field. So I was there and as soon as the game started, I started and, when the game finished, I finished. That was it. I was the announcer for the whole game and I never stopped talking. Minter was experimenting with the transmitter and it was a good chance to test it out, with me talking all the time. I don't know how it went out or how long they stayed on the air or how much they put on the air. I never heard anything about it. So that was the only thing I ever had to do with the station, directly.[47]

"The Silver Box" Broadcast

Bain brought off another ambitious project on November 24, when CFRC aired the second night of the Queen's Dramatic Club's production of John Galsworthy's The Silver Box from the stage of Convocation Hall in Old Arts.[48] Thwaites was the operator.[8]

Prof. J.A. Roy and Mr Charles W. Gates were in charge of production and the principals were John V. Mills, Miss Vera Skinner and Miss Rose Gourlay. S. Frank Ryan (who was also the club's business manager) played "Roper" and Miss Kathleen Whitton was cast as "Wheeler," a maid servant.[49] Miss Whitton did not, however, appear in the play, the part being assumed by Miss Mildred Tape.[50] The second performance got a good review from The Whig, which said that the acting was excellent and the supporting cast strong.[51] Afterward, the cast was photographed in costume before trooping off to the Queen's Cafe.[48] That picture, which might show the CFRC microphone, has not turned up.

A Double Studio Broadcast

Two long programs were aired on November 27, for which the log shows only the date, the time (8 - 10.40 pm) and the antenna current (1.6 amps). It was announced as "a vocal and instrumental recital from the Fleming Hall studio" followed by a dance program at 9.15.[52]

> Last Friday night station C.F.R.C. went on the air with a big double studio programme. At eight o'clock sharp, the click of the microphone was heard and Mr. Thwaites announced the opening number, a piano solo by Mr. C.Y. Hopkins. This was followed by a soprano solo by Mrs. Ashton, the appreciation of which was shown by the numerous phone calls that were received. Mrs. H.P. McGrath of the Kingston Orchestra played several delightful piano numbers, which also received very favourable comment. The radio association are grateful to those artists who so kindly lent their services.
>
> The second part of the programme began at 8:46 when Messrs. Stevens, Phillips and Ashworth came on the air with a burst of melody and mirth, giving the listeners an earful of the latest jazz. They were good. We had no idea there was so much talent latent in the college. This trio has promised to play again in the near future. If you have a favourite number you would like played, address a note to C.F.R.C. and drop it in at the college post office.[53]

Two days later, on a Sunday, Donald J. McDonald (Elec. '26) was the CFRC operator in a public concert broadcast from the Red Room in New Arts (Kingston Hall). The performers were Miss Kathleen Elliott (vocalist), Mrs C.F. Gummer (piano) and H.E. Faver (violin).[54]

In the new year, aside from the three studio programs mentioned, CFRC broadcast Queen's-Varsity hockey on January 15, Queen's-Varsity basketball on January 30 and two women's basketball games: on January 30 when Queen's decisively defeated Ottawa CI and on February 6 when Renfrew CI won 36 to 28.

Beginning with the studio concert on January 8, they managed to keep the antenna current up over 4.0 amperes, except for February 5. There is no indication of what the problem was, but Doug Geiger summarized the situation diplomatically: "During this session a considerable amount of transmitting was done with a 7-watt oscillator, which covered a range of about 3 miles."[55]

A record crowd at Grant Hall on the evening of February 5 saw J. Alex Edmison, E. Russell Smith and John Lansbury of Queen's take the affirmative and defeat the Imperial Debating Team on the subject "Resolved that this House supports the establishment of a base at Singapore." The Queen's Collegians supplied music before and after the debate. Prof. Bain operated for CFRC and there was trouble with the transmission, necessitating a test the next day.[8]

The last log entry before CFRC closed down for the summer of 1926 was the Kingston vs. Sons of Ireland Junior Championship hockey game from Jock Harty Arena on March 12. Minter operated and D.R. McLeod announced. There was actually one more broadcast, a studio concert on March 31, that was not entered in the log. Mrs W. Ashton, Mr Mayer and Master Robert P. Burke sang solos, Mr Birdsall played the piano and Miss Hilda Friendship whistled "Songs of Love" and "Melody in F." Miss Pearl Nesbitt assisted at the piano.[56]

The estimates for E.E. and Radio for 1925-26 are shown in the minutes of the Board of Trustees, and include $500 to pay a radio operator plus $350 for broadcasting services.[57]

Kathleen (Whitton) and Frank Ryan

Kathleen Ryan (Arts '26) has always talked about being taken in to CFRC by her future husband, S. Frank Ryan (Arts '27), in 1924, to see the station shortly after it was born on a kitchen table in the basement of Fleming Hall. After 55 years she couldn't remember many details of the room or about whether there was broadcasting as well as listening going on there. She does remember clearly, however, seeing Science students with headsets, sitting at a kitchen table and listening to an upright rig, likely the De Forest Inter Panel set, and probably to KDKA.

In the middle and late 1970s, Mrs Ryan gave $35,000 to CFRC for upgrading the station. This donation was the inspiration and the seed money for CFRC's "Go Stereo" project.

> **Kathleen Ryan:** I was one of the co-eds in 1924 who had a slight introduction to radio at Queen's by going in with Frank Ryan, who was really in Economics. He saw a great future in radio as an advertising medium. He didn't know very much about advertising, but he was a showman. He was a born showman, an Irishman who could step dance and play the violin and do a bit of singing. So, he was all set for it, and he quoted the great President of the Canadian National Railway, Sir Henry Thornton. Sir Henry believed firmly in radio, and I guess he was the first person in Canada who ran a commercial station (owned by the CNR). Then one night, here in the Chateau Laurier, in Ottawa, the CNR let some religious sect on the air and "Tut-tut" Shields of the Jarvis Street Baptist Church in Toronto called somebody else a thief, a liar and a rogue. And Sir Henry Thornton then decided that he would never have any religion (on the air). But Frank saw the great future of advertising and that's why I was introduced to the kitchen table in Fleming Hall.
>
> I'm sure that the first time I went in, that kitchen table was right in the middle of a basement room. It was not a remarkably sophisticated set-up, but I believe they moved it over to one side and (set it on a

lab) counter. And I think they could broadcast out of that room. There was no studio. As I recall there were just two units, and I don't know whether one was broadcasting and one was receiving. I recall something about a crystal set. The microphone looked more like an old-fashioned telephone receiver and you had to turn it if somebody else wanted to speak too. You couldn't speak back and forth - you had to turn it. Really, they were like disciples, this small group working on radio.

It was restricted to a small group on campus. It wasn't very well publicized on the campus. A few people knew about it, all Science men, except for Frank, and there was one other person who used to have the same idea, but it was more for drama. They felt it would be a great way to put on drama. Oh, there was somebody else who wanted to read poetry. She wrote a poem. Something about "Pines, pines, oh beautiful pines, I twine my hand in your needles" or something. But, it wasn't a very good poem and she was discouraged from reading it. It was explained very carefully that women's voices were high and shrill and the carrier waves wouldn't take them, and we would be very squeaky. You never heard anything like the Science men rationalizing, for women, to get so-and-so OUT! I don't remember any other girl there but this one who wrote a poem and was going to read it. I never met anyone from Arts in there either. In fact, I don't recall more than half a dozen there - or maybe I came in with Frank in off-hours.

If you happened to come in, you'd find out that the Science men had pinched all the headsets, listening to something that was being brought in. Two people could listen to radio if they had on earphones, and there was no mortal way of getting those earphones off the listeners once they got them on. You might just as well go home. Nobody else could hear anything. Particularly women. Not really entertaining for the rest of the people in the room.

It was really Frank who was interested in broadcasting. I had not much interest in it. I just went along to watch where he was going, into the basement at Fleming Hall. (Which is a piece of advice I give to all co-eds: Watch them!) Frank took psychology, and he was closely associated with Prof. Humphrey, an extraordinary psychologist who liked all these mysterious things like radio and sound. He put glasses on his little girl so that it would turn the world upside down - this was before the present government - and she was supposed to make her way around in a normal way. They talked a great deal about the psychology of this new type of communication and how far it could go. Frank was sure that through radio or a development of it, some day (it would be) your newspaper in your living room. I don't know whether he really expected quite the horror we have of commercials. I like commercials, but not the ugly ones. Frank was absolutely keen on advertising and he was determined that he was going to own his own radio station one day.

And then, the first year he was out of Queen's and went into advertising (with Cockfield, Brown Advertising Agency, Toronto), Frank...brought out a survey, largely based on Economics 24A... and made the great report on radio as an advertising medium... and showed that you could make a lot of money out of advertising. At the same time, Sir John Aird was bringing out his elegant Royal Commission Report saying that the air didn't belong to the individual and that there was no way advertising could pay for radio in Canada. And here was poor little Frank saying you could make a million bucks, or it was like a licence to print money to have a radio station. And the Aird Report won. I don't think he quite visualized the CBC. And Sir Henry Thornton certainly didn't think advertising was going to be an atmospheric billboard, but he believed firmly in it.

The very first big program that Frank put on was the "Campbell's Soup Popularity Hour" with Percy Faith's Orchestra. He put Percy on the air, made a millionaire out of him and Campbell's sold lots and lots of soup. And I've always thought it started on the kitchen table in Fleming Hall. So after that, Frank bought a radio station with Ted Rogers. Frank, with Harry Sedgewick, Malcolm Campbell and somebody whose name I forget, went into radio in Windsor, and sold a lot of advertising to give music to the 180,000 people who worked on the all-night assembly lines in Detroit. So that this was a very, very good venture. Then a former Chancellor of Queen's University (who was Chairman of the government's radio regulatory body) was going to close all private stations, but he thought he'd start with CKLW because it had gone ahead so much that the CBC had no audience. So Frank went out to The Company of Adventurers of England Trading into Hudson's Bay and put in wind-chargers for radio communication to their exterior posts. Someone in the US said that Frank had been through a suburb of Hell.

Frank came through and established his radio station in Ottawa (CFRA), not without some difficulty and a great deal of debt and then tried to serve the community, the farms and the small towns and the villages, just as to-day...they're picking up this old local trend. Do I well remember, every week we read 27 weeklies and summarized the news, such as how the bees were doing and how the chipmunks were carrying their tails and where the earthworms were - that was one of the ways you told the weather. Then Frank wanted frequency modulation (FM) and...then he got a 50,000 watt station.

When Frank flew the big tower from the AM station in Ottawa, away down the river and up over NCC-land and dropped it on a foundation on the top of a mountain in Camp Fortune, with an inch and a quarter tolerance, for his FM station...it came right back here...because of what happened in 1924, '25, '26. It really was a joyous career for Frank and it started right here in Queen's. It came back to the kitchen table in the basement of Fleming Hall where CFRC was born.[58]

Dr W.R. Jaffrey's Enquiry

Dr Jaffrey was supposed to have been kept apprised of the use being made of his donation for the development of the radio station, but this was not done. He wrote to Dr W.E. McNeill in May 1926, annoyed that he was not being kept in touch with the work that was done using the money that he had given to the radio station.[59] The Principal's secretary instructed J.M. MacDonnell, Chairman of the Board of Trustees, to get the details from Jemmett.

CFRC Broadcasts 1926-27

Bain resigned his Assistant Professorship after the 1925-26 session, whereupon Doug Geiger returned from doing research at Bell Telephone in Montreal to replace him and to resume direction of the radio station.[23,60] Under Geiger, studio broadcasts were discontinued "due to difficulties in the operation of the station," but there were some innovations.[23] Geiger and his assistant H.J.D. Minter tested the equipment intermittently for many hours on September 30 and October 1, prior to their first football broadcast on the 2nd, an exhibition match in which Queen's defeated Ottawa Senators by 7 to 3. They broadcast a second pre-season game from the stadium, against the Camp Borden team, on October 9. After three successive Dominion Championships, 1922-24, followed by an Intercollegiate title, hopes were very high for the 1926 team. The season opener, played in Toronto on October 16, was not broadcast, so the Journal posted the CN Telegraph press bulletins in the rear of Douglas Library. The two regular season games, against McGill on October 23, and Varsity on November 13, were handled for CFRC by Geiger and Thwaites and by Geiger, respectively.[8]

The Queen's-Varsity Game Broadcast over CFCA

Although CFRC could be picked up in Toronto, CFCA, the Toronto Star station, hired a telephone line from Kingston, sent down its own announcer and aired the Queen's - Varsity match on November 13, "the mightiest football struggle of the season," in a joint broadcast with CHIC, CKNC and CKCL. Toronto heard "a thrilling play-by-play report of the Queen's-Varsity struggle direct from the Richardson stadium at Kingston. The listeners could even see 'Boo Hoo', the Queen's bear mascot, playing with a tiny fox terrier, as well as a graphic detailed account of the game."[61]

The sky was sullen and threatening that cold November day, and the lighting was poor. Moreover, the field was too slippery for the teams to risk any open play, and after three quarters of great defensive football there was no score. In the last stanza, Queen's capitalized on two Varsity mistakes for a 2 to 1 lead. With four minutes remaining, a Batstone kick was returned to the Varsity 35. A Snyder to Trimble end run got clear of the Queen's tacklers, opening the way to the goal line. With the victory galloping away, Pee Wee Chantler, far off to the left of the play, raced madly across the field, caught Trimble by the panties, hauled him down on the Queen's 15, and saved the day with "a forlorn hope tackle." A mighty cheer burst forth and 9,000 fans went wild. Queen's held on to defeat Varsity 3 to 1, creating a 3-way tie for first place. The Whig called it "the game of a lifetime - the greatest game of football that has ever been played on a Kingston gridiron."[62]

The CFCA announcer that day was the young Foster Hewitt, and one of his jobs as part of the small broadcast crew was to help with the setting up. The best vantage point for their broadcast was the steeply sloping tin roof of Richardson Stadium, more than 50 feet above the field. The only way up was an iron ladder on the outside of the north tower, and all of the equipment, including the amplifier and 50-pound auto batteries, had to be carried up. It took five trips and two and a half hours for Foster and a helper to get everything in place and secured against sliding down the slippery slope. Foster himself was lowered to the edge of the roof by wires looped around his middle, under his arms and around his neck. In that precarious spot he could brace his feet against a flagpole and the eavestrough. He sat there in his huge chinchilla coat with a blanket spread over his knees, a soap box in his lap and the big, round microphone on the box. When he got too excited calling the game and leaned over the precipice, his crew would shout to him and tug on the wire.

While the game was taking a good turn for the Tricolor in the last quarter, the weather was taking a bad turn for Foster Hewitt. It began to rain, and with the bitterly cold wind from the lake, the rain turned to hail. By the end of the game, the announcer was frozen fast to the roof of Richardson Stadium. He couldn't move from the spot, and he was stiff with cold besides. The men on the crest of the roof yanked on the lines, tore him from the spot, and dragged him up the icy incline. Foster Hewitt left the seat of his chinchilla coat behind on the roof of Richardson Stadium.[63]

Ronald Ede: My father worked at Queen's and was present and assisting Foster Hewitt when he made his famous broadcast of the football game from the roof of old Richardson Stadium. Foster said they tied a cable around him - my father said it was a rope. They were afraid Foster would be blown off the roof. He was quite a brave man in many

Foster Hewitt broadcasting the Queen's-Varsity football game of November 13, 1926, from the roof of the old Richardson Stadium. Note the round microphone sitting on a box on Hewitt's lap, the wires extending up the slope of the roof and the auto battery sitting near the crest. By the end of the game, Hewitt was frozen to the spot and had to be torn free and dragged up the incline.

-courtesy Queen's Alumni

ways. I rememeber the old stadium, the door and the ladder leading to the roof. It was a long way to the ground from where he did his broadcast.[64]

One of those old-fashioned panoramic strip photographs was taken of the field around half-time in the game, and it shows a minute Foster Hewitt sitting on the roof of Richardson Stadium. One can make out the soap box and microphone on his lap, the cables trailing up the sloping roof and at least one car battery. Three figures are standing just at the crest of the roof. Ron Ede identified the man at the left as his father and the man at the right as Alfie Pierce.[64]

All that is known for sure about the CFRC broadcast of the Queen's-Varsity tussle is that Geiger was the operator, because his name in his handwriting is in the log.[8] Whether Foster Hewitt's voice was relayed directly to Toronto or by way of Fleming Hall, so that CFRC could broadcast a feed, is not recorded.

Out of spite over the loss, the Varsity said: "Last week there was a big re-union of the Queen's alumni down at the burg, and the most prominent figures were Dr. Soandso, the oldest living graduate, and Harry Lee Batstone, the oldest living undergraduate."[65]

CFRC Carries the Rugby Championship from Toronto

Varsity eliminated McGill the next week and then defeated the Tricolor in Toronto for the Intercollegiate Championship. The play-by-play of that final game was fed by Bell line from Toronto and was broadcast over CFRC - the first time that CFRC had ever arranged, and paid, to receive the long distance feed of a program. "Many Kingston people tuned in on Saturday's Queen's-Varsity struggle and say that the results obtained (were) most realistic. Foster Hewitt, who broadcasts for the Toronto Star, is said by Kingston people who heard him on Saturday to be in a class by himself."[66]

The First Convocation Broadcast

Geiger operated on November 11, a four-hour broadcast of dance music from the Armistice Ball at Grant Hall. Minter took the controls the next evening, when CFRC transmitted the Queen's Fall Convocation for the first time. Members of the Board of Trustees, of City Council and RMC shared the platform with the Chancellor, the Rt Hon Sir Robert Borden, a host of professors, Principal Taylor and the Registrar. The last gentleman was greeted by the traditional tossing of coins from the galleries of Grant Hall. Lord Willingdon, Canada's new Governor General, received an honorary Doctor of Laws, as did Sir James Aikins, former Lieutenant Governor of Manitoba, and Sir Clifford Sifton, a member of Laurier's first cabinet. All three addressed Convocation and were heard over CFRC. Another highlight of the ceremonies was the granting of a B.Comm. to Harry Batstone.[67] One wonders what that program must have sounded like, for the taciturn Queen's convocation ceremony has never lent itself particularly well to radio broadcasting.

Earlier that day Lord Willingdon had unveiled the large windows in the Memorial Chapel in Douglas Library, dedicated to the Queen's dead in the Great War.

Broadcast of Provincial Election Returns, December 1, 1926

The election in Kingston looked like a 3-sided Tory race, until Mayor Angrove dropped out. It was most fiercely contested. W.F. Nickle, former Attorney General in the Ferguson government, running as an Independent "Dry" for the Ontario Temperance Act, was supported in speeches by Sir George Foster and Charlotte Whitton. Alderman Thomas Ashmore Kidd stood for the Ferguson "Wet" Conservatives and government control of liquor; that is, for "True Temperance."

Normally The Daily British Whig and the Kingston Daily Standard brought the Dominion and provincial election returns to their own party faithful, assembled in front of the respective newspaper offices. On this very election day, the two Kingston dailies amalgamated into one paper, The Kingston Whig-Standard.[68] Since The Whig offices were in the disarray of reorganization, one non-partisan set of local returns was bulletined in the front window of the old Standard offices for supporters of both parties. Provincial returns, received from the Canadian Press wires, were projected by magic lantern onto a screen across Princess Street. Arrangements were also made for CFRC "to broadcast summaries of results in important constituencies, under the auspices of The Whig-Standard newspaper" on 267.7 metres.[69] The radio service "made a big hit with the radio fans. The station came in very clearly and the results were easily heard."[70] Doug Geiger operated for CFRC that evening. No record survives of who announced or whether the broadcast was done in the CFRC studios or remotely from the Standard offices.

The provincial election was the first time that CFRC got involved with anything but a sporting event outside the University, and the first time that the radio station had appealed to a much wider audience. Suddenly the Queen's station was in the public ear and in the public mind. Over the next ten years the Kingston business community became increasingly interested in having a local commercial radio station to promote regional interests and events. A few local industries later supported CFRC's bringing remote football broadcasts to Kingston. Gradually the Queen's authorities softened their stance toward a commercial arrangement for CFRC, until finally, in 1936, the University and The Whig-Standard co-operated to put CFRC on a commercial basis.

Queen's University Public Lectures on CFRC

Queen's mounted two sets of Monday afternoon public lectures in Convocation Hall in 1926-27, a set of three on English literature followed by six on science. After the first lecture of the literature series, Prof.

A.H. Carr, the Director of Extension, announced that the two subsequent lectures would be broadcast over CFRC, and that all extra-mural students were "invited to listen-in and to tell friends who have radios to hear Queen's on the air."[71] Prof. Wilhelmina Gordon lectured on "Some Elizabethans" on November 29, and Dr George Herbert Clarke spoke on "The Poetry of Robert Browning" on December 6. Doug Geiger operated. The science lectures began after the turn of the year, but were not transmitted. Someone may have argued their inappropriateness to a radio audience, since they were illustrated with lantern slides.

Between January 14 and March 2, when the station shut down for the summer, CFRC broadcast weekly, but only sports: Senior Intercollegiate men's hockey, senior men's basketball, two RMC and other hockey games, plus the Ladies' Intercollegiate Basketball Tournament, February 24 to 26, 1927. Queen's lost to Varsity in the final game.

The New Radio Room and the Mark IV Transmitter

During the summer of 1927, the Department of E.E. re-arranged its laboratory space in Fleming Hall to make room for the large Course I. The only way to make enough space was to convert the three rooms plus half the hallway in the west end of the basement into one large laboratory. The Radio Room had to go. The radio station was moved

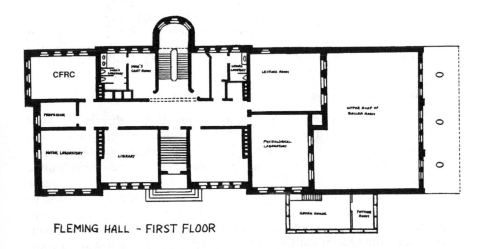

FLEMING HALL - FIRST FLOOR

Plan of the first floor of Fleming Hall. In 1927, the radio station was moved upstairs to the Standards Laboratory in the north-west corner of the first floor to make room for a new laboratory course in the basement. The radio set was re-designed and re-built by Douglas Geiger, and this is where Harold Stewart was photographed with the Mark IV around 1929. After the fire in 1933, Prof. Stewart re-built the radio station in the north-west corner of the second floor, where the transmitters remained until February 1990.
-from a diagram in Queen's Calendar

upstairs to the first floor Standards Laboratory, and Standards was shunted across the hall.[72] Geiger wrote:

> In view of contemplated changes in the laboratories in the basement of the building, it was seen that the broadcasting set would have to be moved from the room in the basement, in which it had been housed since the inception of the station, to a room on the first floor, half of which could be used as a studio. At the same time it was decided that it would be well to re-design and rebuild the set, in order to incorporate in it the advances made in the science since the construction of the original set. New and larger tubes were ordered from Mullard's in England; a new filter, capable of completely smoothing a current of one-half ampere at 3,000 volts, was designed; and the necessary wire for winding and iron for the core ordered. Upon the receipt of the new tubes their characteristic curves were checked against those supplied by the manufacturers, and the design of the set was proceeded with. The new set was ready for the first of the 1927 rugby contests, the Queen's-Argos exhibition game on October 8.[23]

In fact, the work of re-designing and re-building the CFRC transmitter was done in the fall and winter of 1926-27, as Geiger later recalled.[73] Prof. Jemmett's requisition books show that he placed the order for new tubes with Mullard on November 5, 1926, well in advance of the move upstairs.[74] He purchased three VO/350 transmitting valves, three M/3B modulating valves, two 0/30A, two 0/10 transmitting valves and two O.R.A. "B" receiving valves. One month later he ordered the WD-12 tubes for Geiger's new microphone amplifier.[75]

Douglas Geiger and the Trans-Canada Broadcast

Doug Geiger did not stay at Queen's in the summer of 1927, but worked as an engineer in the Transmission Division of General Engineering for Bell Telephone. He was in charge of all lines and equipment west of Winnipeg for the trans-Canada Diamond Jubilee Broadcast from Ottawa, July 1, 1927. Some 20 radio stations from coast to coast in Canada were involved in the hookup, but CFRC was not one of them. CFRC had no affiliation or connection with the CNR network, which formed the basis of the Jubilee system. The broadcast included the official opening of the re-constructed Parliament Buidings and the dedication of the new Victory Tower (now called the Peace Tower) by Prime Minister W.L. Mackenzie King, plus the first ringing of the new Carillon in the Tower by young Percival Price. Governor General Viscount Willingdon delivered a message to the more than ten milliom people of Canada from the King, actress Margaret Anglin read Bliss Carman's poem "Dominion Day, 1927," military bands played and school choirs sang. Portions of the broadcast were recorded from the air by Herbert Berliner of the Compo Company in Lachine, and issued on 78-rpm discs.

<u>Douglas William Geiger</u>: Douglas George Geiger was involved in the first cross-Canada radio broadcast, using all-Canadian facilities, on July 1, 1927. His experience with radio dated back to his student days. At Queen's University in 1922, he and D.M. Jemmett built an experimental broadcasting station, 9BT, during Geiger's final year in Mechanical Engineering. The station later became CFRC, and he continued working on it during 1923, while he completed a degree in Electrical Engineering, and in 1923-24, when he worked at Queen's as a demonstrator. After working for the Bell Telephone Company from April 1924 to September 1926, he returned to Queen's as a lecturer, and during the winter of '26 and '27 he re-designed and re-built CFRC.

At the end of April 1927, he went to Montreal to work for the Bell Telephone Company on a cross-Canada broadcast project.

His boss was John Clark. There were engineering meetings in Ottawa. It was decided the project could be carried out and Geiger was sent to Sudbury, Fort William, Calgary and Edmonton to check out the railroad dispatch circuits. He returned to Montreal in early June and then returned to the west to start lining up the circuits and to adjust the equipment. Standard four-wire repeaters were used. They boosted the signal and were battery- operated. The equipment was put together in May and early June by Northern Electric, who got them from Western Electric in the United States. A repeater station was set up every hundred miles along the railroad line. Engineers from Montreal manned the repeater stations, and Geiger trained them. At that time, the railway had voice dispatched talking circuits between Sudbury and Winnipeg, and between Calgary and Vancouver. Each night, the railroad switched to telegraph dispatch to give the circuits over for testing for the radio broadcast. Geiger worked from 12 midnight to 7 am for three or four weeks with the engineers at the repeater stations, sending tones and testing portions of the circuit. The CPR Superintendent of Western Lines, Mr Neil, came in every night and watched. A battery of telephone operators reported to Geiger about any problems.

There were two routes set up for the broadcast. One via CPR north of Lake Superior, and an alternate one via Chicago. The radio stations were all independent at the time, but would hook into the system to broadcast the cross-Canada program. Final testing of this circuit was never completed before the actual broadcast, because there was always trouble of some kind along the route, especially in the mountains. Final testing was being done early in the morning of July 1st when it was discovered that someone had felled a tree to put up a radio broadcast transmitter, but at the same time felled the line that would carry the broadcast. Mr Neil ordered out a light engine and caboose from west of Calgary and men were sent to repair the line in time for the actual broadcast. There were two broadcast periods on July 1, 1927, one in the morning and one in the evening. The content was speeches, music, etc. Two thunderstorms during the broadcast period, one in Windsor and one in Sudbury, meant that they had to put in plugs to avoid the thunderstorms. The broadcasts were completed successfully.

In late July of 1927, Bell Telephone invited Geiger to rejoin the Company. He was already committed to Queen's for 1927-28, but arranged to start with the Bell in the spring of 1928, and remained until his retirement in 1965.[73]

Harold Stewart on the Mark IV Transmitter

Prof. Stewart published a description of the Mark IV transmitter in an article in 1938, and made some additional comments on it in an interview in 1982. He began by recalling the location of the new Radio Room in Fleming Hall:

> **Harold Stewart**: That's one flight up the steps as you go in the back door, and in the far west end, in the right hand corner.
>
> The transmitter at this time still used a modulated oscillator coupled to the antenna by inductive coupling. The oscillator tube was a Mullard VO/350 with a 240-watt plate (anode) dissipation rating, operated in a shunt-fed Hartley circuit with 3,000 volts DC from an Esco motor generator set on the plate. That's the motor generator given by Dr Jaffrey. The filter was in three sections with a 40-henry choke and had a 2.75 microfarad condenser in each section. This is the filter that Doug Jemmett had designed from the theory he'd acquired at MIT and so on.
>
> Heising constant current modulation was used. Heising, incidentally, was one of the long line of radio developers that came up through Bell Labs. This is now commonly called plate modulation. The modulator was a Mullard M/3B, rated at 150 watts, and the modulation choke was the high-voltage winding of a small distribution-type transformer. Two stages of audio amplification preceded the modulator. These stages were impedance-coupled with transformer-input. Grid bias was supplied by batteries in the audio stages and by a grid leak in the oscillator. All filaments were operated from storage batteries in the E.E. laboratories. The oscillator-tank condenser was oil-filled to prevent flash-over. They thought they needed to do that. No record of the exact power input to the oscillator is available, but it was approximately 600 watts. Assuming an efficiency of 60%, the antenna carrier power would be about 360 watts.
>
> Now, by this time, good double-button carbon microphones were available and the Northern Electric type 387-W was in use at CFRC. While the 387-W has long since been superceded by newer types with better characteristics, it was a great improvement over the ordinary telephone microphone and it imparted something like quality to the transmission. Its disadvantage was its rather low output compared to the ordinary single-button type, which was used in the ordinary telephone. This made it necessary to use an amplifier at the pick-up point. This amplifier, using three type WD-12 low-current, battery-operated triodes in a resistance-coupled circuit, was designed and built by Mr Geiger. It brought the output of the 387-W from about 40 db below six milliwatts to about zero db, or six milliwatts. The latter level was and still is (as of 1938) required for satisfactory transmission between a remote pick-up point and the transmitter.
>
> The question might be asked, "Why not feed the microphone output directly into the line to the station and then amplify it to the level necessary to operate the transmitter?" In a transmission line of more than a few hundred feet, there will be a noise voltage which may easily be equal to or greater than the microphone signal. This noise is largely due to power-line induction and to cross-talk from neighbouring telephone circuits. Amplification at the station end of the

Harold Huton Stewart (Elec. '26) at Douglas Geiger's Mark IV transmitter, when it was built over a laboratory bench on the first floor, west end of Fleming Hall, circa 1929. The primitive design suggests that this is the original 9BT transmitter of 1922 updated by the addition of new tubes and some modern components. The open layout offered the operator no protection from the innards which carried 5,000 volts! Note the three big glass-enveloped Mullard tubes, two M.3B modulators to the left and the V0/350 final on the right. The antenna coupling coil is above Stewart's head, Jemmett's microphone amplifier at centre left and the log book is on the table, centre.
-courtesy Prof. Harold H. Stewart

line will boost the noise to the same extent as the signal, and the noise cannot be filtered out without losing part of the desired signal. However, if the signal is fed into the line at a comparatively high level, it will ordinarily be so much stronger than the noise that the latter will be unnoticed.[76]

Geiger's Transmitter Blueprint, 1927

<u>John Fraser</u>: Basically, this transmitter is a modulated oscillator with shunt modulation. Heising modulation. The oscillator is a shunt-fed Hartley.

The small tubes, the "30" and the type "R", are just the speech amplifier.The modulator consists of two triode tubes - M/3B's - in parallel, so they would have to be class "A" modulators. The B+ or the HT, whichever term you care to use, is supplied by a 3,000-volt motor generator in a three section pi filter for the modulators and the modulated oscilltor, and the B+ and the bias for the speech amplifier are supplied by batteries. The antenna coupling is by simple coupling coil - at the extreme right, the top of the two inductors. The lower coil is the oscillator inductor and the filament tap is movable to adjust the amount of feedback. It's a very, very basic transmitter, about as basic as you can

The Mullard M.3B was a medium power modulating valve with molybdenum anode, designed to be used in circuits with the choke control system of modulation. It was 370 mm long, 127 mm in diameter and was supported and insulated in an adjustable holder. The cost was £9:0:0.

The Mullard VO/350 was a medium power transmitting valve, usable on all wavelengths down to 100 metres. It was 470 mm long, 155 mm in diameter and was supported and insulated in an adjustable holder. The cost was £11:0:0.
-courtesy Ian Nicholson, Archivist, Mullard, Ltd.

get with a tube-type transmitter. Very, very early technology.

It's an antique, really. Even when this drawing was done, it was old technology. I would have thought that they'd be getting away from modulated oscillators by 1928, going to a Master Oscillator - Power Amplifier (MOPA) type to improve the stability. I don't know when the push-pull class "B" modulator stages came in, but the efficiency would be much higher. With straight choke modulation you have to have class "A" modulators, and when you get into high power that's very, very wasteful of DC power, because the efficiency is so low. This is very near the start of tube-type transmitters, the type of technology that would have been used in the original tube voice transmitters.

It's so basic. You've got your signal coming in from the microphone at centre left, where it says "input." You've got your potentiometer here for your volume control (GR #214-500w Pot.), and goes into your input transformer at its immediate right, which is just a matching transformer to the grid of the type "R" tube. The small tubes, the "30A" and the type "R" tube are the speech amplifier. It's amplified there, passed through this inductive coupling here (SN 49 Ret) to the grids of the M/3B modulator tubes, which are in parallel, shunt fed. The DC current is being supplied from the motor generator. It flows through the Modulation Reactor, to the right of the M/3B's. An inductor tends to resist a change in the current flowing through it, so that when these two modulator tubes are driven to a higher current, the result is that the Reactor tends to resist that change of current and the voltage at the common point then changes between the two MA's and the Reactor.

Now, the oscilltor is a simple Hartley. The plate current flows in the top part of the lower coil (at centre, far right), so part of the RF plate current flows in the lower coil. The coil is also coupled to the grid through its bottom part. The two are coupled together because they're part of the same coil, so that part of the RF energy in the plate circuit

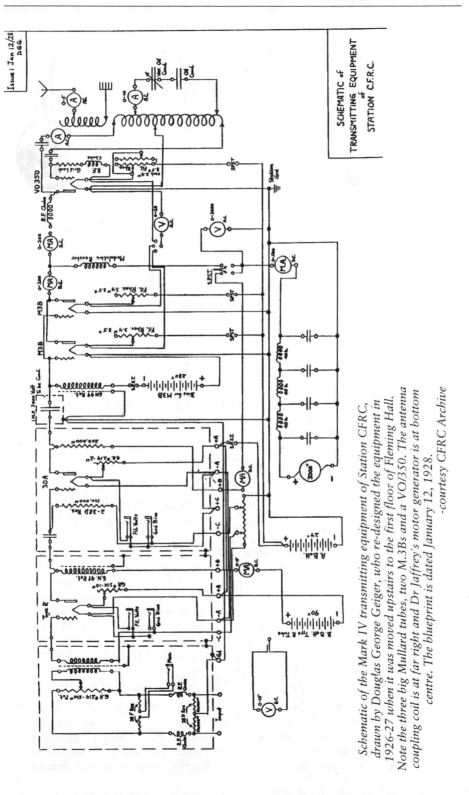

Schematic of the Mark IV transmitting equipment of Station CFRC, drawn by Douglas George Geiger, who re-designed the equipment in 1926-27 when it was moved upstairs to the first floor of Fleming Hall. Note the three big Mullard tubes, two M.3Bs and a VO/350. The antenna coupling coil is at far right and Dr Jaffrey's motor generator is at bottom centre. The blueprint is dated January 12, 1928.

-courtesy CFRC Archive

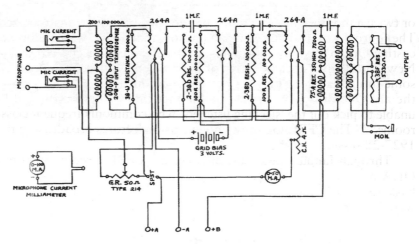

Schematic of microphone amplifier for station CFRC, undated. Prof. Stewart was prepared to swear an affidavit that this blueprint shows part of Geiger's original transmitter.

-courtesy Prof. H.H. Stewart

is fed back to the grid, and that maintains the oscillation. When the voltage at the top of the Modulation Reactor is varied by the varying current drawn by the modulators, the amount of RF power generated in this VO/350 oscillator varies in sympathy with it because its plate voltage is being varied. That's what gives you your amplitude oscillation. Simple as that.[77]

CFRC Broadcasts in 1927-28

There is no evidence in the CFRC log that the new Mark IV transmitter was tested prior to its first broadcast, an exhibition football game between the Tricolor and Toronto Argos on October 8, 1927. Doug Geiger operated.[8]

Fall Convocation was aired on Friday evening, October 21, from Grant Hall. The Journal took exception to The Whig's headline "Carrots and Live Hens Hurled from Gallery": "One would be given to understand that the contents of all the truck gardens and hen-houses of Kingston and vicinity had been commandeered for the occasion. So far as we have been able to ascertain, one carrot and one hen, along with a shower of beans, were all that was 'hurled' from the gallery. The hen was meant as an expression of welcome to Dr J.B. Reynolds, President of the Ontario Agricultural College at Guelph."[78]

CFRC broadcast the three Senior Intercollegiate football games played at Richardson Stadium that fall. Geiger operated the victories over Varsity and McGill on October 22 and November 12, and Minter for the Dominion Championship game on the 19th, which Queen's lost to Hamilton Tigers. "Pep" Leadlay played for the Tigers.

CFRC was planning to produce "special evening programmes once or twice a month" that autumn,[79] but nothing of the sort materialized. The Journal was also nurturing an illusion, that CFRC was still broadcasting Queen's extension lectures. Its article, on a Queen's parallel to New York University's "College of the Air" on WOR, concluded: "At some future date, one may yet be able to obtain a Queen's degree over the air. One drawback would be the fact that professors would be unable to pick out the 'sleeping listeners' who commonly frequent classrooms."[80] The CFRC log shows no extension lectures broadcast in the 1927-28 session.

Through January and most of February, CFRC aired nothing but OHA hockey from the Jock Harty Arena and Senior Intercollegiate basketball from the gymnasium. Newspaper reports of the reception were glowing:

> Reception was again of a high order on Wednesday night and an abundance of real good programmes was heard. Perhaps the most interesting to Kingston fans was the broadcasting by CFRC, the Queen's University station, of the Kingston-Queen's hockey game direct from the Harty Arena. There was no previous announcement that this was going to be done and it was a very pleasant surprise to tune in on the radio and get the results of this closely contested affair. Two Americam stations, as usual, were jamming but with very fine tuning CFRC could be brought in loud and clear.[81]

> Queen's University station CFRC was again on the air last night with the Kingston-Queen's hockey game. After a short test yesterday afternoon, the station was heard in much better form and last night's broadcasting could hardly be improved on. The sounds of the whistle, bell and the crowd at the Harty Arena could all be heard and the ones sitting at home had half the fun of being at the game itself. The announcer of the game deserves praise for the remarkable play by play report given, no action being missed at any time which is something to be said when broadcasting a fast game like hockey. It is hoped that CFRC will be with the radio fans again before the season closes.[82]

That short afternoon test was not logged. It was probably a follow-up to the intermittent failure of the microphone amplifier during the Queen's-Kingston OHA game on February 6.

"Ike Sutton listened in on the radio report of the Brockville-Queen's encounter. Mr. Sutton perched himself behind the radio announcer, rested his tired eyes, particularly one of them, by regarding the ceiling of the rink, instead of the dazzling spectacle on the ice, and listened with an expression on his pan which more closely approached high glee than rapt attention."[83] Sutton, a footballer and basketballer, had recently received a celebrated "shiner." Incidentally, the exemplary CFRC announcer is unidentified.

The following season, hockey broadcasts were discontinued because the atmosphere came across too well: "The reason that hockey is not broadcast is the fact that the cheering of the crowds drowned out the announcer's voice, and until a special box is made in the Arena, broadcasting of hockey will be abandoned. This year they are thinking of building a compartment on top of the stadium to facilitate broadcasting of rugby games. If this is done no longer will Toronto scribes complain of being held up by ropes and graphically describe...hair raising perils of broad-casting on the roof of the stadium."[84]

Kingston Choral Society Broadcasts

The Kingston Choral Society made its debut appearance on CFRC that winter in a 90-minute recital billed as "Part Songs and a Cantata." The program consisted of typical songs by Elgar, Mendelssohn, Bishop, Edward German and Brahms, plus Sir Frederick Bridge's cantata "The Inchcape Rock." CFRC's log for February 20 does not reveal whether this was a studio or a remote concert. The Whig noted: "Last night's reception was well nigh perfect. The concert was under the direction of Dr (C.F.) Gummer and Mrs A.R.B. Williamson acted as accompanist, as Mrs Minton C. Johnson was out of the city."[85]

Dr Gummer and his Kingston Choral Society returned for a second 90-minute broadcast recital on March 26. They put on a slightly different program, this time featuring as the major work Mendelssohn's Violin Concerto in e minor, with Mrs Goldie Bartels-Geiger as soloist.

> Static in abundance marked last night's radio reception and it was an impossibility to enjoy any program that was being broadcasted outside of Kingston. According to old time radio fans, Monday night's static was the heaviest in a long time.
>
> Queen's University station CFRC broadcast a splendid program by the Kingston Choral Society but this was marred at times by the noise. By cutting down the power on the receiving set, it was possible to hear the excellent singing with a minimum of noise but it was too bad that weather conditions were not better for the radio fans. CFRC proved to be Kingston's salvation as far as radio reception went; the announcements, program and reception from the station itself being of the highest order.[86]

The operator for the two choral recitals was Doug Geiger, and the second was his final broadcast at the CFRC controls. He left Queen's to rejoin the Bell Telephone Company in Montreal at the end of the 1927-28 academic year. Prof. Jemmett requisitioned Geiger's last pay cheque on March 23, 1928, on No. 5602, since Geiger was leaving for Montreal on April 1. This was also Minter's last session at CFRC, at $75 a term.

> **Freeman (Casey) Waugh**: I was born in Kingston in December, 1909. When I was a very young man, in my early teens, I once sang over this station. The station was upstairs in Fleming Hall at that time. It consisted of one room that had nothing in it but a piano and a microphone, and possibly a table. That's all I can remember about it. I'm not quite sure of the year, but I think it was 1925 or '26 that I sang over this station.[87]

The CFRC log offers no help in dating this event, but it was likely after the move to the first floor and the installation of the Mark IV transmitter.

Meanwhile, Canadian commercial radio was making huge technical leaps. The first AC batteryless radio station, CFRB in Toronto, opened on February 17, 1927, on 219 metres, operated by the Rogers Batteryless Radio Company and the QRS Music Roll Company.[88] CFRC continued to be dependent on batteries into the late 1930s.

The Satin Finish Broadcast on CKGW, Toronto

Kingston's business community was becoming increasingly interested in promoting the city to tourists through radio broadcasting. It was understood that the University would agree to the use of CFRC for such a purpose, subject to the consent of the Board of Trustees, but no one went so far as to make a concrete proposal.[89] In the fall, one city businessman tried a first little flirtation with CKGW, the "Cheerio Station," in Toronto. Mr W.C. Dover of the Chamber of Commerce and Grant McLachlan, local representative of the Satin Finish Hardwood Flooring Company, put plans together to get a feature talk on Kingston, as a tourist and commercial area, aired on a musical program by the Satin Finish Dance Orchestra.[90] The "first Kingston Night on the Air" marked Kingston as "the pioneer Ontario city to use radio to make its attractions known." They hoped that Kingston's attractions were heard all over the continent, and trusted that good results would follow. Static and KDKA interfered with some local reception of CKGW that night on 312 metres.[91]

S.C. Morgan Takes over Direction of CFRC

Stanley Chapin Morgan (Elec. '16) succeeded Geiger as Station Director of CFRC, and was assisted in the work by William Gordon Richardson (Elec. '26), who took over Minter's position. Morgan came from Elgin, Ontario, and was trained in the School of Mining under Gill and Henderson. At the time of his appointment to Queen's, in October 1928, he had been at the University of Alberta, Edmonton, for seven years and was then Assistant Professor. That year he was on leave of absence at the California Institute of Technology, Pasadena, earning his M.Sc. in E.E.[92]

CFRC Broadcasts in 1928-29

Stan Morgan immediately took over as the operator of the first four CFRC broadcasts in the fall term, all rugby football games from the stadium (October 6, 13 and 20 and November 3). The middle two transmissions were double-headers. All log entries are in Morgan's hand, and no credit to the announcer(s) was recorded.

Prof. A.H. Carr, Director of Extension , announced a series of three 5 o'clock Monday lectures on the ancient world for Convocation Hall. Col. K.L. Stevenson spoke on "Mounds of Babylon" on November 19, 1928, Prof. H.L. Tracy on "Greek Stagecraft" on November 26 and Prof. R.O. Jolliffe on "Virgil" on December 3.[93] These CFRC broadcasts were not publicized, so the only record of their transmission exists in the CFRC log.[8] The Journal had articles on the first two lectures, but neither mentioned CFRC's role.

In the winter term, CFRC broadcast just two hockey games. W.G. Richardson operated the January 11 game against Preston, Queen's Senior OHA debut. Morgan operated the Varsity game on January 23, when he must certainly have noticed that the announcer could barely be heard above the din of the game, and promptly abandoned the effort. Basketball games were cut too. In fact, nothing more was broadcast over CFRC from January 23 until the beginning of the 1929-30 fall term. The Journal took no notice of this.[84]

A further manifestation of the local interest in a commercial radio station for Kingston, perhaps stemming from the CKGW broadcast, occurred when the Kingston Chamber of Commerce applied to the Executive Committee of the Queen's Board of Trustees "for the occasional use of the Queen's Broadcasting facilities to be used for advertising purposes. The Secretary stated that Professor Jemmett estimated that $40.00 an hour would be a proper charge. It was agreed to grant the request of the Chamber of Commerce with the understanding that programmes should be submitted for approval to the University authorities."[94]

CFRC Broadcasts in 1929-30

The Official List of Radio Stations of Canada, published January 1, 1930, shows that CFRC's wavelength was increased on that date from 267.7 to 322.6 metres, for a frequency of 930 kilocycles, an optimum spot right in the middle of the AM dial.

Harold Stewart (Elec. '26) arrived back at his Alma Mater in the fall of 1929 to take up the post of Lecturer in Electrical and Mechanical Engineering. He had been dissatisfied with life in industry. When J.G. Burley told him that Prof. Jemmett was looking for a lecturer, he applied and got the job.[95] For the first while he had nothing to do with CFRC, but by the late winter of 1930 he was a salaried operator, and was soon very deeply involved. In 1929-30 George Ketiladze (Elec. '29) was engaged as Morgan's paid assistant at CFRC.[96]

Queen's Sports

Bell was instructed to install the private telephone line from the stadium to Fleming Hall, as usual, by requisition No. 13661 on September 21. Perhaps the order was cancelled, for no regular league football games were broadcast over CFRC that autumn. Then, on requisition No. 13674, November 12, Jemmett ordered Bell to quickly install "1 pair wires Stadium to Fleming Hall, bridged at exchange & telephones at each end as in former years. Location in stadium is centre top of press box." CFRC broadcast the Intercollegiate Championship game against Varsity, November 16, as well as Queen's loss to "Pep" Leadlay and the Hamilton Tigers in the Eastern Canada Final on November 23. Ketiladze operated the first game alone and worked in company with Morgan on the second. The legendary "Ga" Mungovan (Arts '30) made his debut as a CFRC sports announcer on the 16th, perched at a temporary location on top of the press box right under the roof of Richardson Stadium.[97]

Henry Grattan (Ga) Mungovan had starred as a footballer with the Toronto Argonauts but, after a severe injury to his knee, came to Queen's in October 1927 to get an education. Ga showed himself a "wonderful ball-carrier" against McGill on October 15, looking "like the find of the season," and later substituted for quarterback Ike Sutton for a few plays in the McGill game of November 12, 1927: "He made yards on two successive plunges... leaving the audience with a fair idea of what he could do if he had the regulation allowance of knees."[98] Mungovan was not in the 1929 Tricolor lineup, but he did substitute as utility man from time to time when his knee was not acting up. In 1930 he quarterbacked the Tricolor to the Intercollegiate Championship.

Later he coached Intermediate football, Arts football, KCI football and Senior hockey, was Sports Editor of the Journal and successful Arts-Levana-Theology candidate for First Vice-President in the 1930-31 Alma Mater Society elections.[99]

Ga is still remembered by Queensies and non-Queensies for his colourful language and for one particular football broadcast. Probably his only football broadcast. Three variants of the story were collected, from Jean Campbell (Mrs Stuart) Henderson (Arts '31), Freeman C. (Casey) Waugh (Arts '31 and '34) and Orton Donnelly, so it has to have a foundation in fact.

> Jean Henderson: I remember years ago that Henry Grattan Mungovan was broadcasting one of the Queen's games. It would be about 1928. The game was in a very crucial point and someone dropped the ball, and the announcer said, "Jesus Christ! He's dropped the ball!" It was really shouted into the microphone, and Queen's was dropped from the air. Everything went dead after that - just stillness, and I don't think the rest of the game was broadcast. There were things in the Queen's paper about it.[100]

> Freeman Waugh: There were several occasions during the history of this station, when it distinguished itself. But to my way of thinking, its

greatest moment came during the broadcast of a football game at George Richardson Stadium - the old one. The announcer was an old friend of mine by the name of Ga Mungovan from Toronto. He was a football player himself, but he'd been injured and couldn't play, so they gave him the job of broadcasting the game. Ga's language was not always as mellifluous as it might have been, and during one exciting play there was a fumble - I think it was Varsity fumbled the ball. A Queen's man picked it up and headed for the goal line with it. He was quite a speedster, whoever he was. And Ga was heard to say, distinctly, "Look at that son of a bitch run!"[87]

<u>Orton Donnelly</u>: I remember one incident, broadcasting from the roof, I believe. The broadcaster was Ga Mungovan, who had been a Queen's quarterback and had volunteered for the job on the roof. Queen's had tried one of their well-known end runs, which terminated with the player dropping the ball. Ga's account of the play went something like this: "Queen's quarterback has the ball, he passes back to one of the backs, the back then passes to another back. At the termination of the end run he passes to a third back who was now well down on the field. Oh! He dropped the ball, goddam it," or a remark which was immediately censored. And I believe that ended his career as a broadcaster.[101]

Orton's youngest brother, Douglas, has a fourth variant of this Mungovan story: "In a football game against Varsity, a Queen's back broke away with a pass or interception, late in the game with a close point spread, `...going 10, 20, 30 yards - Jesus Christ, he dropped the ball!' CLICK!"[102]

The stories do point to the game with Varsity on the 16th rather than to the less exciting loss to the Tigers the next week. And Mungovan did make his debut as a CFRC broadcaster in that Varsity game. The Journal made no reference to the nub of the stories, yet the article on that match had a slightly sarcastic and obscurantist air about it:

> Some little criticism of the local broadcast has been voiced in Kingston. One man on Lower Albert Street complained that it was too loud...Letters have been pouring in this week to the station and to Mr. Mungovan from all parts of the province, expressing entire satisfaction with the way in which the big game came over the air...Toronto folk liked the judicial impartiality with which Mr. Mungovan had the good taste to use in describing the play of the two teams even in the most exciting moments. Only once did excitement get the upper hand, and then it was when Gilmore uncorked his great run at half-time - "Ten, fifteen, twenty - twenty-five - thirty yards!" There was an excuse for it... According to his best friends, Ga made only one slip. A friend in Windsor who stood outside a tobacco store all afternoon to hear it, wrote in that once when Ga turned to his helper behind the "mike," a muffled "Who in h - - - was that?" came over the air. Not bad, considering.[103]

The only major Queen's bobble early in that Varsity game happened in the first quarter. "Blurp" Stuart, a great plunging middle, broke through into the clear 50 yards from the goal line and was well on his way, when he tripped.[104] The CFRC log book does not indicate that the broadcast was interrupted at any point or cut short. A newspaper squib verifies, however, that it was interrupted, discreetly avoiding any reference to profanity. So there is circumstantial evidence that Stuart's stumble in this game was the moment in question:

> Saturday's main afternoon event to radio fans was the broadcasting of the Queen's-Varsity game over CFRC. This came in fairly good, except at the start, when station troubles apparently stopped broadcasting.[105]

According to the log, CFRC did not transmit any hockey or basketball games in the winter of 1929-30.

Convocations and Extension Lectures

George Ketiladze operated for the broadcast of the 1929 Fall Convocation in Grant Hall. This was an important convocation, with the granting of ten honorary degrees, the installation of Mr James A. Richardson as the new Chancellor and featuring a speech by the British High Commissioner. It was also the last Queen's Convocation for retiring Principal R. Bruce Taylor and for Dr J.C. Connell, who was being succeeded as Dean of Medicine by Dr Frederick Etherington. Students were exhorted "to refrain from throwing beans, coppers and mothballs upon the visitors and audience on the floor of the hall."[106]

CFRC also broadcast the annual meeting of the Queen's Theological Alumni from Convocation Hall, October 28 and 29, including the induction of the Rev J.M. Shaw as Professor of Systematic Theology plus ensuing addresses, and the first Chancellor's Lecture the second day, by the Rev Dr James Moffatt of Union Theological Seminary.[107] The two following afternoons, Ketiladze operated CFRC for two public lectures from Convocation Hall. Dr Jacob Viner, Department of Economics, University of Chicago, spoke on "The Reparations Problem."[108] Once again the station's batteries got a good work-out, four days running.

An Extension Lecture series began in January 1930, on outstanding figures of the 19th century. All of the lectures were broadcast over CFRC, but that fact was published only once over the course of the series, and in the Journal rather than in the city paper. The schedule was as follows:[109]

Jan. 13 - "Queen Victoria", Principal R. Bruce Taylor
Jan. 20 - "Bismarck", Hugh Sutherland, M.A.
Jan. 27 - "Lord Elgin", Prof. Duncan McArthur
Feb. 3 - "Darwin", Prof. Alexander Macphail
Feb. 10 - "Cecil Rhodes", Prof. Norman Rogers
Feb. 17 - "Pasteur", Dr. J.L. Austin
Feb. 24 - "Disraeli", Principal H.A. Kent
Mar. 3 - "Cardinal Newman", Prof. Nathaniel Micklem

388 IN THE SHADOW OF THE SHIELD

On February 25, Harold Stewart operated a 25-minute program on CFRC advertising the Queen's College Frolic. At a guess, someone must have done a skit or two from the show for the unseen audience.

The Queen's College Frolic

<u>Kathleen Ryan</u>: The College Frolic deserved to be broadcast, because it was good humour. They had a quartet that used to sing, "Veni, Vidi, Vici, Bolsheviki, Means the ship of state is leaky." Then, we had another one. J.A. Roy was Professor of English, a very jolly Scotsman with a lovely accent. He was noted around campus as being a right jolly fellow. So, they had a little skit in honour of him, and "Pete" Dolin (Arts '24) calls up on the phone: "Prof. Roy, this is the Kingston station calling." "Yes sir, what do you want?" "Well, Professor Roy, there's a case of books from Scotland down here for you, and it's leaking all over the platform." And that's what College Frolic was like. It was very high. It lasted a long time at night, but it was very, very good.[58]

<u>Freeman Waugh</u>: For many years at Queen's there was an annual show put on. It was called the "Queen's Frolic." I took part in it in 1928 or '29, I forget which, the first time. In 1930 the name was changed to the "Queen's Review," and during that year we had a rather outstanding group of talented people - present company excepted. The Frolics and the Reviews consisted of intermittent skits and musical numbers, some of which were of questionable taste in both categories. The low-light of the 1930 Queen's Review was a song, written by an old friend of mine named George Ketiladze. He was a White Russian who had come to Canada with his younger brother a few years before to make his fame and fortune. I don't know how well he did in the fortune part, but he became quite famous. He was a huge man, and, without knowing the first thing about wrestling, he won the Intercollegiate Heavyweight Championship two or three times in succession. He was also a gifted pianist, although he was a lousy composer. He hired himself out as an entertainer to help put himself through Science. Not only did he play the piano, he also was quite an accomplished magician. The song that I sang at the 1930 Review was ineptly titled "I Want You," and it was a real stinker. The cover on the song stated, with no modesty whatever, "Successfully introduced by Freeman Waugh." As far as I know, there were three copies sold of this song: George bought one, the fellow who wrote the words bought one, and I bought one. I've long since lost my copy.[87]

George Ketiladze (Elec. '29)

George Valerian Ketiladze came from Tiflis (Tbilisi) in the Georgian SSR, arriving in Dartmouth, NS, without credentials, aboard a lumber ship. During his detention by immigration, he was able to earn money by his entertaining. George came to Queen's in the fall of 1925 with the help of a Mr Williams, a musician and concert director, who wrote to Dr A.L. Clark, Dean of Science. George was admitted "without records."

*George Valerian Ketiladze
(Sc. '29), pianist, composer,
magician, Inter-University
Heavyweight Wrestling
Champion, CFRC operator and
entertainer par excellence. He
was the Big Man on Campus
and is remembered fondly by
his contemporaries.
Courtesy Prof. H.H. Stewart*

Kathleen Ryan: Some of the people who came to the basement where the station was: Ketiladze was a gifted refugee from Georgia, in Russia, not...He was a bear of a man. A huge man and a very attractive personality, with a keen ear for music, and he was quite a definite fan of radio broadcasting. He was a sort of a legend around the University. He lived hard. Ketiladze was such a big, poweful man, and he could pound the daylights out of the piano, without touching it much, you know. He had a light touch, but he was fairly powerful and very, very musical. He enjoyed life, and we had one big joke about late at night, he was standing in front of Grant Hall looking up at the clock. And somebody said, "What are you doing?" He says, "I'm weighing myself." I know it sounds corny, but it was very funny at the time because he was such a distinguished person.[58]

There is an interesting variant of that story. Somebody had walked off with George's coat, scarf, hat and rubbers after the Actors' Ball in 1926, and he wrote: "On my way home while enquiring about my coat I met a gentleman whom I saw at the dance. He was standing before a mail box. He had dropped a nickel into the box and was looking at the town clock to see how much he weighed. He didn't know anything about my coat."[110]

Jean Henderson: The first time I met George Ketiladze was in this house on Union Street. We had a piano in the living room and he played beautifully. He had such huge hands! We all asked him why he never played football because he was so big and strong. And he had his hands insured. He'd been on a ship in the Mediterranean and wanted to come to America or Canada. He came to Canada and he didn't have a sponsor and if you didn't have a sponsor you couldn't get in. George sat down at the piano and played. And this man heard him and said, "Anybody that can play like that will never be a liability to Canada," and he sponsored him. The year I was down here first, in 1927, he sponsored his brother Valerian. George had some knowledge of English - they were from Tiflis in the Georgian province - and Valerian came over and didn't know any English. We would talk to him, forgetting that he couldn't speak, and we thought maybe he could understand. Valerian would just smile at you and George would say, "Speak, you dirty foreigner! Speak!" So I'm sure Valerian learned to speak, because by the end of the year he was doing very nicely. George was very gracious with his talents, though.[100]

Stuart Henderson: He put on a concert at Ban Righ Hall for about an hour and a half or two hours of all his own compositions and it was just magnificent. His piano playing was just absolutely beautiful. He wrote the music for the College Frolic of 1929-30. It was my pleasure to sing in a quartet and Harold Sprott (Arts '32) wrote the words for it. George became quite a talented and well-respected musician in the professional field, and moved into the United States, over around New Jersey - New York, and then all of a sudden his life started to go downhill. He got involved in too fast a group.[111]

George was leader of the "home brew" Queen's Collegians Orchestra in 1926[112] and starred as composer, musician and magician in the College Frolics from 1926. Around 1928 he played piano for Sid Fox in the Venetian Garden Restaurant above the Capitol Cafe, across from Reid's Furniture. If they had a disagreement, little Sid would tell George that he was going to take him outside and whup him. According to Sid Fox, George was a real heavy piano player, and loud.[113] In 1927 and 1928 George was also Intercollegiate heavyweight wrestling champion. Known as the Wrestling Pianist or the Georgian Mountain Lion, he tipped the scales like a carload of cement.

Freeman Waugh: George, although he played the piano beautifully as a soloist, was not the best accompanist in the world. He insisted on sticking arpeggios in all over the place that had nothing to do with the piece that was being sung. I asked him to be my accompanist once at a concert at St Andrew's Church Hall in Kingston, and he agreed. When my turn to perform was announced, I was standing on the platform at the front of the hall, George was seated at the piano on the ground floor of the hall - the piano was on the floor; George was sitting on a piano stool. We started into the song, and much to my surprise and embarrassment, George insisted on so many frills and furbelows that I couldn't keep track of the tune. So I held up my hand and stopped him and said, "George, if you want to play a solo, you go ahead and then we'll do my song." I sat down on the edge of the platform, George played a piece of something and I got back up to my feet and said, "Now we'll try it without the frills, if you don't mind." He was really chastened.

In 1932, I took a year out of university to try to earn a little money in the life insurance business. I had to go to Toronto to take a course, and while I was there I lived with George, who was playing the piano for a living at CFRB in Toronto. While I was there, Duke Ellington's band came to town and was playing at Shea's Hippodrome. Somewhere down through the years, George had met Duke Ellington, and he invited him back to the apartment after the show one night. Also present were a couple of other pretty good musicians, Blaine Mathe, who used to play the violin in the "Happy Gang" and a wonderful Russian basso named Yascha Vilovitsky. I sat in a corner of the room the whole night listening to these characters performing their art. It was a wonderful experience. I had another interesting musical experience during my stay in Toronto that year. George had a nightly assignment to play the piano at CFRB, and he persuaded Andrew Allan, who was the announcer and also I think the Station Manager, to let me sing on one of the late night spots. I don't think that I did very well, and looking back on it now, I shudder to think of one of the songs that

I sang, which was entitled, "That's Why Darkies Were Born." It was a good piece of music, but the sentiments were no hell.

I suppose I knew George Ketiladze about as well as anybody at Queen's, but he was not an easy man to know. He was friendly, but I always got the impression that he was thinking a lot that he wasn't saying. He was so big and so powerful that it was much better to have him as a friend than not. He had a fine mind, did extremely well at his studies, although I don't think he ever put them to very great use. That's partly because of the time in which he graduated, in the middle of the dirty 30s.[87]

George Ketiladze taught at Ottawa Technical School in the late 1930s, then became a successful showman, humourist, musician, magician and actor, known professionally as George Jason. He started with the USO in 1944 and later appeared on Broadway, on radio and on television. Two of his appearances on Jack Parr's "Tonight Show" are preserved in the NBC Archives.[114] George died suddenly in 1959 in Saginaw, Michigan, while on tour with his one-man show.[115]

The Whig-Standard Dominion Election Broadcast

After the last extension lecture in the spring of 1930, CFRC closed down, but not for the entire summer. The log shows that the transmitter was fired up at 7.00 pm on July 28, 1930, for a Whig-Standard broadcast of the Dominion election results, the first time that CFRC had ever been officially on the air outside of a school term and the first time anyone was permitted to hire the Queen's station for a limited "commercial" broadcast.

The initiative came from the newspaper. William J. Coyle (Arts '19), Advertising Manager of The Whig, wrote to the Registrar, Dr W.E. McNeill, on June 30, 1930, suggesting the possibility of linking up with CFRC to broadcast non-partisan reports of the Dominion election. The Whig would have seven telephones in service for enquiries plus the usual lantern projecting the results onto a wall in Market Square. But they felt the time had come to enter the broadcasting field, since the public was no longer willing to wait until the next day for the results. Coyle proposed, as part of the deal, to advertise the Queen's radio service very prominently for several nights in The Whig. CFRC's engineers would do the operating, but the microphone would be manned by newspaper people in The Whig offices and would be linked to Fleming Hall by telephone line.[116]

Prof. Jemmett wrote back to Acting Principal J.C. Connell on July 10, confirming a conversation of the 8th, giving specifications of the station and quoting a rate of $40 an hour for the use of CFRC on election night: "The present set is battery operated and its time limit is five hours, which would do for the Whig's purposes on the 28th."[117] The station was tested on the 25th, when a woman heard her name announced over the air as the winner of a Kingston merchants' Red Hot Sale contest.[118] The CFRC log does not record that test. Two thousand assembled in front of The Whig on the night of July 28 to see the election results projected onto the wall of the Bank of Montreal and to hear the announcements and review of the results, "by a close student of politics" (probably Bill Coyle), over a public address system linked up with the radio. Gen. Dr A.E. Ross defeated Mr Halliday as part of R.B. Bennett's rout of the King government. Ross was carried on the shoulders of his supporters up to the radio room on the top floor of The Whig, where he addressed the radio audience and the assembled crowd. Halliday thanked his supporters as well, by means of CFRC.[119]

Dick Cohen: At the time they broadcast the first election results, shortly after I graduated, Prof. Jemmett phoned me and wanted to know whether I'd be present. I remember all I did was bring a radio with me. I don't think I was even in the studio. I was in a room near the station, monitoring the news as it came through, and that was the extent of it. I received a cheque for a nominal amount afterwards, which I really didn't expect. I didn't want to take it, but Prof. Jemmett insisted, so I did.

CFRC's motor generator set had a commutator ripple since it was not well filtered. So they got a ringing sound, and you could tell that it was Queen's on the air.[120]

William J. Coyle: I was working for The Whig-Standard and trying to do all the promotional things I could. For one election we hired lanterns and slides, and threw shots on the Bank of Montreal wall, on a big screen erected there, from The Whig-Standard building, so that people were able to read it. Now, there was no voice involved. Later on, I hit on this idea of broadcasting an election, so I made arrangements for a telephone connection and I went to see Dr Connell in his big home down on the waterfront - he was Acting Principal. So that would pretty well nail the time - it was between two principalships. He allowed the use of the radio station, and Prof. Jemmett did some of the advising. He said we'd wear their batteries out and we'd wear out some of their machinery - it might cost $50. So I paid $50 for the whole evening rental of CFRC. That meant we had a live microphone on the third floor of The Whig-Standard building. The news came in over a printing machine, the way news did in the newspaper offices. The boys would run upstairs with it. Without any editing we had to read those items and pad them out as best we could. It must have been a terrible broadcast. When we ran out of material, we played records on a machine beside us, to fill in the time, and that must have been pretty bad listening. But it was surprising the number of people who called in from Cataraqui and villages around here - they were quite surprised. It was one of the first times they had ever heard the election returns right on their local radio. And that was CFRC!

I had first worked for the Kingston Standard before the amalgamation, which took place in 1926. I was away in western Canada from '26 till '29, and in '29 I came back to The Kingston Whig-Standard. For the newspaper I did everything from being a reporter to the advertising side and then to the business side of it in about a 40-year period with them. Of course, it was in that period that the radio business was becoming prominent.

I can tell one story about early radio. About 1929 or '30 the World Series was on, and I happened to be quite interested in baseball.

We had a contraption at <u>The Whig</u> that was put in the front window, and a man standing behind it, by moving pegs, could show a ball game in progress. He could show a runner on the bases and move these things around as the game proceeded. Again, the news came over telegraph wires from whatever park it was in, to Toronto. The Canadian Press then sent it to Kingston and it had to be brought from our downstairs office to the front window. So, we might have been two to four minutes behind the actual play going on in New York or wherever it was. This day, we had a very close game coming to an end and a thousand or more people out in the Market Square looking at this thing and following what was going on. And a person came running along King Street, and he said, "You damn fools, it's all over." He had been listening to a radio station about a block away and that 3 or 4 minutes interval - he knew what had happened. It was all over and he was having a great time. So that was the last time we ever showed it on this mechanical screen using telegraphic news, because radio had grown up and replaced it.[121]

Origins of the Mark V Transmitter

The CFRC Mark IV transmitter in 1930 was still basically the old transmitter of 1923, or even the original one of 1922, with modifications. The latest changes, made in 1929, were in the audio part of the transmitter, where the speech-input stages were redesigned to use newer tubes.[26] Now that commercial interest in CFRC was being expressed, the authorities were apparently discussing upgrading the broadcasting equipment and extending the hours of service. Replying to Dr Connell, following <u>The Whig</u>'s approach about co-operation in covering the 1930 election, Prof. Jemmett supplied additional information:

> Confirming our conversation of the 8th...the wave length has recently been changed to 322.6 metres (930 Kilocycles) and the average effective range is 150-200 miles...The design of the station is good and if any increase in output is contemplated, the changes would be in the oscillator, the aerial and the source of high voltage (3,000 volts) direct current. The transmitter and amplifiers up to the modulator would require no change. The cost would be any thing up to say $5,000 - $10,000, depending on the increase desired. The present outfit is very largely home made and would hardly do for any regular and frequent service...The use of the station could be largely increased without greatly increasing the operating charges, except for wages. Two men are usually required, but for a broadcast from outside the University Grounds three would be necessary.

He added two lines post scriptum: "There should be no trouble about getting permission to extend our operating schedule. Something depends on what action the Government may take on the Aird report."[117]

Prof. Jemmett made enquiries over the next month and wrote to Dr Connell in early August:

> With regard to our broadcasting station C.F.R.C. I have written to the Northern Electric Co. and the Canadian Marconi. The Northern makes nothing between a 1 and 5 K.W. set. The Government insists that a 5 K.W. set must be 25 miles outside the city which means that the cost would be prohibitive for us and of no use for class purposes. The one K.W. is no more than our present station could be run at if we added to it.
>
> The Canadian Marconi would not commit themselves to a guess at the price and want full details which I am unable to give at this stage. The Northern are sending their guess in a few days and I have asked Prof. Morgan to take it to you when it comes as I expect to be out of town. Their agent was here and gave me his own guess of $25,000 for a 5 K.W. station not including the building. He says that the old installation at Moncton (C.N.R.A. a 5 K.W. set) is for sale and is sending me the price. He thinks it might be $5,000. This would have to go to Napanee or Tichborne for installation.
>
> To duplicate our present equipment would cost at least $1,500 and it seems to me that this is the cheapest way out, especially as we do not know what action (if any) the Dominion Govt. will take on the Aird Commission report.[122]

Jemmett wrote again on September 24, with prices to put CFRC "in condition to work regularly without interruption." A new high-voltage generator would cost $525, a microphone and speech amplifier $600 and a reserve stock of tubes an additional $500.[123] So, at this point it appears that they had pretty well settled on duplicating the Mark IV equipment, as a back-up, instead of upgrading or buying a second-hand transmitter. Two days later, Dr Connell submitted the report requesting $1,625 of the Executive Committee of the Board of Trustees, and appended his own "opinion that the station should be put in the best possible condition and should be used to a greater extent than it has been in the past."[124] Jemmett had just ordered an armature for type F23 Dynamotor: primary 110 volts, 20 amps; secondary 1,000 volts, 0.4 amps. CFRC probably needed a new motor generator set too.[125]

The University was convinced that CFRC was useful and was doing some good, since "special lectures and addresses...whether delivered by members of staff or by distinguished speakers from outside...are clearly heard in the Kingston area even though speakers have poor voices and are unable to reach the back rows of the hall. Moreover, the University is greatly pleased to know that large numbers of people listen regularly to informative lectures of an hour's length. The popularizing features of commercial broadcasts are not desired by listeners with intellectual interests."[126]

Announcement was made by the Board of Trustees, October 25, that further expenditure on CFRC was endorsed to duplicate the vital parts of the existing equipment, as insurance against breakdown, rather than to increase the power of the transmitter. They granted $1,125 as a capital expenditure for CFRC plus an additional equipment grant of $500 for tubes. "According to Professor D.M. Jemmett, the present station is up-to-date in broadcasting apparatus, and any further addition of equipment for the increase of power would be contrary to the government regulations which govern radio stations...The intention of the University is to use the radio broadcast for lectures of public interest, and other events at the University of an entertaining and instructive nature to the public at large. The station will not be commercialized, but its use by business firms of the city who wish to provide a program or broadcast some special event of importance, as did The Whig-Standard giving the general election results this year, will be permitted."[127]

The Master Plan Changes: Mark V Is Built

<u>Harold Stewart</u>: I graduated from Queen's in E.E. in 1926, and after three years in industry, helping to build motors and transformers and that sort of stuff, I was offered a job in the staff at Queen's. This was a joint appointment, half in Electrical and half in Mechanical, a lecturer at the magnificent sum of $1,500 per annum. Of course, in those days, you were not expected to stay around in the summertime. That was from late September to the end of examinations, then good-bye. And you could go and get another job then. And up until 1929 or '30, you could.

Prof. Morgan took charge of the station in 1928 and was assisted by Mr W.G. Richardson of Science '26, my own year, who was then a demonstrator in the Department and, as of 1938, was on the engineering staff of the CBC. The following year Mr Richardson left and Mr George Ketiladze, Science '29, became the operator under Prof. Morgan. In this year I also returned to Queen's as Lecturer in the Electrical Department.

In the day when I came back here, it was hard to label anything as a studio. The station consisted of one room. Half of it was walled off for a transmitter room, the other half was a studio. I think we hung some drapes around there to reduce some of the bouncing - the echo. And, the man who turned the transmitter on and adjusted the thing probably made the initial announcement. Unless there was something very special. Now, let me recall - when George Ketiladze was the operator, Stan Morgan was in charge of the station. Maybe Stan made the opening announcement, such as "This is CFRC, Queen's University, in Kingston, Ontario" and so on. And then he would introduce the speaker. Now, I don't think we had any record players of any consequence at that point. We might have had something you could plug in for a test, you know, but practically all the programs would be a special speaker on some special subject that they thought the public would be interested in, or convocation or a sporting event. We had wires going over to Grant Hall...to the stadium and maybe to the arena.

Well, when I came on the staff I didn't have anything directly to do with CFRC. I was a lecturer in E.E., and I watched with interest what people were doing at the station, although I didn't go in. I just listened and I talked to people. I had been, and still was an amateur radio operator with my own set at home. I lived here in Kingston since 1913 actually, and so I was a bit curious. Well, by and by I must have sort of wormed myself in there somehow and got an operating job. And they didn't run very many hours a week - maybe one or two programs a week. I can't even remember now, but they were rather infrequent. One of the operators at that time was a chap named George Ketiladze, who was a graduate in E.E., and he was on a Master's program. George was given the assignment of developing the theory of the screen-grid tube, which was just coming on the market then, and there had been no amount of theory published. Well, that was a large assignment, and to make the story short, George never did master it and never did write it. But, anyway, I became an operator and I learned something during that year, '29-'30, that the people around here seemed to like to listen to CFRC. There wasn't much else to listen to, I guess.

When I came there was a functioning transmitter, a big Mullard triode which acted as an oscillator and another big one which acted as a modulator. And there were a couple of smaller tubes ahead of that as a speech amplifier to drive the thing. That's what they had on the air when I came in.

The University people offered some money to the Department to improve the transmitter - there didn't seem to be much interest in actually modifying it. They were going to build a duplicate, so they'd have a fall-back in an emergency. Well, I had been a ham radio operator, and I knew what was going on in the field. I read the magazines and so on, and it seemed to me to be a waste of money to duplicate something that was already out of date. So I said, "Let's build a new one." "Okay. You're it. You build it."

So, why did I build it, or why did we build it? Well, the existing transmitter was a self-excited oscillator type, and with Heising modulation, one tube modulating another. You couldn't get very high modulation percentage - maybe forty. Fifty at the most. And, I suppose it had about maybe 300 watts output. They got good reports from around the country on it, but I thought, let's bring it up to date. So, we built a crystal-controlled set, right out on what's called a breadboard layout. You just nail the stuff down to a table top in effect or up on a board behind it. So we built a crystal-controlled job and used high-level modulation. In the previous set, the modulator and the modulated tube were operated at the same voltage. Well, if you re-arrange the circuits so that the modulator operates at, say 500 volts, and the modulated tube operates at maybe 250 to 300, then you can develop enough signal voltage swing to completely modulate the amplifier that's carrying the RF fed through from the crystal oscillator. Well, we did that anyway, and I think this was finished some time in '30, '31, and we went on the air in a late night broadcast and got, oh, reports from all over the country. Oklahoma, I guess.[128]

By this time AC-operated receivers were coming into general use and broadcasting was developing rapidly with the increasing number of listeners. Transmitter development had not lagged far behind and the master-oscillator power-amplifier (MOPA) type with crystal-controlled oscillator and 100% modulation capability was making its appearance in many stations. In the fall of 1930 I was given permission to build such a transmitter...The crystal was ground for 930 kilocycles...

The new transmitter was completed in the spring of 1931. Its line-up was as follows: a type-210 triode crystal oscillator, a type-865 screen-grid first buffer and a type-210 second buffer. The modulated Class "C" amplifier was also a type-210 operating with 300 volts on the plate, through a voltage-dropping resistor, from a 500-volt motor generator set, which supplied all of the low-power stages. The modulator was a type-250 with a plate voltage of 500. The audio-power output from the modulator was 6 watts and the plate input to the modulated amplifier was 12 watts. Thus the modulator, without overloading, could completely modulate the Class "C" amplifier. This power was used to drive a linear amplifier using two of the Mullard VO/350's in push-pull, supplied by the 3,000-volt motor generator set. Thus available equipment was used to the greatest possible extent consistent with modern design. This transmitter put 200 watts of carrier power into the antenna. No reasonably priced modulation meter was available at this time but overload or excessive modulation could be prevented by keeping the audio voltage below the point where the DC plate current in any stage started to vary appreciably with modulation. The antenna and counterpoise were not changed at this time.[26]

When we got crystal control, we had to stick right to the assigned frequency. We were allowed plus or minus 50 cycles, I think, and we built a home-made monitor to check this. This was a General Radio design, but we built it ouselves because it was cheaper. The principle of the monitor would be simply set up another oscillator which would have a frequency one kilocycle away, and you would feed these two signals in through two detectors - each signal will go through a tuned circuit. One was tuned to 1,050 cycles and the other to 950, so if the beat frequency between them came through at 1,000 cycles, it would be equidistant from the two resonant circuits, and the meter would read "zero." If it deviated one way or another, the meter would go to one side or the other. Anyway, we thought we were okay.

One time I heard a thousand cycle note coming through in the phones, and I thought, "We've got a station in Syracuse that's off frequency, because we're okay." But at the same time we got word from Ottawa that we were off frequency. I protested, but they found out that we actually were. What was happening, I guess, was that we were off 500 cycles and our circuits couldn't respond to 500 cycles. But the second harmonic looked like 1,000 cycles, which was what was supposed to be coming down. So the meter read okay. Later, the limits were changed to +/- 20 cycles on AM.[128]

Purchases for the Building of Mark V

Beginning with a flurry on November 18, 1930, Prof. Jemmett charged certain purchases to "Special Radio Station" or "Special Radio Broadcasting": three VO/350 and two DO/40A transmitting valves from Mullard, variable condensers, fixed and variable resistors, grid leaks, a type 585-M2 microphone transformer, a type 585-D audio frequency transformer, milliammeters and a voltmeter. A motor generator set was ordered from the Lancashire Dynamo and Motor Company of

Canada, Montreal, for $535 net, on December 5. Two weeks later he bought a Type R-600-A transmitter (microphone) and a Type R-1-D transmitter mounting, for $141.40, on requisition No. 12607. Purchases continued into early 1931, with a Type TMU 300A transmitting condenser (0.0003 capacity, 7500 volts) and radio tubes (types X250, X210, X280 and X245).

The requisition book shows that Harold Stewart was paid three monthly installments of $150 from the Special Radio account and then two from E.E. over the period April 11 through September 9, 1931, perhaps for work on refining the Mark V. He even worked on Labour Day. In 1931-32, Morgan was receiving $300 a year, in two instalments, for his regular work at CFRC and Stewart $150.

CFRC Broadcasts in 1930-31

In the meantime, while Mark V was under construction, broadcasting continued with the Mark IV in the fall of 1930. The CFRC season began with Convocation and the installation of the new Principal, Dr William Hamilton Fyfe.[129] Among those receiving honorary LL.D.'s that afternoon were the Premier of Ontario, Hon G. Howard Ferguson, Dr Louis Stephen St Laurent, President of the Canadian Bar Association, Gen. Sir Arthur Currie, Principal of McGill, and Dr. Robert Charles Wallace, President of the University of Alberta. George Ketiladze operated.

The two regular league rugby games played in Kingston were aired on October 25 and November 1, plus the victory over McGill two weeks later for the Intercollegiate Championship. Bill Coyle (Arts '19) of The Whig-Standard and Wilf Charland (Arts '32) handled the announcing duties very capably indeed.[130]

> William J. Coyle: There was some dissatisfaction with Toronto broadcasting. The local fans felt that they were too partial to Varsity, so a couple of us did a broadcast. Perhaps we were too partial to Queen's, but we did a gane for them, and somehow I dropped out after that. I wasn't in the radio business, and I've forgotten just how I got involved in that. Probably through the newspaper side of it. But, I remember doing one for sure, one Queen's football game which we won, but I can't tell you exactly what year. I can still see them making those plays - of course, we had no forward pass in existence. Two bucks and a kick. But Queen's developed a lateral pass that went out to probably two or to three people, and I think it was Leadlay was the last man to get it and he was such a fast runner that he could just skirt the end and go for yards and yards. So, that was the type of play that was the most famous with them.[121]

Principal Fyfe had observed that Canadians had a diseased appetite for speeches.[129] A long series of special lectures was fed to the word-hungry Kingston public that fall, but the Journal made no mention of the contribution of CFRC:

Oct. 27 - "International Peace at the Present Time", Sir
George Foster (on the League of Nations and
World Court)

Oct. 29 - "The Church and Labour", Tom Moore
(President of Canadian Trades and Labour
Congress, delivered to the 3rd Theological
Alumni Conference)

Oct. 31 - "Imperium et Libertas", Dr. R.S. Conway

Nov. 3 - "Virgil's Creative Art", Dr. R.S. Conway
(celebrating the bimillennium of Virgil's birth)

Nov. 6 - "Education and World Peace", Rev. Stanley
Russell

Nov. 18 - "Charles Lamb", Dr. T.R. Glover

Nov. 19 - "Diet in History", Dr. T.R. Glover

Nov. 20 - "Our Debt to Greece", Dr. T.R. Glover

Nov. 25 - "The Conflict of Morality and Convention",
Dr. Herbert I. Stewart

Nov. 26 - "The Puritanism of Bernard Shaw", Dr. H.I.
Stewart

CFRC broadcast little except spoken word in the spring of 1931. The
Journal announced that Senior B hockey and one of the B.W.F. Assaults
might be transmitted, but the log shows no sports programs. A Queen's-
McMaster debate aired on February 20, with Queen's supporting the res-
olution "that this house endorses the stand taken by the Canadian del-
egation at the recent Imperial Conference."[131] The extension lectures
returned to the airwaves Mondays at 5 pm in January with a series on
"World Figures Past and Present." The first, Prof. A.E. Prince speaking
on "Mustafa Kemal," on January 19, was supposed to be broadcast,[132]
but it was not logged. There followed:

Jan. 26 - "William Jennings Bryan", Prof. W.A.
Mackintosh

Feb. 2 - "Sir William Osler", Dr. James Millar

Feb. 9 - "George Bernard Shaw", Principal W.H. Fyfe[1]

Feb. 16 - "Thomas Carlyle", Dr. W.E. McNeill

Feb. 23 - "Gandhi", Prof. Norman Rogers

Mar. 2 - "European Affairs", Dr. E.B. Schmitt

Mar. 9 - "Foch", Prof. Alexander Macphail

Mar. 16 - "Lenin", Prof. T. Callander

The last lecture of the series from Convocation Hall was the occasion of the first logged use of Harold Stewart's Mark V transmitter.

The first complete CFRC Proces Verbal, dated April 1, 1931, turned
up in Prof. Stewart's papers. Following a 6.55 pm sign-on, it logged
recorded popular music, a program transcription (Canada on Parade)
and two studio concerts (a string quartet and a soprano). At 10.01
CFRC carried "The Neilson Hour" originating from CKNC, Toronto,

---

1. "G. B. Shaw is an ally of the Angels, no matter what mince meat he may
make of the churchmen."

and at 11.00 pm signed off the air. They tested the transmitter briefly after midnight and closed down again at 00.12 am, April 2. There is no corroboration of this broadcast in The Whig or in the Journal. Unfortunately, CKNC did not send its post-6 pm schedule to the Toronto newspapers, so it cannot be verified whether "The Neilson Hour" went to air that night.

A test broadcast was sent out at 4 am on April 2, 1931, to determine the range and quality of the new transmitter. They dedicated the program to the Newark News Radio Club in New Jersey, which had indicated interest in the test. Other people around Ontario and Quebec were also asked to tune in. The program featured George Ketiladze playing a recital of popular songs and his own transcriptions and Bill Bothwell (Arts '33) singing songs and novelty numbers. Wilf Charland announced "and put a touch of humour to the proceedings," and S.C. Morgan and Harold Stewart operated.[133] The test was a great success. Nearly a hundred letters arrived at CFRC, from Lachine and Windsor, Maryland, Kentucky, Iowa, Nebraska and Oklahoma City, Oklahoma, about 1,300 miles away, congratulating CFRC on the clarity, power and program quality. Someone in Newark, NJ, wrote: "Your program came over great, with a great amount of volume, and very clear; in fact, at first I thought it was one of the local stations testing, and was almost stunned when your announcer gave the call letters, CFRC."[134]

Before shutting down for the summer, CFRC broadcast the annual Girl Guide Service from Grant Hall on May 24, following their church parade. Principal Fyfe conducted the service and delivered the keynote address. "Miss L. Walker presided at the piano for the musical part of the programme, Messrs. G.A. Radcliffe and James Neilson assisting with violins."[135]

Changes in the Radio Regulations

CFRC's licence for 1931-32 from the Radio Branch of the Department of Marine, Ottawa, dated April 1, 1931, contained a minor operational change. The antenna input between local sunrise and sunset was cut from 500 to 250 watts, and between local sunset and sunrise to only 50 watts. This may reflect the increased effective power of the new transmitter.[136]

Four pages of new typed-in regulations were added to the licence, pending Parliament's action on the Aird Commission Report. The station's transmitter had to be adjustable and operational on any frequency between 520 and 1,500 kilocycles. The allotted frequency had to be absolutely adhered to or the licence could be suspended. The station had to have an approved type of frequency indicator, as described here by Prof. Stewart, and unless it had automatic control, hourly frequency checks were mandatory. Schedules and program descriptions had to be filed with the Department. The logs had to be kept in much greater detail, and signed by the operator. CFRC did keep a proper Proces Verbal from January, 1932, but no one signed it. Operators had to prove that they were British subjects. Phonograph records were prohibited between 7.30 pm and midnight, and pre-recorded programs were restricted to 30 minutes between those times (not to exceed 3.5 hours per week). These transcriptions had to be produced in Canada and were not to be repeated. Direct advertising was not allowed except with permission of the Minister, though indirect advertising was permitted at any time.

CFRC Broadcasts in 1931-32

Prof. Jemmett's requisition No. 12738, October 13, was an order on Bell to "connect Fleming Hall to the Stadium above the press box as in former years." It turned out that the wire had been left in and CFRC still had the sets which Bell had forgotten to collect. The new transmitter went into regular service on the 17th, transmitting the first of three football games, then a series of lectures by Dr S. Mack Eastman on the League of Nations,[137] and the evening lectures from the Third Theological Alumni Conference. This year, the Journal and The Whig repeatedly announced that the lectures would be broadcast over CFRC, and the University heard from listeners who were very pleased with what they had been able to hear.[138] Dr T.Z. Koo of Pekin, Vice-Chairman of the World Student Christian Movement, was heard over CFRC on January 7, talking on "Manchuria and World Peace." Most of an Extension Lecture Series was broadcast in January and February,[139] with Harold Stewart operating.[140]

> Jan. 18 - "The Gold Standard", Prof. F.A. Knox
> Jan. 25 - "Medical History from Hogarth's Prints"
>   (illustrated), Dr. L.J. Austin
> Feb. 1 - "History of Application of Power to
>   Transportation", Prof. L.M. Arkley
> Feb. 8 - "Masefield - The Poet Laureate", Prof. G.H.
>   Clarke
> Feb. 15 - "The Manchurian Conflict", Dr. Mackintosh
>   Bell
> Feb. 22 - "Bacteria and the Age of Man", Prof. G.B.
>   Reed
> Feb. 29 - "The St. Lawrence Seaway", Prof. Duncan A.
>   McArthur

CFRC carried Dr L.J. (Blimy) Austin's lecture, the slides being shown to the audience in Convocation Hall only after the broadcast. Prof. Arkley's lecture depended on his lantern slides and so could not be broadcast.[141]

The Masonic Temple Charity Concert

Lorne Richardson, Worshipful Master of the Ancient St John's Masonic Lodge, organized a charity concert at the Masonic Temple for the evening of February 3, 1932, to aid the unemployment relief work of the Kingston Branch of the Red Cross Society. In January, the local Red Cross had served 2,400 meals to single men each week. Mr W.H.

The Masonic Temple, Wellington and Johnson Streets, location of the CFRC charity concert remote broadcast, February 3, 1932.
-photograph by Arthur Zimmerman

Herrington of Herrington and Slater, past Worshipful Master of the lodge, propopsed and arranged for the program to be broadcast over CFRC.[142] A technician from CKNC, Toronto, was on hand with some needed equipment, after CFRC had expressed doubt that its equipment could handle the large volume of sound that would be produced in the hall.[143] The man from CKNC reportedly said that CFRC's station and equipment "was quite as good as any which he had brought with him but there was not quite enough of it to pick up the choruses by the Temple Choir." The handwriting in the station log suggests that Morgan operated at the studio. Jemmett's requisition book shows that he did not order a Bell line to Fleming Hall from the Temple, corner of Johnson and Wellington streets, for one of the conditions of the broadcast was that Queen's incur no expense.[144]

The concert featured the Temple Male Choir directed by Mr Herbert W. Hartshorn, Mrs Lottie Sanders Wyatt (soprano), Mr James Saunders (baritone), Mr W. Adams (piccolo), Mr John A. Percival (baritone), Mr Arnold Spencer (violin), Mr John Spencer (reading), Master Billie Sharpe (boy soprano), Mrs Edna Davison Burton (soprano), and Mr R.R.F. Harvey (organ). Miss Lois Baker, Mr John Taylor, Mr S.A. Salsbury and Mrs A.R.B. Williamson accompanied.

> The Masonic Temple...was filled to capacity and it is estimated that several thousands listened-in to the broadcast throughout the city and district...Messrs. W.Y. Mills and Duncan McArthur were the radio announcers during the evening and they reported over the air on every subscription that was telephoned in to the four phones at the Temple, which were receiving the donations. Some were phoned from Ottawa and other distant points.[142]

Announcements about subscriptions were made after each number. They also received many telephone calls asking for encores, and although the concert was already very long, the log shows that a soprano solo and a number by the choir were added. The charity concert raised about $1,200 for unemployment relief.

Lois Baker Rich: I remember the Masonic charity concert. It was a very gala affair. The hall was filled with Masonic men and their wives - everybody dressed in their evening clothes, and it was a very beautiful evening. I played for Lottie Sanders Wyatt, who sang "That's Why Darkies Were Born" and "Trees," and also I played for Arnold Spencer, who was a student at Queen's and a very fine violinist. There were many other Kingston artists on the program - and the outstanding Temple Male Choir under Mr Herbert Hartshorn. I remember it was broadcast by CFRC radio, and it was a thrilling moment because we hadn't done any broadcasting up until then - the microphones and the wires all over the place. It went very smoothly, I remember that. There were no hitches. Nothing went wrong.[145]

The Kiwanis Welfare Fund Broadcast

Following upon the success of the Masonic Concert, Mr Herrington wrote to Principal Fyfe requesting the use of the broadcast facilities on the evening of April 22 for a concert at KCVI auditorium in support of charitable work in Kingston.[146] Dr Fyfe was glad to give permission as long as the project did not involve the University in any expense.[144]

Station CKNC of Toronto assisted once again with the remote set-up. Announcers W.H. Herrington and H.B. Muir kept the interest high throughout the long concert, exhorting listeners to call in and then reading the names of donors over the air. The RCHA Band and Hopkirk's Dance Orchestra answered requests. Also on the program were vocalists Mrs Burton, W.P. Black, Mrs Wyatt, Robert Johnson, Miss Etta Dowler, accompanied by Miss Lois Baker and Miss Lenore Black, Miss Anna Corrigan assisted by Miss M. Tierney and boy soprano Master William Sharp with R.R.F. Harvey at the piano. Violinist Miss Muriel Arbuckle, duo pianists the Misses Lois Baker and Agnes Brebner, Kiwanis pianist Charlie Smith and the vocal quartet of William Eva, Herbert W. Hartshorn, James Saunders and Sid Salsbury also participated.[147] It has not been discovered how CFRC became associated with CKNC, nor why the Toronto station was willing to come to the assistance of Queen's radio.

Another Approach by Commercial Interests

From the mid-1920s through the mid-1930s, pressure was steadily increasing upon Queen's to consider the commercial possibilities for its radio station. In the summer of 1932, Principal Fyfe received a communication from G. McLeod Coates, representing Essex Broadcasters, Limited, of Windsor and Detroit, and CBS-affiliated radio station CKOC. Essex was seeking a market for a medium sized station and had selected Kingston. Essex was interested in operating a phantom station daily from its own studios in Kingston, using the CFRC transmitting facilities. Phantom stations owned no physical equipment, but operated with their own licenced call during the off-hours of another radio station. This would save Essex applying to the Department of Marine and would turn a profit for the University. Someone wrote edu-

cational on the letter, and that was likely the gist of Dr Fyfe's reply.[148]
Turning down the Essex proposal may not have been an easy decision,
since Vice-Principal and Treasurer Dr W.E. McNeill would soon have to
announce rigid economies so that Queen's could live within its means
in that period of falling revenues.[149] The University had just completed
a new gymnasium and a new science building, planned before the Great
Depression struck, and money was scarce.

CFRC Discontinues Sports Broadcasting

At this point it was decided not to broadcast Queen's sports any more
over CFRC. The authorities felt that availablity over the radio was contributing to reduced attendance at and revenue from the games. The
Journal countered that the radio served to popularize sports and to
eventually increase attendance by its advertising them, that the poor
crowds really reflected abnormal business conditions and the high price
of admission. A young man had to pay nearly a day's wages to attend
a game with a lady friend: "The putting on the air of these contests, while
aiding the University by advertising value, adds greatly to its prestige and
can only result in building up an unsectional spirit of good will."[150]

CFRC Broadcasts in 1932-33

No sports was broadcast over CFRC in 1932-33. The session began,
according to the new CFRC log book, on October 24 with the evening
addresses at the Fourth Theological Alumni Conference, and continued
with a series of educational lectures in the first term and the Monday
afternoon Extension Lectures in the second.

S.C. Morgan and his crew logged another early morning DX test
broadcast, at 200 watts, beginnng at 3.40 on December 27. Interspersed
among "a well arranged selection of classical and modern music" on
gramophone records, Morgan gave short discourses on CFRC, Queen's
and Kingston and vicinity. The gramophone was rented from the Lindsay
Company on Princess Street. They may have added their own Northern
Electric Type T-4A tone arm and reproducer, requisitioned December 1
on No. 21049. The program, supervised by Morgan and operated by
Harold Stewart, signed off at 4.58 am. They received reception reports
from 20 states and 3 provinces, as far away as Vancouver, BC, and
Jefferson, Oregon, and they were quite gratified.[151] Stewart was still
hoping to reach Australia in these DX tests, but they never did.

The only purely musical broadcast that year was of the Science
Formal with Billy Bissett's Orchestra from the New Gymnasium,
February 10: "It was a poor night for broadcasting, but local recep-
tion of CFRC wasn't too rusty...the announcers were good, especially Les
Williams...a written script would have helped them a lot...Professor
Morgan kept saying Kingston, Ontario, Canada. Why not drop the
'Canada'?" Everybody mis-pronounced Bissett, accenting the last instead
of the first syllable. Oil Thigh and the Levana yell went over big, but the
Queen's and Engineers' yells were cut off the air. "Sid Parkes (Chem. Eng.

'33) speaking '- and the Engineering students have a wild - er - wide reputation etc.'...certainly hope that this broadcasting of formals becomes
an established custom."[152]

One evening a CFRC announcer gave the station letters and introduced George Wallace, President of the Queen's Political and Debating
Union who, in turn, introduced a broadcast debate from Convocation
Hall. Mr E.H. Gilmour and Mr John Parker, for Queen's P&DU, took
the affirmative on "Resolved that this house approves Japan's activities
in Manchuria" and won. The CFRC log says February 23, but the
Journal reported the victory and the broadcast on the 21st.[153]

The last logged program using Mark V was on April 4, 1933, when
Rabbi Maurice Eisendrath of Holy Blossom Synagogue, Toronto, spoke
under the auspices of the International Relations Club of Queen's on
"Danger Zones of the Present Day."[154] A few days later Jemmett finally ordered, on requisition No. 21074, "One 'Green Flyer' governor-
controlled two speed phonograph motor and turn table for 110 volt
Direct Current operation, with 12" turntable and automatic stop
#2A101, catalog #53." Wholesale Radio Service Company in New
York City was the supplier and the cost was $15.40. Two speeds indi-
cate that he was intending it to be able to play the new 33.3 rpm program transcription records.

Fleming Hall was closed in May, after the school term finished, and
James Hamilton, the janitor, went on vacation, though the staff still
had access to the building. In those days the core of Fleming Hall was
largely of wooden construction, with interior brick firewalls. Some
called it a firetrap. The Board of Trustees was scheduled to discuss fire-
proofing the building on June 6, 1933, in the hope that renovations
could be completed before the students returned.

The Fleming Hall Fire

Albert Stover, the campus night watchman, was making his second
round of the night, scheduled to check Fleming Hall around 2 am, June
6. Coming out of the University Avenue side of Ontario Hall, he heard
the sound of breaking glass. Imagining burglars were breaking into the
library, he ran around between Ontario Hall and Douglas Library and
noticed flames coming from two windows on the north side of Fleming
Hall. Stover ran to the nearest telephone, in Carruthers Hall, and called
the fire department at 1.42 am. At almost the same time, alarms were
turned in at box 72 at University and Alice streets and at box 711 near
the Medical Building. A few minutes later the first and second floors of
Fleming were blazing from end to end. The fire quickly spread up to and
over the roof, and the upper sections of the building became a roaring
furnace.

Kingston's Fire Department responded quickly, but the flames had
made considerable progress. Stover showed the KFD the various hydrants
in the area and then phoned some University officials. The firemen
started deploying the first of 12 lines and 4,200 feet of hose but, until

the pumper trucks got working, the water pressure was low and the streams couldn't reach the flames. Fire Chief Armstrong denied that. Not long after the firemen arrived, at 2.15, the roof caved in, and the fire flared up even higher through the open centre of the building.

> Flames shot high above the roof casting a red reflection for a great distance around the University. It was easy to distinguish anyone in the hundreds of people who had gathered to witness the fire, so brightly illuminated was the surrounding campus. The reflection of the fire could be seen in the sky for several miles.[155]

Other campus buildings appeared not to be in immediate danger because, fortunately, there was no wind. To be on the safe side, the firemen kept the roof of Carruthers Hall soaked with water. Around 3.45 am the fire was brought under control, but the last hoses and trucks weren't gone until after 8.30.

The first and second floors of Fleming Hall were almost completely gutted, the tops of the exterior stone walls were severely damaged and the interior brick walls were finished. CFRC, at the north-west end of the first floor, was destroyed, and so were the classrooms and laboratories on that floor. Offices and classrooms on the second floor were a total loss, though some of the books in the library "escaped with a scorching." The basement was largely untouched by the flames and the electrical equipment there suffered damage from water and smoke only. Except

The burnt-out shell of Fleming Hall later in the day of the fire, June 6, 1933. The sloping roof was destroyed and replaced by a less expensive flat one. CFRC, at the north-west end of the first floor, was a total loss and was re-built by Prof. Stewart on the second floor, right above the old room.

The rear view of the burnt-out Fleming Hall shows the old turret, containing the entrance door and staircase. The station had been behind the middle row of windows at the far right, and was rebuilt one floor up.
-photographs by Arthur Wehman

that the main storage battery was destroyed. Total damage was estimated at $140,000.

Early indications were that the fire had started in the vicinity of the janitor's room, on the first floor to the left of the north entrance and stairway. The fire took a very peculiar course, leading to speculation that it had started in two places. Apparently it swept up through the building on both sides of a solid brick firewall through which there was no "unprotected" communication. Firewalls guarding the main part of the building's interior also did not stop the spread of the flames. The Deputy Fire Marshal investigated and determined that the fire was caused by an electrical short-circuit in the ceiling at the east end of the basement hall.[156]

Elizabeth Allan: I can remember listening to CFRC before the fire - I still listen to CFRC for good classical music. We had an old Silver Marshall, the old upright radio, when we lived on King Street. I can remember...standing on Union Street in the middle of the night and watching poor Fleming Hall go up in flames. And it was hot! Fleming Hall was on the other side of the then tennis courts, and quite a ways from Union Street. But I can remember standing with my mother and dad and the flames were really, really bad, and the heat was terrific. Even that distance to Union Street. And I kept thinking, "Oh dear! No more radio station." The newspaper account of it said that they could read the newspaper on Union Street at the height of the fire. It was really a very, very strong fire.[157]

Harold Stewart: I got a telephone call in the morning from my friend Harold Pollock, who, in a doleful voice said that Fleming Hall had burnt out. The stuff in the basement, the power machinery, was all soaked with the water from the fire hoses, and the radio gear was just...finished...all frizzled up. So that was a write-off. We didn't do anything for a while. They had to take time to clean out the building and we had to find some other quarters that year.[128]

Maud Jemmett: Douglas looked out and saw the sky was all red and it looked to him just about where the building was. So, he said he must go out and at about 1 or 2 in the morning he drove down and found that it was his building. The fire was still going on, but he got his books out. He went in and came back smothered in smoke. He had to have a complete bath before he went off again. It was terrible. A lot of his things were all covered in smoke, but anyhow he got them out and filled the car with them, and used the books as they were. They were better than none at all. He didn't lose very much - nothing that was really important.

Well, when I looked in the car, I found he'd heaped up his smokey books inside, and I'd just cleaned the whole car and put in a pair of perfectly clean pale blue velvet curtains. Of course, he didn't bother. He just put in his books that were far more important than any curtains of mine, and the whole place was all smokey and dirty again. I was very annoyed with him, filling up the car with all his dirty old smokey books, just out of the building. Before we could use the car, I had to take all the things out again and reclean the whole car after I'd just relined the whole thing. It certainly upset my cleaning business - the car was smothered in smoke. The children were ashamed to ride in the old thing anyhow. They called it "Biscuit Can."[158]

The original CFRC log book and some of Prof. Jemmett's surviving textbooks still show indelible streaks of soot under their front covers from that fire.

Harold Stewart: Doug Jemmett's car, at the time of the Fleming Hall fire in '33, was an old Essex, and when it needed painting I think Doug painted it. It was later known to the students as the "Yellow Peril," and a year or so later he finally got rid of that and got a new Dodge. The name "Yellow Peril" may be related to one of Doug's hobbies. He was a terrific reader of history, and I think in my final year, around '26, he talked to the class one time about the Yellow Peril.[128]

The day after the fire, dozens of unemployed men gathered around the burned-out building looking for work in the clean-up. The $204,000 insurance would cover reconstruction, but only to the former condition. Modern fireproof construction had to be incorporated throughout Fleming Hall, and so economies had to be made. For instance, the sloping roof was replaced by a flat one, and changes were made in the building's layout.

Immediately after the fire the University began clearing away the debris and removing the unsafe walls in the interior of the building. A squad of students and recent graduates under the direction of Mr. James Bews, superintendent of buildings, spent three weeks at this work. During this time plans for reconstruction were being prepared and on July 6 the contract for rebuilding was let. Meanwhile, as soon as the machinery was removed from the building, the electrical and mechanical engineering staffs commenced repairing the apparatus that could be salvaged. The reconstruction contract stipulated that the basement be completed by August 1...All the laboratories and drafting rooms will be fully equipped and ready for occupation by the opening of the fall term...On the first floor a radio laboratory will occupy all of the west end...The Queen's broadcasting station will now be housed in the northwest corner of the second floor, formerly used for drafting.[159]

The Mark VI Transmitter

From this point, Prof. Morgan took over the requisition book, drawing on a new "Electrical Contents" account. On requisition No. 25633, August 26, he started by buying two pair of Type C Baldwin headphones, three cystals and three Leeds dust-proof crystal holders.[160] Until October 3, delivery was made to Ontario Hall. There followed an electric clock from Kinnear & d'Esterre, aluminum sheeting, slate panels, three

Bruno Model RV-3 velocity microphone, with transformer, cable and universal swivel joint, serial #396. It was ordered by Prof. S.C. Morgan, Engineer-in-Charge of CFRC, from Bruno Laboratories, 20-26 West 22nd Street, New York City, on September 7, 1933. The cost was $33.00.
-photograph by Arthur Zimmerman

distribution transformers, input, interstage and output transformers, a National AGS-X short wave receiver with coils, $2,000 worth of radio equipment (the list now missing), a Kenyon laboratory standard 2A3 amplifier and tubes, a VARIAC adjustable transformer and two Interphone sets (a cradle type and a bar type). From Bruno Laboratories, New York, he ordered a model RV-3 ribbon microphone with transformer and universal swivel joint, and a cheaper RV-2 ribbon microphone plus transformer. Tubes ordered were two type 865 and four type 864 from Hygrade Sylvania.

Before beginning the reconstruction of CFRC, Dr Fyfe consulted Col. W. Arthur Steel, technical advisor of the new Canadian Radio Broadcasting Commission (CRBC) in Ottawa. Col. Steel wrote back on October 2 approving the new construction as proposed, with the exception of a Weston modulation meter.[161]

Changes of Frequency

On October 14, Morgan had to re-order the crystals because CFRC's frequency had been changed from 930 to 915 kilocycles. The requisition reads "2 - 50 cycle crystals ground for 915 K.C. with 150-200 volts on plate, operating temperature of crystal to be as near 50 degrees as possible."

Harold Stewart worked on rebuilding the station through October and November and it was expected to be completed and begin operations in late November or early December. The Mark VI transmitter was capable of nearly 100% modulation, permitting a voice to be boosted to twice its power. Mark VI was installed in a room of its own in the north-west corner of the second floor, while the generators were kept in the first floor Radio Laboratory so that students could experiment with them in off-hours. A new aerial system was set up as an experiment, possibly a single-wire similar to the "T" that was still hanging between Fleming and Ontario Halls as late as February 1990, except that this one was end-fed or L-type. Wires were laid from Fleming to Convocation and Grant Halls. The cost of reconstruction was $5,000 to $6,000, and the completely modernized facilities included a state-of-the-art short-wave transmitter and receiver.[162]

The Assistant Deputy Minister of Marine notified Queen's, in a letter dated November 6, 1933, that the frequency of CFRC was being changed again, this time from 915 to 1510 kilocycles. Principal Fyfe grumbled that this was off the end of the dial of most radios, in the "graveyard area."[161] Hector Charlesworth, Chairman of the CRBC, argued back that 1510 was one of the best clear channels in North America, exclusive to Canada and easy to find at the top of the dial.[163] At that point Harold Stewart was just on the point of completing the new transmitter. The inauguration of the station had to be delayed by several weeks while all sorts of things were altered. The new crystals were returned to the US supplier on November 27 to be reground to 1510 and 1509 kilocycles. The aerial also had to be changed and the transmitter altered in design.[164] When the first broadcast was made the transmitter was still a temporary set-up.

On the day of the first logged broadcast with Mark VI, January 8, 1934, Stan Morgan ordered "1 loose-leaf notebook with filler, for log book for Sta CFRC (Required by Can. Radio Br. Comm.)." That set the fund back $1.25.

Harold Stewart: In 1933 the Fleming Hall fire destroyed most of the transmitter, and (Harold Stewart), at that time station operator, undertook to rebuild it. The same electrical design, with minor changes, was used but the transmitter was built up in a rack form and was located in a special room on the second floor of Fleming Hall, where it now stands. The output was now limited by licence to 100 watts. A considerable amount of the work was done by Mr H.S. Pollock and Mr F.C. Lawson, both of Sc. '32, and by Mr W.G. Richardson, Sc. '26. The use of the same design as in 1931 may seem strange since high-level modulation, using Class B audio power, was coming to the fore at this time, but considering our spare equipment still available, it was not worth while to make a change.

The Canadian Radio Broadcasting Commission had come into being by this time and had taken control of broadcasting. Again the frequency of CFRC was changed, this time from 930 to 915 kilocycles (a split channel, a kind of a bastard frequency now).

The use of a split channel with the possibility of 5,000-cycle beat notes with adjacent channel stations did not seem reasonable and the shift was protested. The protest was useless. New regulations required all stations to maintain their assigned frequencies within plus or minus 50 cycles. This meant not only crystal control but temperature control of the crystal itself, and careful adjustment of the oscillator. This regulation was entirely justified because of the increasing number of broadcasting stations, many of which shared common channels. Had stations been allowed to wander 400 or 500 cycles from their assigned frequencies, satisfactory reception would have been impossible.

At this time, cathode-ray tubes were becoming reasonable in price and a complete Cossor oscillograph with a 5-inch tube was obtained to indicate modulation and distortion. To check the carrier frequency, a General Radio-type frequency monitor was built and is still in use (as of 1938). This type of monitor uses a temperature-controlled crystal oscillator, displaced 1,000 cycles from the transmitter carrier frequency. Any change in carrier frequency produces a change in the frequency difference. The two signals are fed to a detector and the output contains a signal of 1,000 cycles plus or minus the frequency shift. This is applied to a bridge made up of two rectifiers and two parallel resonant circuits. One circuit resonates at 900 cycles and the other at 1,100 cycles. A zero-centre DC-microammeter is the indicator. If the applied voltage is at 1,000 cycles, there is no indication. If the frequency difference is greater or less than 1,000, the bridge is unbalanced and the meter is deflected in a direction dependent on the sign of the shift.

When the transmitter was nearly finished, the frequency had to be shifted again, this time to 1510 kilocycles.[26]

The Official List of Radio Stations of Canada does not show the brief allocation to CFRC of the split frequency of 915 kilocycles (327.7 metres). The Official List of November 1, 1933, does show that the Radio Branch, Department of Marine, had assigned 1510 kilocycles (198.7 metres) to CFRC, a drastic change in the wrong direction.

Harold Stewart: By 1934, we had our transmitter up on the second floor, and the studio was a 2 x 4 framework set up out in the hall with some curtains draped around it to keep down the echo from the long terrazzo hall. Terrazzo at that time because Fleming Hall had burned out in '33 and been rebuilt. By the fall of '34, I don't think we had any turntables worth mentioning.[26]

The Canadian Radio Broadcasting Commission

The consequence of the Aird Royal Commission Report plus pressure by Graham Spry's Canadian Radio League,[165] was the Canadian Broadcasting Act of 1932, introduced by R.B. Bennett's Conservatives. This set up the 5-station CNR Radio Network as the basis of a country-wide public service broadcast system and the Canadian Radio Broadcasting Commission to run it and to regulate Canadian broadcasting. The CRBC was composed of three Commissioners: Hector Charlesworth, the editor of Saturday Night as Chairman; engineer Thomas Maher, a director of CHRC, Quebec City, as Vice-Chairman; and Lt.-Col. W. Arthur Steel of the National Research Council as Technical Advisor, with R.P. Landry as Secretary. The Commission set the tone for Canadian broadcasting, a mixed system teetering somewhere between the controlled British system and the largely unregulated situation in the US. When the CRBC gave way to the CBC in 1936, the Commission owned 7 stations and had 2 networks, of 36 stations in the east and 25 in the west, and reached about 75% of Canadians.[166]

Prof. E.A. Corbett, Department of Extension, University of Alberta at Edmonton, and Chairman of the Central Committee on Radio Broadcasting, had argued for educational broadcasting in the hearings before the CRBC was set up, perhaps on the model of his university's station, CKUA. He informed the universities by circular letter, in September 1933, that the CRBC was anxious to co-operate in their educational work and was interested in a series of inter-university radio debates over its network beginning in October.[167] Dr Fyfe offered Queen's co-operation, but was later surprised to learn that a series of radio lectures was scheduled, excluding Queen's.[161] He had tried to follow the proper course, through the Commission, and suggested that Corbett consult W.J. Dunlop (Arts '12) of his committee, who "could no doubt vouch to you for our respectability." Corbett felt that a possible reason was that CFRC was not connected with the CRBC network. He wrote to Dr Bovey and the Committee to suggest that a Queen's series of lectures might be arranged and that the CRBC should supply the lines.[168]He complained of difficulty in dealing with the CRBC and had hopes for the new Director of Programs for Ontario and Quebec, Ernest L. Bushnell.

W.J. Dunlop, Director of Extension, University of Toronto, had also asked Commission officials why Queen's was excluded and was informed that Kingston had no radio station. Dunlop was a staunch advocate for Queen's and promised to insist that it be given three lectures in the new year, even if the speakers had to talk from Toronto or Ottawa.[169] At this point Dr Fyfe asked S.C. Morgan for the cost of linking CFRC with the Bell or CNR lines, and was informed that two loops would be needed, one for broadcasting and one for control. All of the other necessary equipment for pick-up was at hand.[170] Although it turned out to be too late to include Queen's in the debates, or in the current lecture series until late March, Fyfe was rather pleased that he had been able to stake a claim for Queen's with the CRBC.[171] After discussion with Profs. McNeill, Jemmett and Morgan and Mr Columbus Hanley of the local CNR office, he wrote to Charlesworth asking permission to install the loops at Queen's expense.[172] The CRBC would not foot the bill. Another letter followed in late January 1934 stating, "We will now go ahead in laying down the loop and shall then be at your service for any purposes of educational broadcasting that you may have in hand."[173] Queen's and CFRC were setting out on a new path that would determine the role of the radio station for years to come.

CFRC Broadcasts in the Spring of 1934

> Yesterday afternoon (January 8) saw the fruit of four months untiring effort when H.H. Stewart of the Physics Department succeeded in putting the extension lecture by Dr. W.H. Fyfe "on the air" over the Queen's station CFRC. Mr. Stewart has been working since September last on the new Queen's station and though it is not completed yet, yesterday's broadcast was eloquent of the success of a merely temporary set-up.
>
> The mechanism has been put together here in Kingston, and though the new station has no advantage over last year's station as far as power is concerned, there are other features about the new set up which makes for clearer transmission. The wavelength of the present station is 1510 kilocycles and it is operated on a power of 200 watts.
>
> Notable amongst the distinctive features of the transmitter now in Fleming Hall is an extremely delicate instrument for control of the frequency.

The object of the instrument is to regulate the temperature of a piece of quartz to within 1/10 of a degree Centigrade. This is done thermostatically.

To ensure that the quartz is not affected by the temperature of the outside atmosphere, the quartz is housed in a box of Balsa wood which is a bad conductor of heat. Within this box are the heating elements and the thermostat. Within this outer box is an aluminum case enclosing a smaller box of balsa wood which contains the quartz. The aluminum protects the inner chamber from too rapid changes in temperature as the thermostat switches the current off and on.

The reception of the broadcast of the inaugural lecture given by the Principal yesterday in Convocation Hall was good, and numerous listeners declared that they believed that this transmitting from CFRC represented a great advance in clarity over any previous broadcast from the Queen's station.[174]

This was the first of a long series of Extension and other lectures broadcast that winter and spring:

Jan. 8 - "An Introduction to the Series of Men and Manners of Stuart England", Principal W.H. Fyfe
Jan. 15 - "Archbishop Laud", Prof. P.G.C. Campbell
Jan. 22 - "Donne, Poet and Divine", Prof. H. Alexander
Jan. 29 - "Hobbes, the Bad Man of British Philosophy" Prof. G. Vlastos
Feb. 12 - "Cromwell and Rupert, Roundhead and Cavalier", Prof. A.E. Prince
Feb. 19 - "The London of Pepys and Wren", Mr. E.C. Kyte
Feb. 26 - "Newton", Prof. Alexander Macphail
Mar. 5 - "Milton", Dr. W.E. McNeill
Mar. 29 - "Economic Planning", Prof. W.A. Mackintosh
Apr. 5 - "Why Read Milton?", Dr. W.E. McNeill
Apr. 9 - "My Trip to Germany", Rabbi M. Eisendrath
May 24 - "Experiences in Bird Banding, or Do the Same Birds Return?", Mr. R.O. Merriman

Mr Dunlop kept his promise, and Queen's was assigned two slots in the CRBC Thursday evening university lecture series over the eastern network, March 29 and April 5, 1934.[175] The loops to the CNR outer station on Montreal Street were installed in time to take the feed from CFRC to the network. Each of these lectures began at 8.15 and lasted 15 minutes. The Whig's radio page did not list them, but the Toronto Daily Star showed Toronto station CKNC carrying "University Lecture" at 8.15.[176] Future Principal William Archibald Mackintosh was thus the first person to broadcast from CFRC to the CRBC network.

DX Test, January 29, 1934

The radio boys got to play with the transmitter early one Sunday morning from 5.29 to 6.30, when they sent out a program of recorded music and talks. S.C. Morgan acted as announcer and spoke on "Station CFRC: Construction and Operation" and "Queen's University." Author Mr C.S. Lundy (Arts '27) gave vent to clever monologues on "Kingston as a Tourist Centre" and "Standard Time." For music, the crew selected Paderewski playing his "Minuet in g" and a trasherie of songs including "I'd Love to Meet that Old Sweetheart of Mine," "Paradise," "Always," "Listen to the German Band," "Alcoholic Blues," "On the Riviera" and "Barcelona." Over a hundred reception reports came in, from as far as Alabama, Georgia, Minneapolis and Regina.[177]

Whig-CFRC Broadcast of the Provincial Election

CFRC co-operated with The Whig again to bring the returns of the Ontario Provincial elections to the listeners-in. Large crowds assembled in front of The Whig on King Street, as in the past, to see the results projected on lantern slides, and to listen to the CFRC broadcast over the public address system.

> The outstanding feature of The Whig-Standard (election) service was the broadcast arranged in co-operation with Queen's University and put on the air over the University's own station, CFRC. All arrangements were made through Professor S.C. Morgan, a memeber of the Science Faculty. The broadcast was made from the third story of The Whig-Standard building and amplified at Queen's University. Much appreciation has been expressed for this service...Between broadcasts music was provided by the "Mauna Loa Serenaders", composed of Ed Slater, Hawaiian guitar; "Hub" Joyner, Spanish guitar; Ted Bullock, ukelele; Oakey Freeman, vocalist. Musical selections were also rendered by electrical transcription...The first announcement that a Liberal landslide was predicted was greeted with a loud roar of approval which swept over the crowd in waves, opposed by a few "boo's"...a large number of spectators and listeners remained in front of the office until "God Save the King" was played at midnight... Nothing daunted by the defeat of the local Liberal candidate, a loyal group of supporters formed a parade after midnight and marched up and down Princess Street bearing a banner with the inscription "Mitchell F. Hepburn, our next Premier".[178]

The CFRC log for June 15 records the election broadcast in detail, complete with times of the reports, "Music by Whig-Standard Trio," and speeches by Col. Kidd, Conservative candidate, by Howard Kelly, Col. Bawden and Mr Rupert Davies.

CFRC Facilities in 1934

A questionnaire from the CRBC, dated April 26, 1934, details CFRC's antenna, transmitter, studio and operations.[179] The guyed mast on top of Ontario Hall was 120 feet high and the other end, terminating at Fleming Hall, was 70 feet above the ground. Only the buildings insulated the antenna from the ground. The aerial was a half-wave horizontal, 325 feet long, with a radio frequency transmission line between the transmitter and the aerial. It had a counterpoise and there were buried plates for lightning protection only. The transmitter still had two Mullard VO/350 tubes in the last radio frequency stage, with a rated plate dissipation of 240 watts each. There was 200 watts input to the final stage of the transmitter (class B linear): 3,000 volts x 200 mA = 600 watts; approximate efficiency 33%, therefore output = 200 watts. The frequency stability of the carrier wave was maintained within 50 cycles by a temperature-controlled crystal oscillator, operating with a power input of about one watt and isolated by two buffer stages from the modulated stage. Percentage modulation of the carrier, rated at 95% on program peaks, was monitored with a cathode ray oscilloscope with Cossor power supply and sweep circuit. The radio frequency monitor was still under construction.

CFRC had no regular studios. All programs originated in Convocation Hall, and the velour stage curtains there sufficed as acoustical treatment. The microphones were a Leach 600-A and a 387-W Northern Electric. Velocity mikes were on hand and were awaiting new speech amplifiers. The speech input equipment, both studio and remote, was supplied as sketches (missing), and the speech input power supply was battery operated. The amplifier used for all pick-ups was 3 stages of 64-A's, resistance coupled. They had one turntable, American made, with a Northern Electric T-4A reproducer, used for test purposes only, and a National AGS-X receiver.

Harold Stewart was listed as having 15 years of amateur radio experience and an Amateur Experimental Certificate, but no commercial operating experience. In the summer of 1930 he had worked 5 months with Northern Electric Broadcasting. His notebook, "Experimental & Repair Work at CFRC," details his labours with Mark VI and with the frequency meter from September 7, 1934.[180] He was having his friend Keith MacKinnon, at the CRBC laboratories in Ottawa, do the frequency checks for him. One of these frequency checks survives among Stewart's papers, dated May 28, 1934.[181]

Prof. Jemmett submitted a requisition to Maclachlan Lumber on November 15, 1934, along with a sketch of a frame for a broadcast booth.[182] Mrs Laura Shales, 290 Princess Street, re-modelled the old velour drapes of the Queen's Players for the new radio studio.[183]

CFRC Broadcasts in 1934-35

CFRC sent three lectures to the network in the fall, beginning with the first broadcast of the term on October 18, 1934, when Mr E.C. Kyte, the librarian, spoke on "Some Remarkable Bibles." Harold Stewart noted "Talking booth set up in hall outside transmitting room door. Booth covered with old faculty players curtains. Echo reduced somewhat but not eliminated. Not noticeable on transmitter output unless looked for."[181] He later recalled: "E.C. Kyte was the Chief Librarian. He made an early broadcast from a temporary booth that we had set up in Fleming Hall, a 2 x 4 framework with fabric drapes hung around it. That was the studio, and Kyte and the announcer were in this cage. I think his talk was on 'Some Early Bibles'."[128]

Principal Fyfe gave the third of a series of lectures to the 32-station national network under the auspices of the League of Nations Society of Canada on November 15, on "Education and Peace." The booth was again set up in the hall outside the door. The announcer's script for that 15-minute talk survives in the Queen's Archives, and concludes with the mandatory "This is the Canadian Radio Commission." Since programs to the network were forbidden to mention the name of the announcer or the station of origin over the air, neither is identified in the script. The speech was too short, the operator at the CNR outer station cut out a piece of the closing announcement, and the whole thing finished 90 seconds too soon.[180]

Finally, on December 16, Dr Gregory Vlastos, Department of Philosophy, spoke for 20 minutes on a 36-station Canada-wide hook-up on the program "Canadian Institute of Public Affairs." He was introduced by Alex J. Grant (Arts '35), Queen's representative on the National Council of the new Canadian Federation of Youth, and his subject was "Youth's Opportunities in the Present Crisis."[184] The announcer's script for this program must have been approved by the Commission, but the network part signed off with:

> "The Canadian Institute of Public Affairs" is presented each Sunday at this same time over a coast-to-coast network. Today's program has originated at Queen's University, Kingston, Canada, and has been a feature of the Canadian Radio Broadcasting Commission.
>   Your announcer S.C.M.
>   This is the Canadian Radio Commission.

The script concluded, after the break with the network, with the first documented CFRC station break:

> This is station C.F.R.C., the broadcasting station of Queen's University, Kingston, Canada.

CFRC carried two football games that autumn, both remotes from Varsity Stadium in Toronto. The first, on October 27, was made possible by the generosity of Hemlock Park Dairy and the Monarch Battery Manufacturing Company. W.G. Richardson, of the E.E. Department, was the announcer and Harry (Red) Foster did the half-time resumé.[8,185] A phonograph was connected for stand-by, but it was not needed since the transmission was quite satisfactory.[180]

The Senior Intercollegiate Championship game on November 17 was the second remote, and Vice-Principal McNeill arranged for the raising of $120 for the Bell line and pick-up. Contributions were obtained from Monarch Battery, Jackson Press, Hotel La Salle, Joseph Abramsky and Son, the Superior Tea Room, the Bank of Montreal, Hanson and Edgar Ltd., C. Livingston & Bro., local societies and probably from Queen's itself. Technical preparations were rushed through the morning of the game and Red Foster announced for CFRC and for the Varsity Stadium public address system.[180,186] Ted Reeve's Tricolor defeated Varsity 8 to 7 for the Championship. For that transmission, Stewart first used a cathode ray-oscilloscope with a sweeps circuit to visually monitor the broadcast, and the transmission was fine.[180]

CFRC also did a brief broadcast in support of "Happiness Week," a sort of pre-Christmas spirit booster in the depths of the Great Depression. The Chamber of Commerce organized the week in co-operation with churches, merchants, charitable and relief organizations, the press, service clubs and veterans, to "reap a harvest of joy," to create temporary jobs and to collect usable clothing, bedding, dishes and furniture for the needy. Stan C. Morgan acted as announcer and Dr W.E. McNeill introduced A.J. Meiklejohn (Arts '98) and W.Y. Mills, who spoke for the Chamber of Commerce.[187]

The Christmas Day Empire Broadcast

In 1934, for the third successive year, the BBC broadcast a special Christmas Day "Empire Exchange," and for the first time CFRC joined the CRBC network to bring it to Kingston. The program began with the bells of the Church of the Nativity in Bethlehem playing over a hymn, and went in succession to bells in India, Ottawa, New Zealand, Ireland, and St Paul's and Big Ben in London. This was followed by "live" messages from ordinary people all over the Empire, concluding with a spoken message from His Majesty King George V from Sandringham and the synchronized mass singing of the National Anthem by choirs around the world.[188]

The signal went by underground cable from Broadcasting House in London to the station at Daventry, where it was transmitted omnidirectionally by GSB (31.55 metres), GSG (16.86 metres) and G5SW (using GSD's frequency of 25.53 metres). The program was "picked up by the receiving station at Yamacheiche, Quebec, and relayed to Montreal which served as the distributing point for the CRBC." Harold Stewart noted that he picked up GSD with good strength on the AGS-X receiver using about 10 feet of lamp cord as antenna.[180]

The Christmas Day Empire broadcast had a deeper significance for CFRC, as explained in the announcer's script from the CFRC extro:

> Perhaps you will be interested to know that Sta. CFRC
> has secured permission to join the Canadian Radio

Broadcasting Commission network at any time and broadcast any program being handled by that network. In the near future CFRC hope to be able to broadcast at least one program per week coming from the Radio Comm. If we have reason to feel that CFRC is serving this region by this procedure even more may be done.

Therefore, we will appreciate it very much if you will write to Sta. CFRC, Kingston, Can., telling what you think of the idea, offering any suggestions and criticisms that may occur to you. These will be considered carefully and confidently and if the response is sufficiently vigorous we may have something to announce on our next broadcast which will be given, Jan. 3, 9:00 p.m., when Miss Hilda Laird of the Moderns Dept. of Queen's University, will speak on "Careers for Women" over this station and a Radio Comm. network.

Sta. CFRC wishes you all a very jolly Christmas and a bright and prosperous New Year.

Goodmorning.[189]

This announcement came as a consequence of a letter received from Hector Charlesworth in response to the request of W.E. McNeill of November 26, that CFRC be placed on the CRBC network. The Commission had no objection to CFRC's taking a feed, but the Commisssion had to be notified, in writing, a week in advance of each program to be broadcast so that wire line orders could be issued. Unless CFRC took over 90% of their programs, no simpler scheme could be adopted.[190] All through 1935, however, CFRC was on the list to receive CRBC memoranda and circular letters on network routine and regulations. In fact, a letter from Commissioner Steel is in the CFRC archive, dated April 10, 1934, authorizing CFRC to carry the Royal York Hotel luncheon music by Rex Battle and his Orchestra, daily except Sunday, from 1.00 to 1.30 pm.

On December 24, CFRC finally tried out the new Bruno RV-3 velocity microphone in an unlogged air test. Harold Stewart noted: "We used the old speech amplifier with the one-stage 12A amplifier added to it. This gave high enough level to completely modulate the transmitter...Mike and speech amp. set up in Prof. Rutledge's office. The advantages of having this office for a studio have been discussed several times and it seems probable that we may get it in the near future."[180]

Dr Hilda Laird, former Dean of Women, spoke to the network on January 3 from the booth in the hallway of Fleming, rather than from Dr Rutledge's office. A certain amount of echo was produced as a result, but a monitoring "receiver could not be operated on account of possibility of feed-back. Reports indicate that xmtr was considerably under modulated." That problem was solved by making modifications a few days later.[180] Dr Laird spoke for 15 minutes on "Careers for Women" and her talk was the first of five in the CRBC's University Lecture Series.[191]

Dr Hilda Laird: I was Dean of Women at Queen's for some years and later in the German Department, starting as Lecturer and climbing up to be Head of the Department. To speak over the Queen's Radio was an important event for me, so the occasion stands out very clearly in my memory. I was happy to have a chance to talk about opportunities of employment for women because I felt very strongly that those opportunities were far too limited at that time. A girl really only had a choice between teaching, nursing and stenography. It was a particular satisfaction to me to get my ideas across to a wider audience, and apparently I succeeded, for a week or so later I received a letter of appreciation from somewhere in the Northwest Territories, telling me how clearly I had been heard. And that was, of course, a compliment to the Queen's Radio too.[192]

There were two sets of Extension lectures aired that spring, two in the "Charles Lamb Centenary Lectures" on January 14 and 21, followed by a series of five called "Man and His Changing World."[8,193] Queen's and CFRC took part in the inter-university debating series when Jack T. Weir (Arts '35) and E.T. Sherwood (Arts '35) took on Toronto Varsity in the first round of the Ontario contest on the topic "Resolved that there is as much scope for individuality in industry under government as under restricted competition." Weir and Sherwood upheld the negative from the makeshift booth in the hallway. Harold Stewart connected 3 sets of headphones in series, for the announcer and the speakers, and an extra control loop.[180] The Toronto team of Arnold Smith and Sol Rae spoke from Toronto station CKNC and the whole thing was mixed in Toronto by the CRBC and fed back to CFRC and to the network. The judges listened from their respective homes in Napanee, Oshawa and Belleville and gave the verdict to Toronto.[194] Weir and Sherwood found radio not an ideal medium for debating and challenged the Toronto team to a face-to-face re-match. The CRBC script has a pencilled note showing that the local broadcast ended with a Post Office announcement about tying letters in bundles.

The only other CFRC broadcast that spring was of an afternoon meeting of the Alma Mater Society in Grant Hall. The occasion was the official re-opening of Grant Hall after renovations financed by the alumni. Dr Fyfe accepted the Hall on behalf of the University. Alma Mater Society President Don Bews (Meds '35) introduced the Rt Hon R.B. Bennett, Prime Minister of Canada and the successor to Prof. O.D. Skelton as Rector of Queen's, who delivered the Rectorial Address to the student body and graduates. William Lawson Grant, former Queen's professor, Principal of Upper Canada College and son of Principal G.M. Grant, after whom the Hall was named, was to have been there, but he died a couple of weeks before the event.[195]

The Royal Canadian Horse Artillery Band played before the Grant Hall ceremony and may have been heard over the air briefly. Someone wrote a letter to <u>The Whig</u> a few weeks later asking why CFRC could not broadcast that excellent band once in a while, since US bands were on the air almost every day.[196] An unsigned draft reply exists, explaining that in these hard times CFRC had no money to build modern studio facilities with proper acoustics for music: "Only recently an attempt was made by CFRC to broadcast a band concert arising in Grant Hall. The results were entirely unsatisfactory, the listeners by means of their radio receivers being unable to recognize the piece of music being rendered."[197]

Agnes Macphail, Goodridge Roberts and Julian Huxley spoke at Queen's that month but were not broadcast. In May, Harold Stewart completed a monitor for checking the output quality, for use in connection with the Cossor oscillograph. He also changed the antenna coupling so that the antenna current could be raised from 2.9 to 3.5 amps, and built a high-gain speech amplifier for use with the RV-3 ribbon mike. The amplifier design was from Radio News, an RCA manual and his own ideas - "apparently a poor mixture" because it was unstable at full gain.[180]

> Robert Donnelly: I heard the old CFRC...when they used to broadcast lectures and music and things like that. Mainly phono records. This was in the early '30s, before The Whig-Standard took it over. We used to listen to it quite often. In fact, we used to swear at it because it had a heterodyne down on top of WSYR, the Syracuse station. Most sets in town couldn't get around the 600-700 closed cycle band in those days. Cluttered, but CFRC was sitting on top. It was their second harmonic image coming through the set. So, they used to bounce through there and people'd swear at them every once in a while.[198]

Dr W.E. McNeill wanted very much for CFRC to broadcast the Sunday afternoon New York Philharmonic concerts with Arturo Toscanini, from the CRBC network. Accordingly, he wrote to Senator Rupert Davies of The Whig, asking his opinion and suggesting that these broadcasts would measure the enthusiasm of the citizens for such a service from Queen's. The Managing Editor, Mr Muir, responded that The Whig felt that this should be entirely a Queen's undertaking, but that the newspaper would be glad to promote the service in the Saturday edition, without charge, and might perhaps publish an editorial commending the broadcasts.[199] To Dr McNeill's proposal of a Whig-Standard news broadcast as part of the program, Mr Muir explained that there was no news wire service on Sunday. If the concert became a permanent feature on CFRC, the newspaper could arrange for a news summary or a news commentary as part of the afternoon's programming. Nothing came of the proposal to broadcast the Philharmonic until the next year.

CFRC Broadcasts in 1935-36

CFRC's first broadcast in the autumn term was the Dominion election on October 14, again broadcast in co-operation with The Whig, which advised that CFRC could be found "just off the end of the dial on most sets that stop at 1500."[200] Morgan supervised the installation and the broadcast at The Whig, opting for the first remote use of the new velocity ribbon microphone and the new high-gain speech amplifier.[180] The

Donnelly Radio Service supplied the public address system and the lantern came from the YMCA.[201] Mackenzie King's Liberals swept the country and Prof. Norman McLeod Rogers took the Kingston seat from the Tories for the first time since 1908, defeating Dr A.E. Ross.[202] It turned out to be significant for CFRC that Prof. Rogers was appointed Minister of Labour in the King cabinet.

Two weeks later CFRC carried four evening speeches from the Theological Alumni Conference,[203] and two weeks after that the Intercollegiate Football Championship from Toronto, with Robert Allan Sheppard (Elec. '35) announcing. At half-time, CFRC played recorded music. The Tricolor voted to challenge for the Dominion Championship and Red Foster announced the play-by-play for CFRC from Hamilton on November 23, when Queen's lost to the Tigers. The Jackson Press, Kingston, paid the expenses of both broadcasts, around $125 each.[204]

Arts '38 presented the "Sophomore Soiree" on November 22 in Grant Hall, music by Cuth Knowlton's Orchestra, and featuring the latest singing sensation, Bill Lamb, with Chuck Saunders, Bill Christmas and Cuth himself assisting with the vocal duties.[205] CFRC was to broadcast half an hour of variety, skits and dance music from the Soiree, from 10.00 to 10.30 as the "Doyle's Bread Hour." This can not have been direct advertising, because the licence would not permit it. W.A. (Bill) Neville and E.R. James announced for CFRC, and the broadcast lasted 40 minutes.[8]

The third annual DX test broadcast went out from 3.58 to 5.02 am, November 24, dedicated to the Canadian DX Relay Club of Goderich, Ontario. Mr Leslie Scourfield (Sci. '42), the Club's representative at Queen's, announced for CFRC. They played over a dozen records, did fading and static tests and "DX taps." Stan Morgan was in charge of the broadcast and Harold Stewart operated the transmitter.[8,206]

CFRC was part of the CRBC hook-up on Christmas Day morning for the Empire "This Great Family" broadcast. Announced as an elaborate production, it did not rival the 1934 broadcast for technical daring. After the ringing of the Bethlehem bells, a scripture reading and the bells of Big Ben in London, there was a series of switchovers to bring Christmas greetings from around the world. The program's climax was a brief message from the King, titled "His Great Family," and then the National Anthem was sung by choirs in England, Scotland and Wales.[8,207] This was to be the King's last broadcast to the Empire, for he was in the last stages of a terminal illness.

The winter term on CFRC featured a series of six Monday afternoon Extension Department lectures, inspired by the bimillennial celebration of Horace's birth, on the general topic "Some Aspects of the Classical Tradition."[208] There were three additional lectures broadcast: "The Ethiopian Conflict," by Miss Elizabeth MacCallum (January 24); "An American Looks at the British Empire," by Prof. Howard Robinson of Ohio State (February 12); and "The Byrd Antarctic Expedition," by

Capt. Innes-Taylor (March 10).[8]

Queen's was again part of the CRBC Inter-University Debates series, first against McMaster on February 7 when R.W. Young and J.G. Brown took the affirmative on "Resolved that in Canada today democracy is triumphant." The McMaster side broadcast from Hamilton, and the mixing for the broadcast feed was probably done in Toronto. Queen's defeated McMaster and then took on Western on February 14 on "Resolved that university students should refrain from political activity." That was the last debate of the year on CFRC, so Borden Spears and Robert Ford must have won the decision over Queen's.[8]

King George did not survive a month beyond his last radio broadcast.[209] It was the close of an era for the Empire - Rudyard Kipling and Dame Clara Butt, two great imperial symbols, died the same week. At 8.00 am on January 28, CFRC was connected with the CRBC network for the "broadcast of the funeral services of His Late Majesty, King George V, at Windsor Chapel." The next morning CFRC and the network carried a special memorial program produced in Canada. Vice-Principal McNeill understood that the CRBC had by then assumed the expense of a permanent Bell line connecting CFRC through the CNR with the network, but Queen's was billed $28.23 for a week-long contract covering the memorial programs and the McMaster debate. Stan Morgan sent the bill to the Commission and the CRBC paid it.[210]

From March 1 through April 26, CFRC joined the network 12 times, first when King Edward VIII spoke briefly to the Empire from Broadcasting House in London.[211,212] Eight of those occasions, March 8 to April 26, were for the 22nd to 28th Sunday afternoon concerts by the New York Philharmonic-Symphony Orchestra from Carnegie Hall, conducted by Arturo Toscanini and Hans Lange. Dr McNeill had got his wish. These were Toscanini's last appearances with the Philharmonic as principal conductor. CFRC should have carried the March 1 concert as well, but there was a delay in getting some necessary permission from the CRBC.[212]

CFRC took three other programs off the line from the Commission network. One was an OHA Junior "B" hockey game between the Kingston Red Indians and Toronto St Michael's on March 17, Foster Hewitt announcing, and provided courtesy of The Whig. The broadcast originated from CKCL. In those days, Maple Leaf Gardens did not permit hockey broadcasting to begin until 9 o'clock, so listeners usually missed the first period.[213] Moments into the broadcast, a resistor burnt out and the CFRC transmitter went down for forty minutes: "Two radio men, one in evening dress, lay on their backs and wrestled with tubes, transformers, fuses and other gadgets until the trouble was finally located...an insignificant little piece of metal and porcelain, no bigger than a cigarette...This was the first time that trouble had ever been experienced with that particular part of the transmitter, and being so unusual it was hard to locate... When the broadcast stopped about five thousand people thought that the best thing to do was to telephone and ask what had happened...Bell had three extra operators on duty (and had called in three more)...The flood of calls tied up the entire exchange and some people were unable to even get central...The whole board lighted up at once."[214]

> Robert Donnelly: I remember one time they burnt out in the middle of a game. They had a telephone link with Toronto and we were all set to listen to it. They burnt out a resistor, and didn't get on the air until the second period. The fellow came on apologizing and said, "And we burnt out a little wee gadget. It looks like a pencil. It looks just like a lead pencil." We said, "Great! Shut up and get on with the game!"[198]

CFRC also piped in an Easter program of music from Toronto on the evening of April 12 and a concert by the Chicago Symphony on April 18.

W.H. Fyfe left Queen's in late March, 1936, to take up his new appointment as Principal of the University of Aberdeen. He was therefore not present at the unveiling of his portrait at the 95th Convocation on May 6. The Vice-Principal spoke before and J.M. Macdonnell after the presentation. CFRC aired the entire ceremony, its first Queen's Spring Convocation broadcast. The station returned to the air that night with the music of the Convocation Dance featuring Jack Telgmann's Orchestra. The last broadcast of CFRC as an independent University station was a program on May 27 in support of the Kingston General Hospital's "Sick Children's Fund." W.H. Herrington, Secretary of the special committee, was the announcer, Pearl Nesbitt led the Victoria School Choir, and several children recited and played solos. All of the participants and their musical numbers were listed in the newspaper report of the concert.[215]

Epilogue to Independence

From some time in 1934, CFRC was in effect an affiliate of the CRBC, although confirmatory documentation has not been found. CFRC was permitted to join the network with a week's notice, in writing, and negotiations were carried on to get the CRBC to pay for a permanent hook-up so that CFRC could carry any programs desired at no cost. The loose affiliation continued through 1935. It was quite clear to many that Kingston needed a full-time commercial radio outlet connected with the CRBC network. The airways of the continent were so saturated with signals that Kingston listeners were pretty well cut off from the rest of Canadian radio. Because the city lay in the radio shadow of the Canadian Shield, Kingstonians had great difficulty sorting out radio signals from Toronto and Montreal. The Whig-Standard was interested in an arrangement with the University to operate CFRC as a commercial affiliate, but things changed drastically when the Commission declined to negotiate with CFRC to continue the arrangement into

1936. It looked as though they had been beaten out by some unknown group. Principal Fyfe, Vice-Principal McNeill and Senator Rupert Davies, in consultation with Profs. Jemmett, Morgan and Stewart, and of course with the CRBC and the Radio Branch of the Department of Marine, eventually came to an agreement whereby CFRC would operate with a commercial licence in association with The Whig-Standard. This arrangement would produce a profound change in the philosophy, the methods of operation, in the length of the broadcasting day and in the sound of CFRC. The joint commercial operation of CFRC lasted for six full years, from 1936 to 1942, during which time CFRC came to be beloved by the people of Kingston and vicinity. They called it "the Kingston station" or "the Kingston station from Queen's." During the Oral History Project phase of this study, many people of that generation said that they held Queen's University in very high esteem for what it brought to Kingston through its radio station. CFRC was Kingston's only voice connection with the rest of Canada. CFRC was their listening post on the world, and it was very important to them.

Post-Script on the CNR Radio Network

In June 1923, Sir Henry Thornton of the CNR began masterminding construction of a transcontinental chain of radio stations for the purpose of amusing the railway passengers over the long boring stretches of their journeys, entertaining the guests at its many hotels, and promoting Canada and train travel to the general public. The stations were to air the same program simultaneously and be close enough geographically that when a train left the sphere of one CNR station, it would be entering that of another and the service would be continuous. The CNR built CNRO in Ottawa, CRNA Moncton and CNRV Vancouver, and secured licences for phantoms in Montreal (CNRM), Toronto (CNRT), Winnipeg (CNRW) Halifax, Strathburn, Quebec, Saskatoon, Regina, Red Deer, Edmonton and Calgary. By the time of the Dominion Day transcontinental broadcast in 1927, CN Telegraphs had not yet completed the carrier current line across the country, and had to rent circuits south of the border. The network was completed in December 1929, and carried three hours a week. The CNR broadcast plays, music, school programs, farm reports and concerts - the Toronto Symphony season was broadcast nationally in 1929, a year before CBS signed up the New York Philharmonic. The network never really fulfilled the dream of a transcontinental radio service for trains, but it did unite Canada in a new and different way.

Conservative forces maintained a powerful campaign against the involvement of government-supported agencies in Canadian broadcasting, and it was only by a combination of flukes that public broadcasting survived the determined onslaught. In the wake of the report of the 1932 Standing Committee on Railways and Shipping, which concerned itself with the financial excesses of the CNR, Sir Henry Thornton resigned as Chairman. But with the Aird Royal Commission Report plus pressure from the Radio League, R.B. Bennett's Conservatives brought in the Broadcasting Act of 1932, whereby the CRBC took over the stations and phantoms of the CNR network. These became the basis of the new CRBC system, which in 1936 became the Canadian Broadcasting Corporation.[216]

Kenneth McIntyre worked at CNR repeater points and handled the line transmissions for the radio network from late in 1927. In 1936, he was posted to Kingston's Outer station and was in charge of feeding CFRC the network signals from the CN lines.

Kenneth McIntyre: I used to work handling the program network after I transferred from the Operating Department to the Telegraph Department of the Canadian National Railways in the fall of 1927. I was appointed to a repeater point on the CN at Foleyet, Ontario (424 miles west of Toronto), when the CN had set up their first carrier current telegraph system, getting ten Morse circuits out of a pair of wires. Also, what we called a low pass, which cut off at 2,800 cycles, and it was used a little later on as a long-distance telephone between the CN offices in Toronto and Winnipeg. It operated between 9 am and 5 pm five days a week, from 9 am till about 1 pm on Saturdays, and it was idle all day Sunday. This particular telephone circuit was the base from which the CN started to have a circuit for radio broadcasting. About February, the CN started to experiment with setting up a network of stations, one from Ottawa, using the dinner and dance music of the Chateau Laurier Hotel Orchestra. It was aired in Ottawa over station CNRO. It was carried to Montreal on CN facilities and aired over CRNM and fed on to Toronto, where it was put on the air on CNRT, and then fed to Winnipeg (through manned repeater points at Parry Sound, Capreol, Foleyet, Hornepayne, Jellicoe, Port Arthur and Fort Frances), where it was aired on CNRW. It was just a test program and it wasn't on any regular basis. We might have one to-day and it might be a week or so later before they had thought of some new wrinkle to try out.

During that same winter, we were practicing. Along the pole line of the CNR between Toronto and Winnipeg there's a dispatcher's telephone which they use to give the train orders to the trains. Also the freight and the passenger trains all carried portable telephone sets, so that if they got into difficulty between stations they could hook them up to this one telephone circuit and call the dispatcher and report their troubles.

I remember quite distictly some time in March - it was about 25 below - that an "extra" freight train west of Port Arthur had a derailment and the conductor put his phone up on what he thought was the dispatcher's telephone line. But unfortunately, it was up on the pair of wires that we were using for a broadcast circuit. And he heard this music and he couldn't understand where it was coming from. His train dispatcher was in Port Arthur, and he said, "Hello dispatcher, extra so-and-so, whatever the engine number was, at such-and-such a milage," and the music kept coming. Well, he spoke a little louder the second time, and the train conductor thought, "Well, this dispatcher must have a radio in the office in Port Arthur, and is not listening to his telephone but is listening to the radio." So, he started to use some pretty foul language, and all this monologue was going out over CNRW in Winnipeg, before the operator in Winnipeg cut the circuit. We used our Morse circuit and had our operator in Port Arthur get on the broadcast circuit and tell the fellow that he was on the wrong circuit - to get off of it - and he explained to him where it should be on the pole line. And as soon as he explained to him that he was on the wrong circuit and told him what was happening, he immediately got off the line, and then we continued the broadcast again. I think that has been written up and remembered by quite a number of people.

When I was working at Foleyet, they had a radio operator in the observation car on the Transcontinental between Toronto and Winnepeg. He had a uniform very much like the trainman used, and a cap with "Radio" on it or something. They had no speakers, but there was headphones. It worked pretty well coming out of Toronto till they got up to where they couldn't get a Toronto station, probably around Parry Sound. But the train left at 11.30 at night, and nobody was going to get up at that time of the night and listen to the radio. In the morning, the train would be leaving Capreol about 6 am, and the radio operator couldn't get anything but noise until he got near Winnipeg the next morning. Well, as he got near Winnipeg, then he would get some program again, so it didn't prove to be too satisfactory. I went to Foleyet in October 1927, and it was on then. I don't know just when it started.

Along about 1929 or so, the CN put in a different carrier current system that increased from 10 circuits on a pair of wires to 36 circuits, and a broadcast circuit that went from zero up to about 4,800 cycles. That was what we used for the Canadian Radio Commission and also for the CBC's AM network until 1963, when the Bell Telephone took over the network. From January 1, 1963, the CN established an FM network, using a bare pair of wires, between Ottawa, Montreal and Toronto, the band going up to about 12 kilocycles with the gains that we had.[217]

NOTES

1. Prof. Jemmett's requisition book, requisition number 2460, April 5, 1924.
2. "License to use Radio", DMF, commencing April 1, 1924 and terminating March 31, 1925; QUA, University Secretary's Office, Coll. 1197, Box 1, Agreements and Leases, folder 26.
3. "Science '14", Old McGill Annual, 1914; McGill University Archives.
   "Ontario - Ottawa Valley", Dictionary of Graduates, McGill University, p. 331, 1946; ibid.
   "All Trains Bringing Students", DBW, Sept. 25, 1922, p. 2.
   "Staff Appointments", QJ, Nov. 3, 1922, p. 3.
   Finance and Estates Committee, Sept. 22, 1922, in W.F. Nickle Papers; QUA, Coll. 1027, Box 2, File Folder #56.
4. "Radio to Play Important Part", QJ, Oct. 27, 1924, p. 1.
5. Prof. Jemmett's requisition book, requisition number 2487, Sept. 24, 1924.
6. Ibid., requisition No. 2486, Sept. 24, 1924.
7. Ibid., requisition number 2490, Sept. 25, 1924.
8. CFRC log book, 1923-1932; QUA, A Arch 3707, Box 2.
9. "Listening In", KDS/DSJ, Oct. 25, 1924. p. 9.
10. "Good Radio Service", DBW, Feb. 2, 1925, p. 11.
11. "Listening In", KDS/DSJ, Nov. 29, 1924, p. 11.
12. "Generally", ibid., Jan. 31, 1925.
13. D.R. McLeod, announcer, C.F.R.C., to Editor (dated January 29, 1925), QJ, Jan. 30, 1925, p. 1.
14. "Arts At Home: Programme", ibid., Jan. 20, 1925, p. 4.
15. CFRC log book, Jan. 30, 1925; QUA.
   "With the Amateurs", KDS/DSJ, Feb. 7, 1925.
16 "Queen's Defeat McGill Cagers", QJ, Feb. 3, 1925, p. 1.
17. CFRC log book, Feb. 4, 1925; QUA.
   "Science: Demand Again Exceeds", QJ Jan. 27, 1925, p. 6.
18. "B.C.L.'s", KDS/DSJ, Feb. 14, 1925, p. 7.
19. Exchange of letters, J.E. Macpherson, Bell Telephone, Ottawa, and Principal R. Bruce Taylor, Queen's, Dec. 12 and 14, 1924; QUA, Coll. 1250, Box 19, Principal's Office, Series #1, Subject Files R-S, Radio Station CFRC, 1923-1952.
20. "Radio University", QJ, Jan. 30, 1925, p. 1.
   Ibid., Feb. 6, 1925, p. 4.
21. B.K. Sandwell was succeeded by Dr. George Herbert Clarke (QJ, Oct. 2, 1925). Sandwell founded the Canadian Civil Liberties Association and was editor of Saturday Night from 1932.
22. Prof. Jemmett's requisition book, requisitions number 2515 and 2516, Dec. 9, 1924.
23. D.G. Geiger, "Radio Broadcasting at Queen's", QR 2, Jan. 1928, p. 9.
24. "The Amateurs: Local News", KDS/DSJ, Jan. 24, 1925.
25. "Queen's Broadcasting Station", KDS, May 2, 1925, p. 13.
26. H.H. Stewart, "The Origin and Development of CFRC", PESQU 28, Sept. 1938, p. 30.
   Personal communication from Prof. H.H. Stewart, Aug. 10, 1982.
27. Prof. Jemmett's requisition book, requisition number 6811, May 28, 1925.
28. "Queen's Electrical Equipment", KDS, May 16, 1925, p. 20.
   "Queen's Radio Works Better", KDS, Nov. 3, 1925, p. 11.
   "Prof. Alexander Delivers", QJ, Nov. 26, 1926, p. 1.
29. "Annual Report", QJ, March 20, 1925, p. 4.
   "Listening In", KDS/DSJ, March 28, 1925.

30. Prof. Jemmett's requisition book, requisition number 6860, Sept. 4, 1925.
31. Ibid., requisitions number 6863 and 6864, Sept. 12, 1925.
32. Ibid., requisition number 6894, Oct. 23, 1925.
33. Ibid., requisition number 6897, Oct. 30, 1925.
34. "Harry John Duncan Minter, Faculty of Science," Queen's University Yearbook, 1925, p. 106.
    D.G. Geiger, Radio Broadcasting at Queen's, QR 2 #1, Jan. 1928, p. 9.
    "H.J.D. Minter", QR 4 #4, April 1930, p. 152.
35. "Radio Ripples Show Static", QJ, Nov. 3, 1925, p. 1.
36. "Programme for Science '26", ibid., Nov. 3, 1925, p. 6.
37. "Local Orchestra Broadcast", DBW, Nov. 5, 1925, p. 14.
38. "Radio", QJ, Nov. 6, 1925, p. 1.
39. "Radio", ibid., Nov. 13, 1925, p. 2.
40. Personal communication from Dr. Clarence Y. Hopkins, Ottawa, April 24, 1989.
41. Personal communication from Keith Abbott MacKinnon, Ottawa, Nov. 7, 1982.
42. The CFRC log shows that the following took part in the broadcast concerts:
    John P. La Flair (Mech. '25)
    Stuart Kent Harper (B. Comm. '29)
    Clarence Yardley Hopkins (B.A. '24)
    Mr. Madron
    William Winston Orr (Elec. '25)
    Miss Sanders
    Miss Pearl Seal
    Melville Charles Thurling (Elec. '27)
    Verne Egbert Roy Zufelt (B.A. '29)
43. "Lecture at Queen's" KDS, Nov. 13, 1925, p. 7.
44. "Hall Good for Radio", ibid.,, Nov. 13, 1925.
    "Broadcast Programme", ibid., Nov. 19, 1925, p. 5.
45. "C.F.R.C. Requires Concert Talent", QJ, Nov. 20, 1925, p. 1.
46. "Medical Dance Programme", QJ, Nov. 6, 1925, p. 7.
    "Annual Medical Dance", ibid., Nov. 17, 1925, p. 3.
47. Personal communications from Keith A. MacKinnon, Ottawa, Nov. 7, 1982; Aug. 15, 1987.
48. "Dramatic Club Scores Triumph", QJ, Nov. 27, 1925, p. 1.
49. "Exceptionally Brilliant Cast", ibid., Nov. 6, 1925, p. 1.
50. "Queen's Dramatic Club", DBW, Nov. 24, 1925, p. 2.
51. "Excellent Acting", ibid., Nov. 25, 1925, p. 2.
52. "Radio Notice", QJ, Nov. 27, 1925, p. 6.
53. "Radio", ibid., Dec. 1, 1925, p. 1.
54. "Coming Events", ibid., Nov. 27, 1925, p 2.
55. D.G. Geiger, "Radio Broadcasting at Queen's", QR 2, Jan. 1928, p. 9.
56. "To Broadcast Concert", DBW, March 31, 1926, p. 11.
    "Queen's Radio Station Concert", ibid., April 1, 1926, p. 13.
57. Minutes, Queen's Board of Trustees: Estimates for 1925-26, pp. 253 and 257; QUA.
58. Personal communication from Mrs. Kathleen (Whitton) Ryan, Ottawa, Nov. 7, 1982.
    Speech by Mrs. Kathleen (Whitton) Ryan at the CFRC 55th Anniversary Dinner, Queen's University, Oct. 29, 1977.
    NOTE: A Jehovah's Witness called Hector Charlesworth, Chairman of the CRBC, "a liar, thief, Judas and polecat, fit to associate only with the clergy", after he had called their American leader, Judge Rutherford, "a heavy jowled flannelmouth".

59. Secretary to the Principal, Queen's University, to J.S. Macdonnell, Chairman of the Board of Trustees, May 3, 1926; QUA, Coll 1250, Series 1, Box 19, Principal's Office, Subject Files R-S, Radio Station CFRC, 1923-52.
60. "Queen's Notes", DBW, Sept. 27, 1926, p. 9.
61. "CFCA", Toronto Daily Star, Nov. 12, 1926, p. 30.
   "Radio Fans `See'", ibid., Nov. 15, 1926, p. 7.
62. "Queen's Created", DBW, Nov. 15, 1926, p. 8.
63. "Radio Announcer Describes", KWS, Oct. 12, 1928, p. 17.
   Bill McNeill and Morris Wolfe, The Birth of Radio in Canada: Signing On, Doubleday Canada, Toronto, 1982, p. 82.
   Foster Hewitt, Foster Hewitt: His Own Story, Ryerson, Toronto, 1967.
64. Personal communication from Ronald Ede, Kingston, March 26, 1986.
65. "Champus Cat", The Varsity, Nov. 15, 1926, p. 2.
66. "Heard Game by Radio", DBW, Nov. 29, 1926, p. 12.
67. "Queen's LL.D. to Lord Willingdon", QJ, Nov. 16, 1926, p. 1.
   "Broadcast Speeches", DBW, Nov. 13, 1926, p. 16.
68. The new President of the Whig-Standard was W.R. Givens. W. Rupert Davies was Vice-President and Editor, H.B. Muir was the Managing Director and Dr. Bruce Hopkins and Mr. A.J. Thomson were Directors (KWS, Dec. 4, 1926, p. 1.)
69. "Election Returns", DBW, Nov. 29, 1926, p. 1.
70. "Large Crowds Hear Results", KWS, Dec. 2, 1926, p. 2.
71. "Queen's University Public Lectures", QJ, Nov. 19, 1926, p. 2.
   "Prof. Alexander Delivers Lecture", ibid., Nov. 26, 1926, p. 2.
72. "Further Changes in Fleming", QR 1 #5, Oct. 1927, p. 158.
73. Personal communication from Dr. Douglas William Geiger, Kingston, Sept. 6, 1982, based on a conversation with his father on June 30, 1977.
74. Prof. Jemmett's requisition book, requisition number 1554, Nov. 5, 1926.
75. Ibid., requisition number 1565, Dec. 4, 1926.
76. H.H. Stewart, "The Origin and Development of CFRC", PESQU 28, Sept. 1938, p. 30.
   Personal communication from Prof. H.H. Stewart, Kingston, Aug. 10, 1982.
77. Personal communication from John Fraser, Queen's E.F., Nov. 2, 1989.
78. "Friday Night's Outrages", QJ, Oct. 25, 1927, p. 2.
79. "Queen's Broadcasting", QR 1 #6, Nov. 1927, p. 195.
80. "Air College", QJ, Nov. 9, 1927, p. 1.
81. "Radio Reception", KWS, Jan. 12, 1928, p. 2.
82. "Radio Reception", ibid., Feb. 9, 1928, p. 16.
83. "The Spectator", QJ, Jan. 24, 1928, p. 6.
84. "Radio Station at Queen's", QJ, Feb. 26, 1929, p. 1.
85. "Choral Society Program", KWS, Feb. 21, 1928, p. 6.
86. "Radio Reception", ibid., March 27, 1928, p. 6.
87. Personal communication from Freeman C. Waugh, Kingston, Dec. 8, 1982.
88. "Will Open the First", KWS Feb. 17, 1927, p. 2.
89. "Suggested Kingston Might", ibid., Feb. 17, 1928, p. 16.
90. "Kingston Will Be On the Air", ibid., Aug. 29, 1928, p. 3.
   "Radio Program", ibid., Sept. 10, 1928, p. 2.
   "Kingston Broadcast", ibid., Sept. 12, 1928, p. 2.
91. "Attractions of City", ibid., Sept. 13, 1928, p. 2.
   "Radio Reception", ibid., p. 2.
92. "Personalities", ibid., June 27, 1936, p. 5.
93. "Dr. Carr Announces", QJ, Nov. 16, 1928, p. 4.
94. "Broadcasting City Programmes", Executive Committee, Board of Trustees of Queen's University, Oct. 5, 1928; QUA.

95. Personal communication from Joseph Gerald Burley, Toronto, June 4, 1989.
96. Prof. Jemmett's requisition book, No. 13719, authorizing cheques for S.C. Morgan and G.S. Ketiladze, April 4, 1930.
    Ibid., No. 13726, authorizing mailing of final salary cheques to H.H. Stewart and G.S. Ketiladze, April 14, 1930.
97. "Sideline Notes", KWS, Nov. 18, 1929, p. 2.
98. "Sport Jottings", QJ, Oct. 18, 1927, p. 6.
    "Queen's Sensational Play", ibid., Nov. 15, 1927, p. 1.
99. "The Concensus?", ibid., Oct. 11, 1929, p. 6.
    "The Concensus?", ibid., March 14, 1930, p. 6.
    "Opposing Parties Present", ibid., Oct. 14, 1930, p. 1.
    "Meds-Science", ibid., Oct. 17, 1930, pp. 1 and 3.
100. Personal communication from Jean Campbell Henderson, Bedford, Ohio, Oct. 16, 1982.
101. Personal communication from Orton Donnelly, Hay Bay, Ont., Aug. 21, 1982.
102. Personal communication from Douglas Donnelly, Toronto, March 31, 1989.
103. "Mungovan Makes Debut", QJ, Nov. 22, 1929, p. 4.
104. "Varsity Lucky to Score", KWS, Nov. 18, 1929, p. 2.
105. "Radio Reception", ibid., Nov. 18, 1929, p. 2.
106. "Installation of Chancellor", QJ, Oct. 11, 1929, p. 1.
107. "Theol. Alumni", ibid., Oct. 25, 1929, p. 1.
108. "Noted Economist", ibid., Oct. 29, 1929, p. 1.
    "Prominent Economist", ibid., Nov. 5, 1929, p. 1.
109. "Local Staff to Describe", ibid., Jan. 10, 1930, p. 1.
110. G. Ketiladze to the Editor, ibid., Feb. 26, 1926, p. 2.
111. Personal communication from Stuart Henderson (Arts '32; son of Professor E.W. Henderson), Bedford, Ohio, Oct. 16, 1982.
112. George Ketiladze, pianist and director; "Ducky" LaFrance, violin; Colclough, traps; Joe Howard, banjo; Dunn, saxophone; Kennedy, clarinet; Friskin, sousaphone. Knox N.W. Williams (Arts '26) was the manager of the Queen's Collegians.
113. Personal communication from Sid Fox, Kingston, 1982.
114. George Jason appeared on the "Tonight" Show with Jack Parr twice in 1957 and four times in 1958. N.B.C. has copies of the shows of Jan. 31, 1957, and Feb. 28, 1958.
115. Personal communication from Lewis Donovan Clark (Mining '29), Lyttleton, Colorado, 1982.
116. W.J. Coyle to Dr. W.E. McNeill, June 30, 1930; QUA, Coll. 1250, Box 19, Principal's Office, Series #1, Subject Files R-S, Radio Station CFRC, 1923-1952.
117. Prof. D.M. Jemmett to Principal J.C. Connell, July 10, 1930; QUA, ibid.
118. "Heard on the Radio", KWS, July 26, 1930, p. 16.
119. "Prepare Greatest Election", ibid., July 24, 1930, p. 3.
    "Thousands Enjoyed Election", ibid., July 29, 1930, p. 2.
120. Personal communication from Harold Cohen, Kingston, Sept. 29, 1982 and April 1, 1991.
    Prof. Jemmett's requisition book, No. 13746 for cheques for $5 each to H.A. Cohen and W.J. Henderson, July 29, 1930.
121. Personal communication from William J. Coyle, Kingston, Sept. 2, 1982.
122. D.M. Jemmett to Acting Principal J.C. Connell, Aug. 8, 1930; QUA, Coll. 1250, Box 19, Principal's Office, Series #1.
123. D.M. Jemmett to J.C. Connell, Sept. 24, 1930; ibid.

Prof. Jemmett's requisition No. 13747 was for an armature for Type F23 Dynamotor: primary 110 volts, 20 amps; secondary 1000 volts, 0.4 amps. Likely CFRC needed a new motor generator set.

124. J.C. Connell to Executive Committee, Board of Trustees, Queen's University, Sept. 26, 1930; ibid.

125. Prof. Jemmett's requisition book, No. 13747, July 29, 1930.

126. Note by Carleton Stanley, 21/6/32, attached to Dr. Connell's letter of Sept. 26, Report, to the Executive Committee.

127. Report of Executive Committee, Board of Trustees, Queen's University, Oct. 25, 1930; QUA.
"Radio Station's Equipment", KWS, Nov. 3, 1930, p. 12.

128. Personal communications from Prof. Harold Stewart, Kingston, Aug. 10 and Oct. 20, 1982.

129. "Convocation Programme", QJ, Oct. 24, 1930, p. 1.
"Inaugural Address of William Hamilton Fyfe on the Occasion of his Installation...", October 24, 1930; QUA.

130. "Touch Lines", ibid., Nov. 7, 1930, p. 6.

131. "Queen's Team Will Oppose", ibid., Feb. 20, 1931, p. 1.
"Queen's Debaters Lose", ibid., Feb. 24, 1931, p. 1.

132. "Univ. Begins 1931 Broadcasts", ibid., Jan. 20, 1931, p. 1.

133. "Daybreak Test Put On", KWS, April 2, 1931, p. 2.

134. "Test Broadcast", ibid., April 8, 1931, p. 16.

135. "Annual Service", ibid., May 26, 1931, p. 12.

136. "Licence to use Radio" #30, private commercial broadcasting licence issued to CFRC by the Department of Marine, Dominion of Canada, Ottawa, April 1, 140; QUA, Secretary's Office, Coll. 1197, Box 1, Agreements and Leases, Folder 26.
OLRSC, Department of Marine, Radio Branch, Jan. 1, 1932.

137. "League of Nations", QJ, Oct. 27, 1931, p. 3.

138. "Queen's Theological", ibid., Oct. 27, 1931, p. 1.
"Hon. I Tokugawa", ibid., Oct. 30, 1931, p. 1.
"Lectures of Alumni Conference Air", ibid., p. 1.

139. "Extension Lecture Series", ibid., Jan. 8, 1932, p. 1.
"Extension Lecture Schedule", ibid., Jan. 12, 1932, p. 1.

140. Memorandum from Prof. Harold Stewart, identifying his handwriting in the CFRC log; Aug. 24, 1982.

141. "Professor L.M. Arkley", QJ, Jan. 28, 1932, p. 1.

142. "Programme Is Prepared", KWS, Feb. 2, 1932, p. 3.
"Masons Raise Large Sum", ibid., Feb. 4, 1932, p. 3.

143. "Toronto Station to Help", ibid., Jan. 29, 1932, p. 2.

144. W.H. Fyfe to W.H. Herrington, April 2, 1932; QUA, Coll. 1250, Box 19, Principal's Office, Series #1, Subject Files R-S, Radio Station CFRC, 1923-52.

145. Personal communication from Lois Baker Rich, Kingston, Oct. 20, 1982.

146. W.H. Herrington to W.H. Fyfe, April 1, 1932; QUA, Coll. 1250, Box 19, Principal's Office, Series #1.

147. "Kiwanis Radio Concert", KWS, April 23, 1932, p. 2.

148. G.M. Coates, Essex Broadcasters, Ltd,, Windsor, Ont., to W.H. Fyfe, Aug. 20, 1932; QUA, Coll. 1250, Box 19, Principal's Office, Series #1.

149. "Queen's to Live within Its Means", QJ, Oct. 4, 1932, p. 1.

150. "Editorial", ibid., Oct. 12, 1932, p. 2.

151. "Proces Verbal of Radio Broadcasting Station CFRC, at Kingston, Frequency 930 Kilocycles, Oct. 24, 1932 to Dec. 25, 1935"; QUA, A Arch 3707, Box 2.
"Station CFRC Heard", KWS, Jan. 5, 1933, p. 3.

"Queen's Radio Station", QR, 7 #1, Jan. 1933, p. 15.
152. "Some Impressions", QJ, Feb. 14, 1933, p. 7.
153. "Winning Queen's Debaters", ibid., Feb. 21, 1933, p. 1.
154. "Treaty Must be Revised", KWS, April 5, 1933, p. 2.
155. "Fire Destroys Fleming Hall", ibid., June 6, 1933, p. 1.
156. "Fire Destroys Fleming Hall", ibid., June 6, 1933, p. 1.
     "Watchman's Story", ibid., p. 2.
     "The Fleming Hall Fire", QR, 7 #6, Aug. 1933, p. 165.
157. Personal communication from Elizabeth Black Allan, Kingston, Oct. 15,
     1982.
158. Personal communication from Maud Martineau Jemmett, Kingston, Oct.
     5, 1982.
159. "The Fleming Hall Fire", QR, 7 #6, Aug. 1933, p. 165.
160. The order was for one Y-cut, 160-metre crystal (1860 Kc.), one X-cut 80-
     metre crystal (3625 Kc.) and one X-cut 40- metre crystal (7500 Kc.).
161. W.H. Fyfe to Hector Charlesworth, Chairman of the CRBC, Nov. 18,
     1933; QUA, Coll. 1250, Box 19, Principal's Office, Series #1, Subject
     Files R-S, Radio Station CFRC, 1923-1952.
162. "CFRC, Queen's Broadcasting", QJ, Nov. 3, 1933, p. 1.
163. Hector Charlesworth, Chairman of the C.R.B.C., to W.H. Fyfe, Nov. 22,
     1933; QUA, Coll. 1250, Box 19, Principal's Office, Series #1, Subject
     Files R-S, Radio Station CFRC, 1923-1952.
164. "Queen's Radio Station Again Ready", QR, 7 #9, Dec. 1933, p. 272.
     "Broadcasting Is Resumed", ibid., 8 #2, Feb. 1934, p. 42.
165. Graham Spry, "A Case for National Broadcasting", QQ 38 #1, Winter
     1931, p. 151.
166. Sandy Stewart, From Coast to Coast: A Personal History of Radio in
     Canada, CBC Enterprises, Toronto, 1985.
     Frank Foster, Broadcasting Policy Development, Franfost
     Communications, Ltd., Ottawa, 1982.
167. E.A. Corbett, Chairman, Central Committee on Radio Broadcasting,
     Department of Extension, University of Alberta, Edmonton, to W.H.
     Fyfe, Sept. 21, 1033; QUA, Coll. 1250, Box 19, Principal's Office,
     Series #1, Subject Files R-S.
168. E.A. Corbett to W.H. Fyfe, Nov. 29, 1933; ibid.
169. W.J. Dunlop, Director of Extension and Publicity, University of Toronto,
     to W.H. Fyfe, Dec. 5, 1933; ibid.
     Ibid., Dec. 12, 1933.
170. Letters between W.H. Fyfe and S.C. Morgan, Dec. 14 and 19, 1933; ibid.
171. W.H. Fyfe to Mr. W.J. Dunlop, Dec. 13, 1933, and Jan. 13, 1934; ibid.
     W.J. Dunlop to W.H. Fyfe, Jan. 12, 1934; ibid.
172. "Broadcasting Prospects", memo from W.H. Fyfe, Jan. 13, 1934; ibid.
     W.H. Fyfe to Hector Charlesworth, Jan. 18, 1934; ibid.
173. W.H. Fyfe to Hector Charlesworth, Jan. 27, 1934; ibid.
174. "Broadcasts Begun by Queen's", QJ, Jan. 9, 1934, p. 1.
175. E.W. Jackson, Supervisor of Station Relations, C.R.B.C., Ottawa, to W.H.
     Fyfe, March 6, 1934; QUA, Coll. 1250, Box 19.
     W.H. Fyfe to E.W. Jackson, March 8, 1934; ibid.
176. "Thursday's Programs", Toronto Daily Star, March 28, 1934, p. 32; ibid.,
     April 4, 1934, p. 30.
177. "Toronto Listens", QJ, Jan. 30, 1934, p. 1.
     "Queen's Station Was Heard". KWS, Feb. 7, 1934, p. 2.
     QR 8, March 1934, p. 82.
178. "Whig-Standard Service", KWS, June 20, 1934, p. 3.
179. Canadian Radio Broadcasting Commission, Ottawa, Questionnaire,
     Station CFRC, April 26, 1934; Harold Stewart's papers.

180. "Experimental & Repair Work at CFRC", a Technical Supplies Department notebook kept by H.H. Stewart, Sept. 7, 1934, to July 13, 1937; QUA.
181. Keith A. MacKinnon, C.R.B.C., Ottawa, to H.H. Stewart, May 28, 1934; Harold Stewart's papers.
182. Prof. Jemmett's requisition book, requisition No. 30920, Oct. 15, 1934.
183. Ibid., requisition No. 30933, Nov. 30, 1934.
184. "Philosophy Professor", QJ, Dec. 7, 1934, p. 1.
    Announcer's script, Dec. 16, 1934; CFRC donation to QUA.
185. "Queen's-Varsity Game", KWS, Oct. 26, 1934, p. 3.
    "Queen's vs. Varsity Football Broadcast", ibid. p. 9.
186. "To Be Broadcast", ibid., Nov. 17, 1934, p. 1.
    "Local Merchants Backed", QJ, Nov. 20, 1934, p. 1.
187. "Happiness Week Starts", KWS, Nov. 30, 1934, p. 3.
    "Happiness Week", ibid., Dec. 4, 1934, p. 12.
    "Happiness Week Broadcast", ibid., Dec. 6, 1934, p. 9.
188. "Empire Broadcast", KWS, Dec. 22, 1934, p. 2.
    "By the Radio Editor", Toronto Globe, Dec. 21, 1934, p. 18.
189. CFRC script for "Christmas Empire Program", Dec. 25, 1934; CFRC donation to QUA.
190. Hector CHarlesworth, C.R.B.C., Ottawa, to Dr. W.E. McNeill, Dec. 3, 1934; QUA, Coll. 1250, Box 19, Principal's Office, Series #1, Subject Files R-S, Radio Station CFRC, 1923-1952.
191. "To Speak over Radio", KWS, Jan. 3, 1935, p. 12.
    "Miss Laird Delivers Address", QJ, Jan. 8, 1934, p. 1.
192. Personal communication from Dr. Hilda Laird, Toronto, Oct. 7, 1982.
193. "Prof. Roy to Speak", QJ, Jan. 11, 1935, p. 1.
    "City and District", KWS, Jan. 19, 1935, p. 12.
194. "Weir and Sherwood", QJ, Dec. 7, 1934, p. 1.
    "First Radio Debate", ibid., Jan. 15, 1935, p. 6.
    "Sherwood and Weir Lose", ibid., Jan. 27, 1935, p. 1.
195. "Official Re-Opening", ibid., Feb. 1, 1935, p. 1.
    "Principal Grant", ibid., Feb. 5, 1935, p. 1.
    "Renovated Grant Hall", ibid., Feb. 22, 1935, p. 1.
196. "Wants R.C.H.A. to Broadcast", KWS, March 15, 1935, p. 3.
197. Unpublished letter to Editor, Kingston Whig-Standard, March, 1935; QUA, Coll. 1250, Box 19, Principal's Office, Series #1, Subject Files R-S, Radio Station CFRC, 1923-1952.
198. Personal communication from Robert Donnelly, Brantford and Kingston, June 30, 1986.
199. H.B. Muir, Managing Director, Kingston Whig-Standard, to Dr. W.E. McNeill, Feb. 25, 1935; Harold Stewart's papers.
200. "Radio Broadcast of Election", KWS, Oct. 14, 1935, p. 1.
201. "Whig-Standard Supplied", ibid., Oct. 15, 1935, p. 3.
202. "Norman Rogers Elected", QJ, Oct. 15, 1935, p. 1.
203. "Series of Lectures", ibid., Oct. 26, 1935, p. 1.
204. "Game Broadcast", ibid., Nov. 15, 1935, p. 1.
    "The Big Broadcast", ibid., Nov. 19, 1935, p. 4.
    Dr. McNeill to F. Newman of Jackson Press, Dec. 18, 1935; QUA, Coll. 1250, Box 19, Principal's Office, Series #1, Subject Files R-S, Radio Station CFRC, 1923-1952.
205. "Kuth to Broadcast", QJ, Nov. 15, 1935, p. 8.
    "Programe Is Varied", ibid., Nov. 22, 1935, p. 8.
206. "Test Broadcast by CFRC", ibid., Nov. 29, 1935, p. 8.
207. Queen's Station on An Empire Hook-up", KWS, Dec. , 1935.

208. "Winter Term Extension Lectures", QJ, Dec. 3, 1935, p. 1.
209. "King George Is Dead", KWS Extra Edition, Jan. 20, 1936, p. 1.
210. "Program Supply Service Contract", Bell Telephone Co. of Canada to Queen's; QUA, Coll. 1250, Box 19, Principal's Office, Series #1, Subject Files R-S, Radio Station CFRC, 1923-1952.
    Undated note re interview with Mr. MacGregor; ibid.
211. "Queen's Station to Measure", KWS, Feb. 28, 1936, p. 3.
212. "The Queen's Broadcast", ibid., Feb. 29, 1936, p. 14.
213. "Broadcast of Game Tonight", ibid., March 17, 1936, p. 8.
214. "Radio Resister", ibid., March 18, 1936, p. 9.
215. "Broadcast Was Greatly Enjoyed", KWS, May 28, 1936, p. 3.
216. Frank W. Peers, The Politics of Canadian Broadcasting: 1920-1951, University of Toronto Press, Toronto, 1969.
217. Personal communication from Kenneth McIntyre, Kingston, Jan. 24, 1983.

CHAPTER 11

CFRC announcer Thomas J. Warner at the RV-3 Bruno velocity microphone in small Studio "B" in Fleming Hall, circa 1937, during CFRC's commercial years. Alex. J. McDonald (Elec. '36) is operating the RCA OP-4 mixing board behind the glass in the control room.
-courtesy CFRC Archives

CFRC's Commercial Period, 1936-1942

All through 1935, CFRC dabbled with the CRBC, giving a few feeds to the network and receiving a few. CFRC went on the air only twenty times that year, so was still essentially experimental equipment of the E.E. Department. The Queen's administration was interested in a non-commercial community service broadcasting role for CFRC, with programming mainly from the CRBC, but the expenses involved made this project impossible. The University's revenues were severely depleted. Investment returns were down and government grants had been cut.

In fact, for the past couple of years, Queen's had paid the $50 annual licence fee to the Department of Marine in two instalments. Queen's felt that the radio project could go ahead if the Commission would assume the cost of the rental of the loop to the CNR Outer Station as well as the hourly line charge. The University would pay for the maintenance of its transmitter.

When Queen's applied to the CRBC for relief from these charges in 1935, they were told that the contracts were always made at the beginning of the year. Accordingly, Principal Fyfe wrote to the Chairman, Hector Charlesworth, early in December 1935, enquiring whether the Commission would assume the expense of the loop and the hourly line charge for 1936. He also asked whether CFRC might have permission to broadcast the Sunday concerts of the New York Philharmonic and other items of interest from the network, since interference ruined the reception from Toronto stations.[1] Mr Charlesworth replied that this was impossible since a new station had just been licenced in Kingston on the understanding that it would be the CRBC's local affiliate.[2] G.W. Richardson of the CRBC in Ottawa had privately urged Dr Fyfe to submit CFRC's application quickly, because he knew of this rival commercial interest, and he was surprised that the Principal had asked so much of the Commission at such a critical time.[3]

Through early January 1936, there was a spate of meetings and quietly desperate letters. Senator W. Rupert Davies, partner with H.B. Muir in The Whig-Standard, wrote to Dr Norman Rogers, Minister of Labour, after discussing the situation with Dr McNeill. The Whig had nurtured hopes of sharing in the future development of CFRC, and was disturbed that the principals behind this new radio scheme were unknown, construction was not even begun, and there appeared to be no financial backing. The Senator did not wish to oppose the licence openly but thought it looked like a stock-jobbing proposition.[4] He let on to McNeill that he had been able to hold up the arrangements, and also told him that certain stations in the Maritimes were actually paid by the Commission to carry its programs.[5] About the same time, the Principal wrote to arrange a meeting with Hon C.D. Howe, Minister of Marine, enquiring about rumoured changes to the Radio Commission and offering Queen's help in serving Ontario on the air.[6]

Dr McNeill also wrote to Dr Rogers, to explain the dilemma. Queen's didn't mind a commercial station in Kingston, but wanted CFRC to serve Kingston educationally by carrying certain Commission programs. To that end, he hoped that the government would cover CFRC's costs; that is, the line and loop charges plus salary for two operators..."At a very small cost the government might have the advantage of a broadcasting station here without the disdavantage of commercialism in connection with it."[7] Rogers assured him that he would take up the questions with the Department of Marine "to see if something can be done to give Kingston better radio service than it is getting at the present time."[8] G.W. Richardson had been visited by Senator Davies and was in correspondence with McNeill too, attempting to give Queen's the benefit of his good advice.

The high level heat may have helped bring the mysterious principals in the new Kingston radio venture out into the open. They revealed themselves in a letter, dated January 27. Radio producer B. Howard Bedford, of Chatham, and a prospective partner were interested in buying the CFRC licence, preferably without the Queen's equipment. They would permit the University to retain certain periods of time for its broadcasts, but wanted Queen's valuable franchise "which is pregnant with opportunities commercially to those with a competent knowledge of the business of commercial broadcasting...such a station in the city of Kingston, under aggressive management, could be made a decidedly paying venture."[9] Obviously, Bedford's group had not yet secured its licence from the CRBC. Dr McNeill's reply was terse: "Radio Station CFRC is part of the teaching equipment of Queen's University and we should not consider entering into any arrangement with any commercial company."[10]

The crisis passed as quickly and mysteriously as it had arisen. Just how this came about is not clear, since not all of the documentation survives. Before the end of January, Morgan received a letter from the CRBC dealing with local loops, and another from E.L. Bushnell,[11] and Queen's started making plans for CFRC once again. On February 4, the Principal consulted with Jemmett and Morgan and came up with a list of ideas for improving the University's radio service. Since regulations forbade increasing the power to 1,000 watts within the city limits, they would seek permission to step it up from 100 to 500 watts, at a projected cost of $3,000. Harold Stewart thought that operation between 5.30 and 10 pm daily would require three operators at a total annual salary of $3,000. Since the service would carry mostly Commission progams, it followed that the government should bear all of these costs.[12]

G.W. Richardson advised Dr McNeill that the loop from the CNR station had to be made permanent, that Queen's needed a real broadcasting studio and that Queen's should recognize the value to itself and to its educational aims of a CRBC connection.[13] McNeill countered that Convocation Hall was a perfectly good studio for speakers and that only lectures for the network originated at CFRC. The University expected that all other programs would originate from the Commission, so a studio would be unneccessary. Whenever a studio were needed, a room in Fleming Hall could be fitted up. If the Commission would assume the charges arising from the line transmission of its programs, and allow CFRC to increase its output to 500 watts, a program for Queen's extramurals might even be possible. Hon C.D. Howe was interested in assisting in these activities.[14]

Agreement with the Whig-Standard

The rival dispatched, contacts with the newspaper were now firmly established. A sheet in Queen's Archives titled "Proposals re Broadcasting" seems to be a preliminary agreement between Queen's and The Whig. They proposed asking the Minister of Marine not to licence a commercial station in Kingston, but to authorize boosting CFRC to 500 watts. The $3,000 capital cost of stepping up the power was to be divided equally between Queen's and The Whig. CFRC would take some or all Commission afternoon and evening broadcasts, would air talks on topics of general interest, and would broadcast lectures for extramural students. The Commission would be asked to pay the $3,000 annual salaries, and The Whig would supply material for a daily news program, be responsible for any local advertising "so that Queen's may not be directly associated with this," pay Queen's 50% of the gross advertising revenue and announce the University's educational radio talks in the paper.[15] They were trying to compromise between a commercial and an educational service, and it is not clear just what they had in mind.

For Senator Davies' part, he was hoping that this commercial agreement with Queen's would protect his newspaper's advertising revenue. He did not expect to make any money by this broadcasting arrangement, but he was certainly prepared to lose a little money to this rather than a great deal to a local fully commercial radio station.

The Principal, the Vice-Principal and Messrs Davies and Muir met in the Principal's office on the morning of February 19 and came to a verbal agreement for presentation to C.D. Howe. Muir drew these up in a letter to the Principal. The Whig agreed:

1. To pay the University $1500, 50% of the cost of your estimate of stepping up broadcasting station to 500 watts.
2. We agree to supply the necessary material for a news broadcast to be given between one and three o'clock in the afternoon.
3. We agree to assume complete responsibility for, and control of all local advertising to be broadcast from the University station. We are agreeable to the announcing of such advertising as from The WhigStandard station (this in order to dis-associate the advertising from Queen's University over the air). We agree to pay to the University 50% of the gross revenue from such advertising.
4. We agree to announce, in The Whig-Standard, educational lectures given by Queen's professors as a part of the regular station program, to turn the attention of the greatest possible number of people throughout this district to Queen's station, the entertaining and educational programs furnished therefrom.[16]

Dr McNeill insisted, as part of the agreement, that CFRC arrange to carry the Sunday broadcasts of the New York Philharmonic-Symphony, and that steps be taken to pick up the Metropolitan Opera on Saturday afternoons. For the first four years of the commercial operation of CFRC, the partners did not feel they needed and did not have a signed agreement.

There followed a long period of public silence on the fate of the

Queen's radio station. During this time CFRC was re-modelled according to CRBC specifications. The licence was amended in this period too, to authorize a private commercial broadcasting service. It should be noted that the licence continued to be issued to the E.E. Department of the University, and was never granted to the commercial partnership, as such.

Arthur L. Davies: My entire career in Kingston has been devoted to the newspaper business, first of all with The British Whig, which was purchased by my father in 1925, and later with The Whig-Standard which came into being on December 1, 1926. I served on The Whig as a reporter, and became City Editor of The Whig-Standard when it began publication. After amalgamation, the paper was published from The Whig Building on King Street. The Bank of Montreal immediately across from it made an excellent curtain for us to throw lantern slides giving details of election returns. Later, in the early '30s it was suggested, possibly by Bill Coyle the Advertsing Manager, that we should utilize the facilities of the Queen's radio station to broadcast election returns to people who didn't feel like coming downtown. My recollection is that we negotiated with Prof. Jemmett, who was in charge of the Electrical Engineering Department. A portable booth was erected on the third floor of The Whig-Standard building, and equipment placed in that booth. Mr Coyle broadcast the returns which were prepared on the floor below by the staff of the paper. This was well received by the people in Kingston, and I believe was continued for several elections afterward.

In the mid-30s The Whig and the University radio station formed some kind of a partnership. I believe this was instituted by Harry B. Muir, who was an equal partner with my father in The Whig-Standard. Mr Muir, the Managing Director, was interested in increasing the advertising facilities that were available in Kingston. He foresaw the role that radio was going to play in advertising and which it was playing in some larger cities, and approached the University with a suggestion that The Whig-Standard would attempt to solicit advertising for the radio station, but that the radio station would certainly remain in the control of the University and that Queen's would control all the programs. I had no part in the negotiations, but I believe they were carried on through Dr J.C. Connell, who was Acting Principal of Queen's at the time. Mr Muir urged that daily broadcasting would provide the University with an opportunity to present cultural talks and music on a regular schedule, adding still another facet to its educational program. It would open up opportunities for Queen's students to participate in a new and expanding field of employment and would, hopefully, provide additional resources for experimental work in the broadcasting field. After considerable negotiation, it was agreed that Queen's and The Whig-Standard would jointly apply to the CRBC for a commercial broadcasting licence for CFRC. The newspaper would have a free hand to solicit paid advertising to finance the station, any profits after expenses to be divided equally. Queen's was to have complete control of the program content. The commercial licence was granted, I believe, because the Commission was glad to get an outlet for its programs in the Kingston area without spending any money. Daily broadcasting began on June 29, 1936.

It was necessary to set up a sort of separate organization to look after the advertising and the man selected by The Whig-Standard to be Advertising Manager was Major James Annand. He had office space at the newspaper and his broadcasting was done from studios at Queen's. As well as soliciting the advertising, he announced the news, particularly at noon. Major Annand sometimes would get a little involved in his broadcast and in order to correct it he would say "As you were!" and then start over again. He also had some pronunciations which were not always recognized by the audience. For example, he insisted on referring to the Pope of that time as "Pope Pee-us," and while this may have been correct Latin pronunciation, it seemed strange to Kingston listeners.[17]

Broadcasting Continues

At the end of February, the University announced that it would broadcast a number of Commission programs as a test, beginning on March 1 with a re-transmission of King Edward's address followed by a live Philharmonic broadcast. Listeners were asked to telephone their comments to the station at 2057J or 2633J, or to send a postcard: "If the response should indicate that the station is of real service to the local listeners, the University might be encouraged to develop it and increase the number of broadcast hours."[18] Not a very scientific survey, but permission to carry the Philharmonic was delayed and so Kingston missed pianist Ray Lev's radio debut that Sunday. The survey was re-announced for the following week with stronger exhortations to listeners to make their opinions known, as this was "important to the radio development of this part of Eastern Ontario."[19] On March 8, Kingston heard Toscanini conduct Beethoven's 8th and 9th Symphonies from New York. The Kingston-St Michael's hockey game in March provided an interesting sample of data, with the Bell switchboard flooded and letters praising the station's clarity and power from as far away as New York, Kirkland Lake and Sault-Ste Marie.[20]

About that time, a new operator took charge of the CNR outer station and the CRBC broadcasting circuits in Kingston:

Kenneth McIntyre: I was moved from Capreol to Kingston on the 29th of March, 1936, where I continued to work handling the CRBC network. On Sunday afternoons we fed CFRC the New York Philharmonic from 2 until 4 o'clock each week. That was the only program that we fed to CFRC until The Whig-Standard took over the operation of CFRC. It came in from Niagara Falls, New York, over the Bell facilities to the CN in Toronto, and then was fed west to Vancouver as well as to the east coast. It was one of the most difficult that I ever had to handle, because I don't know a note of music. But I watched the levels on my volume indicator - some of it was very high and some of it was very low - and we had to watch to maintain the same level. Sometimes you'd line up and it would be clear. If it started to rain, the loss increased and you'd have to bring the gain up a little bit. So, what we used to do was get Toronto to call peaks for us and we watched our volume indicators and if we were peaking the same as he was, then we were all right. We were all set the same, for +6. You fed the stations at 0 level into a 600 ohm termination, but we carried it at +6.

At that time, when we fed CFRC just that program, we had one control circuit - which meant a telephone at each end, and we rang each other - and one program circuit. But when CFRC was going to feed the network, we had to establish what we called a "send loop" or circuit. You didn't send and receive on the same circuit. For the regular feed to any station that's going to transmit as well as receive, you've got to have a control circuit, and you've got to have a "send" and a "receive."[21]

CFRC assisted the Sick Children's Fund of the Kingston General Hospital with a special broadcast on May 27. W.H. Herrington was announcer and the program featured several appeals for funds, talks and stories for childen, the Victoria School Choir under Pearl Nesbitt, the children of Miss Hayward's class, Victoria School, plus recitations and instrumental solos. Reception was excellent.[22]

Purchases for the New Station

Aside from the regular purchase of new tubes, resistors, condensers and the odd transformer, the requisition book shows little activity until September 20, 1935, when Morgan bought a relatively inexpensive Bruno microphone and an adjustable floor stand with a heavy base for the old Bruno RV-3 ribbon microphone, serial #396. On January 8,

The RCA OP-4 portable 3-channel broadcast mixing board and amplifier, serial #1050. Three channel pots at lower left, master attenuator at lower right, V-U meter at upper right, milliammeter (DC volts) at upper left. Main power switch, upper centre left.
-photograph by Arthur Zimmerman

1936, he bought a state-of-the-art wide range SR-80 Amperite studio model microphone and stand, and Prof. Jemmett bought a Du Mont model 1150 electronic switch (serial #171) on March 23.

Doug Jemmett renewed the licence on April 3, 1936, with requisition No. 29509, for $25. There was no indication on the requisition that this was to be for a private commercial licence. Purchases began in earnest on April 29, with a model A12 Jensen "High Fidelity" speaker, a Variac Type 200-CU, volume controls, a model 665 type 2 Weston Selective Analyzer (serial #3471) and other components. The first of two RCA Type UZ-4210 transcription turntables was ordered on May 8, and the other on June 3. Perhaps they had thought at first that they could manage with just one turntable. Later they bought a fader for the pick-up, more Variacs, plate transformers and filter transformer rectifiers, receptacles, cables, plugs and steel cabinets.

The major expenditure was on an RCA Type OP-4 portable broadcast amplifier (serial #1050), recommended by the CRBC. It was to operate on AC and needed a power supply. The OP-4 and the type 54-A power supply were ordered May 20 on requisition No. 36968, for $804. Harold Stewart had to drive the OP-4 back to RCA in Montreal on June 15 for correction of some problem incurred in shipping. They bought two Hamilton-Sangamo electric clocks, with red sweep second hand, from Kinnear and d'Esterre and then an RCA Type 50A inductor microphone (serial #1438) with cannon plug and table stand for use with the OP-4 board.

On May 26, Col. Steel sent a blueprint drawn up by his engineering department, "Suggested Speech Input Equipment for Station - CFRC." He appended a list of the equipment indicated thereon (amplifiers, attenuators, announce key, a power unit, pads, repeat coils, jack panel and rack) plus costs, "suggested as one suitable means for meeting the general requirements of such a station."[23]

Renovations to CFRC

James Bews and his crew started work on the new studios in Fleming Hall on June 18, re-building them to CRBC specifications. Again, the CRBC sent a blueprint detailing the acoustic treatment of the studios.[24] For the inside walls they specified Zono acoustical plaster over 3/4" gypsum board, then an air space, 1/2" ten-test, another air space and 3/4" hair felt in the middle of the wall. They wanted 4" of rock wool above a fly screen, ten-test and an air space as a false ceiling, and on the floor a carpet over Ozite or ten-test.

The CRBC blueprint called for the Fleming Hall library, in the south-west corner, to be converted into one large studio with a waiting room all along its east side. Prof. Rutledge's office was between the proposed studio and the transmitter room and he kindly moved to permit the installation of a control room. The CRBC plans were altered, however. The projected big studio was cut into two. The north side, next to the control room, became small studio "B" and the south end

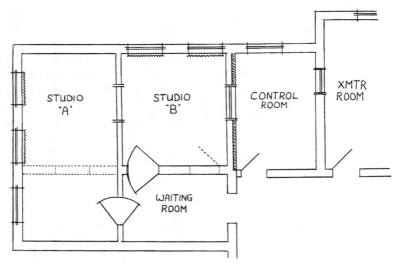

Floorplan of the CFRC studios in Fleming Hall, from a blueprint drawn by the Engineering Department of the Canadian Radio Broadcasting Commission and sent to CFRC by Commissioner and Technical Advisor Lt-Col W. Arthur Steel in the spring of 1936. The blueprint called for the construction of one studio and a large waiting room along its east side, but the plan was altered to give CFRC the long music studio "A" along the south wall of Fleming Hall at the expense of half the waiting room, and a booth studio, "B," next to the control room.
-Prof. Stewart's papers

of the waiting room was incorporated into a long studio "A," along the south side of the building. Studio "A" was large enough to hold a ten-piece orchestra, and the small studio "B," between the large studio and the control room, was for solo work, speeches, news and small drama groups. The windows of the two studios and the control room were in a line, allowing the operator to see through "B" into "A."[25,26] The transmitter room remained in the north-west corner of the building, so CFRC now occupied the entire west end of the second floor of Fleming Hall.

> The control room and studios are treated acoustically. In the ceiling there is a space of four inches filled with rock wool, which is held in place by metal screening. There is another air space below this and then a decorative fabric.
> Between the studios the partition is specially built, using insulating board, air spaces, ordinary and acoustic plaster, and 3/4 inch felt. The acoustic plaster is full of air spaces which pocket and hold the sound to stop refraction. The walls of the studio are lined to a height of four feet from the floor with burlap and the partitions are set in felt.
> Double doors lead out of the studios. These are constructed of heavy wood and insulating board with an

air space between them.

The specifications for the new station call for the use of flannel around the windows of the studios, but for the present, at least, these will be hung with monk-wool. The floors will be carpeted with a thick carpet.

The observation windows are about four feet by three feet in size and are constructed with three panes of glass in each, with an air space between each pane. The panes are set in rubber to avoid the possibility of vibration.

A loud speaker in the control room brings the programs to the man at the controls and there will be an engineer and a control man in charge of every program. These men are skilled in their work, which requires the greatest accuracy. Network programs end at twenty seconds to the hour and the next one begins exactly on the hour. This leaves the staff in the local station just twenty seconds to break from the hook-up, announce their station, and pick up the network again for the succeeding program.[25]

Harold Stewart: When the decision to operate on a daily schedule was made, much work had to be done. We had a transmitter, remote pick-up equipment and two fairly good velocity microphones, but that was all. Studios, speech-input equipment, turntables, and another microphone had to be acquired...

New speech-input equipment was necessary to permit changing the program source without loss of time. Due to lack of funds, it was necessary to compromise here, and an RCA type-OP-4 portable amplifier was installed to handle all programs. The OP-4 is a three-stage, high-gain amplifier, incorporating a three-channel mixer. But three channels were not sufficient to accommodate all program sources, so a control panel had to be made up to handle all input circuits. This made it possible to carry on studio auditions using the main amplifier, while the network program was being fed directly to the transmitter. Loud speakers were placed in the studios as well as in the control room. Program switches automatically controlled microphones, signal lights and loud speakers in the studios.

Both motor-generator sets supplying the transmitter were operated from the main 120-volt battery in the electrical laboratory. Daily operation with this source of power was not practical for several reasons. Furthermore, the noise produced by the motor-generators was not pleasant and would have been annoying to anyone in the radio laboratory or in the transmitter room above. Two rectifiers were therefore installed and a flexibility of voltage-control even better than that provided by the motor-generators was obtained by the use of General Radio variacs. A variac is a continuously-variable-ratio auto-transformer. The motor-generator sets were available in case of rectifier breakdown or AC power failure.[26]

For the first few months, the operator's area in the control room was makeshift. With the construction of some tables in September and October, the classic cockpit was assembled. The operator sat in the well of a U-shaped group of tables in front of the window looking directly through the small studio and into the big studio beyond. Mr Salisbury built a special carrying case for the OP-4 with suitcase handles and snaps, and this was placed beneath the window in front of the operator. To prevent overloading, the operator had to keep his right hand on the master volume control potentiometer, on the right front of the OP-4. On either side of the OP-4, almost forward of the operator, were the two transcription turntables. These had to be close to the operator since they were slow in getting up to playing speed and needed to be given a "hot" start. To do this, the operator held the cued record stationary over the spinning mat and turntable with his free hand until the signal came to start the music, at which point he released the disc. The operator's microphone was to the right and the box of RCA "Shadowgraph" needles was to the left of the OP-4. One play wore a stylus to a chisel edge so it had to be changed for each record side. To the operator's right was the auxiliary control board, a vital expansion to the three-channel OP-4, and the oscilloscope.[27]

CFRC was to continue broadcasting at 100 watts on 1510 kilocycles (198.7 metres), a local clear and an exclusive Canadian channel, shared only with CKCR in Kitchener. CFRC's area of coverage was 65 miles daytime and 40 miles at night: from Brockville to Belleville and north to Perth and Smiths Falls.

Prof. S.C. Morgan was to be Station Director, and Asst. Prof. H.H. Stewart the Engineer-in-Charge of CFRC. Both would continue to teach at Queen's, the radio duties to be "supplementary."

The CFRC Piano

Willis and Company of Montreal, manufacturers of pianos and organs and Canadian agent for Knabe and Mason & Hamlin, wrote to Dr McNeill early in 1935 to offer a deal on a grand piano for the renovated Grant Hall. In reply, Willis was told that Mrs Etherington's grand piano was available for all University purposes. When Queen's needed a piano for the CFRC studio in June 1936, Prof. Morgan turned to Willis and rented a Knabe grand at $15 per month, with option to purchase. The piano was shipped on June 23, complete with a bench and a triangle dolly. Norman Butcher, of Kingston, was in charge of setting up and tuning the piano at CFRC.[28]

In January 1937, on Mr Butcher's advice, Dr McNeill negotiated the purchase of this special "artist-used" Knabe piano (#107072) for $800 cash, less $105 in rent already paid, less $95 worth of advertising on CFRC. The original price had been $1,600. The Knabe served the station for many years. In 1982 it was in the common room of the Morris Hall residence, minus its original black sheen and in sad condition. It was open to unsupervised use, and had recently been set on fire by a student. The sound board was severely damaged and the felts were worn to the wood. The Knabe was then being moved every week to the Quiet Pub and back.

The Program Director/Advertising Manager

Major James Annand was hired by The Whig as Program Director and Advertising Manager of CFRC. He had been an actor in England for many years with Sir George Alexander's Company and with the F.R. Benson Shakespeare Company. Later he had worked under George Bernard Shaw and Granville Barker at the Court Theatre in London, and had appeared with Dame Ellen Terry. He had also been in the moving picture business in England until hard times struck. After he came to Canada, he acted and produced in radio in Toronto. He was in the CRBC's "Up to the Minute," and played the Scottish curator "Old Mac" in Don Henshaw's Royal Ontario Museum radio series "Forgotten Footsteps," over the CRBC.[29]

Major James Annand, Program Director and Advertising Manager of CFRC, 1936-1942. He was an English actor who had worked with Ellen Terry and George Bernard Shaw. He could do every job in the theatre, including making costumes, such as the one he is wearing in this photograph taken from a postcard.
-courtesy Bonnie Nichols

On May 15, Annand issued CFRC's advertising rates, based on 30,000 area families (120,000 individuals) and 20,000 area radio sets. Programming hours were noon to 2 and 6 to 11 pm week-days and Saturday, and 7 to 11 pm on Sundays. Daytime program rates were $25 for one hour or $21.75 for an hour per week for a year. Evening rates were $35 and $29.75, respectively. Spot announcements, limited to 100 words, were $3 or six for $16. CFRC then boasted a library of 78 and 33.3 rpm transcription records and continuities from NBC, featuring Nat Shilkret's Orchestra, Rosario Bourdon and his Orchestra, Xavier Cugat, The Dreamers, Richard Liebert at the organ, Jack, June and Jimmy, comedy sketches, mysteries and dramas, adventure, education and travel shows, religious music and cowboy songs. For a 15-minute transcription program using these discs, the charge was $8 for the spot and $2.50 for use of the transcriptions. High-class live local talent was also available for programs.[30]

Thomas Joseph Warner's name was not mentioned in the early publicity about the commercial operation, but he was probably on staff from the start to assist Major Annand as part-time announcer. Tom Warner was later engaged by the University as a full-time announcer.

Some time in the summer, Mr John Donald Stewart (Arts '29, recently returned from postgraduate work in Germany) was engaged as operator.[31] "In September, 1936, the increase in operating time and the pressure of teaching duties...made it necessary to obtain another operator. Mr Alex McDonald (Sci. '36) was engaged, and...from time to time assisted in the construction of various new parts of the station."[26]

The First Commercial Broadcast

The announcement that Queen's and <u>The Whig-Standard</u> would co-operate to bring Kingston a daily commercial and cultural radio service was held off until the morning of June 24, after most of the preparations had been completed:

> Queen's University, for some years, have had in mind the daily operation of CFRC, primarily to make available to the people of all Eastern Ontario the lectures which Queen's professors have given, on increasing demand, in Picton, Renfrew, Brockville, Belleville and many other centres throughout Eastern Ontario.
>
> The announcement states that the opening of Station CFRC, in daily service, would not have been possible without the co-operation of The Whig-Standard.
>
> The Radio Commission, in discussing the advantages of regular daily service of CFRC, referred to what is generally known here, that Kingston seems to be the centre of an aerial pocket in which reception of radio from the outside is not as dependable as it should be for the purpose of entertainment, education and commercially sponsored programs.
>
> Therefore, to round out the complete service of Station CFRC, times will be made available to national and local advertisers.[32]

The inaugural program was scheduled for June 29, when the Commission welcomed CFRC to the network with a special dedicatory production and a salute to Kingston. At that time, the CRBC was programming just six hours a day to the network, and was trying to expand its hours.

CFRC signed on at 6.45 pm (EDST) with the results of the Puzzleword Contest No. 1, five minutes of recorded music sponsored by Rexall druggists Jury and Peacock, followed by the Kingston Merchants' Goodwill Gift Draw, a special event arranged in connection with the opening of the new commercial station. At 7.00 Joseph Abramsky and Sons presented 15 minutes of "fun and music," featuring transcribed music and a speech by Senator Frankenstein Fishface, Presidential candidate of the Pussyfooter's Party. Abramsky offered prizes for the 20 best letters commenting on the program and extended an invitation to amateur entertainers to appear on future Abramsky broadcasts.[33] At 7.15 there was a quarter hour of music compliments of the Hemlock Park Dairy, including "a special announcement re cottage cheese."[34] The next half hour came courtesy of E. Wray van Luven Auto Sales. CFRC joined the network at 8.00 for "The Story of Old Kingston", a series of dramatic sequences written by Andrew Allan, produced in Toronto by Rupert Lucas, starring "an able cast of leading radio actors" and narrated by Charles Jennings and Robert (Bud) Walker. Percy Faith and his Orchestra supplied the musical decor for the drama.[35]

At 8.30, CFRC fed the network with its own celebration. The

Chalmers Church Choir, under J. Arthur Craig, opened singing "God Save the King." J.M. Macdonnell, Chairman of the Queen's Board of Trustees, spoke briefly on the history of the radio station and then H.B. Muir of The Whig talked about the new venture and the opening. Lottie Sanders-Wyatt followed with a soprano solo and then the CFRC Players (Mima Cook, Katherine Symington, Joan Sterritt and George Tottenam) put on a playlet titled "The Otter," adapted from a story by H.H. Monro (better known as Saki). The choir added a chorus, accompanied by Lois Baker, and Alphonsus Patrick (Phonse) McCue concluded proceedings with a tenor solo. CFRC left the network at 9.00 for a locally produced half hour program presented by the Empire Life Insurance Company, and rejoined the CRBC for "Rhythm River" at 9.30, "With Banners Flying" at 10.00, "Wilderness Adventure" at 10.30 and "Canadian Press News Bulletin" at 10.45. After this heady beginning to a new kind of radio service for Kingston, CFRC signed off the air that night at 11.00.[36]

> **Phonse McCue**: I can recall when the Station Manager called me up and asked me to sing on the first CFRC broadcast on the Canadian Radio Commission. I was quite well known at that time, and Major Annand had heard me at a concert. I was really thrilled to think that they'd ask a person without any radio experience to broadcast from coast to coast.
>
> They had a grand piano there, and we were all in one big studio with chairs along two sides of it. As our turn came to go to the microphone, we were told by Major Annand to go up and do our number. I did "The Hills of Home" - Lenore Black was the accompanist - and I thought I did it fairly well. I had practiced and practiced and practiced for this thing because I thought it was something I'd never do again. These people I was on with were real pro's, and I was awfully glad that I was on a program to open up CFRC. That's something I'll always remember.[37]

> **Lareta Lamoureux**: Alphonse McCue. He had a gorgeous voice.[38]

> **Lenore Black**: I was on the program and I played for Alphonse McCue. I'd been playing for him and every place he sang, he took me as accompanist. He had the most beautiful voice in the world, I think. He was just unique - no one quite like him. A beautiful voice and he's a beautiful person. Major Annand was the MC. It wasn't a big studio, and the piano was right in the centre.[39]

The CFRC Program Schedule

The first day of local broadcasting, June 30, was kind of preliminary and formless. They signed on at 5.55 pm and played transcription discs, interspersed with appeals for comments on reception quality, a "funny story" at 6.38 and another at 6.46. They did a frequency check every 15 minutes. CFRC joined the network from 7.00 until sign-off, for Florian Nelles, soprano; Rex Battle and his Royal York Hotel Orchestra; "Musical Tours"; "Edgar Herring's Novelties"; "Reve de Valse"; Alfred

Wallenstein's Sinfonietta; "Tom Swift and his Electric Rifle" on "Mystery House"; "Serenade to Summer" directed by Jack Arthur; "Sunshine and Deep Shade" directed by Isaac Mamott; Alex Lajoie and his Chez-Maurice Orchestra; finally the Canadian Press News Bulletin at 10.45 and sign-off at 11.00.

By July 2, there were tentative titles for the two locally produced programs: "From Soup to Nuts" from 6.00 to 6.30 pm, followed by "Concert Hall of the Air" until 7.00, and then hook-up with the CRBC. The next day the 6.00 o'clock program was changed to "The Diner's Hour," but it continued to be known as "From Soup to Nuts" by The Whig and by the public. On Friday, July 3, live guests, the McCue Quartet, were added to the transcription discs on "The Diner's Hour."[40] On the 5th they welcomed George A. Patterson, accompanied by Hazel Robinson, and on the 6th Mrs Frances Crawford accompanied by Pearl Nesbitt. In the next month Mrs Lottie Sanders Wyatt, Dr Harold Angrove, George Patterson, Phonse McCue's Harmoneers Quartet, Phyllis Jerome, James Bankier, Donald Mackenzie, Vina Sharpe, Gladys Laurel, and Chuck and Hal with their guitars appeared, midst electrical transcriptions. The surviving log shows that in the first ten months of commercial operation, CFRC presented about 120 local guest artists.

Phonse McCue: I was back every week for a couple of years. I did 15 minutes - maybe three or four songs. During my regular programs, called "Phonse McCue Sings," Kay Grimshaw was my accompanist. She was a really gifted accompanist. That went on for perhaps two years, weekly, and I got no pay. But, I enjoyed doing it - something I'd never done before. One time we had the McCue Quartet, made up of Johnny North and my brothers Len and Wilf and myself. We were singing "The Song of the Islands" and we forgot the words, and every one of us started singing different words. We didn't sing the song through at all with the correct words. We asked people afterwards who were listening to the broadcast if they had heard it, and they didn't notice anything wrong with it. We laughed at that time and time again.[37]

John Donald Stewart: We turned the transmitter on if we were on that shift and announced, or turned it off if we could, if we were on the late shift. And, of course, we had to adjust the volume to the proper level. I remember my brother was in once and reminded me that I was letting the voice level get too low. You might as well have it up nearly to 100% modulation, otherwise you're losing some of the advantage of your power. We had no facilities for cutting discs. We just played commercial 16 inch discs. I remember that amateur artists from the city were there quite frequently. Once I was asked to audition some of these people and I refused because I didn't trust my own judgement on these matters and it wouldn't have been fair to these people to have me judge them. They may have sounded bad to me and they may have been good to somebody else. So I begged off that job.[41]

"Owing to the demand for time for sponsored programs," CFRC expanded its schedule on July 7.[42] They signed on at 11.55 am for

"Organ Reveries," transcribed, "On the March" at 12.15, "Home Sweet Home" for Canada Starch at 12.30 and "Luncheon Hour" at 12.45. The station signed off and shut down the transmitter at 12.59, and returned to the air at 5.55 pm for "The Diner's Hour." "Concert Hall" was cut to 15 minutes so that "Novelties from the Night Club" could go into the schedule at 6.45.

The first locally produced news program, "Canadian Press and Local News Bulletin," appeared on July 9, 12.35 to 12.45 pm. Major Annand arrogated all news reading to himself. The next week, "On the March" replaced "Organ Reveries" and the 12.15 slot was filled by "The Voice of Wisdom," a transcribed program advertising "Gastronox": "...A lady of great experience in psychology, numerology and astrology answers questions for clients regarding things personal, financial, matrimonial, etc. Her advice has been asked for by thousands of people in various difficulties...whether you are a believer in clairvoyant powers or not. This noted psychologist has been consulted by many of the well-known movie stars...there are many more people who ask her advice than are prepared to admit it."[43]

> John Donald Stewart: The advertisements were pre-recorded, as I recall it, or sometimes the announcer would give them. We gave some brief ones, like "Such and such a time, Bulova Watch Time." And I remember Tom announcing "Listen for her theme song. It always identifies her program." That was the "Voice of Wisdom," some woman selling stomach remedies.[41]

Gastronox, a product of the Knox Chemical Company, was the Rolls Royce of indigestion remedies, and came with a very heavy advertising campaign.

The Whig published a letter on July 10, from a citizen who very much appreciated the excellent new radio service from Queen's. This letter sounded the first note of the massive chorus of good feeling that was welling up in the community toward CFRC and the University, a feeling that was still warm in the hearts of Kingstonians when the Oral History Project interviews were recorded in the early 1980s:

> CFRC has come to mean something to Kingston and Eastern Ontario. My work puts me in contact with the people from the country districts. They are all delighted with the service and have a kindly feeling towards Queen's and The Whig-Standard...The noon news broadcast you started two days ago, giving world, national and local news, is very welcome in our house.[44]

The Saturday format at the beginning was pretty much the same as week-days, except that the 12.15 slot was a transcribed feature called "The Dreamer's Program." At first, CFRC signed on at 6.55 pm Sundays and joined the network from 7.00 until 11.00. Then the "Sacred Music Program," featuring local artists, came in at 6.00 pm from Sunday, October 18 until January 31, sponsored by the Radio Division of

Frigidaire Sales and Service of Kingston.

Major Annand had signed up Gastonox and Canada Starch, but the log shows that at this point Kingston businesses weren't as eager to promote their products on the radio as they had previously let on. Most of the programs were sustaining; that is, they were unsponsored. CFRC sold the odd spot announcement to the local Maher Shoe Store, the Chamber of Commerce, Donalda Beauty Salon and the Merchants' Red Hot Sale. Then Dodd's Medicine Company, Empire Life and Kelvinator began to buy time. The Major added the Johnson and Ward Market Quotations at 12.45 and cooked up a tune competition on "The Diner's Program" to get public interest aroused. The successor to "On the March," called "Noontide Melodies," included a transcribed serial about the adventures of the Parker family, "one of the jolliest programs that has been on the air...One is really behind the times without the latest excitements in the life of the Parkers."[45]

CRBC programming wasn't exactly a thrill a minute either. Programs were generally 15 to 30 minutes and featured a lot of live music, short documentaries and brief dramas: "This Week in History," "Perennials from Seed," a talk by Sir Edward Beatty on swimming, "Les Cavaliers de Lasalle," "Chasing Shadows," "Stars of Tomorrow," "Twilight Echoes," "I Cover the Waterfront," "Tribute to Song," "Mystery House," "Jass Nocturne," "Band Box Revue," the Georgian Singers, Bert Anstice and his Mountain Boys, Luigi Romanelli and his King Edward Hotel Orchestra and "Rhythm Rebels." At 10.30 pm the network always carried a quarter hour live dance band remote, with Mart Kenney and his Western Gentlemen, Louis Guenette and his Chateau Frontenac Dance Orchestra, or Lloyd Huntley and his Mount Royal Hotel Orchestra. Saturdays at 7.30 pm they featured Bert Pearl playing the piano and singing in his pre-Happy Gang days. CFRC carried the CRBC feed of the unveiling of the Vimy Memorial by King Edward VIII and all of the attendant speeches and ceremony beginning at 9.05 am, Sunday, July 26, 1936.[46]

Early CFRC Remote Broadcasts

Harold Stewart and his crew had to be out with the remote equipment early on the morning of August 17 to begin a day's broadcasting from the Rexall Streamlined Train. Mr Liggett and the United Drug Company's board of directors held court around the continent that summer on a spiffy royal blue train with a white racing stripe. They brought display cars for the public and put on a sort of convention and pep rally for the local dealers at every stop. The train was wired and equipped for broadcasting - they sent all of the technical details ahead - except that CFRC would have to bring a remote amplifier and an operator. A second man would be needed to work the system of microphone plugs as the broadcast moved down the train.[47] CFRC did three programs from railside that day, 8.25 to 9.30 am, 12.00 to 12.15 pm and 6.00 to 6.59. The Whig received a bill for $16.96 for pick-up equipment and for the services of an operator.

Two days later, The Whig owed another $16.15 ($10 for equipment, $6.15 for an operator) to CFRC for a remote political broadcast from Stirling, Ontario. The station signed on at 3.30 for an hour-long speech by Ontario Premier Mitchell Hepburn.

> **Harold Stewart**: I was sent as an operator to a place called Stirling, up north of Belleville. We had a political broadcast to do, and the speaker was Mitch Hepburn. I don't know if this was pre-arranged or not, but Mitch gets up on a manure spreader and starts to speak. He says, "Ladies and gentleman, this is the first time I have ever addressed an audience from a Tory platform." And somebody in the crowd yells, "Start 'er up Mitch! She's loaded!"[48]

CFRC remained on the air into the afternoon five times in late September and early October for the New York World Series, Giants versus Yankees, picked up from the CRBC. For the first time in years, CFRC carried a Queen's regular season football game, from Varsity Stadium in Toronto, on Saturday, October 17. It came in down the CNR network, so must have been special to CFRC. The station log mentions no sponsors.

The New Antenna

Since 1931, CFRC had had a horizontal, half-wave, end-fed antenna. This radiated a wave with a large high-angle component that was reflected from the ionosphere at considerable distances. This sky wave constantly shifts in angle as the ionosphere fluctuates, and reception at night had been reported from as far away as Australia. What produces local reception is the energy radiated along the ground, but the old aerial produced only a small ground wave. Therefore, from the standpoint of local coverage, the old aerial was poor because the power was wasted on the large sky wave.

Prof. Morgan secured a quotation from the Steel Company of Canada, Montreal, for the erection of a guyed mast of two 30-foot pieces of 4" pipe welded together. It would cost $119.[49] On July 21, Morgan issued a requisition to have the blacksmith construct antenna equipment and a mast, as designed by Mr Bews. Around September, 1936, the old system was replaced by a single wire "T" between the old mast on Ontario Hall and a new 70-foot mast on Fleming Hall. The horizontal portion was 110 feet high, the lead-in nearly vertical. A counterpoise, a small aerial substituting as a ground, was slung under the big aerial near the earth, because of the difficulty of obtaining a good ground in that soil. Right away a great improvement in local reception was noted, because the new aerial produced a large ground wave and a small sky wave. A Stewart-Warner car receiver, roughly calibrated as a field-strength meter, showed an increase in field strength locally of more than two to one. With the previous antenna, fading set in about five miles from Kingston, where the sky wave and ground wave could interact to cancel the signal. With the new antenna there was no appreciable fading 25 miles from the station.[26,50]

The log shows that James Annand and George Tottenham put on a little play, "The Canaan's Cook," 6.13 to 6.24 pm on "The Diner's Program," August 21. Phonse McCue also sang three songs, and there were spots for Empire Life, Dodd's Medicines and the Kinsmen's Club. Dr Frank Llewellyn Harrison played a 15-minute piano recital on CFRC on September 29 at 7.00, just before they went to the network. The next day Sid Fox and his band broadcast in the same slot, and on the 1st and 8th, CFRC signed on at 5.10 so that Cuth Knowlton could give a half hour sponsored program from the large studio.

Cuth Knowlton had played percussion in pit orchestras in the vaudeville and silent film days, then with one of Goldkette's touring Orange Blossom Bands, with the Wyn Phillip Band in Quebec and at the Chateau Laurier for the British Economic Conference throughout the summer of 1932. He played winters at the Roy-York Cafe in Kingston from about 1932-37, and on a City Council contract from 1933-36 with his own band at Lake Ontario Park. During the War, he played in Robert Farnon's AEF Band in England, and often sat in with Glenn Miller.

Cuth Knowlton: One day I thought, why not try to get a commercial thing out of radio, so I went over to the radio station at Queen's and asked them if I could sell a program on the air, could it be broadcast? When it was straightened out, they said sure. So I went down to Steacy's Limited, one of our bigger department stores, and talked to Mr Noble Steacy and his brother Herb, and when I laid it out for them as best I could, why they thought it was a good idea. I made the arrangements with them and then I went over and told Queen's. I was supposed to see Major Annand personally at The Whig about this, and

Cuth Knowlton's Orange Blossom Orchestra at Lake Ontario Park, circa 1933.
                                        -courtesy Cuth Knowlton

when I did, why he eventually confirmed it but admitted he should have done a little research on his own and got some kind of a commercial going. I set up a program and rehearsed it for a little while and eventually went on the air. I believe Tom Warner announced the Steacy show for us. The studio was very, very small and to put a ten piece band in there was quite a chore.[51]

The New Principal, Dr Robert Charles Wallace

Dr R.C. Wallace, former President of the University of Alberta at Edmonton, became the eleventh Principal of Queen's University. This choice would be most significant in the future of CFRC, as Dr Wallace had had connections with the Radio League in Alberta, had been E.A. Corbett's boss at CKUA, and was very enthusiastic about the educational possibilities of radio. His installation was broadcast over CFRC on the afternoon of October 9, 1936. Dr Duncan McArthur, Deputy Minister of Education, delivered one of the congratulatory addresses. That evening CFRC broke away from the network for about an hour to broadcast the speeches given at the dinner in the new Principal's honour.[52]

Queen's Board of Trustees set up a Radio Committee in October to take charge of CFRC's educational broadcasting. The original members were Messrs Ernest Cockburn Kyte, Librarian, and A.E. Currie, Professors S.C. Morgan, H.L. Tracy and J.A. Roy, Drs James Miller, G.H. Ettinger and W.E. McNeill.[53] No records exist of their deliberations, but individual submissions survive. Prof. Roy offered the services of the Faculty Players and observed "Only the best that the University can give should be sent out on the 'Wireless'. Every sort of fare - fare that can be had from Toronto - should not be sent out. Miscellaneous programs are better done elsewhere. Unless circumspection be used, you will very soon lose the sympathy and consequently the support of your special audience of listeners."[54] Prof. F.Ll. Harrison proposed four series of concerts, both live and on records.[55] The Committee set up a series of daily programs to be given by the faculty, 7.30 - 7.45 pm, beginning November 16. Monday talks were devoted to literature, Tuesday to science and medicine, Wednesday to social sciences, Thursday to art and music and Friday to modern books.[56]

From January 1937, the talks began at 7.15, but returned to 7.30 in the fall. Naturally, Mr Kyte delivered most of the Friday talks, and the Resident Artist Andre Bieler and the Resident Musician Frank Harrison handled most Thursdays. (Bieler and Harrison were made Assistant Professors in October, 1941).

Charles A. Millar: Every night in prime time University staff members came and gave talks on various subjects, usually their own vocation. Sometimes they were interesting and sometimes they were not too interesting, especially to the operator. One day somebody who came to give a talk wasn't too sure of all his subject matter so he had a handful of notes. He arranged them all in proper sequence on a music stand beside his chair. It was summertime, the window was open and a gust of wind came up and his papers fluttered all over. Anyway, he carried on in a valiant manner and finished the broadcast. Then, a lot of faculty members weren't too punctual. I mean, they got there on the right day, but in broadcasting you've got to be there in the right minute and second. We were feeding a talk by Mr Kyte to the CBC National Network and it was coming up to time. A couple of minutes before broadcast time and no man to give the talk. We ran to the window to see if anybody was coming across the campus from Douglas Library, and no Mr Kyte. It got to be a matter of less than a minute and we were really worried because we didn't know what we were going to feed the network. What do you do? Feed them music? And we scrambled around. But eventually Mr Kyte lumbered up the stairs and arrived in what he thought was ample time to speak to the people of the nation.[57]

Alex McDonald: Dr McNeill, the Vice-Principal, who was an English teacher, read poetry on the University quarter hour series, and it was beautiful. Really wonderful. The way he read it, anybody could appreciate it. E.C. Kyte was on there, and Hilda Laird, and then there was the University Trio. Prof. Gummer was in it, playing the flute, and somebody played the violin - three professors and they were very good.[58]

Lareta Lamoureux: I recall when we first got our radio. My sister bought it for $30. A little, round box with scrollwork on it. A dollar a week, on time - and there was only one radio station, CFRC, Kingston, by Queen's. The news was the big item. My father'd park his rocking chair by the radio, turn it on, and sit there, his hand beside his ear, listening to the news. Then he'd turn the switch off - save the power! We didn't have the authority to turn it on or off when we felt like it. Goodness sakes! No way! Tubes would blow and you'd have to pay for them!

The radio from Queen's opened up a lifeline of information and music and so on. It wouldn't matter whether it was in town or not, or how corny it was. It was something coming through the air into our house! For us to listen to, eh? A scholar or something talking in our house! It was fascinating. There was a lot of stuff that was way over my head. A lot of the music was over my head, but the fact that it was coming into our house...When we were growing up, Queen's was the high palace of Kingston. To me, Queen's was Kingston.

And later on the soaps came on and that was another ritual in the house, that we had to be very quiet for Ma Perkins. A few years later, on come the Happy Gang. That was another big thing in the household. It came over Kingston station, brought in from Toronto - there was no other station. It was exciting, with the radio on. You didn't make any noise in the house when the radio was on, because it was fascinating. Well, you weren't allowed to anyhow. Dad wouldn't allow it if it was turned on.[38]

The students were to supposed to produce a radio program of their own, "The University Hour," a full hour per week showcasing campus organizations and talent.[59] The project never got out of the planning stage, since the students were too busy to bring off such a co-ordinated and sustained effort.

New Programs and Sponsors

The New York Philharmonic-Symphony Orchestra broadcasts, from CBS, resumed on Sunday, November 8, 1936, and filled the 3.00 - 4.30 pm slot during the season for the whole of CFRC's commercial phase. CFRC first carried the Metropolitan Opera on Saturday, January 16, 1937 - Wagner's "Die Walkure," with the fabled Vienna cast, Lotte Lehmann, Lauritz Melchior, Emanuel List and Friedrich Schorr, conducted by Artur Bodanzky.[60]

The CFRC log includes all of the spots sold by The Whig, and indicates that until around September 1936, local businesses did not chance spending for promotion on the new medium. Gradually they began to buy spot advertising: Doretta Apparel Shop, Lewis Optical, Lucille's Ladies Wear, Frigidaire Sales and Service, Price's Dairy, J. Swift Coal Company, Ward and Hamilton Drugs, D.A. Shaw Ltd., Grand Cafe, Capitol Theatre, Shoe Krafts, Arthey Optometrists, R. Moat and Company, Kinnear and d'Esterre, Jackson Metivier Ladies Wear, Ford V-8 Motor Sales, Austin's Drug Store, Frank Robb Beauty Parlour, Boyd Electric, Golden Lion Grocery, McCullough Shoe Store, Treadgold, Tweddle Hatchery, Star Fruit and Candy Store, Bethel Church and New-Ray Hair Waving.

Hemlock Park Dairy was the first local business to have its own weekly series, a transcription feature called "The General Store," which ran Thursdays at 6.45 from September 17. Empire Shoe Rebuilders sponsored some sort of a canned mystery series. Joseph Abramsky and Sons was next with a detective series titled "Nemesis Incorporated." That ran at 6.45 Mondays and Thursdays until early December, when a music program took over the slot.

The national brands came in as well: Silk Stocking Tea, Mecca Ointment, Carter's Liver Pills, Dr Chase's Medicine Company, Bulova Watch with their time signal, Rexall Drugstores, Bell Telephone, Toronto Star Weekly, Red and White Stores, Dr Jackson, CPR Telegraphs, Imperial Tobacco, Sherwin-Williams Paint and the National System of Bakeries. "Belle and Martha" replaced "Home Sweet Home" for Canada Starch in November.

Dr Chase's Medicine Company sponsored "The Family Doctor," the first Canadian soap opera, written, produced and directed by Andrew Allan. It ran every Monday, Wednesday and Friday at 1.00 and starred Judith Evelyn with Frank Peddie as her father.[61]

The Whig sponsored "Neighbourhood Night" on November 12 in Grant Hall. 1,600 were admitted free and 1,000 more were turned away. The program was MC'd by W.J. Coyle of The Whig, assisted by Major Annand and Tom Warner, and featured "Red" Newman of "The Dumbells" accompanied by Kathleen Grimshaw, xylophoniste Bessie Entwistle, vocalists Edythe Hayman, Dr Harold Angrove and Fred Potts, artistic whistler Hilda Friendship, guitarist Norman Smith, banjoist George McAuley, Al Smith and his Shanty Boys and Sid Fox's Orchestra. Arthur L. Davies moderated a general knowledge contest for high school pupils and, as a special surprise, Senator Davies led a community sing-song! CFRC carried the program from 8.15 to 10.45.[62]

Patent Medicine Advertising on CFRC

Queen's agreement with The Whig stipluated that all advertising sold for CFRC had to be acceptable to the University. After a few months of commercial operation, the Faculty of Medicine complained about the advertising of patent medicines. As far as the listeners were concerned, the University was endorsing these products, and therefore so was the Faculty of Medicine. The CBC's only rules in the matter were that the scripts for patent medicine spots had to be approved by the Department of National Health. The Queen's Journal medical column published the recipe for Dodd's Kidney Pills, concluding, "It seems too bad that those of us now in Medicine will graduate from an institution that lends itself to the advertisement of patent medicines."[63] Principal Wallace conferred with Mr Muir, and they agreed that those particular advertising contracts not be renewed. Moreover, if a satisfactory arrangement with the clients could be reached, all patent medicine contracts should be abrogated immediately.[64] The matter seemed to have been settled amicably, but it rankled with the newspaper, and was a constant source of irritation. Patent medicines were the biggest single source of revenue for CFRC, and cutting them off meant jeopardizing the commercial success of the partnership.

The Canadian Broadcasting Act, 1936

During the 1935 Dominion election, the Liberals were furious over attacks upon them aired by the Conservatives over the CRBC, in the guise of unsponsored folksy chats, the infamous "Uncle Sage" series. The Liberals felt that the CRBC was guilty of lapses of judgement and control, had shown too many administrative weaknesses, and had to go. On regaining office, Prime Minister King convened another special parliamentary committee on broadcasting.[65] It reported on May 26, 1936, recommending repeal of the Broadcasting Act of 1932 and replacement of the CRBC with a crown corporation which would be independent of Parliament. All radio broadcasting in Canada would be completely under the control of the new corporation, and there would be co-operation between the national system and the private stations.

At this point, Alan B. Plaunt of the Radio League lobbied Mr King intensively, and the Prime Minister finally accepted the principles of the old Aird Report. The Canadian Broadcasting Act of 1936 made the new Canadian Broadcasting Corporation both broadcaster, in competition with the commercial interests, and the regulator of all broadcasting in Canada. Unlike the CRBC, its policy making and regulating were separated from its broadcasting function. The Board of Governors set policy for the whole country, and a Manager headed the CBC radio network. The Board would recommend the appointment of the CBC executive and would buffer it from external pressure. Incidentally, two members of Mr King's government at the time were Dr Norman Rogers, Minister of Labour, and Charles Dunning, Minister of Finance.

The Canadian Broadcasting Corporation was officially born on Sunday, November 1, 1936, at midnight. According to the CFRC log, Queen's first linked up with the new CBC at 6.30 pm that day for "Dr H.L. Stewart Reviews the News."

The new Department of Transport (DOT) and the CBC Board of Governors both came into existence on November 2, 1936. Leonard W. Brockington, KC, former City Solicitor for Calgary and counsel for the North West Grain Dealers' Association, was named Chairman of the CBC Board and René Morin of Montreal, Director General of the Trust General du Canada and former Liberal MP, the Vice-Chairman. Members included Col. Wilfrid Bovey, Director of Extramural Relations for McGill; J.W. Godfrey, lawyer from Halifax; Mrs. Nellie McClung of Victoria, writer and leader in the women's rights movement; N.L. Nathanson, Toronto, President of Famous Players and Managing Director of Canada Paramount Corporation; General Victor Odlum, a Vancouver bond broker and early supporter of the Canadian Radio League; Alan B. Plaunt, Ottawa, Secretary of the Radio League; and Rev. Alexandre Vachon, Professor of Chemistry at Laval. The Board unanimously recommended Major W.E. Gladstone Murray, a Canadian who was the BBC's Director of Public Relations, as General Manager and Dr Augustin Frigon, Professor of E.E., Director of l'Ecole Polytechnique in Montreal and a member of the Aird Commission, as Assistant General Manager of the CBC.[66]

L.W. Brockington spoke over the new CBC network on November 4, 1936, to officially inaugurate the new corporation and review the radio situation in Canada.[67]

The Abdication of King Edward VIII

The station logs show that CFRC joined the CBC on the afternoon of Friday, December 4, 1936, for a speech by Lord Elston in London, England. Prime Minister Stanley Baldwin had just broken open a story that had been festering for weeks when he announced to the Commons that the Government would not bring in legislation to permit the King to marry Mrs Simpson while ensuring the present succession. At 11.27 am on the 7th, the CBC hooked up with London again for Baldwin's statement to the House that the King would be given more time to come to his decision. On the morning of the 10th, CFRC carried Edward's abdication speech and Mr Baldwin's statement, and then at 1.00 the statements of Prime Minister Mackenzie King and Mr Ernest Lapointe, Minister of Justice. The next day, CFRC brought to Kingston Edward's farewell broadcast to the Empire before he left for exile, and the day after that the proclamation of King George VI from London. These radio connections were now possible at very short notice, so that the freshest news could be brought instantaneously from overseas from the mouths of the actors themselves. The whole of Canada hung on every word of the royal crisis as they never could before, and never will again.

Lareta Lamoureux: He came on and read his speech that he could not go on the throne without the woman he loved at his side... yeah. Everybody had their ears to the radio for that. That's for sure. It was a sad thing too. I liked her, Mrs Simpson.[38]

Queen's Convocation Broadcast

On May 7, 1937, CFRC broadcast the convocation ceremony from Grant Hall and sent it through the CBC lines to CRCO, Ottawa. The commentator was W.A. Neville, editor of the Queen's Journal. The broadcast followed the plan used in the special convocation for Governor General Lord Tweedsmuir the previous November 7.[68]

Bill Neville: In my student days, 1934-38, I was interested in everything that had to do with communication - the art as opposed to the science. My main outlet was the Queen's Journal, and I worked my way up from freshman reporter to editor-in-chief for 1937-38. I even thought I could be a good radio broadcaster. Through my good friend Alex McDonald I got to know Prof. Morgan and I did a few announcing assignments for CFRC.

As a Kingstonian, it was natural that I should remain in Kingston between the exams in April and the commencement of my summer job early in June. Thus the Prof. and I cooked up the idea of broadcasting the 1937 convocation - they could laureate graduates from all faculties in one afternoon in those days - on a modest network with an Ottawa station. Queen's was heavily weighted with Ottawa natives before Carleton existed, so we gave the broadcast some local colour for the Ottawa listeners. We set up a broadcast booth - primitive and makeshift - in the east balcony of Grant Hall overlooking the stage, and from there we broadcast two hours of the proceedings. It wasn't a very professional performance, but as an experiment it provided some good lessons in remote broadcasting. The only feedback I had was from a few Ottawa friends who said that it was nice for relatives of grads who hadn't been able to go to Kingston.[69]

Changing CFRC's Parameters to Increase Coverage

The new Principal was interested in bringing Queen's educational broadcasting to a much wider audience, as his former charge, the University of Alberta, was doing. Some time prior to the fall of 1936, Prof. Stewart had mentioned to Dr Wallace that a short wave broadcasting station would give Queen's wider coverage at minimal cost, and accordingly, Dr McNeill wrote to Stewart enquiring about expense, coverage and reception quality.[70] In reply, he learned that a 100- to 500-watt home-made short wave transmitter and antenna would cost $1,200-$1,500 and would be all right for speech but not for music. The CRBC was then operating one of these facilities, CRCX, at 100 watts from Bowmanville, and it was heard all over northern Ontario. A 50-metre short wave transmitter could have a 500 mile radius in daytime, but after dark its coverage would be erratic. The signal might skip right over Napanee and Belleville and be received in Winnipeg or Vancouver. Moreover, special receivers were required for short wave and the service would have to be strictly non-commercial. That precluded simulcasting with CFRC-AM. Stewart thought that, all things considered, short wave was not a very attractive solution.

Another way out would be to seek higher power for CFRC or a lower frequency channel than 1510, either of which would also increase CFRC's coverage. To cover Ontario as thoroughly as the University of Alberta covered the prairie would cost Queen's at least $500,000.[71]

Wallace then wrote to Leonard W. Brockington, Chairman of the CBC, enquiring whether the audience for Queen's valuable educational radio work, as well as for its broadcasts of CBC's evening programming, could be extended by increasing CFRC's power to 500 watts or by decreasing its wavelength. Both Belleville and Brockville were on the fringes of CFRC's signal.[72] A copy of the letter was passed on to General Manager Murray, who promised to include the matter in the CBC's current survey.[73] Wallace received an encouraging reply to a letter to Major Murray, asking the Corporation's assistance in finding a way to enlarge the dissemination of cultural offerings from Queen's staff through CFRC. He noted that reception of CFRC was not reliable beyond the immediate vicinity of the city, a radius of about 6 miles. Increasing the radius of coverage to 100 miles could be accomplished by changing power and/or wavelength.[74]

Major Murray enclosed with his reply a memorandum from the CBC's Chief Engineer, who defined the problem: at 100 watts, CFRC had to be on a local shared channel above 1200 kilocycles, while an increase to a kilowatt would permit a move to a regional frequency below 1200. The transmitter, however, would then have to be located outside the city limits, the location to depend on field strength surveys. He estimated the cost of refitting, including a vertical radiator, to be about $35,000, and suggested that Queen's apply to the Department of Transport to increase the power to one kilowatt. A suitable frequency could then be dealt with by the Technical Committee on Broadcasting after the North American Conference in Cuba.[75]

Wallace, after consultation, wrote back to Murray requesting a meeting to explore the possibility of increasing power without moving the transmitter. Queen's could simply not afford to spend $35,000 on CFRC.[76] Prof. Morgan estimated the cost of constructing a kilowatt transmitter and installing it off campus at $15,000. Annual operating costs, including power, employees, taxes, rental of loops, telephone and insurance, would be an additional $10,000.[77] A few weeks later, Prof. Stewart got some estimates on the cost of erecting a 250- to 300-foot four-sided self-supporting steel antenna tower.[78]

On May 6, Wallace made formal application to C.D. Howe, Minister of Transport, for permission to increase CFRC's power to a kilowatt and to erect a vertical radiator.[79] The Chief of Air Services, DOT, referred it to the Technical Committee and advised that "the question of the re-location of the transmitter on an appropriate site which will be satisfactory to the Department will then be taken up with you."[80] Dr Wallace wrote to Major Murray on June 23 to see whether any progress had been made, and offered the Corporation the co-operation of CFRC in their experimental work in educational broadcasting. He was assured that the application was favourably thought of, but there was some question of frequency. In any case, the application had first to go to the CBC/DOT Joint Technical Committee and then to the August meeting of the Board of Governors.[81]

By mid-September there was still no answer, and Wallace wrote to Murray again. This time, he also asked whether payment might be made to stations for carrying the CBC service.[82] It turned out that the matter of increased power had been considered, but since there was no suitable frequency available in the Kingston area, approval was withheld. Moreover, the CBC was not in a financial position to pay any more stations for carrying the service, and that whole matter was under review.[83] The Principal kept reminding Murray about Queen's request for more power, pending any decisions about re-allocation of North American frequencies at the conference in Cuba.[84]

Finally, the DOT decision arrived, dated December 13, 1937, turning down the application of May 6.[85]

In anticipation of permission being granted to go to a kilowatt, Harold Stewart had spent most of the summer of 1937 "preparing detailed designs for an entirely new station to use a 300-foot steel tower as a radiator."[26] Stewart remained hopeful that CFRC would be granted higher power within two years. The refusal was not final. He had reason to believe that the situation would change after the upcoming Canada-US agreement was concluded. Stewart spoke to Vice-Principal McNeill on February 4, 1938, about improvements to the control room, compatible with the new 1000 watt transmitter to be built, and submitted his cost estimate, $2,150, in writing a week later. He even offered to build the equipment himself:

> **Harold Stewart**: ...in a broadcast transmitter, the average modulation must be kept at such a level that sudden increases in sound intensity do not overload the transmitter and produce serious distortion. This means that the effectiveness of the transmitter is being held down to maintain good quality. If the operator could react fast enough to a sudden increase in sound volume before the microphone, he could safely operate at a higher average level of modulation, and still be able to prevent sudden peaks from causing serious distortion in the output...Now a limiting amplifier...does just this. It limits the peaks so that a higher average modulation may be maintained. The net result is that the effectiveness of the transmitter's power is somewhat more than doubled. A 100 watt transmitter on this basis will be about as effective as a 250 watt transmitter on the ordinary basis. This will not double our range, unfortunately, but will produce a noticeable improvement in service...As our carrier power will not be changed, our present licence will still be effective.

A weak voice in the studio has to be amplified to a greater degree than a strong one. With our present equipment this increases the background noise considerably. With the equipment proposed a different system of mixing and control is used...The advantage is to decidedly reduce noise level (hiss and hum) on studio programs...It will provide better operating facilities and greater flexibility in control equipment...This is the accepted method in high quality transmission.

Transmission characteristic measuring equipment...requires special explanation. With increasing complexity of transmitter circuits, a reliable and accurate means for quickly analyzing transmitter characteristics becomes a necessity. It is possible for progressive deterioration in quality to set in, in one or more amplifiers and finally become noticeable to the ear, unless checked and corrected in its early stages. At this stage the deterioration is apparent to the outside listener as well as to the operator. Accurate measurements make detection possible in the early stages...Where the transmitter is a matter of interest to electrical engineering students, this equipment is valuable from the teaching point of view...Since it is not necessarily an integral part of the transmitter and not in continuous use, it will be available at times to the radio laboratory for experimental work on amplifiers and transmission equipment.

I might mention here that I have discussed this equipment with C.B.C. engineers and I learned that on testing one set they immediately ordered five sets for their system...The time required will be about six to eight weeks.[86]

The matter of turntables came up about this time. Major Annand complained to Dr McNeill that one turntable had never worked properly.[87] Stewart affirmed that one had given trouble, that cleaning and greasing the bearings had helped, but serious pitch variation was now showing up. He had opposed buying these RCA turntables in the first place and now suggested that new ones were needed.[88] After further problems, he made some changes to both tables, but the real solution was a new type of motor without mechanical speed reducer.[89]

By the spring of 1938, up-grading of the control room was in progress:

The increasing revenue from commercial programs has made it possible to purchase an RCA transmitter-measuring set which will permit us to extend the experimental work on the application of negative feed-back to the transmitter. A limiting amplifier is being constructed to reduce excessive modulation peaks and to permit the use of a higher-average percentage of modulation. This will be approximately equivalent to doubling the transmitter power. A cathode-ray oscillograph, using an RCA one-inch tube is being incorporated as a modulation indicator in one of the new speech-input cabinets in the control room. This small tube does not permit very accurate determination of modulation percentage but it does give the operator a good idea of what is going on in the transmitter. A larger five-inch oscillograph is available for more accurate measurements.

New turn-tables with heavy-duty synchronous motors using the pole-changing method to shift from 78 to 33 1/3 rpm, will be installed to eliminate the "wows" which occur with the present equipment. New speech-input equipment will increase program-handling facilities. As an example...a game in Richardson Stadium...sent to Montreal or Toronto over Bell Telephone or Canadian National wires. This has to be amplified at the stadium and a carefully controlled level of about zero db fed to the station. This level is not high enough to put out on the network because of loss in line equalizers. It therefore has to be stepped up to about +10 db in another amplifier and this level will be under the control of the station operator, who must be able to listen at times on this program. Continuous visual indication is given by a meter...The transmitter is taking a program from the network at this time. A check must be kept on both programs and levels watched at all times. A flexible speech-input system is therefore...absolutely necessary.[26]

James Annand, as Advertising Director, received notification from a client that the station could not be heard beyond 45 miles radius. The sponsor had mounted a club membership campaign in connection with "The Plainsman" radio serial and a breakfast product and, after four weeks, Kingston's response was very poor. He was requesting proof of coverage. The sponsor's policy was to eliminate non-productive stations, and CFRC was put on probation. A week later, CFRC was instructed to cancel the series and to ship all remaining transcription discs on to the next station in rotation.[90] Annand informed Dr McNeill that even 500 watts would help and, failing that, a lower frequency.[91]

A long letter to Gladstone Murray soon left the Principal's office, detailing CFRC's recent problem with sponsors, and the need for higher power - at least 500 watts - both to satisfy customers and to carry on Queen's educational work. A carbon copy went to Hon Dr Norman Rogers, MP and Rector of Queen's. Murray counselled patience, since the channel re-assignments approved in the Cuban conference were still being worked out.[92]

Dr Wallace renewed the Queen's application for one kilowatt power in a letter to the Hon C.D. Howe, dated March 12, 1938: "At the present time...about 63% of the time on the air is given to the re-broadcasting of the Canadian Broadcasting Corporation programmes to the great advantage of the people of this district and vicinity. It is well known that this area, because of some special conditions probably not fully understood, is a difficult area from the standpoint of radio reception. For that reason our station provides a general service apart altogether from the work which is initiated at the station itself."[93] Mr Howe presented the application to the CBC, but the situation was unchanged. The Technical Committee was unable to find a channel for a kilowatt, but there was hope of a clear channel for a Class II station of that power in Kingston under the provisions of the Havana Agreement.[94] Patience was again necessary.

CFRC Radio Personalities

Wilma and Louise, the Jackson Sisters, first appear in the CFRC log on January 7, 1937, mixed in with transcriptions on "The Diner's Program." They sang "Nobody's Darling But Mine," "The Dear Old Sunny South By the Sea" and "Night Time in Nevada." Bell Telephone, Jackson Metivier and Red and White Stores each had a spot on the show.

> **Louise Wittish**: We used sing on a CFRC program called "From Soup to Nuts." I always wanted a guitar, so for my twelfth birthday my parents bought me one, and an hour after I got it I could play. Then Wilma and I used to try and sing. We used to practice without the guitar after we went to bed. One of the songs that we used to sing so much was "I Want to Go Back to My Little Shack," and Wilma here was trying to get the tenor into it. It seemed we never got past those two lines, and Mum used to open the stairway door and say, "For heaven's sake, I wish you'd **go** back!" That was a couple of years before we got on radio.
>
> Our dad had a sand pit, and one day he brought a contractor named Walter O'Donnell into the house to get warm, because it was about 45 below zero outside. So, while he was in there, Daddy got us to sing for him. And he thought we were real good, so he brought us in to the radio station to have an audition. And they must have thought we were all right too, because they wanted us to come back, on a program called "From Soup to Nuts," from 6.15 to 6.45. We always said we were the nuts, but I don't know whether we were or not. I think we were on three or four times. It was just different people eh, coming in singing. Two other girls who used to be on it quite often were Shirley and Thelma Bryan. They used to bring their tap dancing mat and tap dance.
>
> The first song we sang was "Neath the Maple on the Hill." We'll always remember that. The first time we sang I played the guitar, but after I think that Uncle Josh backed us up, or it was Al Smith. We used to send for Bradley Kincaid's song books. He used to sing on WSM, Nashville. That's where we learned our songs out of. Our announcer was Tom Warner. Very well remember him - handsome man.[95]

Uncle Josh and his Nephews had a regular Saturday afternoon slot on CFRC. In 1938 they had noon to 12.35, and later they were shifted to the 1.00 to 1.30 slot. Uncle Josh specialized in country corn, seasoned with light humour and a real inventive spirit. It was almost a situation comedy at times, recounting their adventures as they travelled to gigs. If their soft-cut demonstration record is any indication of what they could do, they were world-class artists.

Erwin Wendholt: I had a small connection with Uncle Josh and his Nephews when they used to play on CFRC back in 1937. A very good friend of mine, Frank Doyle, sang regularly with them and played guitar and mouth organ and he was very good at it. The odd time when he couldn't be there, he used to delegate either me or my cousin Austin to fill in for him. Uncle Josh was Phil Nichols, who played accordion. Ralph (Chuck) Clark played guitar and sang, "Sparky" Spencer was great on piano, Cliff Parslow sang - the others I don't recall. A couple of times Billy Christmas played with them - a trumpeter. I've forgotten how this worked in with their music because it was mostly hill billy or cowboy music - ballad and comedy songs. They were a great bunch of kidders. They would wait til you were halfway through a song - perhaps you had the words propped up in front of you - and somebody'd whip the words away, and there you were stuck halfway through a verse, sans words. And little bits of business, like putting in sound effects with water glasses and so on, accompanying the songs. A lot of the songs were - this was at the end of the Depression years - hobo songs, about chaps sleeping in barns and looking for work and riding the freight trains across the country.

Phil "Uncle Josh" Nichols at Lake Ontario Park, circa 1940. "Uncle Josh and His Nephews" specialized in "Kountry Korn" and they were excellent artists.
-courtesy Bonnie Nichols

Tom Warner was master of ceremonies - quite a handsome chap, as I remember, and very helpful. He always put you at your ease. Tom would make you feel quite relaxed and introduce you to the microphone. That's about all there was in the little studio we used. The walls were of cork combination, painted drab green, and the microphone standing in the middle of the floor, and your group around that. Step up and do your little solos, and fill in a little background for the rest as they were singing.[96]

Cuth Knowlton: Around '37, '38, my full band was set up in Grant Hall - 12 men in the band - and we were playing away, everybody dancing and having fun. All of a sudden somebody from Queen's radio station come waltzing in and putting up microphones in front of the band. I just was bewildered. Well, it fooled me. I didn't think we were going to broadcast. I hadn't made any program or anything else, but that's the way it was.[51]

John D. Stewart's Famous Misadventure

<u>Harold Stewart</u>: I'll tell a story about my brother. He was the operator one time and he got himself locked in. He forgot his key and couldn't get out. So he put an announcement on the air asking somebody to phone me. They did, so I went over and let him out. I don't know whether he'd like me telling this or not.[48]

<u>John Donald Stewart</u>: I came back from studying in Germany in the spring or summer of 1936, quite a bit in debt, and jobs were scarce then. But I was fortunate that my brother was in technical charge of CFRC and they needed a couple of operators to go into commercial operation in collaboration with <u>The Whig-Standard</u>, and I was able to get a part-time job operating there, which came in very handy to me. I took turns with Alex McDonald operating the station. We used to amuse ourselves by mimicking some of the ads, like the one for Carter's Little Liver Pills - "They do the work of calomel, but have no mercury in them" - and making irreverent twists in the names of some of the programs. Alex was afraid that he would say one of these on the air. Tom Warner had a good clear voice, but Major Annand had more of trained voice - he was a professional actor. He wasn't so good at ad libbing.

Part of the time we'd operate for <u>The Whig-Standard</u> people to give the news, and then we'd play records and announce various people. In the evenings we'd switch to the CBC and carry the network program. Then we'd have to sign off at night around 10.30 or so. It was an easy job and I could get some reading done there or we could have some friends come and visit us in the evening, and it was no problem. One night I'd signed off, and then I wanted to get into the transmitter room where my coat was and my keys were, the telephone was and the switch for turning off the transmitter. But the janitor had

John Donald Stewart at the OP-4 board in the CFRC control room, circa 1937, seen from Studio "B." An oscilloscope and the auxiliary switching panel are at his right, and the Mark VI transmitter is visible through the window behind his head.

-courtesy Prof. Harold Stewart

been showing some people around the building that evening and he'd locked the transmitter room with my keys in there. What was still worse was that Fleming Hall at that time had a lock on the outside door that could be locked from the outside, so that you couldn't get out of the building unless you had a key. And I didn't have a key. So, I didn't know what the dickens to do. I was trying to think of something to do. I tried to get the attention of my friend Ron Bradfield, who worked across in Ontario Hall, but I couldn't attract any attention. Finally it came to me that, well, I'll have to try putting out an appeal on the air. And I did, asking anybody that heard me to please get in touch with my brother. Fortunately, somebody was twiddling his dial and he heard me and he got in touch with my brother, who rescued me.[41]

Major James Annand

Joan (Annand) Henderson: My father was an actor in London in his early days, with George Alexander, Ellen Terry and Henry Irving. Later he was Stage Manager at St James' Theatre in London, working with Bernard Shaw. The theatre was his first love. He was in some of the very early films of the Hepworth Company at the Ealing Studios in London. He was in the London Scottish for many years, and was gassed in the the First War in the first gas attack. After that he did war correspondence work. He enjoyed army life - he was that kind of person who was able to adjust to almost any kind of situation and make the best of it. He never used the "Major" himself. He always identified himself just as Mr Annand, but somehow people picked it up and he was always called "Major" as a sort of a nickname. About 1920, things were pretty bad with the movies, and actors began to move out. Ronald Colman went to the States and my father came to Canada to start a new life, and he went into business here for a while. He joined the 48th Highlanders and did a lot of work with the Shakespeare Society in Toronto - he always kept up his connections with the theatre. Then he moved into radio in a wonderful series on the CRBC, called "Forgotten Footsteps."

At Queen's he had a straightforward job and the station was just a very simple, basic set-up. One difficulty for him was that Queen's didn't allow patent medicines to be advertised, and he could have made an awful lot of money on that. My father used to tell one story against himself, which he always enjoyed. He announced that on Wolfe Island, next Saturday afternoon, there would be a peasant shoot. Instead of a pheasant shoot. That just broke him up, and he couldn't go on for quite a while until he explained that it was a pheasant shoot, and not a peasant shoot. They used students as operators in the summertime, and my brother sometimes ran the engineering end of things.

Dad sang on CFRC. He had studied singing in England for a long time and was very interested in it. He just loved to sing - everything from the old Scottish songs to opera. My mother played the piano, and every night at home I can remember that an hour of music was just part of the evening. Everything was beginning in those days and if you had a talent, you used it. I was a student of music at that time at the Royal Conservatory in Toronto and I think I gave a little recital of songs on CFRC once or twice.

He was a very interesting man and interested in a great number of things. He was a book-binder, a bee keeper, interested in poetry and singing. He could make costumes - if he was putting on a play, he'd sit down at the sewing machine and design the most beautiful velvet costumes for Elizabethan plays - make the scenery, paint it, take us camping, play tennis, go boating. He could mend things around the house, he could tell you something about almost anything. On top of all that, he was a very comforting person to be with. People liked him and liked to be with him. Even if things went wrong on a program or with a play, he seemed to manage to smooth it all over and not get everybody upset. He was a very gentle man, and a real gentleman.

People in Kingston would remember Dad from his work with the soldiers during World War II. He noticed that the boys had nothing to do on their days off and Sundays and arranged for the YMCA to open up to entertain them.

We left in 1942, and moved on to St Catharines where he was Manager of CKTB. From there we moved back to Toronto, where he did the thing he is remembered most for - he was on the CBC in a children's radio series called "Maggie Muggins." Dad played the role of Mr McGarrity, the gardener that Maggie turned to in time of trouble, the old retainer who kept life smooth. He also did some roles in various drama shows and eventually some production work with the CBC.[97]

Lenore Black: I played the piano on CFRC quite often. A Knabe grand piano, a beautiful thing, and I think it was new. I had a 15-minute program of light classical music once a week from the studio, and another time on the organ from Grant Hall, sponsored by Abramsky - for about ten weeks - and I probably got $3. And I played for singers - for Alphonse McCue, as you know, and Major Annand. He only did one number that I remember, and that was "The Cobbler's Song." He had rather a light tenor voice. And I played for his daughter Joan too. One number I played on the organ over in Grant Hall was a popular number then, called "Rosalie, My Darling," and he said "Rose-alley." I don't think he ever got my name right. It was either Eleanor or Leonora. And if he made a boo-boo on the news, he always said "As you were."[39]

Harold Stewart: Major Annand was the Program Director in the days when we operated commercially in our alliance with The Whig-Standard. He was an old Shakespearean actor, and a very competent one, I believe. The thing that I remember most about him is when he stumbled, he'd say, "Harrumph. As you were!" So, that became a famous expression around the place.[48]

Alphonse McCue: Major Annand was a real English gentleman and real precise. I remember quite a few times he made a mistake and he'd say, "As you were!" and then he'd go on broadcasting. That was a joke with everyone on the station, because every broadcast he'd make a mistake of some kind. It was a well-run station and everyone just loved Major Annand. You could stand and talk to him for an hour if you wanted to. He'd listen to you.[37]

Bogart Trumpour: I recall at least one instance when he was giving the news - and Major Annand always had a pronounced English accent - he came on with "Friday last the annual Wolfe Island Peasant Shoot was held...Oho, Oho, I say! As you were! As you were! Har, kaff, kaff! It should have been Pheasant Shoot, you know!" This was typical of CFRC and some of his broadcasts in those days.[98]

Dr William Angus: When the Anguses came to Kingston, in 1937, CFRC was the one and only radio station in Kingston, doing commercials, news and so on, and their chief announcer was Major James Ananand. Whenever Annand made a mistake broadcasting, which he did do, he'd say "Oh! As you were," and then correct himself. He didn't bother to find out how to pronounce place names particularly. He would tell us of something that had happened on this side of the Atlantic in Okla-hama. There was Pudget Sound out on the Pacific coast, the Pot-ah-mac River near Washington, DC, and the Apple-a-chin Mountains. But then came the War in '39 and the German army with its Blitzkrieg just over-running place after place in Poland. Consequently in the news there were a good many names of Polish cities. He struggled with them at first and then he said, "Ach! I give up! It is another town. I can't pronounce those names." And that was that. He finally left Kingston and went to Toronto, getting mixed up with production there. Sandy Webster told me about about him. Sandy was a Queen's grad, who went from Queen's to Lorne Greene's school, and he ran into James Annand on a television program called "Razzle Dazzle," and Annand was still the character he had been.[99]

Thomas J. Warner

Cuth Knowlton: Tommy was an awfully nice fellow. A very good-looking man. A heck of a good fellow.[51]

Phonse McCue: I knew Tom Warner - I talked to him a lot after the broadcast was over - and he was a wonderful man and a good singer. He was a perfectionist, really. I never heard him make a mistake in his broadcasting. Never. He had beautiful English and a good clear voice so that you couldn't mistake what he was saying. He was sort of a mysterious person, and no one got to know him too well. He used to go over to Kay Grimshaw's and talk to her dad quite often. He was pretty deep. I remember he went to war, and when he came back he went off to some other town and I never saw him again.[37]

Thomas J. Warner

Charles Millar: Tom Warner was well known. He was a musician of sorts, I guess, but his main work was arranging for musical groups to get their act into broadcast form as to timing and the way the songs followed one another. Tom was always associated with musical programs, because there were women in them, usually. Now I don't say that disparagingly, but he was one of those kind of guys that just seemed to be fluttering around while there were women in the studio. Yeah, he was a singer of sorts. We had a lot to do with him, but he remained more or less a man of mystery as far as I was concerned.[57]

Tom Warner in Studio "B," playing the station chimes (Deagan Station Chimes, W14103, Req. No. 37028, Oct. 31, 1936) into the Bruno RV-3 ribbon microphone (serial #396, Req. No. 27324, Sept. 7, 1933). The No. 8 music stand on the table was ordered from Warmington's Music Store, 351 Princess Street (Req. No. 37029, Oct. 31, 1936). No one remembers the tune that was played on the CFRC chimes.
-courtesy Bonnie Nichols

<u>Alex McDonald</u>: Tom was a singer, and a pretty fair judge of singers. We had a lot of local people coming around for auditions, and Tommy was a pretty good judge of their ability.

When we would be talking back and forth, when we were on the air, some of his jokes were pretty weak, and I would say "Oh, no! Not again!" "Isn't that awful?" or something like that. The jokes weren't scripted. Scripts were provided for the musical programs that we got from the RCA subscription - "Thesaurus" - these were big 16" discs, but the scripts for the local programs were done at <u>The Whig-Standard</u>. I guess maybe Tommy did some of them and the Major did some of them.[58]

<u>Kenneth McIntyre</u>: I remember listening to Major Annand reading the stock quotations and this day he said he couldn't make out whether it was 14 2/3 or 14 8/3.

Mr Thomas Warner and Mr Millar handled the announcing duties for <u>The Whig-Standard</u> at CFRC. I met Mr Warner one day when I went over to visit CFRC on the second floor of Fleming Hall. They were figuring out how many seconds a thing would take and I thought he was a little bit on the worrying or jittery side, that 20 seconds wasn't going to be enough, that he'd have to speak a little faster or something. I only met him that one time, but I remember speaking with him on the phone different times from our office when we would be discussing something on the network.[21]

<u>Agnes (Morrissey) Neale</u>: I was a student of Kay Grimshaw's, and through her I came to play piano for Tom Warner's early morning programs on CFRC. '36 to '38, around in there. I was probably 10, 11, 12 years old - just a kid - and it was really exciting to be on the radio. I remember getting up very early to get there, because I had to walk. We had no car. It was like a religious program, and he sang quite a bit - a very nice tenor voice - and did readings. It was only about 15 minutes long. As a little girl I admired him because he was handsome and rather dapper - I thought he was a very nice man, someone you admired. Maybe a special feeling that you have, the little girl for the grown man. He sang a lot, not just on the radio, but throughout Kingston. He was quite in demand, really, as an MC and soloist.[100]

Both the Major and Tom Warner hosted shows at Lake Ontario Park and community sing-songs around Kingston.

An Expanded Schedule

Stanley Chapin Morgan left to join the faculty of the University of British Columbia in Vancouver in the summer of 1937. Harold Stewart then rose to Engineer-in-Charge of CFRC. As his assistant he had a former student, Harold Stockwell Pollock.(Elec. '32)

There was no fanfare in the newspaper when CFRC filled in its broadcasting schedule between 1.15 and 5.30 pm, on Monday, October 4, 1937. A short article appeared in The Whig that day, and that was all, probably because all of the new material came from the CBC network. Everything was shifted around and the first day's afternoon schedule looked like this:

12.00 The Boy and Girl Friend
12.15 (to be announced)
12.30 Canadian Press and local News Bulletin
12.45 The Ogilivies (Ogilvie Royal Chefs)
 1.00 As the Moments Fly
 1.15 The Charm School
 1.30 Rex Battle and his Orchestra
 2.00 London Calling
 3.45 The Modern Home
 4.00 Piano Musings
 4.15 Gypsy Violin
 4.30 The Belgrade Stakes
 4.45 Fred Bass Presides
 5.00 Melodious Moments
 5.15 (to be announced)
 5.30 The Story Teller's House
 5.45 Military Parade
 6.00 Canadian Press News

The next Monday, CFRC filled in an hour in the morning with local productions. This time The Whig made an announcement:

> CFRC will commence its daily broadcasting schedule on Monday, October 11, at 8.00 a.m., E.S.T., and every week-day thereafter. The station will be on the air for one hour in the morning (from 8.00 to 9.00 a.m.). "The Canadian Press News" will be presented at 8.00 a.m., followed by "Morning Light", a sacred program at 8.15 a.m. This short devotional service will be in charge of members of the local Ministerial Association. From 8.30 to 9.00 a.m. there will be the program "Keeping an Eye on the Time."[101]

About this time, October 1937, John Donald Stewart left CFRC to join the Mathematics Department of Queen's University.

The first National Hockey League game presented on CFRC was the Toronto-New York Americans match from Maple Leaf Gardens on Saturday, November 6, 1937. The broadcasts started at 9.00 pm, around the beginning of the second period, in order to protect the box office receipts. CFRC carried Foster Hewitt's NHL broadcast from the CBC every Saturday night of the season after that.

Bert Pearl, a young CBC studio pianist, was assigned in June 1937 to hire some musicians to put together a cheery half hour summer replacement musical variety program for the local CBC Toronto station. He rounded up radio veteran Kathleen Stokes, organ, Blaine Mathe, violinist with the Toronto Symphony and Robert Farnon, trumpet. Hugh Bartlett was their announcer, and George Temple produced. The program was very good, and it soon went to the CBC network, where it stayed for about 30 years. Eddie Allen, the accordion-playing tenor, came along in 1938. The other familiar members of the Gang, Cliff McKay, Jimmy Namaro and Joe Niosi were added between 1943 and 1945, Lou Snider in 1947 and Bobby Gimby in 1951.

Blaine Mathe always started the show by knocking twice on his violin:

> Who's there?...It's the Happy Gang!...Wellllll, come
>   onnnnnnn in!
> Keep happy with the Happy Gang!
> Keep happy, start your day with a bang!
> A happy hello to you from the boys and Kay Stokes,
> We hope you like our music, our songs and our jokes -
>   yuk, yuk, yuk!
> Keep happy in the Happy Gang way,
> Keep healthy with Palmolive each day,
> 'Cause if you're happy, and healthy,
> The heck with being wealthy,
> So keep happy with the Happy Gang!

CFRC picked up the Happy Gang on Monday, November 22, 1937, and in those days it ran from 1.00 to 1.30, after The Ogilvies and before Rex Battle and his Orchestra.

The Whig published a detailed schedule of "Kingston CFRC Programs" every day. In February-March 1938, highlights were: Morning Light (devotional; CFRC; 8.15 am); Keep an Eye on the Time (musical; CFRC; 8.30 am); Big Sister (playlet; 11.30 am); Home Folks' Frolic (musical; 11.45 am); The Boy and Girl Friend (musical; 12.00 noon); The Ogilvies (musical; CBC, 12.45 pm); The Happy Gang (musical; CBC, 1.00 pm); The Backstage Wife (playlet; 1.30 pm); London Calling (BBC Empire transmission, 2.00 pm); The Hughes Reel (commentary; 4.30 pm); Treasure Island (5.15 pm); CFRC Variety Program (6.45 pm); Queen's Radio Program (talk; 7.00 pm); Major Bill (children; CBC; 7.15 pm); "Forsaking All Others", with Bette Davis on Radio Theatre (Monday, 9.00 pm); "Big Town" with Edgar G. Robinson (Tuesday, 8.00 pm); Al

Jolson Show (Tuesday, 8.30 pm); One Man's Family (Wednesday, 8.00 pm); "Musical Comedy Memories" from Hollywood with Eddie Cantor and Deanna Durbin (Wednesday, 8.30 pm); Rudy Vallee and his Connecticut Yankees (Thursday, 8.00 pm); CBC Dramatic Hour (Thursday, 9.00 pm); Hollywood Hotel (Friday, 9.00 pm).[102] The soap operas did not come along until the middle of that summer.

On Saturdays in 1938, CFRC carried the Metropolitan Opera at 2.00 pm, NHL hockey at 9.00 and the NBC Symphony at 10.30 pm. Sign-on Sundays was at 11.00 am, for a service from Grant Hall or Chalmers Church; Ken Soble's Amateurs from Toronto at 12.30; Radio City Music Hall of the Air at 1.00; Romance of Sacred Song at 2.00; the New York Philharmonic-Symphony at 3.00; Tudor Manor at 5.00; Dr H.L. Stewart's news commentary at 6.30; Jack Benny and Mary Livingston at 7.00; Canadian Mosaic (musical) at 7.30; Edgar Bergen and Charlie McCarthy at 8.00; and CBC Music Hour at 9.00 pm.

Behind-the-Scenes Personalities at CFRC

<u>Chuck Millar</u>: Alex McDonald was a student in E.E. and later on went with the CBC. He turned out many a good broadcast while he was at the controls of CFRC.[57]

<u>Alex McDonald</u>: I was Science '36, and I worked in CFRC from September 1936 to September 1938, when I left to join the CBC in Toronto as technician and was replaced by Chuck Millar. I was announcer-operator and split the technical duties with Don Stewart. We made station calls and Bulova spot announcements. At first CFRC was on the air sort of three times a day - in the morning for the news and some music, then we signed off, came back on at noon for the news, then signed off again. We returned around five o'clock for the local news and then joined the CBC network. In 1937 CFRC came on in the afternoons with the Happy Gang and the soap operas, and continued on more or less through the supper hour till after the CBC National News.

The control console was an RCA OP-4, and the power supply was underneath. The OP-4 was a three-channel mixer with a master control. Behind the technician was an additional mixer that Prof. Stewart built up to increase the flexibility of the mixer, so that we could switch more microphones through. Also, by means of the auxiliary mixer, we could carry out auditions in the studios and still patch programs through to the transmitter. We were still pretty limited in switching. When I joined the CBC, they had a number of these OP-4's. We used to do our remotes around Toronto with them, and they were killers - they used three great big 135 volt "B" batteries, and they had room for a couple of mikes and cables. And then your amplifiers. You see, in Toronto they still had 25 cycle power - the lights used to flicker - and this was a problem. Oh, the weight of that equipment...that's why I'm so short.

When I joined the Queen's station, the power supply for the transmitter tubes - the plate voltages - was batteries. The batteries were down in the basement, so when we came in we had to go down and

Student operator Alexander J. McDonald (Elec. '36) at the 3-channel RCA OP-4 mixing board in the control room of CFRC, circa 1938. Note the large transcription turntables on either side of the OP-4 and the pile of 16" RCA Thesaurus electrical transcription records in the foreground. An extra switching panel is to his right and, on the table at the extreme right, is the telephone connecting the control room with the CBC line operator at the CNR Outer Station on Montreal Street. McDonald's right hand is on the master attenuator of the OP-4, and a knob on the table, just under his left hand, was for switching the output of each turntable into the OP-4 board.

-courtesy Queen's Archives

switch the batteries over from "charge" to switch the power up to the transmitter room, and then go up and put the transmitter on the air. Put the filaments on first of all, and when they warmed up then we had to switch on the low high-voltage and then the high high-voltage. At night, when we would take the transmitter off the air, we had to do it in reverse, step by step, and finish up by going downstairs to the lab to put the batteries on "charge." Sometimes you'd go into the lab and switch on the light and there'd be a fellow and his girl sitting on the windowsill. They had no place else to go. Harold Stewart re-built the high-voltage power supplies for the transmitter around '38, so he didn't have to worry about charging the batteries. He was also going to build a new transmitter.

When the CBC programming came in in the evening, I think we could arrange to have it patched in ahead of time, and it was a matter of fading out the studio channel and bringing in the CBC on another channel. The CBC had to go through the OP-4 console to our transmitter because we had to have the local announcements done from the same console. If there was a run-over on a local program, the technician or announcer/operator would break in, fade out the local show and "We now join the network of the Canadian Broadcasting Corporation for the program such and such." The telephone to the outside was in the transmitter room, which was behind me, so that its ringing wouldn't bother us. The telephone in the studio was to the network so that we could talk to the repeater attendant at the Kingston Outer Station. If we lost the program or they didn't put it on, we could speak to them - "Oh, breakdown you know, west of Denver." If it came from Hollywood, the problem was always west of Denver.

It seems to me that we played only 33.3 rpm RCA Thesaurus discs - our library was limited to music on these 16" discs. We weren't allowed to play commercial records in those days. We also used to get a fair number of 15-minute commercial programs on 16" discs. They'd be sent in to The Whig and Major Annand would bring them up to the studio. They'd have a script with them and the timing.

The turntables were on either side of the operator. You had to reach over and "slip" the disc on the mat while the table was running - you'd have your hand on the disc and then you'd let it go. There were weights on either side of the pick-up and it was counterbalanced to 2.5-3.0 ounces head pressure - terribly heavy. It used replaceable steel needles and we changed needles every time we played a record. You just made a chisel out of the needle after one play, particularly when you played a shellac disc, and if you played another record with it you'd ruin the damn thing. We were very limited as far as microphones were concerned. I think we had only about four microphones, one or two ribbons. We had an RCA 50A inductor, and another dynamic, and a Bruno velocity - a pretty old microphone.

A fellow by the name of George Young, from the Maritimes, did a trans-Canada series of open air sing-songs, and he did a program from Macdonald Park here in Kingston. I remember being in the studio operating that, and it was fed from our studios to the network through the CNR Outer Station.[58]

Prof. Stewart commented on the CFRC power supply in 1938: "Filament batteries are kept on floating charge during operating hours, and are trickle-charged when the station is shut down. The temperature-controlled crystal oscillator runs twenty-four hours a day, seven days a week, this being necessary to maintain constant transmitter frequency."[26]

Charles A. Millar was and still is a very active "ham" operator in Kingston. He was hired on at CFRC some time in 1937, probably replacing John D. Stewart, and was appointed operator on October 2, 1937.[103]

Chuck Millar: My hiring with Queen's came about because I had some sort of a family alliance with Harold Stewart, and he knew that I was interested in radio, and probably would want to go on and make a lifetime job out of it. So, on that basis, I was hired to help out with the operation of the Queen's station. It concerned primarily seeing that the records were ready for broadcasting and for filling in - in their proper sequence, and the right kind of music for the time of

Chuck Millar announce-operating a program at the microphone and RCA OP-4 board in the CFRC control room in Fleming Hall in the late 1930s. A 16" transcription disc is cued or playing on the far turntable. The photograph was taken before Prof. Stewart succeeded in getting a cradle-type telephone for CFRC, for a candlestick model is visible at the left of the OP-4. Millar's left hand is over the turntable output switching knob.

-courtesy Queen's Archives

day - receiving anybody that was going to do a broadcast, showing the facilities where they were to do their broadcasting and seeing that they had an adequate amount to do for the time that was allocated to them.

We were offered the tremendous sum of $80 a month. What it worked out to on an hourly basis is hard to say because our hours were from 8 o'clock in the morning until the station shut down at night. Anyway, it was pretty good in those days.

Another memory is our old friend Nicky Nicholson, the hockey player, delivering parcels of records from the company that supplied them to broadcast stations. We'd have to sign for Nicky and he'd leave us the records. The chief orchestra that we subscribed to with these records was directed by Rosario Bourdon, playing old standards. These were 16", with as many as 12 or 15 pieces of music recorded on the same disc. There used to be a program that we took from NBC called "Matinee," and the chap on that would come on and say that he would have the symphony-type program played by this certain orchestra...Ransom Myles Sherman. Boy, some of the things he got away with on that afternoon program would shake you today!

Once in a while the control staff was pressed into service to act as announcer, especially in off hours, and sometimes we got to the exalted position of reading the news. We had very wonderful tutors, because W.E. McNeill listened very, very attentively to the way the words were pronounced on local news broadcasts. We used to regularly get letters from his office telling us that we had mis-pronounced words in the news text. He would spell out with hyphens and underlines in red ink the way it was to be pronounced from now on. But, one day, a letter came over from his office in which the word that was mis-pronounced was mis-spelled. I underlined the word and circled it, put "mis-spelled" beside it and sent it back to his office. That was the end of those helpful notes from W.E. McNeill on news broadcasting. I wouldn't have the nerve to do it today.[57]

<u>Arthur L. Davies</u>: Dr W.E. McNeill, Vice-Principal and former distinguished member of the English Faculty, had a long association with the Queen's radio station and he performed a function for the station which is not performed for many radio stations now. He listened carefully to broadcasts, especially the news, and he kept a little notebook. If any announcer made a mistake in pronunciation or grammar, Dr McNeill noted it and brought it to the attention of the announcer. The CFRC announcers enjoyed, or endured, this special advantage.[17]

The Cradle Telephone Story

<u>Harold Stewart</u>: In the 1930s, when we were operating commercially, the Bell brought out their so-called cradle phone, a hand set that hung up on a bracket on the wall in front of you. Well, I heard about this. I said, "Dr McNeill, this is just what we need because the operator needs three hands to run this job, and with this telephone he can just manage to carry on." "No sir! That's a prestige item, and when the first one of those comes into this University, it'll go into the Principal's office. The next one will go in my office. After that we'll think about it." I said, "Look, it costs only 25 cents a month more. I'll

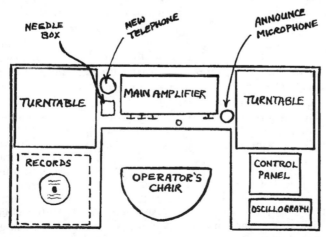

The operator's cockpit in the CFRC control room in Fleming Hall. This diagram was sent to Dr W.E. McNeill by Prof. Stewart on January 10, 1938, to illustrate his argument for the installation of a cradle telephone in the control room.
-from a drawing in Prof. Stewart's papers

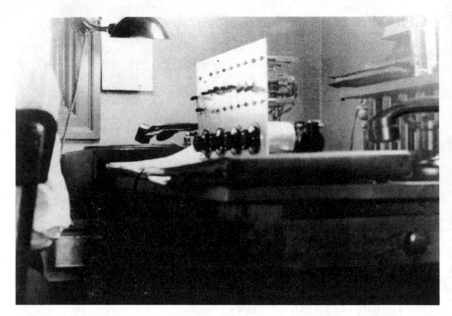

CFRC control room detail to the right of the operator, Fleming Hall, circa 1937.
To the right of the chair is the transcription turntable and its RCA type VX-
4210 tone arm (Req. No. 36959, May 8, 1936), the auxiliary switching board
and the telephone connected to the CNR Outer Station.
-courtesy CFRC Archives

pay for it." "Oh, no. We can't have that."...Billy McNeill said himself
that although I'm not a Queen's man born nor bred, yet, when I die,
there'll be a Queen's man dead. Doug Jemmett said that Billy McNeill
would commit murder for Queen's. He kept the University going in
hard times. I don't remember hearing of anyone being fired and he
didn't cut any salaries.[48,104]

The telephone incident is well documented. Prof. Stewart wrote to
Dr McNeill on January 6, 1938 to inform him that the only telephone
in CFRC was back in the transmitter room. This was an extension of the
stenographer's phone in the Fleming Hall library, and as such caused
problems for the station staff after office hours and in the summer. He
asked for a private line with a cradle phone for the CFRC control room
at a total cost of $8.05 ($3.95 + $4.10) a month.[105] Special circuitry
would prevent the phone ringing during control room announcements.
McNeill agreed to a private line for CFRC, but did not see the need
for a cradle model: "We have only one in the whole University and
that is in the Principal's Office."[106] Stewart countered with a diagram of
the operator's area in the control room and an explanation of how having a two-handed phone could result in overloading and signal distortion, since the operator had always to keep one hand on the master
volume control.[107] McNeill stuck to his principles and CFRC got a candlestick phone for $3.95 a month.

CFRC Productions and Scripts, 1938

<u>Verna Saunders-Kerrison</u>: I played the piano on CFRC - half an hour recital sometimes. I also sang a program or two, accompanying myself. They just telephoned and asked me - they knew me as a teacher, I presume. I did one of the first commercial programs for Abramsky and Sons. They paid me $5, and that was a big fee in those years. I was very proud. There was one little room you went in and you talked with the operator. Then they put you all alone in this other room behind a big glass, and they looked through at you and made motions, put up their hand when you were to start, and so on. You weren't to make any other noises or turn any pages, because that all sounded. They were particular about that sort of thing. They'd always say the piano was just tuned. I didn't have to do any trial run of it - we just went straight on.

I had between 90 and 100 pupils every week in both piano and singing and tried to promote my pupils more than myself - I always sent pupils around to sing at church affairs and had pupils perform on later programs: Jack Bearance on the piano, and vocal programs particularly by Pearl Johnson.[108]

The CBC released a plan at the beginning of February 1938, to broadcast a series originating in Canadian universities.[109] Queen's had not yet been scheduled but enthusiasm on campus ran high. The Committee for the Queen's Radio Program set February 17 for the inaugural program, to air over CFRC before a live audience in Grant Hall. If the talent unearthed were good enough and if the students gave sufficient support, the Queen's Radio Review might become a regular CFRC feature. A second program was scheduled for March 10.

Dorothy (Redeker) Wilson (Arts '40) kept a copy of her announcer's script for that half hour variety show. It featured the Glee Club, directed by Dr Harrison; Maurice Chepesiuk (Meds '43), football player, pianist, vocalist and accordion "flash" played a medley on his "stomach pump"; Bill Gummer (Arts '40), the Union piano pounder, beat out his swing specialties in the groove; the Nelson Eddy of the football field, Hugh Sampson (Mining '39), sang; the Coed Trio, Connie Deuel, Jean Merriam and Mary Uren swung on down with some current favourites, accompanied by Bill Gummer; the silver-toned baritone of the boxing ring, George Silver,

In 1938 the CBC planned to broadcast a series of programs originating in Canadian universities and, in order to unearth Queen's talent, "The Committee for the Queen's Radio Program" aired a live concert over CFRC from Grant Hall on February 17. Here the Co-ed Trio, Mary Uren, Connie Deuel and Jean Merriam, swing on down with the current favourite "Bei Mir Bist Du Schoen," accompanied by Bill Gummer.
-courtesy Dorothy (Redeker) Wilson

crooned; and Gerry Chernoff (Arts '38) of the Drama Guild whacked out the hilarious monologue "In a Shell Hole." Art Parmiter (Arts '39) and Roy Loken (Arts '42) produced.[110]

Although the test program had a number of rough edges, the stars of the first production were retained and more talent was expected for the second show. It did not materialize. Queen's Radio Review was never scheduled for the CBC, but McGill's "Red and White Review" was broadcast coast to coast on March 10.[111]

Florence Daly had sung over CFRC and the Rhythm Racketeers had a regular Saturday night quarter hour in 1937, but they became real stars of stage and radio after Ken Soble's Amateur Hour came to town on April 8, 1938.[112] Eighty people applied to the Chamber of Commerce to audition, and the preliminaries cut the number to 18 for the big show at KCVI auditorium, under the auspices of the Kingston Workers' Association. Judges W.J. Coyle, P.G. Otten and Tom Warner gave first place and an audition on the Ken Soble Amateur Hour in Toronto to Art Macdonald's Rhythm Racketeers. Flossie Cricket, who imitated Martha Raye, was second, and a comedy dance team from Carleton Place came third. Flossie Cricket is Florence (Daly) Courneya, of Kingston.

Florence Courneya: Ken Soble's Amateur Hour came to Kingston, and that was a big deal in those days. At this point I would have been 14 or 15 - I wasn't very old. It was suggested I enter the Amateur Hour, and I was chosen as one of the finalists. Bill McUen's group, the Rhythm Racketeers, was also chosen. After the contest, I was called over to CFRC and they talked to me about having a program. This was quite a thrill at that young age. So, every Sunday night I had a program, called "Songs Just for You," and it was sponsored by Crown Dairy. My theme song was "My Blue Heaven." The song that I did in the amateur contest, and the first song that I sang on CFRC, was "Pennies From Heaven." Tom Warner was there as the head of the announcing staff. Tom and Chuck were finished masters - both excellent.

I can see Major Annand quite clearly. He was quite a big man, rather stoutish, but I do not recall that he did any

A broadcast in CFRC's big Studio "A," circa 1938. The announcer is Charles Millar, the vocalist is "Flossie Cricket" and Kathleen Grimshaw is at the CFRC Knabe. The microphone is an Amperite SR80n studio model, ordered on January 8, 1936.
-courtesy Mrs F. Courneya

announcing. He was in and out and around and about, but when I was there he wasn't around a lot. Of course, I was there only on Sunday nights for my broadcasts. Chuck was ideal for a person who was inclined to be nervous, as I was at the time, because he could put you at ease and his nice, slow, easy manner was perfect. He had a good sense of humour, and if he thought you were nervous, he'd always have some little thing to throw in that would just sort of settle you. He'd give you a little wink on the side and say, "Get to it. You know you can do it." I found Tom the same way - he had a little deeper, different voice - but Chuck was the better. I thought an awful lot of Chuck. They were very good to me, just like a little family group when we did our broadcasting. Kay Grimshaw and George McAuley were the back-up people. It was a very nice, easy station to be in. No hassles. When CKWS started it was so much larger and had lost the nice little friendly appeal that you had at CFRC.

We chose our own programs. I followed the Lucky Strike Hit Parade, and took much of my programming from the hit parade pieces and of course from anyone I liked to hear sing, like Bing Crosby. Between Kay Grimshaw and myself, we selected the pieces we wanted. Then Kay would phone them in to Tom or Chuck, and they would work intros around them which were kind of cute. They didn't just announce the number. They created a little story which brought in the intro of the number and it was quite well done, really.

Tom Warner was the director in the studio, and I don't know whether Tom wrote up all the introductions, but it would all be ready when I got there about 15 minutes before air time. All I would do is have a copy of the songs I would sing, and they would have all the little in-between passages. It usually started out with the theme and two songs, and then Kay and George would play a piece in between - George was excellent on the banjo - and it would end up with probably two more songs. Sometimes Chuck would tell a little story and throw in the advertising for Crown Dairy and we'd end up with "See you next Sunday night," and that was about it. It was very well received. I don't recall that there was any other program as steady as mine. And it was sponsored, but I didn't receive any pay for it. It was strictly on a voluntary basis. I suppose they thought that you should be happy to even be asked. It was a good experience, and I wouldn't have gone on to the things I had later if I hadn't done that.

I also believe that Jack Elder had something to do with broadcasting at CFRC. He was very widely known as a Scottish singer. Oh, he was just tremendous, a great singer of Scottish songs. And Phonse McCue was just a marvellous singer. We did a couple of duets here and there, and he had a great voice. I went on until CKWS came into being, and then I went there as a steady program and did singing, women's interviewing, worked in the library and different things.[113]

<u>Art Macdonald:</u> In the late 1930s the Rhythm Racketeers, our small group of amateur musicians, won a contest and the first prize was that we were given a weekly show on CFRC. We carried that on for about two years, 15 minutes every Saturday night, MC'd by Tom Warner, and he was a very great help to us. At first the group was primarily rhythm - very little melody involved - and we played mostly fairly fast numbers. The kind that would get your toes tapping. Later on,

with a vocalist, we moved into slower and ballad tunes. There was Vin Frasso, guitar, Bill McUen on banjo, my washboard, Allie Atkins on accordion, plus Audrey as vocalist. After we played the first couple of shows, Tom Warner suggested that the sound was simply too thin for transmission, and that we needed a heavier bass. He suggested a piano, so we found Brian Brick, a well-known musician around Kingston, who then became our pianist (and a very good one).

Tom took a great interest in us. The first time we ever met him, and Kingston knew him, was when he MC'd a band concert out at Macdonald Park, and he made quite an impact there. From then he became very active in the entertainment life of Kingston. If there was an MC needed, or any sort of public presentation to be made, it was Tom Warner. I don't know how many hours he spent per day on the air, but it was a lot, and he seemed to do most everything else at the station.

We just went into a building at Queen's University - it didn't have any signs outside saying that it was a radio station. Somebody told us where the door was and up the stairs we went. There seemed to be only about two rooms, the very small studio where we performed and the control room with a small window where you could see who-ever handled the technical side of things. That's just about all there was to it. Life at the station in those days was very informal. We'd watch the clock and at the appointed time a light would go on and somebody would point a finger at us - either Tom Warner or the lad behind the glass - and off we would go. We were never paid, and since it was a Saturday afternoon and most everyone worked and we didn't have cars, we used to walk. I'd carry my washboard gadget. We'd walk along Barrie Street with our instruments, play the show, turn around and walk all the way home again and wonder how we did.[114]

Audrey (McUen) Macdonald: When we were broadcasting, my husband, now, introduced me after they played a number, and he said, "Now, from the sublime to the ridiculous." And, despite that, we got married.[115]

Art Macdonald: One of the things that stemmed from our appearance on CFRC was the opportunity to take part in another amateur show, which led to an appearance on the Ken Soble Amateur Hour, then being nationally broadcast from Toronto. It was a write-in vote. We won for that week, and were asked to appear in the finals. We placed second, and the winner at that time was Jack Kane, a clarinetist who later became a band leader.[114]

The Rhythm Racketeers returned to CFRC, victorious from Ken Soble's show, on The Variety Programme, July 3, 1938.[116]

CFRC Variety Program Scripts (March 1941)

Good evening ladies and gentlemen. I hope you were expecting us, because we've come to stay for the next 15 minutes. Of course, we won't be idle because Florence Daly, that's our featured singer tonite, has a couple of brace of songs, and George McAuley the banjo player is slated for solo along about the half way mark, and no matter what's going on, Kathleen Grimshaw will be busy trying to make the piano accompaniments fit. So it does shape up like a fairly large evening.

Coming events surely do cast their shadows before them, because Monday is the day of days for the Irish, and our program tonite is nearly 100 percent in their honor.

Back to the vocal portion of the program now as Florence tells the musical story of the Daughter of Mother McCree.

Its always with a little bit of reluctance, that we say goodbye, but the hope of the cast tonight is, that having been with you for the past 15 minutes, we might have brought back a few fond memories of that storied island of Eire. Florence Daly sang for you, George McAuley was heard on the banjo, and Kathleen Grimshaw guided the whole affair from the keyboard and it originated in the university studios of the station.[117]

The neighbors are raving/ The men on the street/ Turn in their tracks and beat a retreat/ The reason, its simple, its easy to see/ Geo McAuley all ready to play "Sympathy."

Always a little visionary/ And now, with a wave of the hand/ Florence really does a nice job/ on "Make Believe Island."

Well, we have to move along now to make room for more visitors to your fireside, so we will bow ourselves out, by saying that Florence Daly, Geo McAuley, Kathleen Grimshaw and myself all appreciated the hospitality of your house and we'll invite ourselves back again soon. This program originated in our University studios.[117]

The late 1930's was a boom time for soap operas on commercial radio, and CFRC carried a bunch from the mid summer of 1938. Weekday afternoons you could tune in at 1510 on your dial at 3 o'clock and hear a solid hour of sentimental, super-serious, slippery, sudsy soaps: The Story of Mary Marlin, Ma Perkins, Pepper Young's Family and The Guiding Light. By the next year CFRC had The Road of Life, The Story of Dr Susan, The Story of Mary Marlin and One Man's Family, but spread out over the day. Some members of the Queen's faculty felt that broadcasting soap operas was out of keeping with the high tone of this august educational institution.

Arthur L. Davies: I recall a story which was told to me by Col. Cam Jones, who was a student at Queen's during the time that the University and The Whig-Standard were doing commercial broadcasting. I believe the year was 1937. Cam Jones was running to a class at Fleming Hall, and as he dashed up the stairs he found Prof. Sandy Macphail standing outside the door of the studio of CFRC, apparently listening to a program that was going on. Prof. Macphail seized him by the arm and held him firmly, and said, "Jones, listen to that! It's like having a whorehouse in a church."[17]

By 1939, CFRC's local production was being supplanted by material from the CBC, including the popular programs from the American networks: Don McNeil's Breakfast Club (from Chicago; 9.05 am weekdays); Fibber McGee and Molly (Tuesdays at 9.30 pm); The Aldrich Family (Thursdays at 9.00 pm) and Vic and Sade (Fridays at 4.30 pm). The CBC supplied Don Messer and his New Brunswick Lumberjacks (Saturdays at 6.00 and later 8.00 pm) and the Ontario Farm Broadcast (week-days at 1.30 pm).

<u>Charles A. Millar</u>: The broadcasting was broken up in blocks. We would do an hour around 8 am with a news broadcast and fill it out with music until 9 with requests. As time went on, the hours were filled in between 8 and noon to start with, then in the afternoons the soap operas made their appearance and eventually the whole day became filled. We were concerned mostly with things of interest to the authorities at Queen's who held the franchise for the station. They had us broadcast things that maybe weren't of general interest, but certainly were of interest to the people on the staff at Queen's. Convocation, for one thing. About once a year the theologians came to town for meetings in Convocation Hall and parts were broadcast, especially if a nationally known speaker were on the agenda. We did one broadcast from a new house - when a builder develops a certain

The first CBC Radio mobile unit parked outside the entrance to CFRC, circa 1937. Alex McDonald remembers that this mobile unit had a transmitter in it when he joined the CBC in 1938.
-photograph by E.V. Lord, Picton
-gift of Bill Fitsell, Kingston Whig-Standard

area, he'll build a model house and invite the general public. I know we did one of these from a house up on Hill Street, which at that time seemed to be pretty well out in the sticks.[57]

A CFRC Recording in Convocation Hall

Jack Telgmann treasured a recording of his dance orchestra, relayed by CFRC from Convocation Hall and cut onto a 16" acetate disc at CBO, Ottawa, on March 23, 1938.

Jack Telgmann: Back in the '30s I had a dance band here. We did the odd dance in Grant Hall, and CFRC would broadcast the programs. We played the popular pieces of the day and a lot of classical music in dance rhythm. My father thought that was a terrible thing to do, but it did give a lot of people an appreciation for the classics even though it was in a popular style. Prior to '36 there was a very loud orchestra playing at the La Salle Hotel, so I decided to do something entirely different. I cut out all the brass and used two violins (Jim Reading and his brother-in-law who is now Dr Amodeo), piano-accordion, three saxophones, a very quiet rhythm section and it caught on right away because it was so different and so quiet. Then we got the contract at the La Salle Hotel. After two years a lot of other people did the same thing, so I had to change again, at the start of the swing era. I switched over completely, cut out all of the nice soft music and got into the big band stuff - real deep swing. I had four brass, four reeds (saxophones) and four rhythm, a violin and two girl singers. I did all of the arranging and just stood up in front with the stick.

The recording relayed from Convocation Hall in 1938 was made as a try-out for the Bayer Aspirin Program, which would have gone coast-to-coast. In those days, most programs emanated from Toronto or Montreal, which meant that almost every program sounded like the one just before because, of course, they were all the same musicians. So Bayer got the idea of moving it out of Toronto or Montreal. I imported musicians to Kingston, at quite some expense, and started this dance band here so that we had the musicians here and could employ them. This was a build-up for this program which we hoped to get. The audition had to be recorded. There was no studio or recording facility in Kingston. So we rigged up a studio on the stage of Convocation Hall, pulled the curtains, hung heavy blankets where the backdrops were, to deaden the sound, and then we sent it from there over the Bell line to Ottawa and it was recorded at CBO on a huge record. It lost a lot of fidelity over the line. In 1939 Bayer finally got around to making a decision on the program and we got the contract. But by that time the War had started and my orchestra disbanded, so there was nothing we could do about that contract. I'm quite proud of the arranging that I did for the program and I enjoy listening to it every once in a while. It's still quite modern - all the big bands coming back now are playing much the same type of music.[118]

Bill Neville: CFRC broadcast Jack Telgmann's dance music live from Grant Hall, and on at least one occasion I was the announcer. The Telgmann orchestra was certainly superior to the other groups in the Kingston area. When Jack felt that his ensemble had reached a high standard of performance, he decided to make a recording of a proposed half hour show to submit to the CBC. Most CBC programs had to have catchy alliterative titles, so Jack came up with "Tempos by Telgmann." We did the recording in Convocation Hall, with Alex McDonald as recording engineer. I remember the heavy drawn stage curtains and the dim light by which to read my lines. The musicians had lights on their stands. About 17 pieces plus two female vocalists plus me as announcer - all jammed into this tiny space. The piano was a problem, but the real culprit was the percussionist. Rather than the usual set of traps, Telgmann's man had a set of tympani plus chimes and other symphonic novelties, and he needed a lot of room. After much shuffling and sorting, Jack arranged us in an almost full circle around the mike with him, as conductor, in the opening. The two girls and I were in the centre, with very little elbow room. We did a little routine as each one needed to get close to the mike. I had barely enough standing room to do my part and move out of the way for the singers, who used the same mike. It was hot, loud and pulsating. Anne Cavin was petite and effervescent and gave me a bad time during the show, playing on my nervousness by making faces and threatening to tickle me just when my turn at the mike was approaching.

In July 1939, when I was attending summer school at NYU, I heard Tommy Dorsey at the Pennsylvania Hotel. He was great, but I thought at the time that there wasn't all that much difference between his band and Jack Telgmann's. Unlike Alex, I never pursued my interest in broadcasting. I came to Winnipeg in 1938 and spent many years in advertising, where I had an indirect connection, first as a copywriter and later as an advertising manager, with radio and TV.[69]

The Roosevelt Convocation

It was a great day for the University when the President of the United States arrived to receive an honorary degree from Queen's, to deliver an important foreign policy statement and to open the new international bridge at Ivy Lea. His private train from Niagara Falls pulled in to the CNR Outer Station on Montreal Street at 10 am on August 18, 1938. There he met Lieut.-Gov. Matthews, the Rt Hon W.L. Mackenzie King, Principal Wallace, the Rector Norman McL. Rogers and other dignitaries, and the PWOR Band played. The gates of Richardson stadium swung closed at 10.30, and the procession to the Stadium began at 10.45. Thousands lined the route, as the cavalcade glided down Montreal Street to Brock, along Sydenham to Union to Alfred, and into the stadium at 11.00 sharp.

Dr Wallace presented President Roosevelt to Chancellor J.A. Richardson for laureation. Dr McNeill placed a black silk hood, bordered with blue, on the President's shoulders, and the Chancellor presented Dr Roosevelt to the audience.[119]

Verna Saunders-Kerrison: My mother and I saw him in the George Richardson Stadium - you had to have reserved tickets to go in. It was a lovely day and there was a huge crowd there. They brought him in, someone supporting him on either side. He was very proud of the way he could get around. He didn't want people to think he was so badly crippled. He managed to stand up while he was speaking, with someone holding each arm.[108]

__Charles Millar__: This was a big day for the University, the first time that any privately owned radio station was able to feed every network in the US and Canada simultaneously. And you can well imagine that there was a nervous operator at the switchboard that day. That was me! My finger was kind of nervous when the time was coming up. I watched the clock and knew I had to punch in on time, and that every major network on the whole continent was listening. And I could imagine all kinds of squeaks and howls and everything else. But it went very, very well indeed - I think faultlessly. So, that was quite a feather in the cap of CFRC to have been able to do that on such a big occasion.

Of course, tremendous plans had to be laid because here was one of the first men in the world, more or less a house guest at Queen's University, and they had to make special provision for him because he was lame. They had quite a discussion on a new lectern - quite an expenditure. They got a special one. They also had to make some other arrangements for him so that he would not be in discomfort during the time of the broadcast. Steps were a trouble to him so they built ramps in all the locations where he would be.

He was to be presented with a degree and he had to have the proper attire, including a gown and mortarboard. The most humorous part of it all - and this is purely hearsay, but I think it's close to the truth - is that quite a hurrah was raised because he put it in his suitcase and took it home. They didn't know how to get the gown or the mortarboard back.[57]

Special convocation at Richardson Stadium, August 18, 1938. From left: President F.D. Roosevelt, Col. Watson, Rev Dr J.R. Watts, Principal Wallace, Chancellor J.A. Richardson.
-courtesy A.E. O'Kane, A.B., and Queen's Archives

Alex McDonald: I did the Roosevelt broadcast for the CBC from the centre of the stadium - it was outdoors and the stage was set up in the playing field. I was the technician and I've forgotten whether the University or the CBC provided the mixer and the microphones. It must have been CBC, but I did the mixing. A fellow by the name of Bill Little, who had graduated from Queen's about 1930, was the Regional Engineer of the CBC for Ontario. Monty Wherry was the Chief Operator in the Toronto studios, and they supervised and I did the mixing. That summer I was still with CFRC, and it was a month or so after that I did go to CBC.

Canadians were very surprised to see how crippled and handicapped Roosevelt was. He could hardly move. I remember him saying in his broadcast that, if Canada were attacked, the United States would not stand idly by. Afterwards, he and Mackenzie King had lunch at Ban Righ Hall and they went from there to open the Ivy Lea Bridge. I was at Ivy Lea, but the CBC did it.[58]

Harold Stewart: I recall the special convocation put on in the stadium back in 1938 for President Roosevelt, and he gave quite a stirring address to the assembled crowd. In the course of his remarks he referred to principalities and powers - Biblical quotation - and he said, "Well now, we know that Queen's University is not a principality, but it certainly is a power." Big cheers from the crowd. And a curious outcome of the convocation is the story I think I heard from Herb Hamilton, that a special gown was given to the President for the occasion, that he was to wear, and he took it for granted that he could take this home with him, free. And I understand he was quite annoyed when he was asked to pay for it.

I think we put on a broadcast for the opening of the Ivy Lea Bridge, but I can't recall anything special about it. There were some dry speeches, nothing very stirring. I think Mackenzie King was there and Roosevelt. The usual platitudes about hands across the border and stuff.[48]

Kenneth McIntyre: While I was on vacation during August 1938, the Ivy Lea Bridge was opened by the then Prime Minister Mackenzie King and President Roosevelt, and after the ceremony at the Bridge, President Roosevelt came to Kingston. Shetler, the man that relieved me, noticed the precautions that they were taking at the railway station in Kingston. There were men with rifles on the top of the water tank, which was just behind the station in the yard where the cars were parked.[21]

Lareta Lamoureux: When Roosevelt was here, going down to open the International Bridge, where you're going across the bridge to Barriefield they had a stand there for dignitaries as he was driving by. The photographer was on the stand, and I was waiting for the cavalcade across the road with Bernie, and Bob in the stroller. The photographer snapped the picture and later I happened to be walking down Princess Street past Meyers Studio, looked in the window and there's the President and there's me in the background, with my hand up waving at him. Just as plain... Yes, I saw him. My picture was taken with him!...We should have bought that picture, because it would have been something. It cost a dollar, but we didn't have no money then.[38]

The Radio Committee

The Committee reviewed CFRC's quarter hour educational programming for the 1938-39 academic year in April 1939.[120] They had arranged 110 presentations by 68 staff members, 2 students and 40 outside speakers and musicians, including four special series of lectures. There had also been 25 remote broadcasts, among them three convocations, four Sunday evening services, six University Sunday services, a Summer School concert, the Armistice Day service and a number of speeches.

Mr D.F. Toppin proposed to the Committee a new program for CFRC, The Community Forum, a weekly 15- or 30-minute transmission on timely topics, "To provide a local medium for the exchange of ideas and services which might contribute to the betterment of the community", to "create a greater interest in basic (national and international) problems" and "to contribute substantially to general education and essentially to an intellectual democracy." The topic would be announced a week in advance, and the format would be a brief talk, interview or statement, followed by a question period using listeners' letters. Listeners with ideas might be interviewed. The program could also present qualified young people seeking employment, or underprivileged people needing assistance. It could review current books, suggest further reading and announce important local meetings. The Community Forum never did go to air.

Convocation Broadcasts

Charles Millar: On a broadcast you've got to have something going all the time - somebody talking or a commentator to fill in. In Convocation, somebody stumbles up the stairs and there's blank. Just blank. Well, what do you do? So we went to Principal Wallace and asked, not to change the program, but for more continuity to it, like a master of ceremonies. Eventually they came to see our point of view, that we had to start on time, run things on time and be finished on time. They gave us a broadcast booth in Grant Hall, where we could have a commentator that knew the convocation service and could tell us "The next voice you'll hear will be so and so, having received their degree..." The very fact that we got a partitioned-off place in Grant Hall was something, because that was sacred ground! Geez, one time we went in there, the pillars were all painted fire red - I think this was the doing of Mrs Angus. Oh, geez, that was something! It was the dullest place in the world - you couldn't see in there when you first went in. Now bright red pillars hit you right between the eyes and there was gold leaf all over the place. Later, the Anguses tried their best to make gentlemen out of the staff at CFRC and there was bugger-all chance of that. They wanted us to speak with an accent - right off the farm! Brush off our boots going in in the morning, that was all we gave to their idea of what we should be.[57]

A Confession

Charles Millar: We had some famous characters associated with the station. A lot of listeners remember our Manager, Major Annand, with his famous saying of "As you were" when he made a mistake. But of course you need to have some background on this. In the early days the news broadcasts were done locally and he was the one that did them, usually at 8 am, noon and 6 pm. More of the public would be listening around news time so they would put in more than one spot announcement. We operators had to assemble Major Annand's newscasts for him. The news came in over a ticker tape, in printed form, and you had to tear off strips and paste them on paper in the right sequence. And the Major would come in and very quickly look it over. Sometimes he'd look it over, sometimes he wasn't there in time. He'd come in at the last moment and we'd have the news on his desk for him in front of the microphone, ready for him to read.

But, as time went on, we got some rather sadistic ideas about humour. We could fix this fellow if we didn't put the sentences in the right sequence. So, this we used to do with quite some regularity, and he would get into the worst boxes, because there wouldn't be any cohesion between the end of the sentence where he stopped and the next sentence where he was to start again. He'd run into this and then he would say "As you were." And it became quite a general saying. We did this for quite a while before he became aware of it. Then, of course, he came a little earlier and read over the news and was able to correct these roadblocks, and the fun stopped.

I can see him yet, reading our paste-up and glancing ahead before he said it, and he'd see the next line wasn't going to fit...We wouldn't last very long today, I don't think. He and Tom seemed to share the announcing, but I don't remember Tom ever doing newscasts. Major Annand seemed to think that was the big thing and he did every one. Once in a while he'd get caught some place so one of us would have to do it.

He was quite a colourful character and he introduced a lot of local thought into the broadcasting, which sometimes met with enthusiasm by the University group. But not always. He was a great man for Gilbert and Sullivan and music hall variety. It didn't go so well with the people from the colonies out here. They weren't always in love with what he wanted to broadcast, but he was the major force and we had to listen and do what we were told.

Major Annand used to tell us that he had friends in the musical world - all over the place. Being young in mind we had all kinds of ideas of counts from Italy and France. We had good friends in the CPR telegraph office, and we'd get telegraph forms and type out fictitious messages and leave them on Major Annand's desk. That somebody was going to be passing through Kingston and they'd like to see him, and we'd sign Count Somebody's name. He'd read them, go into raptures over this, and come and show them to us.

Of course the equipment wasn't always that reliable either. Once in a while you'd have to put the odd piece of stick under a relay that wanted to fall out and put you off the air. It added a nice personal touch to it anyway.[57]

Bruce Riggs: My father was Minister at First Baptist Church in Kingston, 1939-1941. One morning we were sitting around the breakfast table listening to the news on CFRC. Then the announcer said that Rev Earl H. Riggs would be taking the Meditation of the Day shortly, but he had not yet arrived at the station. Hearing this, my father exclaimed, "I had forgotten all about having to take this morning's meditation!" Immediately he was out the door, and shortly we heard him on CFRC after a very hurried bicycle ride - we didn't have a car - from the corner of Sydenham and Johnson to the Queen's campus. I do not expect the other listeners could tell, but both Mom and myself knew that he was very much out of puff.[121]

The Mark VII Transmitter

In early 1939, Harold Stewart was concerned that his old Mark VI was on the air over 15 hours a day, allowing little time for maintenance or modification of the transmitter and even less for use in teaching the two radio courses or for research and experimentation. None of that could be done before the 11.15 pm sign-off. The Mark VI could break down at any time, and that might cause a sponsor to cancel, with consequent loss of revenue. He felt responsible to prevent such an occurence, and besides, going to higher power would require new equipment. He therefore recommended to Dr McNeill the construction of a new transmitter, for operation at 100 to 250 watts, to cost about $1,000. The commercial equivalent would cost $4,500 to $6,900. They also needed to spend $560 for a new frequency monitor to replace the home-

Prof. Harold Stewart with the Mark VI transmitter in May 1938. The Mark VI was of the same electrical design as the Mark V built in 1930-31. The point at which the antenna feed wire went through the wall, the plate at top centre, is still discernable in the transmitter room as a slight depression in the wall.
-courtesy Queen's Archives

*Power level indicator, type 586-B, serial #780, manufactured by the
General Radio Company, Cambridge, Mass. This was listed in Prof.
Stewart's inventory as #291, and is part of the Mark VI, at right, just below
Prof. Stewart's right elbow. (previous photo)*
-photograph by Arthur Zimmerman

built one of 1933. The old monitor had proven inadequate recently - the
DOT had caught them 60 cycles off frequency because of a faulty crystal and oven.[122]

Harold Stewart: The first CFRC transmitters involved modulation at
low power levels - about 5 watts - and then amplification of the resul-
tant modulated wave. In 1939 we went for a completely new high
level type modulation where you modulated the final output amplifier. Modulation at high level, at the output stage directly. We used
that new one, that I built with the help of people around, until CKLC
gave us their used RCA transmitter in 1961. We hoped to get a licence
to use it at 1,000 watts, but actually we got only 250. When the commercial business finished the students could operate on that one and
they put negative feedback on it for one thing.[123]

Harry W. Beardsell: I had been a radio amateur before coming to the
station and I obtained my Commercial Radio Operator's licence and
then my first job was with CFRC. I am still very active in amateur
radio and have many contacts with Chuck Millar on the air. I operated
the control room and also did announcing for recorded or network-
fed programs. Cecil Richards, Chuck and I were then full-time operators and Gladstone Fiddes was on a part-time basis.

   Harold Stewart was the Chief Engineer and designed and supervised the building of the new transmitter, which was of the latest technology at the time and therefore more efficient than the old one. We
all had helped in building the new transmitter - Cecil Richards,
Gladstone Fiddes, Chuck Millar, Harold Stewart and myself. This is
the reason that photo was taken. It was put into use and was still
there when I left in April 1940.[124]

   The new Mark VII employed a Bliley "A" cut crystal oscillator,
type 802 oscillator tube, type 802 first buffer tube and type 805 second
buffer tube, high level plate modulation, with two 805's in the last
audio stage, two 805's in the modulated stage, 125 watts each, and
used the AC mains as the primary source of power, with storage batteries
and a motor generator as emergency power supply. Both low and high
power plates were supplied by rectified AC.

   CFRC ordinarily used 5 db of compression, and transmitter distortion at 100% modulation was 2.9% at 400 cycles. The frequency

This photograph was taken in the CFRC control room, probably in the winter of 1938-39, just after the CFRC crew had completed building the new one kilowatt Mark VII transmitter. Because of its location in a populated area, the Mark VII was put into service in 1939 at 100 watts.
From left: Cecil D. Richards (operator), Harry W. Beardsell (operator), Charles A. Millar (announcer), Gladstone Fiddes (operator; Meds '40) and Prof. Harold Stewart (Elec. '26).

-courtesy Harry Beardsell

response of the new transmitter was +/- 2 db , 40 to 10,000 cycles, and the transmitter noise level below 100% modulation was 65 db. They used a General Radio frequency monitor, a cathode-ray oscillograph as modulation monitor and a monitoring receiver as audio monitor, using a diode rectifier and A.F. amplifier. CFRC had a General Electric type K60 broadcast receiver and a National AGS-X high frequency receiver.

The numbers for antenna height were slightly different from before, and did not entirely jibe with what the diagram shows. CFRC had two home-made steel masts, painted according to DOT regulations. One was 60 feet above the roof of Ontario Hall and the other 70 feet above Fleming Hall, giving an antenna height of 120 feet. The masts were about 170 feet apart, and had no lights on them. The accompanying diagram shows that they were still using a counterpoise.[125]

The new 100-250 watt machine was put into service on Sunday, November 26, 1939, and had to be operated below capacity. The licenced power output, taken as 50% of the 200 watts plate input to the final stage, was 100 watts. Tests showed Mark VII to be superior in some respects to the equivalent commercial product, and it performed far better than Mark VI.

The 4th year E.E. class was assigned the project of re-designing

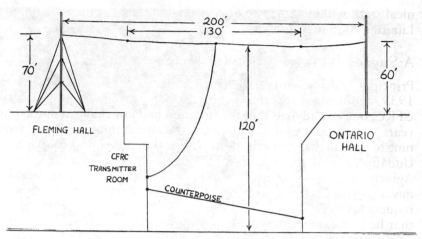

*Dimensions of the CFRC T-antenna, not drawn to scale, sent to the DOT by
Prof. Stewart in December 1940. The counterpoise was still in use.*
                                                        -Prof. Stewart's papers

and re-building the old Mark VI transmitter, which was kept in reserve
for emergency use.[126] The blueprint still exists for that modified Mark
VI, and Prof. Stewart explained: "It was built by a group of my students
back in 1940. The J.A.J. on the print is James A. Jarvis, who graduated in 1940 and went to the RCN and RN early in that year. The basic
transmitter is essentially the same as one I built in 1931 (Mark V), with
a few modifications."[127] The re-built Mark VI was also intended for use
in the radio course, in training men needed for the war effort. The DOT
specified that an Experimental Licence had to be secured for Mark VI,
except that, during the national emergency, the DOT was not issuing
these "unless the operation of radio apparatus is essential for the development of equipment or for the instruction of men for national service." In that case the licence would authorize the use of certain high frequencies and operation into a non-radiating dummy antenna load.[128]

Prof. Stewart and Harold S. Pollock developed a volume-limiting amplifier with negative feedback - compression with feedback - which prevented overmodulation and increased the effective power of CFRC by
about 100%: "It is well known, that if a portion of the output voltage
of an amplifier be fed to the input circuit in phase opposition to the
incoming signal, amplifier gain, and the distortion and noise, for a
given signal output, are reduced. By letting the signal voltage automatically control the percentage of feedback in the output or high level
stage of the amplifier, limiting action is secured with a lower noise level
and with a lower value of distortion than is possible in previously developed types."[129] It was able to handle all ordinary peaks smoothly, with
no thumps as the compression kicked in. They published a paper on their
amplifier after it went into service at CFRC. The idea was later developed by General Electric and the Columbia Broadcasting System tech-

nical staff, without credit to Stewart and Pollock, as the type BA-5-A Limiting Amplifier.[130]

A Successful Application for Increased Power

Principal Wallace wrote to General Manager Murray on March 15, 1939, again enquiring about the possibility of increased power for CFRC, but was advised that the situation had not changed in the past year: "Your application is brought to the attention of the Board from time to time so that it may be dealt with at the earliest possible date."[131] Undaunted, Wallace turned to C.D. Howe: "Now that the Havana Agreement has become effective, we should be very pleased to get permission to have our station increased from 100 Watts to 250 Watts if the request for 1000 Watts is still not possible."[132] The Minister replied that he had explored the situation thoroughly and concluded "that even with the Havana Agreement, an increase for your station is quite impossible. I regret that this is so because I appreciate quite fully the splendid work which your station is doing and were it not for the physical obstacles, the increase would be granted with little delay."[133] Harold Stewart did not understand the meaning of "physical obstacles."

Ten days later, Howe advised that he had learned that there was no technical objection to an increase to 250 watts on the present frequency, provided the US agreed. He suggested withdrawing the application for 1,000 watts and submitting a new one for 250 watts on 1510 kilocycles, which was to become 1490 when the Havana Agreement took effect in December. The application went in to Howe on May 6, 1939, "in the interests of education by radio and the wider distribution of C.B.C. programmes,"[134] and he forwarded it to the Joint Technical Committee, which reported to the Board of Governors. The Board would then advise the Minister of Transport, all in accordance with Section 24 part 1 of the Broadcasting Act.[135]

The decision was relayed on August 15. The Board of Governors had advised the Minister to approve an increase of power for CFRC: "This is subject to the relocation of your transmitter on a site not nearer than two miles from the centre of population of the city of Kingston, in order to avoid undue blanketing of the reception of outside stations on adjacent channels...The site, the height of the proposed towers and the make and type of the proposed transmitter must be approved by the Department before work is started on the new station."[136]

Dr Wallace protested that the current situation was little changed because the increases in power and radius of coverage were small. The University would prefer to implement the change on the present location for two reasons: "1. Our channel is at the high frequency end of the broadcast band and no stations of any importance received in Kingston would be unduly affected by increase in our power. 2. It is important for the training of our students in radio research that the equipment be very easily available to the staff and senior students. That would naturally not be the case if the greater part of the equipment had to be moved out of town."[137]

Mr Rush, Controller of Radio at DOT, was not sympathetic: "The regulations concerning the site of Broadcasting stations have been adopted in consultation with the Canadian Broadcasting Corporation and are being strictly adhered to."[138] Queen's would not move the transmitter, so the struggle for increased power and/or a lower frequency for CFRC was finally over and lost.

Prof. Stewart was still, however, enquiring about installing a vertical radiator on the roof of Fleming Hall in March 1940.[139] For some reason, in August 1940, Major Annand wrote to the CBC's Chief Engineer, Mr Olive, informing him that CFRC was prepared to move out of town and broadcast at 250 watts.[140] This was certainly not the case and the circumstances of the writing are not known.

CFRC at 1490 Kilocycles

The North American Regional Broadcasting Agreement went into effect at 3 am (EST), March 29, 1941. Sixty-nine of the 85 Canadian stations, including CFRC, had to change frequency to a new assigned channel. CFRC went to 1490 kilocycles or 201.2 metres,[141] and had to replace the transmitter crystal and the general radio monitor crystal.[142]

CFRC Personnel, 1939-40

Prof. Stewart wrote to Dr McNeill on August 18, 1939, to suggest a holiday and relief operating schedule for CRFC. He named 5 employees of the station, G. Fiddes, T. Warner, C.A. Millar, C. Richards and Major Annand, who was no longer on the air. There was also Mr Beardsell, who would assist with the operating.[143]

> **Charles Millar:** Gladstone Fiddes was another who used to operate the controls of CFRC. Some may remember that his father was the Minister of Princess Street United Church. Gladstone attended Queen's and had a part-time job in the radio station. One morning I came into work and saw this strange man in the transmitter room with a broom, sweeping up the floor. I didn't recognize him as being one of the operators. It turned out that it was Gladstone's father, Rev Mr Fiddes. So at least we had some class on the cleaning that day. But you never knew what to expect when you came in to work at CFRC in the morning.[57]

> **Rev Dr Victor H. Fiddes:** Gladstone was a radio amateur, and held the licence for VE3CX when we were living in Guelph from 1929-1933. I do not know the years exactly when Glad was active at CFRC, but he lived in Kingston from 1933 until graduating from Medicine in 1940...My niece recalled that her mother, Dorothy Wilson (Arts '37), while engaged to Glad, was teaching at a remote country school away north of Parham and feeling isolated. My brother used to close the night broadcast with a song which was of general appeal but directly addressed to her so she would know he was thinking of her in her isolation.[144]

<u>Charles Millar</u>: There was a chap that lived out in the country, on the Sydenham Road, and it was his job to get to the station about 7.30, quarter to eight in the morning to get the transmitter warmed up and to get the microphones and everything in place to start the day's broadcast. He very seldom made it on time and it usually was past 8, when the news broadcast was to be on, and somebody would have to fill in for him. He decided to get around this problem by moving into the city, right across the road from Fleming Hall on Union Street. This would get around his lateness because he could still stay in bed until the same time and get to work on time. But when he got on Union Street, his lateness was more evident than ever. That was Cecil Richards.[57]

<u>Harold Stewart</u>: Cecil Richards was a transmitter operator. He'd turn the transmitter on and make the odd station break. Cecil was quite a character. He was clever enough, but he was a lazy dog. He didn't like the idea of coming in in the morning early and turning the transmitter on a half hour before sign-on. We had to get the crystal warmed up well ahead of time that so its frequency would be within tolerance by the time we went on the air. And this didn't appeal to him. So he rigged up an alarm clock as a time switch to turn the thing on, and it would have been fine, only the thing failed one time when Cecil was supposed to have been on duty. I think W.E. McNeill was listening at that time and got him on the phone. So we all got in trouble.[48]

<u>Charles Millar</u>: Another operator at CFRC that went on in radio was Harry W. Beardsell, who later went with Canadian Pacific Airways and finished out his time as a navigational officer with CP on the far east run. He now lives in Vancouver and devises means for paraplegics to do common everyday chores, especially operating amateur radio equipment. Harry has been in the forefront of devising and manufacturing these appliances for them.

One of the other people associated with CFRC was Kathleen Grimshaw, a very prominent music teacher here in town. She was the cornerstone of all the local broadcasts, accompanying singers, and she was considered on the staff because she was there to fill in if it was necessary and play the Knabe grand. George McAuley played banjo in one of the local bands, and the day before he was sent overseas, he was married to our Kathleen. He was killed overseas (at Normandy).[145] Kathleen figured in everything that was musical around the station for years and years. She seemed to be able to fit in her playing with any kind of an artist that showed up.[57]

The Royal Tour

The Royal Couple pulled in to the Kingston CNR Outer Station on Montreal Street on the evening of May 20. CFRC broadcast the arrival "for the benefit of those unable to witness the Royal Progress through the city."[146] The cavalcade came down Alfred to Mack to Frontenac Street, around Victoria Park onto Brock, down Barrie to Johnson, onto Sydenham and past the Court House, down Barrie again to Stuart, up University to Union and along to Richardson Stadium. They wheeled once around the stadium, where 10,000 school children cheered their heads off, then along King Street to Market Square. In front of City Hall 2,500 cheered from special stands and a band played. The cavalcade proceeded down to Ontario Street, over the La Salle Causeway and to Fort Henry.

> **Lareta Lamoureux**: I saw Queen Elizabeth and King George, going up Montreal Street to the Outer Station. Sure did.[38]

> **Kenneth McIntyre**: In 1939, King George VI and Queen Elizabeth landed at Peggy's Cove, New Brunswick (sic.), and he gave his first speech outside of England. We (CN operators) had to use earphones. The King stuttered, but he did very well - three words, pause, three words, pause, and I heard a feminine voice saying, "Bertie, you're doing well"...oh, so faint.[21]

THE CFRC LOGS, 1938-1942

The development of the CFRC program schedule through 1938 and 1939 is contained in the daily listings published in The Whig-Standard. Of course, the newspaper listings tell little of the local personalities featured, make no reference to the CFRC announcer and operator, and tell nothing of commercials aired or of pre-emptions for special programs or news bulletins. For these details, one must go to the station log.

Naturally, CFRC broadcast all of the famous speeches by Prime Ministers Chamberlain and Churchill during the crises leading into the Second World War. These were relayed from the CBC network, and were registered in the CFRC log. The author examined the faint carbon copies of the 1938-42 logs in 1982, before they were given to the Queen's University Archives for safe-keeping, and was able to make out a good many of these historic entries. Queen's Archives, however, felt that the logs were largely illegible after April 29, 1937, and rather than return them to CFRC, destroyed them. Since many of the historic broadcasts in those times were unscheduled and were not included in the daily newspaper listings, the proof of their airing over CFRC is lost except to fragile memory.[147]

The Queen's Drama Guild and its famous President, H. Lorne Greene (Arts '37), probably appeared on CFRC during this period, but the surviving log is not helpful in confirming that. And, contrary to legend, the record does not show that Lorne Greene ever acted as an announcer on Queen's radio. He got his first professional voice training after he was picked out of a Drama Guild production by a talent agent from New York and was sent on a scholarship to the Neighborhood Playhouse. Lorne Greene is remembered as "The Voice of Doom" on the CBC news in the early years of the War, but Prof. Stewart pointed out that he was actually called "The Voice of Impending Doom."

Tom Warner Leaves for the War

> **Florence Courneya**: I recall very vividly the day that they had a little party for Tom when he went into the Army. The staff was all called in

and a few people connected with it, and they had this nice party and a cake for Tom going away. I think everybody fell in love with Tom Warner. He was handsome and he was really a nice person. He made you feel so at home. I don't think anyone was dry-eyed when we left that day because everyone hated to say good-bye to Tom.[113]

<u>Cuth Knowlton</u>: When wartime came, Tom joined up with the army, and I did too, of course, in 1939. I took an ammunition examiner's course at Halifax - one of twelve of us in Canada - and we were all sent overseas together. I was only over there a little while at the camp at Borden where there was thousands of men, and I ran into Tom. He was a lieutenant then. Tommy was going to go for captain, and he couldn't get it unless he had this ammunition course down. I gave my books and drawings of ammunition, shells and things to Tom. He did get his captaincy, but I never got my notes back. I used to meet him at the Officers' Mess at Borden when I was playing with my own band, before I went with Robert Farnon's AEF Band. There were three AEF bands in England, Glen Miller from the States, Robert Farnon from Canada and George Melachrino from Britain. Of course, we talked a lot about Kingston and CFRC.[51]

Bits of Verse by "Maple"

Miss Bessie M. Comer lived at 80 Division Street in a little house with "Maple Home" in gold letters on the transom over the front door. She always signed her verses "Maple" and some of her poetry was read over CFRC around 1939 and 1940 on the Shut-In Program. Once she was a guest on the morning program, Watching the Clock, and read her poem "Birds in Winter." For Tom Warner's birthday, October 28, 1939, Kay Grimshaw read the following tribute by Maple over CFRC:

"To Tom"

There was a young man by the name of "Tom"
Who was always so happy and gay.
We heard his voice over CFRC,
But now he has gone away.

To join the men of his Country and King
Who have taken a pledge to be true;
To uphold the right, and fight against wrong,
And ever their duty do.

But we think of him still, and "Best Wishes" we send,
And our prayers will follow him too,
And his friends of the air all want him to know
"To-day we're remembering you".[148]

The same day, Gladstone Fiddes presented Maple's tribute to

"The New M.C."

The new Master of Ceremonies, long known as "Chuck"
At the station of CFRC,

Is a very fine "Millar" - Yes (What?) didn't you know?
Well, that is his name, you see.

And every morning at half past eight
He's on the air with a song,
And a joke or two while "Watching the Clock"
As he sends your greetings along.

He's a funny old fellow, with a "chuck"le and grin,
Who is doing his bit each day
To bring the sunshine into your home(s),
And drive your cares away.

So here's to you, "Chuck", and our voices we raise
In singing the "toast" anew:
"For he's a jolly good fellow" - that's right!
And we all mean You.

Reports on CFRC for 1939 and 1940

As of December 31, 1938, CFRC still had a deficit of about $3,000 over the whole period from May 8, 1936, but the projection for 1939 was for a $1,000+ surplus, even after capital expenditures. The surplus would go into a depreciation reserve and toward technical improvements.[149]

The Vice-Principal and Treasurer, Dr McNeill, reported in May 1940, that the income from CFRC had been significantly increased in 1939, owing to advertising revenues received through the CBC. The total revenue was $11,114.96. Annual expenses were under $5,000, for maintenance and for the four Queen's employees, supervisor Prof. Stewart, announcer Charles Millar, and operators Cecil Richards and Ephraim Diamond. The University had been able to build up a reserve of $17,020.47, enough to build a new station outside the city. That was expected to add $3,000 to annual costs.[150]

A draft for an inspection report in December 1940, lists Cecil D. Richards, Allen Ephraim Diamond (Elec. '43), Richard Samuel Rettie (Physics '41) and Peter Theodore Demos (Physics '41) as the CFRC operators.[125] None had an operator's certificate.

Dr McNeill complained in writing to Prof. Stewart that he had tuned in CFRC at three minutes to eight on November 12, 1940, at 8.01 and again around 8.05 am "and then as there were no signs of life I went to Toronto for my news...I called up the station about nine o'clock and Mr Richards confessed to me that he was the guilty person. He said that he had never been late before and he quite understood that there must be no repetition of his failure to get the station on the air sharply on time." Then, on Saturday the operator left his post so that "the performers were obliged to run overtime in order that there might be something on the air." He also cited Mr Demos' or Mr Diamond's failure to get the Chalmers Church service on the air on time the previous Sunday. Three cases of carelessness in three days! "I wish that you would impress upon the staff that this sort of thing cannot be tolerated and there will be immediate dismissal for anyone who is at fault again. We cannot afford that our station should get a reputation for carelessness. We must be dependable to the second."[151]

Charles Millar: A.E. Diamond (Elec. '43; known as "Ace") came originally from Brockville, and as a student was a part-time operator of CFRC. He was recently Chairman of Cadillac-Fairview Corporation.[57]

A.E. Diamond worked at CFRC from the spring of 1940 until the commercial operation closed down in 1942, part-time during the school year and full-time during the summer. The money he earned got him through engineering at Queen's. Chuck Millar recalled that "Ace" used to bring salami sandwiches to work, and the CFRC staff made him keep them inside a record case because they couldn't stand the smell.

A.E. Diamond: I started at Queen's in September 1939. The War had just broken out, and I came up there with just enough money to pay for my tuition and books. So I had to get a job right away. I worked for a while in a restaurant waiting on tables and then I worked at DOK House - originally a frat house, then it became a boarding house. I waited on tables and got my food there and enough money to pay for my room. Around March or April 1940, I found out that the radio station needed an operator part time. So I was interviewed by the University and by Major Annand of The Whig, and I got a job in the summertime working full time, about 43 hours a week. I was paid $20 a week, with the understanding that when school started in the fall, that would be reduced to half-time. Cec Richards, who was the other major operator, would take up some of that time and the rest would be picked up by another student. So I had this very good arrangement where I had enough money coming in during the summer to pay for my room and board and tuition for the coming year, with a little help from my father, and during the school season I was able to get enough money to support myself.

When I told him that my name was Ephraim, Chuck says, "That's not a good name for you. From now on your name is Ace." He was the one that started that, and I used to get all kinds of names around there, like Ace of Diamonds. But that "Ace" never stuck very far.

The beauty of working at the radio station as an operator during the school year was that we were on the network a significant part of the time, particularly evenings, and I was able to do my homework and studying while on the network and still look after station breaks and any emergencies that might come up. From 7.00 o'clock on, the operator was the only person at the station. Almost every night at the end of the news, which normally finished between 11.15 and 11.30, there would be some type of announcement that came to us through The Whig-Standard. Either a funeral announcement or some current event or a lost child or something like that. And that announcement was usually the responsibility of the operator closing the station. I was the main student operator because I was there for the summer. In the summer there was just Cecil Richards and me.

I enjoyed the work very much, for a lot of reasons. First of all it was interesting - because of the people. Chuck Millar was a wonderful fellow with a great, expansive sense of humour, and a real good guy to work for. Very easy to live with and I really enjoyed his tactfulness. I assumed that he was my boss, although there were times when Major Annand felt he was my boss.

Major Annand was a bit of a fuss-pot - a very decent fellow, but very British in all his attitudes toward life. Very much the British aristocrat. I think I had been working there only a very short time, and he was giving the news. Pope Pius had been quite ill, but he kept referring to him as Pope Pee-us. There had been some calls to the station indicating he was making an error and Major Annand stopped in the middle of the news and gave a long explanation of why it was Pope Pee-us and not Pope Pie-us.

Dr McNeill, a former Professor of English, was a stickler on pronunciation, and he would call Major Annand occasionally, but Chuck Millar regularly, if there was any transgression. At that time the most popular of the soaps was Super Suds, but Chuck was never allowed to pronounce it as "Sooper Suds." He had to say "Siu-per Suds," and he used to practice "Siu-per Suds, Siu-per Suds, Siu-per Suds" to make sure that he didn't transgress and get into "Sooper Suds." We used to kid around a little bit about Dr McNeill, but he was a very, very imposing guy and we really were quite frightened of his phone calls. So, Chuck used to toe the line very, very much on pronunciation.

We had a very fine musical library. I loved music, and it wasn't unusual on a quiet evening during the summer, when I didn't have studies and there wasn't much doing, to have one or two of my friends come up and we would listen to music. I used to monitor the network program through our cathode-ray tube. So, we used to have some entertaining evenings playing music - with interruptions, of course, for the station breaks.

I used to live in Mrs Adams' boarding house at 273 Earl Street with a couple of other students. One of them was a jovial Irishman named Harry Timms, now a real estate broker in Toronto. We became pretty good friends, and I see him occasionally. He got a lot of joy out of life, and was up to the odd practical joke when it suited him, but at the same time, Harry was also a great womanizer. He had dates all over the place. He'd be with one young lady for a month or two and then shift on to somebody else and pursue her for a while. We used to kid him a lot - of course, he was very popular too, and was able to attract some of the nicest ladies around. I'd come off a bit of a joke from him and I wanted to get even, so one night after the news, I came on with "I have a special announcement to make. If Mr Harry Timms is listening in, would he please contact his wife immediately." I then signed off the station and started to think about what was going to happen as a result of this misdemeanor of mine. First of all, I wondered whether Dr McNeill or Chuck would ever find out what it meant. For some reason or other it was never questioned by them. My major concern was what would be Harry's reaction to this. Well, I walked around the house for half an hour, until the house became dark. Maybe ten after twelve I gradually let myself in. The place was dark. I gently walked up the stairs, went into my room and was about to close the door when, from behind, two very strong hands seized me in a very strong grip, picked me up and threw me out the window. Now, the window of my room looked out over the sloping porch roof, and it was kind of slippery. So, there I was, hanging on by my fingernails to keep from falling and Harry was inside the room laughing like hell. He wouldn't let me in for 10 or 15 minutes. We had a good time laughing about it. There were no other repercussions, and he was able to convince his girlfriend that it was just a big practical joke.

My lectures were mornings and, the occasional afternoons when I didn't have a lab, I would be operating at the radio station. The Battle of Britain took place and it wasn't unusual for a number of people to crowd around us in the operating room to listen to the news about what was happening. We were the place on campus where that information could be obtained immediately through the news. If any student was in Fleming Hall, they would just drop in at the radio station to pick up the news program.

They used to have this morning program at 9 o'clock, with Kay Grimshaw and Charlie. It was like a little Happy Gang show. They would read local things, Chuck would wisecrack and Kay would sing. They once got me in to sing with them, and it wasn't very successful. They never asked me again.

It was a well-engineered station. In those days to have a piece of electronic equipment work without interruption day in and day out was quite a feat. I can't recall there ever being a shutdown except once or twice when we had an electrical storm. If there was lightning you might get a trip of the transmitter, and you'd have to re-start it. That happened very rarely. We never had any problem with the transmitter at all. None. No faults in it, and that's saying something. Most of the time we were on the network, which always came through to us by telephone line, and we'd simply connect the network into the station through the patch panel. You used to throw the attenuator down, put the jacks in with the attenuator down, then you'd bring up the attenuator to the required level. Oh, the record cuing was dead on - we could cue within three seconds without any problem at all. Sometimes five seconds, but that was all.

The salami story? Well, I used to be a great lover of Jewish delicatessen - pastrami, corned beef, salami - and even in high school I'd bring in sandwiches and put them in my locker. Of course, people could smell them from outside. At the radio station I was kidded about the smell from the salami sandwiches. I liked the smell, but some people found it offensive. So they made me put them in one of the record storage cabinets...like a filing cabinet.

Harold Stewart, who was my professor in Electrical, was one of the nicest men I'd ever met. He was easy to be with - a real gentleman in terms of how he treated his students. If there was a student who didn't catch up quite as quickly as the others, he would be very gentle and not try to make any cute remarks about him, like some of the professors did. He was very patient, regardless of your level of progress in his work. I had a great admiration for him.[152]

Peter T. Demos: Though my stint at the station was brief, the memory is a very warm one. Of course, Harold Stewart stands out most strongly, both because of the courses he taught us and his generous and good-humoured spirit. In enabling some of us to be involved with CFRC, he added a rare dimension of realism to our education. I recall the clarity of Chuck Millar's diction and the enviable ease with which he spoke. His patience in putting up with our blunders, especially when we were being broken in, was nothing short of saintly. The friendship of fellow operators was also an important part of the experience, particularly that of Eph Diamond, with whom I shared many hours of philosophizing about the affairs and future of mankind.

I remember the announcements of lost articles and animals that we had to make, as well as the announcements of deaths - the last invariably following the playing of a few bars of Sibelius' "Valse Triste" and, as we were instructed to do, with the proper respect and dignity of voice. I once got a call from the Kingston Penitentiary asking us to announce that two prisoners had just escaped. Just as I opened the mike to do this I caught myself thinking that it might be a prank call. It turned out to be real enough, when I called back to check.

One stormy night, a nearby lightning strike induced a strong surge of current through the station's antenna, and caused an arc that set fire to the plastic cylinder carrying the transmitter's main inductance coil. I was alone and a bit dismayed at first, but inspection showed that the damage was limited to the coil. By luck, a plastic tube of the right size was at hand and with some consultation, which must have been with Harold, the station was back on the air in an hour or so. But the real fun was yet to come. I thought it would be smart to confirm that our signal was really going out, and where better to call at 10 pm than the Student Union. Well, it was a mistake! The razzing from the succession of unbelieving guys who came on the phone was one thing, but the calls kept coming till sign-off time. Modern computer networks have nothing on the student grapevine that was at work that night.

An embarrassing recollection is of a coincidence during a regular BBC transmission, which came in just before an 8 o'clock program hosted by the Lucky Strike Tobacco Company. The latter was on a 33.3-rpm record that I had to cue-in beforehand by listening to it over the station monitor. You selected monitor and transmitter channels by plugging patch cords into appropriate but look-alike sockets. These occasionally got mixed up, especially by new operators, so that what was to be monitored would go on the air over the current program. The BBC announced that incendiary bombs had been dropped on London. Then the Lucky Strike show, which I cued on top of the BBC program, began with the theme song, "Light up and listen, in your favourite easy chair..." Not many phone calls came in, but those that did were very clear.[153]

William J. Wilson: I was a part-time operator at CFRC during 1941-42 while I was a student at Queen's. I took some of the afternoon, evening and weekend shifts. Chuck Millar was the station announcer and Bas Brosso was the operator. The weekend shifts in winter were the worst because they used to turn down the heat in Fleming Hall after the Saturday morning classes to save coal. By Sunday afternoon it would be really cold. We carried the CBC network most of the time. Each weekday afternoon there was a 15-minute period when CBC used the network to advise stations of changes in their schedule for the next day. We had to fill with recorded music and listen to the network to get the corrections. I was never given any guidance as to what to play, so had to make up a program from our very skimpy library. We had one or two drawers of 16" 33.3-rpm Thesaurus and Standard Program records with ready-made notes. Since the station was not a regular subscriber, they were pretty old and scratchy.

Bas liked to play the more modern selections, but when we had to make the occasional funeral notice announcement, he would precede it with a sombre classical selection, follow it with the same kind of piece and then get the more modern stuff back on the air as soon as he could. I recall a half hour religious program on Sundays, put on by a local church. Its format was typical of the time - hymns sung by a small group around the piano in the large studio, Bible readings, a short talk by a clergyman and an announcement or two. Also there was the regular morning show master-minded by Chuck. I walked into the station one morning before class and got roped into playing a part in a funny story Chuck wanted to tell. We had two rehearsals in the big studio while a record was playing, and Chuck complained bitterly that I was going to blow the whole thing with my flat delivery. When we performed on air I poured on the expression and Chuck was both surprised and pleased with the result. Gales of laughter in the control room!

There was a group of three gain controls on the lower left front of the OP-4 which controlled inputs from the network and two studio microphones. On the lower right was the master gain to control the overall output to the transmitter. It had a couple of meters - the volume meter on the right was used to keep the program feed to the transmitter at a constant level. There was a control in the middle of the operating desk in front of the OP-4 that one turned to the left to bring up the left turntable and to the right for the right. On my right, mounted in a cabinet rack, was a monitor amplifier, a speaker, a jack field and a row of switches. A tiny oscilloscope eye staring out served as a modulation monitor and ran continuously.

The "no-name" transmitter was housed in two cabinet racks, each about six feet high and maybe two feet deep. I think the wider left rack held the power amplifier and the antenna matching unit. The antenna was a T-type and the ground was a simple counterpoise under it. Prof. Stewart always kept a close eye on the technical performance. I know, because I tested him once by moving one of the transmitter controls a whisper, and found it back to its original setting the next day.[154]

Basil Brosso was hired to replace Cecil Richards. After an interview about things technical with Prof. Stewart he was given a voice test by Dr W.E. McNeill. Early one September morning Brosso couldn't get any output from the transmitter and, looking out the window, noticed that some students had hung a stuffed toy animal on the counterpoise. He thinks that Sparky Spencer must have worked in a shipyard because every Saturday morning, after Uncle Josh and his crew left the station, he had to clean the blackened piano keys with carbon tetrachloride.[155]

The Thin Edge of the Wedge

Queen's and The Whig operated in partnership in CFRC, more or less amicably, for almost four years without a written agreement. In the summer of 1939 The Whig's auditors asked that a formal agreement be drawn up "to complete The Whig-Standard records." Arthur L. Davies discussed the matter with Dr McNeill. Relations were cordial and the University had no objection. The newspaper did not wish to change anything in the formal agreement, though it did wish to have the document signed by the end of the year.[156]

Several undated drafts of the formal agreement exist. One, probably written after May 29, 1940, set out the University's and The Whig's responsibilities in the operation of the radio station. The University was to operate CFRC during agreed hours at its own cost, including engineer, operator and announcer, upkeep and improvement. The Whig handled all advertising copy and continuity, supplied all recorded features, a news service, and newspaper publicity for CFRC. They hired the Program Director and his staff, and paid the University half of the gross revenue from advertising every month. In order to terminate the agreement, a year's notice in writing was stipulated, to be waived if both parties agreed. The subsidiary sections dealt with the working arrangements which were currently in effect, plus the charges for remotes and arrangements for free-time broadcasts. Major Annand was confirmed as Program Director and Advertising Manager, Charles Millar as Announcer and Assistant Program Director and Harold Stewart as Engineer. A Committee of Management was struck, consisting of the Principal, the Vice-Principal and Messrs W.R. and A.L. Davies, and at a meeting on June 7, R.C. Wallace was made Chairman and W.E. McNeill the Secretary. The agreement was signed on December 9, 1940.[157]

By May 1941, CFRC was doing very little of its own studio production. There was a 15-minute religious program every morning, 15 or 30 minutes of local artists performing popular material in the mid morning and short local programs at noon and in the evening. The 15-minute talks by Queen's staff continued five nights a week, and from time to time at irregular hours CFRC broadcast lectures, convocations and other University functions. Every second Sunday they carried a service from a local church. Although CFRC was on the air 105 hours a week, nearly 90% of its programming was taken off the CBC network. On the one hand this was good, since CFRC was getting a share of CBC's advertising revenue - the annual gross advertising revenue was about $26,000. On the other hand, certain advertising was rejected by the University, and CFRC's 100 watts gave a reliable coverage of roughly 20 miles. This was not very attractive to national or to local advertisers.[158]

The Splitting Begins

About this time, Dr Wallace received a long letter from W. Rupert Davies, President of The Whig, who had thought the situation over and had decided that the present set-up was becoming increasingly inadequate for Kingston's needs. Radio was changing rapidly, and was no longer an experimental medium. A strong organization was applying for a commercial radio licence in Belleville, and was likely to do so in Brockville too. With its self-imposed limitations, CFRC would soon lose a great deal of its business volume to the aggressive commercial stations. At the present time, CFRC was not attractive to national advertisers. Mr Davies had just got a long-awaited radio licence for his _Evening Examiner_ in Peterborough and wished to link stations in Peterborough and Kingston in order to compete successfully in Eastern Ontario:

> As radio stations become more and more competitive with newspapers in the dissemination of news and advertising, the more imperative it becomes for newspapers to acquire radio stations wherever they can in order to safeguard themselves against changing conditions...With increased radio competition in Eastern Ontario becoming more certain, and as Queen's University does not wish to make full commercial use of CFRC, I have been wondering if you would consider withdrawing from commercial radio broadcasting and disposing of your interest in the commercial radio broadcasting licence which you hold in this district to me for a consideration to be agreed upon. In this connection, it should be possible to secure for Queen's an experimental licence for teaching purposes and also to arrange for Queen's to have broadcasting time on the Kingston station...CFRC, as at present restricted and operated, could not hope to continue to do the volume of business it is doing now against aggressive commercial stations in Brockville or Belleville or both...You will realize that as my life's savings are invested in a newspaper, I cannot sit idly by while changing conditions and new inventions are vitally affecting these investments.[159]

Senator Davies had heeded the advice of newspaper magnate Victor Sifton to take on a partner in radio who knew something about that business, and approached Roy Thomson of Timmins, who had a reputation for making money out of small newspapers and small radio stations. Thomson would own 49% of the Peterborough radio operation, and run it through Jack Kent Cooke. Kingston was to be the second station in the partnership, but it was to be independently owned by Allied Broadcasting Corporation.[160]

In a conference of the Committee of Management on April 29, 1941, the University agreed that some change was necessary, but countered that CFRC was an essential teaching tool, an important outlet for University functions and its educational aims, and the University stood to lose a most excellent source of revenue.

A much longer letter from Mr Davies followed on May 1, re-iterating why The Whig needed control of the commercial licence in Kingston. He pointed out that in 1936, "when Queen's lost its radio licence because of its non-availability for commercial use," he "helped substantially to secure for it a commercial licence." CFRC was now losing ground in advertising with respect to other commercial stations, and with the new Federal budget, there would be a 40% corporation tax on Queen's profits from CFRC. His own total profits from CFRC since 1936 had amounted to only $1,340. Meanwhile, Canada's newspapers were annually losing over two million dollars in advertising revenue to radio. This was very unsatisfactory. Mr Davies wanted to use the Queen's licence to present programs "of a very high standard" and made Queen's "a very generous offer in view of the returns which our efforts have already brought to the University." He proposed to lease the licence for $5,000 a year, with the option to buy at the end of the fourth year for an additional $5,000. He declined a Queen's feeler to invest in his venture, advising them that the risk was more than they would normally assume. His one kilowatt station would be available to the University for broadcast at certain times, and would co-operate with E.E. in demonstrating its equipment to students. He intended to make application at once.[161] The Principal reported this proposed sale of the radio licence to the Executive Committee of the Board of Trustees on May 3.[158]

Dr Wallace was advised that CFRC might get 250 watts if it moved the transmitter out of town, but a channel for 1,000 watts would be difficult to find. This would require a directional antenna array costing $20,000-$30,000.[162] Other advice was to go for a lease and secure an operating contract from the newspaper.[163] The Principal and Mr Davies went to Ottawa to see Commander C.P. Edwards of the Radio Division, DOT, about Mr Davies' proposal for a 1,000 watt station on 960 kilocycles. The CBC's policy was that there should be only one radio station in Kingston, but Mr Edwards was sympathetic to the University's retaining its present licence for a limited-use low-power experimental and educational station.[164]

Senator Davies wrote again on May 24, urging a decision from Queen's, since the CBC Board would meet on June 9 for the last time before the fall. He had heard from his CFRC advertising agent in Toronto that, since the change to 1490 kilocycles, the word was out that reception of CFRC had deteriorated, and he feared that advertisers would hesitate to purchase time. He enclosed a letter which he thought should be sent to the Radio Branch at once requesting that CFRC's power be increased to 1000 watts on 960 kilocycles, and that the call letters be changed to CKWS. The transmitter would be located on the northern edge of Wolfe Island. Also that Queen's be granted a new non-commercial licence on 1490 kilocycles with a power of 50 watts. E.A. Laport of RCA was preparing the technical brief on CKWS.[165]

The Principal drafted a letter which merely supported Mr Davies' application, provided that 1490 kilocycles at 100 watts be retained by Queen's for experimental work and educational broadcasting.[166] Dr McNeill warned that this approach would not fulfill the verbal agreement with Mr Davies, under which Queen's would acquire the new licence and Mr Davies would lease to buy.[167] A letter very similar to what Mr Davies had suggested was sent to the Radio Branch on June 4, including "Under the arrangement which is proposed Queen's University would transfer the 1000 watt licence to the Kingston Whig-Standard to be operated as a commercial station."[168] Dr McNeill was still concerned that Mr Davies would not be liable for the payment of the

$25,000 to the University if he were granted a licence for CKWS at 960 kilocycles. There was no contract, and The Whig's application for and receiving of a licence was entirely independent of Queen's University.

About that time, a letter was received from a member of the Special Committee of the Executive of Queen's Trustees, noting that CFRC was in an exceptional position. Kingston is in a radio "pocket." The poor reception of outside Canadian stations means that a Kingston station has its very own area to serve with very little competition from other Canadians. With some aggressive selling and 1,000 watts, giving a primary coverage of 100,000 persons and a secondary of 170,000, CFRC could boost its advertising rates very substantially. He therefore advised that Mr Davies' offer was not good enough.[169] Dr Wallace replied that the newspaper's offer was not unreasonable. Queen's was not in a position to sustain a competitive commercial radio operation. Besides, they might then be taxed on their profits and there might be further clashes between the needs of a commercial operation and the main purposes of a university.[170]

The Department of Transport, in its acknowledgement of receipt of the application, noted that no transfer of licence was needed, since Queen's proposed to continue operating CFRC as a university station. The application for one kilowatt on 960 would be considered as coming from The Whig-Standard.[171] Dr Wallace did not immediately distinguish between the transferring of Queen's commercial rights and transferring of the licence to the newspaper.[171,172] Incidentally, war conditions now made it essential that operators be British subjects of certifiably good character and, to protect the realm, the transmitter had to be disabled when unattended. The station had secret instructions from the DOT that, upon directives from the Eastern Air Command, the transmitter be shut down immediately and without comment on the air.[173] Another secret order forbade the broadcast of plain language weather reports.[174]

Senator Davies appeared before the CBC on June 9, 1941, and reported to Dr Wallace that the Board would not grant a licence for 1,000 watts if Queen's operated the station. They did not believe that a university should be involved in commercial radio, and required assurance from Mr Davies that he was going to operate this proposed new station. The CBC's Technical Department was left to find a frequency for a 1000 watt station.[175] After telephone negotiations with Dr McNeill, Senator Davies wrote a letter on July 6 confirming his proposal to pay Queen's $25,000, provided that he received a 1000 watt commercial licence. At that point, the kilowatt licence seemed unlikely, and if a 250 watt licence were granted, something else would have to be negotiated. Queen's would be retaining its licence for CFRC, so Mr Davies could not pay them for that. Perhaps he would offer $10,000 for the commercial business and goodwill, plus $3,000 a year for five years if Queen's guaranteed to keep out of Kingston commercial radio for that period, "and indefinitely."[176]

The final agreement was approved by the fall meeting of the Board of Trustees, in Douglas Library on October 18, and Queen's radio rights were "sold" to The Whig-Standard. In spite of the fact that the University could not actually sell its licence or its radio rights to Mr Davies, he did honour his offer and pay a consideration to Queen's for getting out of the commercial radio business. The University would discontinue commercial operations on the date that CKWS commenced operations and would carry no programs originating outside of Kingston. It would operate CFRC "to satisfy the educational requirements of the University" and would not compete with CKWS for a full ten years. The Corporation agreed to pay Queen's $10,000 on or before the starting date of CKWS, $3,000 a year for the next five years and $100 a year for the following five years. The agreement expired on December 31, 1951.

Over the next few months, the University sought some modifications. They wanted permission to broadcast the concerts of the New York Philharmonic or the matinees of the Metropolitan Opera whenever CKWS was unable to do so, and Senator Davies gave his informal approval, but refused to incorporate it into the agreement. He had no objection to the University's carrying the occasional educational broadcast, but did not want the competition of regular educational features. He did offer to let Queen's broadcast over CKWS, as it would give Queen's a larger audience and would help CKWS at the same time.[177]

Dr Wallace explained to the CBC Supervisor of Station Relations that CFRC would broadcast a 15-minute educational talk each evening plus appropriate university functions. He expected that the Queen's station would be on the air an average of an hour a day after CKWS became active.[178]

Radio Mechanics Course for the RCAF

A 13-week course on the fundamentals of radio for 150 RCAF radio technician recruits was given at Queen's beginning on June 16, 1941, part of a Canada-wide program at universities to train 2,500 men for the war effort. Queen's stood to make a lot of money on the deal. Prof. Stewart was in charge of the Queen's course, and other instructors included Drs Norman Miller, H.W. Harkness and B.W. Sargent, plus Harold S. Pollock and A.E. Diamond. The recruits were housed in the gymnasium and took their meals in the Student Union.[179]

CFRC is mentioned in Sea Flight: A Fleet Air Arm Pilot's Story by Hugh Popham, in the section about British Navy air training in Kingston during the Second War. One spring, one of the men arranged to broadcast about half a dozen programs called "Fleet Air Arm Half Hour," from the single tiny studio at the University.[180]

CFRC Gives Way to CKWS

> Harold Stewart: We carried on daily broadasting until 1942, when it bcame fairly obvious to all that we could not go on selling Lydia Pinkham's Pink Pills on a university station. But The Whig-Standard wanted the revenue, so we parted company - amicably. And all this time it was obvious, of course, that more power was a desirable thing. We

designed a new transmitter for a kilowatt and we were even planning how high the tower would be. We had our good friend Keith MacKinnon, who is a consultant on antennas, design an antenna for us and all this while he was trying to get a channel. The two things that stood in our way - we'd have to find a channel where we could operate and we'd have to move out of town. We were hoping that that could be done, but it never was. We never got the licence, and when The Whig-Standard station CKWS opened up in the fall of 1942, they invited Dr McNeill, our Vice-Principal, to make some comment. And he said, "For years, the University has been trying to get a licence for more power and a better channel. And we were never able to do this. It remained for the opulent Mr Davies to succeed where the University had failed." And he says, "Verily, the meek may inherit the earth. But not yet." Well, I'm not sure how Mr Davies took that. I guess he took it all right. Nobody got murdered over it anyway. After that, the station activity subsided very quickly and substantially. We went back to week-end operation with student volunteers.[48,181]

CFRC went off the air as a commercial station at 6.59 pm on August 31, 1942, with the words "Thanks for listening and good-bye." Listeners were told to re-tune their radios to the new station and CKWS signed on as the CBC affiliate in Kingston at 7.00 pm on 960 kilocycles and 1,000 watts, with an organ concert and speeches.[182] That day Allied Broadcasting paid Queen's $10,000 as the first installment of the agreement.[183]

Characteristically, W.E. McNeill sent a note to Allied Broadcasting on Saturday, January 2, 1943, reminding them that the first annual payment of $3,000 to the University had fallen due on December 31, 1942.[184]

Ironing Out the Bugs

CFRC went on the air only twice in 1942 after relinquishing its commercial licence; first to broadcast Convocation, and then to air the opening address of the annual Theological Conference. Allied objected in writing, since Queen's had failed to obtain the necessary permission from Allied. The Directors realized that no slight was intended.[185] Drs Wallace[186] and McNeill[187] pointed out that the restriction was only upon programs originating outside Kingston, that the University had not agreed to restrict its own educational services. The two programs in question had originated on the campus.

On January 14, Arthur Davies informed the Principal that Allied's directors had agreed the day before "that the University is at liberty to use the facilities of CFRC to broadcast any program originating in the City of Kingston that is non-commercial in nature and which is an extension of the educational facilities of Queen's University. The directors have accepted the University's interpretation of the agreement on this matter." He was also instructed to offer Queen's the use of CKWS for a 15-minute educational program once a week at no cost, the time slot, content and length of run to be worked out later.[188]

From the Whig's Point of View

<u>Arthur L. Davies</u>: As time went on, a certain amount of friction began to arise between the University faculty and <u>The Whig-Standard</u> over the type of advertising that was being put over the station. This applied particularly to the Medical Faculty and the patent medicines that were being advertised on the station. On the other hand, <u>The Whig</u> began to chafe a bit about the programs that the University thought were good and would attract an audience, but which <u>The Whig</u> thought were perhaps a little over the heads of the average listeners. The newspaper felt it unfair that it was taxed on any profits it derived from the radio station, but Queen's was exempt from taxation. The War was beginning to siphon off staff and replacements were difficult to obtain. The limited range of CFRC was being criticized by residents of neighbouring townships, who could receive some American stations but couldn't receive this Canadian station.

In any event, in 1941 Roy H. Thomson was beginning his empire up in Timmins, where he had several radio stations, and he wrote to my father suggesting that he come to Kingston to discuss radio broadcasting in our area. He arrived accompanied by a general manager of radio, Jack Kent Cooke. They met with my father and myself and spent the day pointing out what they saw as flaws in our broadcasting arrangement with Queen's, and the advantages of forming a partnership with them. We were impressed. Later that year I went to Timmins to inspect their operation and reported back favourably. Mr H.B. Muir had died in January 1939, and my father had subsequently purchased the Muir family interest in <u>The Whig-Standard</u> and the <u>Peterborough Examiner</u>. Thus it was my father who entered into negotiations with Dr R.C. Wallace, then Principal of Queen's, and Dr W.E. McNeill, Treasurer of the University, to dissolve the broadcasting partnership. Dr McNeill was a devoted executive of the University and he was not going to let <u>The Whig-Standard</u> off the hook financially without some real concessions. I have forgotten exactly what those concessions were, but they were substantial and eventually were agreed to by both the University and <u>The Whig-Standard</u>. Queen's felt it should be compensated generously for giving up its interest in the commercial broadcasting licence granted by the CBC. Both my father and Mr Thomson, who had formed the Allied Broadcasting Company, agreed and a figure was arrived at based on a projection of commercial profits for CFRC over a period of years. Queen's agreed to withdraw entirely from commercial broadcasting, and to support Allied Broadcasting Company in an application for a commercial radio broadcasting licence for the Kingston area. The Commission was sympathetic to the application because it wished to increase the coverage for its programs far beyond the range of the Queen's transmitter and at no capital cost to itself.

It was decided to build an addition to <u>The Whig-Standard</u> building to house studios and other facilities, and land was leased on Wolfe Island, where a powerful transmitter was installed together with an array of signal towers to send a strong signal north, east and west, but to protect a radio station being built at Watertown. This all took time, because due to the war, practically all this equipment was allocated by priority. However, in mid-summer 1942, the new station was ready for preliminary tests and the date for the formal opening was set for August 31. The opening of CKWS was to be something special and my father, who was interested in cultural affairs, thought that it would be fitting if we had a well-known Canadian artist take a leading part. Sir Ernest MacMillan of Toronto, noted Canadian organist, conductor and composer was asked to come to Kingston and give an organ recital as the feature of the opening of CKWS. Of course, we had to procure an organ, and we were able to get the use of the organ at Sydenham Street United Church. This meant a remote hook-up with the station - something which we hadn't experienced before, as this was the first broadcast. The remote hook-up was installed and tested and declared to be in first-class shape. Sir Ernest took his place at the console of the Sydenham Street United Church organ and played his program. Unfortunately, this first remote broadcast, arranged by the staff of CKWS, failed to reach the transmitting station on Wolfe Island. None of it was heard by anyone listening in because the remote hook-up just wasn't working. The speakers who spoke from the studio at The Whig-Standard office were heard perfectly well. The invited guests at the studios in the Whig-Standard Building were enthusiastic because they heard the program beautifully, but our unseen audience did not receive it. Perhaps a few members of the technical staff of the Queen's station had a quiet smile.

When The Whig entered into contact with Queen's to operate the radio station, there were two strong stipulations made by Dr W.E. McNeill. One was that the New York Philharmonic program every Sunday afternoon must be carried and the other was that the broadcast of the opera every Saturday afternoon must be carried. In the negotiations just before we gave up association with CFRC, and The Whig began operating CKWS, the stipulation that these two broadcasts must be carried by CKWS was also put in the agreement, for how many years or for what length of time this was to take place. I do recall when a time came when CKWS felt that it didn't get many listeners for the opera broadcast, and that it interfered with football broadcasts. It was decided to stop the opera broadcasts and this brought about a considerable amount of criticism from citizens who liked to hear the opera broadcasts, and I remember particularly Professor Lower was a leading critic of this action by the station.[17]

Kenneth McIntyre: Regarding the time that we cut over from CFRC transmission to CKWS, in 1942, the CN had their own loops to CFRC, and we just had three loops, so that when tCKWS was going to start on the network, our CNR superintendent in Toronto wrote a letter instructing me that at two o'clock in the afternoon of that day, I was to cut the feed to CFRC and swing the loops over from CFRC to The Whig-Standard Building on the corner of Clarence and King streets. And I notified Mr Millar to the effect that they were going to be without any feed from 2 until 7. Dr McNeill called me a short time before I was to do it, and wondered if there was any way that we could maintain the feed continuously to CFRC up until 7 o'clock when CKWS would take over. He explained to me that they were expecting President Roosevelt to speak, and that it would be important that his speech would be carried. I read my instructions from my superintendent to him, and I said, "I must do what I'm told. That's what I'm paid for." Well, Dr McNeill was a pretty convincing man, and what I should have done was said, "Well, Dr McNeill, would you call our superintendent in Toronto and request that?" And I didn't. I made that mistake.

Anyway, I thought, after listening to Dr McNeill and his convincing words that maybe we might be able to do some little switching on our own. If CKWS was not going to feed the network, we could cut the telephone control circuit over to CKWS at 2 o'clock and the arrangement with Chuck Millar was that we would use just the local city telephone if we needed it, and if CFRC would accept the network on the "send loop," that I would feed CFRC the "send loop." When he moved over onto it, I would cut the feed off the "receive loop" and we would cut the "receive loop" over to CKWS. Mr Millar complied and we had our lineman at the pole on King and Clarence streets cut the telephone over. It worked fine. But, when we attempted to cut over the "receive loop," and CKWS and I tried it, it was unbalanced. I was in contact with the lineman, and said, "Cut the 'receive loop' back to CFRC again," and when he said he'd cut it back, it worked perfectly to CFRC. So, I got Mr Millar on the city phone and asked if he'd move back onto his "receive loop," and I cut the feed off the "send loop," cut the send loop into CKWS and it worked. We didn't tell CKWS what we did, but we just fed CKWS that evening on the "send loop," and when the lineman and I checked it out the next day, he found **his** mistake. He put the "receive loop" back, regular, and CKWS had to change over. We tested them all out and they were all fine.[21]

Kevin Nagle: Major Annand is the one person who really sticks in my mind. I remember him very clearly doing the six o'clock news, and he'd come on the air and say, "This is Major Annand with the day's news." Not the smooth, polished type of announcer that you have to-day. He had a very clipped English accent.[189]

Charles Millar: Of course, they left the CFRC transmitter behind and had their own erected on Wolfe Island with a radio link from the studios on King Street, because at that time the phone lines weren't adequate to take the programming. That link to Wolfe Island is a story in itself, and it was out more often than it was on. Eventually it got to be pretty reliable. The big thing came when the telephone company opened up a series of lines to Wolfe Island and gave us separate ones to feed the programs on, and then our troubles were over. The transmitter was put there on purpose to protect stations on adjoining frequencies in the US, and most of the transmission was headed in a generally northerly direction so that it would cover the Kingston area completely, and not bother our neighbours to the south.

At the studios of CKWS on King Street, we had a great supply of would-be announcers come to us from the five or six other stations operated in the loop by the syndicate that owned CKWS. One of these was Mickey Carlton, who had some odd ways with him. One of them was that if had trouble with his car, he'd bring the part to work with him, and get it fixed after his hours were over. On this particular morning he brought a tire that had given him trouble, and he rolled it up the stairs - the studios were on the second floor of the Whig-

Standard Building - and Mickey brought the tire up so it wouldn't be taken out of his car. He parked it on the little landing on top of the stairs, but it wasn't too securely fixed. So, when Mickey turned his back to go to his office, the tire took off, rolled down the stairs, through the plate glass door, right across the street and bounced off the wall of the Bank of Montreal building across the way. The thing that made it interesting was that Senator Davies happened to be coming to work and he didn't even have to open the door. It was smashed open for him, and he needed an explanation of that tire going across King Street.[57]

<u>Kevin Nagle</u>: You went in the front door of the Whig-Standard Building on King Street and then you turned left into CKWS. Roy Hoffstetter was the Manager, Charlie Millar was the Program Director - he was a good teacher - and Florence Daly was in charge of the huge library of records. We had a soft-cut recorder for anything that was to be delayed, and there was no possible way to edit those.

CKWS was operated by Northern Broadcasting at that time, and I remember one time Roy Thomson came to Kingston and I was earning something like $17.50 a week. I foolishly asked to see him, and I was ushered into the Station Manager's office. He was short and stout with very thick glasses and he was imposing. Mr Thomson asked, "What would you like to see me about?" I said I'd like to discuss a raise. He spent about an hour explaining to me that really I should be very thankful that I had the opportunity to work for his organization, that it was a great training school and that I should be paying him for the privilege of working there to acquire the experience I was getting. It always struck me that the man who even at that time was a very, very wealthy person would spend an hour of his very valuable time explaining to me why I shouldn't get this very small raise. Even $2.50 a week raise would have been wonderful. He was a very forceful man. You always bettered yourself by moving. At that time, any move was a pay increase. That's one of the reasons why there was such a big turnover in staff at CKWS, along with the company policy of no stars. You'd get to a certain point and that's it! You could sit at that point the rest of your life, or you could leave and go somewhere, and that always meant more money. It's a great mover - a great incentive.

To get the program from the King Street studio over to the transmitter on Wolfe Island there were no quality telephone lines available, so they had a 40.2 megahertz AM transmitter that had been a "Cone of Silence" marker transmitter at an airport.[190] That broadcast the signal to Wolfe Island, where it was picked up on a receiver and then fed in to the main transmitter on 960 kilohertz. Later that link was replaced by an FM transmitter, and I think CKWS was probably the first commercial FM station in the country on the air. It was only 15 watts, but it was heard by short wave listeners all over the world.

People used to phone and say, "Could I speak to Tommy Dorsey?" Because of these good quality transcription records, they thought the musicians were actually there! Once in a while, you were feeling kind of hairy, you'd say, "Yeah. Just a minute," and get one of the other guys to get on the phone and be Tommy Dorsey or whoever.[189]

Post Script

In Steve Cutway's first years as Station Manager of CFRC, Queen's Comptroller called him up to say that he had found a pocket of $80,000 in the name of CFRC. This was apparently the original $25,500 paid to Queen's by Allied Broadcasting, plus accrued interest, untouched all those years![191]

Other Radio Licences Held by Queen's University

Prof. Stewart wrote to the Radio Division of the DOT in May 1939, enquiring about a licence for "a high frequency relay broadcast transmitter for remote pick-up work" at CFRC.[192] This was for a special transmitter, operating well above the broadcast band at 38,600 kilocycles/second, to send CFRC's own remote broadcasts to its studios. The DOT consulted with the Joint Technical Committee of the CBC, and advised that such a transmitter was to be used only "where wire circuits are inaccessible or impracticable." A private commercial licence had to be obtained, and the annual fee was $10.[193] The application form was submitted after July 1940, and Private Commercial Station licence No. 878 for VD2Z, 20 watts, A3 type of emission, was issued on September 11.[194] The call was changed in 1949 to CJY 40.[195] The licence for broadcast pick-up transmitter CJY 40 was cancelled by the DOT on March 14, 1957, and the ten dollar fee for 1956-57 returned to the University.[196]

With his request for the renewal of VD2Z in 1941, Stewart also requested renewal of the licence for a portable receiver used as the CFRC field strength measuring equipment. The DOT had no trace of such a licence, and requested completion of an application.[197] The licence, dated June 28, 1941, authorizes Queen's "to establish and operate a Commercial Receiving Station in a car operated by Professor H.H. Stewart, in the Kingston, Ont., area for field strength measurements in connection with broadcasting station CFRC."[198]

The Queen's Amateur Radio Association, 1935-1939

Interest in amateur radio on campus was revived in 1935 by Bill Motherwell, who organized "The Radio Club" and was its first President. Prof. Harold Stewart was Honorary President, K.R. Eland was Vice-President and Warren Raynor (Mech. '39) the Secretary-Treasurer. In the first year they operated under Motherwell's call, VE3BY, on the top floor of Old Arts.[199] They tried to get a transmitter for their licenced members and gave code practice sessions for those who wanted to learn. By 1937 they had their own call, VE3VX, a 40-watt CW transmitter with antennae for 40 and 80 metres, and a "shack" in the basement of the old Students' Union. They had contacted nearly 500 amateur stations over a radius of more than 1,000 miles, and were expecting to get into 'phone operation in 1938.

With the outbreak of war in the fall of 1939, all amateur radio operation was banned. The club had to dismantle most of its equipment.[200] Among its executive members in its five years of existence were L.A. Milton, James A. Jarvis (Elec. '40), L.G. Askwith and Ernest J. Wiggins (Chem. Eng. '38).

The Electrical Engineering Club took over the radio shack in 1943, and in October 1945 they were awaiting return of VE3VX from the DOT. Over the six years of inactivity, most of the old equipment had disappeared, so a transmitter was hastily built with Prof. Stewart's assistance, and an aerial was strung from the Students' Union to the Gymnasium. They were ready to operate by November 8, but they had to wait some months until the restrictions were lifted on their frequencies.[201] In the winter of 1946-47 the transmitter was re-built and a small communications receiver purchased, and the next fall the Radio Club was re-organized as a special branch of the E.E. Club.[202]

When Prof. James L. Mason (Eng. Phys.'69) was first associated with the Radio Club as a student in 1965-66, it had about 30 or 40 members. At that time, it met in the clubrooms on the second floor of Clark Hall, and used a Halicrafters HT 37 transmitter and a General Radio HRO 50 receiver. It also had a 2-metre transmitter, modified from a General Electric strip line taxi or police set. The club used to phone-patch traffic so that students could call home by radio at no charge, and participate in contest operation. Around 1972-73 the club moved to Goodwin Hall, where the antenna remains, and in the early 1980s its space was required so it moved to Abramsky Hall and took along the dipole antenna. Dr Mason became the licence holder for VE3VX in 1978. Although he has had no connection with the club since the early 1980s, he continues to renew the licence. He has heard nothing of the Radio Club since 1986, and believes that it has gone defunct again.[203]

NOTES

1. Dr. Fyfe to Hector Charlesworth, C.R.B.C., Dec. 6, 1935; QUA, Coll. 1250, Box 19, Principal's Office, Series #1, Subject Files R-S, Radio Station CFRC, 1923-1952.
2. Hector Charlesworth to Dr. W.H. Fyfe, Dec. 12, 1935; ibid.
3. G.W. Richardson to Dr. W.E. McNeill, Dec. 17, 1935; ibid.
4. W. Rupert Davies to Hon. Norman McLeod Rogers, Ottawa, Jan. 3, 1936; ibid.
5. Dr. W.E. McNeill to W.G. Richardson, Jan. 4, 1936; ibid.
6. W.H. Fyfe to Hon. C.D. Howe, Ottawa, Jan. 4, 1936; ibid.
7. W.E. McNeill to Hon. Norman McL. Rogers, Jan. 8, 1936; ibid.
8. Hon. Norman Rogers to Dr. W.E. McNeill, Jan. 13, 1936.
9. B. Howard Bedford, Chatham, to the Secretary, Queen's University, Jan. 27, 1936; ibid.
10. W.E. McNeill to B. Howard Bedford, Jan. 29, 1936; ibid.
11. Prof. S.C. Morgan to Dr. W.H. Fyfe, Jan. 29, 1936; ibid.
12. "Broadcasting, February 4, 1936: Information secured from Professors Jemmett and Morgan"; ibid.
13. "Notes from G.W. Richardson re: Radio Broadcasting from Queen's", undated; ibid.
14. Dr. W.E. McNeill to G.W. Richardson, Feb. 17, 1936; ibid.
15. "Proposals re Broadcasting", undated; ibid.
16. H.B. Muir to Dr. W.E. McNeill, Feb. 19, 1936; ibid.
17. Personal communication from Arthur L. Davies, Kingston, Sept. 2, 14 and 30, 1982.
18. "Queen's Station to Measure", KWS, Feb. 28, 1936, p. 3.
19. "The Queen's Broadcast", ibid., Feb. 29, 1936, p. 14.
   "Queen's Radio Station", ibid., March 7, 1936, p. 3.
20. "Queen's Radio Station Conducts Local Survey", QR 10 #4, April, 1936, p. 103.
21. Personal communication from Kenneth McIntyre, Kingston, Jan. 24, 1983.
22. "Broadcast Was Greatly Enjoyed", KWS, May 28, 1936, p. 3.
23. Memorandum, H.M. Smith for Colonel Steel, C.R.B.C., to CFRC, May 26, 1936 (file no. 3-1-20); Prof. Stewart's papers.
24. "Details of Acoustical Treatment in Studio, C.F.R.C., Queen's University", C.R.B.C. Engineering Department, May 28, 1936; Prof. Stewart's papers.
25. "Great Care Taken with Acoustics", KWS, June 29, 1936, p. 3.
26. Harold H. Stewart, "The Origin and Development of CFRC", PESQU 28, Sept. 1938, p. 30.
27. Harold H. Stewart to Dr. W.E. McNeill, Jan. 10, 1938; Prof. Stewart's papers.
28. CFRC Piano file: QUA.
29. "Personalities", KWS, June 27, 1936, p. 5.
   Bill McNeil and Morris Wolfe, The Birth of Radio in Canada: Signing On, Doubleday, Toronto, 1982, p. 215.
30. "Rates Effective May 15, 1936: Broadcasting Station CFRC", QUA, Coll. Box 19, Principal's Office, CFRC.
31. "Queen's Radio Station", QR 10 #6, Aug. 1936, p. 171.
32. Queen's Radio Will Be Operated", KWS, June 24, 1936, p. 3.
33. "Tune in Tonight", ibid., June 29, 1936, p. 10.
34. "City and District: Special Broadcast", ibid., p. 14.
35. "Queen's Radio Station", ibid., June 25, 1936, p. 3.

36. "Local Choir to Sing", ibid., June 26, 1936, p. 3.
37. Personal communication from Alphonsus Patrick McCue, Kingston, Oct. 12, 1982.
38. Personal communication from Lareta Lamoureux, Kingston, Sept. 3, 1982.
39. Personal communication from Lenore Black, Kingston, Oct. 15, 1982.
40. "CFRC Programs: More Local Talent", KWS, July 3, 1936, p. 11.
41. Personal communication from John Donald Stewart, Ottawa, Nov. 7, 1982.
42. "CFRC on Air at Noon on Tuesday", KWS, July 6, 1936, p. 1.
43. "Has Uncanny Powers", ibid., July 25, 1936, p. 12.
44. To Editor from "We Feed the Body", ibid., July 10, 1936, p. 4.
45. "Listen in at Noon", ibid., July 25, 1936, p. 12.
46. "Vimy Unveiling", ibid., July 22, 1936, p. 2.
47. "Crowds Throng Streamlined", ibid., Aug. 17, 1936, p. 3.
    United Drug Company to Chief Engineer, CFRC, Aug. 17, 1936; Prof. Stewart's papers.
48. Personal communication from Harold Stewart, Kingston, Aug. 10, 1982.
49. H.M. Mahwhinney, Steel Company of Canada, Montreal, to S.C. Morgan, June 30, 1936; Prof. Stewart's papers.
50. "New Aerial on CFRC", KWS, Sept. 4, 1936, p. 3.
51. Personal communication from Cuth Knowlton, Kingston, Oct. 27, 1982.
52. "The Kilocycler Says", QJ, Oct. 2, 1936, p. 7.
53. R.C. Wallace to E.C. Kyte, Oct. 15, 1936; QUA, Coll. 1250, Box 19, Principal's Office, Series #1, CFRC, 1923-1952.
54. James A. Roy to R.C. Wallace, Oct. 22, 1936; ibid.
55. Frank Ll. Harrison to R.C. Wallace, Nov. 2, 1936; ibid.
56. "Schedule of talks by members of staff"; ibid.
57. Personal communication from Charles A. Millar, Kingston, Sept. 3, 1982.
58. Personal communication from Alexander John McDonald (Elec. '36), Victoria, Oct. 25, 1986.
59. "The Kilocycler Says", QJ, Oct. 2, 1936, p. 7.
60. W.H. Seltsam, Metropolitan Opera Annals: A Chronicle of Artists and Performances, H.W. Wilson Co., and Metropolitan Opera Guild, New York, 1947, p. 605.
61. Andrew Allan, Andrew Allan: A Self Portrait, MacMillan, Toronto, 1974, p. 70.
62. "'Red' Newman Wants", KWS, Nov. 10, 1036, p. 3.
    "Neighbourhood Night", ibid., Nov. 11, 1936, p. 3.
    "Neighbourhood Night", ibid., Nov. 13, 1936, p. 2.
63. "The Campuscope", QJ, Nov. 24, 1936, p. 2.
64. Dr. R.C. Wallace to Gladstone Murray, CBC, Dec. 8, 1936; QUA, Coll. 1250, Box 19, Principal's Office, series #1, Radio Station CFRC, 1923-52.
65. "Special Committee", KWS, March 18, 1936, p. 3.
66. "Government Announces", ibid., Sept. 12, 1936, p. 1.
    Frank W. Peers, The Politics of Canadian Broadcasting: 1920-1951, University of Toronto Press, 1969.
67. Leonard W. Brockington, "Canada Is On the Air", Saturday Night 52 #2, Nov. 14, 1936, p. 4.
68. "CFRC: A Progressive Station", QJ, May 7, 1937, p. 4.
    "Ceremony to be Broadcast Today", ibid., p. 7.
69. Personal communication from William A. Neville, Winnipeg, Oct. 2 and Nov. 7, 1989.
70. Dr. W.E. McNeill to Prof. H.H. Stewart, Oct. 8, 1936; Prof. Stewart's papers.
71. Prof. H.H. Stewart to Dr. W.E. McNeill, Oct. 8, 1936; Prof. Stewart's papers.
72. R.C. Wallace to Leonard W. Brockington, Nov. 10, 1936; QUA, Coll.

1250, Box 19, Principal's Office, CFRC, 1923-1952.

73. Gladstone Murray to R.C. Wallace, Nov. 19, 1936; ibid.
74. R.C. Wallace to Major Gladstone Murray, March 3, 1937; ibid.
75. Gladstone Murray to R.C. Wallace, March 11, 1937; ibid.
    Memorandum from CBC Chief Engineer, March 10, 1937; ibid.
76. R.C. Wallace to Major Gladstone Murray, March 23, 1937; ibid.
77. Prof. S.C. Morgan to R.C. Wallace, April 19, 1937; ibid.
78. Dominion Bridge Company, Lachine, to H.H. Stewart, May 6, 1937; ibid.
    Ajax Engineers, Toronto, to H.H. Stewart, May 7, 1937; ibid.
79. R.C. Wallace to Hon. Mr. C.D. Howe, Minister of Transport, Ottawa,
    May 6, 1937; QUA, Coll. 1250, Box 19, Principal's Office, Series #1,
    Radio Statiuon CFRC.
80. C.P. Edwards, Chief of Air Services, D.O.T., Ottawa, to R.C. Wallace,
    May 14, 1937; Prof. Stewart's papers.
81. Donald Manson, Chief Executive Assistant, CBC, Ottawa, to R.C.
    Wallace, June 23, 1937; ibid.
    Gladstone Murray to R.C. Wallace, June 29, 1937; ibid.
82. R.C. Wallace to Major Gladstone Murray, Sept. 16, 1937; ibid.
83. Gladstone Murray to R.C. Wallace, Sept. 20, 1937; ibid.
84. R.C. Wallace to Gladstone Murray, Oct. 28, 1937; ibid.
85. G.C.W. Browne, Assistant Controller of Radio for Chief of Air Services,
    Department of Transport, Ottawa, to Director, Queen's University,
    Dec. 13, 1937; ibid.
86. H.H. Stewart to Dr. W.E. McNeill, Feb. 11, 1938; Prof. Stewart's papers.
87. Dr. W.E. McNeill to H.H. Stewart, Feb. 19, 1938; ibid.
88. H.H. Stewart to Dr. W.E. McNeill, Feb. 24, 1938; ibid.
89. H.H. Stewart to Dr. W.E. McNeill, March 23, 1938; ibid.
90. L.C. Powell, Cockfield Brown and Co., Toronto, to CFRC, Dec. 15, 1937;
    QUA, Coll. 1250, Box 19, Principal's Office.
    C.N. telegram: Herbert, All-Canada Radio Facilities, Toronto, to
    Major James Annand, Dec. 21, 1937; ibid.
91. James Annand to Dr. W.E. McNeill, Dec. 22, 1937; ibid.
92. R.C. Wallace to Gladstone Murray, Dec. 28, 1937; ibid.
    Gladstone Murray to R.C. Wallace, Dec. 31, 1937; ibid.
93. R.C. Wallace to Hon. C.D. Howe, Minister of Transport, March 12, 1938,
    c.c. to Major Murray; ibid.
94. Hon. C.D. Howe to R.C. Wallace, March 29, 1938; ibid.
95. Personal communication from Louise Wittish and Wilma Nicol, Kingston,
    March 2, 1983.
96. Personal communication from Erwin Wendholt, Kingston, March 2 and 8,
    1983.
97. Personal communication from Joan (Annand) Henderson, Mississauga,
    Oct. 9, 1983.
    "Farewell Tendered Annands", KWS, April 4, 1942, p. 3.
98. Personal communication from Bogart Trumpour, Kingston, Sept. 20, 1982.
99. Personal communication from Dr. William Angus, Kingston, Aug. 25,
    1982.
100. Personal communication from Agnes (Morrissey) Neale, Ottawa, Nov. 17,
    1982.
101. "CFRC Program Highlights", KWS, Oct. 9, 1937, p. 5.
102. "Radio Programs: Kingston CFRC Programs", Kingston Whig-Standard 12
    #74-80, 5, Saturday, February 26 to Saturday, March 5, 1938.
103. R.C. Wallace to Charles A. Millar, Oct. 2, 1937; QUA, Coll. 1250, Box
    19, Principal's Office, Series #1, CFRC.
104. "Days of Austerity Recalled", KWS, Nov. 21, 1973, p. 28.
105. H.H. Stewart to Dr. W.E. McNeill, Jan. 6, 1938; Prof. Stewart's papers.

106. Dr. W.E. McNeill to H.H. Stewart, Jan. 7, 1938; ibid.
107. H.H. Stewart to Dr. W.E. McNeill, Jan. 10, 1938; ibid.
108. Personal communication from Verna Saunders-Kerrison, Kingston, Oct. 19, 1082.
109. "CBC Releases Broadcast Plans", QJ, Feb. 1, 1938, p. 8.
110. Script for "Queen's Radio Review", courtesy Dorothy Wilson, North Vancouver, B.C., Oct. 7, 1989.
     "Fine Radio Talent Revealed", QJ, Feb. 8, 1938, p. 1.
     "Date Set for Radio", ibid., Feb. 11, 1938, p. 6.
     "Students Present Program", ibid., Feb. 15, 1938, p. 1.
111. "More Talent Expected", QJ, Feb. 22, 1938, p. 1.
     "McGill on Air", ibid., March 8, 1938, p. 8.
112. "80 Have Entered", KWS, April 6, 1938, p. 12.
     "Amateur Show of K.W.A.", ibid., April 9, 1938, p. 3.
113. Personal communication from Florence (Daly) Courneya, Kingston, Oct. 13, 1982.
114. Personal communication from Art Macdonald, Ottawa, Nov. 7, 1982.
115. Personal communication from Audrey (McUen) Macdonald, Ottawa, Nov. 7, 1982.
116. "CFRC Program Highlights", KWS, July 2, 1938, p. 5.
117. Scripts courtesy of Florence (Daly) Courneya, Kingston.
118. Personal communication from Jack Telgmann, Kingston, Oct. 4, 1982.
119. "President Roosevelt's Address", QR 12 #7, Oct. 1938, p. 197.
     "President Roosevelt Honoured", ibid., p. 199.
120. "Radio Programmes, 1938-39", R.M. Winter, Secretary, Radio Committee, April 18, 1939; Prof. Stewart's papers.
121. Personal communication from Bruce Riggs, Queen's Archives, Kingston, Aug. 30, 1989.
122. H.H. Stewart to Dr. W.E. McNeill, Feb. 10, 1939; Prof. Stewart's papers.
123. Personal communication from Prof. Harold H. Stewart, Kingston, June 23, 1986.
124. Personal communication from Harry W. Beardsell, Vancouver, May 24 and July 17, 1989.
125. "Data for Inspection Report", 1940-41 licence year, Dec. 1940; Prof. Stewart's papers.
     Notebook containing complete calculations and specifications for Mark VII; ibid.
126. H.H. Stewart to R.C. Wallace, March 29, 1940; QUA, Coll. 1250, Box 19, Principal's Office, CFRC.
     "New Transmitter Built", QR 13 #9, Dec. 1939, p. 266.
127. H.H. Stewart to E.A. DeCoste, National Museum of Science and Technology, Ottawa, Nov. 24, 1983.
128. H.H. Stewart to J.W. Bain, Chief Engineer, Radio Division, D.O.T., Ottawa, Dec. 12, 1940; Prof. Stewart's papers.
     Walter A. Rush, Air Services, Radio Division, D.O.T., Ottawa, to H.H. Stewart, Dec. 31, 1940; ibid.
129. H.H. Stewart to Dr. J.A. Gray, Secretary, Science Research Committee, Queen's University, March 17, 1939; Prof. Stewart's papers.
130. H.H. Stewart and H.S. Pollock, "Compression with Feedback", Electronics 13 # 2, Feb. 1940, p. 19.
     R.C. Wallace to H.H. Stewart, April 9, 1948; Prof. Stewart's papers.
     H.H. Stewart to Canadian General Electric, Toronto, April 22, 1948; ibid.
     J.G. Smart, Electronics Section, Toronto District Office, Canadian G.E., to H.H. Stewart, May 4, 1948, ibid.
     N.J. Peterson, Broadcast Equipment Sales, General Electric, Syracuse, New

York, to H.H. Stewart, June 23, 1948; ibid.

131. Gladstone Murray to R.C. Wallace, March 17, 1939; QUA, Coll. 1250, Box 19, Principal's Office, Series #1, CFRC.

132. R.C. Wallace to Hon. C.D. Howe, April 22, 1939; ibid.

133. Hon. C.D. Howe to R.C. Wallace, April 24, 1939; ibid.

134. Hon. C.D. Howe to R.C. Wallace, May 4, 1939; ibid.
R.C. Wallace to Hon. C.D. Howe, May 6, 1939; ibid.

135. Walter A. Rush, Controller of Radio, Air Services, D.O.T., to R.C. Wallace, May 11, 1939; ibid.

136. G.C.W. Browne, Assistant Controller of Radio, Air Services, D.O.T., Ottawa, to Manager, CFRC, Aug. 15, 1939; ibid.

137. R.C. Wallace to G.C.W. Browne, Aug. 31, 1939; ibid.

138. Walter A. Rush to R.C. Wallace, Sept. 13, 1939; ibid.

139. H.H. Stewart to K.A. Mackinnon, Transmission and Development, CBC, Ottawa, March 17, 1940; cited in letter below.
W.G. Richardson, Transmission and Development Department, Engineering Division, CBC, Ottawa, to H.H. Stewart, March 20, 1940; Prof. Stewart's papers.

140. W.G. Richardson, CBC, Ottawa, to H.H. Stewart, Aug. 8, 1940; ibid.

14 1. Walter A. Rush to Manager, CFRC, Feb, 10, 1941; ibid.
"Notice to All Standard Broadcasting Station Licensees and Suppliers of Frequency Control Equipment", D.O.T., undated; ibid.

142. H.H. Stewart to Walter A. Rush, April 8, 1941; ibid.

143. H.H. Stewart to Dr. W.E. McNeill, August 18, 1939; ibid.

144. Personal communications from Rev. Dr. Victor H. Fiddes, Niagara Falls, Nov. 1, 1988 and Aug. 25, 1989.

145. "G. McAuley Killed in Action", KWS, July 14, 1944p. 1.

146. "CFRC Program Highlights", ibid., May 20, 1939, p. 5.

147. "General Description" in Finding Aid to "Records of Queen's Radio Station CFRC", #3707; QUA.

148. Manuscript, Bits of Verse by "Maple", courtesy Mrs. George T. White, Kingston.

149. "Radio Station", memo from Dr. W.E. McNeill to Dr. R.C. Wallace, Jan. 26, 1939; QUA, Coll. 1250, Box 19, Principal's Office, Series #1, CFRC.

150. "Radio Station CFRC", Queen's Board of Trustees, May 8, 1940, p. 60; QUA, Coll. 1227, Vol. 8.

151. Dr. W.E. McNeill to H.H. Stewart, Nov. 13, 1940; Prof. Stewart's papers.

152. Personal communication from Allen Ephraim Diamond, Toronto, Oct. 20, 1989.

153. Personal communication from Dr. Peter Theodore Demos (Physics '41), Belmont, Mass., March 11, 1990.

154. Personal communications from William J. Wilson (Sci. '45), Ottawa, Oct. 6 and Nov. 15, 1989.
Mr. Wilson retired around 1979 from the Department of Communications, where he was responsible for the regulation of all non-broadcasting radio stations in Canada and for the technical regulation of all broadcasting stations and CATV systems.

155. Personal communication from Basil Brosso, Kingston, March 20, 1991.

156. Arthur L. Davies, General Manager, The Whig-Standard, to R.C. Wallace, Nov. 25, 1939; QUA, Coll. 1250, Box 19, Principal's Office, Series #1, CFRC.

157. "Agreement between Queen's University and Kingston Whig-Standard re CFRC", Dec. 9, 1940; ibid.

158. "The Story and Problem of Radio Station CFRC", Report of the Executive Committee of the Board of Trustees, Queen's University, May 3, 1941; QUA, Coll. 1227, Vol. 8, p. 129.

159. W. Rupert Davies to R.C. Wallace, April 16, 1941; QUA, Coll. 1250, Box 19, Principal's Office, Series #1, CFRC.
160. Russell Braddon, <u>Roy Thomson of Fleet Street</u>, Collins, Toronto, 1965. "Peterborough Radio Station", KWS April 2, 1942, p. 3.
    By the end of 1942, Northern Broadcasting Management was operating: CKWS, Kingston; CKGB, Timmins; CFCH, North Bay; CKVD, Val D'Or; CHEX, Peterborough; CJKL, Kirkland Lake; CKRN, Rouyn-Noranda.
161. W. Rupert Davies to R.C. Wallace, May 1, 1941; QUA, Coll. 1250, Box 19, Principal's Office, Series #1, CFRC.
162. Alex E. McRae, Ottawa, to R.C. Wallace, May 5, 1941; ibid.
163. R.D. Harkness, Montreal, to R.C. Wallace, May 2, 6 and 10, 1941; ibid.
164. "Purchase of Queen's Radio Rights", Board of Trustees, May 17, 1941; QUA, Coll. 1227, Vol. 8, p. 116.
165. W. Rupert Davies to R.C. Wallace, including draft of a letter to the Radio Branch, D.O.T., Ottawa, May 24, 1941; ibid.
166. Draft of a letter to Radio Branch, D.O.T., Ottawa, from R.C. Wallace, May 26, 1941; ibid.
167. Dr. W.E. McNeill to R.C. Wallace, May 27, 1941; ibid.
168. R.C. Wallace to Radio Branch, D.O.T., June 4, 1941; ibid.
169. John Irwin, Montreal, to R.C. Wallace, June 3, 1941; ibid.
170. R.C. Wallace to John Irwin, June 10, 1941; ibid.
171. Walter A. Rush to R.C. Wallace, June 7, 1941; ibid.
172. R.C. Wallace to Donald Cameron, Director, Department of Extension, University of Alberta, Edmonton, June 26, 1941; ibid.
173. <u>SECRET</u> letter, Walter A. Rush, Controller of Radio, D.O.T., Ottawa, to Queen's University, Jan. 29, 1942; ibid.
174. <u>SECRET</u> letter, Walter A. Rush, Controller of Radio, Chief Radio Censor, Radio Division, Air Services, D.O.T., Ottawa, to Queen's University, Oct. 29, 1942; ibid.
175. W.R. Davies to R.C. Wallace, June 25, 1941; QUA, Executive Committee, Board of Trustees, July 5, 1941, Coll. 1227, Vol. 8, p. 150.
176. W.R. Davies to R.C. Wallace, July 6, 1941; ibid., p. 151.
177. "Sale of Queen's Radio Station: Proposed Agreement Between Queen's University and the Allied Broadcasting Corporation Limited", Dec. 13, 1941, drawn up by Dr. W.E. McNeill and presented to Executive Committee, Board of Trustees, Jan. 15, 1942; QUA, Coll. 1227, Vol. 8, page 201.
    "Radio Rights Sold", QR <u>15</u> #8, Nov. 1941, p. 225.
178. R.C. Wallace to J.R. Radford, Supervisor of Station Relations, CBC, Toronto, July 25, 1942; QUA, Coll 1250, Box 19, Principal's Office, Series #1, CFRC.
179. "100 Radio Technicians", QR <u>15</u> #4, April 1941, p. 101.
    "To Train 150 Men", ibid., <u>15</u> #5, May 1941, p. 127.
    "R.C.A.F. Radio", ibid., <u>15</u> #6, Aug. 1941, p. 158.
    "Opening talk to RCAF Radio Mechanics `Ab Initio' Class Starting at Queen's June 23/41, H.H. Stewart"; Prof. Stewart's papers.
180. Hugh Popham, <u>Sea Flight: A Fleet Air Arm Pilot's Story</u>, Wm. Kimber and Co., London, 1954, p. 36. (Thanks to John MacMillan).
181. "Queen's Station Ceases", QR <u>16</u> #7, Oct. 1942, p. 203.
    Herbert J. Hamilton, <u>Queen's, Queen's, Queen's</u>, McGill-Queen's Press, 1981, p. 295.
    Dr. McNeill's words were: "But gradually it became clear that the plan could not long continue. The University station was diverted from its main task of experimenting and teaching; <u>The Whig-Standard</u> had too carefully to temper its advertising contracts to the ethics and aesthetics

of high-minded professors; both parties were frustrated by the limited reach of a 100-watts transmitter. The soaring <u>Whig-Standard</u> felt like Pegasus trying to be a rocking-horse. So the partnership has been dissolved in complete good-will. From to-night, one station which had meagrely served two masters became two stations richly serving a community. Mr. W.R. Davies, the very powerful, resourceful and persuasive President of CKWS, has got a 1000-watts transmitter and a better wave length - advantages for which the University had striven in vain for many years. The meek may ultimately inherit the earth, but not yet."

182. "CKWS Opens Tonight", KWS, Aug. 31, 1942, p. 2.
"Brilliant Program Opens", ibid., Sept. 1, 1942, p. 2.
183. "Report of the Executive Committee of the Board of Trustees of Queen's University: Allied Broadcasting Corporation", Sept. 5, 1942; QUA, Coll. 1227, Vol. 8.
184. Dr. W.E. McNeill to Allied Broadcasting Corp., Jan. 2, 1943; QUA, Coll. 1250, Box 19, Principal's Office, Series #1, CFRC.
185. Arthur L. Davies, Managing-Director, CKWS, to R.C. Wallace, Dec. 15, 1942; ibid.
186. R.C. Wallace to Arthur L. Davies, Dec. 17, 1942; ibid.
187. W.E. McNeill to Arthur L. Davies, Dec. 18, 1942; ibid.
188. Arthur L. Davies to R.C. Wallace, Jan. 14, 1943; ibid.
189. Personal communication from Kevin Nagle, Kingston, Nov. 18, 1982.
190. "Radio Beam Equipment in Use", KWS, Aug. 29, 1942, p. 14.
191. Personal communication from Steve Cutway, CFRC Station Manager, July 21, 1985.
192. H.H. Stewart to Radio Division, D.O.T., Ottawa, May 9, 1939; Prof. Stewart's papers.
193. G.C.W. Browne, Assistant Controller of Radio, D.O.T., Ottawa, to H.H. Stewart, Aug. 19, 1939; ibid.
Walter A. Rush to H.H. Stewart, July 19, 1940; ibid.
194. Walter A. Rush to H.H. Stewart, Sept. 11, 1940; ibid.
195. "Licence to Use Radio" No. 705, Call Sign CJY40, 38.6 Mc/s, at 20 watts, April 1, 1949, D.O.T., Ottawa; ibid.
196. W.B. "Buster" Doubleday, Officer-in-Charge, Air Services, Telecommunications Division, D.O.T., Kingston, to H.H. Stewart, March 14, 1957; ibid.
197. Walter A. Rush to H.H. Stewart, May 15, 1941; ibid.
198. "Licence to Use Radio" No. 49, Department of Transport, Ottawa, June 28, 1941; ibid.
Walter A. Rush to H.H. Stewart, July 9, 1941; ibid.
199. "Radio Club", PESQU <u>28</u>, Sept. 1938, p. 71.
200. "Radio Club", ibid., <u>29</u>, 1940, p. 87.
201. "Opening Meeting for E.E.C.", QJ, Oct. 16, 1945, p. 6.
"Queen's Amateurs", ibid., Nov. 16, 1945, p. 1.
202. "Radio Club to Meet", ibid., Nov. 25, 1947, p. 4.
203. Personal communication from Dr. James L. Mason, Kingston, Aug. 21, 1989.

CHAPTER 12

An unidentified student operator at the OP-4 board in Fleming Hall, circa 1950. By this time the cockpit had been re-designed. The turntables were new and were set back from the console.
-Queen's Tricolor, 1950

Summer Radio Institute to Student Radio Club, 1942-57

<u>Harold Stewart</u>: As I have mentioned, the operation subsided very quickly and substantially after <u>The Whig-Standard</u> agreement was terminated. We went back to week-end broadcasts with student volunteers - the students were faithful - they operated. After the War, the Anguses, along with the people from the CBC, organized a Summer Radio Institute which went for several years, and many people who were later prominent took part in this and it was very well run, as you might expect the Anguses to do. After that we subsided again, until about 1957, when Arthur Lower and a few people looked at CFRC and said, in effect, "There's an idle resource. You're not really justifying your existence. We should get organized for something more extensive than what you're doing." So they formed a committee, and I think that committee was the origin of the Radio Advisory Subcommittee.

> We were given an FM transmitter by Westinghouse, specially through the influence of our good friend A.C. Monteith (Elec. '23), known as "Monty." He gave us a one kilowatt FM transmitter about 1952, and we put it on the air in '54.[1]

In late 1942, after the break-up of the commercial partnership, CFRC broadcast a Queen's convocation and the opening address at the annual Theological Conference. No CFRC logs exist from the period 1943-45 to indicate the broadcasting that went on nor the names of the student volunteers involved. In fact, no station logs were kept. J.R. Radford, the Supervisor of Station Relations, CBC, wrote to Dr McNeill in 1945 noting that since the establishment of CKWS, his office had not received any logs from CFRC. Regulations required submission of a true copy of the log of each broadcast.[2] Prof. Stewart replied that an exact log was not made. He submitted one culled from memory which showed 4 test broadasts in March 1943 (6th, 13th, 20th and 27th) and 5 in March 1944 (13th, 18th, 20th, 25th and 27th). Stewart operated the programs "for benefit of Electrical Engineering students," and all consisted of recorded music, test tones and station announcements. Six went from around 9 am to noon, and the others were early afternoon transmissions.[3]

Prof. Stewart's Home-Built FM Transmitter

Harold Stewart: I was actually building an FM transmitter while the Summer Institute was going on, on the Armstrong design, but I didn't get it to a satisfactory stage because I had to carry on with my regular job. I had read about what Edwin Armstrong was doing, and in '44 I had been down to Columbia University and talked with some of his assistants. It was actually a personal hobby done on campus in my spare time, in the summertime.[4]

Stewart began planning his experimental FM transmitter before the Summer Radio Institute was conceived. A radio inspector, Mr Cox, had assured him that an experimental FM licence could be obtained without negotiation, so on August 23, 1943, Stewart sent a commercial order to the Department of Transport (DOT), enclosing $10 for a licence.[5] Walter A. Rush, Controller of Radio in the Radio Division, Air Services, sent by return mail a proper application form, and advised that the licence fee was $5.[6] The formal application was submitted on September 9, for operation in the 42-50 megacycle band:

> **Antenna Data Given Taken from Data on Old Station**
> **Proposed Xmtr Details Given:**
> Armstrong System. 100 Kc Xtal Osc.
> 6ST7 Multipliers and Auxiliary Tubes.
> 807 Amp.   810 Output Stage.
> Antenna Power - 250 Watts Max.[7]

The Joint Technical Committee and the Board of Governors of the CBC reviewed the application and the Board recommended to the

Minister that it be refused on the grounds that such licences were then being granted only to radio manufacturers for the experimental development of receivers. They would not deal further with it.[8]

Dr Wallace mentioned the matter to R.S. Lambert, Supervisor of Educational Broadcasts for the CBC. Lambert was very concerned because he was hoping that the CBC would make use of FM for educational purposes immediately after the War. He was not clear on whether Queen's application was for educational broadcasting or for engineering experiments.[9] Lambert discussed the situation with Dr Augustin Frigon, General Manager of the CBC, and with G.W. Olive, CBC's Chief Engineer, and requested a copy of the application.[10] Mr Olive then advised Dr Wallace through R.S. Lambert to renew the application, indicating clearly that it was for an experimental station, that it would be operated into a dummy antenna load until authorization could be given to radiate, and including its purpose plus the frequencies requested. Lambert forwarded a memo to the Joint Committee, to be presented when the renewed application came forward.[11]

Stewart supplied the required details to Wallace on February 3, 1944, and the renewed application was submitted on the 7th:

> Its purpose would be to give Electrical and Physics students an opportunity of working with F.M. and to provide opportunity for development in a field not yet fully exploited. It would, of course, if the University so desires, be used for educational broadcasts when the local public has acquired the necessary receivers. The transmitter would be operated into a dummy antenna until authority was given to use a regular antenna. If the transmitter is to be received by any standard receivers it would have to operate in the standard F.M. band between 42 and 50 megacycles. This would be very desirable. Proposed power output is 250 watts.[12]

The application was forwarded by G.W. Olive to the licencing authority, the Radio Division of Air Services, DOT, but the Assistant Controller of Radio advised that experimental FM licences had been suspended for the duration of the War, except to manufacturers developing equipment for the war effort.[13] Stewart could not assure a direct application of his project to the war effort and, as they had for years operated transmitters into dummy loads at Queen's without a licence, he saw no need to pursue the matter with the DOT.[14] Wallace, however, wrote back: "F.M. is becoming so significant that any research work that can be done by us may have real importance."[15] The DOT accepted the point and was prepared to recommend that the Minister grant an experimental licence provided that the equipment were solely for instruction and were connected to a dummy load to confine the radiation to the campus area.[16] Wallace re-applied on March 25, stating the special conditions, and the notice of the granting of licence No. 58 was dated April 6, 1944, authorizing a 250-watt station for 1944-45, pending receipt of the $5 licence fee.[17]

Licence No. 58 for the experimental FM station, call VE9BH, was dated April 3, 1944. Power was not to exceed 250 watts and frequency was not specified. The station was to be "utilized in conjunction with a 'dummy' non-radiating antenna only," and only for "the practical demonstration of radio equipment for the instruction of students."[18] VE9BH was renewed yearly by the Department of Reconstruction and Supply and then by the Minister of Transport under the same restrictions through March 31, 1953.

The Queen's Summer Radio Institute

Dr R.C. Wallace had been a very active supporter of the Canadian Radio League in its struggle to promote the Aird Report and establish a national public broadcasting service in the early 1930s. At the time he was President of the University of Alberta, which pioneered in education through radio with its station, CKUA. His colleague Dr E.A. Corbett, Director of the Department of Extension which managed CKUA, was also on the executive of the Radio League and, from 1936, was Director of the Canadian Association for Adult Education. As President of the Association of Canadian Clubs, Wallace must have been heavily canvassed by its National Secretary, Graham Spry. So Wallace was well-connected in public and educational radio circles.[19]

Wallace was appointed Chairman of the National Advisory Council on School Broadcasting in the spring of 1944, and he was very interested in getting Queen's involved in the CBC's expanding educational service. He thought that Queen's could contribute scripts, programs and research on audience response to broadcasting, and might eventually branch into teaching of the radio arts. The beginnings of Queen's expertise, he figured, would arise out of the activities of the Queen's Drama Guild and those of the Division of Drama of the Summer School of Fine Arts. Accordingly, Dr Wallace and Dr G.B. Harrison, Head of English, arranged through the Rockefeller Foundation to send their Director of Drama, Dr William Angus, to New York City for a 3-week course in radio production.

> **Willam Angus:** When the Anguses first came to Kingston in 1937, CFRC was the one and only radio station in Kingston doing commercials, news and so on.
>
> It was after the announcement in the Queen's Alumni Review of CFRC becoming stereo. I didn't exactly like what the article in the Review said, and I wrote a letter saying, "It was easy enough for each generation of students, anchored as they are in the present, to have the illusion that history begins with them." Then I made the point that CFRC owed a great deal to earlier generations, going back to contributions made by the Queen's Drama Guild Radio Workshop and the Summer Radio Institute of 1945-48. In '45 it was a pioneer - the only summer radio institute done by any college in the country. In fact, a short time after our successful first year, Lorne Greene, a Queen's graduate and also an instructor for a week in the Summer Radio Institute, started up his Academy of Radio Arts in Toronto. That drew

people to him because his course was the usual length of a university course, whereas ours was only for two months in the summer.

Nevertheless, there was impetus at Queen's back as early as 1943, 1944. Particularly 1944. I was still a member of the English Department until 1947, when Drama got divorced from English. G.B. Harrison was my boss, and I think he had known R.S. Lambert, Supervisor of Educational Broadcasts for the CBC, in England before Lambert came over here. I think he was sort of under a cloud when he came - some famous case to do with something that Lambert had said.

Anyway, G.B. was a Shakespeare authority - in fact, he edited all the Shakespeare plays put out by Penguin - so he was interested in getting Shakespeare produced at Queen's. I was Director of Drama and he got me with the student Drama Guild to produce first of all <u>Twelfth Night</u> in the second term of 1943-44. Robertson Davies came to see it. He was not too much impressed. He thought Malvolio was right good - I could see it his way. It's not easy to produce. So, G.B. thought, "Oh, well, let's go for the top the next time," so in November it was <u>Hamlet</u>. And then Shakespeare every November for a few years in Convocation Hall over in the Old Arts Building. Then G.B. had the idea that we should get this thing on radio. So, he said, "Let's put the last act of <u>Hamlet</u> on a record and we'll approach the CBC with it to see if we can't get them to do something for us." It must have been late 1943 or early 1944. So, G.B. proposed to the CBC that we should produce something experimental as a sample of what we could do. And they said, "Well, fine. We'll send down a director for you," and he said, "Oh, that isn't necessary. We've got a director at Queen's." We had a director for stage production, but neither G.B. nor I knew anything about producing radio, never having done it. While CFRC was still commercial, before 1942, there was the Queen's Quarter Hour, when members of staff...oh, everybody...was called upon more or less in rotation to do his 15-minute broadcast. I did several. I read poetry, I gave 15-minute talks - we all did. So that we had a bit of experience broadcasting in the use of the mike, but the actual production and so on, we didn't know a thing about that. Oh, dear!

So, G.B. had us experiment making a radio version of the last act of <u>Hamlet</u>, with sound effects, etc., and we put it on a disc. I can't remember where we recorded it - it was something that G.B. Harrison was more in charge of than I was. I was a jack-of-all-trades, and I also did whatever sound effects were required. I was pretty busy directing and doing the sword play of the duel in the fifth act with a couple of forks or something. Oh, gee, it was awful. That wasn't radio at all - one man doesn't do it all. The result was terrible. Andrew Allan and R.S. Lambert and others listened to it. I was told, when Andrew Allan heard this fifth act of <u>Hamlet</u> he just couldn't believe it! It was absolutely incredible to him. It was like a farmer seeing a camel - there's no such animal! It was awful! Gaw, it wasn't radio, it wasn't anything. It was just wrong. The only thing Andrew Allan liked about the recording was the music that went with it. That was the "Dead March" from <u>Faust</u>. There was a German refugee student who chose the music.

And so it became necessary for someone at Queen's to learn how to do radio production of one kind and another. Talks, music, drama and so on, and I was sent down to New York for a month in 1944 to a sort of school that advertised in the <u>New York Times</u>, based in the Rockefeller Center, where NBC is. In that month I didn't really learn very much. It wasn't too good a school, really.[20]

The Radio Workshop of the Queen's Drama Guild

That summer of 1944 the Division of Drama of the Summer School of Fine Arts had its first playwrights enrolled, and offered its first prizes for playwriting. The Queen's Radio Workshop was also established to realize the writers' work. Three radio plays submitted that summer were produced by the Radio Workshop and broadcast over CKWS under the direction of Dr Angus and Mr C.B. Rittenhouse: <u>The Bridge</u> by Nora Mooney of Winnipeg; <u>Underground</u> by Queen's student Kenneth Phin (Meds '44); and <u>City Music</u> by Lenore Osborne of Montreal. The Workshop broadcast two more plays over CKWS in September, in aid of the Community Chest: Ken Phin's <u>My Light is Spent</u> and <u>Jonas Hanway</u> by Nora Mooney.

The executive of the Queen's Drama Guild, a student club, took a critical step in the history of radio at Queen's when they decided to add radio production to their activities in 1944-45, by incorporating the Queen's Radio Workshop as one of the Guild's departments.[21]

The Summer Radio Institute Takes Shape

Principal Wallace wrote to R.S. Lambert on October 16, 1944, seeking co-operation with the CBC over Queen's plans to found a Summer Radio Institute at Queen's.[22] A conference was arranged at the CBC offices in Toronto, for 10.00 am, Thursday, November 2, with Drs Wallace, Angus and Harrison representing Queen's.[23] For the CBC there was Ernest L. Bushnell, Director General of Programs; Andrew Allan, Supervisor of Drama; W.H. Brodie, Supervisor of Broadcast Language; R.S. Lambert, Supervisor of Educational Broadcasts; and M.L. Poole, CBC Regional Engineer. After Mr Bushnell took the chair, Wallace explained that the proposed institute was an experiment of the drama section of the Summer School of Fine Arts. It would give elementary training in the radio arts: announcing, acting, script and continuity writing, production and control room technique. Queen's had facilities, an experimental FM station and studio equipment; or they might ask to use CKWS for practical work. There would be no academic credit at first, and students would not be encouraged to expect a job in radio. The costs would be met from the fees of the 30 or 40 students.

The CBC people felt that interest would be high, and that the demand for highly qualified people in radio, FM and television would grow. Although the tone of the notes is cool and cautious, Bushnell offered the CBC's co-operation and suggested that Wallace discuss the matter with Dr Frigon. Andrew Allan was skeptical, since two UBC summer school courses in script writing had produced no good prospects.[24]

Augustin Frigon, General Manager of the CBC, met Wallace at luncheon at the Chateau Laurier on the 28th, and offered generous assistance by CBC officers in getting the institute launched. Once the curriculum had been blocked out, he invited Wallace to request particular people for short visits.[25] Early in December, William Angus went to Toronto to discuss the details of the CBC's role with R.S. Lambert and W.H. (Steve) Brodie. Bushnell could not be present. These gentlemen were not satisfied that Angus and Rittenhouse could teach the practical aspects of radio, not having had first-hand experience. Angus shared these doubts,[26] and it was later suggested that having Aurèle Séguin, Directeur de Radio College of the CBC, in an advisory capacity would strengthen the practical side of the teaching.[27] Some time in this period it was decided to send Bill Angus back to New York City for more training in radio production and pedagogy.

<u>William Angus</u>: We arranged for a meetimg with CBC mandarins in Toronto to approach them with the idea of teaching radio in a summer course, a summer radio workshop. It was G.B. Harrison's initial proposal, and he got in touch with his friend R.S. Lambert and through him with others.

G.B. Harrison and I and Principal Wallace went to Toronto in November to confer with the CBC mandarins at 55 York Street. There were a number of key people there: Ernie Bushnell, R.S. Lambert, the famous Andrew Allan and W.H. (Steve) Brodie, who was in charge of all announcers. Our purpose, we thought, was to try to convince them that they should co-operate with us in putting on a summer course in radio. When we got there, they turned the tables on us, and they were eager to have us put on a course, when we thought we were going to sell the idea to them. So the thing was on. It was all set up that we'd have this thing called the Summer Radio Institute. That name was attached to it thanks to Lambert's idea. He had got in touch with NBC people down in the States, and they had a Radio Institute, so we had to have a Radio Institute. I didn't like the term. I would have preferred "Workshop," but it didn't matter.

They were so much in favour of it, G.B. Harrison and R.C. Wallace recommended that the Rockefeller Foundation grant me a fellowship to go down to New York for the second semester to learn how to teach radio arts. I was given a leave of absence from the first of February until the first of June to go to New York City again to take courses at Columbia University and New York University. I was to learn as much as I possibly could about radio production and particularly the teaching - the pedagogy and method of teaching radio arts to students. And that was a very productive and very intensive period. While I was down there, Aurèle Séguin was supposed to come down and meet me, and consult with me. I didn't see him until he came back up here to Kingston, but that didn't matter. We got to calling him Yehudi, the man who was never there. He was a very nice, affable person, really. He was wonderful.

I took afternoon and evening non-credit courses at Columbia and New York University. The Columbia courses were conducted by members of staff of NBC, and we met in studios of NBC at Rockefeller Center. The CBS courses were down at NYU, near Washington Square, in the Greenwich Village part of the town, in charge of a fellow by the name of Wallace House, who taught basic and advanced radio acting. I got an awful lot out of him. He had come from Toronto originally, out of Canadian radio, so we got along beautifully.

There were courses in sound effects, in music for radio, radio talks, radio interviews, dramatic writing, production and acting. All those subjects were separate courses, which prepared me for running an institute. I spent a great deal of my spare time in the library, the control rooms and studios of CBS. Much more than NBC. Thanks to the Rockefeller Foundation fellowship, I got a pass that let me into the studios of CBS and NBC and even to sit with the director in the control room, and hear him commenting to me without using talkback to talk to the actors. And I got a great deal out of that. Of the two I found CBS far, far better for my purposes than NBC, in spite of R.S. Lambert, who was favourable to NBC. NBC was not so co-operative and rather formal. The reputation that NBC had was that they were rather controlled by the legal department. But NYU and CBS, you had the feeling of a sort of a large family. It was co-operative, much less formal, and we got a great deal out of it.

I modelled the Summer Radio Institute very much on what NYU was doing because they knew how and it was very good. It wasn't like NBC's set-up, with this one and that one coming in just to give a lecture or demonstration. It was curricular with a calendar, and you could take various courses conducted by a permanent staff. So, I got a great deal out of CBS, notes and experience, because you also got practical training. You got to do the actual sound effects and other operations that could be handled. My spare time was spent pretty much in the studios and control rooms of CBS, up on 52nd Street. I got to know a number of the people there, and I got a great deal out of going to see the rehearsals and then stay for the broadcast that immediately followed. It was very helpful. And not only that - whenever I went to a studio to see a broadcast, I saw to it that I got a copy of the script that they did, whatever it was. It might be a soap opera, a drama, an interview, a comedy, a talk - oh, just the miscellaneous variety that was broadcast by the networks. And by the end of a week I'd have a bundle and send them up here to my wife to keep until I got back.

Some of the rehearsals were very interesting, especially Burl Ives. He wouldn't stay out in the studio; he'd come into the control room and he was a sort of a practical joker, a cut-up. Not when he went on the air, but during rehearsals. He was quite a character. So was Fred Allen. Several times during reheasals I wrote down what revisions he seemed to have put into the script. Some of them were bawdy! They didn't go out on the air. The actors would see me sitting there, and they had an idea that I was representing the sponsor.

Besides that - and I give myself credit for this - I went down Madison Avenue from one advertising agency to another, B.B.D. & O., Young & Rubicam and so on, that had the accounts of the various sponsors of the programs, and told them what I wanted and why. They wouldn't be broadcast, but would be used as specimens of this sort of thing for students for the purpose of teaching, and eventually I got - oh, Lord - two fairly big, wide-drawer filing cabinets full of scripts, each in its folder. Somewhere around 600 scripts altogether.

All sorts of things. Fred Allen, Jack Benny. Oh, and I got quite a lot of soap operas. The students appreciated them and used them a good deal as models. In fact, we lost some of the Fred Allen scripts, particularly. There were sticky fingers among the students.

I had quite a number of sessions with the Chief Engineer down at NYU, picking his brains and finding out what facilities were necessary for a radio workshop at a university - the equipment for the kind of broadcasts we'd be doing, for an up-to-date radio station, which CFRC really didn't have then. In fact, CFRC was then confined to small studio "B," the control room and the transmitter room in Fleming Hall. I was insisting that we needed the larger studio "A," the big corner room. And it was not an easy job to get that. R.C. Wallace didn't manage to get it, because it was being held onto by a very prestigious, important, senior research professor, the Director of Applied Science Research, Joe Gray. He had appropriated studio "A" for his research lab, and he wasn't going to let it go. No sir! Not even R.C. Wallace was going to get it from him. So, my wife went to him and we managed to get the bigger studio thanks to her good offices. R.C. Wallace said to my wife, "You're certainly a diplomat." So we got the two studios, which we needed for the full program and all the instruction that we wanted to do. We had things going sometimes in both studios, and switched from one to the other in the control room.

Down in New York, seeing what equipment they had at NYU and talking with the engineer, I found out what we needed in the way of two turntables and how you managed to hook them up and switch from one to the other - there was a good deal of highly technical instructions and that was passed on to Harold Stewart, who was Station Manager. My wife had to convey this information to him and he had to convey it to Dr McNeill, the Treasurer, and we managed to get the facilities that were necessary to be an up-to-date radio station for the first summer of the Institute.

We bought disc recording equipment, which they hadn't had, and we had a time getting it, because of restrictions on that kind of material during the War. It was hard to come by during the War - it was the Armed Services that were cornering the market. Down in New York I couldn't get a recorder. We worked to get it through Montreal and we had a terrible time. We appealed to Aurèle Séguin in Montreal to see if he could get recording equipment. We eventually did get it.

I have a sort of record of the whole period down there, by means of letters that I wrote to my wife, sometimes every day, but I never went more than two days without writing a letter. As a result of what I'd done picking brains down there, I could even send back a floorplan sketch showing what equipment we needed in Fleming Hall for the courses in the Summer School. I'm looking at a thing here that has what would be in the big studio, in studio "B," what we'd need in the control room and, of course, transmission facilities beyond that. It required two turntables with a console between them and - there's a good deal here, now that I look at it again, that I can't exactly interpret now. It was all pretty technical, coming from this Chief Engineer in New York, and made sense to Harold Stewart. It says here we need, "A table and chair in studio 'A' and a mixer on the table, with three or four, preferably four, channels. We need earphones on the output of the mixer, or pre-amplifier. Also from the mixer, a cable with a female plug to connect with a cable with a male plug which connects with the input of the cutter of the recording apparatus. Mikes in the big studio plug into mixer..." and so on. A detailed indication of what we needed, so that we could at times be recording something out of studio "A," while something else was going on in studio "B." All this equipment made sense to Harold Stewart, and we managed to get it. Whatever was done in the radio station was under his management. He was always a radio station manager throughout the year.

Oh boy! Oh! God, I had to do more damn things when I was in New York, and my wife up here. I had to write her letters what she had to do, to see the person in the Extension Department, because Summer School was an Extension Department thing, to contact R.C. Wallace, G.B. Harrison and Harold Stewart and...oh, gosh! She was my deputy, doing things, and she had to deal with Professor Joe Gray. And Doug Jemmett was apprehensive. He went about the whole building there, putting labels on the doors, that this was Electrical Engineering lab number so-and-so and so-forth. He was afraid that the camel might get more than his nose into the tent and that these radio people would take over. And it became absolutely clear that we had no intention of that at all, and Harold Stewart and I got along beautifully. Doug was Head of the E.E. Department then and not really the radio person that Harold Stewart had become. The studio remained thereafter for use by winter students as well.

Then also, I notice in another letter, I sent up a list of what was necessary for sound effects: for walking on a boardwalk, for walking on gravel, for rain, for wheels of a coach, and what to do for horses' hooves, all that sort of thing. And also a collection of sound effect and musical records. We had to get all of that.

R.C. Wallace was quite in favour of radio. He said that it was as important a thing as the invention of type for printing, because here was a medium of communication of the present day. I corresponded back and forth with Wallace when I was down in New York and he was always very supportive, very much in favour of it.[20]

Preparing for the First Session

Dr Angus drew up a detailed prospectus of the first summer of the Institute, to run from July 3 to August 16, 1945. Basic tuition would be $50. The regular staff would be William Angus and Charles B. Rittenhouse, M.A. (Supervisor of Speech Training and Dramatics, Montreal Protestant Board of School Commissioners, and Associate Director of Queen's Summer Theatre), who had written and produced many plays for radio. W.H. Brodie and Andrew Allan would be the visiting staff. Brodie and staff would teach "Speech for Radio," including announcing, pronunciation, style and microphone technique, with practice, recording and criticism. Rittenhouse would teach "Writing for Radio," in seminars and individual conferences, by analyzing professional scripts and listening to CBC programs. Angus and staff would give "Introduction to Radio Production Directing." There would also be special guest lectures by R.C. Wallace, Dr Frigon, M.L. Poole, E.L. Bushnell, Alice Frick (script editor, CBC Drama), Neil Morrison

(Supervisor of Talks and Public Affairs, CBC), R.S. Lambert, Lloyd Moore (Manager of CFRB, Toronto), Jack Radford (Supervisor of Station Relations, CBC), Dr Arthur L. Phelps (University of Manitoba), Dr R.A. Chipman (Director of Research in Short Wave, NRC), Dan MacArthur (Chief News Editor, CBC) and Andrew Allan himself.[28] More meetings followed and L.M. Baudet and Rupert Caplan (Supervisor of Production, CBC) were added to the list. Letters of invitation were dispatched in early March.[29]

Almost from the beginning, it appeared that Andrew Allan would not come to Queen's. He had summer commitments in the west and preferred to vacation in British Columbia. Besides, he was concerned about his fitness as a teacher. He might be impatient with amateur performers and with inadequate facilities "which would make impossible his usual meticulous and exacting methods which he has applied to his professional productions. He said, however, that if Mr Bushnell assigned him to Queen's he would have no choice."[28] He would also be floundering without his own technicians and sound men. He suggested that Esse W. Ljungh of Winnipeg would do the job better. There were also concerns that the CBC people would be resentful of being ordered to go to Queen's, that they would be distrustful of amateurs and nervous about trying to be academics. It was felt that the CBC staff should be housed among the students or with Queen's faculty so that they would feel comfortable and part of the whole operation.[30] There was a lot of problem finding late replacements for both the radio and the drama courses: Rupert Caplan for Andrew Allan; Marguerite Carlson of Hackettstown, NY, for Charles Rittenhouse in Drama; Elspeth Chisholm for Neil Morrison. CBC producer Kay Stevenson and Edgar Stone of the CBC's Commercial Department were also added to the faculty.

The Presto Y-4 Disc Recorder

Mr Poole had stated at the first meeting with the CBC that the Institute would need recording and playback equipment.[24] Dr McNeill placed an order for a Presto Y-4 recorder with a Montreal firm in early April. Queen's had AA1 priority. Two weeks later, Wallace asked Angus in New York to get in touch with the Presto Recording Corporation in the US to see what was happening. The order had to be approved by the War Production Board in Washington, and Angus suggested that Queen's, the CBC and Canada's diplomatic representative in Washington should bombard the Board with appeals if they were to get delivery by July 1.[31] Letters did no good,[32] so Angus suggested approaching the Ministry of Pensions and Health, stressing the possible contribution to the rehabilitation and education of veterans.[33] The Deputy Minister of Veterans' Affairs advised that the Canadian Army had to wait 11 months for a Presto.[34] There was a recorder at the Vimy Barracks, but it would have to be connected by wire with Queen's, as it could not be moved. Aurèle Séguin offered another solution: V.E. Day had passed and a recorder at CBC Quebec City might be released to Queen's on loan.[35] The CBC eventually lent the Institute a Presto K-8 and another recorder, a micro-

*The CFRC Presto Y-4 disc
recording machine.
At the first meeting with the CBC
in late 1944, the organizers of the
Summer Radio Institute were told
that they would need recording
and playback equipment. In those
days before tape recording,
instantaneous reference recordings
were cut with a stylus into a soft
acetate material coating an
aluminum or glass disc.
Dr W.E. McNeill placed the order
for the Y-4 in April 1945.
-photograph by Arthur Zimmerman*

phone, a pre-amplifier and mixer, recordings, films and booklets. The
Institute's own Presto Y-4 was expected to arrive in December 1945.

The Summer Radio Institute's First Summer

Over one hundred people applied to the first Queen's Summer Radio
Institute. The committee of Angus, Harrison, Miss Kathleen Healey
(Publicity Manager for the Summer School), Lambert and Brodie selected 55 candidates, of whom 4 dropped out. Among them were 21 teachers, 12 from private radio and 5 veterans. Fourteen CBC staffers came,
including Dr Frigon, and several from private radio and academia,
including Lloyd Moore and Howard Milsom of CFRB and Dr R.A.
Chipman of the National Research Council.

The course stressed practice rather than theory, and students received
individual help from their instructors. Each one got experience in operating the board, sound effects, acting and directing. That year 45 students
each produced a program that was recorded, which involved writing or
selecting a script, casting from auditions, choosing a production staff,
directing rehearsals and finally cutting the record. Three 30-minute
student dramas were broadcast over CFRC and one was later broadcast
from Montreal over the CBC Trans Canada Network. Ralph Purser
(Elec. '47) operated the transmitter for the Institute and once gave a
short piano recital, broadcast remote from Grant Hall, to assist a student assignment.[36]

<u>William Angus</u>: So, the first summer was arranged very much with the
co-operation of the CBC. I was the Director and Aurèle Séguin of
Radio-Canada in Montreal was appointed the Associate Director.
We had W.H. Brodie, who was in charge of all the announcers; we had
Elspeth Chisholm, who did talk shows, as Public Affairs; we had their
Chief Engineer from Toronto, and a person who was in charge of the
music programs. But Andrew Allan - no. He didn't want to risk his
prestige and associate with inexperienced, raw non-professionals. So,
he wouldn't have anything to do with it, and in his place was Esse
Ljungh, directing productions of radio drama, who stayed with us

also through that first summer. Besides that, I got people from private stations, Howard Milsom, who did the Buckingham Theatre, to direct the broadcasting of a half-hour play. And we got Rupert Caplan from Montreal. The kids called him Uncle Rupert. With the co-operation of Augustin Frigon, who was the head man of CBC, they supplied us with a whole series of CBC people that first summer, to come for one day, a few days, a week, to help with the instruction. W.H. Brodie himself came for a week; we got drama producer Kay Stevenson for a broadcast. She produced R.S. Lambert's school programs, and so she was given to us for a weekend that summer. R.S. Lambert was with us for a week, and Neil Morrison, the talks person sent Elspeth Chisholm. If we were teaching radio production in a Radio Institute, we had to give them the whole thing.

The first year was especially good - 52 students in all and they were not only university students. The people who enroled in the Summer Radio Institute weren't doing it just for a hobby or for the information. Some of these people were going back to stations they'd come from, and others were hoping to get into radio somewhere, a private station or the CBC.

They didn't all go into radio. Dorothy Cotter (1946), Barbara Monture (1945), Larry Palef (1947) made the CBC - Larry was a Queen's winter session student as well - and Florence Fraser (1946). Edie Shindman (Arts '49) - she did go to CBC and was assistant to Kate Aitken in women's programs. There was quite a good number of people from private commercial stations, from Halifax to Winnipeg. There was even a woman from Colgate University who had written a book about radio and came up in 1945 to learn some more - Florence Felten French. A good many of them were veterans, home from the War, that looked at radio as a career to get into. My wife took a course too that first summer, and Terry French - then he started CKLC, "Limestone City."

It began in the forenoon and went through the afternoon and sometimes into the evening, with rehearsals and so on. On the second floor of Fleming Hall. There were some lectures, but most of it was workshop. Once a week that summer there was a Queen's radio hour on CFRC. There would be "Kingston Chronicle," which was news, and there would be some music and an interview. As Dr McNeill would say, "They might as well be speaking into a lead pipe. There'll be nobody listening."

At the Institute I did dramatic writing and dramatic production - the writing and production of radio plays. Elspeth Chisholm did talks and public affairs, and we'd have somebody for a week doing announcing. It was Steve Brodie the first year, teaching announcing. The second year it was Gilbert Harding - oh, boy, was he a character! A well-known notorious character. W. Bruce Adams from Central Tech in Toronto took the engineering course the first year, and then he became our engineer the next three summers.

You know that 8 typed pages will be 15 minutes, usually. You get so that you get pretty efficient at timing, and in production you do it with a stopwatch. In a drama program the director marks his script with a tick for 15 seconds and two ticks for 30 seconds, and so on, minute by minute during rehearsal and during production, so that he's got the thing accurately timed. You don't have to do that on the

Students and staff of the first Summer Radio Institute at Queen's University in the summer of 1945.

<u>*Back Row*</u>: *Francoise Rouleau, Pauline M. Spooner, office secretary, Professor H.H. Stewart, Dorothea C. Graves, Marion G. Bell, Clive Taylor (recorder), Stuart MacFarlane, W. Bruce Adams (Instructor in Radio, Central Technical School, Toronto), John G. Sutherland, Kenneth W. Grant, Steven B. (Bud) Hayward, Douglas T. Manning, Donald P. Delaney, Bernard (Sigh) Blumer.*
<u>*Third Row*</u>: *Elizabeth G. Saint, Louise M. Wyatt, Miriam A. Cameron, Martha Harrower, Jean Lorroway, Sadie Miller, Cecile A. Chenier, L. Sylvia James, Marion G. Northcott*
<u>*Second Row*</u>: *Eleanor G. (Alfie) Beatty, Florence Felton French, Agnes (Nan) Husband, Harry M. Grant, Grace S. Bartholomew, Barbara A. Monture, Judith A. Bainard, Eleanor L. Swan, Vernadelle Tibbles, Robert Morin, Frances R. Silcox, Fern A. Rahmel, Chris G. Hansen.*
<u>*Front Row*</u>: *Maxine E. Sutherland, Adelle Seeley, Shirley M. MacKellar, Marily D. (Buttercup) Plottel, Rupert V. McCabe, Joyce Taylor, Dr William Angus, Aurèle Séguin (Director of Radio College, CBC), L. Maurice Lacasse, Constance H. Denny, Father Matthew G. Meehan, Russell J. Eastcott.*
<u>*Absent*</u>: *Jean K. Evans, Margaret A. Shortliffe, Duncan A. Chisholm, Harry Furniss, Garth D. Gowan, Robert A. Kennedy, Pierre Saucier, Hideo Shigei.*

<u>**INSTRUCTORS**</u>: *Dr Augustin Frigon, W.H. Brodie, M.L. Poole, Ernest L. Bushnell, Alice Frick, Elspeth Chisholm, R.S. Lambert, Kay Stevenson, J.M. Beaudet, J.R. Radford, Arthur L. Phelps, Edgar Stone, D.C. MacArthur, Rupert Caplan.*
-courtesy Dr William Angus, Mrs Eleanor Foster and Verna (Tibbles) Vowles

---

stage. Things were done live - down in New York in '45 I also observed some of the TV productions, right in the studio. CBS studio was down in Central Square Station - a great, big barn of a place up above the concourse. Oh, I bought records down in the States too, where they were more easily got. I got quite a library of sound effects down in New York, and also a collection of mood music. Then we found a company in Toronto that supplied that sort of thing.

We got going and we recorded what we did. Something would be rehearsed, whether it was a talk, or news or drama, it got recorded and played back to the kids so that they could hear it. Not only those that did the broadcast, but the whole lot. Aurèle Séguin, who was Director of Radio-Canada in Montreal, he wouldn't touch a disc recorder. No sir! Because you had to learn how and you had to get proficient, and he wasn't going to risk making a mistake. If it could go wrong he didn't want anything to do with it, so he wouldn't do any recording. I had to do that. We met in studio "B" and studio "A" and down in the library.

They sent Rita Greer Allen, the wife of Robert Allen, both very important CBC people - well, free-lance more than anything else - to be with us for a day to see what the deuce we did. They had a CBC program, Canadian Round Up, a sort of documentary program. Oh, she was quite impressed. Quite a number of people in Toronto, like Andrew Allan and some others, thought that this is Kingston, not Toronto, and you can't expect much of it. A typical Toronto parochial attitude toward Kingston. Boy, it was an eye-opener to this Rita Greer Allen, that this course was really quite good, that there was something going on and something for the students to be involved in from the time they came in the morning right through the whole morning and afternoon. They were shifted from dealing with somebody on the subject of news to somebody on music programs to sound effects to radio drama and radio drama writing. I lectured on that in the library. In the control room and the two studios there would be something else going on, while in another classroom somebody was doing something with a group. They got divided into groups that worked together - one would have their turn on producing a radio play, while another group was involved with something else. They were all recorded, and you might have 3 or 4 discs to play back for the whole group to hear. It was a thing that needed a great deal of organization, which I learned down at NYU - a time for this, a time for that and it had to be worked out away ahead of time for the curriculum for the whole thing right through the whole period. It was quite a task, but it turned out to be well done. We were complimented on it by Wallace and by Dr McNeill, and by H.L. Tracy, who was Director of Summer Session. The first two summers were quite good. A camaraderie developed among the students. There was one woman who wanted to make a kind of club of it.[20]

In his report on the first Radio Institute, Angus made many recommendations for the improvement of facilities for teaching, including converting the two studios into one large one to improve microphone placement and acoustics, a piano in the large studio, a non-reverberating room for disc playback, another small announce studio for practice and for cutting records, new turntables or at least overhaul of the old motors and lighter pick ups, two modern microphones, such as RCA ribbon velocity 44 BX or 77C, and enlarged sound-proof windows between the control room and the studio. There was a monitor for the large studio in the control room and talk-back to it, but the small studio had neither. A perennial problem was that the two studios could not be handled simultaneously by the single control room. The capital cost might be $2,000.[37] D.G. McKinstry, CBC's Chief Architect, was invited down to look at the studios and suggest what remodelling should be done.[38] There is no evidence that anything was done, but Harold Stewart bought two microphones.

Summer Radio Institute, 1946

For the second session, July 3 to August 16, 1946, they expected Aurèle Séguin, W.H. Brodie, Elspeth Chisholm, W. Bruce Adams, Mrs. Elsie Park Gowan, Lorne Greene, R.S. Lambert, Howard Milsom, Rupert Caplan, Kay Stevenson and Michael Barkway, Chief Representative of the BBC in Canada.[39] They also sought J. Frank Willis from the CBC.[40] There would be a junior course for a maximum of 50 students and a senior course for 10. Lorne Greene offered a $400 tuition scholarship to a graduate who would take the full course in 1946-47 at his Academy in Toronto.[41] Again, the CBC lent equipment to the Institute through our old friend Alex J. McDonald, who was now with the Corporation.[42]

In the end, Séguin did not come, nor did Brodie, who was replaced by Gilbert Harding, the BBC representative to the CBC: "Mr Greene says that Mr Harding is very good 'in announce work'."[43] Lorne Greene was unable to attend.[44] The staff consisted of Bill Angus, Gilbert Harding, Elspeth Chisholm, Bruce Adams, Prof. Henry Alexander of Queen's, Kay Stevenson, Howard Milsom, Michael Barkway, R.S. Lambert and Rupert Caplan. The Institute attracted only 36 students. A notable feature of this summer was that they broadcast an hour-long program over CFRC each Monday, Wednesday and Saturday evening for 3 weeks.[45] In one of these broadcasts, on July 19, R.C. Wallace delivered a short address on "The Functioning of Radio as an Agent for Good or Evil."[46] A visitor's view of a day at the Institute, including a few moments of a rehearsal with Rupert Caplan, appeared in the Queen's Review.[47]

<u>William Angus</u>: Some of the visiting firemen who came to the Summer Radio Institute to help us out with the direction of a play that was broadcast, or for a week to teach radio announcing, were characters.

Rupert Caplan directs, from left: George Hickey, Cornwall; Clyde Douglas, Halifax; Bernard Trotter, Kingston; Shirley Elkin, Montreal. Summer Radio Institute, 1946.
-courtesy Queen's Archives

Characters! Rupert Caplan came every summer over a weekend to direct the broadcast of a play. People like that came just for a day or so, and we managed to give them accommodation at our house on Collingwood Street - room and board while they were here - and Rupert was quite a character. A fellow with quite a powerful, strong, resonant voice - it would fill up a room. He was rather upsetting really, because he was such an exuberant, buoyant fellow. He'd come with a bottle of Irish whisky and he and my father consumed it - my father wasn't too good at drinking, but anyway - and we had a lovely cocker spaniel that we called "Scamp." Rupert, with his strong, booming, Lorne Greene sort of voice would yell at the dog, "Scamporo! Scamporovich! Scampolino!" and so on, and the poor dog was upset too. There was a dish that my wife cooked and served to him and he said, oh, he loved it so. He was going to beat up his wife when he got back home and have her make the same thing.

The first time Caplan came, my daughter was only 14 years old and, of course, he had remarks to make about her. Oh, he'd make remarks about anything. The next summer he came again at his usual time, and the first thing he says to my 15 year old daughter in his loud, bellowing voice, "Oh, aren't you married yet?"

And the other fellow that was memorable, the other character, was Gilbert Harding, who came to us from the BBC in England for a week. He did a job that W.H. Brodie had done and Lorne Greene had done, teaching announcing. In fact, he applied for the job, and I got a letter from R.S. Lambert saying that this man had applied and he sort of warned me in a very subtle way about this man who was coming. In effect saying, don't blame the CBC. This fellow came and doused himself with what we called "Repel," which was to keep the mosquitoes away. He was on the way to becoming alcoholic - he had been for some time. In fact, he came from England the way they used to send remittance men out to the colonies. When his term was up in Toronto, the next place they put him was Egypt. They didn't want him back in England. He was broadcasting one night from Portsmouth Harbour when the Royal Navy fleet was doing something rather special, and eventually he says, "Ah, the fleet is all lit up, and so am I." That didn't go down too well.

One night at our house, he came into the living room with a bottle of whisky and a bottle of wine. The wine was for my wife. She maybe had just a glass and after he'd finished the whisky, he also finished the wine. And he could hold it too, and he'd regale us with stories. The only one I remember now was of his two aunts that were whacky. Oh, they were crazy! He told of a time that they went to the funeral of a friend of theirs who had died, and as the casket was being lowered down into the grave, one aunt went forward and shook her finger at the casket and said, "You promised to mend my fence, and you never did it!"

He would go to lunch, when he was here, to the LaSalle Hotel which was still in existence then, and he was driven to the hotel by Bruce Adams, a Torontonian who was on my staff as technician. He actually gave a few lectures on technical matters to the students. This fellow was very sedate and very proper. He's driving the car down Princess Street to get to Bagot and his companion sitting beside him leans out of the car and yells at a little kid that's on the sidewalk, saying, "Get off the road, you filthy little beast!" And there were other things like that which were typical of him that occurred as they went to lunch at the LaSalle Hotel. Then one day at the hotel after they'd gone in and sat at a table quite a distance from the door, they ordered their meal, and friend Harding said he didn't want gravy on his potatoes. I think that was it. Unfortunately, the potatoes came from the kitchen with gravy on them. Uuuuh! Oh! He stood up and just blasted them on that and stalked out of the dining room. And poor little Bruce, who was his chauffeur, had to tag along with him.

Well, he was here only a week, and the week immediately after that Bruce was driving down Princess Street alone in his car, and he said he would see people on the sidewalk, nudging one another and pointing at his car. That was the car that the fellow had been in that yelled at people, and so on. Oh, Harding was very entertaining, but he was darn good at his job.[20]

Summer Radio Institute, 1947

For the 1947 session, the Institute secured the services of Elsie Park Gowan, CBC Edmonton, for dramatic writing, W. Bruce Adams, Elspeth

Chisholm for talks, Lorne Greene for announcing, R.S. Lambert, Rupert Caplan and Kay Stevenson of the CBC, Michael Barkway of the BBC and Howard Milsom of CFRB. The CBC again lent equipment, scripts and program transcriptions.

Student enrolment was "rather disappointingly low" in 1947, but the "quantity and quality of production was remarkably good...It was generally felt that the fewer students was not due either to a lack of demand for a course like this or to its inadequacy but, rather, to a deficiency in its publicity." Minor revisions were made to increase course flexibility and to improve the quality of the product, and the fee was raised from $55 to $60 for 1948.[48]

Students taking the Queen's University Summer Radio Institute Course gain practical experience through broadcasting over the Queen's University radio station CFRC, probably 1947.
Left to right: Nancy Burns (Ogdensburg, NY), Mary V. Jamieson (Sarnia) and John Jordan (CJIC, Sault Ste. Marie).
-courtesy Dr William Angus

Larry Palef: I was taking a pre-Meds course at the time, rather heavy with comparative anatomy and cutting up cats, which really didn't appeal to me. I saw this ad outside Douglas Library for people to audition for Hamlet. Sylvio Narazzano, the film director, had left Queen's and consequently there was the role of Laertes open and available to anyone who could pass the audition. I auditioned with several others for G.B. Harrison and Dr William Angus, and they were delighted with the reading and that was the beginning of my downfall. I was great as Laertes, as far as the audition went and as the dress rehearsal went. But when it came time to go on, I wasn't available. I came down with a fever of 104 and ended up with viral pneumonia and in Kingston General Hospital for a long time. Doc Angus tried to get me out the first week or so, when the play opened, and I just was too ill to make it. Bernard Trotter subbed for me with palmed notes - did a very quick study and I understand did an excellent job. So, I had to withdraw for the year, because I wasn't able to come back in time to write my finals.

As a result of that, I decided that radio was for me. It was much easier - you didn't have to memorize at that time and I felt I would enjoy it more. Consequently, I took the summer course that Dr Angus was offering - I think it was the summer of 1947 - and it was most enjoyable. There was Lorne Greene, who allowed me to go to his Academy of Radio Arts after the course free of charge on sort of a scholarship, there was Rupert Caplan, Elspeth Chisholm from the International Service of the CBC, and several others. It was a good foundation in broadcasting that I received at that time.[49]

Queen's University Summer Radio Institute, 1947.
From left: Elsie Park Gowan, Larry Palef, Dr William Angus.
-courtesy Dr William Angus

<u>Elspeth Chisholm</u>: Alice Frick and I were hired together at CBC – many women were picked out across the country by Neil Morrison's predecessor, Hugh Whitney Morrison, during the war – Hugh packed as many women as he could find, who looked promising, into jobs that men were leaving. We were part producers, but mainly we were national program planners. I had people from all over the country do the legends and yarns of their regions on Canadian Yarns, and Margaret Angus was one of MY broadcasters. I found myself out of a job for six weeks one summer, and the only solid thing I had to look forward to was the Queen's Summer School. When I got a letter from Bill Angus telling me what my duties would be, it included training announcers, most of whom were male. CBC did have female announcers. Madelaine Charlebois was an announcer all during the war from CBO, Ottawa. So I had to get books and bone up hastily on something I'd never done, because I was a natural. I had never had any voice training, never wanted any. Indeed, I sounded too rafeened for Gloria Harron, who was producing Trans-Canada Matinee. I had to teach these people, and I remember telling them that the way the BBC announcer got those tones was to raise the back of his palate. Altogether it was a beautiful experience, very intense, a lot of studio work. Bill was great, but very demanding, very much into everything. He was with us a lot in class. We flattered ourselves that we did a better job than CKWS by putting very good programs on the air.Bernard Trotter did a thing for a weekly show called something like Kingston Chronicle. I took to Bernard; after that Bernard got a job with the CBC in Winnipeg. Then I was hired by CBC Overseas Service in Montreal, and my supervisor seemed to me to always take the credit when I did anything good and give me the blame when I did anything not-so-good. When he left for higher and greater things, I went to our Chief Supervisor and said, "I don't want anybody else like that again. Ever. I'm Assistant Head and should be in line" – but it was a tradition

On a hot summer day, Elspeth Chisholm (left) explains a fine point about talks broadcasting to two students of the Summer Radio Institute.
-courtesy Elspeth Chisholm

that women were not made heads of sub-departments of the CBC. You could get so far and no further. But I said, "There's Bernard Trotter in Winnipeg doing his apprenticeship...Couldn't we have him in?" They will never understand why I FLEW down the corridor and greeted Bernard, kissed him on both cheeks and welcomed him when he came. So, you see, there I was a king-maker, and he went on to greater and greater things.

Bernard Trotter: I'm specially interested in CFRC because I spent 15 years of my life in broadcasting with the CBC, and I can truthfully say that I got started at CFRC when I was a graduate student here in 1945-46. I went to the Summer Radio Institute in 1946 and got very interested in broadcasting. As a result, Lorne Greene offered me a scholarship at his Academy, beginning that fall. That's really how I got going. One of his objects in coming down to Queen's and offering the tuition scholarship was that he thought it was important that people with university background got into broadcasting. Most had come out of journalism or whatever.

I can't remember whether I did anything on CFRC during the winter of 1945-46, when I was working on my M.A. I know that I did get involved in a drama broadcast at CKWS. I was interested in acting. I guess I was beginning to think about what work I was going to do, and my M.A. work convinced me that I didn't want to do any more academic work for awhile. Journalism attracted me and radio drama too. Perhaps the availability of the course and just a kind of vague interest that maybe there was an interesting career there, led me to take the course. It was so easy to do because I was living at home and could just plug in. I guess I was going to finish my M.A., until the Lorne Greene opportunity came along. It was all serendipitous, mainly.

The Summer Institute was a unique thing at the time. The biggest name who turned up there was Gilbert Harding, who came as a lecturer. He was the BBC's assistant representative in Canada, and a completely eccentric character whose lectures were very funny. Harding was also very colourful in his speech, and told us a lot about the BBC - not always from a flattering point of view. But since the BBC was the image of the experienced public broadcaster at that time, it was very valuable. He told about his work with them during the War. He would go out and do man-in-the-street broadcasts about events of the day, which was a pretty elaborate business in those days. You had to have a sound truck with a disc recording machine. At one point Mrs Roosevelt was visiting Britain and he went out to find out how Britons were reacting to this. He stopped one very genteel lady and said, "Are you aware, Madam, that Mrs Roosevelt is visiting our country?" And she said, "Oh, yes!" He said, "What do you think is the purpose of her visit?" And the lady said, "Why, of course, to have intercourse with the American troops."

I remember Howard Milsom, who came to do some productions with us, and to teach us radio dramatic production. He was the producer of Buckingham Theatre, a quite successful Canadian commercial drama series. The other major outsider was Elspeth Chisholm, who gave the course on talks - the art of talking on the radio - and writing, in order to be able to talk effectively. She was a marvellous broadcaster herself, a commentator on the CBC and then a producer with the International Service. She was a very good teacher. She would tell us the theory of it, and then we'd have to do it, and then she'd tear apart what we had done, both in terms of the writing style and the delivery. All in all, it was an exceedingly good course and a lot of fun. There were several people in that summer course who went on to careers in broadcasting: Drew Crossan became a major variety person in CBC television, and I worked with the English section of the CBC International Service in Montreal.

The Lorne Greene Academy was really quite fantastic. Up to that time there had been no broadcast training facility in Canada at all. Everybody learned on the job. He felt it was necessary to raise the standards, particularly in the private sector. There were a lot of veterans, both men and women, with educational gratuities which would pay their fees at approved institutions. He set up the Academy in an old house on Jarvis Street in Toronto, right across from the CBC, and he got marvellous people as instructors - Mavor Moore taught drama, Andrew Allan taught production, Lister Sinclair taught writing, there was a speech man and Lancaster for sound effects. It was just like a regular school, with classes in a five day a week timetable. Production classes were the most fun, because that's where it all came together - a lot of practical work because everybody in the class had to produce at least one play, which was recorded using our classmates as actors. We had to write a full length play, news, commercials. It was a very stimulating year and we worked very, very hard, usually from 9 am until very late at night. I worked so hard that I got ill and had to take a year off. I finished my M.A. and then went to work for the CBC in 1948.[50]

William Angus: Lorne Greene left Queen's in the spring of 1937, and I came in the fall, so I can't claim him as one of my boys. However, he worked for me in the Summer Radio Institute one summer. While he was a student here, he was a member of a group that was very active in drama: Robertson Davies, Lorne Greene, Arthur (Suds) Sutherland and a fourth. Sutherland later started up "International Players," that played in Kingston and upstate New York in the summers and in Toronto in the winter, including Barbara Hamilton, Russ Waller and others.

For another anecdote, Fred Euringer with his car was assigned to go to the airport in Toronto to meet Lorne Greene and bring him here in 1971, when Lorne got his honorary degree. And there was Fred at the airport reception area, and here comes Lorne Greene, and people coming over to him and saying, "Hey, Ben! Ben Cartwright!" I had the job of putting the hood on Lorne, when he came here.[20]

Summer Radio Institute, 1948

The enrolment fell again in 1948, to 19. On staff were: W.H. Brodie, CBC Supervisor of Broadcast Language for announcing (in place of Lorne Greene); Helen James, CBC Talks and Public Affairs; Esse W. Ljungh, CBC Radio Drama; Rupert Caplan; Elspeth Chisholm, CBC International Service; Dr Glen Shortliffe, who was experienced in news analysis and "opinion" broadcasting; Joseph Schull and Hugh Kemp for dramatic writing; Mrs Margaret Angus for non-dramatic programs; and W. Bruce Adams for technical.[51]

Class in radio writing on the lawn in front of Fleming Hall, July 1949. Elisabeth (Elsie) Park Gowan, visiting professor from the University of Alberta, Edmonton, is at right, and Dr William Angus is at rear centre.
-courtesy Dr William Angus

<u>Esse W. Ljungh</u>: Harry J. Boyle was the Program Director at the CBC in Toronto and Andrew Allan, Rupert Caplan, J. Frank Willis and I - each of us had our own specialties in dramatic production and there was always a certain kind of rivalry in those days. I did the international plays, and was very independent. I could select a play and go ahead and do it. There was no question about it. I valued the freedom that you had at the CBC, and the confidence they had in you. My assignment at the Institute was in the production end of it, teaching people where to stand at a microphone, how to deal with pronunciation and various other things along that line.[52]

Shakespeare Adaptations for the CBC

<u>William Angus</u>: Evidently Lambert was satisfied that we'd done a good job with the Institute, and he approached me with a job at the end of the first year to do the radio adaptation of <u>Julius Caesar</u>, for the Education Department of the CBC. I took it on and we had a series of letters thereafter, because Lambert was rather a nitpicker. I thought I'd produced a radio drama - of course Shakespeare wrote drama - and Lambert didn't want it a drama. He wanted just a reading, but we compromised eventually. We did it, act by act, one act each week in January-February 1946. G.B. Harrison, the Shakespeare authority, did a five minute commentary to introduce each program and then came the production of my adaptation of each act.

And the first one was very successful. <u>Julius Caesar</u> was entered in the Ohio State Competition and, fortunately, got first prize. A letter to me from Lambert: "You'll be interested to know that the judges at the 1946 exhibition announced on May 4th that a First award had been made in the classification 'School Broadcasts for Junior and Senior High School' to our recent presentation of Shakespeare's 'Julius

Caesar'...For a rich and brilliant portrayal of Shakespearean literature, executed on a superbly high level of production, an example for
other education programs to follow, both in the school and out."[20]

Kay Stevenson was the producer. Bud Knapp played Julius Caesar,
Tommy Tweed was Cassius, Frank Peddie was Brutus, Douglas Master
was Mark Antony, and the rest of the cast included Bernard Braden, Alan
King, Lister Sinclair as Casca, Ruth Springford as Calpurnia, Alice Hill,
Heldy Rennie, Frank Perry and Herb Gott. The music was composed and
conducted by Lucio Agostini and Gordon Keeble announced.

> <u>William Angus</u>: The next year R.S. Lambert decided that G.B. and I
> should do <u>Macbeth</u>. And <u>Macbeth</u> is a jinx show, on stage or in any
> way. Professionals will tell you that. Something goes wrong somehow.
> I produced <u>Macbeth</u> on the stage here, and in the first night of the
> duel between Macduff and Macbeth, Macduff - darn fool he was -
> didn't follow the swordplay routine worked out by an expert from
> RMC. He messed it up and got it on his wrist. Damn near broke his
> wrist! That was the second one to be done by the CBC with my adap-
> tations, and it didn't go well. It was not a success. No fault of G.B.
> Harrison's or mine. The fault was in Toronto really. First of all, it was
> not well cast I would say, when I found out who was cast and I heard
> the first broadcast of it. The wrong people had been selected. It didn't
> go well. Probably Kay Stevenson was the director, but the actors acted
> up rather much. I heard that a good many of the actors were in the con-
> trol room arguing with the director a good deal of the time. It wasn't
> good, so they decided if they continued doing that sort of thing, they'd
> have the adaptations made by somebody in Toronto who could actu-
> ally be there in the control room - to see to it that they did his script
> properly. So, I lost out. I didn't get much of a fee on that either.
> If I remember correctly, I could have carried on and done what-
> ever they did the next year, if I would go to Toronto to be in the con-
> trol room along with their director, to answer questions and settle
> arguments. But I just didn't want to be in Toronto at all, so I didn't
> have the job after <u>Macbeth</u>.[20]

The End of the Summer Radio Institute

The Queen's Summer School Committee, Principal Wallace, Dean W.A.
Mackintosh, Miss Jean Royce, H.K. Hutton and Miss K. Healey, discussed the Radio Institute at a meeting on December 31, 1948: "It was
the general opinion that it should be dropped until there is more demand
for such an Institute. It was felt that a Radio Workshop in connection
with the Drama Course might be worked out."[53] Accordingly, the Radio
and Drama Departments were combined in the 1949 Summer School,
under William Angus, and the course in radio was reduced to Drama S2,
a Radio Workshop. William Angus and his wife Margaret gave Drama
S2, "A practical introductory course in writing for radio and radio production with much individual work, personal conference, rehearsal and
actual broadcast of programmes. This course does not offer degree
credit but may be taken for a Certificate or as a Refresher Course."
The fee for the course was reduced as well, to $15.[54]

Eileen Green, now Mrs Fleming, was recommended by her high school principal for a job at the Department of Extension at Queen's, and was exempted from her grade 13 finals so that she could start work on May 9, 1949. Mr Hutton, the Director, enrolled her in two Summer School courses, Public Speaking with Arnold Edinborough and the Radio Workshop. Mr Hutton had some connection with rural and farm groups and needed someone to do speaking engagements and radio broadcasts on behalf of his Department.

Eileen remembers spending several hours in each course, every day for a month or six weeks, in a small attic-like room in Old Arts. Her classmates were mostly teachers and librarians. The Radio Workshop was all practical work - students reading parts from scripts in front of a stand-up microphone. William Angus would correct and critique the students, and Margaret Angus would critique the reading. They did not do any broadcasting, but they did cut records. After the courses were finished, Eileen was given the job of compiling the minutes from Rural Farm Forum meetings into a program which she broadcast over CKWS one morning a week.[55]

William Angus: We tried it for 4 summers. 52 the first summer, 38 the second summer, it was down to maybe less than 20 in the third, and there were no more than a dozen in the fourth year. Of course, immediately after the War, there were quite a number of veterans among those first 52, who had got grants from the government. Year by year and progressively it declined in enrolment. Well, for two very good reasons. One reason was that it was just a summer course, and Lorne Greene had found out how to do it and had set up his Academy of Radio Arts in Toronto, which was September through May I think, a regular whole school year. He always was a smart fellow. He was with us for a whole week one year teaching announcing. The exercises he had to drill on were commercials - selling things. He might have seen what the deuce our course was like. I don't know whether that had any influence on him or not. Anyway, when he started up his whole school year of the Academy - it must have been after our second summer - that's where the veterans and all of this went. The second is that the market was rather saturated. There weren't employment opportunities so much as there had been immediately after the War. Our enrolment just dwindled and it was decided that that was enough.

After the first year, I didn't have Aurèle Séguin any more - didn't need an Associate Director - but I had Elspeth Chisholm full time and she was very, very good on talks and interviews, public affairs and that sort of thing. We had her every summer. And from the second summer on we had Bruce Adams as engineer. Salaried people. We heard of and heard from Augustin Frigon, but we didn't see him. He didn't come here. And every year I got for a week-end somebody from the CBC or from CFRB's Buckingham Theatre to do a dramatic production. Well, some of the pupils might be going to a private commercial station rather than CBC, and they should find out something about that. Another thing that I had that the kids appreciated very much - I'd get the log, a day's broadcast at CBS.

The CBC gave us a great deal of assistance in the personnel. That would have cost a good deal. The personnel didn't cost us anything in fee or salary. They were salaried CBC people. I think Lambert had talked them into it, to get a University doing something for the school broadcasts, something of benefit to him - writers and performers. In spite of Andrew Allan.[56] One exercise Lambert had us do was a program that, if it was good enough, would be broadcast on one of his school programs. The whole class was set the assignment of doing a half hour script on "Edmonton, the Gateway to the North." Whatever came out of it, it was used by Lambert. The next year it was R.L. Stevenson's <u>The Black Arrow</u> for school broadcast. He thought there might be Canadian writers produced here. Very few went into the CBC. Drew Crossan was another one, Larry Palef, Bernard Trotter, Ernie Mutimer, "Tippy-toe" Douglas from CBC Halifax - he always stood up on tippy-toes when he was at the mike - Edie Shindman worked for Kate Aitken, and Michael Roth, the magician, who had Razzle Dazzle on CBC-TV, a children's program.

The installations necessary for the courses given there in '45 provided CFRC with what was needed then to be an up-to-date radio station. What was done then carried on in the regular session, thanks to the engineers who had the CFRC Club, and the Drama Guild's Radio Workshop. The Institute did a great deal for CFRC, to make it what it has become.

So, the filing cabinets were moved out of Fleming Hall and came down to beside my desk in the Old Arts building after the Summer Radio Institute folded, which it did, after 1948.

After the Radio Institute finished, we continued with the Radio Workshop in the winter session. Wallace felt that it was sort of an extension of Drama. After the first year there was a so-called Queen's Drama Guild Radio Workshop. The Drama Guild did stage performances - oh Lord, that goes back to the Queen's Dramatic Club started by Rev Dr Dyde in 1899. I can't remember that I did much with it - the students took over. The E.E. students were quite eager to get into radio and did a good job with it. The Radio Workshop didn't do very much in the way of drama, but they did do popular and classical music programs, talks and news - the students could handle it. My son was sports broadcaster for Queen's Radio Workshop regular session. Ted Bond, heck, he played on the stage too. He was Marchbanks in Shaw's <u>Candida</u> - very good too. Bob de Pencier, on the engineering faculty, acted in <u>Romeo and Juliet</u>. Grant Sampson was in a couple of plays. Sid Penstone was our electrician for two or three years, and was he a wizard! Whew! Of course, he still is. He did the lighting for the stage productions and he was indispensable. Bernard Trotter is another one - his father was Head of the History Department. In fact he was CBC man in London for a while, and he met his wife at Lorne Greene's Academy.[20]

The Drama Guild Radio Workshop Revives CFRC

With the success of the first Summer Radio Institute and the enlarging of general radio expertise among the Queen's student body, the Radio Workshop of the Queen's Drama Guild wanted to try broadcasting over CFRC. William Angus, the Guild's faculty advisor, spoke to the members about the Radio Institute at the organizational meeting on October 11, 1945, and the facilities of CFRC were put at the disposal of the Workshop. The Radio Workshop announced that it would broadcast as often as programs could be prepared.[57] The first effort was the play Pigs Is Pigs, adapted by Dr Angus from a short story and re-worked during his radio course in New York. Auditions took place on October 25 and the play went on the air on Hallowe'en night.[58]

William Angus: The author of Pigs Is Pigs was Ellis Parker Butler, and I was rather addicted to him somewhat, oh, for a number of years. He wrote quite a number of humorous little stories which appeared in a magazine month after month. Pigs Is Pigs, of course, is his best known - sort of classic - about Mike Flannery, who is an express agent in a small railway station somewhere in New Jersey, who received a crate of guinea pigs - a couple of guinea pigs - and Mr Morehouse, to whom they were assigned, came to collect his shipment and said they were pets. No, no. To Flannery, these things were pigs. "Pigs is pigs, and pigs is not pets," and so the rate was different. It was going to cost the man who wanted the pigs a good deal more to take them as pigs than it would as pets, and he refused. So it goes, and eventually you've got more than two pigs and quite a good deal of very comic intercourse.

I wanted to write...something. I never was really very creative. My forte was much more interpretation than creation - my wife's the creative writer - but I could manage to do an adaptation - take somebody else's thing and make something of it - and I settled on Pigs Is Pigs, took it down to New York with me and got advice on it down there for revision. I was called on to direct it as my exercise in the advanced dramatic production course at NBC. Somebody in this course was connected with a small station in Brooklyn and it got done there, but it has never been published. The only things of mine that were broadcast of any consequence was Pigs Is Pigs by the Buckingham Theatre, and the two Shakespeare adaptations.

However, I did get a fee from Buckingham Theatre for my adaptation of Pigs Is Pigs, around February '46. I played in that - I was Flannery, the Irishman that insisted that pigs is pigs. It's a damn clever script - it was a clever short story. We had to get permission from Ellis Parker Butler's widow to broadcast it. I managed to put that on and direct it also on CFRC. It was easy enough doing. I didn't have to do any sound effects or any control room stuff at all. We had to rehearse with the students, getting us all doing it properly and then put it on the air.

Also something that was broadcast in New York but nowhere else, and only on a small local station, was a thing that I called Daily Nightmare. That was a very free adaptation of a Chekhoff one-act farce, about a commuter who goes from his home every morning to his job in the city. He has to go by train and back, and he's beset by all sorts of neighbours to pick up for them in the city and bring home whatever he could manage to get. It just drives him up the wall eventually. I have him called down to breakfast and several members of the family asking him to get this and that and the other thing. Then he goes out to get to his train, and we hear the train coming, and a neighbour asking him to get something for her, and you hear the footsteps of the two of them running along while she's after him to do that. We have a scene in the women's lingerie department of a store, and the salesgirl says, "Let me show you what I've got in a girdle" and he says, "Oh, please, I'm a married man!" It's quite a good farce, but it required a rather large cast and didn't manage to get produced here. Rupert Caplan was interested in it, if it could be cut. It would be much more economical to have fewer people, and we just couldn't do that.[20]

The broadcast of Pigs Is Pigs over CFRC took place at 7.30 pm, Wednesday, October 31, 1945, and it marked the beginning of the revival of CFRC as a campus radio station. This was the first time in years that there had been a real program aired over the Queen's station. Barbara Monture (Levana '48) produced for the Drama Guild Radio Workshop, Dr Angus played Flannery, Frank Hoffer (Arts '47) was Mr Morehouse, and the cast included Lorne Brown and Ed Somppi.[58] The broadcast lasted half an hour.

The next week, the Workshop went on the air at 7.15 pm with 15 minutes of recorded music preceding Charles Tazwell's play J. Smith and Wife. This experimental work, without sound effects or dramatic tension, was guest directed by Dr Glen Shortliffe of the French Department, and featured Ed Somppi, Joan Connor, Stan Slorance and Ernest Smeathers. Garth Gunter announced.[59]

On the third Wednesday evening they inserted into the music segment a 5-minute digest of and commentary on campus and Canadian University Press news, through the good offices of the Journal. Alan D. Gray gathered the material and wrote the script. Garth Gunter read it. The play was The First Dance by John Latouche. Betty Potts, Maryal Edwards, Barbara Monture, Sylvia Shaver, Lawrence Palef, Jack Fitzpatrick, Sandy Webster (Arts '49) and Monroe Scott starred. Bernard Trotter and Dorothy Arron handled the sound effects.[60] The transmitter blew a tube at the climax of the play and CFRC went off the air for the duration.[61]

About this time, the Drama Guild was preparing Shakespeare's Hamlet for Convocation Hall. As a promotion, the cast aired a "dehydrated" version with narrative - a K-ration Shakespeare - beginning at 7.10 pm. Douglas Dale (Arts '47) was Hamlet, Frank Hoffer was King Claudius, Glen Wilms the ghost, Sandy Webster played Polonius, Larry Palef was Laertes, Mary Black (Arts '46) was Queen Gertrude, Marg Matheson (Arts '47) was Ophelia and Archie Malloch (Arts '47) was Horatio. Gordon Robertson (Arts '48) and Don Heap played the gravediggers and Stan Slorance narrated.[62] Five days later Larry Palef fell ill and was replaced by Bernard Trotter.

After a few successful brief broadcasts, certain students believed that they could now handle the whole radio enterprise by themselves. An anonymous letter to the editor of the Journal insisted that there was no need of a faculty supervisor for CFRC, that CFRC be dissociated from the Drama Guild and that a radio society with student executive be incorporated by the Alma Mater Society. This would "give students the opportunity to exhibit their many and varied talents," as at UBC.[63] Fortunately for the development of CFRC, this was not a popular opinion among the students.

Changes were coming in any case. The Workshop adopted a new policy of varying its productions - there would be a round table discussion the next week on union security, chaired by Prof. Clarence H. Curtis, and a musical variety program after that.[62] A new constitution was drawn up and a major reorganization instituted on Sunday, November 25, at a meeting at the Angus home. To work more efficiently, the Workshop divided itself into several production departments. Brockwell P. Mordy, President of the Workshop, was made the student Manager, responsible for co-ordinating the work of the departments. Barbara Monture was put in charge of programs, Alan D. Gray of newscasts, Kay Barclay of features, Joan Connor of music and Bernard Trotter of sound effects. Joan Connor and Archie Malloch were responsible for continuity writing and Jack Sutherland (Chem. Eng. '47) and Harvey Sheffield (Elec. '45) were in charge of technical operations. The Workshop also decided to broadcast for a full hour each week.[64]

The first hour-long program introduced a new quiz show, modelled on "Information Please," that was going to run every second or third week in the next term. Ian Campbell emceed "What D' Y' Know?" Kate Macdonnell, Dave Carlyle, Herb Lawler and Mark Stern were the experts. One of the questions was the old gag about the bells in Grant Hall Tower - there aren't any. This was followed by 30 minutes of recorded selections from Gilbert and Sullivan's H.M.S. Pinafore, the Glee Club's upcoming production. The last broadcast of the term, on December 5, featured a quarter hour of Christmas Carols with a male quartet led by Ed Somppi.[65]

Dr McNeill was "a good deal disturbed" about the Wednesday night broadcasts over CFRC. He felt that most of what was going on had nothing to do with satisfying the educational requirements of the University, but was in direct competition for audience with CKWS. There was a danger that Allied Broadcasting might refuse to make its annual payment to Queen's. Moreover, on the basis of the letter to the editor,[63] he was afraid that students considered CFRC an open facility, like the gymnasium or the library. Prof. Stewart was the Station Manager, in charge of the physical plant, but "Dr. Angus has no direct interest in much of what is going on and cannot have a great deal of control over the students." Prof. Jemmett had complained to him of the misuse of privileges - he had found the floor of the studio covered with cigarette butts.[66] Dr McNeill's fretful brow must have been soothed because the Wednesday night broadcasts continued. Bill Angus had, in fact, final responsibility for all programs broadcast by the Radio Workshop.

The Directors of Allied Broadcasting were far from upset by the student efforts. In fact, at the beginning of the 1946-47 session, Arthur Davies and Roy Thomson made a verbal proposal to Dr Wallace to help CFRC broadcast public service and educational programs for 5 or 6 hours every day.[67] Prof. Stewart conferred with the Principal about the offer and then submitted a letter estimating expenses, based on the current one or two hours a week on the air and the six weeks of the Summer Institute. The cost would be $1,000 a year, including licence and routine maintenance, but excluding replacement of large items. He and Jemmett agreed that CFRC really needed new control room equipment, a new frequency monitor and general modernization. This would cost about $6,500, and a full-time operator would come to another $1,500-1,800 annually. Attached to the letter is a tentative schedule showing 11 CBC network programs and 3 of transcribed music, from 9.45 am to 8.00 pm.[68] The Allied proposal was renewed the next spring.[67]

In the winter-spring term of 1946, CFRC was on the air each Wednesday from January 16 through February 27. The first week was "What D' Y' Know?", emceed by Jack Houck (Meds '49), and The Fourth Man, directed by Barbara Monture.[69] The cast was Jim Barker, Lorne Brown, Don Heap, Frank Hoffer and Sandy Webster. The next week Alan Gray read the news, Dr Glen Shortliffe commented on the news, Bill Crowe, jive virtuoso of Science '47, played piano, and the Workshop mounted a 15-minute play, The Silver Coronet. Recorded music filled the empty spaces.[70] The third week it was news, recorded jive, Dave Boyes' Meds '49 Octet and the quiz.[71] A short drama, Sherrill, opened the program on February 13, and then there was recorded music, news and the Glee Club stars (Jean Graham, Verna McClure, Hope Ross, Murray Gill, Bob Osborne, Gordon Robertson and Charles Blancer) doing selections from H.M.S. Pinafore.[72] A format began to take shape the next week: news at 7.00; a round table on "The Enigma of Russia"; Meds Octet from 7.30 - 7.45 and the comedy Little Red Riding Hood, with Marg Matheson, Polly Sheppard, Lorne Brown, Bernard Trotter and Sandy Webster.[73] By the close of that season, the Workshop had involved more than 50 students in its broadcasts.

The 1946 Tricolor reported: "One of the features this year was a series of quiz programs...(with) quizmaster Ian Campbell...The key men behind the scenes, who make the broadcasts possible - the strong silent men of the Science faculty...Sandy Webster was the Workshop's star announcer."[74]

William Angus: Sandy Webster, of course, was a Queen's grad who got into radio partly through Queen's. I think he did some things in the Drama Guild Workshop, but then he was the first one to get the scholarship that Lorne Greene offered to one person from Queen's to go and get instruction at his Academy of Radio Arts in Toronto, tuition free. Sandy was the first one, and actually he came through his term with Lorne Greene's school with first prize in acting. His radio career started in Hamilton. He had a long, hard row to hoe for a while there, and then he got into CBC doing a very good show called "The Craigs." It was a family of farmers who broadcast a daily program that was for farmers. There was one time Sandy and rest of the cast went to some sort of county agricultural fair, and people came up to him and asked for advice on some farming problem, which Sandy couldn't answer at all. Because he wasn't a farmer. All he knew was how to read the script.[20]

A.R. (Sandy) Webster: I arrived at Queen's in the fall of 1945 with the first wave of veterans, and it was not very long after I arrived there that Dr Angus got me interested in the Drama Guild, and I was one of the performers in the 1945 Drama Guild production of <u>Hamlet</u>. I think it was in the spring of 1946 that Dr Angus first introduced me to CFRC, which at that time had rather limited studio facilities on the second floor of Fleming Hall. I remember that there were a number of us undergraduate students working with Dr Angus, and we produced maybe two or three hours of planned programming one night each week. The Program Director in that first year that I worked with CFRC was Barbara Monture, who had been a student with Dr Angus at one of his Summer Radio Workshops. He had had associated with him some pretty impressive people, Rupert Caplan and Esse W. Ljungh, outstanding CBC producers of drama during that golden age, and Joseph Schull, a writer who later wrote some wonderful and exciting things for CBC Wednesday Night, which I had the privilege of playing.

As I say, Barbara was the Program Director in my first year at CFRC, and I had the rather splendid title of Chief Announcer, which meant that I got to read campus-related news and promote student functions, the Drama Guild and the aquacade and interviewing people sometimes, and introducing people. I recall quite vividly huddling with Bernard Trotter in the corridor outside the studio door. I think Bernard was going to do a kind of current events talk. I was getting my introduction straight and he was getting our lines of communication generally straight. Somebody was finishing up a program at that very moment in the studio, and I remember opening the door very carefully and slowly and getting a signal from Cam Searle, the operator, so that Bernie and I tip-toed into the studio, through the hazards and around, and slid in behind the microphone while somebody else was sliding out, and Bernard got introduced. In one of the succeeding years, I became the Program Director and Larry Palef was the Chief Announcer. Larry, of course, has since become very well known in Toronto as a CBC staff and a free-lance announcer.

One of the things that was for me a most thrilling and invigorating experience was performing in a radio play which Dr Angus directed, a comedy called <u>Pis Is Pigs</u>. I can't recall now just what it was I did in that, whether it was the station master or the man who owned the pigs, or just one of the pigs. I do know that all of us involved with that program had a great time, and it was for me a first experience of radio drama on a professional level, where you started at a certain time and finished at a certain time, and Dr Angus was holding the stopwatch on it. We had music and sound effects which we had great fun inventing when we couldn't get the required effect off disc.

At the time, I had no idea that I would later be a radio actor in Toronto, earning a good part of my living that way. The engineering boys, who were very much interested in CFRC, also did some broadcsting and I think it was mostly popular musical and dance recordings that they did. Bill Sutherland (Sci '47) told me he remembered hearing some of the broadcasts that we had done, and the music that the engineers were broadcasting. A younger brother of his, who was at Queen's at the same time, had a real interest in jazz and eventually did an hour a week broadcasting jazz records. And I remember listening to that in my final year, and developed a taste for it. Wyatt McLean, another Arts 49er, reminded me that another person who had done some announcing back in those days was Ken Phin, who was in Medicine.[75]

Larry Palef: Sandy Webster was the Program Director of CFRC at the time - who also played Polonius in Hamlet - and Doc Angus figured that we needed more announcers on staff at CFRC, so Sandy "hired" me as well - I passed the audition there - and it was fun being Chief Announcer. We did some drama shows, and that was live radio - no tape to edit. I soon went from acting to announcing parts and also "hired" Maurice Halperin of Timmins, who was in Commerce at Queen's, to do some of the announcing as well.

I've got to thank Queen's, Doc Angus and his wife for allowing me the opportunity of really forming the foundation for a career with the CBC. I guess, without that pneumonia, I just wouldn't be here at CBC right now.[49]

The Radio Workshop, 1946-47

The new constitution provided that the Drama Guild appoint, on the recommendation of Dr Angus, a five member executive of the Radio Workshop. For 1946-47, Sandy Webster was re-appointed Chief Announcer, Edith Shindman became Student Dramatic Producer, Bryce Seggie the Director of Features and Ted Burkholder the Director of Music. The executive then chose one of their group, Sandy Webster, as the Program Director. Sandy had had previous radio experience with CFAR, Flin Flon.[76] The E.E. Department was responsible for the technical operation of CFRC, and they selected Ralph B. Purser (Elec. '47) as Chief Engineer. John Robert Wright, Campbell Leach (Cam) Searle, Ronald E. McKay and Donald M. Atkinson, all Electrical '47, assisted.[77]

The 1946-47 session opened with a disc recording session, the first ever by the Workshop at CFRC, of Nora Mooney's Jonas Hanway, directed by Bill Angus. The disc got air-play over CKWS on October 3 in aid of the Kingston Community Chest.[78] Weekly broadcasts resumed on the 16th at 7.00 pm with a regular feature, "Campus Roundup" with Jim Kirk (Chem. Eng. '50) - he had received some seasoning at CKFI, Fort Frances. Then there were interviews and Lovely Is the Rose, a play written by Joan Connor at the Summer Radio Institute. Betty Potts, Debby Pierce and Max Cohen starred. Transmitter problems bothered listeners that first evening.[79]

The Radio Workshop did two other plays from the Summer Institute: Lenore Osborne's City Music (with Barbara Bews, Joan Bowra, Bob Joyce and Bud Morden), and Drew Crossan's Final Concerto (Barbara Monture, Doug Dale and Bill Hodge). During Education Week they aired Utterly Fantastic (Willie Dowler, Cliff Morris, Bill Hodge, Betty Thurston and Sandy Webster). They also recorded Jonah and the Whale for a delayed broadcast. Edith Shindman directed this fable, "dramatized in a negro dialect," with Larry Palef, Mike Rodney, Bill Bauer, Bud Morden and Sandy Webster.[80] Maryal Edwards, Shirley Geiger and others took care of sound effects and the turntables were handled by Alex Davidson and Joan Hamilton. Announcers that term included Larry Palef, Cliff Morris, Monroe Scott, Don Boyes, Bryce Seggie, Helen Gougeon, Bud Morden and the deep-voiced Maurice Halperin.[74] Hi Bialik interviewed Drama Guild members, and Mike Nelles, Leslie McNaughton, S.T. Ringer, Dean A.V. Douglas and Prof. Roy provided commentaries.

In the winter term the Workshop went from weekly to bimonthly production, still Wednesdays from 7.00 to 8.00 pm.[81] The Journal noted only three Workshop broadcasts. On January 29, Jim Kirk's Campus Roundup started off, Bryce Seggie interviewed Jerry Barclay, Levana President, and Miss Honor Ince directed Larry V. Thornton and Larry Palef in her Radio Institute adaptation of The Pipe Smokers. Mrs William Angus debuted on the Radio Workshop telling one of her stories, The Prince Passed By, from the CBC series Canadian Yarns. CKWS hadn't picked up that Trans Canada Network series, so Kingston had missed hearing how the future Edward VII bypassed Kingston on his 1860 tour because of local sectarian strife. Ted Burkholder closed the program with recorded music.[82] The next broadcast was on February 12, and the last featured a remote hook-up to Grant Hall for a poetry reading by W.C. Lougheed (Arts '49) to piano music by Harris Arbique (Arts '48) and a brief recital by Mr Arbique. Bryce Seggie interviewed Dr P.G.C. Campbell of the French Department, the oldest active Queen's professor, and the SCM speaker, Rev K.H. Ting, of Shanghai, wrapped it up speaking on "China as a World Power."[83]

The Radio Workshop, 1947-48

Workshop broadcasting was almost weekly in 1947-48. The Program Director, Larry Palef, had won the 1947 Summer Radio Institute scholarship to the Lorne Greene Academy. Maurice Halperin was Chief Announcer, Edith Shindman contined as Drama Producer and Ted Burkholder as Music Director. Jim Kirk was News and Florence Fraser (Arts '51), formerly of the CBC Central News Room in Toronto, Features Director.[84] Auditions on Friday, October 3,[85] preceded the first broadcast the next Wednesday: The Check Book by Edith Marks of the Summer Institute (Marg MacGregor, Betty McRae, Don Beavis, Don Nixon, Jim Fogo), plus the Meds '49 Octet. A fortnight later they produced Mrs Angus' play Royal Welcome, a re-working of The Prince Passed By.[86] The Workshop program of November 5 was a 4 1/2 hour experiment, featuring News Roundup, a round-table discussion organized by Flo Fraser on "The Utility of Philosophy," Norman Corwin's play Untitled (John Cowan, Sandy Webster, Larry Palef, Don Beavis, Jack Riddell, Willie Dowler and Maurice Halperin), recorded music and a BBC recorded play. The engineers were also given an hour, in which they presented music by the local King Trio, Stu Jenness and Doug (Ziggy) Creighton.[87] The last show of 1947 reverted to one hour and presented an Arts versus Science quiz, introduced by Flo Fraser and MC'd by Ken Phin.[88]

Some time in 1947, the CFRC engineers pulled a classic radio boner. They were broadcasting a prominent ecclesiast to the Theological Alumni banquet: "Back in the control room the more pagan engineers were playing some Ellington records...just as the climax of the sermon was reached, the startled theologs heard, coming in behind the speaker, the climax of 'Main Stem'. Somehow the engineers' record had fed back and had been broadcast with the sermon. It is rumoured that one of the engineers was forced to take theology as a peace offering."[89]

A proposal by Hugh S. Jackson (Elec. '48), in the CFRC file in Queen's Archives in the 1948 sequence,[90] probably dates from the fall of 1947. He listed all of the things that CFRC could broadcast, and someone wrote comments under each: sports (CKWS might object), pep rallies (sports and rallies are not programs that can be examined in advance for content and quality), dramatics (being done Wednesday nights), dance band pickups (CKWS and the Musicians' Union might object), classical and popular recorded music (being done Wednesday nights, but playing records may not fulfil the educational function of CFRC), special events such as the Eleanor Roosevelt AMS lecture, and features, including Queen's news casts, commentary, surveys and participation shows (all this has been done Wednesdays). The proposal was "to operate CFRC on a regular schedule, afternoons and evenings, Friday, Saturday and Sunday, plus air time desired by Dr Angus and the Radio Workshop. Technical personnel to be provided by members of the Electrical Engineering club who are interested, under the supervision of Professor H.H. Stewart. Program department to be made up of all Queen's students who show ability and interest."

The sub-scripter wondered whether this wouldn't make excessive demands on Profs. Stewart and Angus. Quality control, planning and checking program content would be a great deal of work. Besides, "Even the students' enthusiasm, put to pragmatic test, would probably wane." Clearly, this proposal by an engineer heralded the next expansion of CFRC's schedule, which was brought about by the engineers.

The Engineers' Friday Night Broadcasts

The strong, silent Electrical Engineers came on the air at 9.30 pm, Thursday, October 23, 1947, with a half hour of dance music, "Club '48," hosted by Don Armstrong (Elec. '48). At 10.00 they cut over to Hugh Jackson (Sci. '48) at the Science '48 Ball in Grant Hall for the music of the Eric-James Band. Then Doug Creighton hosted "The 10.30 Express," from records. Operators were Cam Searle (Elec. '47), Don Moore (Elec. '45) and Art Bossert (Elec. '48). The remote gave them an excuse to gauge student interest in dance music programming on CFRC, and the students wanted more.[91] The Electricals' hour of music on the November 5 Workshop proved their competence,[87] and they were permitted an 8.00 to 11.00 pm series of pop music shows on Friday, November 14, 21 and 28. The last included four half hours of recorded music, two 15-minute studio productions - Stu Jenness' (Sci. '48) "Piano and Patter" and Don Beavis' (Arts '50) "Solid Syncopations" - and a half hour remote from the Science '50 "Gambler's Gambol" in Grant Hall, with the Eric-James Band.[92] A further venture that has the flavour of Electricals was the first Queen's "Sports Night," broadcast by CFRC on Saturday, December 6. Jim Kirk called a Queen's-Varsity water polo game and a Queen's-Ottawa basketball game. Then they cut to the Science Soph-Frosh Dance in Grant Hall and finally over to the music of Ivor Edwards at "Sports Night."[93]

1948 opened with broadcasts of Eleanor Roosevelt's Alma Mater Society lecture and a special convocation on January 8.[94] Commentary was by Jim Kirk, and the engineers tape recorded both.[95,96] Anyone interested was welcome to drop into the CFRC studios to look at or hear an excised tape strip containing Mrs Roosevelt's cough.[96]

CFRC's schedule was now expanded to two nights a week, the Radio Workshop on Wednesdays and the Electricals Friday nights.

The engineers, Gord Henderson, Art Bossert, Bob Burnett (Sc. '48), Jack Harold, Jim Kirk, Stu Jenness and Doug Creighton, under Hugh Jackson and Cam Searle, were not overseen by the Workshop.[96] Friday shows were pop-oriented, and often included intercollegiate hockey games or dance remotes - Boyd Valleau at the Arts Formal, Hal MacFarlane at the Levana Formal - and local talent such as pianists Stu Jenness, Jack Harold, Don Beavis and Dave McQueen, vocalist Donna Scott or Doug (Ziggy) Creighton and his jazz crew. They aired

Meds '49 Octet at the CFRC microphones, Studio "B," February 16, 1948. From left: Mel Shaw, George Stone, Wilf Roy, Stan Morill, Dave Boyes, Gord Louden, Del Blaine, Bill Simmons.
-courtesy Dr W.A. Roy

a Sunday Hour Service from Grant Hall,[95] two Saturday night hockey games[97] and their own recording of the "Campus Frolics of 1948."[98] The engineers also started a phone-in record request program that was very popular.[99] The Friday productions got far more coverage in the Queen's Journal than the Workshop's did.

The first Workshop program, January 14, 1948, featured an unmentionably horrible ghost story (Betty McRae, Connie Wilson, Henry Balker, Doug Dale, Maurice Schwartz, George Toller and Sandy Webster), directed by Larry Palef. Flo Fraser interviewed veterans' wives on campus.[100] In February, Mary Harrison hosted a new program of thumbnail sketches of campus personalities.[101] The last Wednesday night featured a 3 1/2 hour extravaganza: an hour of transcribed classical music with Ted Burkholder; Campus Roundup; Rev A.M. Laverty speaking on "Education for Citizenship"; the play <u>Two Gentlemen from Soho</u> (Gord Robertson, Harry Threapleton, Michael Roth, Dorothy Bradley, Joan Bowra, Art Todd and Bert Caldwell); and a comedy skit called "How a Program Should Not be Aired" or "It Can Happen Here" (Don Beavis, John Cowan, Flo Fraser and Larry Palef).[102]

<u>W. (James) E. Kirk</u>: I was a member of the <u>Journal</u> staff, and since I had had experience in radio, became the weekly news commentator. I held that post for my four years at Queen's and, as news announcer, I became a member of the Drama Guild. In the first year we did a bunch of remotes - dances in Grant Hall, wrestling and hockey. I believe I am the first to broadcast an intercollegiate water polo game. For hockey we had a rickety platform in the rafters of Jock Harty Arena. One of the Electrical club was along to handle the physical hookup of a pre amplifier to a phone line back to the studio. There was a venerable, almost perpetual referee with a full head of white hair known as "Snowball." If he made a contentious call there would be derogatory shouts and chants of "Snowball! Snowball!" which irritated him, and then tennis balls often appeared from nowhere and bounded about the arena. Although I didn't do it and voiced my strong disapproval, many of the tennis balls originated on the platform high in the rafters. The remotes from Grant Hall started with the year dances - an hour segment featuring the music of the orchestra and sometimes the vocalist - and we would con the organizers into contributing a few dollars to the CFRC record fund in exchange for publicity. I did several as the announcer. A highlight was when we persuaded Ted White (Sci. '48) to sing "You Made Me Love You" in a very suggestive rendition.

For the broadcast of the granting of an honorary degree to Eleanor Roosevelt, an Electrical and I were in the projection booth high at the back of Grant Hall. I did a commentary until we switched to a microphone on the stage for the presentation and oration. One of my all-time bloopers occurred at that time. I was commenting on the academic procession and said, "...and it made a very colourful scene there on the stage with all of the various gowns going up." We showed one of the neophytes a piece of recording tape which we claimed was Mrs Roosevelt's cough which we had edited out. Once I had to be at a formal dance in Belleville at broadcast time, so I sat at the microphone dressed in my tux, did the news into a recording machine and rushed out to catch the train. One of the early delayed news broadcasts. We had a lot of fun with the sound effects and the sound effect recordings which we patched together for our presentation of <u>Jonah and the Whale</u>. I also recall the Arts v/s Science quiz. We were all astounded by the erudite answers of Len Harvey (Chem. '50), who came up with game winning scores.[103]

<u>Florence Fraser McHugh</u>: I was with the Summer Radio Institute in 1946, and then I started Queen's in 1947. I think I got involved with CFRC in 1947-48 because I had been involved with the Summer Institute and I had also worked for CBC for about a year before I

came down to Queen's. So it seemed kind of natural for me to start working with CFRC. Then I went out to UBC on an exchange, and when I came back I was again very much involved with the radio station in 1949-50. In that year, I was Features Editor and I was also in the Drama Guild. As the person in charge of features, I was supposed to arrange for other people to do stories, but quite often you couldn't nail somebody to do it. So you'd end up doing it yourself and you'd be on maybe two or three times a week and maybe more. We worked quite hard and quite steadily at it.

I know I did some interviews - the most interesting one was with Eleanor Roosevelt, when she was up here getting her honorary degree in a January convocation in 1948. She was a fascinating, charming lady and I really enjoyed that very much. I guess I talked to her for about maybe 20 minutes or so, and it was broadcast. The questions were fairly light. We talked about women's rights, which she was very much involved with at the time, and about all her fan mail. There was one funny thing at the end, and I know it's on a soft-cut disc in the CFRC archives. I got so excited about the whole thing and so involved that I forgot that I had made copious notes, which were on my lap. And I started to laugh or something and this whole pile went crashing down on the floor. And that sort of ended the interview right there. There were so many people who wanted to talk to her, but I think I was the only one who got her on disc. I know Doc Angus was there with me, involved with the technical things. I think it was done in Grant Hall, so it must have been some sort of portable equipment. That evening I had talked to people I had formerly known at CBC, and they had me do an item on CBC News Roundup from CKWS, with a story on Mrs Roosevelt.

I remember Lou Tepper and Larry Palef. Because of my work on CFRC and also with the Drama Guild, I got the Lorne Greene scholarship and went to his Academy of Radio Arts the following year, 1950-51. Larry had gone on the same scholarship. And Bern Trotter. I later worked for three years on staff at CBC and used to see Larry Palef all the time.[104]

CFRC Broadcasts, 1948-49

During the summer of 1948, "the wire men in Science '49... completely re-built the control room. Despite the fact that they alone know how to run it now, it is as up to date as science and money will permit...They have a first-class wire recorder, a Presto disc recorder, reliable remote control broadcasting equipment as well as two very modern and impressive studios complete with awe-inspiring clocks. An 'on the air' sign has not been forgotten either."[89]

The Radio Workshop probably tried too hard in 1947-48 to provide very high-toned fare on CFRC, because the policy was modified in the next academic year:

> Unlike last year, the workshop plans to bring their programs down to the level of their listeners. Edification and education will be partially subservient to variety and timeliness. Instead of an extensive and almost

monopolistic diet of classical music, a history of jazz
series is being planned. For this the workshop has lined
up a very extensive collection of the best that records
can give. Gone too will be the scratchiness and fuzzi-
ness of last year's discs. Long plays of a re-hash nature are
to be replaced with interviews, discussions of campus
interest and new and entertaining scripts. The wide
awake "Campus Roundup" is going to be enlarged with
Jim Kirk handling the reins. Furthermore, much stress will
be placed on live talent. Jack Harold is slated to play
some terrific piano with other performers in the offing in
subsequent broadcasts.

To accomplish these things, Producer Don Beavis
is bubbling over with ideas. He proposes lengthening
the (Wednesday) short hour of last year if and when the
quality and quantity of materials suggests.[89]

For the 1948-49 Radio Workshop, Don Beavis (Arts '50) was
Program Director. He had had summer experience at CJBQ, Belleville,
where he improvised "The Madhouse Little Theatre" between discs,
poking fun at commercial radio. After taking-off a horse race, he concluded, "Remember, to play the horses you've got to be a horse." Jim
Kirk was News Editor and Sports, Debbie Pierce was in charge of
drama, Lou Tepper was in charge of features, Shiela Orr handled the
music and Jim Robertson was the continuity man. Radio veterans all.[89]
The new Workshop series began on Wednesday, October 27, 1948,
and the engineers' that Friday.[105] Both series carried on much as before.

Talks and Features, under Lou Tepper, devised a series for the
Workshop called "As You Like It," which widened the possibilities for
student participation. They put out a call for opinions, personal experiences, short stories and poems.[106] The Workshop also promoted the all-
Queen's opera Evangeline, before its premiere by the Glee Club at the
La Salle Ballroom, December 1, 1948. Graham George composed it,
Arnold Edinborough directed and Alan Crofoot sang.[107] Another
Workshop innovation was a limited-enrolment 3-month course in writing for radio, given by Mrs Angus on Sunday afternoons. At that point,
with experience writing CBC scripts behind her, she was the district
reporter for CBC's Canadian Chronicle.[106]

The engineers introduced "Keyboard Kapers" with Jim Baldwin,
Don Beavis or Jack Harold, "Sports on the Air" with Mike Milovick
(Elec. '49), "Request Time" with Bill Greene, Bill Grant, Vern
McCullough or Ken Lloyd, and half an hour called "Levana Time"
with Mary Eleanor Thorburn.[108] It is curious that it was the engineers
who first gave the Queen's women's faculty a radio voice.

In the new year, the Workshop out-spaced the Engineering Faculty
night in the Journal. Lou Tepper came up with another people program, "In this Corner". Students were invited to air their gripes, and the
person targetted would have to reply.[109] They also featured discussions
on "What's wrong with Queen's women?" and "Should political parties
come back onto the campus?", a contest, three plays in a series called

"My City," the scripts courtesy of Rupert Caplan, and some live jazz. The Electricals' regular features were "Slide Rule News," "Keyboard Kapers" and "Sports on the Air." They also broadcast an Intercollegiate Assault and promoted the ISS campaign by inviting Dr Wallace and Dean Douglas to speak.[109,110]

> **Larry Palef**: On one show we had Principal Wallace and we asked him to speak only for 3 1/2 minutes because we had a very tightly timed show that Wednesday night. I was amazed, because he said, "No problem," and he spoke for 3 1/2 minutes without even looking at the clock. That still amazes me to this day.[49]

The Riders of the Southern Trails, 1950.
-courtesy Fred Paquin

On March 4, the Electricals introduced "The Riders of the Southern Trails, a new barn dance combo, who will strut their stuff."[110] Ron Scott and the Riders became a CFRC institution during their few years of existence.

> **Fred Paquin**: I used to play with the Riders of the Southern Trails on CFRC and WS at different times on the weekends. I joined the group in July 1949, and played with them through until '53 or '54, in along there some time. We had Fred (Pappy) Ryan on steel, Clarence (Slim) Barker on bass and vocal, Johnnie Stephenson on violin and vocal, Ron Scott on mandolin, violin, and he did a lot of bluegrass-type vocalizing. And myself, I done vocals and was on rhythm (flat top) guitar. We used to play all the jamborees and we were on tour for a couple of years, several times with Stoney Cooper and Wilma Lee. All one night stands. I was pretty glad to see the end of those. All you done was set up, you'd play, you'd tear down, jump in the sack at the motel, and the next morning, bright and early, back in the car and back on the road again. Hit the next destination and do it all over again - two weeks at a time. One of the announcers we had on CFRC was Louie Tepper. Louie had a very deep-sounding voice, almost like that Lorne Greene "Voice of Doom."[112]

> **William Angus**: Years ago, back in Fleming Hall, Lou Tepper was the station's Chief Announcer. I was usually there as supervisor of the Drama Guild Radio Workshop. There was a local group of country and western musicians, with their guitars and so on, who gave us a program one night. The leader of the band, being introduced by our Chief Announcer, said, "Thank you, Cousin Lou" and then he went into his program. When that was finished, he told the people who might be listening where his band might be playing in the surrounding district in the next week - Sydenham, Glenburnie, Harrowsmith - and after that they came out of the studio and gathered in the transmitter room. They were looking over all the equipment

in there, the dials and switches, and finally a fellow turned to me and he says, "Which one o' them dials tells us how many people are listening?"[20]

CFRC Programming, 1949-50

Lou Tepper, Program Director for the Workshop, announced in the fall that the Radio Workshop was switching its broadcast night to Thursday. At the same time the Workshop introduced a new system of "package" presentations, in which each department's programs was produced by its own crew. This arrangement would give more students a chance to get radio experience and would allow more rehearsal time for each production. The first packaged production was a radio play.[113]

7.00 pm	Ron Scott Sings Folk Songs
7.15 pm	Latest of Popular Hits
7.30 pm	CFRC News
7.45 pm	CFRC Sports
8.00 pm	Judy Ettinger Show
8.15 pm	Jack Harold at the Piano
8.30 pm	In this Corner, with Bill Bauer
9.00 pm	Radio Play
9.30 pm	Interview
9.45 pm	History of Music, with Ted Bond
10.00 pm	Mike Humphries Variety Show
10.30 pm	Classical Music
11.00 pm	Sign Off

"Had" Wilson was Program Director for the engineers. Their innovation for 1949-50 was "My Favourite Record," a request program calculated to solve CFRC's record shortage. A student could lend any record to CFRC for the term, and hear it played on "My Favourite Record" at any time, on request.[114]

William Angus: The Radio Club comes after the involvement of the Drama Guild and students in dramatic production and so on. They sort of phased themselves out and the Radio Club took over. I think the most active ones in that were the engineers. They were very strong and very interested, of course, because it was practical experience for them in a technical field. And the club got quite well organized, and it was very, very good, but it was much more dominated in all of its aspects by the engineers in the succeeding period. Harold Stewart was still Station Manager, and for a time I was considered Program Manager. Anything that was broadcast had to have my okay, but it was actually operated by the E.E. students and they did the announcing and everything else.

I made them take a program off the air one night because I thought it wasn't the sort of thing that would give Queen's a good name. I vetoed it; afterwards I regretted doing it. I called Harold Stewart and said, "Take them off the air. We don't want Queen's broadcasting that sort of thing." It was the most terrible music and remarks and so on that were being made. It was awful. It would make you ashamed for Queen's, you know. It wasn't good for the Queen's image, I felt. I was embarrassed afterwards. I shouldn't have taken it off, really, because maybe, as McNeill said, they were talking into a lead pipe.

In fact, Arthur Lower one time at an Arts Faculty meeting wanted to know what could be done about CFRC. He didn't like their programs, particularly music, and he gave a sample of what he didn't like. And then I spoke up and said Harold Stewart was Manager and I was Program Manager and we didn't like to run the students with too tight a rein - give them some leeway. The engineers even did some drama programs. I didn't do much with it. We did develop an audience. Well, my wife stepped into my job when I went out. She was Director of Radio for 11 years.

Something that Queen's took pride in, including the Principal, was that our university was the only university that had its own station on the campus. There were other universities that did broadcasting, but they had to go to the commercial stations to get time to do their broadcasting. Queen's students and even the Principal were rather proud of the fact that we were the only one that had its own station on the campus.

We had soft-cut discs first. The recordings weren't for a record - they were for rehearsal. We later had a tape recorder in a wooden cabinet, and sizeable - weighed 60 or 70 pounds. What was it called? A Brush Sound Mirror. In fact, when Dr McNeill retired, the students gave him one of these things as a present. Boy, he recorded all kinds of things. He made a hobby of it. The student broadcaster, whether for announcing or whatever, was recorded on tape and then played back to him so that he heard what he sounded like. And then the tape was erased; it wasn't saved. If we wanted something permanent, it was on disc.[20]

Margaret Angus: I started radio work because I was interested in Kingston history and partly because I talk a lot. It was at a party at the house of the Head of the History Department, who was entertaining some visiting CBC broadcasters, and someone said there must be some interesting things about Kingston. I had just been doing some research and found a fascinating story, which I told and before I left the party that evening, the CBC person asked, "Would you be interested in writing that story and telling it on the air?" "Sure!" I was doing some writing at the time, for Saturday Night and for the newspaper. I got a call from them saying we really mean it - can you do it? It was for a program called Canadian Yarns, and I did quite a number for that program. Then I had my own series on the CBC, three different series of eight or nine each, short stories based on historical fact. I broadcast the first couple, but it was much easier writing for an actor to read. It was great fun at first, because I had an incredible number of ideas and I turned out a finished script per week. Finally, it got to where it was a burden, and it was no longer a pleasure, so I dropped it.

After that I did some on-the-spot interviews for the CBC International Service. Then I decided that, having done interviews, two documentaries, and straight news broadcasts - I broadcast for the International Service on the re-opening of Fort Henry (c. 1938) - I thought I would like to try some different sort of writing to prove to myself that I could sell this sort of thing. So, I wrote a couple of radio dramas. The CBC domestic service did one and the International Service did the other. By this time I was involved in doing some other writing and this radio work really got me into directing radio at CFRC, since I had experience in a number of lines that we wanted students to begin to learn - interviewing and writing short scripts.

The way I got really first connected with CFRC was through my husband's work with the Radio Workshop of the student Drama Guild. The Radio Workshop was a small section of the Drama Guild where people interested in broadcasting, especially plays, got a chance to do things. I was an assistant in Drama and as my husband became busier, and as I began to have a better knowledge of radio, I was given the job of working with the Radio Workshop. They broadcast spasmodically at first, then on a regular basis on Thursday nights, as they broadened out to do more than plays - they did the news, interviews, plays and music. Finally it worked up to be a regular Thursday night, when Drama Guilders took over the whole broadcast thing and the Science people did the Friday and Saturday night. But the Workshop, certainly from about 1946 or 1947, was fairly active in radio. The Radio Institute gave a basis of professionalism for broadcasting that there hadn't been before.[115]

Lou Tepper, Chief Announcer and Program Director of CFRC, circa 1950

Lou Tepper: My first connection with CFRC began in the summer of 1948, when I enrolled at the Queen's Summer Radio Institute. That was the last of such Institutes to be given, and that was the year that I graduated from high school and was to begin my first year in Queen's Arts. I'm Arts '51. We had Esse W. Ljungh, director and producer of plays at CBC in the golden age of radio, Bud Knapp, who wrote a great deal for the CBC, W.H. Brodie, Head of Language Broadcast for the CBC - very charming fellow - looked and carried himself like a retired colonel, Dr Angus and Mrs Angus, and there were others. Of course, I was interested in radio, so when the fall came and I started in Arts, I joined the Queen's Drama Guild and the Queen's Radio Workshop, which was part of the Drama Guild, and I became its Program Director.

As Program Director, my responsibilities at the beginning of the term were to find out who was interested in becoming a member of this section of the Drama Guild, checking them out on air to see which of them would be good announcer material. We delegated responsibilities to various people for programming, direction, writing and production. I was more or less co-ordinator - made sure that everything was done, that everybody delivered when he or she was supposed to - and I did a fair amount of on-air work. Mostly announcing, reading poems occasionally and interviewing people. We were overseen by Dr Angus and by Mrs Angus, the faculty advisors. On the technical side, Prof. Stewart was in charge of the engineers, who physically ran the station. Really we broadcast to the campus - we were a campus radio station.

A group of us set out to do some quality broadcasting. We thought we would provide a very nice contrast to what was the current fare on commercial stations, and we provided three evenings of radio broadcasts per week at our height, of interest to the campus. We had campus news, sports, academic things - we were a sort of campus bulletin board - and we had discussion and interview programs with staff and students alike. I remember one interview I did with the football coach Frank Tindall. It was one of the first I had ever done, and I suppose I had done very little preparation for it, thinking that my guest would be talkative, only to discover after we began on air that he was anything but talkative. He would make the perfect witness in court. He said only what he had to say and he spoke with tremendous economy of langauge, and I was really perspiring by the time we finished, to keep it sort of talky and light. Other than that, we had talented students who would sing or play musical instruments. We had a group from the West Indies who sang calypsos, who were very popular. And also Ron Scott and the Riders of the Southern Trails. They did country and western stuff and they had a big following. Ron Scott was a student at Queen's, but I don't think any of the members of his band were students. One of the members of that band, walking into the control room one night before we went on the air, looking at all the dials and knobs, said to me, "Which one of these do you turn to find out how many people are listening?" Well, we didn't have one of those, and I guess they still don't.

One of the things I did was read poems on occasion to a background of appropriate music. That, I understand, had a following. We had radio plays, of course, and I think some of those productions were comparable to the sort of thing that one would hear on commercial radio. Certainly it was a wonderful outlet for people who wanted experience in radio, on air, writing or producing.

We also worked very closely with the Queen's Journal, and a number of the people who were very active in Journal used to do campus news or part of news programs and write occasionally. Don Brittain, the then editor, was very active and he has made a very big name for himself in film documentaries, mainly with the CBC. Bruce Dunlop, who is now a law professor in Toronto, and Don Gordon, now teaching in communications, were both in my year. I remember once introducing him as "And here is Don Gordon, the inside dope...I mean, with the inside dope."

The engineers, after we got the station going again, decided that they would do something. So they took a night or two for themselves. Mostly music - it went longer and longer with phoned in record requests - and that became a very big thing. But that wasn't ours. In 1948 we had perhaps half a dozen records to play on air. We played them over and over and over again, just to have something to play on air. So we developed the bright notion of contacting the various record companies and got from them their various releases in the same manner as the commercial stations. We built up a tremendous library over that three year period, all for free, and people contributed their own records on a gift or loan basis. I remember very vividly sitting with Doc Angus, listening to people who were trying out for announcing spots on the station, and how we worked out a system of grading.

What I recall most about the studio is that it wasn't terribly big and I remember these very heavy velour curtains that covered entire walls. They were excellent really, because they did things for the sound.

Otherwise the sound would have been bouncing off wooden or concrete walls. Certainly it wasn't a large operation. Normally one operator could keep the whole thing going, and it was always an engineering student who operated the station while we were there. Norm McKinney from Toronto had a splendid announcing voice, and he had worked in radio before. His nom-de-radio was George Norman, and I don't know what he ended up doing.

The remotes I was invloved in were graduations. I remember going to a professor, who shall remain nameless, and asking him would he be good enough to let himself be interviewed on the air on his specialty. He listened very politely while I told him what we'd like him to do, looked at me after a short pause and said, "And what are you paying?" When I told him that it was non-profit, that no one got paid and that we had no budget for that sort of thing, he very politely declined. He was the only person who did. Everyone else we asked was quite happy to be of assistance. I interviewed Prof. Bieler once, on some aspect of art in Canada. I think everything we did was live - I don't even think we had recording facilities. Certainly that was before tape.

John Bermingham, who was a year or two behind me at Queen's, was one of the people that Dr Angus and I auditioned for an announce spot. John was very active at CFRC. He's the one person I can tkink of who actually did follow up and make a career of radio. He's the manager of one of the local commercial stations and he's won a number of awards over the years for excellence in his work. David Quance also went with the CBC, but as a writer or in some technical capacity.

There was little, if any, impromptu or unscripted stuff. If we didn't have actual scripts, then at least we had blocked out the main points of what we were going to do. Most of it was rehearsed, and well rehearsed, before we went on the air. Everything that went on the air was rehearsed at least three times. Not newscasts, but certainly plays and anything where timing was crucial and had to fit a certain slot. We tried to do that. We tried to make ourselves as professional as we could. It was a very happy, entertaining and educational sort of thing and there was tremendous camaraderie among the people who worked there.

It was a useful experience for any of us who did it. Certainly has been for me. I was able to use it in a small way, working for the two local radio stations from time to time while I was an undergrad and a short time afterward, just to fill in at Christmas and in the summer holidays.[116]

In 1950, the Drama Guild Radio Workshop was on the air four hours a week. Lou Tepper was Program Director, Doug Creighton was Music Director, Dick Crowther headed Talks and Features, Bruce Davenport was News, producing "Journal on the Air," and Don Brittain was Sports Director. Flo Fraser was Drama Director and Murray Beech operated for the Workshop. The CFRC Staff, the club of engineers who operated the station that year, were K.A. Laidman, H. Eberly, B. Bellerby and F.R. Carruthers. H.S. Wilson was Station Manager and Prof. Stewart the Faculty Consultant.[117]

Charles F. Currey (Elec. '51): I wrote sports for the Journal from 1947 and worked for CFRC, beginning about 1949. The rivalry between the Arts and Science students on control of both the Journal and CFRC intensified in '49 and I left to concentrate my services at CFRC, where my Science friends were. I had a regular 15-minute Friday night sportscast covering intramural and intercollegiate athletics, introduced by a very worn and familiar tune. In 1950 we began doing play-by-play of sports other than football. Hockey was a disaster. I really knew few players, even on the Queen's team. We did only one hockey game. In 1950-51 Don Brittain and I did all the home basketball games, set up in the centre of the gym balcony in a rat's nest of wires and equipment. By the end of a game we would be in a total sweat with excitement and the mental effort of trying to talk the ball into the basket for Queen's. We also did intercollegiate assaults from there. The boxing was particularly exciting. What the fighters lacked in skill and finesse they made up in ferocity! With the blows raining and blood on the mat and the competitors, we even forgot to continue the blow-by-blow account for the listeners.[118]

Others who were active with CFRC in 1950 were George Norman (whose real name was George Norman McKinney), John Bermingham, Ted Bond, Marg Dickerson, Bruce Dunlop, Mike Humphries, Anne Larmour, Doug Timms, Jan Swoboda, Donald R. Gordon and Joan Torgeson.[119]

> **Donald R. Gordon:** With Doug Creighton, Jan Swoboda and others, we faked a studio in Grant Hall Tower, with wind fx, typewriters and other Steve Wilson aspects, such that people actually phoned Grant Hall to reach us. We also did some of the first all-night record programs, mainly because of a gentle lust for student nurses. Often putting the microphone under the announcing table so that we wouldn't spill our brews. We had time, indifference and leeway to learn quite a bit.
>
> I thought of it when I was, as a CBC correspondent in Amman, Jordan, doing a report to the network from the back of an abandoned BBC wireless bus in 1953, surrounded by shaky Arabs and curious cows. It was the same set of circumstances and CFRC training helped quite a bit.[119]

E.J. (Ted) Bond: I came to Queen's as a freshman in 1949, and one of the first things I did was to join the Queen's Drama Guild. I discovered that it had a branch operation called the Radio Workshop, under the direction of Margaret Angus. I had been interested in radio for quite a long time, so I went straight into the Radio Workshop and I didn't do anything with the Drama Guild proper until the year I graduated.

The whole Thursday night thing on CFRC was under the aegis of the Radio Workshop of the Queen's Drama Guild. Apart from working out programs, talking about ideas for programs, rehearsing, doing live radio drama - that was just about it - I also attended extra-curricular classes in radio script writing and production that were given by Margaret Angus, which were part of the Radio Workshop and were very worthwhile.

Lou Tepper was our number one announcer, our answer to Lorne Greene. He was as good an announcer as anything in the CBC, a very impressive announcer indeed, and a very nice guy. Sid Penstone was the operator all the way through, as I recall.

I did programs with the Radio Workshop in 1949-50 and 1950-51 - both ran Thursday nights right through the year. The first was a history of music, called "Music in History," and I wrote it and produced it. I have the script for October 20, 1949. The announcer was Mary Austin and the operator was Murray Beech. John Bermingham (Arts '54) read my rather complicated script, which was mostly cribbed from a book which had the same title as the show. About half of each 15-minute show was talk and about half of it was music. I used all of the resources of the Music Room - I had access to the records there - and I could illustrate just about everything that I wanted to say about the history of music in Europe. LP's started being sold in Canada in 1949, so they were a novelty, and certainly none of the stuff that I wanted to play was on LP. They were all 78's. I also used the piano for simple harmonic examples. John had a lot of trouble with the pronunciation of foreign names, so I had to write them out for him in phonetic spelling. I was playing a medieval piece for the vihuela, and the guitarist was Emilio Pujol. You can imagine how I had to write this out phonetically, and John had difficulty reading it without breaking up. This whole thing started when I wanted to play a madrigal called "Au Joli Jeu," and John says, "What's this about a jolly Jew?" So, I figured we'd better do something.

The following year, 1950-51, I immediately joined up with the Radio Workshop again. That year I did a different kind of show. I've always been interested in records - in fact my earliest memory is of watching a record go around on a turntable, when I was two years old, and I've been a record freak ever since - performances that were unusually fine. So I devised a program called "Accent on Performance." It ran for an hour on Thursday night - it must have, just to accommodate the stuff that I wanted to play. Marg Dickerson and I prepared the script together, and I may have produced it from the studio. It was a scripted program that was supposed to sound unscripted. And it probably sounded like a scripted program that was supposed to sound unscripted.

Marg and I talked back and forth in this scripted casual fashion, and then the especially fine performance was played. Afterwards we said things like, "Well, that was quite something, wasn't it, Marg?" And she would say, "Yeah, that was quite something, Ted." All read from the script, of course. I got a bang out of that program too.

I did a third program in the fall of 1952, which I did with Tamara Lipkowitsch, with her marvellous hushed tones. The program was called "Contrast," and the idea was to set two contrasting pieces of music together and see what it was like. I still have a script.

Margaret Angus was the Director of the Radio Workshop and she was the one who made the final decision about what programs were to go on. She was ultimately responsible for all the program scheduling. She managed the whole affair, pretty well from minute to minute. She was there all the time, and very much in charge. She wouldn't be standing over my shoulder, at least not after the first program - she'd probably be talking to somebody else, working out some arrangements. There were no restrictions that everyone wouldn't take for granted. I don't remember anything that I might want to do that I couldn't do, and certainly there was no air of being under heavy regulation.

I didn't do anything else on CFRC until I came back here as a member of the faculty. In 1966, Margaret Angus invited me to be the host and moderator of a series of six spontaneous discussion programs on the general topic of "Freedom." I did the introductory program all by myself, reading a little piece I'd composed on freedom. Then there was a certain topic up for discussion and I acted as moderator, each time with at least two members of faculty, usually from different departments. That show was taped.[120]

<u>John Bermingham</u>: My interest in CFRC was triggered in my freshman year at Queen's, 1950, by Lou Tepper, Chairman of the Radio Workshop and now a very successful lawyer here. He has a great radio voice, a very deep, resonant voice. One of my first recollections of CFRC in my first year was Lou MCing a live country music show. A local group led by Ron Scott would come into the studio, and Lou used to have a great time emceeing that show.

I was Chairman of the Radio Workshop for two years in the early '50's, following Lou Tepper. We looked after the programming and broadcast operations on Thursday nights at CFRC. We did a variety of programming, the idea was to present as many different kinds of programming as we could. I did one called "Rocking Chair," a blend of jazz and some current pop hits at the tail end of the Big Band Era. Ted Bond (Arts '53) wrote most interesting scripts for "Music in History," and he's still interested in that field and writes the occasional review for <u>The Whig-Standard</u>. We did dramas, Broadway music - you had to scrounge around campus to see if anybody had a copy of <u>Oklahoma</u> or whatever. On Friday and Saturday nights, the engineering people took over, and theirs was a more informal operation. Whoever wanted to do a show came in and did a disc jockey show or whatever. There was always a great battle between the Drama Guild and the engineers to see who got the biggest audience at Queen's. We all boasted that one did better than the other. CFRC and the Drama Guild shared one tape recorder, a Brush Sound Mirror, for rehearsals at the Drama Guild and then taken over to CFRC. We took it out and recorded a Queen's Review, with all original music, in 1951, a show called <u>Dear Susie</u>. It was a big hit on campus, and we played the complete recording on the air. Subsequently somebody had the recordings pressed onto disc.

Gary Smith, who also wrote many humorous articles for the <u>Journal</u>, used to write some comedy scripts for us. One of his dramatic skits had a line, "Have some more wine, my dear," and no one could find a glass of water to pour to make the sound. So Gary Smith ran out and filled a wastepaper basket full of water, and poured it into another one with a great gushing sound. Which of course broke up all the actors, and we immediately went to recorded music for about 15 minutes while we composed ourselves again. Mike Humphries, a leading light in the Drama Guild, starred in <u>Othello</u>, and he also did a fair amount of announcing on CFRC. It was all volunteer student help and there were occasions when someone would get involved in something else or forget, and you'd have to improvise a program very quickly. I used to play the piano a little bit, and one night about three of the people who were supposed to do programs didn't show up. I did a disc jockey program, then ran in and had somebody introduce me as John Cornelius who would now play the piano. I did that for 15 minutes, then ran into another studio and did something else. I'm sure it sounded terrible, but it was kind of exciting and fun.

I was fortunate enough to win the Lorne Greene Scholarship in Radio Arts in my graduating year. It entitled you to a year's free tuition at the Lorne Greene Academy in Toronto. Unfortunately, the year I won it the Lorne Greene Academy closed down, and I think Lorne took off for Hollywood, so I wasn't able to take advantage of it. To-day, I'm President of Bermingham Marketing Limited, which is an advertising agency specializing in broadcast advertising and other forms as well.[121]

Jane Sherman: I started in Notre Dame High School in Kingston, and somebody got the idea that we should be on the air, so we put a program on CKWS for about 3 or 4 months one year - none of us knowing anything. Don Browning, now with the CBC, warned me not to get involved in radio unless I really liked it, because I'd be hooked for the rest of my life. And it's true. By the end of that year, I was Chairman of that, and felt if I was to go on, I needed to know what I was doing. Dr and Mrs Angus were running the Radio Workshop as part of Summer School, so several of us from Notre Dame took that course that summer. They taught us the rudiments of script writing, of drama, and we had a lot of fun producing plays and doing sound effects. The day that I was doing sound effects somebody "died" and I didn't fall correctly the first, second, third or fourth times. I was kind of bent and bruised by the time that afternoon was over. But we learned all that - horses' hooves, slamming doors and crumpling cellophane.

I was in Arts '55 and part of the Radio Workshop Thursday nights at CFRC. There was a student Chairman elected each year for the Radio Workshop. The cast of characters varied each year, and so did the programmimg. We had Dr Margaret Angus as our advisor. She helped to organize the programming, discussed any problems and any new ideas and gave us suggestions. She is a very fine writer and broadcaster in her own right, so it was a great thing for people to have her as a resource person.

The control room work was done by the engineers, people like Sid Penstone, who spent their whole lives in the control room under the counter. The engineers Bob Radford (Elec. '56), Paul Carroll, Paul Hughes and David Spendlove (Elec. '56) put up with what we Artsies didn't know about the radio station. The station was cramped and crowded with a lot of odds and sods of stuff scattered around, if you know old radio stations. A music library out in the hall. The program that I did was called "Leave It to Levana," and Levana was the women's society at Queen's, as distinct from the faculty societies. They didn't admit women to their secret organizations, and so we had our own and very fine organization. The program was the usual news, sports, interviews type of thing, with women on the campus. We talked to cheerleaders, team captains, club executives and people who represented Queen's at debates and so on. We must have filled with music, because I'm not sure we found 15 minutes worth of material each week.

So many times over the years, when I was working either as an amateur or as a professional broadcaster, I had occasion to think back about the things Mrs Angus would say about inflection and making sure that you don't dribble out the ends of your sentences. To remember that you're talking to an individual, not you people out there. I always felt that that stood me in very good stead, and it was a lot of fun. It's interesting, the number of people who have stayed connected with radio or have gone into TV, who got a start at CFRC.[122]

<u>Catherine M. Perkins</u>: Though I wasn't involved through my whole time at Queen's with CFRC, I especially remember 1957. I got involved through the Drama Guild, and the first play I did there was a ghost story, with all these clanking chains, weird sounds and horses galloping off. They had a collection of special effects and I was very impressed. I think it was mostly live, but we could record for critiquing. June Pryce, in our residence, whom I'd gone all through high school with, got involved in the technical side of the station, and used to talk up the club in residence. There are all sorts of nice romances that flourished over in Fleming Hall - she married the technician who used to do her show, now Dr Robert Sanderson. The engineers looked after all the technical side. There was almost a snobbery about it - of course the technical people wouldn't get involved on air, nor would the on-air people touch the machinery. There was a man on the other side of the glass, pointing, and he had nothing to do with you, didn't know your name and didn't want to. It was an engineers' club, we thought.

One of my favourite recollections of CFRC involved the St Andrew's Exchange Student of our time, Ian MacGregor, a very distinguished lawyer in Toronto these days. He did a classics show on a Friday night, and Friday night was carousing night at Queen's. Ian got carried away, but he showed up for his show, and the rolling R's that were part of his normal speech were very pronounced that night. He was introducing Schubert's "Unfinished" Symphony, and he had done quite a formal presentation. Then, almost as an afterthought, he said, "You understand, of course, why the symphony is unfinished. Because Schubert died of a terrible dose." I would like to have seen Mrs Angus' face when that came over the air. In those days, we didn't even think "dose," let alone say it on a public broadcast.

They used to go broadcasting live from hockey games, and you'd see the little team from CFRC in the old Jock Harty Arena, where the psychology building is now. There they'd be, all bundled up, with their telephones connected to the studio and little mikes. Jim O'Grady, Sports Editor for the <u>Journal,</u> was also announcer for hockey. It was alternative radio even then. They kept jazz going before the commercial stations realized it was a good thing again. And classics and plays. It taught us a lot. I think CFRC was more of a teaching tool, part of education, even than it is now. Your programs are complex and colourful, but the University was smaller, a lot more of us came from non-urban areas. The teaching part of CFRC was considerable.[123]

The Kingston Board of Education's Director of Education, Mr Ritter, asked Principal Wallace whether CFRC might be able to carry the CBC's national school programs two hours a day. CKWS carried only the Friday programs.[124] Prof. Stewart estimated it would cost Queen's $6 a week, if there were no line charges and if CKWS assumed the $2 an hour for the operator.[125] Nothing came of this matter.

The original program logs had to be sent to the CBC every day, and duplicates have not been found at Queen's. A sample schedule in the Journal from the spring of 1951 shows the sort of programming that was being done on CFRC:

Thursday evening
7.00 pm Sign on
         Show Time (Theme: Swedish Rhapsody
         - David Rose) Music fron "Annie Get Your
         Gun" (MGM Records)
7.30   News with Pete Braden
7.45   Inside Story - Murray Mackay interviews
         interesting campus figures
8.00   Light Classics - Mike Humphries, announcer
8.30   Talking with Tepper; talks with heads of
         campus organizations
8.45   Jazz from Town Hall (Jazz at the
         Philharmonic)
9.00   Drama; written by students
9.30   Rocking Chair - John Bermingham (Theme:
         Rocking Chair; Claude Thornhill, Vaughn
         Monroe, Spike Jones, Doris Day)
10.00   Sign off; God Save the King

Friday
5.30 pm Sign on; warm up with popular music
5.45   Pinnochio
6.00   Pinto Pete (western music)
6.15   UNESCO World Review; report by Bob
         Wright
6.30   Supper Serenade (Theme: The Continental
         - Shearing)
7.00   Sports - Paul Toune
7.15   News - Bob Radford
7.30   Jazz Time (U.S. Armed Forces Radio Service)
8.00   Madness and Merriment (novelty popular)
8.30   Levana Time (Theme: String of Pearls -
         Glenn Miller)
9.00   Yale Glee Club (Columbia L.P.)
9.30   1490 Classics (Theme: Meditation - Boston
         Pops)
10.00   Request Time
11.00   Dance Time (Nick Seiler, remote from Grant
         Hall)

11.30	Studio X (drama)
11.45	Cool off (Good-bye - Glenn Miller); God Save the King
Saturday	
5.30 pm	Sign on; warm up with popular music
5.45	Pinnochio
6.00	Pinto Pete (western music)
6.15	Supper Serenade (light classics and musicals)
7.00	Remote broadcast from Students' Union, featuring Dan McRae and Sid Penstone, duets
7.30	Jazz Club (Jimmy Dorsey)
8.00	Waltz Time (Henri Rene's Orchestra)
8.30	Intercollegiate basketball from the gym
10.30	Lean Back and Listen (popular music; Dorsey, David Rose, Dinah Shore, Frank Sinatra, Glenn Miller)
11.00	Call It Anything (Oscar Peterson, Harry James, Mario Lanza, Ray Noble, Dick Jergens, Gene Krupa)
11.30	Guess What (novelty popular) Sign off (Good-Bye - Glenn Miller); God Save the King

H.R. Cavanagh was the Manager for the engineers in 1950-51, and his executive was J.C. Wood, E.R. Bruce, R.D. Colvin, G.G.F. May and S.B. Upper.[126] The next year the CFRC radio station executive consisted of J.C. Wood, Station Director, Harry Brien, Doug Entwistle, Bob Wright and Dick Cowper, and John Bermingham became the Workshop Program Director.[127]

The old Presto disc recorder was supplemented in late 1953 by a model PT6-VAH Magnecord Voyager tape deck, a single case (7.75 x 17.75 x 19.25 inches), portable (42 lbs), dual speed (7.5 and 15 ips) monophonic machine taking up to 7" reels.[128] By this time the CFRC record library had grown to some 2,000 discs.[129]

A typical Radio Workshop Thursday evening in 1953 showed that the standards and the general tone were still high:[130]

6.50 pm	Sign on and warm up
7.00	Campus News, with Pete Handley
7.10	Sports Interviews; Claude Root interviews campus sport celebrities
7.20	Leave It to Levana, with Jane Sherman
7.30	Relaxing with Music, Paul Weston recordings with Walter Masters
7.45	Talent Time; Male quartet: John Brownlie, Derrick Best, Roger Ingall and Will Friskin
8.00	Discussion: On the subsidization of university sports
8.30	Around the Turntable; jazz recordings with

	Fred Flynn
9.00	The Music Room; Beethoven and Stravinsky
9.30	Dramatic Moments: Shaw's "Candida"
10.00	Mixing with Mike: Mike Humphries plays popular records
10.30	Sign off

At some point the heavy discipline at the station was relaxed on Friday and Saturday nights to see if common sense would prevail among the engineers. It didn't. The door was always left open, horseplay in the studios increased, guests wandered around unsupervised and people who had been drinking were sick in the station. The students lost the use of the station for six weeks and were allowed to resume on probation. A list of regulations was posted, including: A member of the executive would be in charge each night; The operator was second in command, in charge of conduct in the studio and of the aired product; The door must be locked at all times and a guard posted; Guests must be sponsored and supervised by a member; Profanity will not be tolerated in the studios; Membership cards were issued, subject to cancellation if four consecutive meetings were missed; An active member had to attend three meetings a month.[131] There was no such problem with the members of the Radio Workshop.

Over the Christmas holidays, 1951, a group of students under Prof. Stewart's direction completely re-built and renovated the control room. Sidney Penstone, of Winnipeg, did the re-wiring.

Sidney R. Penstone: I was in Science '52 and '53 when I first got involved in the station, and that would have been in my third year, back about 1951. I'm Elec. '55, and I returned to Queen's for a Master's degree, worked in Ottawa till '63 and came back onto faculty. I'm now Associate Dean in Applied Science.

My group arrived at Queen's in the late '40s, and we saw the tail end of the veterans. Science 48 1/2 was still here the year that I arrived. A lot of the fellows that we knew in Science '49, '50 and '51 were still veterans, so we saw a different face of Queen's than later on when the age profile started dropping back to be mainly high school kids. It was a pretty responsible bunch at that time, and if you wanted to get involved in the radio station, you went up there shaking in your shoes, because these were real men that were operating the radio station. Our impression was, if we were real good and did all sorts of dirty jobs for a while, someone might let us near the radio station. It wasn't a democratic organization at all.

The programs produced by the Drama Guild on Thursday evenings sounded like real radio. We engineers went on on Friday and Saturday nights, whenever people felt they had an opportunity to do it, and some of it sounded like real radio and some of it sounded like real fun. There was no official organization to that engineers' side of it. You had to show that you would participate, that you had something to add to the activity, that you'd work very hard at it. It was run in a very dictator-like atmosphere. There were a few people who decided who was going to get on the air and who wasn't. Very much the same as the Drama Guild, but they did it in a very nice manner, with auditions. And things were very seldom arranged beforehand. Some people on the engineering nights would, in a sense, have a slot and what they did in that slot was pretty much up to them, as long as they didn't do anything that would get us in trouble with the Principal. We had a feeling that retribution would be swift if a student stepped out of line. I don't know how true they were, but there were always stories about how someone had done something the previous year and had been expelled instantly.

There was some friendly rivalry between the engineers and the Drama Guild. The Drama Guild didn't do any operating. All of the operation of the station was handled by engineers, almost without exception Electricals. Harold Stewart was the staff member who really looked after things, but he left the day-to-day operations to this group of third and mostly fourth year Electricals, who had a key to the building - a marvellous thing - and operated the radio station. We were green at that time - we just stood around in awe and watched. We'd come in Thursday nights sometimes and if we were very quiet, we could watch the Drama Guild people - Lou Tepper and later Johnny Bermingham, who sounded like real announcers on the air.

A group in Science '53 was the main bunch that I got involved with. Dan McRae (Elec. '53) was a great big lanky fellow. Well over six feet. Very musical. Played the piano, the banjo, knew jazz inside and out. He and I played piano, just very loosely - he played exceptionally good bass because he knew his jazz chords very well - and once or twice we were inveigled into playing on Friday night. One thing led to another and we started doing programs - he had a marvellous collection of old jazz records and did a jazz show. Dan was a first class operator. His hands were so enormous that he could pick up a 78 from the top with one hand, turn it and literally slam it down on the turntable with great precision. He prided himself on how fast he could flip a 78. Except once in a while he would miss and the record would be shattered by the spindle. Bob Radford (Elec. '56) was more involved in management. His father ran CFJR, Brockville, so he had a background as a radio boy and knew how it was done commercially. So, the people who did programs were mainly those who knew something about their own kind of music.

Oh, we did take-offs on other peoples' shows, and we had a lot of people come on and do something that we knew they could do well. We'd heard them do it over in the Students' Union or at a Grant Hall dance or show, and we'd invite them over. I think we were a little more flexible than the Drama Guild, which had a more structured format. If we got good listener response, we asked them to come back and do it again. We had a schedule with regular shows, but we had room in it for special events. We had a couple of fellows who played the piano rather well, Beavis and Pope, who used to play in the Union and at dances, if their arms were twisted. We did interviews. We did good radio as well as Saturday Night Live radio.

We had bad evenings when people did things that were not good radio. We also had some very fun things. We were struck by how well Orson Welles' Martian invasion had worked, so one of our dramatic highlights was the night we burned Grant Hall down. We actually did have it scripted. We dragged out some sound effects, but most of it was done by people yelling and screaming and being crowd noise. It succeeded beyond our wildest expectations, because many people hop around on the dial, and by the time they tuned in we were broadcasting the noise of fire engines and crowds milling around. We got all kinds of phone calls from people in the city, and all kinds of people drove down to Grant Hall to see it burn down. We were pretty proud of that and felt that we had succeeded.

About that time there was a program on commercial radio, a girl with a very dreamy, sexy voice used to do a record show. That sort of thing just wouldn't appear on a Drama Guild Thursday night. Lynne Goldman, in Arts and Science, had a voice something like that, so she came on the engineers' night and did a show like that. Which worked pretty well. We sort of did things as we went along. George Lake, one of the engineers, liked playing quiet jazz, and got a fairly regular show going. He had quite a low voice, and a nice, easy way of talking in between records. Gradually we began hearing from co-eds. Who is this guy with the sexy voice? Unintentionally, we ended up with a male sexy voice as well as a female sexy voice on the air.

Most of the programs were quite personal. One thing the station never did was run a bunch of LP's from one end to the other. Announcing was what everybody wanted to do and everybody wanted to do something on the air that was uniquely theirs. Doug Entwistle (Elec. '52) was the only guy who had LP's, and he did a classics program from his own or borrowed LP's. The station had virtually no LP's. We had GE cartridges so we could play 33's, but our tables couldn't play 45's. When we knew someone was going to have a program with LP's, we'd actually change the cartridge in the arm on one of the turntables. Eventually they insisted that we put better arms in before anyone else would let us play their LP's - we were tracking up in the ounces! We bought another set of arms with lighter cartridges because there's no way people's LP's would have survived that sort of treatment.

One night Doug came out with "Sending 100 watts of surging power tippy-toeing right to the edge of the Kingston Traffic Circle," which was about the limit of our AM signal in those days.

I have a rather foggy memory of what the physical layout was like. I remember that the equipment was pretty sketchy. For our main console we used the OP-4, an RCA portable amplifier designed for doing remotes. It sat in a box, vertically, and had only three channels on it. Harold Stewart built the pre-amps that I saw on the turntables. These were great, big tables that handled the old 16" transcription records. A turntable like that takes a full 15 seconds to get up to speed, so the way you cued them was by holding the rubber mat. You sat the record on the rubber mat, and the mat sat on top of a felt surface on top of the big turntable, and you cued the record by spinning the mat back and forth and you held the mat while the turntable ran. Then you let it go. These huge turntables made an enormous rumbling noise inside the control room, groaning and scuffing away as you held the mat back. So we never did broadcast from the control room. We always had to have an operator in the control room and an announcer in the studio.

We also did remotes, and I can't remember how the devil we did them. We must have had some kind of a portable unit. The one I recall was battery-powered and very old, but worked quite well. We used to have the Bell put in the line every time we wanted it - twisted brown wire that we used to see coming into the end of Fleming Hall from across the campus. We used to worry why we had to pay Bell so much money when we could have run a wire across the road from the gym ourselves. But we never did that. We were told that if we did, Bell would cut the University telephone service off. We used to do broadcasts from Grant Hall. But I can't recall too many hockey events. We had a pair of wires hanging out from the balcony and that was the connection. Bell had left them in and every time we went over we'd pick them up, but we'd have to pay Bell for that particular day to put the grasshopper or put the fuse in somewhere. We didn't keep the lines in all of the time. That would have cost us too much.

I don't know when we got thinking that we needed a new console. But we persuaded Harold Stewart that a bunch of us in Electrical should design some new equipment for the control room. Budgets were kind of hard to come by in those days. To us it was a mystery where the radio station got its money, but it seemed like a major event if you could spend $100. That was a pretty staggering amount of money. A fair number of people got involved in the re-design of the equipment, but I don't know where the design even came from! We decided to build our own console. We looked at catalogues of what real, professional consoles looked like, and decided that we'd take apart our existing equipment and put it into a nice, big steel console. We'd take the pre-amps out of the turntables and put them into this box, we'd take the guts out of our OP-4 amplifier and we'd drop it into this console, and we'd put all kinds of switches and keys on it. The OP-4 had the amplifiers and the gain control - there were three mike channels and a master. Then all the switching for the microphones, lights and so on were put in a separate box - a black box which was nothing but a row of keys. They were largely unlabelled and that was part of the mystery of operating a radio station. You had to know which key did what. And then over behind us was the other half of the system, a patch panel with jacks. Certain jacks had to be in all the time or nothing went over the air. When you patched in remotes, you had to break into one of these channels and steal it back from the studio, because there were only three channels on the main amplifier.

When we decided to build things, we were going to start right from scratch. We bought sheets of steel from Mullard and Lumb, had them cut by torch to the shape we wanted down at McLaughlin Hall, welded by Seth Brown, and spray painted it so it looked just like a very expensive RCA or Northern Electric console, complete with pedestals on the end, a sloping front and a lid that opened up on a piano hinge. It's funny the amount of looking around we had to do to find somebody who would sell us a piano hinge. That was expensive, and I recall a lot of agonizing over spending the money to put a piano hinge in the top of the console. We had another amplifier, the cue amplifier, hidden underneath the table. When you were cuing up a record, you switched the output of the pre-amp for the turntable over to the cue amplifier, which had totally separate loud speakers to cue it off-air. I think by then I had run afoul of regulations about grades, and was repeating. George Lake, who was in Electrical, took it upon himself to do a real design, in true engineering fashion, of a brand new cue amplifier, using all the new techniques he'd learned in his course. He got Harold Stewart to buy all the parts and he assembled it very carefully. Only one trouble. It oscillated. Howled like a banshee. We finally fixed it by taking the input transformer off and turning it 90 degrees. So I guess it was coupling between the input and output transformer in the same chassis.

This console had every type of switching we'd ever dreamed about. We gradually wired that in, then shut the radio station down over the Christmas holidays and a group of us proceeded to re-wire everything. We had to re-wire the 'ON AIR' lights, all the microphone outlets, all the switching for all the speakers, the monitor system, the cue system - all now went into a single large case. Queen's used to turn the heat off over the holidays, and it got pretty cold up in Fleming Hall. So we brought up the load resistors from the basement, used for the machine experiments, and we brought up the 250-volt DC. In those days there was a panel in the control room and another in the transmitter room which had bare voltages. After all, that was an E.E. lab and everyone knew you had to have voltages. When we re-did the control room, we took them out of there, so we ran wires from the power panel in the transmitter room into the control room to run these big wire racks - about 3 feet square and 6 inches deep with about 20 knife switches along the top to select which of the toaster type wires you had on. We had a couple of these sitting in there to give us heat, and we worked around the clock. It was really dumb, come to think of it. Various people came and helped. I think I was probably the key man in deciding what got wired - I'd done most of the paper design. Georgia Johnson, who I later married, worked with me, soldering. And Bob Radford. There was some panic getting it ready, because the Drama Guild was hoping that nothing would disturb their operation. And it didn't. As I recall, we did pretty well in getting it back on the air. It worked!

We built a wooden platform - that was the origin of the platform in the old control room - to stop people from leaning over the operator's shoulder, poking at things and getting in his way. We felt that by having the operator up on a platform, it sort of separated him from the rest of the world. We got the two big old turntables up on it, and sat the console between them on a wooden shelf. The operator was on a swivel chair so he could move back and forth smoothly between the turntables. What we forgot was there was no means to stop this guy from rolling back off the platform. So, after a few disasters, we put a chain on the bottom of the chair, which prevented the bottom of the chair from going back too far. Still didn't always work. The odd guy would lose his chair - he'd stand up suddenly and the chair would fall over because it would catch on the chain.

We used an awful lot of lever switches, but I can't remember that we bought very many. We were great at gadgetry, and it seemed like so much fun to hook up all these complicated switching systems using lever keys, so that if you pressed this key you automatically locked out that studio and this one, and turned off that speaker, and turned on that light. To us that was the nearest thing to computers in those days. The keys were the ones used by the telephone company, with a little black lever sticking out of the front - one position in the centre, one position up and one down. In the back there's an enormous pile-up of contacts. By using the three positions, you could come up with very complicated operations. These keys were very expensive, but we had boxes of them. I don't think we ever bought any. I think we always scrounged them from Northern Electric. It was a lot of fun taking these apart and re-building your own switches, so that some of these didn't always work. You were always in for some surprises. The good operators knew that you had to hold this particular key slightly to the right as you moved it, because sometimes it had a habit of not closing. A certain amount of skill was involved in operating the control room.

Operators took a test which was very informal. As the older fellows graduated and moved on, we became the hierarchy in the station, and Dan McRae's test was really vicious. The big turntables used synchronous motors, and there was a clutch to engage the turntable. If you were careful when you first turned on the power in the station, you could start these motors idling the wrong way. It was a dirty trick to play on anybody, but you'd start the motor the wrong way, and sit this operator down and he'd put his record on and cue it. It would come to the time and he'd engage the clutch, hold his cued record back, turn it on the air, the record would start up the wrong way and the arm would immediately fly off the record. That was part of the operator's test. Did the guy know what we'd done to him? Did he have enough presence of mind to bail himself out of a situation like that? Or we'd unplug the main microphone in the studio and leave an extra microphone in the corner, plugged in. We were looking for fellows who could think on their feet. We had girl operators in those days - they could cope with this sort of thing.

I can't remember why we used to bug the Drama Guild sometimes, except we found some of them a little pompous. Not the people who ran it. We only gave them our very best operators. No doubt about that - we never gave them an inexperienced operator. But, by giving them our most experienced operators, we also gave them fellows who knew how to scare the daylights out of a green announcer. They might let people in the booth hear things that really weren't going out over the air, while they had things well under control. Or the talkback didn't appear to be working — the operator was mouthing instead of talking. When the announcer said that he couldn't hear the operator, we'd give signs through the glass that we couldn't hear him and we couldn't understand what his problems were.

The Drama Guild operated very regular hours and we knew when they were going to be on the air. Our hours were very flexible. If somebody came back in the Christmas holidays, we went on the air. If no one was around, we didn't. We used to do request shows. At that time Fleming Hall had only one telephone, and it was in the Library almost at the other end of the long hallway. The record library was just outside the studio, so for awhile we used to have somebody answer the phone down in the Library. They would transfer the request up to someone in the studio, who would look it up in the catalogue, get the record and put it on the pile. Finally, unknown to Bell and to Prof. Stewart, we used to run the telephone down the hall on our own wiring and bring it down to the record librarian.

The present people won't remember the AM transmitter that we had. The 250-watt (RCA) job is a late gift from CKLC. Before that, the transmitter that we were using sat in big open steel racks - black wrinkle finish - built from scratch by Harold Stewart and Harold Pollock. You would go to Hammond Manufacturing in Guelph, and order steel rails and panels for the front, modulation transformer and input transformer, and you'd drill holes in these plates and mount tube sockets in them, wire up the tubes and wind your own coils. It was a 100 watt transmitter, had an open back on it - totally unacceptable by modern safety standards - you could kill yourself if you grabbed ahold of it. I know. I burned my finger on it once, by trying to adjust something while it was on the air. Something you were never supposed to do, but we were having some difficulty one night and I foolishly stuck my finger in where I shouldn't. The limiter was a rather interesting one they had built. In fact they wrote a paper on it and it was published in Electronics.[132]

The transmitter was a great big thing and took up as much space as the present one. Most of it was empty space, because there were large air-wound coils and lots of space around the vacuum tubes. And I remember the particular problem we used to have with the modulators. When I went to work for a commercial radio station in 1955, I found that this was traditional, that modulator tubes were pairs and they were always off balance. Every commercial radio station engineer had a boxful of tubes of doubtful or mixed characteristics. They were very expensive. As your modulators went off balance, after you signed off at night, you would paw through the box and try to pair tubes on the basis of scribbled notes on the bases.

Part of the problem was that the modulation transformer couldn't stand any unbalanced DC in it. It was a good quality transformer, but you had to make sure that the push-pull modulators drew equal DC currents, otherwise they'd cause some saturation in the core. One of our main problems was keeping the modulators balanced. The limiter worked all right. RF stages, the output stages, no problems. For open houses we used to hang a fluorescent light on the feeder of the antenna, so the crowd could see the lamp glow on and off when people talked.

There were some old 78 rpm recordings that we used to use for special puroposes. One was the Wilf Carter recording of "When the Iceworm Nests Again." There wasn't very much western or cowboy music played on the station, except by the engineers on Friday or Saturday nights. Gord Dorland (Mech. '53) ran a program of cowboy music, and "Iceworm" became such a popular piece that he'd always get requests for it. It became almost a trademark. Another record that had a special purpose was "Autumn Leaves." When we used to set up the limiter on the AM transmitter, we had to have an operator in the control room keeping the levels right up bang on 100%, or 0 V-U. Dan McRae found that this particular record, if you set it correctly on the first phrase of the music, would go right up to 100% on the rest of the record. So, you could put this record on and go into the transmitter room, and know it was just as if an operator was there. We pulled that one out of circulation and kept it for adjusting the limiter on the old AM transmitter. We also had a sign-off theme called "Good-Bye," that we used to play after the last record of the evening, just before "God Save the King." It was a rather restful record, and over top of

"Good-Bye" we'd give the sign-off information, and maybe thank the Department of Physics and Ontario Hall for holding up the other end of the aerial for us.

Dr Angus bought a Brush Sound Mirror for the Drama Guild, one of the first tape recorders around here, and we experimented with making recordings and putting them on the air. We rapidly learned that good recorders ran faster than the Sound Mirror; you got away from all the wow and flutter. So, we took the regular capstan off and put a smaller capstan on and made our recordings at our own speed, which was somewhat higher than standard. The result was, of course, that none of these tapes could be played on anybody else's machine. Instead of doing everything live, the Drama Guild was starting to experiment by recording ahead of time in the Drama Guild for airing on Thursday nights. That was new for us and pretty sensational. We had no permanent tape installation - we just brought the Sound Mirror over and set it up and patched the phone jack or something into it and brought it in as a remote. What we called "NEMO" line. To this day, I don't know what "NEMO" stood for. We inherited jargon from people who had worked with CBC. We never knew where some of these mystery words came from, but we had "NEMO 1" and "NEMO 2" up on the patch panel. In fact, they were our remote input lines. To get them into the system you patched Grant Hall into "NEMO 1," and then "NEMO 1" came up on a key on the control console. Someone thought that "NEMO" meant "no one" in Latin, but we didn't know what the significance of that was.

Head alignment was a problem with those old Sound Mirrors. You couldn't align the heads - you took them the way they sat. We didn't make much use of tape on the engineering side because we didn't own the tape recorder.[133]

Margaret Angus: There are marvellous things you can do with tape to cut out boo-boo's and make people sound very good. In the days when we got a tape deck and began to use a lot of tape, we had a student in an administrative position come in to give a talk. The boys taped it and worked on it afterwards, and when he heard it on the air he came over to us and said, "Well! It went awfully well, didn't it?" They said, "Yes. It went marvellously. Would you like to hear some of the things we took out?" "Oh, I didn't realize you took anything out." So they played him back a tape they had spliced together of all the ooo's, ahhh's, errr's and it was really hilarious. He never afterwards talked making great pauses like that, without remembering that tape the boys put together. They really took him down a peg, to show him that he wasn't that good after all.[115]

CFRC Goes Simulcast on FM, 91.9 MegaHertz

The original Queen's application for an FM licence was filed by Prof. Stewart on September 9, 1943,[7] and the experimental call VE9BH, for use with a dummy antenna, was granted.[18] At the end of the War, the CBC Board of Governors made recommendations to the Minister about policies for dealing with applications for FM licences, and the Manager of CFRC was asked if there were still interest at Queen's in FM broadcasting.[134]

Harold Stewart was a representative of education on the Canadian Radio Technical Planning Board in 1945, and acted as liason between the Board and the National Advisory Council on School Broadcasting. He wrote to R.S. Lambert, Honorary Secretary of the Council, advising him that Mr W.G. Richardson of the CBC Transmission and Development Department recommended that anyone planning on FM broadcasting should so inform the Radio Branch, DOT, as a Canada-US conference in January would allocate FM channels along the border. Lambert wrote to Principal Wallace "if intentions to use FM radio for educational purposes are not filed now with the Department of Transport, it may be too late to do so after the Conference in January."[135]

Dr Wallace's letter to the DOT was dated ten days later.[136]

Two years later, G.C.W. Browne of the DOT informed Stewart that the CBC Board of Governors was now prepared to accept applications for FM private commercial stations from other than AM licencees. Queen's was on record as desirous of applying for an FM licence whenever they became available.[137] Stewart responded with a request for an application form and information. He explained that the hope was to obtain a private commercial FM licence to do educational broadcasting "in the not too distant future." They would use an FM transmitter which was "still in the initial stages of development."[138] A few years later, Queen's received a one-kilowatt FM transmitter as a gift.

<u>Harold Stewart</u>: The FM transmitter was presented to us by the Westinghouse Corporation from Pittsburgh, and it was free. They gave them to a number of colleges, but I don't think to many Canadian ones. We got ours because we had a friend at court, Alexander C. Monteith (Elec. '23), known as "Monty," who was I think at that time Vice-President of Manufacturing for Westinghouse in Pittsburgh. He made sure that Queen's got this one kilowatt FM mono transmitter about 1952. Monty persuaded Canadian Westinghouse to set up a maintenance fund, which did buy spare tubes and many things for years. Don Maclean, Canadian Westinghouse, was responsible for dispensing the maintenance fund. He was one of our grads, and a good one.

Two students, Sid Penstone (Elec. '55) and a chap named J.D. McGeachy (Mech. '53), who is now a professor in Mechanical Engineering, put up the double-V antenna on the top of Fleming Hall at the risk of life and limb. We didn't really appreciate that it was a hazardous job. They could have fallen, the pulley could have broken, any number of things could have happened. The FM transmitter went on the air in 1954, operating at 91.9 and it had a second harmonic that interfered with local reception of Syracuse TV station channel 8. It was one of the few TV stations that people could get, and our second harmonic landed right in their channel. Now, we didn't have a

A·C·Monteith.
Alexander Crawford Monteith (Elec. '23). "Monty" arranged for Westinghouse to give the 1 kilowatt FM-1 transmitter to Queen's radio station in 1952.
-Queen's Tricolor, 1923

very big second harmonic - it was a commercial transmitter and its harmonic radiation was well within the government specifications. Although we had no legal obligation to protect that distant TV station, as a public relations gesture we thought we'd like to do something to cut that second harmonic down.

We're now into 1956. Sid Penstone had done the job raising the double-V antenna when he was just a student, so now he's graduated and been with the Defence Research Board, and now he comes back to do a post-graduate year. Post-graduate students need a problem, so I gave Sid the job of designing a filter for that transmitter, for his M.Sc. project. Well, Sid spent a good deal of time investigating all of the possibilities, and he came up with a co-axial filter which knocked the second harmonic down by about 100 db, which we could hardly believe. He did a damn good job on it and practically eliminated the second harmonic problem. Sid did other things - he built amplifiers and he'd had experience outside (in radio at Brockville), became an E.E. Professor and he's now Associate Dean, Faculty of Applied Science. A great guy.[1]

The original communication from Alexander C. Monteith, Vice-President of Westinghouse in Pittsburgh, offering to donate an FM transmitter to Queen's, has not been located. What does exist is a letter from Monty to Prof. Jemmett, dated June 20, 1952, advising that the FM-1 transmitter and a set of tubes had just been shipped from Baltimore.[139] Two copies of instruction book 81-210-1 arrived with the equipment, along with drawings that would be useful in getting and adjusting a suitable crystal. The Order Service Department of Westinghouse suggested a Bliley Type MC-7 crystal mounted in a Type TC-92 oven.[140]

Prof. Stewart enquired of the DOT on March 26, 1952, if it would be possible to submit an application to install an FM transmitter as part of Queen's educational service. Transport advised that it was prepared to forward such an application, to include an engineering brief, to the CBC Board of Governors. FM channel #220, 91.9 megacycles/second (mHz), was tentatively reserved for the Kingston area, and could be had pending clearance of the proposal with the FCC in the US.[141]

James L. Hunt at the student-built console in the control room in Fleming Hall, circa 1953. The room was re-designed and re-wired by Sid Penstone (Elec. '55) and a crew over Christmas vacation, 1951. The mixing board was put into a steel case with a piano-hinged lid and the case was painted "Northern Electric Blue." Hunt sits on the small dias designed to protect the operator against meddling hands.
-courtesy James L. Hunt

Specifications of the proposed FM station were sent to G.C.W. Browne on September 19, with a copy to CFRC's consulting radio engineer, Keith A. MacKinnon. Browne confirmed that channel 220 was assigned to Kingston, for operation at 1400 watts radiated from an antenna 100 feet above the ground.[142] The preliminary application for an educational FM private commercial broadcasting station, to operate in conjunction with CFRC and as an extension to it, was filed November 21, 1952.[143] Keith MacKinnon forwarded the required four copies of the engineering brief to the DOT.[144] At the time there was a steel shortage and the Minister of Trade and Commerce would not authorize the purchase of steel for the construction of new radio stations. That restriction did not apply to CFRC, however, and the application was approved by Order in Council P.C. 1953-414 in the spring of 1953. The station, to be called CFRC-FM, would have a nominal power of one kilowatt and would operate on channel 220 or 91.9 megacycles (mHz). Stewart was asked to return Private Commercial Station Licence No. 87 to the office of the Controller of Telecommunications for revision.[145] The licence form authorizing the operation of CFRC at 1490 kilocycles (kHz) as a non-commercial station, from April 1, 1951, to March 31, 1954, was surcharged on March 31, 1953, to permit the operation of CFRC-FM from Fleming Hall at 91.9 megacycles and 1270 watts ERP. The FM licence became effective on March 31, 1953. By its terms, CFRC-FM was required to carry simultaneously all the programs broadcast by CFRC.[146] Separate FM programming would require a formal application for permission and a separate private commercial licence costing $50.

James L. Hunt in Studio "B" announcing a live talent show taking place in Studio "A," circa 1953. Note the RCA RX 88 "bombshell" microphone.
-courtesy James L. Hunt

Keith Geddes (Dewey) Fillmore (Phys. '55) at the "N.E. Blue" steel case console with the piano-hinged lid, circa 1953.
-courtesy James L. Hunt

Willis Hunt and unidentified person doing a jazz program in Studio "B" in Fleming Hall, circa 1953.
-courtesy James L. Hunt

In the same hearings, the Board had a request from Robert S. Grant to authorize a licence for a new AM radio station in Kingston, to operate at 1,000 watts on 1380 kilocycles.[147] The Board felt that there was a place for another private commercial AM station in Kingston, approved the request and CKLC came into existence.

Report on FM-1 and Antenna Problems

By May 31, 1953, the FM-1 was up to specifications, and Dan McRae (Elec. `53) reported on the operating characteristics of the transmitter. Problems with the feedback circuit and RF interference in the distortion analyzer were sorted out, and the defect in the tuning motor drive noted. He advised on the warm-up time of the crystal and of the transmitter, pointed out the warm-up drift in the meters, and the instability of the modulator plate current. If improperly tuned, the frequency centre would shift and even moderate modulation would trip protective relays in the final cathode, and the transmitter would dump. "The transmitter goes off the air with about 140% modulation... Modulation is best set by program material rather than a steady tone due to the pre-emphasis circuits. I had best success using a dixieland jazz long playing recording that was heavily compressed in recording. This gave a fairly constant level with a good spectrum of sound. I usually set the modulation to 80%, and at no time during a varying program of music and speech was the transmitter over-modulated when the control room VU meter was relied upon."[148]

The Andrew Type 1202 multi-V 2-bay side-mounted FM antenna was ordered through Canadian Marconi in Montreal and received in June 1953.[149] In answer to a letter from Mr Browne, dated February 4, 1954, Prof. Stewart was unable to give a definite date for putting the FM transmitter on the air.[150] This was partly owing to delays in shipment of the antenna, Jones MM-707 Micromatch, cables and fittings. Stewart explained that the antenna had been wrongly adjusted at the factory and had to be returned to Marconi twice.[151] In fact, Andrew had mixed up two almost identical antennae in shipping from Chicago and had sent Queen's one for 95.1 mHz.[152] By February 1954, trials had been carried out, but cable fittings were not complete and connections were temporary. In addition, the transmitter's second harmonic at 183.8 mHz was interfering with television channel 8. There was also interference with other channels, due to cross-modulation with CKWS-FM on 96.3 and CKLC-FM on 99.5 mHz in receiver input circuits. Student volunteers were working on the problem, filtering the feeder. The work was progressing, but he had to apply for an extension to June 30, 1954.[148] This was granted,[153] and the licence was renewed on April 1.

Sid Penstone asked Prof. Stewart to obtain the operating conditions, "the ratings, characteristic curves, and typical operation for the final - WL 5736."[154] Stewart sent to Westinghouse on February 12.[155]

Sidney Penstone: The FM came in about 1953, and because we had to carry the same programming as AM, it didn't have as much impact as you might have expected. The FM was kind of impressive when we finally got it in and working. I must have been in my final year, still involved in the station to some extent, but by now I'd learned to keep these things under control.

We were offered one kilowatt, three kilowatts and ten kilowatts by Westinghouse. I remember the way Prof. Stewart made the decision. We took a ruler and measured the doors in the building, to see what size transmitter we could get in without major renovations. And it turned out a kilowatt seemed the most appropriate one. None of the others could have been got in without major surgery on the doors. Now, that's just my memory of it and it may have been only one of the factors that went into it. Hydro wasn't a worry. Shortly afterwards we found that Physical Plant was very concerned that the transmitter would fall through the floor, when they found how much it weighed. So Prof. Stewart had to do some calculations on the loading to satisfy Mr Hinton. Everything that got done, you had to go and talk to Mr Hinton about. If you were on the wrong side of Mr Hinton, you wouldn't get it done. He also had very strong ideas about who was authorized to have keys.

The Westinghouse FM transmitter came in during the summer, when I was off working. Prof. Stewart told me how it came in. They had Bob Purtell come with his trucks and cranes - very ingenious fellows, Purtell and his crew. They looked at the size of the windows and there was no way to get it in through a window, so they dragged it up the stairway, landing by landing. In those days there was a turret on the back of Fleming Hall (where the bridge to the new annex is). At every floor there was a window that looked out from the landing, and similarly there was a window looking out the front. So they put a winch truck out on the front lawn and ran a block and cable out through the window to the front. They put planks up the stairs, and using the leverage over the window sill, dragged it first up one flight towards the front of the building, then they moved the winch truck around to the back of the building for the next landing, put the planks up there, and dragged it up to the next landing.

I was involved in firing up the new transmitter for the first time, and it scared us. Prof. Stewart looked it all over, made sure it had survived and read the manual. I was still a student, and Prof. Stewart was the only one around who could understand that sort of complicated stuff and had a pretty good idea how it should operate. It was

CFRC TRANSMITTER

ELECTRICAL ENGINEERING LABORATORY

II

Window sign on door of CFRC transmitter room, #307, Fleming Hall. -photograph by Arthur Zimmerman

a fairly involved modulator and had a tube of a type we had had no experience with. It was the final with the big cooling fins on the anode. It had a big plug of lead in the middle of the cooling fins as a heat sink, and there was a blower that blew the air right up through the fins so it would never get hot. We thought that would look after itself. And the driver was a small tube, with a tantalum alloy anode. So, we turned it on and started bringing it up, adjusting the meters, and to our horror the driver tube, this little fellow, suddenly got red hot! We knew from the old AM transmitter that if tubes got red hot, we were in trouble - we were doing something wrong. For instance if someone forgot to connect the antenna, our finals would get red hot, give off gas and go gassy after that. So we quickly turned off the switches and had a brief consultation. Prof. Stewart made a point of finding out about that driver and that's perfectly normal. They operate cherry red, the hotter the better, because as the alloy gets hot it picks up gases out of the envelope, and the vacuum is fine running at that sort of temperature. That was our introduction to that type of tube.

So, in 1953 we got the transmitter up and running for the first time, and we immediately got calls from people who said, "I was watching channel 8 from Watertown until you came on. I'm going to phone the Department of Transport and you'd better get off!" At that time most of the reception from the US was fringe, and Watertown was much too far away to be classified as primary coverage area. But people in Kingston had found that by using lots of aerial and turning it the right way and holding their mouth just right, they could watch it. The frequency of channel 8 turns out to be the same as the second harmonic of our FM transmitter, 184 megaHertz. Every time we went on the air, we'd blot out all the channel 8 reception for a range of about a mile and half from the transmitter, all on our north side, with antennas that were facing south toward CFRC and Watertown. Legally we didn't have to do anything about it, but from the point of view of public relations, we thought maybe we should. For many months we went down to quite low power to the point where the calls would trickle off to a low level. Eventually, of course, we had to fix it. At that time I was starting my Master's work, and as my research project I got involved in looking at the methods of reducing second harmonic radiation from FM transmitters. The outcome of that research was an harmonic trap, which was built in the shop in Electrical and was installed in the antenna line on top of the transmitter. We were then able to watch channel 8 on a TV receiver right in the building, so we knew that we'd fixed the problem. However, we still got a few sporadic phone calls, since some of the TV sets being used near the campus were sufficiently out of adjustment or the design of the stages in them was such that they would generate second harmonic in the first tubes inside the receiver itself. At this point we had the DOT make tests, and we satisfied them that our radiated level of harmonic was way below any possible requirement. It's important to point out that the transmitter, as delivered, met all of the FCC and DOT requirements for harmonic radiation. It's just that we were in an environment where people were trying to listen to very, very weak signals from across the border. A few still complained, and our answer to them was that they should call up their TV repair or perhaps they should listen to CFRC instead of watching channel 8.

After I left, the station went through quite a few changes. It got a permanent technician, Jack Harrison, who was part of the Electronic Service Shop - in 1960, about the time the station was moved to Carruthers Hall. He was supposed to provide the necessary maintenance. When I was there, most of the maintenance work on the station was done by the Electrical students, or by a combination of us and the technician in Electrical. I don't recall that we always had a permanent technician. I was it for awhile when I came back as a graduate student. If something had to be replaced, we went to Harold Stewart and he ordered the necesary part.

The radio station wasn't a democratic organization. I remember in my latter years at the station, when I wasn't really active in the station myself, that students involved in running it would come and talk to some of us old-timers - at the ripe old age of 23 by that time - about

AEG Bayly technician disconnects Sid Penstone's filter trap and FM-1 from the antenna feed cable as the new FM stereo transmitter is prepared for service in Fleming Hall, September 16, 1987. The first test broadcast with the new transmitter began at 14:26:00 with the "Triumphal March" from Verdi's "Aida." -photograph by Arthur Zimmerman

the philosophy of running a radio station. One of the arguments at that time was that the radio station had to become democratic, to become open to everybody. In the eyes of many of us, that was the downfall of the radio station. From that point on you ran into all kinds of troubles, when a certain group of people wasn't clearly in charge. If things were not going well on the air, then these people would simply say, "Look, that's not appropriate. You're not going to be on the air any more. We're going to put someone else on." What was worrying us was when you try to become democratic, everybody felt they had a right to be on the air, and that seemed to us to be quite inappropriate for a radio station that tried to project an image of Queen's University. We felt that you had to select who was on the air - the Drama Guild did it and they did it in an undemocratic manner. They looked for people who had some talent and did things properly and according to the rules. The radio station club at that time seemed to be trying to move to a large activity and everybody had a democratic right to get on the air and do whatever they liked. As the station got bigger and bigger, some of the people who had been heavily involved, contributing their labour free, phased themselves out of it and moved on.

I don't really know much about the activity of the station by the middle '50s. It had moved into a different operating style. Certainly we never had any difficulty getting operators. People would work all night if you wanted them to, as long as it was what the people who operated the radio station felt was appropriate. This was a moving group - you never knew quite who it was - a flexible group. We used Mrs Angus as a sounding board sometimes, because we respected her opinion as to what was good on the air and what you could expect people to do. We didn't tolerate sloppy operating or stupid programming on the air. If someone was doing something just plain silly on the air, he would gradually be told that it just wasn't appropriate. Certainly there was no problem with profanity, or with off-colour material, ever. It was a privilege to get on the air and people were willing to be a record filing clerk - for a whole session or a whole year to work your way into the system - to get to where they were felt good enough to select records for one of the programs. It was a small group of keen people - a self-generating group. It had no legal constitution - it developed one - and no president as such. It had a Chief Engineer, who was always Electrical, and I don't remember what structure it had. It just seemed that things were appropriate. Prof. Stewart was the boss as far as we were concerned, and some senior students made sure that we didn't step out of line, and gradually we became senior students and made sure that those beneath us didn't step out of line.

The thought of asking for more broadcast time? We had all we could handle. We didn't want to broadcast any other times. We really wanted to be part of the community here in Kingston and what we said to be thought of as "that's what Queen's is doing," but we didn't do anything official about it. Our listeners were our customers, and we felt that we were representing Queen's. That's why we wouldn't let anybody do something that would be offensive to anybody in Kingston. Can't think that we had any troubles with the local population. They were very much on our side, it seems to me. Queen's to them was a welcome relief from whatever other radio there was in town. Of course the Drama Guild projected a pretty carefully worked out image.[133]

K.W. Greaves: Most of those who belonged to CFRC were Sciencemen, but there were one or two medical bods and even the odd Artsman. We were not a club or a University organization at all, since the station was considered part of the Electrical Department. We were, however, like a club in many ways and used to meet at regular intervals. There were no dues, but everyone was given some sort of job and, most important, access to Fleming Hall and the station itself. We operated the station and provided the programming on Friday and Saturday nights. On Thursday nights the technical operation was also done by CFRC people, but the programming was in the hands of the Drama Guild, a group of mainly Arts students whom we considered rather artsy-fartsy. They tended to wear beards and sandals and take themselves very seriously. They used to put on "drahma" and earnest talk shows about philosophy and communications art. The CFRC people, on the other hand, tended to be a rather scruffy lot, as befitted engineers, and to take a facetious view of life in general and radio communications in particular. Our main forte was the DJ business - playing records and blathering in between - and most of it had no redeeming social significance whatsoever. Our plays tended toward farce and our interviews toward football players. In Studio "A" we had an incredibly ancient, decrepit, low fidelity machine dubbed "the cocktail bar" - a waist-high bench on wheels with three turntables which could be operated simultaneously - for feeding effects records into the mix at appropriate times. We also had a mock door-and-screen for entries and exits, a box of gravel for outdoor footsteps, two coconut shells for horses and crumpled cellophane for frying eggs - all the most modern equipment. Almost all of this stuff came into use in a single show April 1, 1952 or 1953, our famous "Orson Welles broadcast." An excited announcer broke into the scheduled record show to inform the world that Grant Hall was on fire. This was followed by a live broadcast of the fire, including the sounds of crowds, sirens, roaring flames, and crashing walls, together with on-the-scene interviews with those watching the holocaust. This must have surprised those who knew how long it took to set up our remote facilities and how seldom they worked. Quite a few students showed up to see the fire and a number of taxis were seen cruising up and down University Avenue in a puzzled sort of way. Perhaps the most satisfying result was a call right after the show from the Kingston Fire Department which had, of course, been warned beforehand. They congratulated us on the realism of the broadcast and thanked us for giving them three more pumpers than they actually had.

There was Sid Penstone, Bob Radford, Dan McRae, Ernie Jury and George Podolski and my room-mate Jim Hunt (Arts '55), who introduced me to classical music and Dixieland, simultaneously, leading to an enthusiasm which has persisted. It was a fascinating group, all very bright and with a tendency to look upon Authority with a regrettable lack of awe. We discovered that the basement of Fleming gave ready access to the labyrinth of steam tunnels beneath the campus and, at night, virtually every University building, if one was prepared to crawl great distances through dirty, hot tunnels. We removed and hid a bust (of Homer?) - a particular trophy of the Drama Guild - from the English Department. We inverted a large garbage can, painted Engineering yellow, over the top of the flagpole on Old Arts tower. This required the use of lineman's spurs on a cold, dark, windy night on a very thin, whippy pole. I still don't know how they got it down, but it took several weeks. Then there was the night we set off the cannons in the Martello tower at Fort Frederick.[156]

Hugh Lightbody: Doug Frame was Station Manager and I was Program Director of CFRC when it was in Fleming Hall. The biggest differences to-day from the station we worked in are the outstanding studio and control room facilities now and the very highly professional quality of the programming. Where CFRC is now, in the basement of Carruthers Hall, was a Civil Engineering testing lab. A ram used to come down out of the ceiling there - I think they used to test concrete pillars in this area.

While we attempted to maintain some kind of professional standard, our basic requirement was to satisfy the desire of a lot of students who wanted to participate. Talent in our time had little to do with appearing on air, and if you'd heard some of those broadcasts, you'd understand that. There were a few people who were simply natural announcers and broadcasters and they put on some really outstanding shows. Most of us were there principally for our own enjoyment and probably not that of the listening audience. There was Doug McIlreath, and Doug Thompson had been at CJAD, Montreal.

The whole facility was cramped in there in Fleming, including a not terribly extensive record library, mostly 78s. The short hallway to Studio "A" held the record library, and that studio doubled for record auditioning. A nearby classroom was left unlocked for folks to congregate and keep out of the operator's hair. We started getting 45s at that time, and that helped. We even had a few LPs, which we owned, mostly.

I did the basketball broadcasts and Jim O'Grady, a lawyer in Ottawa who became President of the Consumers' Association, did the hockey and some of the colour stuff in basketball. At one time we had two loops to the gym - one control loop and one for broadcasts - but our budgets were slashed dramatically one year, and we had to go with one loop. Which meant that once the broadcast started, there was no hope of communicating with CFRC unless you had a runner. You'd be broadcasting like crazy in the gym but nothing was getting to the transmitter. Sometimes it didn't matter. One night Western was here for a very big basketball game. I had been furnished with the roster - of course, I knew all the Queen's team - and with all the cheering and the colour, the teams came on the floor and we talked up the practice and the statistics, and then the jump, the game started and I looked down at the program. There wasn't a single Western player on the floor who was on the program. All the numbers were different. I had no idea who these people were. That didn't stop the broadcast - it was probably one of the more colourful games. Somebody gave us a legitimate sheet about half time, but I just carried on with whatever name - we tried to give a good ethnic and national flavour to the Western team, and mixed the names up a good deal. That was the emphasis in those times - the fun of participating. A good learning experience as well.

In Fleming Hall, the operator sat there rather grandly at the console - and you felt pretty important when you sat there operating. There was a problem that led to a great loss of decorum. The operator's chair was on wheels and whereas he would like to roll back about 24 inches to stretch his legs, unfortunately the platform that the chair sat on went back only about 18. So every new operator spent part of his time under the tape recorder, because he would roll back and disappear from view. Then with the terrazzo floors, sometimes he was finished for the evening if he hit his head. We had a few problems like that.

You remember Thelma Hunter, Ann Dorland and Doug Thompson, Bob Sanderson, Arnie Matthews, Don Harrison, Graham Skerrit - a tremendous mix of people from different faculties participated. And Ralfe Clench. We knew that Ralfe was very sensitive about the spelling of his name. One of my jobs was putting the program lists into the Journal. We'd meet on Mondays in Fleming Hall and talk about what had happened on the week-end, did we want to make improvements, did anyone have any concern about the way their program was listed? And invariably Ralfe's concern was that Ralfe had been spelled wrong. He liked it to be Ralfe J. Clench, as described. And I suppose that Ralfe never realized that I never once put his name into the list in the proper way. I always intentionally spelled it wrong. And Ralfe even got very sensitive about leaving the "J" out of his name, so one week I submitted the list to the Journal with the middle initial "J" in everybody's name. Of course, the list wasn't necessarily what actually happened on the week-end anyway, but there was space in the Journal and they expected to hear from us.

Feeling versus the Drama Club must have been running quite high in engineering at this time. Jim Bethune was with 3 other actors at a stand-up mike, and they had just got nicely into the drama, when one of our engineers, not making any attempt to keep the noise down, flung the studio door open and walked into the midst of the live presentation. Of course, it startled the bejeebers out of everybody. But then, when he whipped out his lighter and set fire to Bethune's script, the fun really began! The dramatization proceeded more and more rapidly to attempt to finish the script before it became a cinder. The occasion was considered not really very funny by the Drama Guild, although it was one of the highlights as far as I was concerned.

The general noise was always a problem. Soundproofing was not something we had money for. The doors were forever being slammed. Rads got very noisy when the heat came on, if it came on on the week-end, and there was a lot of banging and crashing. We had a so-called audition turntable in studio "A," but you couldn't audition anything there if anyone was on the air in studio "B," because the noise would seep through. One night, during a rather dramatic recitation of poetry in studio "B," somebody put on a recording of a train wreck in studio "A" and gradually increased the power on this thing until frankly the lights in the studio dimmed. And the person who was trying to get through the poetry, J. Gordon Penny, was hurrying to get through the poetry before the building collapsed. I mean, it really sounded like that was going to happen. When that thing crashed, everybody in that studio nearly screamed. The noise built up, the lights dimmed and somebody thought that the transmitter had gone.

One night, I think it was again Penny wanting to read a piece of poetry of some importance. He was understandably concerned about the background noise that often went on during these readings, and we promised him that, even though we didn't understand the poetry, we'd make sure that it was sent out over the air properly. This was the same year Marilyn Bell had made her famous swim across Lake Ontario. The papers were full of it, and some obscure western star made a hideous recording about her swim, called "How Far is She Now?" It was absolutely terrible, and we kept a copy of this for some useful time. It occurred to us that the time was the next time Penny read a piece of poetry. When he gave the signal to the operator to play a piece of classical music that somehow tied in to this poetry reading, while we let the classic go out over the air, we at the same time plugged the damned "How Far is She Now?" into his monitor. We finally drove him out of the radio station, although that hadn't been the intention. These kinds of things were rife.

One year Doug Frame proposed that we do a DX broadcast. There are hobbyists, called DXers, who tune up and down the band in the wee hours trying to listen to distant radio stations. The mythology was that once CFRC's full 100 watts blasting out of Kingston had been picked up in Australia. Which was wholly possible. Through the DX radio club we scheduled on 1490 from 3 to 4 am (January 16, 1956), and theoretically everything was cleared on that channel throughout North America. We talked it up on campus for a long time and were desperately asking people to send a card or call in. After we had done all this work we didn't want to not receive a single call. It was a cold Saturday night after a dance, and we got the girls home. Only 2 or 3 of us turned up at the station in the middle of the night, and we decided to crank it up to 250 watts. Our licence allowed that, but normally we didn't do it for a number of technical reasons. We had the transmitter set up with more tubes in it, turned everything on and nothing worked. There we were, in the middle of the night behind the transmitter, frantically trying to get the thing going. About half-past three we got it on the air somehow. I don't know what we did, but we hit the right wire at the right time. We had only half an hour left but we had it all taped and ran the show through anyway. About 10 minutes into the broadcast we received our first phone call, from a friend of ours who wasn't feeling any pain. And he said, "I just wanted you to know you're coming in loud and clear at 79 Alfred Street." We got about 35 QSL cards for that night, from Nebraska, Maryland and one from Dubuque, Iowa. I think we were interfered with by a station in Havana. When we plotted where we'd heard from, we could see how this station had bulged up into the southern US, so our coverage was affected. We didn't get out of North America, but we got about 1,000 miles.

Mrs. Angus didn't run the station in our time. She ran the Drama Guild and the Thursday night Radio Workshop of the Queen's Drama Guild. They did classical music and live dramas. Friday and Saturday nights were the engineering nights. After we left, they got more serious about the programming, the station became far more professional and its whole orientation changed.

Some people used to come up with some dramatic sign-offs. One of the best ones was by Bob Held (Meds '5?), who was really a very, very good broadcaster. At the time there was a good deal of friendly rivalry between us as a non-commercial university station and the local Kingston stations. Of course, there was tremendous rivalry between the Kingston stations, so if one got a western show you could count on it that the other would have a western show a few days later. We weren't considerd competition, so they didn't really care what we played. But we would try to participate. When CKWS called itself "Radio Kingston," and CKLC called itself "The Top Spot on Your Radio Dial," we found that highly irritating, because they were at 1380 and we were at 1490. I can remember one night Held signing off the station and saying, "You've been listening to CFRC at 1490 kilocycles on your dial. That's 110 kilocycles **ABOVE** the top spot on your radio dial." We had a call from CKLC about that one.

We used to irritate them, but we were friends with them too. Remember Al Boliska and Ron Bertrand - they'd help us out and loan us recordings. John Bermingham, who owned CKLC, was a CFRC alumnus. Our other favourite sign-off was to thank all the people who had particiapted. You thanked your operator, the man operating the transmitter, and Ontario Hall for holding up its end of the antenna. We thought that was nice. We also used to sign off as "Kingston's least powerful AM station, most powerful FM station," to rub a little salt into the wounds.

I wonder if Sid Penstone would mind us telling that he was one of the people who missed his year over the tremendous contribution that he was making to CFRC. After that happened a couple of times, Sid got out of full-time participation, went on to get a Master's and Ph.D. and became a professor in the Electrical Department. I can remember I was getting desperate to graduate from Queen's, and I asked, "How in hell do you get out of this place, Sid? It just seems to be getting heavier and heavier." He said, "Well, I can't give you any specific answers, but I will tell you that the first six years to the Bachelor's is the toughest." I think while he was a student here, he was also the chief engineer for a radio station in Cornwall.[157]

Doug Frame: I joined the station in the fall of 1954, and we had three nights. The engineers had Friday and Saturday nights, from about 6 till around 11 pm, depending on how much sleep you'd had the night before. We weren't normally on on Sundays, but occasionally we would go on impromptu. Other times were all right if it didn't interfere with your work. Some of the former managers of the station had lost their years because they were down there all the time. But it was a very amateurish operation.

I remembering coming in my first day for the first meeting of the club and Sid was there. He took us into the control room and said, "Well, here it is. I built all this stuff, and I'm now going to take you into the lecture room and go over the whole thing." On the board he drew the schematic of the whole control room and everything in it. He had wires and lines that just covered the board from end to end. I sat there and said, "If I have to understand all that, I'm gonna quit!" Sid knew that thing inside out. He designed it, and could fix all aspects of it. I worked with him, usually as the tool carrier, watched what he did, and I learned a lot from him.

I remember those Monday noon meetings to set up the program. Hugh here was the Program Director who ran that and set up the time slots. We had some programs that were on every week regularly and then there were other slots that were sort of open. It was just a matter of saying, "Okay, who wants eight o'clock? He'll take eight. You'll take nine. He'll take the next half hour..." At the time there were about 30 members altogether, not counting the Drama Guild.

The AM transmitter was an old one that Harold Stewart built, and it's not there now. Some of the older parts - the modulation monitors and things like that - are still there. The control console was also built by Sid, and the transcription turntable stands. Behind that was a patch panel. Sitting on a wooden table was a brand new Magnecord tape recorder. It was not only one of the nicest pieces of equipment, it was probably also one of the heaviest, and took about four people to carry. We did hockey and basketball broadcasts from the old Jock Harty Arena and the Gym. We took over the remote amplifier (OP-4), and connected up Bell lines between there and here. We ran our own wires over to the Arena, just behind Jackson Hall. The Gym is across a city street, so we had to get Bell lines into that.

The FM transmitter up there now is the same one we had. It went in around 1952, but was off the air until we got it working around 1954. Sid Penstone actually did most of the work on it and got it

*Don Harrison
(Comm. '57)
operates in
Fleming Hall,
March 1958.
-courtesy
Dr Robert
Sanderson*

working. I worked along with him. When I came it was sitting up there idle, and I asked Sid why it wasn't on the air. He said I turn it up and I got a second harmonic in channel 8, right in the middle of the TV band. Channel 8 was coming from Watertown so people's antennas pointed south, and our FM antenna pointed north. You'd push the button to turn the station on and the telephone would ring within 30 seconds. That's all it took - "Your station's bothering our TV set again!" Sid worked on a second harmonic pipe trap, which is up there yet on top of the transmitter, and was able to cut it down so that at least we could turn on. And we got it up to about 100 watts - about all we could run on the 1,000 watt transmitter. We could run it at low level and get away with it, but every once in a while, when conditions were right, the telephone would go. The Department of Transport checked us out and we were fine and technically clear, and I guess we were concerned about public relations so we sometimes had to turn the power down even more than that.[158]

Bob and June (Pryce) Sanderson: Bob Sanderson was introduced to CFRC by a final year physics student whom he met while studying in the Music Room of the Students' Union. He was looking for someone to take over his CFRC classics program, and Bob inherited that as well as his magnificent room on Clergy Street. Though in Arts, Bob ultimately became Chief Operator because of his skill with the equipment and his knowledge of the patch panel. That he was not in engineering was no reason to exclude him from the post. Bob wrote and performed a 15-minute comedy program, "Anything Goes," every Saturday night, under the name Uncle Igor. June wanted to meet that crazy fellow, so her house-mate, Thelma Hunter, brought her along to meet her remote operator.

To the best of our knowledge, the first female full member was Anne Dorland, who joined in 1953. Her brother Gordon (Mech. '53) introduced her. She brought her friend Ellie Williamson in to share her disc jockey show. Three other women followed - Thelma Hunter, June Pryce and Beverley Phillips - each introduced by the previous one. Those five are the only women that we recall who ever sustained a continuing membership during that time. But males and females were on an absolutely equal basis at CFRC. This was probably unusual for the era.

The Drama Guild used the facilities every Thursday evening and absorbed only a couple of our most senior operators - Bob and Ralfe Clench did all of the operating for the Guild. Our rag-tag club of about equal proportions of Arts and Engineers provided programming Friday and Saturday evenings and technical operations for all three nights. I don't think this club had any name but CFRC. Most members visited the station on both our evenings and spent part of the night. Documenting repairs and keeping good wiring diagrams was a bore - long-time members carried the idiosyncrasies of a part of the station in their heads. In those days the station could effectively be re-wired through the patch panel and most technical problems bypassed temporarily. It was not uncommon to see an operator madly keeping things going with one turntable, the record cueing amplifier bypassed to drive the transmitter, working over the shoulder of someone rooting around in front of him in the guts of the console.

We had a wire recorder, but it was so bad that it wasn't used. Then the Magnecord M-90 tape deck was purchased. Under the lid of our luggable remote OP-4 console was a hand-written note: "This equipment is out of date" and signed "Guglielmo Marconi." To operate a remote, one disconnected the control room and

Ron Elliott keeps on operating while Bob Sanderson roots about under the console. Manfred Ficker, I.C. for the night, looks on anxiously, circa 1958.
-courtesy
Dr Robert Sanderson

studios using patch cords in the panel, and plugged the remote directly into the transmitter. This worked fine except that any loud shriek by the interviewee would cause the AM limiter to kick the transmitter off the air. The operator had to watch the mike levels carefully and anticipate increases in volume. We learned that, after isolating the AM transmitter for a remote, we could patch the station to the FM transmitter and send out a different signal on FM. Since most of the club wasn't very sports minded, when sports was on we'd broadcast classics on FM. It took a long time for someone in authority to catch on to us and point out that the Broadcasting Act made separatecasting very illegal and that we should cease and desist.

Program timimg and station breaks on the hour were not taken too seriously, but absolute reliability was. There was considerable peer pressure for quality. If any real goof went out over the air, there was an instant phone call from some member and the perpetrator received a lot of verbal abuse at the next Monday meeting. There were only two cardinal, non-negotiable rules in the club. First, no swearing or off-colour language on the second floor of Fleming Hall. This stemmed not from any great virtue but from the fact that the studio walls were far from sound proof. You had to be careful because the power in the microphone control circuits held the microphones OFF. If the circuits died, all the microphones would drop open and be live. The second was no alcohol or smell of alcohol on the premises. One night, one of our E.E.'s "In Charge" dropped in on the way to a party. A mickey of liquor fell out of a pocket or package and broke. There was no appeal. This person was out forever.

Each fall the station was given a thorough cleaning - the influx of women probably set a higher standard of cleanliness. During the 1955 cleanup, one of the men was surprised to find most of the AM trans-

Radio Office, Richardson Hall, 1958. From left, Bob Sanderson, Mrs Angus, Ralfe Clench, Grundoon.
-courtesy
Dr Robert Sanderson

mitter missing. An energetic Anne Dorland had removed all the tubes and detachable things and put them to soak in soapy water in the sink of the women's washroom. While they soaked, she was giving the inside of the transmitter a good scrub with soap and water to get rid of all those pesky numbers beside the holes where the "light bulbs" had been. Somehow, by air time, the tubes were dried out and replaced in the correct sockets. The AM transmitter behaved itself better than before.

Ralfe J. Clench (Arts '58), one of the true eccentrics cherished by Queen's and Kingston, hosted an opera broadcast on Saturday afternoons. Hugh Lightbody, the Program Director, repeatedly failed to ensure that Ralfe's name was spelled correctly in the Journal listings. Everyone could recite in a sing-song voice, "Ralfe with an E; Clench, no E; J in the middle."

One of the first things I taught new operators was to reach down and grab two of his chair's three legs and turn them so that two back legs were parallel to and as far as possible from the edge of the platform. If an operator found himself falling backward, he was to fall to the right, against the rack, so that his head wouldn't hit the terrazzo floor. Experienced announcers might surreptitiously spin a turntable backwards when a new operator was not looking. That would start the motor idling backwards and the tone arm would fly off the record as soon as it was released. And everything was controlled by the operator. There was not even a cough switch in the studio for the announcer.

There had been a Bell broadcast line between old Jock Harty Arena and the station for years. However, when we wanted to broadcast from the Coffee Shop in the Students' Union, a new line had to be strung across Union Street. The City pointed out that it was illegal to string a wire across a public road, but what was illegal in the daytime was less so in the middle of the night. One summer, one of our engineers working for Bell had kicked a coil of appropriate wire off a truck near Fleming Hall, so we had plenty of wire. At 3 am, after ensuring the coast was clear, two engineers wearing lineman's spurs shot up the telephone poles on opposite sides of Union Street. The line was strung and our first remote from the Union went off without a hitch. After all, we knew that foregiveness is much easier to obtain than permission.

Grundoon teaches Bob Sanderson how to operate the new state-of-the-art N.E. console in Master Control, Carruthers Hall, circa 1958.
-courtesy Grundoon Sanderson

After the Pryce-Sanderson wedding in August '57, our dog Grundoon became the station mascot. He put the station off the air only once in his puppy years by chewing on a wire under the console. In those days, Fleming Hall was still locked from the inside, so his main duty was to accompany people leaving the building. They would carry him down, unlock the door, attach the keys to his collar and send him back upstairs to his place under the control panel.[159]

Ernie Jury: I was more involved with the technical operation of the station than with programming. Before we signed off at midnight, we played our theme, "String of Pearls." In the classical area there was Paul Carroll, Dave Spendlove and Jim Hunt. Stalwarts on the engineering side were Dan McRae, Bob Radford, Grant Mervyn (Elec. '56), George Lake and Sid Penstone. Our early remotes were basketball broadcasts from the gym, called by Ken McKee, an excellent sportscaster. He could fill in the hollow spots in the game with personal stats: "A rush has been started up the floor by..." and he would throw in, "His scoring average to date this year has been..." and then he'd finish off the play, better than many commercial running commentaries. For a couple of years we used a low power short wave (40 metre) transmitter for our remotes, because we lacked funds to rent Bell lines to the gym. We installed our own lines to Grant Hall, where we had a cubby-hole in the wall on the second floor outside the centre door, and we could string one through the trees to the old Jock Harty Arena. I had my first motorcycle ride on one of my early remotes. A CFRC member owned a machine and I rode behind him with the remote amplifier nestled between us, as I grasped his leather jacket for dear life.

"Coon" Wood felt that all radio stations had to have commercials, so he worked up a few for "National Freight Cigars," inspired by Col. Jemmett's smoking rather pungent cigars in the basement labs. He had a unique station break too. He'd put about a foot of water into a 5 gallon cigarette butt ashcan and say, "This is CFRC, broadcasting from Kingston, only a stone's throw from Lake Ontario. Here, I'll show you. Yes, here's a nice sized throwing stone. I'll throw it...mmmfffff...Now listen." And he'd drop the station keys into the bucket of water. Kerplop! "See? There it is, in the lake."

Jimmy Thicke (Elec. '53) made a cardboard doll with pivoted arms and legs, actuated by a string at the back and he quietly entered the studio and made it dance while Vivienne Sterns was broadcasting Levana News. She had self-control of steel because she never missed a beat, until the red light went off and then she collapsed in a heap of laughter.

One fall my friend John McGeachy (Mech. '53), who was experienced in sailing, took a contract from Prof. Stewart's meagre research grant to paint the masts on Ontario and Fleming Halls and I acted as his ground-man, hoisting him up the pole on a boatswain's chair with the halyard that normally held the antenna up. We were nervous of the mast on Ontario Hall, because the insulators on the three guy wires appeared to be in tension, so that if one broke the whole thing would fall over the edge of the building. Around that time, I went up the pole on Ontario Hall to grease the pulley axle at the top. I knew better than to look down, but I looked up at the moving clouds and had the sensation that the pole was falling over. Believe me, I left my fingerprints up there.

We had the problem of how to get the new FM antenna up the AM mast on Fleming Hall and how to get a man up there to secure the U-bolts to hold it to the mast. During a dry run, we lost control of the halyard that held the AM antenna and it ended hanging halfway up the mast. Fortunately, it had jammed in the block at the top or else the AM antenna would have wound up on the ground. Sid Penstone ran to get a long stick with a hook on the end, used by the Drama Guild to adjust lights, but it would not reach. It was a misty day and the pole was too slippery for shinnying. Then Sid announced that there was a case of beer for anyone who could retrieve the lanyard. John McGeachy stepped forward, shinnied up the 15 or 20 feet and brought it down.

I think I can take the credit for Ralfe Clench's habit of carrying tools with him. He was a freshman up from Hamilton and I was working on the patch panel and was looking for a particular screwdriver or pair of pliers. I remarked to Ralfe that tools are no damn good to you unless you had them at hand when you needed them. A few days later he appeared with a small pouch at his belt and a screwdriver or two and a pair of pliers in it. This pouch grew in size as time went on.

Col. Jemmett had shown the senior E.E. students how to hook up the rotating equipment in the basement to produce 60-cycle power from the large battery in the basement or from the steam-driven generators in the heating plant and feed it up to run the CFRC transmitter. When Hurricane Hazel struck in November '54, power was out in most of Kingston but we continued to run CFRC from internal resources. No other Kingston station was on the air that evening. We got Prof. Stewart's okay, phoned CKWS and ran a crummy twisted-pair line down the hall from the Library phone to the patch panel and we alone carried the CBC National News in the area.[160]

Another Roy Thomson Story

As early as the fall of 1946, Messrs Arthur Davies and Roy Thomson offered to help CFRC "to broadcast public service and educational programs for a reasonable time each day." Allied would supply CFRC with an operator and a program director for 5 hours a day, and would seek to make available some CBC sustaining programs which CKWS could not use. In the 1946 discussions, the Principal felt that CFRC's equipment had deteriorated and needed replacing: "It was suggested by the University representatives that Allied might be disposed to enlarge on its original offer and pay for the necessary replacement equipment or possibly lend equipment to the University." The cost came to several thousand dollars, and the Allied directors declined to put capital into a Queen's asset, finding that it had no spare equipment that would restore CFRC's capabilities. Allied was still interested in a public service function for CFRC, so the original offer to assume the operating costs was renewed.[161] The negotiations must have dragged on and on.

> Harold Stewart: Some time ago, around 1950, before CKLC appeared in Kingston, Roy Thomson came to Queen's and they proposed that CFRC operations should be beefed up, so that it could be said that Kingston had two full-fledged stations operating full time, and therefore did not need another one. This struck me as rather peculiar in a way, so I said to Mr Thomson at an interview in Dr Wallace's office, "Mr Thomson, I thought you were a free enterprise man." "Oh, why certainly I am," he says. "But not that free." Well, maybe we wanted too much to qualify for a full time station. That deal never went through, and obviously CKLC is now a flourishing station. Better that it happened that way. Allied sent a letter to Wallace saying that it was withdrawing the offer of assistance to CFRC until the matter of the licence for the new station were settled, and graciously suggested that the offer would then be renewed.[162]

The Lower Committee Report

By the fall of 1955, after a period when the Radio Workshop's standards were sliding, the Queen's Journal applauded its renewed effort to

> provide something more distinctively representative of a university group...For years the "radio Voice of Queen's" has been characterized, for the most part, by decidedly amateur specimens of that Genus Americanus, the disc jockey. Its program schedules have been barely distinguishable from those of ordinary commercial stations - only the records are older and the voices younger...because of lack of time or of imagination, the over-emphasis on recorded music, generally presented without much inspiration and oriented towards the "pop" field, has continued to be the predominant feature of CFRC programming...Surely a university radio station can draw on resources that will make it much more than a pale imitation of commercial stations.[163]

Several members of the Queen's faculty, notably Prof. A.R.M. Lower, were concerned that the students were abusing their privileges and that CFRC was a wasted resource. An informal committee of staff first met around Christmas, 1956, to consider the future of Queen's

Radio and to try to do something for Queen's by means of the station. Its members, who had either broadcast over the CBC or were vitally interested in radio, were Arthur R.M. Lower, Chairman, Margaret Angus, Secretary, William Angus, Ralfe Clench, H.W. Curran, H.M. Estall, H. Morris Love, David McNaughton, John Meisel, M. Ross, G. Shortliffe and Harold H. Stewart. The committee met 8 times after the Christmas holidays of 1956 and there were 7 meetings of its sub-committees. After "a careful and thorough investigation of the possibilities of broadcasting over radio station CFRC," the committee drew up a 7-page report.

The committee was concerned that this valuable resource, while useful to Electrical Engineering in instruction in the technical side of radio, "was not very useful to the University as a broadcasting medium...the programs generally not being of the type which are to be expected from a university."[164] If properly used, CFRC could render valuable service and good will to Queen's and be an informal extension of extramural work, providing an educational service to the community. It could build a sympathetic audience in Kingston, publicize research at Queen's and be a medium for pronouncements on university policy and needs. The committee pointed to United States universities which used their radio services to "bind themselves to their people closely." Indeed, the committee took as its model the radio service of the University of Wisconsin, which had a budget of $150,000 and its own building.

Prompt action was neccessary if the station were to implement a new policy and extended hours. The report of the Fowler Royal Commission on Broadcasting had just been published and changes might soon be enacted which could interfere with any new policies regarding Queen's Radio.

The committee recommended the appointment of a Director of Radio, to have full control of programming and budget planning at CFRC, on a 2/3 year basis, with a status equivalent to that of an assistant professor. The Director of Radio should be a member of the Faculty of Arts, with all attendant benefits and the possibility of promotion. A board should be constituted to advise the Director of Radio. The committee stressed the advisory function of this board, to consist of the Principal, the Vice-Principal and Treasurer as ex-officio members, two members elected from each faculty, plus student members. The Director would consult the board on matters of general policy, but for details and programming matters, should consult informally with appropriate members of staff or strike standing committees. Margaret Angus was suggested as the first Director of Radio, since she had "had considerable training and experience in the direction of the station as well as important contacts with radio on a national scale through creative contributions to the CBC."

Principal Mackintosh was also concerned about the use which was being made and could be made of CFRC. He encouraged discussion between the Treasurer, Mr Tillotson, and the committee, and agreed that the station needed "more careful thought and better organized administration." He appointed Mrs Margaret Angus as the Program Director and a provisional committee of 13 members to advise on program planning: A.R.M. Lower and John Meisel (Arts), H.M. Love and H.H. Stewart (Science), G. Malcolm Brown and D.L. Wilson (Medicine), D.M. Mathers (Theology), Ralfe Clench and David McNaughton (students), with Principal Mackintosh, Vice-Principal Corry and M.C. Tillotson (ex officio), and Mrs Angus as Secretary. The Senate would later appoint the committee officially, with whatever additions it felt necessary.[165]

> <u>Margaret Angus</u>: Queen's is fortunate in having a radio station and a licence to use the public airwaves. In April, 1957, we presented a report to the Principal. He agreed that better use might be made of the station with someone to administer and co-ordinate long range program planning. Consequently, that summer, radio was set up as a University department, with a budget for supplies, equipment and personnel. And since I had considerable experience in radio, I was appointed Director of Radio Broadcasting. The Committee continues to act in an advisory capacity in matters of general policy, and now has student members also.[166]
>
> I suppose I was considered suitable to be Director of Radio Broadcasting because I had managed the Radio Workshop, and I also had done a lot of writing for radio in every sort of way, and had been a local correspondent for a CBC program called "News Roundup." I used to do interviews for them. That made me eligible for the sort of thing that the University wanted when they decided that they should have a formal appointment as Director of Radio. The actual appointment as Director came about because more and more people were listening to CFRC, and there had been occasional times when what went out over the air was not really approved of by the University. They thought it might be better if there was someone who could not exactly tone them down, but direct them in a better way. Since I had a lot of contact with students, knew the students fairly well, I was moderately acceptable, even to the engineers, and we worked well together. I had a formal appointment as Director of Radio at Queen's from 1957 to 1968.[115]

The Department of Electrical Engineering informed the committee that broadcasting hours had to be restricted because of interference caused to its laboratory work. The committee therefore proposed that the station program:

Thursday 7.00 pm to 11.00 pm
Friday   7.00 pm to 11.00 pm
Saturday 1.00 pm to 11.00 pm
Sunday   9.00 am to 4.00 pm

With the apporoval of the new Director of Radio, the schedule might be extended to midnight or beyond on Thursday, Friday or Saturday. This privilege might be suspended at any time by Prof. Stewart, the Technical Director of CFRC.

The committee also worked out a sample program schedule, incorporating the sort of wide variety of fare that it thought should emanate from a university station: good music of various types, drama, informative and educational discussion with the emphasis on interviews and round tables, direct University publicity and student interests and activities. They also hoped that the CBC might offer progamming to CFRC.

The budget would include a salary for the Director and a secretary. Prof. Stewart suggested that there be put aside something to pay members of the CFRC Club, who operated the stations at examination times and during holidays. The committee also recommended the hiring of a part-time announcer, to be selected by the Director from a pool of students. The total budget suggested in the first year, exclusive of the Director and secretary, was $5,479.00. This might drop to $4,100 once facilities had been set up. Space had still not been found for an office.

When CFRC signed on again on Thursday, October 3, 1957, broadcasting Thursdays and Fridays from 7.00 to 11.00 pm and Saturdays from 1 pm to midnight, a whole new governing structure for the station was being moved into place.[167]

*Bob Sanderson in the CFRC transmitter
room, 1958
Prof. Stewart's Mark VII (left),
Westinghouse FM-1 (right)
-courtesy Dr. Robert Sanderson*

NOTES

1. Personal communication from Prof. Harold Stewart, Kingston, Aug. 10, 1982.
2. W.E. McNeill to H.H. Stewart, Jan. 12, 1945; Prof. Stewart's papers.
3. "CFRC Operations Since the Opening of CKWS", undated (probably after Jan. 12, 1945); ibid.
4. Personal communication from Prof. Harold H. Stewart, Kingston, June 23, 1986.
5. H.H. Stewart to R.C. Wallace, undated; QUA, Coll. 1250, Box 19, Principal's Office, Series 1, Subject Files R-S, Radio Station CFRC, 1923-1952.
6. Walter A. Rush to H.H. Stewart, August 27, 1943; ibid.
7. Stewart's memo to himself, Sept. 9, 1943; Prof. Stewart's papers.
8. Walter A. Rush to R.C. Wallace, Dec. 8, 1943; QUA, Coll. 1250, Box 19, Principal's Office.
9. R.S. Lambert to R.C. Wallace, Dec. 21, 1943; ibid.
10. R.S. Lambert to Dr. R.C. Wallace, Jan. 11, 1945; ibid.
11. R.S. Lambert to R.C. Wallace, Jan. 31, 1944; ibid.
12. H.H. Stewart to R.C. Wallace, Feb. 3, 1944; ibid.
    Letter of application, Dr. R.C. Wallace to G.W. Olive, CBC, Ottawa, Feb. 7, 1944; ibid.
13. G.W. Olive to R.C. Wallace, Feb. 11, 1944; ibid.
    G.C.W. Browne, Assistant Controller of Radio, D.O.T., Ottawa, to R.C. Wallace, March 3, 1944; ibid.
14. Memo, H.H. Stewart to R.C. Wallace, undated; ibid.
15. R.C. Wallace to G.C.W. Browne, March 11, 1944; ibid.
16. G.C.W. Browne to R.C. Wallace, March 21, 1944; ibid.
17. R.C. Wallace to G.C.W. Browne, March 25, 1944; ibid.
    G.C.W. Browne to R.C. Wallace, April 6, 1944; ibid.
    R.C. Wallce to G.C.W. Browne, May 1, 1944; ibid.
18. "Licence to Use Radio", No. 58, call sign VE9BH, Ministry of Munitions and Supply, Ottawa, April 3, 1944.
19. Frank W. Peers, The Politics of Canadian Broadcasting, 1920-1951, University of Toronto Press, 1969, p. 85n.
    Joe McCallum, CKUA and Forty Wondrous Years of Radio, University of Alberta, Edmonton, 1967; Queen's University Special Collections.
    Edward Jordan, "Recollections of CKUA", New Trail, University of Alberta, Summer, 1987, p. 22.
20. Personal communications from Dr. William Angus, Kingston, Aug. 25, 1982; June 18, 1986.
    Gwen Morton Herbst, "The Summer Radio Institute at Queen's", QR 21 #2, Feb. 1947, p. 40.
    NOTE: The Hamlet recording may have been inspired by the CBC. On Jan. 3, 1944, Lambert wrote to Wallace that he had been in touch with Harrison "about the interesting experiments with a dramatic reading from Hamlet by college students that we are planning. This recording will probably be made in the latter part of February"; QUA, Coll. 1252, Box 15, Principal's Files, Series #3.
21. "Radio Workshop Established", QR 18 #7, Oct. 1944, p. 187.
22. R.C. Wallace to R.S. Lambert, Supervisor of Educational Broadcasts, CBC, Toronto, Oct. 16, 1944; QUA, Coll. 1250, Box 19, Principal's Office.

23. R.C. Wallace to William Angus, Oct. 23, 1944; ibid.
24. "Proposed Radio Institute: Notes of a Conference held at 55 York Street, Toronto, on November 2nd, 1944..."; QUA, Coll. 1250, Box 19, Principal's Office, CBC File.
25. R.C. Wallace to R.S. Lambert, Nov. 30, 1944; ibid.
26. E.L. Bushnell, Director General of Programmes, CBC, Toronto, to R.C. Wallace, Dec. 16, 1944; QUA, Coll. 1250, Series #1, Box 21, Principal's Office.
27. R.C. Wallace to Dr. A. Frigon, General Manager, CBC, Ottawa, Jan. 10, 1945; ibid.
28. "The Summer Radio Institute at Queen's University, Kingston, Ontario", enclosure with letter, W. Angus to R.C. Wallace, Dec. 18, 1944; ibid. "Queen's to Offer Summer Radio Institute with the Active Co-operation of C.B.C.", QR 19 #4, April, 1945, p. 102.
29. R.C. Wallace to Rupert Caplan, March 10, 1945; QUA, Coll. 1250, Box 21, Principal's File.
30. Notes by G.B. Harrison, meeting with R.S. Lambert and Andrew Allan, Toronto, March 6, 1945; ibid.
31. W. Angus to R.C. Wallace, April 18, 1945; ibid.
32. R.C. Wallace to H.B. Esterly, Radio and Radar Division, War Production Board, Washington, D.C., April 25, 1945; ibid.
33. W. Angus to R.C. Wallace, April 30, 1945; ibid.
34. R.C. Wallace to W. Angus, May 9, 1945; ibid.
35. W. Angus to R.C. Wallace, May 15, 1945; ibid.
36. Report on the Summer Radio Institute of 1945; QUA, Coll. 1250, Series #1, Summer Radio Institute, Principal's Files, Box 21. Ralph Purser to Nancy Cutway, 1981; CFRC Archives.
37. "Summer Radio Institute - 1945: Facilities"; ibid. W. Angus to R.C. Wallace, Jan. 31, 1946; ibid.
38. R.C. Wallace to D.G. McKinstry, CBC, Montreal, Feb. 15, 1946; ibid.
39. "Memo for Dr. Wallace from Dr. Angus", undated; ibid. "Summer Course in Radio Arts", QR 20 #4, April, 1946, p. 97.
40. R.C. Wallace to A. Davidson Dunton, General Manager, CBC, Ottawa, Jan. 29, 1946; QUA, Coll. 1250, Series #1, Box 21, Summer Radio Institute.
41. R.C. Wallace to W. Angus, May 7, 1946; ibid.
42. W. Angus to Aurele Seguin, June 22, 1946; ibid.
43. W. Angus to R.C. Wallace, June 26, 1946; ibid. Memo, W. Angus to R.C. Wallace, June 28, 1946; ibid.
44. R.C. Wallace to Lorne Greene, July 22 and 31, 1946; ibid.
45. "Report of the Summer School, 1946", Principal's Report, Queen's University, April, 1947, p. 55. "CFRC Broadcast Time Table", QJ, July 19, 1946, p. 6. "Radio Schedule", ibid., July 26, 1946, p. 3
46. Untitled address: QUA, Coll. 1250, Series #1, Principal's Office, Box 21, 4 pp. "Radio for Good or Evil", address by Dr. R.C. Wallce, July 19, 1946; soft-cut acetate disc, CFRC Archives.
47. Gwen Morton Herbst, "The Summer Radio Institute at Queen's", QR 21 #2, Feb. 1947, p. 40.
48. "The Summer Radio Institute of 1947", undated manuscript; QUA, Coll. 1250, Series #1, Principal's Files, Box 21.
49. Personal communication from Larry Palef, Toronto, Oct. 30, 1982.
50. Personal communication from Bernard Trotter, Kingston, Sept. 17, 1982.
51. W. Angus to R.C. Wallace, May 28, 1948; QUA, Coll. 1250, Series #1, Box 21, Summer Radio Institute, Principal's File.

52. Personal communication from Dr. Esse W. Ljungh, Kingston, May 16, 1985.
53. Meeting, Summer School Committee, Friday, Dec. 31, 1948; QUA, Summer School Committee File, Box 4/5, Call No. 3602.
54. Queen's Summer School of the Fine Arts calendar, 1949, p. 14. Summer School and Extramural Courses in Arts, 1949, p. 34.
55. Personal communications from Mrs. Eileen (Green) Fleming, Kingston, May 5, 1986 and Aug. 25, 1989.
56. NOTE: Records show that most CBC people and all those from private radio were paid an honorarium of around $65-75 a week plus living expenses. The total cost of staff in 1945 was some $2,000, and $2,185 in 1947.
57. "Drama Guild Organization", QJ, Oct. 10, 1945, p. 1. "New Radio Institute", ibid., Oct. 23, 1945, p. 1.
58. "Once More Campus Radio", ibid., Oct. 30, 1945, p. 1.
59. "Radio Workshop Will Present", ibid., Nov. 6, 1945, p. 1.
60. "Journal Initiates News", ibid., Nov. 13, 1945, p. 1. "CFRC Featuring "The First Dance"", ibid.
61. "Columnizing the Campus", ibid., Nov. 16, 1945, p. 3.
62. "Radio Group Offers `Hamlet'", ibid., Nov. 20, 1945, p. 1.
63. Letter to the Editor, ibid., Nov. 20, 1945, p. 4.
64. "IRC, PAC, DU Plan Radio Forum", ibid., Nov. 27, 1945, p. 1.
65. "Radio Workshop Begins Series", ibid., Dec. 4, 1945, p. 1.
66. Memorandum, W.E. McNeill to R.C. Wallace, Dec. 10, 1945; QUA, Coll. 1250, Box 19, Principal's Office.
67. Arthur L. Davies to R.C. Wallace, March 21, 1947; Prof. Stewart's papers.
68. H.H. Stewart to R.C. Wallace, Oct. 24, 1946; QUA, Coll. 1250, Box 19, Series #1, Principal's Office.
69. "CFRC to Take to Airwaves", QJ, Jan. 15, 1946, p. 1.
70. "CFRC Featuring", ibid., Jan. 22, 1946, p. 1.
71. "CFRC Featuring Meds '49", ibid., Jan. 29, 1946, p. 1.
72. "CFRC to Feature Glee Club", ibid., Feb. 12, 1946, p. 4.
73. "CFRC Round Table Discussion", ibid., Feb. 19, 1946, p. 1.
74. "CFRC: The Voice of Queen's", Tricolor, 1946, p. 61.
75. Personal communication from Sandy Webster, Toronto, Nov. 2, 1982.
76. "CFRC Completes Year", QJ, March 5, 1946, p. 1. "CFRC - A Survey", ibid., Nov. 26, 1946, p. 6.
77. "Declaration of Secrecy in the Operation of Radio Apparatus", H.H. Stewart to A.G.E. Argue, Radio Inspector, Kingston, March 20, 1947; Prof. Stewart's papers.
78. "Radio Workshop Cuts First Disc", QJ, Oct. 4, 1946, p. 6.
79. "CFRC Resumes Weekly"; ibid., Oct. 16, 1946, p. 1.
80. "Radio Workshop", ibid., Nov. 19, 1946, p. 6. "Final CFRC Broadcast", ibid., Nov. 26, 1946, p. 1.
81. "Inspection Report - Private Commercial Broadcasting Stations", Radio Division, D.O.T., Feb. 13, 1947; Prof. Stewart's papers.
82. "Varied Program on CFRC", QJ, Jan. 28, 1947, p. 1.
83. "CFRC Casts Term Ending", ibid., Feb. 25, 1947, p. 1.
84. "CFRC First Broadcast", ibid., Oct. 7, 1947, p. 1.
85. "Radio Workshop", ibid., Oct. 3, 1947, p. 4.
86. "Royal Welcome Tomorrow", ibid., Oct. 21, 1947, p. 1.
87. "Full Evening's Slate", ibid., Nov. 4, 1947, p. 1.
88. "CFRC Program Featuring Arts", ibid., Nov. 25, 1947, p. 4.
89. "CFRC Cleans House", unknown Queen's publication, Oct. 1948;

CFRC archives.
90. Untitled proposal by H.S. Jackson; QUA, Coll. 1250, Box 19, Principal's Office, Series #1, Radio Station CFRC, 1923-52.
91. "Dance Music Feature", QJ, Oct. 28, 1947, p. 4.
92. "What's When", QJ, Nov. 14, 1947, p. 4; ibid., Nov. 21, 1947, p. 5. "CFRC Pop Program Friday", ibid., Nov. 28, 1947, p. 5.
93. "Premiere of Sports Night", ibid., Dec. 2, 1947, p. 1.
94. "Dr. Eleanor Roosevelt", ibid., Jan. 10, 1948, p. 1.
95. "Station CFRC Inaugurates", ibid., Jan. 13, 1948, p. 5.
96. "Science Radio Amateurs", ibid., Jan. 27, 1948, p. 3.
97. "CFRC on the Air", ibid., Jan. 23, 1948, p. 7. "CFRC on the Air", ibid., Feb. 13, 1948, p. 5.
98. "Campus Frolics on CFRC", ibid., Feb. 24, 1948, p. 4.
99. "CFRC Mike to be at Levana", ibid., Jan. 30, 1948, p. 5.
100. "CFRC Ghost Play Will Spook", ibid., Jan. 13, 1948, p. 5.
101. "CFRC to Sketch Campus", ibid., Feb. 3, 1948, p. 5.
102. "CFRC's Last Program", ibid., Feb. 24, 1948, p. 1.
103. Personal communication from James E. Kirk, International Falls, Minnesota, Oct. 31 and Nov. 16, 1989.
104. Personal communication from Florence Fraser McHugh, Toronto, Oct. 17, 1982.
105. "Tomorrow Night CFRC", QJ, Oct. 26, 1948, p. 1. "Electricals Take to Ether", ibid., Oct. 29, 1945, p. 5.
106. "Radio Workshop Talent", ibid., Nov. 2, 1948, p. 4. "Radio Writing Course", ibid., Oct. 26, 1948, p. 5.
107. "CFRC to Give Gen", ibid., Nov. 23, 1948, p. 1.
108. "CFRC Again Takes", ibid., Nov. 19, 1948, p. 5. "CFRC", ibid., Nov. 26, 1948, p. 7. "CFRC Blurbs", ibid., Dec. 3, 1948, p. 5.
109. "New CFRC Program", ibid., Jan. 25, 1949, p. 1.
110. "Color Night Banquet", ibid., March 4, 1949, p. 5.
111. "200 ISS Collectors Besiege", ibid., Jan. 14, 1949, p. 4.
112. Personal communication from Fred Paquin, Kingston, Oct. 31, 1982.
113. "Radio Workshop Opens Tomorrow", QJ, Oct. 12, 1949, p. 1.
114. "CFRC to Relieve Record", ibid., Oct. 21, 1949, p. 5.
115. Personal communication from Mrs. Margaret Angus, Kingston, Aug. 25, 1982.
116. Personal communication from Lou Tepper, Kingston, Aug. 20, 1982.
117. "CFRC", Tricolor, 1950.
118. Personal communication from Charles F. Currey, Pittsburgh, Penna., Nov. 28, 1989.
119. "Pins Awarded: Radio 1950", Drama Guild Scrapbook; QUA. Donald R. Gordon to Nancy Cutway, undated, circa 1981; CFRC Archives.
120. Personal communication from Dr. E.J. Bond, Kingston, Nov. 19, 1982.
121. Personal communication from John Bermingham, Kingston, Aug. 31, 1982.
122. Personal communication from Jane (Sherman) Kaduck (Arts '55), Kingston, Aug. 24, 1982.
123. Personal communication from Catherine M. Perkins (Arts '58), Kingston, Aug. 24, 1982.
124. R.C. Wallace to H.H. Stewart, Dec. 29, 1950; Prof. Stewart's papers.
125. H.H. Stewart to R.C. Wallace, Jan. 15, 1951; ibid.
126. "CFRC Radio Station", Tricolor, 1951.
127. "CFRC 1490", Tricolor, 1952.

128. George W. Birley, Kingsway Film Equipment Limited, Toronto, to H.H. Stewart, Dec. 14, 1953; Prof. Stewart's papers.
     Queen's requisition #33635, Official Order No. 19647, to Kingsway Film Equipment Limited, Toronto, Dec. 21, 1953; ibid.
129. "CFRC and Radio Workshop", Tricolor, 1954.
130. "CFRC Radio Workshop", QJ, Feb. 5, 1953.
131. "Forewarned is Forearmed", undated list of regulations for conduct at CFRC, c. 1952; Prof. Stewart's papers.
132. H.H. Stewart and H.S. Pollock, "Compression with Feedback", Electronics 13 #2, 19, Feb. 1940.
133. Personal communication from Prof. Sidney R. Penstone, Sept. 21, 1982 and March 16, 1983.
     NOTE:Alex J. McDonald always understood that "NEMO" referred to a remote broadcast, and supplied a reference: Carl G. Dietsch (NBC). "Radio Broadcasting" in "Radio Engineering Handbook (ed. Kieth Kinney, 3rd edition), 1941, p. 264; personal communications, March 6, 1990.
134. G.C.W. Browne, Controller of Radio, Radio Division, Air Services, D.O.T., Ottawa, to Manager, CFRC, Oct. 24, 1945; Prof. Stewart's papers.
135. R.S. Lambert, CBC, Toronto, to R.C. Wallace, Nov. 23, 1945; QUA, Coll. 1250, Box 19, Series #1, Principal's Office.
136. R.C. Wallace to D.O.T., Ottawa, Dec. 3, 1945; ibid.
137. G.C.W. Browne, D.O.T., Ottawa, to H.H. Stewart, Aug. 15, 1947; Prof. Stewart's papers.
138. H.H. Stewart to G.C.W. Browne, Sept. 18, 1947; ibid.
139. A.C. Monteith, Vice-President, Westinghouse Electric Corporation, Pittsburgh, to D.M. Jemmett, June 29, 1952; ibid.
140. From P.E. Higgins, Order Service Department, Westinghouse Electric Corporation, Baltimore, June 20, 1952; ibid.
141. G.C.W. Browne to H.H. Stewart, Sept. 13, 1952; ibid.
142. H.H. Stewart to Mr. F.C. Nixon, Telecommunications Division, Air Services, D.O.T., Ottawa, Sept. 19, 1952; ibid.
     G.C.W. Browne to H.H. Stewart, Oct. 15, 1952; ibid.
143. G.C.W. Browne to H.H. Stewart, Dec. 2, 1952; ibid.
     NOTE: Mr. Browne sent along D.O.T. form AR-5-21 and CBC form 269 for making a formal application, in triplacate. These had to be received before December 8, 1952, to be in time for the Board's next meeting. There was a slight error in the engineering brief, so Keith MacKinnon hand-delivered the revised brief and the applications in triplicate to the D.O.T. on December 8.
144. Keith A. MacKinnon, Consulting Radio Engineer, Ottawa, to H.H. Stewart, Nov. 3, 1952; ibid.
     Keith A. MacKinnon, "Engineering Brief Submitted in Support of the Application of CFRC, Kingston, Ontario, for Authority to Install and Operate a Frequency Modulated Broadcasting Station on the Assignment Channel 220, 1270 watts, 100 feet", Nov. 3, 1952; ibid.
145. G.C.W. Browne to H.H. Stewart, March 31, 1953; ibid.
     H.H. Stewart to G.C.W. Browne, April 13, 1953; ibid.
146. Private Commercial Broadcasting Station Licence No. 87, D.O.T., Ottawa, April 1, 1951, surcharged March 31, 1953; ibid.
147. "Public Announcement No. 66", Board of Governors, Canadian Broadcasting Corporation, Dec. 22, 1952; ibid.
148. Dan McRae, "FM Transmitter", May 31, 1953; ibid.
149. Packing list, sales order 34792, Andrew International Corporation,

Chicago, June 25, 1953; ibid.
150. Draft, H.H. Stewart to G.C.W. Browne, Feb. 12, 1954; ibid.
151. H.H. Stewart to F.R. Robertson, Canadian Marconi, Montreal, Dec. 14, 1953; ibid.
152. Memo, H.H. Stewart to Mr. Winney, Queen's University Financial Services, July 26, 1953; ibid.
    J.W. McLeod, Whitby, Ontario, for Andrew International Corp., Chicago, to H.H. Stewart, Dec. 9, 1953; ibid.
153. G.C.W. Browne to H.H. Stewart, Feb. 25, 1954; ibid.
154. Small notebook leaf from Sid Penstone, Feb. 12, 1954; ibid.
155. Draft, H.H. Stewart to D.D. McLean, Canadian Westinghouse, Hamilton, Feb. 12, 1954; ibid.
156. Personal communication from K.W. Greaves, M.D. (Meds '54), Hamilton, Nov. 10, 1989.
157. Personal communication from Hugh Allan Lightbody (Elec. '57), Toronto, Oct. 17, 1982.
158. Personal communication from John Douglas Frame (Elec. '57), Ottawa, Oct. 17, 1982.
159. Personal communication from Dr. Bob (Arts '57) and June (Pryce) Sanderson (Arts '58), Toronto, Oct. 20, 1990.
160. Personal communication from Ernest J. Jury (Elec. '55), Ottawa, Oct. 22, 1990.
161. Arthur L. Davies, Director, CKWS, to R.C. Wallace, March 21, 1947; QUA, Principal's File, Box 19, Radio CFRC, Whig-Standard Arrangements.
162. Personal communication from H.H. Stewart, Kingston, Oct. 20, 1982.
163. "CFRC Thursday Night", QJ, Oct. 25, 1955, p. 6.
164. "Report of an Informal Committee of Staff on the Subject of the Queen's Radio Station: Its Present and Possible Uses"; undated, probably spring, 1957; gift of Prof. H.H. Stewart.
165. "Re: University Radio Station", Principal W.A. Mackintosh to staff, Sept. 3, 1957; Prof. Stewart's papers.
166. "40 Years of Queen's University on the Air", script read by Mrs. Angus over CFRC, Oct. 27, 1962; QUA, A.ARCH sr 333.
167. "Queen's Radio Sees", QJ, Oct. 1, 1957, p. 1.

EPILOGUE

*Transmitter room, Fleming Hall, 1990, showing the RCA
BTA 250L, a gift of CKLC in 1961 (left), and the
Westinghouse FM-1 (right).*
                    -photograph by Arthur Zimmerman

Epilogue: 1958 to the Launching of Stereo

Margaret Angus: I had a formal appointment as Director of Radio at
Queen's from 1957 to 1968. It was seen as a very lowly job, with the
sort of pay that went with a lowly job. But it was also seen as an important job for the University in its relationship with the public. My
appointment came directly from Principal Mackintosh and I reported
directly to the Principal, which is unusual in a minor administrative post.
Of course, Prof. Harold Stewart was Technical Director, and we had
a Radio Advisory Committee made up of professors of various faculties, who met occasionally and patted us on the back for the good job
we were doing. Once in a while, it would suggest that we might need
a little more money to get some better records and things of this sort.
That was the administrative set-up of CFRC when I started.

Actually, we worked well together, once the students accepted
the fact that if they were going to do regular broadcasting in the

Margaret Angus in new Station Manager's office, Carruthers Hall, 1960.
-courtesy Dr Ronald Elliott

University, the University should have a little bit to say about it. Only very occasionally did a knuckle have to be rapped about a remark that had been made on the air. As they did more broadcasting and as the competition to get a chance to broadcast became keener, the quality went up very decidedly. It was a great satisfaction to have a letter from the Principal saying that he was a regular listener, and thought that the quality had improved immensely. Sometimes we wondered if the technical quality was a little suspect. I remember with absolute horror, starting a broadcast and suddenly having the AM station go off the air. Somebody said, "What you do is go into the transmitter room and hit the transmitter where the 'X' is marked in chalk on the side." Surprisingly, you gave it a little tap and we went back on the air. That was because we were working with equipment that had been so sadly out of date and so changed. Because it was an experimental station, and the engineers were constantly trying something new. What had started out as a wiring system with colour coding got a little haywire because they ran out of a particular colour, and just used another one. Then somebody lost the slip of paper that said what this meant.[1]

CFRC is a unique facility among campus stations, owned and operated by a university rather than by a student council. It is supported entirely out of private contributions and endowment income from private donations and bequests. It receives no public money from any source and is entirely non-commercial. CFRC is a cultural service with an academic dimension, but slightly apart from the main stream of teaching and research, like the Art Centre, the Performing Arts Office, McGill-Queen's Press and Queen's Quarterly, a projection into the wider community. The University sees its purpose as providing an opportunity for students to use their creative and organizational talents in a practical setting and to provide Kingston area listeners with programming alternative to that available on the commercially-operated stations, services which are complementary to rather than competitive with other broadcasters.[2]

When the CFRC Student Radio Club set itself up in 1958, the students decided that its executive would be appointed by the outgoing executive, rather than be elected. The rationale was that the executive would best be able to pick the best and most experienced people to run the club and have input into the running of the station. By 1961-62, there were about 130 members, with some 60 involved in weekly broadcasting. Students were in charge of the transmitters, broadcast operations and announcing. The student Program Manager worked with Mrs Angus in all program planning. The apprenticeship method of training new broadcasters, just showing people how to work the equipment, had given way to formal standardized lessons by senior members. Announcers learned microphone technique, regulations, how to select and organize music and how to do the paperwork. Operators took a more complex course in working the console, tape decks, patching, and cueing records. In the formal operator's test, the examiner might unplug a turntable, scramble the patch field or disable parts of the equipment. The objective was to put the candidate under stress to see if he could handle it. Only senior people with wide technical experience could aspire to being announce-operators.

By the fall of 1958, CFRC's service to Queen's and the community and its public relations value to the University were becoming recognized and it was desired to increase broadcasting to a full 52 weeks a year. The facilities, however, were quite outdated and inadequate for that task. CFRC had lost Studio "A" to an E.E. laboratory, Studio "B" was cramped and had windows which leaked noise, the hallway library was overflowing and vulnerable and the control room was crowded. Records were auditioned on a portable set in a classroom. Mrs Angus' office and the LP collection were across the road in Richardson Hall, and Fleming Hall was locked during broadcast hours. Broadcast hours were limited because the transmitters were right above the E.E. labs and everything had to be shielded from the high frequency fields. In addition, E.E.'s expansion had reached the point where all of CFRC's space might soon be needed. When the Treasurer suggested to Prof. Stewart that CFRC might move, the Radio Advisory Committee struck a subcommittee on October 6, consisting of Margaret Angus and Professors H.M. Love and Harold Stewart, to look into solutions.

The subcommittee decided on a partial move. A complete move would be far too expensive and would involve new technical applications to the DOT, a new antenna - meaning a tower and radial ground system at an off-campus site - a new transmitter and a new technician. Student operation would then be impossible. The old AM T-antenna was protected by the grandfather clause, however, so if the transmitters were left in Fleming Hall but the studios moved to a nearby site, the DOT might permit that shift. The only costs would be for a new Northern Electric R 5420E console, new turntables, intercom, microphones and structural changes in the new location. The 32-hour schedule could then not be expanded, but the number of weeks on air could, and CBC programming might be made available. At that time, all CFRC work was done by E.E.'s technician, John Fraser. The subcommittee also advised hiring a technician to be shared with E.E.[3]

Construction of the new CFRC studios in the former Civil Engineering laboratory in the basement of the John Carruthers Science Hall began in August 1959. The station was expanding from 800 to 2,000 square feet, and would have a Master Control, two studios, technician's room and two libraries of more than 5,000 discs, including

200 LP's. Most of the equipment in Master Control was Northern Electric (N.E.). The new CFRC studios were officially opened by Principal Mackintosh on January 7, 1960.[4]

John A. (Jack) Harrison was hired in 1960, half-time for CFRC and half-time for the servicing of university electronic gear, the beginning of the present Electronic Service Shop. He had been a designer and draughtsman in the Armed Forces, retired at 50, and had just completed two years at CKLC. Electronics was not his field, but he was a radio ham. Jack retired at end of December 1976, and Gary Racine, hired in 1966, succeeded him.[5]

CFRC operated 33 to 34 hours a week, for 20 weeks, with a strict policy of playing about 50% popular music and good quality jazz - "songs which have endured through the years and which are known as 'good music'" - about 50% classical records and no rock 'n' roll, hillbilly, western or current favourites.[6]

The old black monster, the Mark VII AM transmitter of 1939, was starting to show its age, to drift off frequency and fail occasionally. Prof. Stewart enquired of RCA in November 1958, about a new vertical radiator, console, frequency deviation monitor, turntables and pre-amplifier, microphones and a used RCA BTA-250M commercial AM transmitter to operate at 100 watts.[7] Around April 1959, CFRC replaced the failing Micro turntables with high-torque McCurdy CH-12 3-speed transcription models.[8]

Terry and John French, of CKLC, gave their old RCA BTA-250L (Serial 1009) 250-watt transmitter to Queen's after CKLC received permission to double its power. The BTA-250L was installed in the transmitter room on the second floor of Fleming Hall in the spring of 1961. Its output stages had to be modified and Stewart provided the diagrams and Harrison made the changes to bring the BTA-250L on line as a 100-watt transmitter. To bring it up to specification for DOT approval, the output network was adapted to an antenna-counterpoise system from a grounded system and the 810 output tubes changed for 805's. The frequency had to be changed from 1240 to 1490 kHz and tuned, and the crystals ground down or new ones acquired. This took some time. On September 29, 1961, Terry D. French, Managing Director of CKLC, on behalf of the directors and shareholders of St. Lawrence Broadcasting Co. Ltd., presented the BTA-250L to Queen's. The ceremony was broadcast over both CKLC and CFRC.[9] Stewart kept the BTA-250L off the air until DOT approval was granted on Jan 15, 1962.[10]

Some members considered the introduction of a new club Coke machine more momentous in their day-to-day lives than a new transmitter. In 1961 the Richardson Memorial Fund Committee purchased a Steinway semi-grand piano for CFRC's Studio "A."[11]

By October 1962, Master Control boasted a N.E. R5420E console with rack-mounted monitor amplifiers, two McCurdy CH-12 turntables with N.E. control units, and twin rack-mountable Magnecord tape decks. The RCA BTA-250L at 1490 kHz was fed by a limiting amplifier and the Westinghouse Type FM-1 at 91.9 mHz was fed directly, both through underground cables to Fleming Hall. The AM antenna was a top-loaded vertical supported between two guyed steel poles at 120 feet above grade, with a counter-poise. There was a permanent CBC feed through Bell for the Emergency Measures Organization warning device. They had a 4-channel battery-operated N.E. R5460D console for remote use.[12]

A telephone survey of February 1963 showed that 86.6% of students listened to CFRC at least one week-end a month.[13]

The first Queen's football game broadcast by CFRC in modern times was on October 24, 1964, when the Queen's - Varsity match was aired from Toronto. Lawrie Rotenberg was the broadcaster.

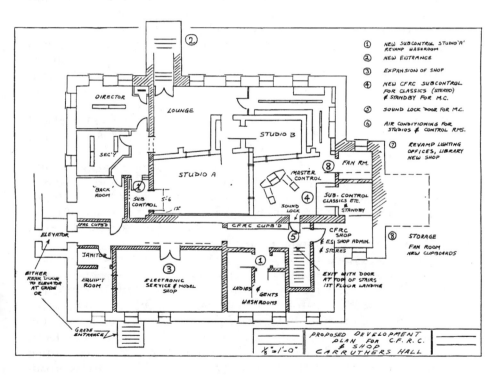

Proposed development plan for radio station CFRC, 1966, in the basement of Carruthers Hall. This was a step in the evolution of the present "doughnut" configuration of the station. The fledgling Electronic Service Shop was located where the lounge is now. Studio "A" was much larger before a piece was cut out to house Subcontrol, and thereby create Control Room 3 and the now vanished Studio "D." In the mid-1970s, the large Studio "B" and the library space were converted into announce booths "B" and "C" and stereo-capable Control Room 2. A notorious trick before alteration was to open the back door to Master Control (the present CR 1) while it was "On Air" and roll a metal ashtray cannister down the back staircase.
-Prof. Stewart's papers

The Board of Broadcast Governors proposed in 1964 that stations with both AM and FM frequencies provide a minimum of 2 hours a day of separate FM broadcasting, with reduced advertising to upgrade the quality and 25% of the time devoted to the arts, letters and science. The commercial operators protested, preferring a move to make FM economically attractive. At a BBG hearing in Toronto, Bernard Trotter, Executive Assistant to the Principal, argued that $3,500 for equipment to permit CFRC to separatecast wasn't in that year's budget,[14] and asked for exemption.[15] Mrs Angus testified that in all other respects CFRC exceeded the current standards, but was willing to add the required 2 separatecast hours after each day's AM-FM schedule. The BBG did not insist upon CFRC's compliance.

In 1966 a second on-air auxiliary control room was added to relieve the demand for production time and studio time for training a growing number of students. Sub-control was made from the old CFRC washroom and stood against Studio "A" in what is now the corridor that passes by the station office. It was run with surplus equipment, the old Red Board portable amplifier.[16] The former men's washroom was then divided into two and refitted.

Margaret Angus, lovingly known as Mrs A. or G.A., for Granny Angus, retired as Director of Radio on August 31, 1968. Her replacement was Andrew Marshall (Arts '66), who had joined as a student in 1962. Under Marshall's regime the stations began to change in direction. He introduced a new image, moving toward the kind of music that students enjoy, like rock and folk-rock, plus a dinner hour show featuring high quality pops. Toward midnight, "The New Sound of Nocturne, the party sound of up-tempo and variety listening." He also talked of plans for separate FM programming and stereo.[17]

   In 1970, Principal Corry approved, in principle, plans to broadcast every day in stereo by 1972. That summer, Prof. Stewart drew up a document concerning the relationship of CFRC to E.E., expressing problems that the Department was having with CFRC's strong radiated fields penetrating Fleming Hall. E.E. could not soon move, so if Queen's decided upon a daily schedule of 12-15 hours, the transmitters and antennae would have to be relocated outside of the city, in accordance with government regulations. Queen's would have to seek higher power and perhaps another channel, and he was concerned about increased costs. If Queen's wished to consolidate responsibility for CRFC under a single head, members of E.E. would not wish to be involved, nor would the current technical personnel.[18] The Order-in-Council of June 4, 1970, caused all discussion and plans to be put aside.

The licence of CFRC-FM was renewed in March 1970 for one year, as the CRTC studied the policy of licencing stations owned by government agencies. Then, on June 4, 1970, the Government of Canada proclaimed in Order-in-Council PC 1970-992 that it was against policy for the CRTC to issue broadcasting licences to government agencies or government-supported institutions. Queen's would have to re-shape CFRC's corporate structure and its relationship to the University if the licences were to continue beyond March 31, 1972. After much agonizing, the policy came under reconsideration and on February 29, 1972, CFRC and three other campus stations were given a reprieve until March 31, 1974.[19] Order-in-Council PC 1972-1569 gave the CRTC a new definition which exempted campus stations and any agency not directly controlled by government.[20]

Engineering consultant A.G. Day wrote to Mr Trotter in December 1971, for Ottawa-Cornwall Broadcasting Ltd., a subsidiary of Bushnell Communications, Ltd. The firm had acquired a television licence for channel 6 in the Kingston/Belleville area, the transmitter to be sited 30 miles from Kingston at Deseronto. Mr Day's analysis showed that CFRC-FM at 91.9 mHz was likely to interfere in Kingston because of the strength of its signal locally and its location at the lower end of the FM band.[21]

Day proposed three solutions. Queen's thought individual residential traps impractical and expensive. Changing frequency to 99.5 would lose CFRC-FM its established audience, unless Bushnell sweetened the deal by donating stereo transmitting equipment, perhaps as a 50th anniversary present to CFRC. Marshall favoured relocation, to Wolfe Island to reduce CFRC's field strength over the city, or co-siting with the Deseronto TV transmitter to make the field strengths of the two equally strong in the immediate area. Either would provide wider coverage and all day operation, allowing extension course broadcasts, as long as CFRC received acceptable maintenance of its equipment.[22]

Prof. Stewart consulted Keith Mackinnon and did a reception experiment at home. The problem was not of great magnitude and his evidence would not yield Queen's much leverage. Mackinnon estimated that CFRC would need 10-20 kW at the Deseronto site, and Stewart figured that the transmitter and aerial might cost $25,000, "a hard bargain to drive." Furthermore Bell Canada would charge $500 per month for 2 equalized 15-kHz channels to link the campus with Deseronto. A microwave link might cost $20,000.[23] On the other hand, changing to 99.5 mHz and doubling power for stereo to the maximum, 3 kW ERP at the Fleming Hall location, would require a 6-bay antenna array, too tall for the 69-foot mast.[24] MacKinnon noted that the DOC considered this sort of interference a problem for the new applicant to solve.[25]

Queen's was offered $10,000 of the $25,000 needed to relocate. The method that was finally chosen to decide was procrastination, and the money was lost. Prof. Stewart thought that perhaps Queen's was too greedy. In fact, there has been no interference problem with channel 6 at all in the last decade as TV tuners have improved.[5,26]

Canadian Content regulations was a big issue back in 1970. The new rules were phased in and CFRC had to have 33% Can-Con.[27]

CFRC did not program rock until 1969, and then it was restricted to the late night "Nocturne" on AM. In early 1970 the Radio Committee addressed the issue of the introduction of rock 'n' roll on CFRC, and all 22 members, without exception, were aghast and dismayed at the prospect. Andrew Marshall defended the new policy, and argued vehemently in favour of it, supported by the student members. The committee did not like it but did not recommend against it.

CFRC applied around January 1969 to separate the two stations, thereby expanding the service: AM to become Queensradio aimed at the campus with student-oriented music, campus news and discussion, while FM would be Fine Arts Radio, featuring concerts and lectures, aimed at the community.[28] Marshall had hoped to split AM from FM in the autumn, when he hoped to introduce rock to AM, but a clause inserted into the licence in 1964 specified simulcasting, so the request was not considered until late in the year. Separatecasting began only in January 1970.[29]

While Andrew Marshall had a vision of FM becoming a Fine Arts Radio, the club saw CFRC as an alternative to CKWS and CKLC in a more experimental sense. There was tension because Andy was perceived as wanting to be high-brow, presenting music for the elite. Later, under Steve Cutway, the elitist sense of FM was replaced by the alternative sense. The emphasis on FM was always on professionalism and the creation of an alternative sound.

Around 1970, there was a plan to develop the University Centre, on grand scale proportions. All student activities and services would be housed there, including CFRC and all the new equipment gradually being acquired to help modernize the studios. CFRC was assigned 2,500 square feet in the proposed centre, and the club spent a great deal of time planning for the best use of that space, until they suddenly learned what was to be directly above, on the next storey of the centre. A bowling alley![30]

The dream of CFRC-FM stereo broadcasting every weekday may date as far back as 1961. The documentary record is very thin on this subject. Most of the club was content with the twin AM-FM voices and had no thought of splitting for stereo. Andrew Marshall was instrumental in moving toward the concept of stereo FM,[31] and he began to acquire equipment that would be stereo-capable. The first stereo tape recorder purchased was a Scully, in 1970.

Andrew Marshall, as Chairman of the CFRC Modifications Committee, reported to Principal Deutsch on June 28, 1973, including a floor plan showing extensive architectual and electrical renovations, new stereo Control Room 2, equipment for the new stereo control facility, including a console, two turntables, a Scully 280, a record/playback and playback cart machine, changes to the heating system to eliminate noise in the studio areas, and partial air conditioning. The total cost was estimated at $24,500.[32] He also proposed installing a new stereo transmitter at an off-campus site.

The "Proposed Basement Floor Plan, Carruthers Hall" blueprint, dated May 22, 1973, shows the basic doughnut layout as it is now. Jack Harrison began searching for a new 8-channel stereo console (3 mic level, 4 medium level, one medium level for remote; with cue, talkback and monitor facilities, and rotary controls) for CR 2 in the fall of 1973.[33] During the summer of 1973, the facilities at CFRC underwent reorganization, extensive renovation and modernization, with carpeting, stereo equipment and air conditioning of the studios for $40,000. By March 1974, the ESS had moved upstairs to Carruthers 103 and the club took possession of a new lounge and office.[34]

In February or March 1974, Marshall decided that it was time to move on, and he resigned as Station Manager. Steve Cutway (Arts '72) received the appointment, beginning July 1. He inherited a station with three control rooms, CR2 still incomplete, four studios and increased library, office and lounge space.[35]

CBC radio programs were available locally on a very limited basis through the affiliate, CKWS, and segments of the public unable to receive CBC Toronto directly were unhappy with this partial service. In the early 1970s, the CBC was seeking to expand its network service in the area and was looking for the strongest possible FM frequency for Radio I, its basic AM service. Queen's had discussions with the CBC on the possibility of affiliating CFRC-FM with CBC Radio and receiving a new transmitter and operational costs for CFRC. The hope was to merge without being submerged. The club was fearful of losing creative control of programming and of decreasing involvement with FM.[36]

Ceremony at the opening of the new CFRC stereo Control Room 2 in the basement of Carruthers Hall, October 27, 1977.
At the microphone, from left, Margaret Angus (former Director of Radio), Kathleen Ryan (CFRC benefactor), Steve Cutway (CFRC Station Manager) and Andrew Davey (Radio Club President). Vice-Principal (Services) H. Morris Love is at rear centre.
-photograph by Arthur Zimmerman

Discussions concluded when CBC decided that it wanted to operate the service itself, and therefore to seek its own FM frequency in the area as a repeater for Radio I.

A new CBC-FM stereo network was to come into existence by the end of 1975 and there was still interest in affiliating CFRC-FM with that, thereby gaining a more powerful frequency, stereo, a 7-day operation and additional capital and operating costs from the CBC, though Queen's would have to be assured of expanded programming opportunity for the CFRC Radio Club in such an operation.[37] Trotter feared the CRTC might make conversion to stereo a condition of FM renewal and hoped that prospects of affiliation with the CBC might avoid this. Radio I received a licence to broadcast at 107.5 MHz, effective July 28, 1975, and was expected to start operations in Kingston by the summer of 1976. Inflation depleted the CBC's capital funds, however, and both the Radio I and stereo projects were slowed.

Steve Cutway approached Prof. Harold Pollock, CFRC's Technical Advisor, about the possibility of daily stereo FM broadcasting after 6.30 pm. E.E. was then conducting research in the broadcast portion of the FM Band, which would be interfered with by CFRC-FM. Relocation of the antenna to the Elrond College building or John Orr Tower was impossible, since the DOC prohibited such installations within city limits. The Vivarium site and the towers of CKWS or CKLC were still possibilities.[38]

The Radio Club's constitution was amended in the spring of 1974. The club chairman had been elected annually from the executive, which was appointed each spring by the Manager and the outgoing executive. The Chairman was usually the AM or the FM Program Director, but this would become an appointed position, like the rest.[38] To this point remote sports broadcasts had been supported by private donations, solicited in an annual canvass on the campus or <u>ad hoc</u> in the cafeteria. Cutway proposed that the AMS include in its 1975 elections a referendum to increase the activity fee by $0.25 for a CFRC Sports Trust Fund and it passed.

Kathleen Ryan addressing the CFRC 55th Anniversary Dinner, October 29, 1977. From left: Kathleen Ryan, Nancy Cutway, Steve Cutway, Harry J. Boyle. -courtesy CFRC Archives

The Manager had reported directly to the Principal, but following a review of responsibilities for cultural activities initiated by the new Principal, Dr Watts, the Manager would henceforth report to the Vice-Principal (Academic).[39]

Principal Ronald L. Watts announced at the re-union dinner in October 1976, that Mrs Kathleen Ryan (Arts '26) would establish a foundation at Queen's, funded to at least $500,000. The Old Medical Building would be renovated with this money to house the Queen's Archives, and CFRC would receive two grants totalling approximately $35,000 to "purchase urgently needed FM equipment. Part of Mrs Ryan's concern for CFRC stems from the fact that her late husband, also a Queen's graduate, was a student broadcaster at CFRC during its initial years of operation."[40]

At the CFRC 55th Anniversary Dinner, October 29, 1977, she gave the station the balance of the $35,000 gift to set up the CFRC Equipment Fund, the seed money and the inspiration for the "Go Stereo" project.

> **Kathleen Ryan**: This is why I'm here to-night, because of what happened here back in 1924-25-26...That's where it all started, on the kitchen table in Fleming Hall...I feel in giving you this cheque, I'm rendering unto Caesar what is Caesar's, and it is really from Frank Ryan's enterprises in recognition of what you have done and what you can do in the future.[41]

NOTES

1. Personal communication from Mrs. Margaret Angus, Kingston, Aug. 25, 1982.
2. CFRC and CFRC-FM Licence renewal application, Oct. 27, 1975; gift of Bernard Trotter.
3. Report of a Sub-committee on Queen's Radio Broadcasting, Jan. 13, 1959; Prof. Stewart's papers.
   CFRC Proposals (hand-written), Jan. 1, 1959; ibid.
4. "New Studios at Queen's for Radio Station CFRC", KWS, Jan. 9, 1960, p. 13.
5. Personal communication from Gary Racine, Kingston, October 22, 1990.
6. "New Studios at Queen's for Radio Station CFRC", KWS, Jan. 9, 1960, p. 13.
   "Open House at CFRC", QJ, Jan. 12, 1960, p. 1.
   Queen's University News, P.R. Dept., Oct. 1962.
7. D.C. Tucker, Manager, RCA Victor, Montreal, to H.H. Stewart, Dec. 19, 1958; Prof. Stewart's papers.
8. R.A. Dunlop, Sales Manager, McCurdy Radio Industries, Toronto, to H.H. Stewart, April 10, 1959; ibid.
9. "CKLC, Kingston: Presents", Canadian Broadcaster 20 #20, Oct. 19, 1961.
   "Transmitter Presented", KWS, Sept. 30, 1961, p. 13.
10. H.H. Stewart to K.A. Mackinnon, c/o CKCK, Regina, Oct. 13, 1961; Prof. Stewart's papers.
    F.G. Nixon, DOT, Ottawa, to Queen's E.E., Jan. 15, 1962; ibid.
    H.H. Stewart to K.A. Mackinnon, Feb. 22, 1962; ibid.
11. Margaret Angus to Advisory Committee on Radio Broadcasting, September, 1961; ibid.
12. Article by D.S. Swain, Queen's Public Relations, from information supplied by Mrs. Angus, memo dated Sept. 28, 1962; QUA.
13. Minutes, Advisory Committee on Radio Broadcasting, Nov. 14, 1963; Prof. Stewart's papers.
14. KWS, Mar. 12, 1964, p. 21.
15. "Queen's Wants Exemption", KWS, Mar. 11, 1964, p. 5.
16. Minutes, Radio Advisory Committee, Jan. 25, 1966; Prof. Stewart's papers.
17. Bernard Trotter to Andrew Marshall, London, Ont., Jan. 22, 1968; QUA, Coll. 3707, Box 2/6, Radio Broadcasting, 1968-69.
    "New Look to CFRC", QJ, Sept. 20, 1968, p. 9.
18. Draft, August 18, 1970; Prof. Stewart's papers.
19. "Radio Licences Extended", KWS, Feb. 29, 1972, p. 13.
    "CFRC Gets Licence", ibid., Mar. 3, 1972, p. 14.
20. Frank Foster, Broadcasting Policy Development, Franfost Communications Ltd., Ottawa, 1982, p. 365.
21. A.G. Day, Ottawa, to Bernard Trotter, Dec. 8, 1971; Prof. Stewart's papers.
22. Memo, H.H. Stewart to B. Trotter, Dec. 16, 1971; ibid.
    Dave Steer, CFRC Chief Engineer, Report on frequency change, Feb. 8, 1972; ibid.
    J.A. Harrison to Bliley Electric Co., Erie, Pa., Feb. 15, 1972; ibid.
    C.A. Allinson, Caldwell A/V Equipment, Scarborough, to J.A. Harrison, Feb. 16, 1972; ibid.
23. H.H. Stewart to K.A. MacKinnon, Dec. 21, 1971; ibid.

24. H.H. Stewart to K.A. MacKinnon, Feb. 2, 1972; ibid.
25. K.A. MacKinnon to H.H. Stewart, Dec. 24, 1971; ibid.
    J.A. Harrison to C.A. Allinson, Caldwell A/V, Sept. 5, 1972, ibid.
26. Personal communication from H.H. Stewart, Kingston, Aug. 10, 1982.
27. Andrew K. Marshall, "A Brief Regarding the Problems Facing Canada's
    Only Educational Radio Station, CFRC, as a Result of the Proposed
    Canadian Content Regulations," c. 1970; CFRC archives.
28. "CFRC Diversifies", QJ, Oct. 9, 1969, p. 3.
29. "CFRC Expands Service", QJ, Jan. 15, 1970, p. 1.
    "Queen's Radio Splits", ibid., p. 2.
30. Personal communication from Steve Cutway, Kingston, March 11, 1983.
31. A.K. Marshall, "Advisory Committee Report to the Senate Committee on
    Fine Arts and Public Lectures," April, 1970; CFRC archives.
32. Andrew K. Marshall to J.J. Deutsch, June 28, 1973; Prof. Stewart's
    papers.
33. J.A. Harrison to radio supply firms, Sept. 6, 1973; ibid.
34. Memo, P.R. Schell to D.T. Arkett, July 27, 1973, ibid.
    Queen's Gazette 6 #12, March 20, 1974.
35. "CFRC Continues to Expand", QJ, Apr. 25, 1974, p. 5.
36. Scott C. Mackenzie, "The CFRC Student Radio Club and CBC
    Affiliation", Jan. 22, 1974; Prof. Stewart's papers.
    "Joining of CFRC with CBC", QJ, May 30, 1974, p. 1.
    "CBC Affiliation", ibid., June 13, 1974, p. 3.
37. Bernard Trotter in Advisory Subcommittee on Radio Broadcasting
    Report, May, 1974; CFRC archives.
38. Minutes, Radio Advisory Committee, Dec. 6, 1974; ibid.
39. Advisory Subcommittee on Radio Broadcasting Annual Report, April 9,
    1975; ibid.
40. Queen's News Dept. Release, Oct. 15, 1976.
41. Kathleen Ryan, speech at CFRC 55th Anniversary Dinner, Queen's, Oct.
    29, 1977; CFRC Archives.

Appendix A: A Chronology, 1980-90

February 1, 1980 - On the advice of the Task Force on the Implementation of CFRC-FM Stereo, Principal Watts announced support in principle for the stereo proposal. The target year was 1982.

Wilf Barkley of the ESS with the AEG Bayly-Telefunken S 3175/21 FM stereo transmitter in place in front of the old Westinghouse FM-1 in Room 307, Fleming Hall, September 16, 1987.
-photograph by Arthur Zimmerman

January, 1982 - CBC Ottawa donated a McCurdy SS4370 stereo audio console, vintage 1967, to CFRC for installation in expanded CR 3. Officially opened in October 1982.

Spring 1982 - The Radio Club added over $72,000 to the transmitter fund, mainly through a "Go Stereo" raffle (October 19, 1980) and a campus-wide student referendum.

March 1982 - The CRTC froze all decisions on FM licencing indefinitely, pending application by the CBC for a series of low power FM relays.

March 7, 1983 - CFRC applied to the CRTC to broadcast in a limited schedule in stereo on 101.9 mHz, with precedent quoted. Not considered.

May 9, 1986 - The CRTC approved licence for CFRC-FM to broadcast daily in stereo at 101.9 FM. Plans to begin in September were stalled when Pittsburgh Township objected to the proposed tower at the Vivarium site. It was then learned that Queen's had plans to sell the land.

November 25, 1986 - Dr John Meisel submitted the report of the Task Force on Campus Radio.

September 16, 1986 - New AEG Bayly stereo FM transmitter arrived and went into service in place of the old Westinghouse FM-1.

April, 1988 - Cantel, Inc., agreed to allow CFRC-FM to co-site on its cellular telephone tower at Highways 15 and 401.

December 15, 1989 - Steve Cutway retired as Station Manager.

February 3, 1990 - CFRC-FM stereo, 101.9, went on the air in a seven day a week schedule and CFRC, 1490 AM, was shut down.

Appendix B: CFRC and SRI Alumni Who Worked in the Media

Dr Margaret Angus - CBC documentary writer
Dr William Angus - adaptations for CBC radio (Ohio State Award, 1946)
James Annand - CBC radio and television actor

John Bermingham - former owner of CKLC, Kingston; Bermingham
    Marketing
Nicole Bérubé - correspondent, Radio-Canada, Toronto
Dr Edward J. Bond - CKWS; record reviewer, Whig-Standard
Phil Brown - part-timer, CKWS, Kingston
Donald Brittain - film maker
Ian Byers - part-timer, CKWS
Paul I. Calback - CKBC, Bathurst, N.B.; CFMK, Kingston
Sally (Sarah) Caudwell - producer, CBC Newsmagazine
Rick Choma - CFLY/CKLC News, Kingston
Jim Clark - CKTN, Trenton
Dorothy Cotter - CBC
William Cowling - part-timer, CKWS
Dave Craig - technician, CKEY, Toronto
Drew Crossan - CBC writer
David Cunningham - former Program Director, CFMK, Kingston
Chris Cuthbert - CBC Sports
Florence Daly (Courneya) - CKWS
Sean Eckford - CKUE, Smiths Falls
Brenda Finlay - CFRN-TV news anchor, Edmonton
Eileen Green (Fleming) - program host, CKWS
Florence Fraser (McHugh) - CBC reporter
Pam Godfrey (Samuels) - CBC correspondent, London, Ontario
Lynne Goldman - CBC, western Canada
Donald Gollan -
Donald Gordon, Jr - CBC correspondent, London, England
Peter Griffiths - newspaper industry
Tom Holden - CBC middle management
Brian Holt - CBC executive
George (Keiladze) Jason - magician, monologist, pianist
Jeffrey Kaufman - Global News, CBC
Paul Kennedy - CBC producer; host of "Ontario Morning"
Ted Kennedy - former Music Director, K97, Edmonton
Rick Leblanc - CBC-TV cameraman
Daniel H. McRae - CBC engineering group, Montreal
Andrew Marshall - freelance broadcaster, magazine publisher
Alexander J. McDonald - retired CBC engineer
Charles Millar - CKWS technical staff, retired
Christian Molard - correspondent, Belgian Radio or Deutsche Welle
Barbara Monture - CBC
Ernest Mutimer - CBC
Larry Palef - retired CBC announcer
Catherine Perkins - Saturday Night; Editor, Queen's Alumni Review
Dave Perrault - Global News (behind the scenes)
David Quance - CBCRobert Radford - CFJR, Brockville
Bob Rews - CBC Current Affairs
William Gordon Richardson - CRBC and CBC management
Shelagh Rogers - CBC announcer and program host

Michael Roth - CBC TV, "Razzle Dazzle"
Frank Ryan - owner of CFRA, Ottawa
Arnie Schwisberg - syndicated radio jazz producer
Jane Sherman (Kaduck) - formerly CKWS and CKWS-TV
Edith Shindman (Houzer) - CBC radio
Jeffrey Simpson - Globe and Mail columnist, CBC commentator
Sally Southey - CBC news correspondent
Charles Taylor - free-lance for CBC in England and Europe; Globe
    and Mail correspondent in China
Doug Thompson - CJAD, Montreal
Bernard Trotter - International Service, CBC; Supervisor, CBC
    Public Affairs, retired
Tim Turnbull - formerly CKEY, Toronto
Thomas J. Warner - General Manager, WJLB, Detroit, retired
Peter Watts - CBC TV Sports, Edmonton; The Sports Network
A.R. "Sandy" Webster - CBC actor
(Martin) Campbell West - News Director, private station,
    Yellowknife, N.W.T.
Scott Whitley - part-time sports broadcaster, CKWS-TV, Kingston
Anna Zee (Zanetti) - CJBQ, Belleville; CFLY, Kingston; Halifax
Arthur Zimmerman - CFMX, Port Hope; interim host of CBC's
    "Mostly Music," Ottawa; part-timer, CKWS News
Abe Zvonkin - radio entrepreneur

INDEX

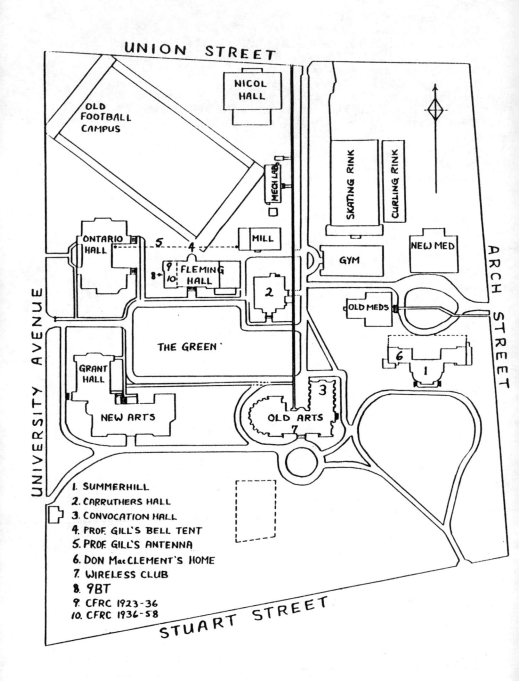

UNION STREET

OLD FOOTBALL CAMPUS

NICOL HALL

SKATING RINK

CURLING RINK

MECH. LAB.

MILL

ONTARIO HALL

5   4

9
8
10

FLEMING HALL

2

GYM

NEW MED

OLD MEDS

THE GREEN

GRANT HALL

6

1

NEW ARTS

3

OLD ARTS

7

UNIVERSITY AVENUE

ARCH STREET

1. SUMMERHILL
2. CARRUTHERS HALL
3. CONVOCATION HALL
4. PROF. GILL'S BELL TENT
5. PROF. GILL'S ANTENNA
6. DON MacCLEMENT'S HOME
7. WIRELESS CLUB
8. 9BT
9. CFRC 1923-36
10. CFRC 1936-58

STUART STREET